CRC DICTIONARY OF AGRICULTURAL SCIENCES

Library of Congress Cataloging-in-Publication Data

Lewis, Robert A. (Robert Alan)
CRC dictionary of agricultural sciences / Robert A. Lewis.
p. cm.
ISBN 0-8493-2327-4 (alk. paper)
1. Agriculture—Dictionaries. I. CRC Press. II. Title.

S411 .L39 2001
630′.3—dc21 2001052483

International Standard Book Number 0-8493-2327-4
Library of Congress Card Number 2001052483
Printed in the United States of America 1 2 3 4 5 6 7 8 9 0
Printed on acid-free paper

CRC DICTIONARY OF AGRICULTURAL SCIENCES

Robert A. Lewis

CRC PRESS

Boca Raton London New York Washington, D.C.

About The Author

Robert A. Lewis, Ph.D., is the author or editor of numerous scientific papers and books written in English and German. Contributions include basic research in physiology, endocrinology, and animal behavior; applied research on air pollution and toxic chemicals; and chemical, biological, and environmental monitoring and specimen banking as applied to human health and environmental effects.

In addition to a long record of basic and applied research, the author has made contributions to contemporary society through outstanding national and international achievements as an inventor and as an academic and government administrator in both the U.S. and Germany. He developed and coordinated several programs of research that were national or international in scope, and he has served as an administrator of governmental organizations and national programs in the U.S. Although an American, he has also served as an academic administrator, director of a research institute, and leader of national programs of research in the Federal Republic of Germany.

Dr. Lewis led a team that selected the National Environmental Research and Assessment Parks of the Federal Republic of Germany and later developed sampling designs and protocols for chemical and biological monitoring and specimen banking to be conducted on a continuing basis.

The author has made a number of technological innovations including invention and codesign of the Zonal Air Pollution System (ZAPS), a research tool that is able to continuously deliver and monitor the direct and indirect effects of gaseous pollutants on agricultural crops and other ecosystems throughout entire growing seasons. Since introduction, this pioneer system (in its original form or with design modifications) has been used in various programs in the U.S., Canada, and Europe to assess the impacts of sulfur dioxide, ozone, and carbon dioxide on various crops including, for example, grasses, broccoli, clover, cotton, tobacco, peas, soybeans, wheat, and even trees. It has also been employed in the National Crop Loss Assessment Program sponsored by the EPA.

Other technological innovations include the design, construction, and application of an air quality monitoring and micrometeorological network in Montana to help assess the effects of air pollution on the range resources of the state.

The author also successfully standardized and applied remote sensing to the assessment of chronic pollution effects on ponderosa pine forests and grasslands in the northcentral Great Plains.

Among many other contributions to agriculture, Dr. Lewis served as a rapporteur at the White House Conference on Technology for a Sustainable Future; The President's Task Force on Water Quality; the American–German Specimen Banking Program; and the Health, Safety and Environmental Task Force for the U.S. Strategic Petroleum Reserve.

Dr. Lewis holds degrees from Ohio State University, Rutgers University, and the University of Washington. His biography appears in various editions of *Who's Who in America*, *Who's Who in Science and Technology*, and various other biographic sources.

A

1080. sodium fluoroacetate.

α. lower case symbol for the Greek letter alpha, *q.v.*

a-, **an-**. prefixes meaning **1:** on, in, at. **2:** negation of the quality expressed by the root; without, lacking, not.

a posteriori. pertaining to a manner or system of reasoning whereby one attempts to determine cause from knowledge of the effect or outcome. *Cf. a priori*.

a priori. pertaining to a manner or system of reasoning whereby one infers an effect or outcome from a notion or theory regarding cause. *Cf. a posteriori*.

A. **1:** absorbance. **2:** ampere.

$\mathbf{A_r}$. relative atomic mass.

A. upper case symbol for the Gr. letter alpha, *q.v.*

A.S.A. acetylsalicylic acid.

A-frame. **1:** a wooden or metal frame that resembles a capital A. Such structures typically support or hold in position heavy loads. **2:** a structure such as a house with steeply angled sides that meet at the top forming the shape of the letter A.

A-scale sound level (A-weighted sound level). a measurement of sound that approximates the sensitivity of the human ear. It is used to note the intensity or annoyance of sounds.

A-weighted sound level. A-scale sound level.

aa. amino acid.

AA. amino acid.

AAA. Agricultural Adjustment Act.

AAF. *N*-2-fluorenylacetamide.

2-AAF. *N*-2-fluorenylacetamide.

AAL. ambient air level.

AAPFCO. Association of American Plant Food Control Officials.

ab. antibody.

Ab. antibody.

ab-. a prefix indicating from, away from, off, derived from.

ABA. abscisic acid.

abacterial. lacking or free from bacteria.

abalone. common name for mollusks of the genus *Haliotis*.

abamectin (ABM; avermectin B_1). an avermectin, *q.v.*

abate. **1:** to diminish, subside, or decrease, as in the case of environmental pollution or the symptoms of an illness or disease. **2a:** to halt, cease, be arrested. **2b:** to bring about a decrease. **2c:** to arrest or cause to cease.

Abate. temephos.

abatement. **1:** decrease, curtailment, reduction, subsidence, attenuation, cessation, or termination of a condition or the intensity of a process (e.g., symptoms of illness, astorm, environmental pollution). the process, methods, or means used to abate a condition or process. **3:** decrease in severity of pain or the severity of a disease or its symptoms. **4:** wrongful entry onto land. **5a:** reduction of a bequest due to a deficiency of assets. **5b:** voiding or reduction of a will or similar instrument. **noise abatement**. the reduction of noise in areas of human activity and habitation to levels that are not disruptive or harmful by technological or regulatory means. **pollution abatement**. the reduction or elimination of pollution by the use of technological or legal measures, or both. **smoke abatement**. the application of pollution abatement measures to the reduction or elimination of smoke in the environment.

abattoir. slaughterhouse.

abaxial. **1:** away from, facing away from, or furthest from the axis. **2a:** pertaining to the surface of a leaf or other lateral organ that is developmentally furthest from the apex of the axis (stem or stalk) that bears it. **2b:** the lower surface of a leaf or other lateral organ. *Cf.* adaxial.

abdomen. *See* belly, venter.

abdomin-, **abdomino-**. prefixes that denote a relationship to the abdomen.

abdominal. of or pertaining to the abdomen.

abdominal cavity. that part of the body cavity (coelom) of vertebrate animals in which the digestive organs are suspended. In most vertebrates, the abdominal cavity is continuous with the thoracic (chest) cavity. In mammals, however, the abdomen is separated from the thoracic cavity by a muscular diaphragm and is bordered caudally by the pelvis.

abdominal dropsy. ascites.

abdomino-. *See* abdomin-

Aberdeen Angus (angus**;** black angus**;** doddie). **1a:** an early-maturing breed of robust, usually black, hornless beef cattle that originated in Aberdeen and Angus counties, Scotland. **1b:** an animal of this breed.

aberrant. **1a:** abnormal; anomalous, deviant; unusual. **1b:** deviating significantly from that which is deemed typical or normal. **2a:** of or pertaining to an abnormality or anomaly. **2b:** pertaining to an abnormal subject or process or to an abnormal property of such (e.g., aberrant behavior). **3:** said of **a:** an ectopic anatomic structure (i.e., one that is not in the usual or expected site) or **b:** a structure (such as a duct or nerve) that does not follow a typical course.

aberration. **1a:** an abnormality or anomaly; irregularity, deviation, divergence. **1b:** oddity, peculiarity; eccentricity. **1c:** an exception to that which is the normal or expected situation or condition. **2a:** abnormal growth or development. **2b:** the result of abnormal growth or development. **chromosome aberration**. any irregularity or abnormality in the number of chromosomes or the structure of a chromosome.

abience. **1:** avoidance or the tendency to avoid or withdraw from the source of a stimulus.

abient. of or pertaining to abience. *Cf.* adient.

Abies (fir). a large genus of North Temperate Zone coniferous evergreen trees (Family Pinaceae). All are conical in aspect, bearing upright cones, flat leaves, and smooth circular leaf scars, which distinguish them from spruce (*Picea*). The branches are stiff and the foliage is aromatic. ***A. balsamea*** (balsam fir**;** Canada balsam**;** eastern fir). an evergreen; the only fir tree (Family Pinaceae) native to northeastern North America where it occurs mostly in coniferous forests, often inpure stands. The foliage, as browse, is important to deer and moose in winter. It is used in the manufacture of many wooden products and as Christmas trees. ***A. lasiocarpa*** (alpine fir**;** Arizona fir**;** mountain balsam**;** Rocky Mountain fir). an ornamental, pyramid-shaped tree of western North America.

abietene. a volatile oil distilled from the resin or balsam of the *Pinus sabiniana* and used in varnishes and lacquers.

abiological. not biological; not concerned with, involving, or produced by living organisms.

abiosis. 1a: the absence of life. **1b:** incompatibility with life. **2:** nonviability. **3:** abiotrophy.

abiotic. **1a:** nonliving, devoid of life. **1b:** not biological in origin. **1c:** pertaining to or characterized by the absence of living organisms or their products; not including living organisms. **2:** of or pertaining to non-living components of the environment (e.g., soil, groundwater, climate). **3a:** incompatible or antagonistic to life. **3b:** disposed to prevent, inhibit, or destroy life; incompatible with life.

abiotic disease (physiogenic disease). A disease, such as a mineral deficiency, that is not caused by a living organism or product of a living organism.

abiotic environment. **1a:** the nonliving environment. **1b:** the nonliving components, collectively, of an ecosystem; the spectrum of physical and chemical factors in any system that do not arise from biological activity. **2:** that part of the environment, in reference to an individual organism, taxon, or living system, that consists of nonbiological factors. *Cf.* biotic environment, abiotic, factor (abiotic).

abiotrophic. of or pertaining to abiotrophy.

abiotrophy. a loss of vigor or deterioration of an organism in the absence of apparent injury. Specific cells and tissues are usually affected.

abirritant. able to soothe or relieve irritation; an abirritant preparation. *See* abirritation.

abirritation. **1a:** weakness. **1b:** absence of tone or strength. **2:** decreased responsiveness to stimuli. *Cf.* asthenia; atony.

ablactate. to wean.

ablactation. the weaning of a child or other mammal.

ablate. **1:** to remove surgically or to deplete or disperse (e.g., by erosion, melting, vaporization). **2:** to be ablated; to undergo ablation.

ablation. **1a:** surgical removal of all or part of an organ. **1b:** removal of harmful materials from the body by mechanical means. **2:** erosion of rock surfaces, snow, or glacial ice by running or falling water, melting, sublimation or evaporation, or calving (breaking off) by running or falling water. **3:** all environmental processes whereby ice or snow erodes or dissipates.

abluent. **1:** cleansing or purifying. **2:** a cleansing or purifying agent such as a detergent.

ablution. a cleansing or washing; bath, bathing.

ABM. abamectin. *See* avermectin.

Abney level. a hand-held level used to determine elevations and angles of slope in percent or degrees.

abnormal. **1a:** aberrant, anomalous; deviant; irregular; deviating from normal, not normal, excessively above or below normal. **1b:** deviating from the accepted, the average, or the usual state, structure, custom, or prescription. **1c:** peculiar, strange, bizarre, unnatural. **2:** beyond the range of conditions considered to be wholesome or healthful. **3:** exceptional, rare.

abnormality. **1a:** the state or quality being abnormal. **1b:** a peculiarity, eccentricity, idiosyncrasy. **1c:** aberration; perversion. **2:** an anomaly; dysfunction, deformity, malformation, or handicap. *Cf.* aberration, abnormity.

abnormity. **1:** *archaic:* abnormality. **2:** deformity. **3:** a monstrosity, especially a grossly deformed neonate or juvenile animal.

abode tick. *Argas persinus.*

abomasal. of or pertaining to the abomasum of ruminants.

abomasitis. inflammation of the abomasum.

abomasum (rennet stomach). the fourth chamber (the "fourth stomach") of the compound stomach of ruminants. It is glandular and thus considered to be the true stomach. The other compartments are the rumen (paunch), reticulum, and omasum, *q.v.* *See also* abomasal disorders, Ruminantia.

aborad. aboral.

aboral (aborad). away from the mouth; in a direction away from the mouth. *Cf.* orad.

abort. **1a:** to end, falter, terminate, conclude, finish; especially to terminate an activity or process prior to completion. **2:** miscarry; to give birth to a dead or nonviable embryo or fetus. **3a:** to cause to miscarry by any means. **3b:** to terminate gestation at an early stage by removing the embryo or fetus from the uterus. **4:** to arrest a disease in its early stages. **5a:** to arrest the growth or development of a plant or animal. **5b:** to cause an animal or organ to remain rudimentary. **5c:** to inhibit the production of seeds or to induce green or immature fruit to drop.

aborticide. abortifacient (def. 2).

abortient. abortifacient (def. 1b).

abortifacient. **1a:** abortive. **1b:** abortient, abortigenic, able to induce abortion. **2:** an agent that induces abortion. Effective abortifacients are used to control reproduction in livestock include estrogens (e.g., in cattle) and prostaglandins (in cattle and horses).

abortigenic. abortifacient (def. 1b).

abortion. **1:** the act or process of aborting. *See also* miscarriage. **2a:** the termination of an activity or procedure before completion. **2b:** anything that fails to realize full development or ceases to progress prior to maturity. **2c:** arrest of a disease at an early stage. **1d:** arrest of the development of a conceptus. **3a:** death of an embryo or fetus in utero. **3b:** miscarriage; expulsion of a dead or nonviable embryo or fetus from the uterus. **3c:** the nonviable being resulting from an abortion. **3d:** premature expulsion from the uterus of an embryo or fetus. **3e:** in domestic animals, expulsion of the fetus any time prior to completion of pregnancy. **4:** the arrest of a disease in its early stages. **5a:** arrested growth or development of a plant or animal. **5b:** prevention of development of a plant, animal, or organ beyond a rudimentary stage. **5c:** inhibition of seed production in plants or premature induction of the dropping of fruit. **6:** brucellosis in domestic animals, especially bovine brucellosis, *q.v.* **7:** any of a number of contagious diseases of domestic animals marked by abortion. **accidental abortion**. that due to accidental injury or trauma. **artificial abortion**. induced abortion. **bovine infectious abortion**. bovine brucellosis. **chlamydial bovine abortion**. epidemic bovine abortion. **chlamydial ovine abortion**. endemic abortion of ewes. **complete abortion**. complete expulsion or extraction of an embryo, fetus, or any gestational product from the reproductive tract. **contagious abortion**. bovine brucellosis. **contagious bovine abortion**. bovine brucellosis. **enzootic abortion of ewes**. an infectious abortion of sheep caused by *Chlamydia psittaci*. **equine virus abortion.** a highly contagious disease of mares that induces abortion. It is caused by equine rhinopneumonitis virus. **foothill abortion of cattle**. epidemic bovine abortion. **incomplete abortion**. that in which the fetus is expelled but part of the placenta remains in the uterus. **incomplete spontaneous abortion**. a naturally occurring spontaneous abortion in which the fetus is expelled but part of the placenta remains in the uterus. **induced abortion** (artificial abortion; induced parturition). artificial termination of gestation at an early stage either by removing the embryo or fetus from the uterus or by use of an abortifacient. **infectious (infected) abortion,** an abortion complicated by infection. **missed abortion**. retention of a dead fetus in utero. **natural abortion**. spontaneous abortion. **necrogenic abortion**. in botany, a rapid necrosis of the tissues of a plant that are adjacent to the point of contact with a pathogen. This tends to deter expansion of the disease process. **spontaneous abortion**. **1a:** a naturally occurring abortion, one that has not been artificially induced. **1b:** a naturally occurring abortion that occurs during the early weeks of gestation (the first 12 weeks in humans). *See also* miscarriage. **vibrionic abortion**. abortion in cattle or sheep caused by *Campylobacter fetus vibrio*.

abortionist. a person who performs abortions, especially one who illegally induces abortions in humans.

abortive. **1a:** unproductive. **1b:** unsuccessful; terminated prematurely; failed. **2:** barren or unproductive of offspring. *See* abort; *Cf.* abortifacient.

abortus. the aborted products of conception, especially an aborted conceptus.

abrade (abrase). **1:** abrase; wear away by friction (e.g., scrape, erode, rub off). **2a:** scour; to scour a surface. **2b:** to scrape away the surface layers as in the removal of the superficial layers of skin or mucosa by scraping. **2:** to constantly irritate; to wear down physically or spiritually.

abrase. abrade.

abrasion. **1a:** a scraping or wearing away by friction of a portion of a surface as in the abrasion of a plowshare by soil particles. **1b:** The physical weathering of a rock surface by running water, glaciers, or windborne particles. **1c:** frictional wear of tools or machinery parts (e.g., abrasion of harrows by rocks and other soil components). **2:** a portion of skin or mucous membrane from which the surface layers have been removed by friction (e.g., by scraping). **3:** a superficial wound (e.g., of the skin) produced by a scrape or scraping.

abrasive. 1. causing abrasion. 2. any substance or material that abrades or is used to abrade.

abrasiveness. that quality of a substance that enables it to abrade or to causes abrasion.

abric acid. a substance in the seeds of *Abrus precatorius* that suppresses digestion. Thus individuals poisoned by abrin may have undigested food in the digestive tract. *See also* abrin.

abrin (toxalbumin; abrus agglutinin). one of

the most poisonous naturally occurring substances known. It is an intensely irritant, extremely toxic, proteinaceous phytotoxin that is inactivated by heating. It is the chief active principle of *Abrus precatorius*. Abrin is actually any of five nearly identical lectins (abrins a-d; mol. wt. 63,000 to 67,000 Da) composed of 2 disulfide-linked polypeptide chains. The A-chain inhibits ribosomal protein synthesis causing cell death; the B-chain binds to the plasma membrane of the cells of the intestinal wall, permitting entry of the A-chain into the cytoplasm. Initial symptoms of intoxication are those of intense gastrointestinal irritation such as vomiting and a bloody diarrhea that develops within 1-3 days following consumption of *Abrus* seeds. As a result, changes in electrolyte balance and plasma composition occur, causing cardiac arrhythmia and cerebral edema with drowsiness, convulsions, coma, and cardiovascular collapse. Further symptoms may include diffuse hemorrhages, hemolytic anemia, oliguria, and fatal uremia with death from heart failure. Digestive processes are inhibited such that autopsy may reveal undigested food. Ulcers occur in the mouth. LD_{50} *i.p.* in laboratory mice is reported at 0.02 mg/kg. Abrin is chemically similar to botulinum toxin. *See* abric acid, *Abrus precatorius*, phytotoxin.

abrin poisoning. abrism.

abrism (abrin poisoning). poisoning by abrin or by the abrin-containing seeds of *Abrus precatorius*. *See* abrin.

abrus agglutinin. abrin.

Abrus. a genus of leguminous plants that produce toxic lectins or phytotoxins. ***A. precatorius*** (Indian liquorice; Jequirity bean; precatory bean, rosary bead; rosary pea, crabseye, prayer bean, love bean, lucky bean; seminole bead; black-eyed Susan). the root of this species is sometimes used as a substitute for liquorice. The seeds can prove fatal if chewed.

abscess. **1:** an inflamed, often swollen, localized collection of pus that involves the displacement or disintegration of tissue with or without necrosis. It may be caused by physical injury as in snakebite, by chemical erosion of tissue (e.g., by bleach or strong acids), or by acute or chronic infection (e.g., a boil or carbuncle). **2:** a cavity formed within firm tissue by liquefactive necrosis.

abscise (absciss). to separate by abscission.

abscisic acid (abscissic acid; [S-(Z,E)]-5-(1-hydroxy-2,6,6-trimethyl-4-oxo-2-cyclohexen-1-yl)-3-methyl-2,4-pentadienoic acid; dormin; abscisin II; ABA). a natural plant growth regulator, $C_{15}H_{20}O_4$, with numerous, usually inhibitory, effects on the growth and development of many green plants. It promotes leaf senescence and abscission of fruit and leaves (fruit drop, leaf drop); regulates seed and bud dormancy, apical dominance, and suppresses flowering of long-day plants held on short days. Its secretion, stimulated by stress due to lack of water, causes the stomata to close during the day.

abscisin II. obsolete name for abscisic acid.

nescent leaves, mature fruit, or unfertilized flowers) from a plant. In this process, a layer of cells (the abscission layer, *q.v.*) that is part of an abscission zone, *q.v.*, at the base of the organ separate from the base of the organ. A protective layer remains. Abscission is controlled chiefly by the abscissic acid which stimulates and auxins which generally inhibit the process. Gibberellins may play a role.

abscission layer. (separation layer; absciss layer). a narrow, distinctive, transverse layer within an abscission zone of a green plant. It varies in thickness and is composed of cells that contain abundant starch and dense cytoplasm but are smaller and differ in shape from bordering cells. Dissolution of cell walls of these cells leaves only the vascular bundles which are ultimately unable to support the stem and abscission of the leaf, flower, or fruit together with some portion of the stalk or stem occurs.

abscission zone. the region at the base of a leaf, flower, stalk of a fruit, or portion of a stem in which an abscission layer develops. The vascular bundles within this zone are typically reduced in diameter; the sclerenchyma is weak or lacking; the collenchyma is absent; and the cytoplasm of some of the parenchyma cells is denser than that in neighboring cells.

absinth. absinthe.

absinthe (absinth). **1:** a commercially important volatile oil produced by *Artemisia*, *q.v.* **2:** a common name of *Artemisia absinthium*. **3:** a highly toxic alcoholic beverage that contains oil of wormwood, anise, and other herbs.

absinthium. *Artemisia absinthium*.

absolute age. *See* age.

absolute temperature scale. Kelvin temperature scale.

absorb. **1:** to take up or take in a substance, especially across membranes that bound tissues, organs, or cells. **2:** to interact with incident radiation such that the energy is attenuated. **3:** to immerse or engage completely, as the mind.

absorbance (A). **1:** the amount of incident light of a particular wavelength that is absorbed by a substance in solution. It is measured spectrophotometrically and can be used, e.g., to determine the concentration of a substance in solution. **2:** a measure of the capacity of a medium to absorb ionizing radiation. It is expressed as the logarithm of the ratio of the intensity of the radiation that enters the medium to that which is transmitted.

absorbate. *See* absorption.

absorbency. the penetration of one substance into another; the capacity to absorb. *See* absorbance.

absorbent. **1:** absorptive, porous, spongy, able to readily absorb or soak up. **2a:** a substance or material that can absorb other substances on contact; absorbefacient (def. 3). **2b:** a substance or material which, on contact with a liquid or gas, preferentially absorbs components with which it has an affinity and which is thereby altered chemically or physically.

absorbifacient. **1:** referring to the facilitation of absorption. **2:** causing or facilitating absorption. **3:** an agent that facilitates absorption.

absorption (uptake). **1a:** the penetration of one substance (the absorbate) into or through the inner structure of another (the absorbent). **1b:** the assimilation or incorporation of one substance into another by capillary action, chemical processes, osmosis, or solvation. **1c:** the transport of molecules across tissues or cell membranes into a living system or system component. **1d:** the process (active or passive) of taking in. **1e:** the uptake into a living system of water, solutes, or other substances by active or passive mechanisms. **2:** the attenuation of energy (e.g., light, sound waves) impinging on a substance or structure. Such diminution indicates that some portion of the energy striking the object is neither transmitted nor reflected. **3:** in the health sciences, the process of movement of a substance from the site of application or exposure to the systemic circulation. *Cf.* absorption of xenobiotics. **4:** the concentration of one's thought upon a single subject, object, or activity to the exclusion of others. **active absorption**. the movement of ions and water into an organism as a result of metabolic processes, against an electrochemical potential gradient. **cellular absorption**. the absorption of a substance through a cell membrane into the interior of a cell. Cellular absorption may occur via passive diffusion, filtration, carrier-mediated transport (active transport and facilitated diffusion), or by endocytosis (phagocytosis and pinocytosis). **cutaneous absorption** (percutaneous absorption). absorption through the skin. **dermal absorption**. the passage of a substance into the interior of a vertebrate animal via the skin. **external absorption**. absorption of substances by the skin and mucous membranes. **mouth (oral) absorption**. absorption through the oral mucosa. Certain xenobiotics, especially alkaloids, can be so absorbed. **parenteral absorption**. absorption other than from the gastrointestinal tract. **passive absorption**. the movement of materials into an organism by diffusion. **pathologic absorption**. the absorption of substances, such as the products of disease, into blood or lymph that would otherwise normally be excreted. *See also* route (route of absorption), absorption of xenobiotics. **percutaneous absorption**. cutaneous absorption. **route of absorption**. **1:** route of exposure. *See* exposure (route of exposure). **2:** occasionally taken to mean the "intestinal route" only. This usage is not recommended.

absorption spectrum. the specific differential pattern of light of different wavelengths absorbed by many organic substances. It is normally represented by a plot of absorbance versus wavelength. The curve, so displayed, is characteristic of a given substance.

absorption terrace. an embankment that slows the surface run-off of water, generally permitting increased absorption by the soil.

absorptive. pertaining to or capable of absorbing; absorbent (def. 1).

abstract of title. a written summary of the essential elements of a title to a particular piece of real property. These include evidence of ownership and any covenants or encumbrances that bear on ownership, use, or disposition of the property.

abundance. **1a:** plenty, profusion. **1b:** excess. **2:** wealth, prosperity. **3:** the total number of individuals of a specified population or type of organism (e.g., a species, race, breed, cultivar) per unit area or volume at a given location or within a specified community, ecosystem, or region. **absolute abundance**. the total number of individuals of a specified type in a designated area, volume, population, or community. **relative abundance**. the abundance of one specified type of organism in a given area, volume, population, or community relative to other specified types or categories of organisms within the same boundaries.

abundant. present in large amounts or in large numbers (plentiful; ample; bountiful; profuse).

abutment. **1:** an adjoining, connecting, or bordering structure (or part of a structure) that accommodates and resists pressure (an arch, beam, strut, or retaining wall). **2:** the point or locus at which abutment (def. 1) occurs. **3:** the act of abutting. *See also* retaining wall.

abutting. adjacent, adjoining, joining, bordering, touching, or connecting; said of pieces of land that adjoin or share a common boundary.

Ac. acetyl.

ACA. Agricultural Credit Association.

Acacia (acacia). a genus of some 50 species of warm-climate leguminous shrubs and trees (Family Fabaceae, formerly Leguminosae), several of which contain toxic or medically important principals. ***A. berlandieri*** (guajillo). a deep-rooted, often spiny, perennial shrub or small tree of the southwestern United States that contains a toxic amine, *N*-methyl-*beta*-phenylethylamine. Consumption of large amounts of the leaves or fruit over a period of months by sheep and goats produces ataxia ("limber leg," "guajillo wobbles") and sometimes death. ***A. catechu***. a source of catechin. ***A. georginae***. a tree of northern Australia, the leaves of which contain concentrations of fluoroacetate that are lethal to cattle and sheep. ***A. greggii*** (cat's claw, catclaw, catclaw acacia). a cyanogenic shrub. ***A. senegal***. an African tree. It is a source of gum arabic.

acacia. **1:** common name for plants of the genus *Acacia*. **2:** an alimentary demulcent used in syrups or gums to lubricate, coat, and protect irritated mucous membranes. As with other demulcents (e.g., egg albumin, gelatin, glycerol, propylene glycol, starch, tragacanth), acacia is used in veterinary practice as a vehicle for insoluble medications or to mask the taste of medications.

acanth-. a prefix meaning spine; spiny, or thorny.

acanthocarpous. of or pertaining to fruit covered with spines.

Acanthocephala (thorny-headed worms). a phylum of elongated cylindrical parasitic worms that have no mouth or digestive tract. They have a spiny, retractile proboscis by which they attach to the intestinal wall of the host. Adults infest the intestines of vertebrates, chiefly wild fish and amphibians. The larvae infest arthropods (the first intermediate hosts), especially insects and crustaceans, and wild fish (chiefly salmonids and marine fish). *See also* Aschelminthes.

acanthor. a larval acanthocephalan. *See* Acanthocephala.

Acanthoscelides obtectus (bean weevil). a pest of growing cowpea, kidney bean, and lima bean; it also attacks most dry, stored beans, peas, cowpeas, and lentils. It is a stout, sturdy beetle, the adult of which feeds on growing beans until the pods are formed. It then lays its eggs in the seed. The larvae burrow into and consume dried beans.

acapnia. a state of reduced carbon dioxide in the circulating blood.

Acari. Acarina.

acarian. acarine.

acariasis (acaridiasis; acarinosis). any disease of plants or animals caused by ticks or mites.

acaricidal. lethal to mites.

acaricide (miticide). **1:** a substance that kills mites. **2:** a substance that is specifically lethal to phytophagous mites. **3:** a pesticide that may be applied, usually as a paste or solution, to domestic animals or humans to kill ectoparasitic mites. **4:** an agent lethal to mites and ticks (acarines).

acarid (acaridan; acarus). **1:** common name of any mite or tick of the family Acari for any member of the order Acarina.

Acarida. Acarina.

Acaridae. a large family of very small mites (Order Acarina) that are rarely more than 0.5 mm in length. They feed by piercing the skin or outer covering of plants or animals. They often infest grain, meal, flour, dried fruits, and meats. Handling of infested foods may result in a severe dermatitis.

acaridan. acarid.

acaridiasis. acariasis.

Acarina (Acari; Acarida). a large order of arachnids that includes mites and ticks; some are microscopic. The body is typically round (sometimes oval, occasionally wormlike), compact, and the abdomen is fused with the cephalothorax, concealing the segmentation of the body. The head is moveable by virtue of its attachment to the body by a hinge. There are four pairs of legs. *See also* mite, tick, Argasidae, Ixodidae.

acarine (acarian). **1:** of, pertaining to, or characteristic of mites and ticks. **2:** of or pertaining to any arachnid of the order Acarina, *q.v.* **3:** a common name of any arachnid of the order Acarina.

acarine disease. **1:** any disease caused by mites or ticks. **2:** a disease of adult bees caused by infestation of the trachea. **3:** Isle of Wight disease.

acarinosis. acariasis.

acarology. the science that treats mites and ticks, especially those that infest living organisms or transmit disease.

acarotoxic. poisonous to mites and ticks.

acarus. acarid.

Acarus. a genus of mites (Family Acaridae). ***A. rhizoglypticus hyacinthi***. an ectoparasite that develops in rotting onions and may cause dermatitis.

acaudal, acaudate. lacking a tail.

acaul-. prefix meaning stemless, lacking stems.

acauline. lacking a stem.

acaulis. lacking a stem.

accelerant. **1:** something that speeds up a process. **2:** chemical accelerator; any chemical used as an accelerant (def. 3). **3a:** a substance that increases the rate of a chemical process (e.g., combustion, the vulcanization of rubber). **3b:** in forestry and in fire prevention and investigation, any material used to initiate or accelerate the spread of a fire or to make a fire more intense. The most common accelerants are flammable liquids. Arsonists often use accelerants in setting fire to a building. **chemical accelerant**. chemical accelerator.

accelerator. **1:** anything (e.g., a substance or a person) that hastens, speeds up, or expedites an action or process (e.g., a chemical reaction, absorption, muscular movement, a behavioral response). **2:** a mechanism (e.g., in an automobile or tractor), usually operated by a foot, that controls the speed of an engine. **3:** accelerant (def. 3). **4:** any substance, muscle, or nerve that speeds up a bodily movement. **5:** a chemical, usually alkaline, that is added to a developer to speed the development of photographic film. **6:** an electrostatic or electromagnetic device such as a cyclotron (particle accelerator, atom smasher) that produces high-energy particles and focuses them on a target. **chemical accelerator** (chemical accellerant). any chemical that increases the rate of a chemical process.

acceptable daily intake (ADI; allowable human daily intake). the amount of a substance (usually a pesticide, drug, or food additive) that can be indefinitely consumed daily without significant lifetime risk to health. It is usually expressed as mg/kg body weight/day.

accessory. **1a:** complementary. **1b:** additional; supplementary. **1c:** accompanying; ancillary, **1d:** secondary.

accessory fruit. *See* fruit.

accessory gland. any small glandular mass

that is similar to and lies near a larger gland.

accident. **1:** an unexpected occurrence, incident, or injury. **2:** an unplanned, unforeseen occurrence, especially during the course of a routine process, activity, or occurrence. **3:** an unexplained result, especially one considered to be due to negligence. **4:** a mishap; misfortune; misadventure. **5:** a harmful event (e.g., injury, damage, or death) resulting from negligence or unexplained causes.

acclimatation. acclimatization.

acclimate. **1:** acclimatize. **2a:** inure, accustom, harden. **2b:** adapt.

acclimation. **1:** a normal phenotypic adjustment of an organism to environmental change, including behavioral, physiological, and morphological changes. **2a:** a gradual and reversible physiological or morphological adjustment of physiology or morphology in response to changing environmental conditions. **2b:** habituation or increased tolerance of an organism to an altered environment. **3:** acclimatization. *Cf.* acclimatization, accommodation, adaptation, adjustment.

acclimatization (acclimatation). **1a:** the process or act of becoming acclimatized. **1b:** the result of acclimatizing. **2:** adaptation; increased tolerance of a species to a pronounced environmental change over several generations. **3:** acclimation; physiological and/or behavioral adjustment of an organism to a new or changing climate. *Cf.* acclimation, accommodation, adaptation, adjustment.

acclimatize (acclimate; harden; inure). **1:** to adapt to a new environment. **2:** to develop increased tolerance to new environmental conditions. **3:** habituate; harden, inure, or toughen as in the hardening-off of plants.

acclivity. **1:** a grade, incline, or slope. **2:** an ascending slope from a point of reference (e.g., from the base of hill). *Cf.* declivity.

acclivous. having a gentle upward slope.

accommodate. to fit or adjust; adapt; conform.

accommodation. **1:** the process, state, or condition of fitting, adjusting, adapting, or conforming. **2:** the capacity of an organism to adapt to changes in the environment. *Cf.* acclimation, acclimatization. **3:** a decrease in the threshold of response to a stimulus with repetitive stimulation. **4:** the adjustment of the curvature of the lens of the eye that permits one to focus the image of an object on the retina at varying distances; it is especially important in enabling close vision. *Cf.* acclimation, acclimatization, adaptation, adjustment, habituation.

accredit. **1:** to endorse or approve. **1b:** to certify, license, or otherwise recognize officially.

accredited. **1:** approved, endorsed, sanctioned. **2:** officially recognized as meeting the essential requirements, as of academic excellence (e.g., an accredited school). **3:** certfied, licensed, authorized, commissioned, or otherwise provided with credentials. **4:** accepted as authoritative or official.

accredited herd. **1:** a herd of dairy cattle certified as being free from tuberculosis, on the basis of tests administered under the direction of the U.S.D.A. **2:** in England, an accredited herd is one that is certified as free from brucellosis.

accrescence. **1:** continuous growth. **2:** accretion.

accrescent. growing continuously; increasing in size with age. Used mainly in reference to plants that continue growth (especially growth of the calyx) after flowering.

accretion. **1a:** accumulation, accrual, or buildup. **1b:** continuous or discontinuous growth (increase in size) by external accumulation of material. **1b:** infrequently used in reference to organic growth: growth and continued development by incorporating (with or without modification) from within. **2a:** deposition of material by sedimentation. **2b:** the gradual build up of new land by aggradation, *q.v.* **3:** recovery or acquisition of flooded land by use of barriers such as dikes or dams and by draining. **3:** accrual of land by an owner through natural processes such as aggradation. **4:** adherence of separate particles. **5:** the result of a process of accretion. **6:** in forestry: incremental growth, i.e., increase in height or diameter as opposed to increase in volume.

accretion borer. a hollow auger used to remove a core of wood from a tree in order to estimate accretion by counting and measuring the width of annual rings.

accretion cutting (accretion thinning). the thinning of trees in a stand or forest to obtain greater girth in those trees that remain.

accretion thinning. accretion cutting.

accretionary. characterized by or produced by accretion.

accretive. of or pertaining to accretion or to growth by accretion.

accrual. **1:** the act or process of accruing, gathering, accumulating, or amassing. **2:** storing or hoarding. **3:** something that accrues (e.g., money), especially something that accrues or has accrued during a specified period.

accrual basis (accrual method). an accounting method that records income when earned and expenses when incurred, regardless of the date of receipt or dispensing of funds. This method is often used to keep farm accounts.

accrual method. accrual basis.

accrue. **1a:** to gather, amass, or accumulate. **1b:** to store, or hoard.

accumulation. **1:** the state or process of accumulating, amassing, concentrating, or collecting something. **2:** the state or condition in which the concentration of a substance within a given system (e.g., a cell) is greater than that in the immediate environment. **3:** in agronomy, the increase of one or more constituents in the soil at a given location due to translocation. Such accumulation can be a residual due to translocation of material out of the site or horizon or can result from imported material. Clay particles and water-soluble substances most frequently accumulate.

accumulator (concentrator). **1:** anything that accumulates. **2a:** accumulator organism, *q.v.*, a living organism that actively concentrates, within its tissues, a particular element or compound to levels above that in its food or environment. Mechanisms vary, but accumulation is often due to an inability of the organism to translocate or further metabolize a particular substance once it has been absorbed. **2b:** accumulator plant, *q.v.* **3:** any device that stores energy, such as a cylinder that contains water under pressure as a source of power. **facultative accumulator**. an organism that acts as an accumulator of a specific environmental chemical under certain circumstances. **obligate accumulator**. an organism that selectively accumulates a specific environmental chemical or class of compounds.

accumulator organism. a living organism that actively concentrates, within its tissues, a particular element or compound to levels above that in its food or environment. Mechanisms vary, but accumulation is often due to an inability of the organism to translocate or further metabolize a particular substance once it has been absorbed. *See also* plant (accumulatory plant), accumulator.

accumulator plant. *See* plant (accumulator plant), accumulator.

accuracy. **1:** freedom from error. **2a:** ideally, the degree of agreement between a calculated or measured value and the true value of a variable. **2b:** in practice, the degree of agreement between a measured value and a value of an accepted (certified) standard. This is usually necessary and accepted because the "true" value remains unknown. *Cf.* precision.

accuracy of selection. the degree of agreement between an animal's true breeding value and a calculated or estimated value.

accurate. **1a:** free from error. **1b:** quantitatively equivalent to or in conformity with a standard, reference, or accepted true value. *Cf.* precise.

-ace. a suffix that means partaking of.

ACE. Agriculture in Concert with the Environment.

acentric. **1:** not centered; lacking a center. **2:** of or pertaining to a chromosome or chromatid that lacks a centromere. *Cf.* acrocentric, metacentric, telocentric.

acephal-. a prefix indicating headlessness, lack of a head.

acephalous. having no head.

acephate (acetylphosphoramidothioic acid *O,S*-dimethyl ester; Ortho 12420; Orthene). a white, water soluble, crystalline solid, $C_4H_{10}NO_3PS$, with a melting point of 72 to 80°C, used as an insecticide against aphids and other foliage pests. It is moderately toxic to humans and laboratory animals by ingestion and is carcinogenic to some laboratory animals. Signs of acute oral intoxication of mallards include ataxia, imbalance, jerkiness, hopping and falling, mild spasms in the legs and feet, immobility, spread wings, and intermittent tremors.

acephatemet. a white, crystalline insecticide, $_3OCH_3SPONH_2$, that is slightly water-soluble and melts at 39 to 41°C. It is used to control cutworms and borers.

acer-. a prefix referring to the genus *Acer*.

Acer (maple). a genus of attractive, long-lived, deciduous trees and shrubs, commonly called maples (Family Aceraceae). Leaves are opposite, often lobed or notched; the flowers are small and born in clusters that produce two-winged fruits termed keys or samaras. Most species are grown from seed; some from cuttings. Members of the genus have a wide variety of uses, e.g., for lumber, as ornamentals or shade trees, for the production of maple syrup and maple cream. ***A. negundo*** (box elder; ashleaf maple; Manitoba maple). a fast-growing, small to medium-sized shade tree with a short trunk, light gray-brown bark, and a rounded crown of light green foliage. The branches are brittle with green twigs and easily broken in storms. Although short-lived, it is used in shelterbelts; the wood is used for paper pulp and the manufacture of woodenware. In the eastern United States, it is often considered a weed tree. ***A. nigrum*** (black maple; black sugar maple; hard maple; rock maple). one of the American sugar maples, it is a large, hardy tree with a rounded, dense crown. It attains a height of about 24 m (80 ft). The sap is sweet and it is often tapped for maple syrup. It is used as a shade tree, and because the wood is both strong and beautiful, it has many other uses including the manufacture of railroad ties, bowling pins, shoe lasts, furniture and veneer, and flooring. ***A. rubrum*** (red maple; scarlet maple; swamp maple). a large tree with a narrow crown; the flowers, fruit, leaf stalks, and autumn foliage are red; the bark is thin, gray, and smooth; twigs are reddish. Its natural range covers most of the eastern United States It is an attractive shade tree. ***A. saccharum***. a large tree with a rounded dense crown; the bark is light gray, becoming rough and deeply furrowed with age. The sugar maple is valuable as a shade tree, for timber, and for its sap which is used to make maple syrup.

acerb-. a prefix meaning harsh, bitter, stinging.

aceroid. Maple-like.

aceros-. needle-like.

acervuli. type of spore-bearing fruiting body.

acet-, aceto-. **1:** prefixes denoting or pertaining to acid. **2:** prefixes that denote the two-carbon skeleton of acetic acid or any moiety of a chemical comprised of such a fragment.

acetabulum. the socket in the pelvic girdle of a tetrapod that receives the ball-shaped head of the femur, forming the hip joint.

2-acetamidofluorene. *N*-2-fluorenylacetamide.

L-α-acetamido-β-mercaptopropionic acid. acetylcysteine.

***p*-acetamidophenol**. acetaminophen.

2-acetaminofluorene. *N*-2-fluorenylacetamide.

acetaminophen (APAP; *N*-(4-hydroxyphenyl) acetamide; 4′-hydroxyacetanilide; *p*-hydroxy-acetanilide; *p*-acetamidophenol; *p*-acetaminophenol; *p*-acetylaminophenol; *N*-acetyl-*p*-aminophenol; paracetamol). a white, odorless, crystalline powder,

$$CH_3CONHC_6H_4OH$$

It is used as a nonprescription medication to help control pain and fever. Its use is similar that of aspirin, but it does not reduce inflammation or swelling, and is less irritating to the stomach than aspirin. It generally does not relieve visceral pain in livestock and other mammals, although it is helpful in controlling some colic attacks for a brief period of time.

***p*-acetaminophenol**. acetaminophen.

acetate replacing factor. thioctic acid.

(acetato)phenylmercury. phenylmercuric acetate.

acetic acid (ethanoic acid; vinegar acid; methane carboxylic acid). a weak, caustic, organic acid, CH_3COOH, with a pungent odor. It is the essential ingredient (4-6%) in vinegar. Reagent grade contains at least 36% acetic acid by weight in aqueous solution. Acetic acid, produced by lactic acid fermentation, occurs in milk and many dairy products, adding to their flavor. Acetic acid is a source of carbon in certain algae such as *Myrmicaria natalensis* and *Chlamydomonas*. It is an active component in the venom of certain ants and the defensive fluid of arachnids such as *Mastigoproctus giganteus*. This acid has numerous research, domestic, and agricultural uses. Acetic acid (4-5%) is the most frequently-used ruminal acidifying agent in the treatment of ruminal stasis, simple indigestion, or acute urea poisoning of ruminants. **glacial acetic acid**. pure acetic acid (at least 99.8%); it is a corrosive, extremely toxic, vesicant acid that is produced by destructive distillation of wood or by the reaction of acetylene and water. Ingestion of small amounts can severely erode the lining of the gastrointestinal tract. Acute exposure in humans and many animals may cause vomiting, diarrhea, bloody feces and urine, and death from cardiovascular failure. Chronic exposure can cause eye irritation, bronchitis, and eroded tooth enamel.

acetic fermentation. the enzymatic production of acetic acid from weak aqueous solutions of ethanol by bacterial enzymes as when vinegar forms from hard cider or wine on sitting.

acetoaminofluorene. *N*-2-fluorenylacetamide.

Acetobacter (vinegar bacteria). a genus of bacteria (Family Nitrobacteriaceae) that grow in beer and wine.

acetogenic bacterium. a bacterium that uses carbonate as a terminal electron acceptor and releases acetic acid as a waste product.

acetonaemia. acetonemia.

acetone (2-propanone; dimethylketone; dimethylketal; methylacetal; β-keto-propane; pyroacetic ether; ketone propane; **propanone; pyroacetic ether)**. the simplest ketone, acetone is a colorless, sweet- smelling, volatile, highly flammable liquid ketone, CH_3COCH_3, used as a solvent. It is miscible with water, ethanol, and most oils. Acetone has many uses in industry, commerce, and research. In agriculture it is used chiefly as a solvent for paint, varnish, and lacquer and cleaning of dry metal machine parts. It irritates eyes and tissues of the respiratory tract. It can damage the skin by dissolving fats in the skin. Inhalation by humans can cause headaches, excitement, bronchial irritation, faintness, and fatigue. Exposure to large amounts can cause unconsciousness. Acetone on the breath of a lactating mammal such as a cow, indicates a failure to oxidize fatty material in the feed, rendering the female deficient in deficient in carbohydrates.

acetonemia (acetonaemia). **1:** relatively high concentrations of ketone bodies in circulating blood, body fluids, and tissues. a special case of hyperketonemia, *q.v.* The condition in humans is initially associated with excitement followed by increasing depression. *See also* ketone body, ketonemia ketosis. **2:** partially synonymous with ketosis, *q.v.* **bovine acetonemia**. bovine ketosis. **ovine acetonemia**. ovine ketosis.

acetonemic (acetonaemic). pertaining to, characterized by, or caused by acetonemia.

3-(α-acetonylbenzyl)-4-hydroxycoumarin. warfarin.

3-(α-acetonylfurfuryl)-4-hydroxycoumarin. coumafuryl.

***o*-acetoxybenzoic acid**. acetylsalicylic acid.

(acetato)trimetaarsenitodicopper. cupric acetoarsenite.

acetoxyphenylmercury. phenylmercuric acetate.

acetyl (Ac; acetyl radical). The radical, $CH_3CO—$; it can be pictured as an acetic acid molecule with the hydroxyl group removed.

acetyl CoA. acetyl coenzyme a.

acetyl coenzyme A (acetyl CoA). an acetyl thioester of coenzyme A produced during the breakdown of carbohydrates as in the tricarboxylic acid cycle.

acetyl radical. acetyl.

acetylacetone (diacetylmethane; 2,4-pentanedione). a mobile, colorless or yellow, sweet-smelling, flammable liquid solvent,

$$CH_3COCH_2OCCH_2$$

This liquid is soluble in slightly acidic

water. The liquid turns brown and forms resins on exposure to light. It is used as a pesticide, fungicide, a paint and varnish drier, and an additive to lubricants. It is mildly irritant to the skin of higher vertebrates.

2-acetylaminofluorene. *N*-2-fluorenylacetamide.

***N*-acetyl-*p*-aminophenol.** acetaminophen.

***p*-acetylaminophenol.** acetaminophen.

acetylation. the introduction of one or more acetyl radicals (CH_3CO) into an organic molecule that contains one or more OH or NH_2 groups. Metabolic acetylations often render poisonous compounds less toxic. In systems where acetylation predominates over deacetylation as in domestic rabbits, there is usually a net excretion of the acetylated substances. Rabbits thus excrete significant amounts of acetylated amines.

acetylcholine (ACh; acetylethanoltrimethylammonium hydroxide). an acetyl ester,

$$(CH_3)_3N(CH_2)_2COOCH_3$$

It is a reversible product of choline and the normal substrate of acetylcholinesterase. It is a neurotransmitter that acts across synapses between adjacent nerve endings in the PNS and CNS, and also across neuromuscular junctions invertebrates and in many invertebrate animals. ACh is released from the nerve terminal in response to a neuronal action potential, transmits the stimulus across the junctional gap to the adjacent neuron or muscle cell, and then rapidly decomposes in a hydrolytic reaction catalyzed by acetylcholinesterase, *q.v.*

acetylcholinesterase inhibitor. anticholinesterase.

acetylcholinesterase (AChE; acetylhydrolase; specific cholinesterase; true cholinesterase; acetylcholine acetylhydrolase). **1:** a carboxylic esterase that is widely distributed throughout the vertebrate nervous system and in many non-nervous tissues such as muscle and erythrocytes. It specifically and reversibly inactivates the neurotransmitter, acetylcholine, by catalyzing its conversion to choline plus acetate in the body. AChE also occurs in many venoms of elapids (snakes of the family Elapidae) and those of the Hydrophidae (sea snakes). It is absent in venoms of the Viperidae and Crotalidae. The toxicity of certain classes of pesticides (e.g., carbamates, organophosphates) is due to inhibition of AChE. **2:** pseudocholinesterase.

acetylcysteine (*N*-acetyl-L-cysteine; L-α-acetamido-β-mercaptopropionic acid; *N*-acetyl-3-mercaptoalanine). an organic chemical used as a mucolytic and in the treatment of acetaminophen poisoning. As a rule, a 10 or 20% solution is administered to domestic animals as an aerosol or by intratracheal instillation to help clear airways of mucoid or purulent airway secretions.

***N*-acetyl-L-cysteine.** acetylcysteine.

acetylene (ethyne). a colorless, highly flammable, asphyxiant gas with an ethereal odor. It is slightly soluble in water, but readily dissolves in ethanol and acetone. It has many uses, e.g., in miner's lamps, the manufacture of acetic acid (*q.v.*), and certain tests such as the acetylene-block assay and the acetylene-reduction assay.

acetylene-block assay. an estimate of denitrification by measuring the amount of nitrous oxide released from acetylene-treated soil.

acetylene-reduction assay. an estimate of nitrogenase activity by measuring the rate of acetylene reduced to ethylene.

acetylethanoltrimethylammonium hydroxide. acetylcholine.

acetylformic acid. pyruvic acid.

acetylhydrolase. acetylcholinesterase.

***N*-acetyl-3-mercaptoalanine**. acetylcysteine.

acetylphosphoramidothioic acid *O,S*-dimethyl ester. acephate.

acetylsalicylic acid (HAsc; ASA; A.S.A.; *o*-acetoxybenzoic acid). the acidic form of aspirin,

$$CH_3COOCC_6H_4COOH.$$

At a pH substantially below the pKa of 3.2 (as in the stomach of most mammals), most of this acid remains in a neutral form and is thus rapidly absorbed. *See* aspirin.

ACH. **1:** air changes per hour. **2:** acetyl choline (ACh is preferred).

achalasia. the inability of a circular muscle (e.g., that of the esophagus or rectum) to relax. This may exert pressure on the muscle, causing an expansion of the lumen above the contracted muscle.

AChE, ACHE. acetylcholinesterase.

achene (akene). a small, dry, indehiscent one-seeded fruit that develops from a simple ovary. The pericarp is a thin wall, attached to the seed at only one point, and does not split open when the seed ripens.

achlorhydria the absence of hydrochloric acid in the gastric juice.

achlorophyllous. lacking chlorophyll.

Acholeplasma. A genus of minute nonmotile bacteria with a flexible cell wall (Order Mycoplasmatales). Saprophytic and parasitic species are known. **A. granularum**. a species that occurs in swine; it is not known to be pathogenic. **A. laidlawii**. a saprophyte that occurs in humus, soil, manure, and sewage. It is the type species of the genus. *Cf. Mycoplasma*.

achondroplasia (achondroplasty; osteosclerosis congenita; fetal rickets in humans). a heritable malformation the skeleton of many animals including humans, cattle, and several breeds of domestic fowl. Cartilage fails to convert normally to bone during development, resulting in dwarfism which is observable at birth. The epiphyses of long bones are chiefly affected and the extremities are thus short. The head may be enlarged and the nose or muzzle flattened. **avian achondroplasia**. a condition known to occur in several breeds of domestic fowl due to the presence of a dominant autosomal gene. **bovine achondroplasia** (bull-dog calf). the newborn calf has the typical facies of achondroplasia with dwarfing and foreshortening of the legs. The head is short and thick, with a jutting lower jaw, a split lip, and a swollen tongue. The fetus is occasionally born live; if so, it usually dies within a few days. **homozygous achondroplasia**. a severe, usually fatal, achondroplasia that affects the offspring of achondroplastic parents.

achondroplastic. of, pertaining to, or characterized by chondroplasia.

achondroplasty. achondroplasia.

Achroia grisella (lesser wax moth). a bee moth (Family Pyralidae, Subfamily Galleriinae). It is a smaller species but with the same type of egg-laying and scavenging behavior as the greater wax moth, *Galleria mellonella*.

achromatic. **1:** achromic; colorless. **2:** able to reflect, transmit, or emit light without separating it into colors. **3:** descriptive of cells or tissues that are difficult to stain.

achromic. colorless.

acicul-. a prefix referring to a needle or needles.

acicular (aciform; aciculate). **1a:** needle-shaped, needle-like. **1b:** having a sharp, needle-like tip. **2:** aciculate.

acicular leaf. a needle-shaped leaf, as the leaf of a pine tree.

aciculate. acicular.

aciculifruticeta. needle-leaved plant.

aciculilignosa. narrow-leaved evergreen and deciduous forest vegetation.

aciculisilva. needle-leaved forest.

acid. **1:** any substance that releases hydrogen ions (H^+; protons) or hydroxonium ions (H_3O^+) in aqueous solution. The pH is thus below 7.0. **2:** a substance that tends to release a proton (a Lowry-Brønsted acid; Brønsted acid). **3:** a substance (e.g., $AlCl_3$, BF_3, SO_3) that accepts an electron pair from a base (a Lewis acid). **4:** a substance that reacts with a base to form a salt. **5:** any substance that dissolves metals with the release of hydrogen gas. **6:** vernacular usage: **a:** any substance that has a sour taste. **b:** LSD. **c:** sour tasting. **7:** said of igneous rocks that have a higher percentage of silica than orthoclase. 8: acidic. note: the major groups of acids include mineral (inorganic) acids and several classes of organic acids: carboxylic acids, dicarboxylic acids, fatty acids (aliphatic and aromatic forms), and amino acids (*See* below). *See also* entries for specific acids. abietinic acid. a crystalline acid $C_{19}H_{28}O_2$, from colophony rosin. There are three isomeric forms. It can stimulate growth of lactic and butyric bacteria.

acid anhydride. *See* anhydride.

acid board. a metal sheet in a wooden frame that is placed over the top of a beehive when the keeper tends it. The top of the board is painted black and the underside is covered by cloth saturated with carbolic acid. Placed on the hive, the heavy acrid fumes force the bees to leave.

acid deposition. acidic air pollutants (chiefly nitrogen oxides and sulfur oxides) deposited in the biogeosphere primarily as nitric acid or sulfuric acid. These materials may be deposited as dry particulates or acidic gases (**dry deposition**) or in rain, snow, or fog (wet deposition or acid precipitation). *Cf.* acid precipitation.

acid detergent fiber (ADF). the quantity of fiber extracted from forages by use of an acidic detergent as a means of appraising the quality of the forage for livestock. The extract includes cellulose, lignin, ADIN, and acid-insoluble ash.

acid detergent insoluble nitrogen (ADIN). dietary protein or nitrogen that is undigestible by livestock.

acid-fast. resistant to decolorization by acids; used most commonly as a descriptor of dyes and histological stains.

acid gland. **1:** oxyntic gland; a gastric gland that secretes the hydrochloric acid component of the gastric juice. **2:** any of a complex of honeybee poison glands. **3:** one of the lateral glands of the body segments of a millipede. It secretes hydrogen cyanide and/or other irritant substances.

acid mine drainage. leachate from pyritic coal mines and spoils that is often acid due to the presence of acidic sulfates (chiefly ferrous sulfate) and to bacterial and chemical oxidation of reduced iron to ferric hydroxide precipitates. Such drainage can kill or render fishes and other aquatic organisms infertile. It can also introduce heavy metals into the water supply and otherwise lower the quality of drinking water. *See also* acid mine water.

acid mine water. mine water that contains free sulfuric acid due to the weathering and leaching of iron pyrites. *See also* acid mine drainage.

acid poisoning. a serious, life-threatening type of corrosive poisoning by ingestion of a strong acid. The victim experiences immediate, severe pain with destruction of the lining of the mouth and upper gastrointestinal tract. This may be accompanied by pain in the affected areas, nausea, vomiting (the vomit is the color of coffee grounds), thirst, shock, circulatory collapse, and death. Rapid death by asphyxia may occur due to edema of the glottis. *See also* alkali poisoning.

acid precipitation. rainfall, snowfall, and other forms of precipitation that contain high concentrations of nitric or sulfuric acids. It results mainly from the reaction of atmospheric sulfur oxides and nitrogen oxides with moisture. These gases arise mainly from the combustion of fossil fuels and often disperse over long distances, producing acidic precipitation that may damage a whole ecosystem in regions that are remote from the source. Water bodies with poor buffering capacity may become acidified to the point where fish and other aquatic biota are killed or population growth is impaired. The acidic waters may release toxic metals (especially aluminum) from bottom sediments, adding significantly to the toxicity of the water. Metal ions may also leach from the soils into affected water bodies. Acid precipitation may also damage terrestrial vegetation by direct effects and indirectly, for example, by leaching nutrients from the soil. *Cf.* acid rain, acid deposition.

acid rain. rainfall with a pH of about 4 or less. It is a form of acid precipitation, *q.v.* Acid rain and other forms of acid precipitation are formed from atmospheric sulfur oxides and nitrogen oxides that are released into the atmosphere chiefly from fossil fuel-burning systems, most importantly coal-fired power plants, metal-processing factories, and motor vehicles. These gases undergo complex reactions in the atmosphere and combine with moisture to form sulfuric or nitric acids. Acid rain is especially injurious to aquatic life, but can damage entire ecosystems (e.g., lakes, forests). *Cf.* acid precipitation, acid deposition.

acid-base balance. **1:** the ratio of acids to bases in a solution. **2:** the normal ratio of acids to bases maintained by a healthy organism with body fluids of normal pH. This ratio is usually maintained within a very narrow range. Disturbances of acid-base balance can have extremely serious impacts on the affected organism. **3:** the maintenance of the acid:base ratio in blood and other body fluids.

acid-base metabolism. the metabolic processes that maintain acids and bases at normal levels within a living organism. *See* acid-base balance.

acidic. **1a:** of or pertaining to an acid. **1b:** **1b:** acid-forming. **1c:** having the properties of an acid. **3:** pertaining to a substance, material, or system (e.g., a habitat) having a pH below 7. *Cf.* alkaline.

acidic amino acid. *See* amino acid.

acidic environment. an environment of low pH and usually poor nutrient content. (e.g., a cranberry bog).

acidic habitat. *See* acidic environment.

acidification. the process of rendering a material acid; the process of acidifying. **soil**

acidification. a process of acidification that occurs slowly during soil formation under natural conditions. This process can be accelerated, for example, especially in poorly buffered soils, by acid precipitation; by the addition of $NH_4{}^+$-forming nitrogen fertilizers; and by the addition of sulfur and reduced sulfur compounds to the soil.

acidify. **1a:** to make acid; to render a material acidic; to convert into an acid. **1b:** to make more acidic. **2:** to become acid. **3:** to make or become sour.

acidism (acidismus). poisoning by exogenous acids. *See* acid poisoning.

acidismus. acidism.

acidity. **1:** the property, state, or degree of being acid or of being sour in taste. **2:** the acid content of a solution or the extent to which it is acid. **3:** the concentration of any chemical species that is titratable by strong base in solution. **4:** an excessively acid quality. **5:** sourness or tartness. **soil acidity**. The hydrogen ion concentration in the soil solution expressed as pH.

acidobiontic. living in an acidic habitat.

acidophil. acidophile.

acidophile (acidophil). **1:** acidophilic. **2a:** an acidophilic substance or tissue, one that tolerates acidic environments. **2b:** an organism that lives successfully in acidic habitats and environments (e.g., cranberries).

acidophilic (acidophilous; oxyphilous). **1:** acid tolerant; tolerant of or thriving in acidic habitats or environments. **2:** descriptive of materials (e.g., cells or tissues) that stain readily with acid stains. *See also* aciduric.

acidophilous. **1:** acidophilic. **2:** descriptive of organisms (e.g., bacteria and green plants, that thrive in an acid soils or other media.

acidophilus. acidophilous (def. 2).

acidophily. the state or condition of being acidophilic.

acidophobe. an organism that is intolerant of acidic environments.

acidophobic. intolerant of acidic environments.

acidophoby. the state or condition of being acidophobic.

acidoses. plural of acidosis.

acidosic. acidotic.

acidosis (plural, acidoses; oxyosis). **1a:** a serious disturbance of the acid-base balance, *q.v.*, within a living organism due to an accumulation of acids in body fluids or to a decrease in alkali reserves. **1b:** a serious increase in the hydrogen ion concentration of blood. This is defined in humans as a reduction of the pH of arterial blood to less than 7.35. In most vertebrate animals, acidosis is a consequence of diabetes, starvation, impairment of liver function, respiratory dysfunction, or failure of the kidneys to efficiently use bicarbonate or excrete hydrogen. In cattle, acidosis may result from excessive intake of concentrate feed. The rumen wall becomes inflamed; the animal becomes progressively dehydrated; death may result in extreme cases. Acidosis is a common cause of death of diarrheic calves. **metabolic acidosis**. acidosis caused by excess acid in, or loss of bicarbonate from, body fluids. It often results from shock or heart failure. **renal tubular acidosis**. that due to failure of the kidney to either eliminate hydrogen or to effectively use bicarbonate. **respiratory acidosis**. that due to a low breathing rate (hypoventilation) resulting in a high concentration of carbon dioxide in the blood and consequent production of large quantities of carbonic acid. This form of acidosis has numerous causes including the use of narcotics, tranquilizers, or other drugs that inhibit breathing reflexes.

acidotic (acidosic). of, pertaining to, having, or characterized by acidosis.

acidulate. to make or to become slightly sour or acid.

acidulation. the process of treating a material to make it slightly sour or acidic, sometimes with a view to increasing its solubility (e.g., the treatment of phosphate rock with a mineral acid).

acidulous. slightly sour or acid.

aciduric. tolerant of acid media or environments that are more acid than the optimum. *See* acidophilic.

aciform. acicular.

acin-. a prefix meaning saber-shaped.

acinar gland (alveolar gland; saccular gland). any saclike gland with a lumen (e.g., the active mammary gland).

acini. plural of acinus.

aciniform. **1:** clustered like grapes; resembling grapes. **2:** acinous.

acinose. acinous.

acinotubular gland. tubuloacinar gland.

acinous (acinose). consisting of acini. *See* acinus.

acinous gland. a gland in which the secretory part has a very small lumen and is shaped like a grape, e.g., the lacrymal gland.

acinus (plural, acini). **1:** one of the small drupelets or berries of an aggregate fruit such as a raspberry. **2:** a berry such as a grape or currant. **3:** the smallest secreting portion of a gland. **4:** in anatomy, a minute rounded lobule.

Acipenser. a genus of sturgeons. *See* Acipenseridae.

Acipenseridae (the sturgeon family). a family of bony fishes that includes the genus *Acipenser*. They are large, ranging in size to about 3.5 m; they occur in large rivers, lakes, and in marine waters where they feed on bottom vegetation. They are characterized by a heterocercal tail, five longitudinal rows of keeled bony plates, a long snout, ventral mouth, and a lack of teeth.

ackee. a common name for *Blighia sapida* and its fruit in Haiti.

ackee poisoning (Jamaican vomiting sickness). an acute and often fatal reaction, mainly among children, to eating the unripe fruit of *Blighia sapida* and perhaps sometimes the ripe aril. The ripe fruit readily becomes rancid and may also be toxic. The ripe aril, properly prepared, is apparently safe. This condition is marked by sudden onset of nausea and violent vomiting with periods of relief that last no more than a few hours, followed by convulsions, coma, and death in most cases. The condition is marked clinically by severe hypoglycemia.

Acleris variana. *See* budworm.

acme. **1:** apex; the highest point attained. **2:** the stage of peak magnitude, strength, or intensity of any process, disease, sign, or symptom. **3:** a period of maximum vigor. **3:** culmination.

acmic. of or pertaining to the acme of a system or process.

aconitase. the enzyme that catalyzes the reversible interconversion of citric acid, *cis*-aconitic acid, and isocitric acid in the tricarboxylic acid cycle (citric acid cycle; Krebs cycle). Aconitase is the target of fluoroacetate, a potent rodenticide that condenses with oxaloacetate to yield an aconitase inhibitor. This results in an accumulation of citric acid and the consequent inhibition of aerobic energy production.

aconite (monkshood; wolfsbane; the "queen mother of poisons"). **1:** any plant of the genus *Aconitum*. **2a:** the dried tuberous root of *Aconitum*, especially *A. napellus* and *A. lycoctonum*, the favored poison of the legendary Greek sorceress, Medea, and the presumed arrow poison of ancient China and possibly of ancient Gaul. Aconite is extremely toxic to many animals other than humans. *See Aconitum* and aconite poisoning. **2b:** the dried root of *A. ferox* (Indian aconite). **3:** an extremely poisonous and rapid-acting mixture of alkaloids extracted from the dried root of *A. napellus*.

aconite poisoning. poisoning by plants of the genus *Aconitum*. The condition in humans and livestock usually runs a severe course; death may occur within a few hours. Domestic cattle salivate heavily, frothing at the mouth, belching, and bloating; they become weak and restless, with an irregular heartbeat. Aconite poisoning of cattle has been confused with larkspur (delphinium) poisoning. The presenting symptoms in humans are usually a tingling and burning sensation about the mouth and throat followed by numbness of the tongue, throat, and face, and constriction of the throat. Common also are nausea, vertigo, vomiting, giddiness, muscular weakness, severe thoracic pain, hypotension, impaired vision and speech, paralysis of the respiratory system, convulsions, and syncope. There is a progressive burning, tingling, and numbness of the entire body with CNS depression and, finally, respiratory arrest and death due to cardiac failure. Terminal stages are marked by severe pain and paralysis of facial muscles. Many victims remain conscious until near death; some complain of yellow-green vision and tinnitus.

Aconitum (aconite; friar's cowl; monkshood; mousebane; wolfbane; wolf's-bane). a genus of more than 100 species of poisonous North Temperate Zone flowering plants (Family Ranunculaceae). Some are violently poisonous. Aconitum was the queen mother of poisons and was the poison of choice by Medea, the fabled Greek sorceress. Several species occur in North America, most notably *A. napellus*. *Aconitum* is closely related to, and pharmacologically similar to, *Delphinium* (larkspurs). The entire plant is toxic; even small amounts of leaves or roots can be fatal if ingested. There is no dependable antidote. A number of monobasic diterpenoid alkaloids have been isolated from various species of *Aconitum*. The most abundant, aconitine, is nearly identical to delphinine; other common examples are picratonitine, aconine, benzoylamine, and neopelline. These alkaloids act rapidly to stimulate and then depress the CNS *See* aconite poisoning. ***A. columbianum*** (monkshood; wolfbane; aconite; western monkshood). a species that occurs at higher elevations in the northern Rocky Mountain and Pacific Coast states of the United States. ***A. ferox*** (Indian aconite; bish; visha; bishma; bikhroot). This species is violently poisonous; A very small dose can be fatal. It is cardiotoxic and was placed on arrow points by natives to hunt big game. ***A. lutescens*** (yellow monkshood). a native North American species found from Idaho to New Mexico. ***A. napellus*** (cultivated aconite; garden monkshood; garden wolfbane). an extremely dangerous plant with attractive violet-blue (sometimes white or mauve) hooded flowers. Widely cultivated in gardens, they readily escape and are often abundant on roadsides, in old fields, and other abandoned sites. This plant contains a number of toxic alkaloids, the chief of which is aconitine. ***A. uncinatum*** (wild monkshood). a species found in the United States from Pennsylvania to Georgia. All parts are toxic, producing symptoms similar to those of *A. napellus*, usually including dysphagia and paresthesia of the lips, mouth, and throat, a colicky abdominal pain, nausea, vomiting, bloody diarrhea, ataxia, convulsions, coma, and respiratory collapse.

acorn. the fruit or seed of trees and shrubs of the genus *Quercus* (oaks), *q.*v., eaten by domesticated grazing animals, cattle, hogs, and goats, and by several kinds of wildlife, both birds and mammals, and to a limited extent by humans. *See also* acorn poisoning, oak poisoning.

acorn fruit (pink nose; blue nose). a sign of a disease of citrus trees caused by the virus, *Citrivir pertinaciae*c. Diseased fruit is acorn-shaped and the stylar end has very thin skin. Serious outbreaks may damage as much as one-half of the fruit of a given tree.

acorn poisoning. a syndrome of livestock due to ingestion of green acorns. The condition is more common in Europe than in North America, occurring only when animals are allowed to forage in oak stands in years when acorns are abundant. Symptoms and lesions are very similar to those of oak poisoning. Milk letdown decreases and the milk may be bitter. *See also* oak poisoning, *Quercus*.

acorn squash. a small squash with a deeply

fluted diamond or acorn-shaped shell.

Acorus calamus (sweet flag; flagroot; calmus; calamus; sweet cane; sweet grass). an iris-like herb (Family Araceae) that produces an oil that is carcinogenic, cardiotoxic, and hepatotoxic. *See also* asarone, oil of calamus, Araceae.

acotyledon. a plant that lacks cotyledons. Such are considered lower in an evolutionary sense than seed plants.

acoustics. the science of sound, its production, transmission, and effects.

ACP. Agricultural Conservation Program.

acquired. **1a:** of or pertaining to a non-heritable disposition, tendency, or trait. **1b:** non-genetic; not inherited. **1c:** acquired, obtained, or accumulated from the external environment. **2:** of or pertaining to a disease that is acquired after birth or hatching in response to external stimuli or environmental factors or agents such as antigens or pathogens.

acquired character. a phenotype of a plant or animal that results from environmental influences and is thus not heritable.

acraldehyde. acrolein.

acre. **1:** an English foot-pound-second (fps) unit of measurement of land area equal to 4.84 x 10^3 square yards, 1/640 of a square mile, 160 square rods, 4.046 x 10^3 square meters, or 0.4047 hectares. **2:** in the vernacular, a land area of unspecified but presumed significant size ("they have their own acre now"). *See also* acres.

acre-foot. the basic measure of agricultural water use, it is the volume of water that would cover one United States or English acre to a depth of one foot. This is equivalent to 43,560 cubic feet or 325,851 U.S. gallons of fresh water.

acre-inch. the volume of water that would cover one U.S. or English acre to a depth of one inch. This is equivalent to 3,630 cubic feet or 27,154.2 U.S. gallons of fresh water.

acreage. **1a:** land area expressed in acres. **1b:** acres collectively. **2:** an approximately one acre plot of land. **3:** an un specified or indefinite amount of land. **failed acreage**. properly planted and managed crops that failed to harvest or were destroyed by a natural disaster. Such acreage is eligible for indemnification if covered by the federal crop insurance program.

acreage allotment. a federal quota system in the United States that limits the area of land on an individual farm that can be planted in a designated type of crop (e.g., cotton, peanuts, rice, tobacco). The acreage allowed is based on previous production and the total acreage deemed necessary to produce adequate supplies of the crop nationally.

acreage control. a control established by any of a number of farm programs with a view to reducing farm output by limiting the acreage that can be planted.

acreage limitation. **1:** in relation to commodity policy, acreage limitation can refer to planting constraints under a set-aside, a paid land diversion, or (formerly) an acreage reduction program. **2:** ownership limitation; ownership entitlement; non-full-cost entitlement. The maximum number of acres that may be irrigated with less than full-cost water from Bureau of Reclamation projects. The acreage limitation for individuals or legal entities that represent 25 persons or less is usually 960 acres. This amount varies, however, in relation to the legal status of a landowner.

Acremonium. a genus of endophytic fungi (Family Moniliaceae, Order Moniliales) that causes eumycotic mycetoma. ***A. coenophialum***. a species that parasitizes *Festuca arundinacea*, *q.v.* It infests tall fescue pastures in the United States, New Zealand, and perhaps elsewhere, causing tall fescue toxicosis (summer syndrome). Among domestic animals, cattle are most often affected. Signs of poisoning include unthriftiness, high fever, heat intolerance, salivation, and seeking of water and shade. *See* fescue poisoning. ***A. loliae***. a species that infests *Lolium perenne*. It causes perennial ryegrass staggers in sheep, cattle, horses, and deer. Signs of the disease include tremors, incoordination, convulsive spasms, and collapse.

acres. **1:** plural of acre. **2a:** an unspecified number of acres. **2a:** acreage; an unspecified or indefinite amount of land, usually farmland or forested land. **harvested acres**. The cropland actually harvested for a particular crop. **planted acres**. the acres planted for a particular crop, whether harvested or not.

acres per animal unit month. the estimated acreage required at a given site, farm or ranch, region, or ecosystem (e.g., rangeland in southeastern Montana) that has forage adequate to sustain one animal unit for one month with suitable management and usage.

acri-. a prefix meaning acrid or sharp.

9-acridinamine. aminacrine.

acrifolia. in botany, sharp-leaved.

acrisol. any of a major soil group. Such soils are strongly weathered and are chiefly tropical in origin. Acrisols have an argic B horizon with an ion exchange capacity of less than 24 $cmol_c kg^{-1}$ clay and a base saturation of less than 50% within a portion of this horizon at a depth of no more than 200 cm.

acritical. not marked by or having a crisis.

acro-. a prefix referring to height or to the top of an organism, entity, or thing.

acrocentric. of or pertaining to a chromosome or chromatid in which the centromere is located near one end of the chromosome, giving the appearance, at metaphase, of a wishbone. *Cf.* acentric, meta centric, telocentric.

Acrocomia. a genus of tropical American feather palms with pinnately arranged leaves and a swollen trunk ringed by blackish spines, pinnately arranged leaves. *See* grugru.

acrodynia. **1:** pain in the extremities or peripheral parts of the body. **2:** erythredema, acrodynic erythema, pink disease, Swift's disease, dermatopolyneuritis; a syndrome chiefly of infant humans associated with the ingestion of mercury. Symptoms in infants and children include erythema of the nose, chest, and extremities; polyneuritis; gastrointestinal problems; and failure to thrive. The disease in adults is marked by anorexia, photophobia, tachycardia, and excessive sweating. **3:** a condition in rats, dogs, and swine due to a deficiency of pyridoxine. The tips of the ears, nose, lips, and feet become red, swollen, and eventually necrotic.

acrolein (2-propenal; acrylic aldehyde; acrylaldehyde; acraldehyde; aqualin; allyl aldehyde). a colorless to light yellow, volatile, asphyxiating, irritant, highly reactive, liquid alkenic aldehyde, $CH_2{=}CHCHO$, that is soluble in water, ethanol, and ether. Acrolein is a highly toxic herbicide and has been used in chemical warfare gases. It is a component of photochemical smog that enters the environment from numerous industrial sources, automotive exhausts, and combustion (e.g., of tobacco, forest fires). Poisoning of terrestrial vertebrates can result from contact, inhalation or ingestion. Acrolein is especially irritating to the skin and mucous membranes and can cause tissue necrosis on contact. The vapor is an intense lacrimator; both liquid and vapor easily damage the eye. The fumes are extremely irritant to the lungs and upper respiratory tract and, if inhaled, can produce an asthmatic reaction, bronchitis, tracheobronchitis, pulmonary edema, pneumonia, emphysema, and nephritis. Symptoms of ingestion are mainly those of severe gastroenteritis.

acropetal. pertaining to the development of a plant upward from the base to the apex such that the oldest tissue or structure is situated at the base (e.g., leaves, flowers). *Cf.* basipetal, centripetal, centrifugal.

acrophytia. plant communities that occur in alpine regions.

acrosome. the thin anterior cap or hook-like prolongation of a spermatozoon. It probably derives from the Golgi apparatus.

acrospire. the spiral plumule of a germinating seed.

acrospore. a spore (e.g., a basidiospore) borne at the extremity of a sporophore.

Acrostichum. a genus of large tropical ferns (Family Polypodiaceae), up to 12 feet in height, that occur mostly in mangrove swamps. the habit is varied, but the sporangia cover the entire lower surface of each frond. They do well under greenhouse conditions if kept in pots with plenty of water. *A. aureum* is the probably the most attractive species having fronds up to 8 feet long.

acrylaldehyde. acrolein.

acrylic aldehyde. acrolein.

acrylonitrile (vinyl cyanide; cyanoethylene; 2-propenenitrile; propenenitrile; AN). a clear, colorless, flammable, highly reactive, and extremely poisonous synthetic liquid cyanide, $CH_2CH{\cdot}CN$. The fumes are toxic and highly explosive in air; combustion of this compound releases extremely poisonous cyanide gases. It has many commercial and industrial uses including the manufacture of dyes, pharmaceuticals, and pesticidal fumigants of grain. It is readily absorbed by contact, inhalation, or ingestion.

Actaea (*Actea*; baneberry; cohosh; necklaceweed; coralberry; herb-Christopher). a genus of mildly to highly toxic perennial herbs (Family Ranunculaceae) that are native to the woodlands of eastern North America or to forests from the Rocky Mountains to the Pacific Coast. There is one European species. The flowers are white and the berries are white or pigmented. Baneberries do well in rock gardens, borders, and in wild gardens. Most are potentially dangerous to humans and livestock (especially sheep). Prolonged contact with a baneberry produces an inflammation of the skin in humans. Symptoms following ingestion may not appear for a considerable time, averaging about 48 hours. Ingestion of small amounts of roots, sap or berries can cause a burning in the stomach, dizziness, and increased pulse rate. Larger amounts may cause nausea, vomiting, bloody diarrhea, an accelerated heart rate, shock, and convulsions. Human fatalities are known. An irritant oil, protoanemonin, is the poisonous principle. Descriptions of some of these plants and the various scientific names applied by different authors can be confusing to one inexperienced with this genus. ***A. alba*** (white baneberry; doll's eyes; white cohosh). the fruit is white. ***A. odorata*** (bitter weed). a species that causes heavy losses of sheep and goats in the southwestern United States. ***A. richardsoni*** (rubber weed). poisonous to sheep. ***A. rubra*** (red baneberry; snakeberry; red cohosh). the fruit is cherry red and mildly poisonous, but distasteful. ***A. spicata*** (black cohosh; black baneberry; red cohosh). the flowers are white or bluish; the berries are purple-black.

Actea. *Actaea*.

ACTH. (adrenocorticotrop(h)in; adrenocorticotrop(h)ic hormone; adrenotropic hormone; corticotrophin). a straight-chain polypeptide hormone secreted by the anterior lobe of the pituitary gland in response to corticotropin-releasing hormone (CRH) and partly (as part of a negative feedback loop) by the concentration of hydrocortisone in the circulating blood. The chief functions of ACTH are promotion of growth and maintenance of the adrenal cortex and stimulation of the synthesis and release of various adrenal corticosteroid hormones (e.g., hydrocortisone, aldosterone) and androgens. It also promotes the breakdown and release of fatty acids from fat cells. ACTH secretion increases during most conditions of stress, including trauma, burns, infection, emotional stress, and hypotension. ACTH is used therapeutically to treat inflammatory disorders.

actin-. a prefix denoting or pertaining to a ray or rays.

Actinea odorata (bitterweed; bitter rubberweed; Colorado rubberweed). an annual herb (Family Compositae) that is most common on overgrazed ranges in the western United States It is poisonous to sheep, which may consume it when preferred forage is limited or absent. Poisoned sheep may lose their appetite, may no longer chew the cud, froth at the mouth, bloat and suffer abdominal pain, a green nasal discharge may be seen.

Actinea odorata. synonym of *Hymenoxys odorata*.

Actinidia chinensis (Chinese gooseberry; kiwi (def. 2); yang tao). a tropical tree orclimbing shrub (Family Actinidiaceae) that is indigenous to China. It is cultivated both in the tropic and temperate zones (most notably in New Zealand) for its edible fruit (kiwi fruit). The ripe fruit is dark brown, about 1 in. (2.5 cm) in diameter; the edible portion is greenish yellow and has a flavor similar to that of the strawberry.

Actinidia polygama (silver vine). a vine of eastern Asia with fragrant white flowers and yellow, edible fruit. The male plant has leaves with silvery-white markings.

actino-, actin-. a prefix meaning ray or beam. In the physical sciences, it relates to radiation; in agriculture and the biological sciences, it usually indicates the presence of raylike structures or being radiate in form.

actinobacillosis(wooden tongue; big jaw; lumpy jaw; cruels). a chronic disease of cattle, sheep, and swine that affects the tongue and throat; fatalities are rare. The infective agent is *Actinobacillus lignieresi*. In cattle, the tongue is chiefly affected; in sheep and swine, the lips, cheeks, and jaws become swollen. Onset is often rapid; affected cattle may refuse feed, drool, and make licking or chewing movements. The root of the tongue becomes hard, swollen, and may develop surface lesions. The tongue is rarely affected in sheep and goats, but pustular lesions develop in the nose, lips, and lower jaw. Extensive scarring that interferes with feeding and breeding may result. Lymph nodes of the neck usually become enlarged. Internal organs may ultimately become affected. In swine, abscesses of the udder may occur. *See also Actinobacillus*, actinymycosis.

actinobacillus. common name bacteria of the genus *Actinobacillus*.

Actinobacillus. a genus of spherical or rod-shaped, aerobic, facultatively anaerobic bacteria (family Brucellaceae) with fermentative metabolism. They are nonmotile, do not form spores, and contain Gram-negative rods and coccal elements. Most are pathogenic to animals. Some species are pathogenic to animals of agricultural interest or importance. ***A. actinoides***. a form that is pathogenic to goats via inhalation and has been isolated from calves afflicted with chronic pneumonia. ***A. actinomycetem comitans***. a species of doubtful taxonomic position. Although it occurs in actinomycotic lesions with actinomycetes, its pathogenicity has not been established. ***A. equuli***. a species that causes suppurating lesions, especially of the joints and kidneys in piglets of domestic swine and foals of horses. It causes an endocarditis in older pigs. ***A. lignieresii***. the type species of the genus, it causes actinobacillosis in cattle and swine and pustular lesions of the skin and lungs in sheep and goats. *A. mallei*. The causative agent of glanders, *q.v.*

actinomorphic (actinomorphous). pertaining to plants (especially flowers) that are radially symmetric. *Cf*. zygomorphic.

actinomorphous. actinomorphic.

Actinomyces (*Ascomyces*; ray fungi). a genus of filamentous, pathogenic, Gram-positive, nonmotile, chemoheterotrophic, non-spore-forming bacteria (Order Actomycetales, Family Actinomycetaceae). Representatives are obligate or facultative anaerobes; metabolism is fermentative, products of which may include acetic, formic, lactic, and succinic acids. The type species, it is the causative agent of actinomycosis in cattle; it may also infect humans and horses. ***A. bovis***. the cause of actinomycosis (bovine actinomycosis or lumpy jaw) in cattle and other animals. The disease is characterized by lumpy, often suppurating tumors, especially about the jaws. ***A. israelii***. the cause of human actinomycosis; it occasionally infects cattle. ***A. micropolyspora faeni***. the causative agent of farmer's lung, a form of aleukia (alimentary toxic aleukia). ***A. naeslundii***. normally occurs in the oral cavity, sometimes infecting cattle and, less often, humans. ***A. naeslundii***. a species that occurs in the oral cavity, sometimes producing periodontal lesions in some animals. Human infections are known. ***A. suis***. this species, and perhaps

A. bovis, causes a primary chronic suppurative, granulomatous mastitis in sows (swine actinomycosis, *q.v.*). It sometimes causes a generalized infection that involves the lungs, kidneys, spleen, and other visceral organs (*See* lumpy jaw). ***A. viscosus***. a species that sometimes occurs in the oral cavity of humans and certain other mammals, especially dogs. It contributes to or causes periodontal disease.

actinomyces. a common name of any of several anaerobic bacteria of the genus *Actinomyces*. *Cf.* actinomycete.

Actinomycetaceae. a family of mold-like bacteria (Order Actinomycetales) that are nonmotile, filamentous, branching, Gram-positive, and mostly anaerobic forms. They are widely distributed in soils and bottom sediments of rivers and lakes. Many strains produce antibiotics (e.g., actinomycin) and many are pathogenic (*Actinomyces spp*. and *Streptomyces spp*.). This group of organisms plays a significant role in decomposing and stabilizing solid wastes by composting.

Actinomycetales. an order of moldlike bacteria comprised of the families Mycobacteriaceae, Actinomycetaceae, Actinoplanaceae, Dermatophilaceae, Micromonosporaceae, Nocardiaceae, and Streptomycetaceae. All lack endospores and may be club-shaped, rod-shaped, or filamentous in form.

actinomycete. **1:** common name of any of a group of branched, filamentous, Gram-positive, mostly anaerobic, nonmotile, fungus-like bacteria (Family Actinomycetaceae; Order Actinomycetales) found in soil and river and lake bottom sediments. Included are bacteria of the genera *Actinomyces* and *Streptomyces*. Many species are pathogenic and many produce antibiotics. **2:** sometimes erroneously applied to any member of the order Actinomycetales.

actinomycin. any of a large family of antibacterial, antifungal, and cytotoxic antibiotic peptides produced by bacteria of the general *Actinomyces* and *Streptomyces*. They have cytostatic and radiomimetic activity. Although useful against Gram-positive bacteria and fungi and certain eukaryotes (e.g., certain cancer cells), actinomycins are highly toxic to humans. Among the more important actinomycins are cactinomycin and dactinomycin. *See also* actinomycosis. **actinomycin A**. an orange-red, highly toxic, heat-stable, antibacterial agent that is soluble in ethanol and ethyl ether. It is produced by *Actinomyces antibioticus* and some species of *Streptomyces*. It blocks transcription in gram-positive bacteria and in eukaryotic cells. **actinomycin A_{IV}**. dactinomycin. **actinomycin B**. an extremely toxic, ethanol-insoluble, antibacterial agent, with properties otherwise similar to those of actinomycin A. It is not used clinically. **actinomycin C**. cactinomycin. **ACC**. actinomycin C. *See* cactinomycin. **actinomycin C_1**. dactinomycin. **actinomycin D**. dactinomycin. *See also* cactinomycin. **Actinomycin D®**. dactinomycin. **actinomycin I_1**. dactinomycin. **actinomycin X_1**. dactinomycin.

actinomycosis (actinophytosis; lumpy jaw; big jaw; clams; clyers; wooden tongue). a type of chronic suppurative disease (sometimes mistaken for actinobacillosis) that most commonly affects humans, horses, cattle, swine, and dogs. It usually affects the face and neck with the formation of abscesses, or pus-filled cavities beneath the skin. These abscesses spread rapidly and form channels that discharge a yellow granular pus on the surface of the skin. The disease occurs when certain species of *Actinomyces* that normally occur in the mouth and pharynx invade tissues where they can cause chronic, pustular abscesses or granulomas and eventual tissue damage. Actinomycosis also sometimes affects visceral organs such as the lungs, kidneys, and spleen.

The pelvic region may be affected, especially in women that use certain kinds of intrauterine devices. There are four main types of the disease: cervicofacial, abdominal, thoracic, and generalized. Fatalities are rare. *Cf.* lumpy jaw. *See also* coccidiomycosis, corky scab of potato.). **abdominal actinomycosis**. a form of the disease that sometimes occurs following appendicitis or any perforation of the large intestine or stomach. **bovine actinomycosis** (lumpy jaw; big jaw). a chronic infectious disease of cattle and swine that may persist for up to a year. The infective agent is *Actinomyces bovis*. Lesions commonly occur in the mandible. The disease is characterized by difficult breathing, swollen gums and loose teeth, bony growths through the tissues and skin, foul-smelling abscesses, and enlarged turbinated bones (i.e., those of the nasal concha in the walls of the nasal cavity) that contain pus. Treatment consists of the oral or intravenous administration of iodide compounds and surgical drainage of the affected areas. The disease is rarely fatal. *See also* lumpy jaw. **cervicofacical actinomycosis**. a type that usually results from bacterial invasion of the soft tissues of the mouth, throat, and neck following infection of a tooth or tonsil. **generalized actinomycosis**. a mycosis that involves the skin, brain, liver, and urogenital system. **human actinomycosis**. seen most often in farmers and others who live in rural settings, it is caused by *A. israeli* and *A. hominia*, actinomycetes that normally occur in the mouth and intestines. The disease usually affects the cervicofacial area, abdomen, or thorax. **swine actinomycosis**. a disease similar to that of bovine actinomycosis, *q.v.*, but swine may also suffer a tumorlike enlargement of the udders. **thoracic actinomycosis**. a form of the disease that may occur if matter containing pus moves down the throat and bacteria are aspirated into the lungs. *See also* Actinomyces, actinobacillosis.

actinomycotic. of or pertaining to actinomycosis.

actinophytosis. botryomycosis.

Actinopterygii (ray-finned fishes). a subclass of the Osteichthyes that contains the majority of living bony fishes. The fins comprise a web of skin strengthened by nearly parallel rows of horny rays. The scales in most living species are reduced or absent, the jaw suspension is hyostylic, and there is usually a swim bladder that serves as a hydrostatic organ. There are three subdivisions: the Chondrostei, Holostei, and Teleostei. *Cf.* Choanichthyes.

actinotoxemia. radiation sickness. *See* sickness.

action. **1a:** motion, movement. **1b:** performance. **1c:** an act; a thing done; a step taken; deed. **2:** observed or overt behavior. **3:** a legal or judicial proceeding to protect a right, prevent wrong, or publish a public offense or violation of law. **4:** a morbid process (pathology). **antagonistic action**. the opposition of a chemical or muscle to the action, respectively, of another chemical or muscle. **bacteriocidal action**. any process that destroys bacteria. **bacteriostatic action**. an action or process that retards growth of bacteria. **biologic action**. an action that has biological consequences. **cumulative action**. the usually sudden appearance of a biologic, pharmacologic, or toxic effect by a stimulus following several subthreshold exposures. **drug action**. the action of a drug on an organism or on various organ-systems. **specific action**. a unique action produced by a substance on a particular organism or part of such organism. **synergistic action**. the production, by a substance or stimulus, of an enhanced effect (i.e., greater than additive) when applied together with another specific substance (stimulus) or type of substance.

action level. 1: in general, the concentration of a chemical residue in food or feed above which adverse effects are possible and corrective action is indicated. **2:** the concentration in the environment of a hazardous substance that is high enough to warrant action. **3:** that concentration of a pesticide residue in food or animal feeds, set by the Environmental Protection Agency, that require the Food and Drug Administration and USDA to take enforcement action. The residues covered by this term (inadvertent residues) are those that are present by virtue of prior legal use or accidental contamination.

action system. a behavior pattern in an organism.

actionable. subject to or affording grounds for a governmental action or a lawsuit.

actionable fire. a fire that violates existing laws, regulations, or otherwise affords grounds for an action or a lawsuit.

actium. a rocky shore biotic community.

activate. 1: to make active; to cause to function; to trigger, actuate, stimulate, or mobilize. **1a:** to render more reactive or easier to activate. **2a:** to render a substance capable of reacting, or to make it more active. **2b:** to convert an inactive or relatively inactive chemical into a reactive form or into a different substance. **2c:** to speed up reaction by various means such as heating. **2d:** to increase the activity of specific parts of a molecule by insertion of suitable neighboring groups. **3:** to aerate sewage in order to accelerate the decomposition of organic matter. **4:** to induce radioactivity by bombardment with neutrons or other types of radiation. *See also* activation, activator.

activated carbon (activated charcoal; active carbon). a highly adsorbent form of carbon produced by heating wood, charcoal, or other organic matter (e.g., coconut shells, bones) to a temperature of 800-900°C (1,472-15652°F) in the absence of air (and in the presence of steam). Adsorbed gases are thus driven off, yielding a highly porous material. Activated carbon is very effective in decolorizing and deodorizing materials, in controlling flavor in milk, and removing impurities and toxic substances from liquids or gases (e.g., as in gas masks). It is used in veterinary practice as a GI tract protectant to bind irritants or caustic chemicals, thus preventing or reducing tissue irritation or erosion. Activated carbon is also used to remove odors and toxic substances from gaseous emissions, and in waste treatment to remove dissolved organic matter from wastewater by adsorption.

activated charcoal. activated carbon.

activated ergosterol. vitamin D_2.

activation. 1: the process of rendering active. **biochemical activation. 1:** activation by means of a biochemical process. **2:** metabolic activation. **biological activation** (bioactivation). **1:** activation by means of a biological process. **2:** metabolic activation. **enzyme activation.** the conversion of a preenzyme into an active enzyme, usually catalyzed by a kinase or another preenzyme. **metabolic activation** (biochemical activation; biological activation; physiological activation). the enzymatic conversion of a foreign or endogenous substance to a more reactive form. Activation reactions of toxicological significance are usually monooxygenations that yield electrophiles that may react with nucleophilic moieties on endogenous proteins or DNA. **multistep metabolic activation.** activation, e.g., of carcinogens and certain other toxic chemicals in which a substance undergoes a series of transformations before generating the ultimate metabolites. **physiological activation.** metabolic activation.

activator. 1a: something that activates;

stimulus, trigger. **1b:** any substance that renders another substance (e.g., a pesticide) active or more active (i.e., capable of entering into its normal reactions or causing its normal effects). **2:** a substance that provokes a specific activity (e.g., an inducer of specific hormone secretion). *See also* activation, proactivator. **compost activator**. a chemical added to a compost pile to increase the rate of decomposition of the plant material in the pile.

active. **1a:** energetic. **1b:** productive of or capable of action or movement. **1c:** in action, in motion, or performing a function. **2:** having a tendency to act. **3:** capable of acting in a specific manner. **biologically active** (bioactive). pertaining to or denoting the capacity to act on and elicit a response from living cells, tissues, organs, or organisms. *See also* biologically active agent.

active agent. *See* active ingredient.

active carbon. activated carbon.

active ingredient (active agent). **1:** that constituent of a product that produces the intended effect with normal usage. **2a:** the component that kills, inhibits the activity of, repels, mitigates, or otherwise controls target pests, excluding plant regulators, defoliants, or desiccants. In the United States, pesticides are regulated by the Environmental Protection Agency chiefly on the basis of active ingredients. **2b:** in the case of a plant regulator, an ingredientthat physiologically alters the rate of growth, maturation, or behavior of ornamental crop plants. **2c:** in the case of a defoliant, an ingredient that causes the leaves or foliage to abscise. **2d:** in the case of a desiccant, an ingredient that artificially accelerates the drying of plant tissue.

active process. any process driven, at least in part, by metabolic energy. *Cf.* passive process.

active transport. the transport of a substance across a biological membrane against a concentration gradient or an electrochemical gradient. This process depends upon availability of specific carrier proteins and is driven by metabolic energy. *See also* diffusion.

activity. **1a:** motion, movement, animation, or any process associated with the flow of energy. **1b:** the state or quality of being active. **biological activity** (bioactivity). **1:** the behavior of a tissue, organ, or living organism as reflected in action or movement. **2a:** the capacity of an exogenous or endogenous agent (e.g., a medication, a poison) to directly affect one or more biological processes. **2b**: the specific action of any agent on a biological system or process. **enzyme activity**. the specific action of an enzyme on a chemical reaction or metabolic process.

actophile. an organism that lives and thrives on rocky seashores.

actophilous. of or pertaining to actophiles.

actophily. the condition or circumstance whereby an organism lives and thrives on rocky seashores.

actual. a commodity that is available for delivery. *Cf.* future.

actual production history (APH). a measure of an individual farmer's annual production of a commodity during a multi-year period. When the actual production changes by more than a designated percentage from the norm established by the APH, a producer enrolled in the crop insurance program becomes eligible for an indemnity payment.

actual use. the actual use of an area (in animal units or animal unit months) by livestock and/or wildlife as opposed to the recommended use.

Acuaria spiralis. a nematode parasite of gallinaceous (e.g., domestic fowl, pheasants, turkey) and other birds. It infests the esophagus, proventriculus, and sometimes the intestines.

aculeat-. prefix denoting or pertaining to stings or prickles; stinging.

Aculeata. a division of the insect order Hymenoptera comprised of those forms in which the ovipositor is modified as a sting. Included are bees, ants, and true wasps.

aculeate. **1a:** having sharp points, prickles, or a sting. **1b:** pointed, stinging. **1c:** incisive. **2:** denoting, having the characteristics of, or belonging to the Aculeata (a division of the insect order Hymenoptera).

acuminat-. prefix meaning tapering, pointed.

acuminate. tapering gradually to a pointed tip or apex.

acut-. prefix meaning acute, sharp.

acute. **1:** sharp, keen, pointed. **2b:** critical, grave, serious. **3:** descriptive of a disease that is transient or of short duration; not chronic. **4:** intense; severe, marked by sharpness or severity; critical. **5a:** of or pertaining to a single, relatively brief, usually intense, exposure to a pathogen, toxicant or other biologically active agent. **5b:** the development of signs or symptoms following an acute exposure (see 5a) to a biologically active agent. **6:** of or pertaining to a stimulus intense enough to induce a rapid response. **7:** astute, discerning, shrewd.

acute condition (acute disease**;** acute disorder). any condition characterized by rapid onset, sharp rise, short duration, and pronounced or severe symptoms (often including pain, high fever, dyspnea). *Cf.* chronic condition.

acyclic. **1:** not cyclic, as an acyclic flower. **2:** of or pertaining to a chemical compound that does not contain a closed chain or ring of atoms (e.g., as in benzene or cyclohexane).

ad-. prefix meaning to, on, onto.

ad lib. abbreviation for ad libitum.

ad libitum (*ad lib.*) freely, at pleasure, as desired. Commonly used to indicate continual availability of feed to an individiual or group of animals.

ad valorem. in proportion to the value; it usually denotes a tariff against property.

ad valorem duty. a tariff assessed as a fixed percentage of the value of an imported commodity or product as stated in the invoice. **specific duty**. a tariff assessed as a charge on each unit or specified quantity of an imported item commodity or product.

ad valorem tax. a tax assessed on the basis of the value of the property, commodity, or item to be taxed.

adactylia. adactyly.

adactylism. adactyly.

adactyly (adactylia; adactylism). congenital absence of fingers or toes. In Holstein cattle, it is due to the presence of autosomal recessive genes.

Adansonia digitata (baobab**;** cream-of-tattar tree). a massive tree (Family Bombacaceae) that is indigenous to tropical Africa and cultivated in India. It bears a large edible fruit ("monkey's bread"); the fibers are cloth.

adaptation (adaption**;** adaptive adaption). **1:** any process by which an organism adjusts to environmental stress. **2a:** a process of natural selection that results in improved fitness of an organism or population to a particular environment; **2b:** the fitness of an organism to a particular environment. **3:** an inherited phenotype (structure or trait), resulting from natural selection, that enhances survival or reproductive success of an organism or population in a specific environment, environmental variable, or activity. **4a:** a process by which an organism or sense organ ceases to respond to a continuing or repeated stimulus of a particular type when the intensity of the stimulus remains constant. **4b:** adjustment of the pupil and retina to changes in illumination. *Cf.* accommodation. **5:** the dynamic adjustment of an organ or organism to a constantly changing environment. Such adaptation depends upon biophysical, biochemical, physiological, and behavioral responses. **sensory adaptation**. a reduction in the frequency of response (impulse generation) of a receptor following a repeated or continuing stimulus of constant intensity; habituation.

adaptedness. the degree of adaptation to a habitat, niche, or to the environment more generally considered.

adaption. adaptation.

adaptive. **1a:** of, pertaining to, descriptive of, or contributing to adaptation. **1b:** able to adapt.

adaptive adaption. adaptation.

adaptive behavior. behavior that improves an animal's fitness under a given set of environmental circumstances. Thus, animals structurally adapted to fly do so under appropriate circumstances and most most animals avoid many natural environmental hazards. *See* adaptation.

adaptive radiation. The evolution, by natural selection, of new life forms (taxa) that occupy previously unexploited ecological niches.

adaptive value. the comparative fitness, *q.v.*, of different genotypes in a given environment; the survival and reproductive value of one genotype relative to other genotypes in the population. One can give numerous examples. As one example, numerous plants have developed toxins, presumably in some cases in response to herbivory. Such toxins are often much less toxic to herbivores that have co-adapted to the evolving plant toxins than to other animals.

adaptiveness, adaptability. the ability of an organism to adjust to new or changing stimuli, experiences, or conditions.

adaxial. **1a:** toward, facing toward, or nearest the axis. **2a:** of or pertaining to the surface of a leaf or other lateral organ that is developmentally nearest to the apex of the axis (stem or stalk) that bears it. **2b:** the upper surface of a leaf or other lateral organ. *Cf.* abaxial.

ADC. Animal Damage Control Program (now designated the Wildlife Services Program), *q.v.*

adder. **1:** part of the common name of any of a variety of venomous snakes, especially of the family Viperidae. **2:** any of several harmless North American snakes such as *Heterodon contortrix* (puff adder; hognose snake; blowing adder; sand viper; spreading adder) and *Lampropeltis* (milksnakes), so-called because they are sometimes seen in dairy barns preying on mice. **3:** *Vipera berus*. **4:** any of a number of venomous species in which "adder" is included as part of the common name. **death adder**. *Acanthophis spp.* **true adder**. any snake of the genus *Vipera*, especially *Vipera berus*, or sometimes any snake of the family Viperidae.

additive. **1:** of or pertaining to addition. **2:**

cumulative; produced or characterized by addition. **3:** referring to the ability to add or be added; cumulative. **4a:** something added, as the addition of one substance to another. **4b:** any substance not a normal part of a food, animal feed, or other material, which is added in order to alter or enhance its natural properties, to preserve or extend its life, or to confer new properties (e.g., feed additive, food additive). **5:** a substance which, when added to a petroleum product, confers special properties or enhances the natural properties of the product. *Cf.* synergistic, antagonistic. **feed additive**. a substance added to animal feed such as minerals, antibiotics, nutrients, vitamins, or hormones (*See* def. 3). **food additive**. any substance added directly to a food prior to sale, usually to improve flavor, odor, appearance, or to preserve or extend its useful life. Examples are nitrate, nitrite, sulfite, color (artificial color), butylated hydroxytoluene (BHT), butylated hydroxyanisole (BHA), and flavor (artificial flavor). *See also* bread, contaminant, Delaney Clause.

additive effect. *See* effect.

additivity. The property, characteristic, or state of being additive.

addled egg. **1:** a rotten egg. **2:** an egg in which the yolk and white are intermixed.

adenine (6-aminopurine). a purine base, $C_5H_5N_5$, adenine combines with the sugar ribose to form adenosine, which in turn can be bonded with from one to three phosphoric acid units, yielding the three nucleotides adenosine monophosphate (AMP), adenosine diphosphate (ADP), and adenosine triphosphate (ATP). All of these derivatives perform important functions in cellular metabolism. Adenine is one of four nitrogenous bases that are components of DNA. A modified form of adenosine monophosphate is thought to be a secondary messenger in the propagation of many hormonal stimuli. Adenine is also an integral part of the structure of many coenzymes.

adenitis. **1:** a glandular inflammation. **2:** lymphadenitis.

adeno-, aden-. a prefix denoting or pertaining to glands or glandular tissue.

adenocarcinoma. any of a large number of malignant tumors that arise from epithelium. They contain cells characteristic of, and named for, the tissue from which they arise, although the cancer cells often become so degenerate that specific tissue characteristics are lost. Thus, the tissue of origin of a disseminated adenocarcinoma may be indeterminant. Adenocarcinomas of the ovary and oviduct occur rather frequently in adult domestic fowl and are not known to be transmissible.

adenohypophyseal growth hormone. somatotrophin.

adenohypophysis. anterior lobe of the pituitary gland (*See* pituitary gland).

Adenoidalpharyngeal-conjunctival virus. adenovirus.

adenoma. a gland-like, usually benign, well-circumscribed tumor that does not invade adjacent tissue. *Cf.* adenocarcinoma.

Adenophorasida (Adenophorea; Aphasmida; Aphasmidia). a class of nematodes that includes the genera *Trichuris, Capillaria,* and *Trichinela*, all of which include important parasites of man and domestic animals.

Adenophorea. Adenophorasida.

adenosine. a white, crystalline, water-soluble condensation product (a nucleoside) of adenine and ribose, $C_{10}H_{13}N_5O_4$.

adenosine 5′-dihydrogen monophosphate. AMP.

adenosine diphosphate. ADP.

adenosine monophosphate. AMP.

adenosine 5′-(tetrahydrogen triphosphate. ATP.

adenosine 5′-trihydrogen diphosphate. ADP.

adenosine triphosphate. ATP.

***Adenostoma fasciculatum*.** a small California evergreen shrub (Family Rosaceae) with small heath-like flowers found in the hard chaparral community of California. Together with a number of other species (e.g., *Arctostaphylos spp.*) that make up the hard chaparral community, these plants secrete phenolic compounds such as arbutin and *p*-hydroxy cinnamic acid, which dissolve in rainfall, wash to the ground, and inhibit the growth of grasses (Gramineae) near the plant. Substantial bare areas thus develop within and somewhat beyond the edge of the community. When the chaparral is destroyed by fire, the growth of grasses is initially vigorous, but slows as the chaparral community is reestablished.

Adenoviridae. a family of double-stranded DNA viruses, the adenovirusus. They infect and develop in the nuclei of cells in mammals and birds.

adenovirus (Adenoidalpharyngeal-conjunctival virus; APC virus). any or a group of DNA viruses (Family Adenoviridae) that cause eye and respiratory diseases. **bovine adenovirus.** common name of viruses of the genus *Mastadenovirus*; there are nine known serotypes that cause a mild upper respiratory tract disease in cattle. **porcine adenovirus.** any virus of the genus *Mastadenovirus*. There are four recognized serotypes cause a mild upper respiratory tract disease in swine.

adenylic acid. 1: AMP. **2:** a generic term applied to a group of isomeric nucleotides that contain the adenyl radical, $C_5H_4N_5\cdot$.

adenylic acid. AMP.

ADF. acid detergent fiber.

ADG. average daily gain.

ADH. 1: antidiuretic hormone (*See* vasopressin). **2:** alcohol dehydrogenase.

adherence. 1: obedience, loyalty, devotion. **2:** adhesion (definitions 1 and 2). **3:** the amount of sediment or soil adhering to the skin that could be absorbed. It is usually expressed as surface density (mg/cm^2) (*Cf.* adhesion, def. 4). *Cf.* adhesion.

adhesion. 1: a molecular attraction that holds the surfaces of two bodies together. **2:** a firm solid attachment. **3a:** the union of two body parts (organs or tissues) that are not normally joined, by the formation of fibrous tissue; the site of union of such tissue. **3b:** the tissue that joins two organs or tissues that are not normally joined. **4:** the attraction of molecules in the walls of the interstices of a sediment to water molecules. *Cf.* adherence.

adhesive. 1: adhering, sticky, gummy, clinging. **2:** in physics, denoting or pertaining to the molecular force in the area of contact between unlike bodies that tends to unite them. **3:** a substance (e.g., glue, paste, mucilage) that sticks to surfaces of other substances.

adhesive agent. an adhesive material that causes material such as herbicide sprays to stick to the parts of a plant that are sprayed.

ADI. acceptable daily intake.

adiabat. a line on a thermodynamic chart

indicating the pressure and temperature of a substance undergoing an adiabatic change.

adiabatic. **1a:** occurring without gain or loss of heat. **1b:** of or pertaining to any change or process in which there is no gain or loss of heat. **1c:** an adiabatic process. *Cf.* diabatic, isentropic.

adiabatic chart (adiabatic diagram). a graph used to analyze adiabatic processes.

adiabatic diagram. adiabatic chart.

adiabatic lapse rate. *See* lapse rate.

adiabatic process. **1a:** any process in which there is no net gain or loss of heat. **1b:** any process that proceeds without heat entering or leaving a system. **2:** a system or process operating in isolation. Such a system is ideal or theoretical and does not, in fact, exist.

adiabatic rate. *See* lapse rate.

adiaphoresis. anhidrosis.

adiaspiromycosis. a rare pulmonary mycosis of man, many rodents, and other burrowing or aquatic animals. *Chrysosporium parvum*.

adience. **1:** the tendency to approach the source of a stimulus. **2:** movement toward a stimulus; an approach response. *See also* abience.

adient. positive; having the tendency to approach the source of stimulus.

ADIN. acid detergent insoluble nitrogen.

adipose. **1a:** fat, fatty; consisting of, resembling, or relating to fat.

adipose tissue (fatty tissue). a type of connective tissue comprised of aggregated of fat cells (adipocytes). In mammals, this tissue occurs chiefly in subcutaneous tissue, in the mesenteries, around the kidneys, and in the mediastinum. Fat cells are present in all connective tissue, but adipose tissue is reserved for large masses of fat cells. Adipose tissue serves as a reserve (sometimes the main) energy source of animals; it provides insulation against heat loss and can act as a shock absorber. *See* fat, brown fat.

adiposis. obesity.

adiposity. **1:** adiposis; obesity. 2. unusual accumulation of fat in a tissue, organ, or body part.

adipsia (adipsy). **1a:** absence of thirst. **1b:** complete failure to drink.

adipsy. adipsia.

adjusted base period price. the average market price received by a farmer during the most recent 10 years, divided by the index of average prices received by farmers for all farm products during the sample period. It is used to calculate parity.

adjustment. **1a:** change, alteration. **1b:** correction, alignment, modification. **1c:** shift. **2:** acclimation; a functional, often transitory, response to stimuli whereby an organism adapts to its immediate environment. **3a:** the process by which an organism becomes adjusted to its environment. **3b:** acclimation; a functional, often transitory, response to stimuli by which an organism adapts to its immediate environment. **4:** changes in range management practices due to changing seasons, environment (e.g., drought), or situation. *Cf.* acclimation, acclimatization, adaptation.

adjuvant. **1a:** serving to contribute, aid, facilitate, or augment. **1b:** that which augments, helps, or facilitates. **2:** serving as an adjuvant in medicine or veterinary medicine. **3a:** an excipient that enhances or facilitates the action of another ingredient (usually the principal ingredient) in a formulation. Thus, use of an adjuvant may intensify the toxicity, antigenicity, or therapeutic action of a chemical formulation. For example, the activity of a dithiocarbamate fungicide is increased by the addition of 2-mercaptothiazole. **3b:** a method, drug, or other means that enhances the effectiveness of a medication or of medical treatment. For example, an excipient which, when combined with an antigen and an administered to a human or other animal, enhances and prolongs antibody production. **4:** in agriculture, adjuvants are also used to increase the effectiveness of various agrochemicals (e.g., fertilizers, insecticides).

adlay (Job's tears). a tropical millet grown in parts of southeast Asia, especially the Phillipines. The seeds are unusually large.

administration. **1:** management, direction, or supervision. **2:** giving, dispensing, delivery, as in the administration of a drug.

Administrative Procedure Act (APA). P.L. 79-404 (July 11, 1946), as amended, establishes, for example, the minimum procedural requirements or models for federal agency rule-making and certain types of hearings. Thus, as an example, the APA establishes procedures for informal rulemaking, which may include notice-and-comment requirements, or formal rulemaking, including trial-type hearings. Exemptions from rulemaking requirements are included in the APA. The provisions of this act apply to USDA rulemaking, unless exempted under the provisions of another law.

admissibility. **1:** the state of being admissible. **2:** data or other information that is relevant and proper to be considered in reaching a decision concerning issues raised in a judicial proceeding. Data are admissible by precedent or mutual agreement and do not require defense prior to acceptance.

admissible. pertaining to data or physical evidence deemed relevant and proper for consideration in arriving at a decision in a judicial proceeding.

adnat-. a prefix meaning joined.

adnation. **1:** union or fusion, especially of unlike plant parts, during growth. **2:** growth that results in one side of a plant or plant part adherent to the stem.

adobe. **1:** heavy textured clay soil found in desert basins of the southwestern United States and Mexico. **2:** adobe brick.

adobe brick. a roughly molded, sun-dried brick, usually of large dimensions, made of adobe, silt, and small amounts of straw or grass.

adolescence. the period of life from onset of puberty until maturity; used mainly in reference to *Homo sapiens* and other primates.

adolescent. **1a:** minor; teenager. **1b:** youth; youthful. **1c:** immature, not fully mature. **2:** childish, puerile, sophomoric.

adonidin. any of the cardiotoxic phenanthrene glycosides found in various species of *Adonis*. Effects are similar to those of digitalin.

Adonis (pheasant's eye; bird's eye; false hellebore). a genus of toxic herbs (Family Ranunculaceae). They are native to Europe, Asia, and Africa and have been introduced elsewhere, including the Australian region and North America. They have proven lethal to livestock under field conditions in Australia, New Zealand, and Europe, but incidents are unknown in North America, where they have been introduced as garden plants. The toxic principles are phenanthrene glycosides such as adonidin. ***A. vernalis*** (false hellebore; vernal pheasant's eye; bird's eye). Native to Europe, Asian parts of the former U.S.S.R., and Labrador. Constituents include adonitoxin, acetyladonitoxin, adonivernoside, adonitol, adoniverhith, convallatoxin (strophanthidin), and vernadin. *See* adonidin.

adonitoxin (3-[(6-deoxy-α-L-mannopyranosyl)-oxy]-14,16-dihydroxy-19-oxocard-20-(22)-enolide). a toxic phenanthrene glycoside isolated from *Adonis vernalis*, *q.v.*

ADP (adenosine 5′-trihydrogen diphosphate; adenosine diphosphate). a nucleotide comprised of adenosine and D-ribose to which two phosphate groups are attached, one of which is a high-energy group. ADP is a component of all living cells and is an important coenzyme in numerous biochemical reactions. It undergoes oxidative phosphorylation in mitochondria to yield ATP and is reconstituted by the condensation of adenosine and phosphoric acid. High concentrations of ADP stimulate, and low concentrations inhibit, oxidative phosphorylation. *See also* adenosine monophosphate, adenosine triphosphate, adenosine triphosphatase.

ADR. adverse drug reaction.

adrenal. **1:** at, on, or toward the kidney. **2:** an adrenal gland or separate piece of adrenal gland tissue. **3:** of or pertaining to the adrenal gland.

adrenal body. adrenal gland.

adrenal cortex. the outer portion of the adrenal gland; it secretes various steroid hormones. The outer region of the cortex *zona glomerulosa* produces mineralocorticoids and glucocorticoids; the inner region, the *zona reticularis-fasciculata* secretes mainly cortisol and certain other glucocorticoids.

adrenal gland (adrenal body; suprarenal gland; adrenal, def. 2). either of a pair of glandular, ductless glands in vertebrates. They are situated on or near the kidneys in mammals and are superior to the kidney in primates. Each is comprised of an inner medulla that secretes adrenalin and noradrenalin, and an outer cortex that secretes steroid hormones (corticosteroids). In some vertebrates, the two types of adrenosecretory tissue form separate glands.

adrenal medulla. the inner portion of the adrenal gland; it secretes adrenalin and noradrenalin.

adrenalin. **1:** epinephrine. **2:** a frequently used spelling in the United States for adrenaline. **3:** often the preferred term for epinephrine in physiology and allied sciences such as toxicology. *Cf.* adrenaline.

adrenaline. **1:** epinephrine. **2:** the official British Pharmacopeia name for epinephrine and the preferred usage in the United Kingdom. *Cf.* adrenalin.

adrenergic. **1:** pertaining to or denoting nerve cells, fibers, or nerve terminals of the autonomic nervous system that release adrenalin or noradrenalin as a neurotransmitter. **2a:** pertaining to drug actions similar to those of the sympathetic nervous system. **2b:** a drug or other agent that produces effects similar to those of the sympathetic nervous system.

adrenocortical. pertaining to or denoting the cortex of the adrenal gland or its secretions. *See also* ACTH.

adrenocorticotrop(h)in. ACTH.

adrenocorticotrop(h)ic hormone. ACTH.

adrenogenital syndrome. a condition caused by hypersecretion of, or excessive treat ment, with adrenal androgens. This brings about premature puberty of the male, penile enlargement, hirsutism in adult females, and shortened stature. In congenital cases, the female may be misidentified as a male.

adrenotropic hormone. ACTH.

adsorb. **1:** to adhere as an ultrathin layer on the surface of a solid or fluid without covalent bonding; said of molecules. Thus, for example, exchangeable ions readily adsorb to a surface of clay or humus. **2:** to attract and retain other substances on a surface by means of chemical or physical forces.

adsorbate. a solid, liquid, or gas that is adsorbed, as to the surface of a solid.

adsorbent. **1:** pertaining to or denoting adsorption or the capacity to adsorb. **2:** a solid or liquid (e.g., activated alumina, activated carbon, charcoal, silica gel, metals, water, mercury) that adsorbs other materials.

adsorption. **1:** the adherence of a substance (liquid, gas, or solid) to the surface of another as an ultrathin layer. **2:** a process of attracting and retaining other substances on a surface by chemical or physical forces. **3:** an advanced treatment method in which activated carbon, *q.v.*, is used to remove organic matter from wastewater. **4:** in spectroscopy, the absorption of light energy by a substance. **5:** the binding of bacteriophage particles to specific receptors on a host cell preceding the injection of nucleic acid into the cell. **immune adsorption**. the removal of an antibody from antiserum by a specific antigen or the removal of an antigen by a specific antiserum.

adsorption complex. The materials in a soil, chiefly clay and humus, that readily adsorb small particles.

adult. **1a:** full grown, mature. **1b:** descriptive of a man or woman as opposed to a boy or girl. **2:** normally an animal capable, in principle, of reproduction.

adulterant. **1:** an impurity; an additive or contaminant that is considered to be undesirable. **2:** any chemical impurity or substance that is not allowed by law in a particular food, pesticide, or other formulation. **food adulterant**. an impurity that impairs the quality or bulk of a foodstuff, feces, toxic substances or poisonous organisms, and radioactive fallout.

adulterate. **1:** pollute, contaminate, taint. **2a:** to add an impurity (an adulterant), especially one disallowed by law, to a material (e.g., to food, a feed, a pesticide, a fertilizer) that impairs its strength or quality. **2b:** to add a substance that alters the bulk of a product or dilutes the concentration of active ingredients (e.g., adding water to milk).

adulterated. **1:** said of a food, plant, animal, pesticide formulation, or any substance that contains an adulterant; one from which an active or significant constituent has been wholly or partly removed, or one that falls below a legal standard for purity. *See also* food (adulterated food) **2:** it is used in reference to a pesticide, if **a:** its strength or purity is less than the standard of quality claimed on the label under which it is sold; **b:** any substance has been substituted wholly or in part for the active agent; **c:** any valuable constituent has been partly or wholly extracted.

adulteration. **1:** the act or process of adulterating. *See* adulterate. **2:** a reduction in purity or quality of any substance by the deliberate addition of an ingredient not ordinarily part of that substance, especially where its presence is restricted by law.

advection. **1:** the process of transfer of an atmospheric property by virtue of the mass motion (the velocity field) of the atmosphere. **2:** the process of transfer of water or an aqueous property (e.g., heat or matter) due entirely to mass movement, usually by horizontal currents, of a body of water.

adventitia. the external covering of an organ (e.g., a blood vessel) or other structure, derived from connective tissue.

adventitious. **1:** accidental; unexpected; out of place; occurring in an unusual or unexpected location. **2:** descriptive of a tissue or organ that arises in an abnormal position or at an unusual stage of development. **2a:** secondary. **2b:** developing from secondary meristem. **3:** extraordinary.

adventitious bud. a bud that develops in an unusual location usually due to injury to the plant.

adventive. **1a:** newly arrived; appearing. **1b:** of or pertaining to an organism that is accidentally introduced to a new region and is imperfectly adapted; not native. Such an organism may be more susceptible than native organisms to certain poisonous substances in the region due to a lack of physiological or genetic adaptation.

adventive embryony. a phenomenon, common in *Citrus spp.*, wherein several viable embryos occur in each seed.

adverse. **1a:** detrimental, unfavorable. **1b:** antagonistic, hostile; opposed to; against. **2:** in the biological sciences, adverse usually means detrimental, harmful, damaging, or unfavorable to a living system.

adverse drug effect. *See* effect.

adverse drug reaction (ADR). any of a great variety of dose-dependent and dose-independent drug interactions, *q.v.*, that increase toxicity or result in therapeutic failure of a drug.

adverse effect. *See* effect.

adverse reaction. adverse effect. *See* effect.

adverse weather. in agriculture, adverse weather is generally considered to be weather that is unfavorable for outdoor agricultural operations.

advisory. **1:** having or exercising the authority or capacity to advise. **2:** intended to advise. **3:** a nonregulatory document that communicates risk information to persons who may be responsible for making risk management decisions.

adz (adze). **1** an axe-like tool that is used chiefly to rough dress and square large timbers. It has been largely replaced by machine tools and is no longer commonly used. An adz typically has a steel head which, unlike an axe, has a cutting edge that is mounted at right angles to the handle. Other than this construction varies; the head of some adzes have a single cutting edge while the other side is formed into a hammer or may resemble that of a pick. Some adzes have a head with two back-to-back cutting edges. **2:** to dress or shape (wood) with an adz.

adze. adz.

adzuki bean. a tropical legume (Family Fabaceae) cultivated in China and Japan.

AEC. anion exchange capacity.

aecia. a type of fruiting body.

Aedes. a genus of mosquitos (Family Culicidae). Many species are pests, and a few are disease vectors. ***A. aegypti***. a widely-distributed tropical mosquito that breeds in stagnant water near human habitations. It transmits human yellow fever and dengue. ***A. albopictus***. a large Asian species that transmits dengue and other infectious diseases. It now occurs also in the southern United States

Aelurostrongylus. A common genus of nematodes; lungworms that infest cats. Slugs and land snails are common intermediate hosts, and snail-eating animals such as certain birds, frogs, lizards, and rodents are accidental hosts that also convey these parasites to cats. ***A. abstrusis***. The most widely-distributed lungworm of the cat, it occurs, for example, in North America, Europe, and Australia. Diagnosis is difficult, since specific signs are often absent. Signs may include coughing, difficult breathing (dyspnea), wasting, and pulmonary rales.

aeolation (eolation). erosion of a land surface by wind blown particles.

aeolian (eolian). **1:** pertaining to or caused by winds; windblown. **2:** denoting or pertaining to materials transported by wind (e.g., sand, soil particles). *See* aeolian deposit, loess .

aeolian deposit (eolian deposit). any fine, dry sediment (e.g., desert sand, light volcanic ash, loess) that is deposited following transport by wind, sometimes for extremely long distances, and deposited when the wind dies. *Cf.* alluvial deposit.

aer-, **aero-**. of or pertaining air or the atmosphere.

aerate. **1:** to expose to fresh air; to ventilate or air (as a room). *Cf.* aerify. **2:** to supply (blood) with oxygen by respiration. **3a:** to introduce air into a substance, especially to replace stagnant air in soil with fresh air. **3b:** to introduce air into (a liquid) by stirring, spraying, or by some similar method. **3c:** to supply or impregnate with air. **3d:** to pass a thin layer of cool air over milk to eliminate odors. **4a:** to charge or combine with gas,sometimes carbon dioxide; to cause a gas to bubble through. **4b:** to make effervescent or sparkling.

aerated. subjected to aeration or charged with a gas such as air or carbon dioxide. *See* aerate, aeration.

aerated static pile. a composting system using controlled aeration from a series of perforated pipes that run beneath each pile and connect to a pump that draws or blows air through the piles.

aeration. **1:** ventilation, airing, freshening; replenishment of air, a process that supports and promotes biological degradation of organic materials, for example, in water bodies, soil, compost. The process may be passive, as by direct exposure to air, or active, when air is introduced into a medium by a blower or bubbling device. **2:** the replenishment of oxygen in venous blood in the lungs. **soil aeration**. aeration of soil by the atmosphere, which is usually effected by percolation of rain through the soil. Stagnant air already in the soil is driven out and fresh air is drawn into the soil as the water drains off or is withdrawn by plants.

aerator. **1:** an apparatus for aerating water or other fluids. **2:** a device for introducing air into a bin of wheat or other grain in order to prevent the accumulation of moisture, keeping it free of fungi and insects.

aerial. of, referring to, occurring in, or transported by the air (airborne).

aerial. *See* antenna (def. 1).

aerial bulblet. bulblet.

aerial dusting (crop dusting). the dissemination of pesticides or herbicides as powder over crops from a low-flying airplane.*See* aerial fertilization, aerial seeding.

aerial fertilization. broadcast distribution of fertilizers over agricultural crops or fields from an aircraft. *See also* aerial dusting.

aerial photograph. a photograph of a portion of the Earth's surface taken from an airplane or some other type of airborne platform.

aerial photography. **1:** the taking of aerial photographs. **2:** a special branch of photography in which aerial photographs are taken, usually at regular, spatial intervals, and used to provide information, for example, about land forms, crops, forests, and various structures on the ground.

aerial seeding. broadcast seeding from an aircraft.

aerifier. a machine that digs or punches holes in sod, thus facilitating the free movement of air.

aerify. **1a:** to aerate. **1b:** to infuse or force air into. **2:** to make aeriform; to vaporize; to change into an gaseous state. **3:** to foster free movement of air in sod by punching or cutting holes into the sod and by loosening soil.

aerobe (aerobic organism). an organism, especially a microorganism, that lives and grows (by means of aerobic cellular respiration, *q.v.*) only in the presence of free (molecular) oxygen. **facultative aerobe**. an organism capable of both aerobic and anaerobic growth. **obligate aerobe**. an aerobe that is incapable of functioning in the absence of molecular oxygen.

aerobic (oxybiotic). **1a:** of or pertaining to an aerobe or to processes (e.g., aerobic respiration) that proceed in the presence of molecular oxygen. **1b:** oxybiotic; growing or occurring only in the presence of molecular oxygen. **1c:** pertaining to growth in the presence of molecular oxygen. **2a:** pertaining to an entity or environment that contains molecular oxygen. **2b:** having a continuous supply of molecular oxygen. **3:** not anaerobic; obligately aerobic. **4:** said of soil that is well-drained. **excessively aerobic**. descriptive of a soil or soil horizon that is usually too dry to support adequate plant growth.

aerobic compost. *See* compost.

aerobic organism. aerobe.

aerobic treatment. the decomposition of complex organic compounds in the presence of oxygen by the use of microbes.

aerobiology. the science that treats biologically important constituents of the atmosphere and the factors that influence their behavior. (*e.g,* allergens, microorganisms, pollutants, spores).

aerogen. a gas-producing bacterium.

aerology. **1:** the branch of meteorology involving the observation of the atmosphere by means of balloons, airplanes, etc. **2:** formerly used by the U.S. Navy to mean meteorology.

aerometric. pertaining to measurement of the properties or contaminants of air.

Aeromonas. a genus of Gram-negative, rod-shaped to coccoid bacteria (Family Pseudomonadaceae) that occur individually, paired, clumped, or in chains and are facultatively anaerobic; motile or nonmotile, usually with one terminal flagellum. They occur in water and sewage. Some are pathogenic to fresh water and marine animals. The type species, ***A. hydrophila***, causes red leg disease of frogs.

aerosis. the generation of gas within a tissue or tissues.

aerosol. a suspension of ultramicroscopic liquid or solid particles, dispersed in a gas, that have a negligible falling velocity. The particles often exceed the normal size limits of colloids (ca. 1 nm to 1 μm in diameter). Fog and smoke are naturally occurring aerosols; aerosol sprays produced by humans include inhalants, some insecticides, perfumes, paints, and antiperspirants. **2:** a product, packaged under pressure, that contains materials intended for inhalation, topical application, or introduction into body orifices (*e.g.*, pesticides, various therapeutic drugs, shaving cream). Many of these products are eye or skin irritants, some are seriously toxic, and inhalation is difficult to avoid. **liquid aerosol**. one in which the suspended particles are liquid (e.g., coal tar). **respirable aerosol**. any aerosol with an aerodynamic particle size of less than 10 μm. **solid aerosol**. one in which the suspended particles are solid (e.g., flue dust).

aerosol dispersion. aerosolization.

aerosolization (aerosol dispersion). dispersion, as by spraying, of a liquid aerosol in the form of a fine mist, often for therapeutic purposes, especially via the respiratory passages.

aerotaxis. movement toward or away from air or oxygen. *Cf.* anemotaxis.

aerotropism. growth or movement toward air or oxygen.

Aesculus (buckeye; horsechestnut). a genus of ornamental deciduous shrubs and trees (Family Hippocastanaceae) with compound leaves composed of 5 to 7 palmately arranged leaflets. *A. glabra* (Ohio buckeye) and *A. hippocastanum* (horsechestnut) are the best known species. The seed is a distinctive, shiny-brown nut about 2.5 to 3.0 cm in diameter and similar in appearance to that of the edible chestnut (*Castanea spp.*). The nuts and probably the young growth and sprouts are neurotoxic due to the presence of esculin. These plants can be lethal to humans, livestock, and laboratory animals. The main symptoms are those of gastroenteritis, with inflammation of mucous membranes, depression, debility, and paralysis.

aestatifruticeta. a deciduous broad-leaved bush community of the aestilignosa, *q.v.*

aestatisilvae. a broad-leaved deciduous woodland community of the aestilignosa, *q.v.*

aesthetic value. the importance or value placed on an entity by humans due to the pleasure derived from its beauty.

aestiduralignosa. a mixed evergreen deciduous hardwood forest community.

aestilignosa. deciduous broad-leaved woodland communities (aestatisilvae) or bush communities (aestatifruticeta) of temperate climates that are characterized by recurring periods of mild, cold and damp, and dry weather. *Cf.* aestatifruticeta, aestatisilvae.

aestiv-, estiv-. pertaining to summer.

aestival (estival). of or pertaining to summer. *Cf.* hibernal, vernal.

aestival pond (estival pond). a pond that retains some water throughout the summer or during extended dry periods but freezes completely in winter.

aestivate (estivate). to spend summers in a state of lowered metabolism.

aestivation (estivation). **1:** the manner in which the immature flower parts are folded before expanding with the opening of the bud. **2:** dormancy or torpor of an animal during summer or a dry, hot period. *Protopterus* (lungfish), for example, bury themselves in swamp mud before it dries out in the hot season and emerge at the beginning of the rainy season. *Cf.* hibernation.

aetiologic, aetiological. etiologic, etiological.

aetiologic agent. etiologic agent.

aetiological. etiological.

aetiology. etiology.

AFBF. American Farm Bureau Federation.

afebrile (apyretic). without fever; not feverish.

affidavit. a written statement taken under oath by a legally valid authority.

affin-. a prefix indicating relationship.

affinin (*N*-(2-methylpropyl)-2,6,8-decatrieneamide);*N*-isobutyl-2,6,8-decatrieneamide; *N*-isobutyldeca-*trans*-2-*cis*-6-*trans*-8-trienamide; spilanthol). an insecticidal analgesic lipid amide from *Heliopsis longipes*.

affliction. **1:** an ailment or infirmity. **1b:** sometimes used as a synonym for disease.

affluent. **1a:** thriving. **1b:** wealthy, pros wealthy. **2a:** flowing toward. **3a:** tributary (adj). **3b:** a body of water (e.g., creek, river, canal) that flows into a larger body of water (e.g., river, lake, ocean). **3b:** raw sewage that flows into a waste treatment facility.

afforestation. the establishment of trees on land that has never been forested (e.g., prairie) or has lacked forest cover for a very long time (e.g., forested land cleared during pioneer days).

aflatoxicol (aflatoxin P_1). a mutagenic aflatoxin.

aflatoxicosis (aflatoxin poisoning; turkey X disease). a fungal disease due to ingestion of peanuts, peanut products, soybeans, maize, and other cereal grains, and animal feeds such as groundnut (peanut) meal contaminated by aflatoxins, *q.v.* All domestic animals are susceptible, but growing fowl (e.g., ducklings, turkey poults, pheasant chicks), young swine, pregnant sows, calves, and dogs are most often affected. Suckling mammals are also highly susceptible and may be exposed through milk while nursing. Adult cattle, sheep, and goats are fairly resistant to the more acute forms of the disease, but may respond to long continued exposure. Signs and symptoms vary, for example, as a function of sex, age, species and strain, nutritional status, amount and duration of intake, and the level of contamination of the rations. Ataxia, convulsions, muscle spasms, and a catarrhal inflammation of the intestines, especially of the duodenum, are usual. Common signs also include depression, lack of appetite, reduced rate of growth, loss of condition, bruising; reduced egg production, fertility, and hatchability of fowl; increased mortality. Most outbreaks formerly identified as "moldy corn toxicosis," "hemorrhagic syndrome of poultry," and "aspergillus toxicosis" were caused by aflatoxins in feed.

aflatoxin. a class of mycotoxins synthesized by certain strains of *Aspergillus flavus*, *A. parasiticus*, and related species that infest foods and feeds such as milk, peanuts, peanut meal, peanut butter, corn, cottonseed meal, and dried chili peppers. All are potent toxins, carcinogens, teratogens, and mutagens. They have a pronounced effect on cell-mediated immunity in humans and the domestic fowl. The presence of *Aspergillus* in food or animal feed does not necessarily indicate the presence of aflatoxins because their production depends upon factors such as temperature, moisture, aeration, and substrate. The Food and Drug Administration (FDA) has established action levels for aflatoxins present in food or feed to protect human and animal health. Action is considered when aflatoxin levels exceed 20 ppb for corn and other grains intended for immature mammals and poultry, for dairy animals, or when the destination is unknown; 20 ppb for animal feeds, other than corn or cottonseed meal; 100 ppb for corn and other grains intended for breeding beef cattle, breeding swine, or mature poultry; 200 ppb for corn and other grains intended for finishing swine of 100 pounds or greater and 300 ppb for corn and other grains intended for feedlot beef cattle and for cottonseed meal intended for beef cattle, swine, or poultry. All corn exported from the United States is tested for aflatoxin. Most important are the aflatoxins B, G, and M. **aflatoxins B**. all exhibit blue florescence. **aflatoxin B_1** (2,3,-6aα,9aα-tetrahydro-4-methoxycyclopenta-[c]-furo-[3′,2′:4,5]furo[2,3-h][1]benzopyran-1,11-dione). an extremely powerful liver poison and carcinogen in many animal species. It is extremely toxic to trout, day-old ducklings, turkeys, cattle, rabbits, cats, some laboratory rats, and swine. Dogs, sheep, and guinea pigs are slightly less sensitive; hamsters, some laboratory rats, and the domestic fowl are more resistant; certain strains of laboratory mice appear to be completely resistant. Aflatoxin B_1 is an extremely powerful environmental mutagen and carcinogen and contributes to the development of primary cancer of the liver in humans. **aflatoxin B_2** (2,3,6aα,8,9,9aα-hexahydro-4-methoxycyclopenta[c]furo-[3′,2′:4,5]furo[2,3-h][1]benzopyran-1,11-dione). a generally less toxic derivative of aflatoxin B_1. It is, nevertheless, extremely toxic to day-old ducklings and is also carcinogenic and mutagenic. **aflatoxins G**. all exhibit green fluorescence. **aflatoxin G_1** (3,4,7aα,10-aα-tetrahydro-5-methoxy-1H,12H-furo-[3′,2′:4,5]-furo[2,3-h]pyrano[3,4-c][1]-benzopyran-1,12-dione). a mycotoxin secreted by *Aspergillosus flavus* and *A. parasiticus*, it is extremely poisonous to day-old ducklings and is also carcinogenic and mutagenic. **aflatoxin G_2** (3,4,7aα,9,-10,10aα-hexahydro-5-methoxy-1H,12H-furo[3′,2′:4,5]-furo[2,3h]-pyrano[3,4-c][1]benzopyran-1,12-dione). a less toxic derivative of aflatoxin G_1, it is, however, extremely toxic to ducklings and is also carcinogenic and mutagenic. **aflatoxins M** (milk toxins). extremely toxic hydroxylated derivatives of aflatoxins B. They occur in the milk of cattle fed infested meal. **aflatoxin** M_1 (2,3,6a,9a-tetrahydro-9a-hydroxy-4-methoxycyclopenta[c]furo[2,3-h]-[1]benzopyran-1,11-dione). a carcinogenic, mutagenic compound that is extremely toxic to day-old Pekin ducklings. **aflatoxin M_2** (2,3,6a, 8,9,9a-hexahydro-9a-hydroxy-4-methoxycyclopenta[c]furo-[3′,2′:4,5]-furo[2,3-h][1]benzopyran-1,11-dione). effects are similar to those of aflatoxin M_1. **aflatoxin P_1**. aflatoxicol. *See* Reye's syndrome.

aflatoxin poisoning. aflatoxicosis.

AFP. air-filled porosity.

African. of, pertaining to, originating in, or related in some way to Africa.

African army worm. *See* ajugarin.

African ass. *Equus asinus*.

African milk plant. *Euphorbia lactea*.

African oil palm. *Elaeis guineensis*.

African swine fever (wart hog disease). a highly contagious viral disease of domestic swine. Signs include fever, cyanosis, internal hemorrhage, and usually death.

Africander. **1:** a breed of red cattle with curved horns found in parts of Southern Africa. Originally used as draught animals, Africanders are now used for beef production. They are resistant to heat and to disease. **2:** a breed of sheep.

after-ripening. physiological changes that take place during a period of apparent dormancy of many presumably mature seeds, especially in plants of the family

Rosaceae. The state cannot be broken by environmental conditions normally favorable to germination until these physiological changes (e.g., degradation of inhibitors) have occurred. This assures that germination does not occur prematurely.

afterbirth. a common name for the placenta and associated membranes expelled by way of the vagina soon after birth (about 8-10 minutes in humans).

aftergrowth. regrowth (e.g., of a crop or stand of trees) following a cutting or harvest, yielding a new crop.

aftermath. **1:** grass or other range or pasture forage that grows quickly after grazing or cutting. **2:** the grazing of crop residues in an area following harvest.

Ag. **1:** antigen. **2:** the symbol for silver.

agalactia (agalactosis). **1a:** the absence or near absence of milk in the breast after parturition. **2:** failure to secrete milk following birthing. Agalactia may occur in swine as a result of post-farrowing shock.

agalactorrhea. the lack of secretion or flow of breast milk.

agalactosis. agalactia.

agalactous. pertaining to agalactia or to the absence or decline of breast milk.

agamospermy. any type of apomixis other than vegetative propagation. *See also* amphimixis.

agar (agar agar). a complex gelatinous polysaccharide extracted from an oriental seaweed. It is used as a gelling agent to solidify liquid nutrients in producing a culture medium in which microorganisms or tissue may be grown. It is also used as a stabilizer in foods.

agar agar. agar.

agaric. **1:** any mushroom that bears gills on the underside of the cap. **2:** any mushroom of the family Agaricaceae, especially those of the genus *Agaricus*.

agaricoid. resembling an agaric.

Agaricus (agaric; agaricus). a large genus of mushrooms (Family Agaricaceae), some species of which are edible, others of which are poisonous. Poisoning by *Agaricus spp.* typically produces mild to severe gastrointestinal distress, which usually appears within a few hours following ingestion. Recovery within 1 hour to 2 days is common. ***A. californicus*** (California agaricus). a poisonous species seen most often on urban lawns and parks in California. Signs are mostly those of gastric distress. It is nearly indistinguishable from *A. campestis*, the edible meadow mushroom. ***A. hondensis*** (felt-ringed agaricus). a poisonous, foul-smelling species, usually seen scattered in woodlands from British Columbia to California. ***A. meleagris*** (Western flat-topped agaricus). a foul-smelling species assumed to be poisonous, it grows in lawns, meadows, and along woodland paths in western North America and sometimes in the east. ***A. xanthodermus*** (yellow-foot agaricus). a poisonous mushroom with an unpleasant odor that grows in lawns, along paths and under hedges in urban areas. Effects of ingestion include vomiting and/or diarrhea and usually appear within 2 hours of ingestion and usually persist for a day or two.

agaricus. a common name of mushrooms of the genus *Agaricus*.

Agave. a genus of attractive evergreen plants (Family Agavaceae) that usually have long succulent leaves with terminal and sometimes marginal spines. ***A. lecheguilla*** (lechuguilla). a hepatotoxic plant (Family Amaryllidaceae) that also causes hepato genic photosensitivity in grazing animals. Horses appear never to be affected under natural conditions, and cattle are rarely affected. Sheep and goats, however, are often affected. Grazing on agave can cause sluggishness, progressively intense jaundice with a yellowish discharge from the eyes and nostrils, eventual coma, and death. The urine is often dark, and a purplish area may appear beneath the coronary band of the hoof. Symptoms of photosensitization may or may not appear. Lesions are those of nephritis and severe toxic hepatitis. *See also* phylloerythrin.

age. **1:** the period that has passed since germination, hatching, pupation, birth, or other stage at which an organism is considered to emerge as an individual or, in some cases, as an adult. **2:** the ages of human life. These are the so-called seven ages of man: infancy, childhood, adolescence, maturity, midlife, senescence, and senility. **3a:** to mature or ripen. **3b:** to grow old; to senesce. **4:** to ripen, season, or mellow. **5:** an organism's normal life span. **6:** a period of time characterized in the human mind by the dominance of a significant process or entity. Thus, the Tertiary period is referred to as the Age of Mammals and, more recently, the Industrial Age and the Atomic Age. **6:** any very long period of time (e.g., an eon, epoch, period, era). **absolute age**. age measured in units of time such as years rather in subjective measures such as physiological condition or stage of development.

age and month-of-calving factors. factors such as different ages and months of the year at calving that affect lactation in dairy cattle. They are an aid in standardizing lactation records for genetic assessments.

age at last calving. the age of a cow at her most recent calving date minus the cow's date of birth.

age class. **1a:** a category of individuals within a population of plants or animals that fall within a specified age range (i.e., are of about the same age); a cohort.

age composition. age distribution.

age distribution (age structure; age composition). the number or proportion of individuals within each age class of a population.

age specific death rate. the death rate for a specified age class (e.g., yearlings) or cohort of a population. It is calculated as the number of individuals that die within a specified age class divided by the number of individuals that reach the age class.

age specific fecundity rate. the average number of offspring produced per unit time by individuals in a specified age class or cohort.

age specific survival rate. the number or proportion of individuals that survive to a specified age from an initial age class or cohort.

age structure. age distribution.

aged. **1:** past tense of age. **2:** having lived or existed for a long time; old, of advanced age. **3:** characteristic of old age. **4:** of the age of: a horse aged 20 years. **5:** ripened; brought to maturity or peak of flavor or smoothness (e.g., of wine, cheese. **6:** in physiography, approaching the state of peneplain.

Agency for International Development (AID; United StatesID). an agency of the U.S. Department of State that administers development projects in foreign countries, many of which are agriculturally related.

agent. **1a:** anything that produces an effect; the means, method, vehicle, instrument, force, or substance used to produce an effect. **1b:** that which acts upon a target.

See also stimulus. **aetiologic agent**. etiologic agent. **biological agent**. an agent that is biologically active; one that causes a change in or a response by a sensitive biologic system; one that acts upon a biologic target. **damaging agent**. an agent that is injurious to a living system and without immediate, direct biologic value to the affected system. **etiologic agent**. an agent that acts alone or in conjunction with other such agents to cause disease. **nonselective agent**. nonspecific agent. **nonspecific agent** (nonselective agent). an agent that acts on numerous targets (e.g., hypoxia, cold, a pyrogen). All are selective to some degree. **selective agent** (specific agent). an agent that acts on one or a few targets or on one type of target. **specific agent**. selective agent. **toxic agent**. a chemical substance or physical agent that acts directly to poison a living system at concentrations or in amounts that are not normally encountered by the system.

Agent Blue (herbicide blue). a solution of cacodylic acid (0.37 kg/l) used as a selective herbicide in the short-term control of rice and other food crops. It was employed by the United States army during the Vietnam war as a defoliant.

Agent Orange (herbicide orange). a defoliant herbicide formulation used widely by United States military forces in the Vietnam war to remove forest cover and destroy the crops. It contained equal parts of 2,4-D and 2,4,5-T and was contaminated with trace amounts of TCDD (a dioxin). A large number of birth defects among infants was soon associated with exposure of Vietnamese families to agent Orange. Subsequent to their return to the United States, American veterans who may have been exposed to Agent Orange exhibited an unusually high incidence of cancer and birth defects among their children. *See also* Agent Blue, Agent Purple, Agent White, dioxin, TCDD, 2,4-D, 2,4,5-T.

Agent Purple (herbicide purple). an herbicide used by United States military forces to control vegetation early in the Vietnam conflict, but it was replaced by Agent Orange in 1964. It is a mixture of the *n*-butyl esters of 2,4-D and 2,4,5-T and the isobutyl ester of 2,4,5-T, at concentrations of 0.50, 0.31, and 0.22 kg/l respectively. *See also* Agent Orange, Agent Blue, Agent White, 2,4-D, 2,4,5-T.

Agent White (herbicide white). an herbicide used by the United States military forces to control vegetation in the Vietnam conflict. It is a mixture of the triisopropanolamine salt of 2,4-D and picloram, at respective concentrations of 0.24 and 0.07 kg/l. *See* Agent Orange, Agent Blue, Agent Purple, 2,4-D, 2,4,5-T.

agglomeration. **1:** an accumulation, mass, clump, or cluster usually of similar units; an aggregation (def. 1, *q.v.*). **2:** a step in the granulation of fertilizers whereby small particles agglomerate to form large stable granules.

agglutinable. able to agglutinate; able to respond to an agglutinant by agglutinating.

agglutinant. **1a:** a substance that effects adhesion. **1b:** able to effect adhesion. **2:** agglutinin.

agglutinate. **1a:** to clump together as the result of an antigen—antibody reaction; used mainly in reference to cells (e.g., erythrocytes, pollen) or bacteria. **1b:** to cause to agglutinate (def. 1a). **2:** a clump of cells (e.g., erythrocytes) formed by agglutination.

agglutination. **1:** the clumping or cross-linkage of cells (e.g., erythrocytes or bacteria) or bacteria especially due to the interaction of antigens on the surface with certain antibodies (agglutinins) whereby bridges form between the antigen determi nant sites of different cells. *Cf.* lectin. **2a:** the joining or clumping of solid particles, the surfaces of which are coated with a thin adhesive layer **2b:** the capture of solid particles by impact on a surface coated with adhesive material.

agglutinative. able to cause agglutination. *Cf.* agglutinant.

agglutinin. **1a:** a substance that is able to agglutinate particles. **1b:** an antibody that causes agglutination. *See also* lectin, snakebite, venom (snake venom). **2:** **abrus agglutinin**. abrin. **plant agglutinin** (phytoagglutinin). a lectin of plant origin. *See also* phytohemagglutinin.

agglutinogen. an antigen present on the surface of cells or particles such as erythrocytes that reacts with agglutinin to produce agglutination.

aggradation. **1:** accumulation of sediment in a stream channel, an alluvial fan, or on a flood plain. This can ultimately build up to form a solid surface at grade (i.e., level with the adjacent land). **2:** sediment accumulation on a grade or slope. **3:** establishment of a uniform grade or slope by adding sediment or other material to the channels or floors of water bodies such as streams, ponds, and lakes.

aggreg-. a prefix meaning clustered or aggregated.

aggregate. **1a:** a mixture. **1b:** composite; **1c:** a conglomeration. (e.g., of soils or rocks) **1d:** entire, total, whole. **1e:** to accumulate, gather, or bring together. **1f:** combine or assemble. **2:** aggregation. **3:** a component (e.g., a particle) of a combination, conglomeration, or mixture. **4:** in agronomy, discrete clusters of naturally formed particles (e.g., crumbs, granules, clods, faecal pellets, concretions). **soil aggregate**. discrete clusters of particles formed naturally or artificially in soil. **species aggregate** (collective group). a group of presumably closely-related species that resemble one another morphologically, and are thus difficult to distinguish on this basis. Members of such a group have more in common with each other morphologically than with other members of the genus.

aggregated subarea. a subdivision of a water resource region, based on hydrologic boundaries.

aggregation. **1:** a mixture, combination, conglomeration. **2:** the process of formation of an aggregate. **3a:** a gathering or assemblage. **3b:** a gathering or assemblage of animals due to nonsocial stimuli other than social organization (e.g., the gathering of animals about a water hole or butterflies about a mud puddle). **3b:** a conspecific group of organisms that have a limited degree of coordination and integration. **3d:** the clustering of offspring about a parent. *Cf. See* aggregate (species aggregate).

aging. **1:** the process of growing old. **2:** the gradual, generally irreversible, functional decline of a mature organism as a function of time, with increased probability of organ failures, disability, and death. **3a:** the ripening of meat; the increasing tenderness of meat with age when properly stored. **3b:** the ripening of cheese which, depending on the cheese and method of treatment and storage, may become softer and/or more flavorful.

agist. to take charge of another person's livestock and care for them on one's own land.

agister. one who takes charge of and cares for another persons livestock on his or her own land for a fee.

agistment. the money paid to an agister for the care and feeding of one's livestock.

Agkistrodon (= *Ancistrodon*). a genus of 12

species of venomous snakes (Family Crotalidae) that inhabit North America, South America, and Asia. The head is broad, depressed and distinct from the neck; there is a distinct canthus. They lack rattles and are sometimes mistaken for harmless species. The body is often stout, tapered, and cylindrical, or depressed, with a short to moderately long tail. Signs and symptoms of envenomation by North American moccasins (*A. spp.*) may include pain, discoloration, swelling and edema about the bite, ecchymosis, thirst, nausea, weakness, some hemorrhage, some degree of shock, respiratory distress, and swelling of regional lymph nodes. ***A. bilineatus*** (cantil; Mexican moccasin). an extremely dangerous, brightly colored, chocolate brown to black aquatic pit viper with one thin white line along the canthus rostralis that continues behind the eye and another such line that runs above the upper lip to the corner of the mouth. Juveniles are marked by light-edged crossbands on a lighter ground color; the venter is dark brown with white markings edged with black. This snake reaches a maximum length of more than 1.2 m. It is often aggressive and the bite may cause serious local lesions, although human fatalities are uncommon. The cantil indigenous to the lower regions of the Rio Grande southward to Nicaragua. It is found along stream banks and in swampy areas on either coast of Mexico, from Sonora and Nuevo Leon, southward to the east coast of Nicaragua and the west coast of Guatemala. ***A. contortrix*** (= *A. mokasen*; American copperhead; copperhead; highland moccasin). one of the most common venomous snakes throughout its range from Massachusetts to northern Florida, west through Nebraska, Oklahoma, and west Texas. The head is triangular with a facial pit; the pupil is elliptical; the top of the head is yellowish to coppery red and paler laterally; the body is relatively stout and most subcaudal scales are undivided. It is a russet, orange-brown, or pinkish buff snake with dark brown or reddish crossbands; the belly is pale pink with large black spots or mottling. The tip of the tail is yellow in young snakes, but is dark green, brown, or black in adults. In the eastern United States it occurs most often in rocky sites in low wooded mountains or hills, wet meadows and the edges of swamps; in the southern United States it also occurs in lowland swamps; along the Rio Grande it occurs in thick stands of cane. It is nocturnal in warm weather, diurnal in cool weather. The copperhead often basks on rocks or mudflats and often hibernates in rocky areas on ledges together with various venomous and nonvenomous snakes. Copperheads usually strike reluctantly, and while the bite is painful, it is seldom life-threatening to adult humans and antivenin is rarely needed. The tail vibrates when the snake is angry. ***A. piscivorus*** (cottonmouth; cottonmouth moccasin; American water moccasin; water moccasin; cottonmouth water moccasin). a dangerous, pugnacious, semiaquatic, heavy-bodied snake, up to 1.8 m long. The body is brown or olive-colored with very dark crossbands that surround lighter centers; large specimens are sometimes uniformly black above. The underside is yellow, densely marbled with dark gray or black. The cottonmouth is often mistaken for nonvenomous watersnakes, even though the latter are generally not pugnacious and lack the distinctive elliptical of *A. piscivorus* and other crotalids. The cottonmouth inhabits swamps, shallow ponds and lakes, lagoons, and sluggish waters in the southern and central United States from Dismal Swamp in southeastern Virginia to Illinois, southward to southern Florida, and west to eastern Oklahoma and central Texas. Cottonmouths often threaten or strike when approached. While the bite is seldom fatal to humans, tissue destruction can be severe.

aglucone. aglycone.

aglycone (aglucone; genin). the noncarbohydrate moiety of a glycoside molecule. Operationally, it is any nonsugar resulting from hydrolysis of a glycoside. It accounts largely for the toxicity of toxic glycosides.

Agnatha. the class of vertebrate animals that contains the jawless fish, represented today by cyclostomes (hagfishes and lamprey eels).

agonal. **1:** of or pertaining to death, agony, the time immediately preceding death, or to the moment of death. **2:** pertaining generally to death or agony. **3:** pertaining to terminal disease.

agonic line. a line that passes through points of zero magnetic declination on the surface of the earth.

agonist. **1a:** a contracting muscle whose action is opposed by another muscle (antagonist). **1b:** an agonistic muscle; the muscle ("prime mover") that is directly engaged in the contraction that produces a desired movement.

agonistic. **1:** pertaining to any of various types of animal behavior associated with fighting, whether including conciliatory or appeasement behavior and flight as well as threatening postures or displays, fighting, and other forms of aggression. In the early stages of an encounter which might lead to conflict, an animal often shows signs of both aggression and fright. **2:** pertaining to an agonist.

agonistic behavior. *See* agonistic.

agoraphobia. a fear of open places. This may develop, for example, when an animal has been housed in a stall for a long period of time. Horses that exhibit agoraphobia are often called barn rats or referred to as picketline-bound.

agostadero. a summer pasture (southwestern United States).

agouti. **1:** common name of rodents of the genus *Agouti*. **2:** a coat color typical of many wild mammals. It is consists of alternate light brown or brownish-yellow and dark brown strips of fur.

Agouti. **1:** a genus of rodents (Family Dasyproctidea) of forests and plains ecosystems. They are closely related to the guinea pig, *Cavia*. They can run swiftly and are superficially pig-like in appearance; the body color is agouti (def. 2). They are a reservoir host for *Trypanosoma cruzi*. **2:** the wild type coloration of many rodents. The ground color is a speckled brown or chestnut usually with alternate light and dark grizzled banding.

agr-, agri-, agro-. prefixes indicating or pertaining to fields, farms, or farming.

agrad. a cultivated plant.

agranulocyte. a type of leukocyte (e.g., lymphocytes and monocytes) that contains no distinct cytoplasmic granules. *Cf.* granulocyte.

agrarian. **1:** of or pertaining to land or acreage. **2a:** of, by, pertaining to, or characteristic of farmers, their practices, or way of life. **2b:** pertaining to cultivation or to a cultivated area. **3:** growing wild on agricultural land. **4:** one who favors agrarianism. *Cf.* agrestic.

agrarian zone. the area of land devoted to cultivation in a region.

agrarianism. **1a:** the doctrine of equitable redistribution of landed property. **1b:** a movement or political party that promotes land reform to improve the financial situation of the farmer. **2:** the life style associ ated with an agricultural or rural economy

agrestal (agrestial). growing wild; referring to plants that grow wild on arable land.

agrestial. agrestal.

agrestic. 1a: of or pertaining to agricultural land or to the countryside. **1b:** rural, agrarian, agricultural. **1c:** not urban.

agri-, agro-. prefixes referring to agriculture or the cultivation of crops.

agribusiness. **1:** the entire complex of business and business-related activities that support farming as a productive enterprise, including the provision of warehousing, supplies, and services, including veterinary services; the manufacture and sale of agricultural machinery; the processing, storage, distribution, and sale of farm commodities. Included also are indirect services such as the education and training as by land grant universities, soil and water conservation services, county extension services, and agricultural experiment stations. **2:** any large company or corporate entity that secures its profit from agricultural products.

agric. **1:** a B-horizon of clay and humus in a soil profile that was deposited and formed as a consequence of cultivation. **2:** abbr. for agriculture, agricultural.

agricere. a resinous or wax-like coating of organic matter on soil particles.

agrichemical. agricultural chemical.

agricultural (abbrs.: agrl., agric). **1a:** denoting, pertaining to, having to do with, or characteristic of agriculture. **1b:** pertaining to one engaged in farming as his or her chief occupation. **2:** of, pertaining to, or characteristic of a farmer, or farming, or activities or phenomena related to farming.

agricultural agent. **1:** county agent. **2:** an expert employed by a business to promote agriculture in its sales or marketing territory.

agricultural agent. *See* county agent.

agricultural aircraft. an aircraft designed or adapted for use in agriculture or forestry.

agricultural biochemistry. *See* biochemistry.

agricultural chemical (agrichemical; agrochemical). any manufactured or formulated chemical commonly used in agriculture. This importantly includes algicides, fertilizers, herbicides, insecticides, soil conditioners, fungicides, feed additives, pharmaceuticals, and other chemicals used to protect crops or livestock or to increase their yield or quality. Many of these chemicals or their decomposition products pollute the environment and may accumulate in soils or water bodies or may pass through food chains. Depending on chemical properties (e.g., solubility, volatility, stability), such chemicals may produce off-site as well as on-site pollution through runoff, suspension in the atmosphere by winds, food chain accumulation, or other processes. In some cases, traces may remain on fruit or other products used as foods even after thorough washing. Some pose a danger not only to humans, but also to fish, wildlife, and/or livestock. *See also* agricultural chemistry, agricultural biochemistry, agroecosystem.

agricultural chemistry. the science of chemical composition and behavior of agricultural chemicals and the substances or materials with which they may react. Studies and applications relating to the production, protection, and use of crops, livestock, foods, and other materials produced by agricultural activities are included. This science is also concerned with the chemistry of all processes whereby food and fiber are obtained, stored, distributed, and used. The chemistry of related life processes, chemical genetics, and chemical methods used to increase yields, improve quality of products, and reduce costs are important components of the field. *See also* agricultural chemical.

agricultural commodity. any agricultural goods or product.

agricultural depopulation. movement of people from rural areas.

agricultural diversification. **1:** the production of a variety of crops and/or livestock on a given tract or farm. **2:** a system of farming that facilitates and manages the production of a variety of plants and animals and their products. Well-managed diversification systems afford more stability than monocultures or other specialized systems. *Cf.* agricultural specialization, monoculture.

agricultural economics. the discipline that treats the economic aspects of agricultural production and marketing, including economic impacts on and economic interactions with society, commerce, industry, and government.

agricultural ecosystem. agroecosystem.

agricultural engineer. an engineer that specializes in agricultural engineering.

agricultural engineering. the branch of engineering concerned with improving the effectiveness and efficiency of agricultural ventures and operations; improving the quality, production, and delivery of agricultural commodities to the appropriate markets. Diverse engineering skills and specialties fall under this definition, include the design of soil management and erosion control devices, farm machinery, irrigation systems, water quality and water supply equipment, as well as food processing machinery.

agricultural experiment station. established by the Hatch Act of 1887, an agricultural experiment station was to be affiliated with the land grant college of agriculture in each state. Research undertaken at these stations underpins the curriculum of these colleges as well as outreach to the farming community.

agricultural geologist (agrogeologist). One trained in agricultural geology. *See* geology.

agricultural geology. *See* geology.

agricultural holding. real property (e.g., a farm) that constitutes a basic unit of agricultural production.

agricultural machinery. mechanized equipment used to till, plant, cultivate, or harvest crops.

agricultural pipe drain. a network of perforated or porous pipes placed in a trench lined with gravel or similar material as a means of draining subsoil.

agricultural product. any product produced on a farm, including derivative products.

agricultural region. a large land area defined by a dominant agricultural attribute such as the major crop or type of crop produced or the type of farming employed.

agricultural resource base. natural resources (e.g., soil, water) and environmental conditions (climate) that sustain agricultural practices and the production of a crop.

agricultural science. any of a large array of sciences that treats all aspects of agricultural theory and practice including the selection, breeding, and management of crops and domestic animals. Included are subdisciplines within a very broad range of basic and applied sciences that draw on and applies their knowledge and/or methods to the problems or interests of agriculture, considered broadly. Included are large subdivisions of the life sciences and chemistry including, for example, microbiology, parasitology, veterinary medicine, biochemistry, ecology, poultry husbandry, animal husbandry, agrology, agronomy, agricultural engineering, biotechnology, gene technology, agricultural and resource economics, and waste management.

agricultural specialization. **1:** the production of one (monoculture) or very few crops or types of livestock on a given tract or farm. **2:** a system of farming that optimizes production of specific a type or species of plant or animal. Such systems may be more efficiently managed and operated than a diversified system, but are often less stable. *Cf.* agricultural diversification, monoculture.

agricultural substitute. a commodity that is not produced as a result of farming practices but is manufactured as a substitute for such (e.g., soy protein meat extenders or substitutes, corn-oil margarine).

agricultural treadmill. the production of food at lower and lower prices due to increased efficiency of production with no increase in profits.

agricultural waste. all waste materials produced as a result of agricultural activities, including, for example, residues from the application of pesticides, herbicides, fertilizers and other chemicals, wastewater, manure, bedding material, and any material that is not recovered for reuse.

agricultural zoning. protection of farmland and farming activities from incompatible nonfarm uses through zoning. Agricultural zoning can specify, for example, the uses permitted with an agricultural zone, minimum farm size, the number and type of nonfarm dwellings permitted, and/or the presence and size of a buffer zone to separate farm and nonfarm land.

Agricultural Adjustment Act (AAA) of 1933. an act of Congress that provided economic relief to farmers in the form of payments to farmers for limiting production. Provisions for loans and credit to farmers and farm cooperatives were also included.

Agricultural Conservation Program (ACP). a federal program, administered by the Farm Service Agency, to encourage and assist owners of agricultural lands to conduct conservation practices. In particular, it shares the cost of improving or restoring soil fertility, minimizing erosion by wind and water, and conserving agricultural and natural resources including wildlife.

Agricultural Credit Association (ACA). an institution of the Farm Credit System that has the authority to make loans to agricultural producers, rural homeowners, and certain farm-related businesses.

Agricultural Fair Practices Act of 1967 (P.L. 90-288). a law enacted to protect farmers from retaliation by buyers of their products because the farmers are members of a cooperative. The act allows farmers to file complaints with the USDA, which can then institute court proceedings, if they believe their rights under the law have been violated. Additional bills have been introduced in recent years to give producers more bargaining power under the act.

Agricultural Marketing Service (AMS). an agency of the U. S. Department of Agriculture that sets standards for grades of cotton, tobacco, meat, dairy products, eggs, fruits, and vegetables. It also administers inspection and grading services, market news services, and oversees the administration for federal marketing orders.

Agricultural Quarantine Inspection (AQI). A program administered by the Animal and Plant Health Inspection Service of the U.S.Department of Agriculture. Incoming passengers, luggage, and cargo at U.S. ports of entry are inspected in order to protect United States agriculture from foreign animal and plant pests and diseases.

Agricultural Research Service (ARS). an agency of the USDA that conducts basic, applied, and developmental research on livestock, plants, soil, water and air quality, energy, food safety and quality, nutrition, food processing, storage, and distribution efficiency, non-food agricultural products, and international development.

Agricultural Stabilization and Conservation Service (ASCS). an agency of the USDA that was formerly chiefly responsible for administering the farm commodity price and income support programs and the conservation cost-sharing programs. Its functions were folded into the newly-established Farm Service Agency in 1994.

Agricultural Trade Development and Assistance Act of 1954 (P.L. 83-480; "PL 480"; "Public Law 480"). a public law that aims to expand foreign markets for United States agricultural products, combat hunger, and encourage economic development in developing countries. The act established a federal program that makes United States agricultural commodities available overseas through long-term credit at low interest rates, donation of food commodities to poor nations (*See* Food for Peace Program), and by exchanging agri-
agricultural commodities for local currency.

agriculture (abbr.: agric.). **1a:** the science or art of cultivation of the soil or the culture of plants or animals in water systems and associated activities that may be required to produce, harvest, and market useful crops and/or to raise animals for profit (e.g., farm layout, community relations, compliance with cultural values and relevant laws and regulations, the use of agricultural chemicals such as fertilizers, herbicides, and pesticides, irrigation, and pollution control). **1b:** the art and culture associated with agriculture as defined above. **1c**: agriculture in the broadest sense includes not only the lore and art of farming but also the knowledge and applications gained from scientific disciplines such as agronomy, agroeconomics, agricultural biochemistry, agricultural ecology, agricultural engineering, biotechnology, and waste management. Included in this definition also are industries and agribusinesses that support the production or distribution of agricultural products and services. In addition to foods, agricultural enterprises produce a remarkably wide variety of products including chemicals (e.g., dyes, drugs, alcohols, opiates, fertilizers, soil conditioners, pesticides, herbicides, fungicides, and insecticides), fibers, cultured pearls, herbs, food additives, tobacco, saddle horses, and *Bos taurus*). **2:** farming *per se*. **alternative agriculture** (unconventional agriculture). agriculture that is considered by many to be unconventional. Some of the practices of alternative agriculture are, however, conventional and even essential agricultural practices; others are on their way to becoming conventional. Alternative agriculture systematically employs farming methods aimed at reducing agricultural pollution and fostering sustainability, while improving efficiency and profitability. Management practices are employed that, to the extent possible, make use of natural processes (e.g., nitrogen fixation, nutrient cycles, and predator/parasite-pest relationships). *Cf.* conventional agriculture, sustainable agri culture. **contemporary agriculture**. modern agriculture. **conventional agriculture**. agriculture as practiced by most agriculturalists today. It is comprised of four major subsystems: biological, labor, farm economics, and socioeconomics. The biological system applies knowledge taken from the various biological sciences to the productive management of crops and livestock. The labor component includes the means of achieving productive goals by planning and the integration of human skills, labor, machinery, and external energy. Farm economics deals with all aspects of a farm's economy, including the quantities of crops, livestock, and/or goods produced and associated risks, supplies used, costs of production, pricing, income and profit. The socioeconomic component applies the theory and logic of social science and economics to the problems faced by a particular type of farm or agricultural production. Included are considerations of the costs of production, marketing of products, labor costs and availability, land use rights, machinery, interactions with the sociopolitical environment, research, credit, taxation, and technical assistance. *See also* alternate agriculture, modern agriculture. **industrial agriculture**. agriculture that employs modern equipment and methods for large-scale production, processing, and packaging of agricultural goods. **industrialized agriculture**. agriculture marked by an industrialized structure and organization. One of the most industrialized agricultural sectors in the United States is that of animal production. When such a structure is initiated in a region based on conventional agriculture, conflict among affected parties often arise. Major concerns often include the question of social impacts on rural communities, the question of private rights and their protection, health and environmental risks due to the adoption of new production systems, and nuisance impacts such as the odor from industrialized swine operations. **industrialized animal agriculture**. livestock production, management, and marketing characterized by an industrialized structure, organization, and corresponding efficiencies. **modern agriculture** (contemporary agriculture). agriculture that draws on and includes the major components of conventional agriculture but is moving rapidly beyond conventional methods and tradition. Modern agriculture depends to a high degree on the changing structure and needs of society as a whole. It draws heavily on modern scientific knowledge including bioengineering, genetics and genetic engineering, and especially on ecology. It is characterized by the application of modern ecological knowledge and ecosystems theory to farming. *See also* agricultural ecology, agroecosystem, alternative agriculture. **organic agriculture**. organic farming. **organic agriculture**. *See* organic farming. **preindustrial agriculture**. nonmechanized agriculture, that which is dependent chiefly on human and animal labor including human and animal-powered tools (e.g, plows, harrows), as it was prior to the industrial revolution. Some societies are still in an essentially preindustrial phase of agricultural development. Traditional agriculture, *q.v.*, although practiced also in industrial societies, is a form of preindustrial agriculture. **subsistence agriculture**. subsistence farming. *See* farming. **sustainable agriculture**. a systematic approach to agriculture in which the productive capacity, integrity, and biotic (floral and faunal) diversity of an agricultural enterprise or agroecosystem is maintained more or less indefinitely even in the face of widely varying uncontrolled conditions. It is an integrated system designed to indefinitely satisfy food and fiber needs. Such a system maintains or enhances environmental quality, uses resources efficiently and effectively, uses environmentally friendly technology, maintains the economic viability of the farm, and enhances the quality of life for neighbors, producers, and consumers. Practices employed in sustainable agriculture include crop rotation, animal and green manures, soil and water conserving tillage systems, and integrated pest management. The FACT ACT of 1990 defines sustainable agriculture as an integrated system of plant and animal production practices having a site-specific application that will, over the long term (1) satisfy human needs for food, fiber, and other important agricultural commodities; (2) enhance environmental quality and the natural resource base upon which the agricultural economy depends; (3) make the most efficient use of non-renewable resources and on-farm resources and integrate, where appropriate, natural biological cycles and controls; (4) sustain the economic viability of farm operations; and (e) enhance the quality of life for farmers and others. **traditional agriculture**. agriculture, including preindustrial agriculture, that is based largely on traditional knowledge acquired by local rural people by written and/or oral traditions passed down through many generations. The origins of such traditions are the result of direct interaction of a rural population with the local environment. Traditional farming is based on human and animal labor and certain farming customs, principles, and practices, adapted to the local environment and to the type of farming engaged. These may include control of weeds and pests, the use of multiple cropping patterns that maintain spatial and temporal diversity, efficient use of space and resources, irrigation and water conservation, and replacement of nutrients and the maintenance of soil fertility by means of partially-closed cycles of water, nutrients, energy, and wastes. Soilnutrients may be replenished by adding materials such as manure, urine from livestock, and/or forest litter to the soil, by allowing some fields to remain fallow each season, and by planting legumes and practicing crop rotation. **unconventional agriculture**. alternate agriculture. **village level agriculture**. the use of land and resources to produce crops and livestock intended to be optimal for a village or small community as a whole. Much of the work is manual and resources may be shared, including the use of a commons, *q.v.* *See also* green (def. 1). A mix of modern and traditional agriculture may be practiced depending on the resources available, education of the population, government regulations, and other factors. **wet-rice agriculture**. the cultivation of rice by planting on dry land, transferring the seedlings to a flooded field, and draining the field before harvesting.

Agriculture in Concert with the Environment (ACE). A program of the U.S. Environmental Protection Agency, conducted in cooperation with the Sustainable Agriculture Research and Education (SARE) program administered by the USDA. The program funds research projects that reduce the risk of pollution from pesticides and soluble fertilizers.

agriculturist. **1a:** a person who farms or is otherwise directly engaged in agriculture (e.g., a farmer, a husbandman). **1b:** a person engaged in any aspect of agriculture including, for example, the production of crops, agricultural research, or agricultural education. **2:** one who is trained in the theory or practice of agriculture, including writers, editors, educators, and agricultural scientists and others who engage in agricultural research.

Agrimet. phorate.

Agrimonia eupatoria (meadow grass). an herb (Family Rosaceae), it causes a phototoxic contact dermatitis (meadow grass dermatitis). Sunbathers are most often affected.

agrioecology. agricultural ecology.

agriotype. the wild or ancestral type of a

cultivated plant or domestic farm animal.

agrium. a biotic community on land under the influence of humans, especially on agricultural land.

agrl. abbreviation of agricultural.

agro-. a prefix meaning: **1a:** of, pertaining to, or belonging to productive or agricultural land (e.g., farmland, forested land) or soil. **1b:** pertaining to the cultivation of crops. **2:** agricultural.

agro-city. *See* agro-town.

agroanalogous. jargon denoting the presence of a similar climate and a suitable environment for the production of similar (analogous) crops. *See* analogous, analogous organ.

agro-town (agro-city). a group of collective farms in the former U.S.S.R. that was managed and operated as a unit.

agrobiologist. a specialist in agrobiology.

Agrobacterium. a genus of bacterium (Family Rhizobiaceae) that do not fix free nitrogen. Three of the four species are plant pathogens that produce galls and hairy root.

agrobiologic. of or pertaining to agrobiology.

agrobiology. the science that applies biological knowledge and methods to the study of those processes that affect or are important to agricultural productivity such as plant and animal nutrition, growth, and production, aspects of soil management that affect production, and pest control.

agrochemical. agricultural chemical.

agroclimatic. of or pertaining to relationships between climate and agricultural production, especially with respect to the adaptation of crops to a particular climate.

agroclimatological. of or pertaining to agricultural climatology.

agroclimatology. agricultural climatology. *See* climatology.

agroecology. agricultural ecology. *See* ecology.

agroeconomic. **1:** of or pertaining to the economics of agriculture in general, especially that of a nation's agricultural economy. **2:** pertaining to the financial concerns of farmers as a group. **3:** pertaining to the economics of a particular type of agronomic system.

agroeconomic zone. an agricultural area defined by environmental factors such as climate, soil type, land use, other available resources, and market access. These factors affect the nature of farming, crops produced, and marketing methods.

agroecosystem (agricultural ecosystem). **1:** any of a variety of types of ecosystems produced, to a high degree controlled, and modified through human agricultural and sociopolitical activities. Such systems differ generally from more natural ecosystems in a number of ways. They are, for example, characterized by absence or short-circuiting of natural succession, simplified structure, often a monoculture, regular geometric patterns of crop plants that favor the growth of pest populations, and serious disruptions (e.g., from plowing, burning, harvesting) that are more frequent than in natural systems. Major controlling factors may include intensity and frequency of natural (e.g., drought) and human (e.g., plowing, fire, harvesting) perturbation. Much of the analytical framework is rooted in systems theory. Agroecosystems may appear to be less complex than many natural ecosystems because of their simplified community structure. Such may not be the case as they are strongly influenced by varied and intense interactions with human activities as well as those biotic and abiotic factors that influence natural ecosystems. The characteristics of a given agroecosystem are both site- and time-specific as a function of (1) agricultural practices employed, on the particular plants and animals that are introduced and managed, marketability of products, and social or political interventions, (2) geographic, physiographic, climatological, and other naturally occurring biotic and abiotic environmental factors. Agricultural chemicals (e.g., fertilizers, pesticides) are often employed to sustain productivity and biotic composition of the system or to improve or maintain the commercial value of the system and its products. Many of these chemicals are toxic to humans and other living organisms that may be exposed, for example, due to resuspension of chemicals in air or surface waters, by percolation through soil into groundwater, and/or by transport through food chains. Organisms living outside the ecosystem may thus also be exposed. **2:** an agricultural unit, often a field (e.g., cropland, grassland, orchard), including all living and nonliving components and the constellation of processes and interactions, internal and external, that control features such as nutrient flux, population structure, cycles of growth, development, and cropping, productivity, stability or sustainability, and resilience.

agroecotype. a crop plant adapted to a particular environment, especially an edaphic ecotype, *q.v.*

agroforestry. a generic term used to describe an old, widely practiced land use system in which trees and shrubs are grown simultaneously or sequentially with forage crops and/or livestock with a view to maximizing productivity on a sustained basis. Such systems can be characterized by their structure (e.g., the mix of trees, crops, and animals), sustainability, productivity, and socioeconomic and cultural adaptability. The composition and characteristics of agroforestry systems are site-specific and strongly dependent on the local environment and cultural practices. Cattle and sheep are the livestock most commonly involved in agroforestry.

agrogeologist. agricultural geologist.

agrogeology. agricultural geology. *See* geology.

agrogorod. an set of amalgamated collective farms as in the former Soviet Socialist Republics.

agroindustrialize. **1:** to industrialize the agriculture of a region or especially that of a developing nation. **2:** to convert or organize a particular type of agricultural endeavor or system of production into an agroindustry (e.g., to agroindustrialize poultry production). **3:** to become an agroindustry; to convert a commercial enterprise or industry into an agroindustry. *See also* agriculture (industrial agriculture).

agroindustry. agriculture conducted on a large scale by industrial/commercial organizations that conduct or contribute to enterprises involving industrial agriculture. *See* agriculture.

agrologic. of or pertaining to agrology.

agrologist. a specialist in agrology.

agrology. the agricultural science that treats the origin, composition, structure, analysis, and classification of soils, especially in relation to crop production. *Cf.* agronomy.

agrometeorology. agricultural meteorology.

Agromyza. the type genus of the family Agromyzidae. Included are economically im portant insect pests such as leaf miners and certain gall-producing flies that infest both cultivated and wild plants.

agromyzid. **1a:** of or pertaining to the Agromyzidae. **1b:** common name of any insect that is assigned to the family Agromyzidae.

Agromyzidae. a family of small or minute two-winged flies. The larvae are phytophagous; many are leaf miners on cultivated plants.

agronomic practice. any activity that uses principles or information from the science of agronomy to support, foster, or enhance crop production. Such practices include selection of seed, seedbed preparation, fertilization, use of suitable soil amendments, planting, cultivation, harvesting, crop sequencing or rotation, strip-cropping, and use of cover crops.

agronomic. of, pertaining to, characteristic of, or designed to promote agronomy.

agronomist. a scientist or other specialist in agronomy. **pasture agronomist**. an agronomist that specializes in the study or management grasses that grow in pastures.

agronomy. **1a:** that branch of agriculture devoted to the theory, practice, and scientific management of soil, land, and field-crop production. **1b:** the scientific utilization of agricultural land for the production of crops. *Cf.* agrology.

agrophile. an organism that lives successfully in cultivated soils.

agrophilous. thriving in cultivated soils.

agrophily. the state, condition, or process of living and thriving in cultivated soils.

agropyron mosaic. a viral disease of wetland grasses (*See Agropyron*). The leaves become mottled and necrotic and the plants are stunted

Agropyron (wetland grasses; wheatgrass; quack grass). a genus of wild, widely-distributed, usually perennial, North and South American grasses (Family Gramineae); they favor temperate and cool regions. The spikes are erect and usually bear a single, several-flowered, sessile spikelets. Grasses of this genus are dominant at high elevations in the northern Rocky Mountains, in the short grass prairie of western North and South Dakota, and in mixed grass prairies. They may, at times, become heavily ergotized, posing a hazard to livestock, wild herbivores, and potentially to humans. *See Claviceps*. ***A. dasystachyum*** (hairy-spiked grass). a species that favors dunes and dry open soil. ***A.inerme*** (beardless bluebunch wheatgrass; beardless wheatgrass). a perennial forage plant that grows wild in the western United States It is also used for erosion control. ***A. repens*** (witch grass; couch grass; quick grass; Chiendent (Que.)). an annual grass with slender, creeping rootstocks that are often considered a weedy species (*See* chlorfenac, Eptam, EPTC). ***A. smithii*** (western wheatgrass). a cool season (C3) grass of the shrub steppe, it occurs on prairie and drier soils throughout much of the United States and into Quebec.

Agrostemma (corn cockle). a genus of tall, silky, annual or biennial European herbs (Family Caryophyllaceae) that contain toxic saponins. ***A. githago*** (corn cockle; purple cockle). a common noxious weed of grain fields and sometimes roadsides. It is occasionally cultivated as an ornamental annual. The petals are purplish-red with black spots. The whole plant is poisonous. The seeds may be harvested with winter wheat or rye and, unless carefully separated from the grain, may cause a sometimes fatal poisoning of livestock. Humans who use contaminated flour may also suffer serious poisoning. In mild cases, humans may suffer a sore throat, nausea, dizziness, weakness, acute indigestion with severe stomach pains, and fever. Severely poisoned individuals may also experience delirium, a much reduced breathing rate, sharp pains along the spine, reduced or even arrested breathing, coma, and death. *See* githagism.

Agrostis (bentgrass). a genus of widely distributed stiff-stemmed grasses (Family Gramineae) including species that are important as hay, in pasturage, and as lawn or turf grasses. Most are slender, delicate plants, native to cool climates. They have open or contracted panicles, and each spikelet bears a single small flower. ***A. alba*** (redtop; herd's grass). a wild North American, rapidly growing, somewhat drought-resistant, turf-forming, perennial grass with reddish panicles. Often mixed with clover, redtop is much used for pasture and hay in northeastern North America. Nevertheless, it is sometimes heavily ergotized and can thus be hazardous to livestock and herbivorous wildlife. This grass is also effective in erosion control. *See Claviceps*. ***A. palustris*** (creeping bentgrass). a creeping, moisture-tolerant grass that tolerates a range of soil conditions. It can spread rapidly by stolons. It is a lawn and putting-green grass, European in origin. It is also used to stabilize irrigation and drainage ditches at the waterline. ***A. tenuis*** (colonial bentgrass; creeping turfgrass.). a perennial grass, European in origin, that is not suitable as a pasture grass due to low productivity, but is often used in erosion control and in lawn mixtures.

agrostographer. a specialist in agrostography.

agrostographic. of or pertaining to agrostography.

agrostography. the branch of agrostology that treats the description of grasses.

agrostologic. of or pertaining to agrostology.

agrostologist. a taxonomist who specializes in agrostology.

agrostology. **1:** the division of plant taxonomy that treats grasses and grasslike plants. **2:** graminology.

agrotechnology. biotechnology, *q.v.*, applied to agricultural systems, productivity, marketability, and other concerns.

agrotechny. the branch of agriculture that deals with the conversion of agricultural crops into manufactured products on the farm or in the farm community.

Agrotis ypsilon (black cutworm). a surface cutworm (family Phalaenidae) found throughout much of the United States It is a pest of truck crops.

agrotype. **1:** an agricultural variety, cultivar, or race. Among plants and animals that have been domesticated for a long period of time, however, it may not be possible to establish the race (or even the species) in relation to progenitors or related feral forms. **2:** any of various soil types that support agriculture.

agstone. agricultural limestone, a soil liming material, as opposed to limestone rock and that used in cement manufacture.

AHS. African horse sickness. *See* sickness.

AI. artificial insemination.

AI daughter. a cow deliverd by an artificially inseminated dam.

aid. that which is used to control or guide a horse while riding, e.g., the hands, knees, legs, body weight, and voice.

AID. Agency for International Development.

ailment. illness (def. 2); a condition of mild illness, physical disorder, or chronic disease; roughly synonymous with disease.

air. the mixture of gases that make up the atmosphere, *q.v.*, that surrounds the earth. **ambient air**. **1:** any unconfined portion of the atmosphere; open air; outdoor air. **2:** the air immediately surrounding an organism or other object of interest. **3:** outdoor air. This definition is the one most commonly encountered in air pollution science and in various government regulations. **atmospheric air**. the air that blankets the earth. **indoor air**. air in a confined area such as a home or office building. Indoor air unless monitored and managed will become polluted, sometimes dangerously so. **open air**. ambient air; outdoor air. **outdoor air**. **1:** ambient air. **2:** atmospheric air at a given location. **pure air**. usually taken to be a mixture of gases that contains, when dry, about 78% nitrogen, 20.95% free oxygen, 0.93% argon, 0.03% carbon dioxide, and trace amounts of numerous other components.

air changes per hour (ACH). a depiction of the movement of a volume of air in a specified period of time. Thus, a building has one air change per hour if all of the air in the building is replaced in one hour.

air classification. the separation of mixed wastes by means of a stream of air, whereby light wastes are carried upward and heavy or high density components drop out.

air curtain. a "curtain" of bubbles released from a perforated pipe. Such are used in the containment of oil spills in relatively still waters and are sometimes employed to restrain fish from entering polluted areas of a water body.

air drainage. flow of cold (dense) air down a slope due to gravity; this may be compensated by upward movement of warm air in the same manner.

air dry. to dry or dehumidify material (hay, wood, forage) by exposure to open air.

air layering (Chinese layering). a method of propagating woody plants by rooting branches. The stem to be rooted, without disturbing its attachment to the plant, is incised, and sphagnum moss is wrapped around the cut part and well moistened. The stem is then wrapped tightly with semiporous material such as cheesecloth, plastic, or wire mesh and left until the roots are well-established in the sphagnum moss. The branch is then removed from the parent plant and transplanted.

air plowing. the pumping of air into the deeper, oxygen-poor region of a water body (e.g., a pond or lake) to aerate the water and bottom sediments directly and to promote further aeration by causing the deeper waters and sediments to intermix with oxygen-rich surface waters.

air pollution index. any of a number subjective indices that are taken to be a measure of pollution intensity. Such indices are usually a function of the concentration of one or more ambient air pollutants. The index most widely used in the United States is 10 times the SO_2 concentration plus the CO concentration (in ppm by volume) plus twice the coefficient of haze. Human health is usually considered to be at risk when this index rises to a value of 50 or more. Several similar indices are used locally. *See* air quality index.

air pressure. the force exerted by unconfined, compressed, or partially evacuated air on any surface with which it is in contact. *Cf.* atmospheric pressure.

air quality index. a standardized indicator of air pollution intensity formulated by the United States Environmental Protection

Agency (EPA) for public use. This index is based on a set of numbers derived from a combination of environmental concentrations of ozone, suspended particulates, sulfur dioxide, and carbon dioxide. No pollution is indicated by zero; 100 is standard; 200, alert; 300, warning; 400, emergency; 500, significant harm can be expected. *See* pollution index.

air quality standard. the concentration of a particular air pollutant that may not be exceeded during a specified time in a given geographic area as prescribed by governmental regulations.

air sac mite. *Cytodites nudus*.

air sacs. **1:** extensions of the bronchi (usually of secondary bronchi) of birds that form thin-walled sacs. The air sacs project into the lungs and ramify throughout the body, forming a reservoir. These sacs allow double passage of air through the lungs. That is, the lungs are ventilated during both inspiration and expiration. **2:** small sac-like dilations of the tracheae of many insects that fly rapidly. They increase the ventilation of the system tracheae in these insects.

air tanker (airtanker). a fixed-wing aircraft with the ability to drop water or other fire suppressants or retardants to aid, for example, in extinguishing or controlling forest fires.

air-filled porosity (AFP). the air capacity of a compost.

airborne release. the release of any chemical into the air. This may occur actively and deliberately as when spraying crops with a pesticide. It may also occur passively as when pesticides or other agricultural chemicals are resuspended, e.g., by mechanical disturbance, wind erosion, or evaporation.

airborne. carried or transported by or through the atmosphere.

airsacculitis. an inflammation of the air sacs of birds.

airtanker. air tanker.

airway. **1:** any passage in the respiratory tract through which air passes during breathing (e.g., nasal cavity, mouth, tracheae, bronchi, and bronchioles). **2:** a passage for air in a mine.

aitchbone (edge bone). **1:** the rump bone (floor of the pelvis) of an animal, especially of beef cattle. **2:** the cut of beef that includes this bone.

Aizen Food Blue. FD&C Blue No. 1

Ajuga remota (bugle weed). a European herbaceous plant (Family Laminaceae), the leaves of which contain ajugarins.

ajugarin. any of a class of diterpenes isolated from the leaves of *Ajuga remota* and related plants. Ajugarins I, II, and III are effective antifeedants against the African army worm, a moth larva (Family Noctuidae) that feeds on alfalfa and various grains, sometimes in huge numbers. Ajugarin IV is insecticidal, and ajugarin V appears to be biologically inactive.

akala. *Rubus macraci*.

akee. a common name of *Blighia sapida* and its fruit in the United States and Haiti.

akene. achene.

aki. a common name for *Blighia sapida* and its fruit in Haiti.

akinesia (akinesis). the absence or cessation of voluntary movement, e.g., of an animal.

akinesic. akinetic.

akinetic (akinesic). of or pertaining to akinesia.

Al. symbol for aluminum.

ala tau. a breed of cattle, gray-brown in color, found in Kasakhstan. It serves as both a dairy and beef cow.

alachlor (2-chloro-*N*-(2,6-diethylphenyl)-*N*-(methoxymethyl)acetamide; 2-chloro-2′,6′-di-ethyl-*N*-(methoxymethyl)acetanilide; metachlor; Lasso). a preemergence selective anilide herbicide used chiefly to control weeds in corn and soybean fields. It is absorbed through the roots and inhibits root elongation. This compound persists in the soil for up to 3 months and is a common groundwater pollutant in some areas. Propachlor, diphenamid, and naptalam are herbicides with similar action. Alachlor is readily absorbed through the skin of humans and may produce allergic skin reactions, eye irritation, and liver damage. Some effects associated with alachlor may be largely due to contamination by polychlorinated dibenzodioxins.

Alar. trade name for daminozide.

alarm reaction (alarm response). the sum of all nonspecific responses to the sudden exposure to chemical or physical stimuli to which an organism is not adapted.

alarm response. alarm reaction.

alarm stimulus. a stressor; a stimulus that causes an alarm reaction.

Alaska spruce beetle. *Dendroctonus borealis*.

alat-. a prefix meaning winged or wing-like.

alata. a winged adult insect in taxa (e.g., aphids) that include wingless (aptera) forms. *See also* aptera.

alate. winged, having wings.

alb-, albi-, albo-. prefixes meaning white.

albicaul-. a prefix meaning white-stemmed.

albiflor-. a prefix meaning white-flowered.

albinism (congenital leukoderma; congenital leukopathia). **1:** a congenital deficiency or lack of pigment in the integument (e.g., skin, hair, feathers, eyes), often inherited as an autosomal recessive trait; sometimes only the eyes are affected. The condition results from an abnormality in the mechanism of melanin production. **cutaneous albinism**. a sex-linked, heritable condition marked by patterned loss of skin pigment, often with a white forelock. **ocular albinism**. sex-linked absence of pigment chiefly in the iris, choroid, and retina. **oculocutaneous albinism**. the heritable absence of pigment in the integument and eyes, transmitted by an autosomal recessive gene. **2:** a lack of chlorophyll in the leaves and/or stems of green plants in those tissues where it normally occurs.

albino. **1a:** an animal, the integument (e.g., skin, hair, feathers) of which in most cases is deficient in pigment (but see albinism). The skin of albino vertebrates is white and the eyes are pink. **1b:** a plant that is deficient in chlorophyll.

albomacula-. a prefix meaning white-spotted.

albumen (ovalbumin; egg white; dried egg white; egg albumin). the form of albumin present in commercially prepared egg white. It comprises about 60% of the fresh egg of the domestic fowl. *See also* albumin.

albumin. **1:** any of a class of water-soluble proteins that occur in nearly all animal tissues and in many plants. Albumins coagulate when heating and dissolve in pure water and dilute salt solutions. **2:** the pure protein of egg white (albumen, ovalbumin). *See also* albumen.

albuminous gland. any gland that secretes an albuminous (i.e., water-soluble, proteinaceous) fluid.

albuminuretic. referring to or causing albuminuria.

albuminuria. the presence, in easily detected concentrations, of serum protein (especially serum albumin) in urine. The condition may indicate renal impairment.

alburnum. the combined cambium and phloem cells of a tree trunk. It lies between the bark and sapwood.

Alcaligenes. A genus of Gram-negative, rod-shaped, typically aerobic bacteria (Family Achromobacteraceae). They occur chiefly in the intestinal canal of humans, livestock, and other animals, in dairy products, decaying matter. ***A. viscosum***. a species found in dairy products; milk infested with these bacteria is glutinous or stringy.

alcohol. **1:** any of a class of oxygenated organic compounds that have one or more functional hydroxyl groups bonded to a hydrocarbon chain but not to carbon atoms of an aromatic ring. Phenols are thus excluded from the family of alcohols even though a few phenols (e.g., phenyl methanol have the properties of an alcohol. Most alcohols are toxic. *See* alcohol poisoning. **2:** a vernacular term for beverages that contain ethanol (e.g., beers, wines, and spirits). **3:** a vernacular term for ethanol, *q.v.* **absolute alcohol**. a liquid that contains 99% ethanol and not more than 1% water. **acetic-alcohol**. *See* fixation. **aliphatic alcohol** (alkyl alcohol). an alcohol in which the hydroxyl groups (1 or more) are attached to a carbon atom of a branched or straight-chain aliphatic hydrocarbon. They are classified as primary, secondary, and tertiary based on the location of the substituted carbon atom in the hydrocarbon chain (thus there is primary, secondary, and tertiary butanol). Aliphatic alcohols are widely used as industrial solvents. Ethanol is toxic but is used extensively in beverages for human consumption; the other aliphatic alcohols are more toxic than ethanol. **alkyl alcohol**. aliphatic alcohol. **allyl alcohol** (2-propene-1-ol; 1-propenol-3; vinyl carbinol; olefinic alcohol). a colorless, toxic, strongly irritant liquid,

$$CH_2{=}CHCH_2OH$$

Allyl alcohol has a pungent, mustard-like odor. It is used in the manufacture of war gas, resins, and plasticizers. It is strongly irritating to eyes, mouth, and lungs and very toxic to laboratory rats. **amino alcohol**. alkamine. **α-(aminomethyl)-3,4-dihydroxybenzyl alcohol**. norepinephrine. **amyl alcohol**. **1:** amyl hydrate; a compound, $C_5H_{11}OH$, with eight possible isomers (discounting optically active isomers); six are commercially available (*n*-amyl alcohol, primary; isoamyl alcohol, primary; 2-methyl-1-butanol; 2-pentanol; 3-pentanol; *tert*-pentyl alcohol), all of which are toxic. In addition to the pure isomers, definite mixtures are sold under a variety of names, some of which are identical to that of a pure isomer. **2:** *n*-amyl alcohol, primary. **amyl alcohol, commercial**. a highly flammable aromatic liquid used in low concentrations as a flavoring agent in foods and liquor. It is an irritant of the eyes and upper respiratory tract. Inhalation of the vapor causes violent coughing, and ingestion of even small amounts can cause headache and vertigo. Ingestion of large amounts can be fatal. **amyl alcohol, normal**. *n*-amyl alcohol, primary. **amyl alcohol, primary**. **1:** *n*-amyl alcohol, primary. **2:** a flammable, commercial mixture of primary amyl alcohols that is sold under this name. It contains 60% primary amyl alcohol, 35% 2-methyl-1-butanol, and 5% 3-methyl-1-butanol. ***n*-amyl alcohol, primary** (amyl alcohol normal; amyl alcohol, primary; 1-pentanol; *n*-butyl carbinol; *n*-pentanol; pentyl alcohol; primary amyl alcohol). a commercially available straight-chain primary alcohol and isomer of amyl alcohol. It is a colorless liquid with a mild odor and is moderately toxic by ingestion and skin contact, a moderate fire hazard, and explosive in air if exposed to flames. It is a severe eye, skin, and upper respiratory tract irritant. Symptoms following ingestion of a toxic dose may include headache, nausea, vomiting, delirium, and methemoglobinemia. *See* amyl alcohol. ***dl-sec*-amyl alcohol**. 2-butanol. **anhydrous alcohol**. absolute alcohol. **anise alcohol** (4-methoxybenzenemethanol; *p*-methoxybenzyl alcohol; anisyl alcohol). an extremely toxic organic liquid that is nearly insoluble in water but freely soluble in ethanol and ethyl ether. **anisyl alcohol**. anise alcohol. **aromatic alcohol** (aryl alcohol). any alcohol in which the hydroxyl group is a substituent of an alkyl side chain of an aromatic compound. **aryl alcohol**. an aromatic alcohol. **benzal alcohol**. benzyl alcohol. **benzoyl alcohol**. benzyl alcohol. **benzyl alcohol** (benzenemethanol; benzene carbinol; phenylcarbinol; phenyl-methanol; α-hydroxytoluene; α-hydroxy-toluene; benzal alcohol; benzoyl alcohol; phenol carbinol; phenylmethyl alcohol; α-toluenol). a flammable, white, liquid hydrocarbon, $C_6H_5CH_2OH$, with a slightly sweetish odor and a sharp, burning taste. It is used as a solvent in perfumes, as a local anesthetic, and as a synthetic fruit flavoring for beverages, sweets, and chewing gum. Benzyl alcohol occurs in *Jasminum* (jasmine), *Hyacinthus* (hyacinth), ilangilang oils, Peru and Tolu-balsams, and many other plant-derived oils in which it may also occur as the ester. It is moderately toxic to laboratory mice when ingested, causing respiratory and muscular paralysis, convulsions, and narcosis. It is moderately irritant and corrosive to the skin and severely irritant to the eyes and mucous membranes of humans. Contact may cause inflammation and blistering of the skin and mucous membranes. Ingestion of excessive amounts can cause a burning sensation in the mouth and stomach, vomiting, diarrhea, headache, confusion, CNS depression, and coma. It can cause adverse reactions in persons sensitive to petrochemical derivatives. *Cf.* sodium benzoate. **butyl alcohol**. 1-butanol. **1-butyl alcohol**. 1-butanol. **2-butyl alcohol**. 2-butanol. ***n*-butyl alcohol**. 1-butanol. ***sec*-butyl alcohol**. 2-butanol. ***tert*-butyl alcohol**. *tert*-butanol. **butyl alcohol, secondary**. 2-butanol. **caprylic alcohol**. 1-octanol. **caprylic alcohol, secondary**. 2-octanol. **2-chloroethyl alcohol**. ethylene chlorohydrin. **crotonyl alcohol**. crotyl alcohol. **crotyl alcohol** (2-buten-1-ol; crotonyl alcohol). an extremely toxic, moderately water-soluble liquid,

$$CH_3CH{=}CHCH_2OH$$

It is miscible with ethanol. It is produced by the reduction of crotonaldehyde. **dehydrated alcohol**. ethanol. **denatured alcohol**. alcohol (ethanol) rendered unfit for consumption by the addition of toxic substances. It is used primarily as a solvent. **dihydric alcohol**. dihydroxy alcohol. **dihydroxy alcohol**. an alcohol that contains two functional hydroxyl groups; epinephrine. **3,4-dihydroxy-α-[(methylamino)methyl]benzyl alcohol**. epinephrine. **diluted alcohol** (diluted ethanol). USP alcohol that contains not less than 41% nor more than 42% by weight of ethanol. It is used as a solvent. **ethyl alcohol**. ethanol. **fermentation butyl alcohol**. isobutyl alcohol. **grain alcohol**. ethanol. ***n*-hexyl alcohol**. 1-hexanol. **isoamyl alcohol, primary** (3-methyl-1-butanol; isopentyl alcohol; isobutyl carbinol; 2-methyl-4-butanol). a moderately toxic, commercially available, branched-chain primary alcohol and isomer of amyl alcohol. It is a colorless liquid,

$$(CH_3)_2CHCH_2CH_2OH$$

The odor is disagreeable and it has a pungent. Primary isoamyl alcohol is poisonous by all routes of exposure; the vapor is toxic and irritant. It is a fire and explosion hazard; when heated to decomposition, it releases acrid smoke and fumes. ***sec*-isoamyl alcohol**. a branched chain secondary alcohol and isomer of amyl alcohol; it is not commercially available. **isobutyl alcohol** (2-methyl-1-propanol; isopropylcarbinol; 1-hydroxymethylpropane; isobutanol; fermentation butyl alcohol). a colorless, flammable, moderately toxic liquid. It is a mild irritant and a narcotic at high concentrations. **isopentyl alcohol**. isoamyl alcohol, primary. **isopropyl alcohol** (IPA; dimethylcarbinol; *sec*-propyl alcohol; isopropanol; 2-propanol). a highly toxic, colorless liquid, $(CH_3)_2CHOH$, with a pleasant odor; it is an isomer of propyl alcohol and is more toxic than ethanol. It is used in high volume industrially. Isopropyl alcohol is a CNS depressant with effects, when ingested, that are similar to those of ethanol intoxication: the appearance of being very drunk, but with more extreme effects. Even when used as rubbing alcohol to sponge parts of the body, there is some danger of coma. *Cf.* isopropanol. *See also* isopropyl rubbing alcohol, isopropyl alcohol poisoning. **isopropyl rubbing alcohol**. an aqueous solution, for external use only, that contains 68 to 72%, by volume, of isopropyl alcohol. It is flammable, toxic by ingestion or inhalation, and is used as a rubefacient; it is suitable only for external use. *Cf.* rubbing alcohol. ***p*-methoxy-benzyl alcohol**. anise alcohol. **methyl alcohol**. methanol. **1-methylbutyl alcohol**. 2-pentanol. **2-methylpropyl alcohol**. isobutanol. **monohydric alcohol**. monohydroxy alcohol. **monohydroxy alcohol**. an alcohol that contains only one functional hydroxyl group. **olefinic alcohol**. **1:** an alcohol in which the hydroxyl group is attached to an unsaturated hydrocarbon skeleton. **2:** allyl alcohol. ***tert*-pentyl alcohol** (*tert*-amyl alcohol; 2-methyl-2-butanol; dimethyl ethyl carbinol; ethyl dimethyl carbinol; *tert*-pentanol; amylene hydrate). a commercially available branched-chained tertiary alcohol and isomer of amyl alcohol. It is a volatile, moderately toxic liquid,

$$(CH_3)_2C(OH)CH_2CH_3$$

It has a characteristic odor, a burning taste, and irritates mucous membranes. At high concentrations, it has narcotic effects. It is also a serious fire hazard. **pentyl alcohol**. *n*-amyl alcohol, primary. **phenyl alcohol**. phenol (def. 1). **phenylmethyl alcohol**. benzyl alcohol. **polyhydric alcohol**. polyhydroxy alcohol. **polyhydroxy alcohol**. any alcohol that contains 2 or more functional hydroxyl groups. Such alcohols are widely utilized as heat exchangers, in hydraulic fluids, and in anti-freeze. They are toxic to laboratory animals and humans. **primary alcohol**. an alcohol in which the carbon atom that bears the hydroxyl group is also attached to two hydrogen atoms. **primary amyl alcohol**. *n*-amyl alcohol, primary. ***n*-propyl alcohol**. 1-propanol. ***sec*-propyl alcohol**. isopropyl alcohol. **propyl alcohol, secondary**. 2-propanol. **propylic alcohol**. 1-propanol. **rubbing alcohol**. a 70% aqueous solution of ethanol by volume plus denaturants, or an aqueous solution of isopropyl alcohol (isopropyl rubbing alcohol). It is flammable, toxic, and suitable only for external use. *Cf.* isopropyl rubbing alcohol. **secondary alcohol**. an alcohol in which the carbon atom that bears the hydroxyl group is also bonded to one hydrogen atom and two other groups. **secondary butyl alcohol**. 2-butanol. **secondary propyl alcohol**. 2-propanol. **sugar alcohol**. an alcohol produced by reducing an aldehyde or ketone of a sugar. ***tert*-amyl alcohol**. *tert*-pentyl alcohol. **tertiary alcohol**. an alcohol in which the carbon atom that bears the hydroxyl group is bonded to three other groups, none of which is hydrogen. **thiobutyl alcohol**. *n*-butyl mercaptan. **trihydric alcohol**. trihydroxy alcohol. **trihydroxy alcohol**. an alcohol that contains three functional hydroxyl groups. **wood alcohol**. methanol.

alcohol-alizarin test. a colorimetric test in which alcohol and alizarin are added to milk. If the test mixture produces a lilac-red color, the milk is considered normal; a yellowish-brown color indicates sour milk; violet indicates milk from a cow with mastitis.

alcoholic. **1:** referring to or denoting alcohol, usually ethanol. **2:** pertaining to a substance that has been treated with or contains alcohol, especially ethanol. **2:** a person afflicted by alcoholism.

aldehyde. **1:** any partially oxidized hydrocarbon compound that contains a terminal carbonyl group, C=O. The general formula is RHC=O, in which R is a hydrocarbon group (e.g., CH_3, C_2H_5) that may consist of saturated or unsaturated straight chains, branched chains, or rings. Aldehydes in the atmosphere arise from the exhausts of internal combustion engines and, to some extent, result from the photochemical reactions of hydrocarbons. Many are toxic and most are intensely irritant compounds. The lower aldehydes are highly water soluble and thus readily damage eyes and mucous membranes, especially those of the upper respiratory tract. The higher aldehydes are less water soluble and thus tend to penetrate deeper into the respiratory tract with serious damage to the lungs. **2:** acetaldehyde. *Cf.* ketone.

alder buckthorn. *Rhamnus frangula*.

alder witches' broom. *Taphrina epiphylla*.

alder. common name of any tree or shrub of the genus *Alnus*.

alder. *See Alnus*.

Alderney. a breed of cattle from the Channel Island of Alderney. It appears to be identical to the Guernsey.

aldicarb (2-methyl-2-(methylthio)propanal *O*-[(methylamino)carbonyl]oxime; 2-methyl-2- (methylthio)propionaldehyde *O*-(methylcarbamoyl)oxime; Temik®). a colorless, crystalline carbamate,

$$CH_3SC(CH_3)_2CH{=}NOCONHCH_3$$

The melting point is 100°C; it is sold under the name Temik®. Aldicarb is a very dangerous systemic poison (an anticholinesterase) used as a systemic insecticide, miticide, nematocide, and as the pesticidal component of some household products used to eradicate pests. *See also* carbamate poisoning.

aldose. a monosaccharide such as glucose or ribose that possesses an aldehyde group (-CHO).

aldosterone (electrocortin). a corticosteroid hormone, $C_{21}H_{28}O_5$, secreted by the *zona glomerulosa* of the outer mammalian adrenal cortex and by the adrenal tissue of many other vertebrates. It is a potent regulator of electrolyte (sodium and potassium) balance. Excessive secretion produces hypertension due to salt retention. Aldosterone secretion is normally regulated by angiotensin, but can also be stimulated by ACTH, lowered blood volume, or by insufficiency of sodium in the diet.

aldrin (1,2,3,4,10,10-hexachloro-1,4,4a,5,-8,8a-hexahydro-1,4:5,8-dimethanonaphthalene; HHDN; compound 118; Aldrin®). the designated common name of an insecticide that contains 95% or more of 1,2,3,4,10,10-hexachloro-1,4,4a,5,8,-8a-hexahydro-1,4:5,8-dimethanonaphtha lene, $C_{12}H_8Cl_6$. It is a colorless, water-insoluble, crystalline chlorinated diene. A derivative of chlorinated naphthalene, it is related chemically and toxicologically to dieldrin. It is highly toxic to insects and to most vertebrate animals, including humans, and is carcinogenic in some laboratory animals. As a result, its manufacture and use are now banned in the United States and use is banned or restricted in a number of other countries. Intoxication of mostvertebrate animals can result from absorption via skin, respiratory organs, or by ingestion. *See also* DDT.

Aldrin®. aldrin.

alemtejo. a soft, ripe cheese made from ewe's milk, sometimes in a mixture with goat's milk.

aletophyte. a weed found in mesic habitats.

aleukia. the absence of leukocytes in circulating blood. *Cf.* leukopenia, aleukocytic, aleukocytosis. **alimentary toxic aleukia** (ATA). a mycotoxicosis of livestock and sometimes humans associated with the ingestion of old grain infested with fungi of the genera *Actinomyces, Alternaria, Cladosporium, Fusarium, Mucor, Penicillium, Piptocephalis, rhizopus, Trichoderma, Trichothecium,* or *Verticillium.* The responsible poisons appear to be trichothecenes such as acetyl T-2 toxin and monoacetylnivalenol (fusarenon-X).

aleukocytic. lacking leukocytes.

aleukocytosis. a low proportion of leukocytes in the circulating blood.

aleukocytosis. an absence of or a low proportion of leukocytes in blood. *Cf.* aleukia.

Aleurites (tung tree; tung oil tree; tung nut). a genus of small trees (Family Euphorbaceae) indigenous to China and Japan. They are widely cultivated in the United States, especially in states bordering the Gulf of Mexico. The leaves and seeds are toxic. The toxicity of foliage, in descending order, is *A. fordii*, *A. montana*, *A. moluccana*, and *A. trisperma*. *A. trisperma* is about half as toxic as *A. fordii* to domestic fowl. Ingestion of foliage or untreated tung meal by cattle causes hemorrhagic inflammation with necrotic lesions of the gastrointestinal tract, producing hemorrhagic diarrhea. Ano- rexia, listlessness, and emaciation may result if exposure is prolonged. As little as 3.5 g/kg/day of fresh macerated leaves can produce these symptoms. Symptoms in humans appear in less than 1 hour following ingestion of as little as a single nut and include nausea, severe vomiting, abdominal cramps, weakness, and exhaustion. In more acute cases, shock, dehydration, cyanosis, diminished reflexes, respiratory depression, and death may result.

aleurone. a protein found in the outer layer of the endosperm of cereal seeds. *See* aleurone layer.

aleurone layer. the outer differentiated layer of the endosperm of cereal seeds. The cells (aleuroplasts) contain aleurone grains that form a food reserve; the cells also synthesize an amylase, which catalyses the digestion of starch in the endosperm.

aleuroplast. a plastid that stores protein, often in the form of aleurone grains. *See* aleurone layer.

alexipharmac. alexipharmic.

alexipharmic (def.2). antidote.

alexipharmic, alexipharmical (alexipharmac). **1a:** antidotal; acting as an antidote. **1b:** effective against or counteracting the harmful effects of a poison. *Cf.* alexiteric. **2:** an antidote.

alexiteric. protective or effective against

infection or poison.

alfalfa dwarf. **1a:** a viral disease of alfalfa that causes dwarfing of affected plants. **1b:** a plant affected by this disease.

alfalfa gall midge. *Asphordylia websteri*.

alfalfa leaf meal. a stock feed comprised mainly of ground leafy materials from alfalfa hay or meal. It contains no more than 18% crude fiber and contains few impurities including material from other plants.

alfalfa meal. a stock feed of ground alfalfa hay that contains few impurities and less than 33% crude fiber.

alfalfa-molasses feed. a stock feed comprised of alfalfa meal and cane- or beet molasses, the latter comprising 20 to 40% of the blend. It contributes additional proteins and vitamins to the diet of cattle and sheep.

alfalfa mosaic. alfalfa yellows.

alfalfa yellows (alfalfa mosaic). boron deficiency disease characterized by pale leaflets or bronzed leaves with green midribs, or, in some cases, a stopping of terminal growth. The margins of the leaves may dry up and shrivel.

alfalfa. *See Medicago*, especially *M. sativa*. *See also* lucerne.

alfileria. *Erodium cicutarium*.

alfillerilla. *Erodium cicutarium*.

alfisol. any of a type of soil having a gray to brown A horizon, a large or fairly large amount of bases, and a B horizon comprised of alluvial clay. Such soils form mostly beneath forest or savanna vegetation in climates with at least some seasonal deficit in moisture.

alfombrilla. *Drymaria arenarioides*.

alga (plural, algae). a common name for organisms of three phyla of unicellular and multicellular plants. These are Phaeophyta (brown algae), Rhodophyta (red algae), and Chlorophyta (green algae). All contain chlorophyll and are aquatic or occur in damp habitats. The term was formerly also applied to Cyanobacteria (blue-green bacteria). Algae are primitive eukaryotic, often microscopic, photosynthesizing, nonflowering plants. Some species are colonial. They primarily inhabit natural bodies of water and sometimes soil. They can adversely affect water quality when present in large numbers by reducing or even exhausting the amount of dissolved oxygen in the water.

algae. plural form of alga.

algal. of, pertaining to, resembling, or caused by algae.

algal bloom. *See* bloom.

algal poisoning. poisoning of wildlife, domestic animals, or humans who drink from water bodies that have algal blooms. Onset and course of the condition occur rapidly, often ending in death within 24 hours. Signs and symptoms usually include nausea, vomiting (in many animals), abdominal pain, diarrhea, muscle tremors, prostration, labored breathing, cyanosis, paralysis, and convulsions. Bloody feces and photosensitization are common in cases that do not progress rapidly. *See also* alga, bloom, factor (fast death factor), *Microcystis aeruginosa*.

algicide. a substance (e.g., copper sulfate) used to destroy algae.

algologist. a specialist in algology.

algology (phycology). the scientific study of algae.

ali-, ala-. prefixes meaning wing.

alicyclic. alicyclic hydrocarbon.

alicyclic compound. alicyclic hydrocarbon.

alienated. **1:** distanced; separated; divorced. **2:** descriptive, in the United States, of land owned by one party that is bounded on all sides by land owned by another party.

alighting board. a lip at the entrance of a beehive on which incoming bees can land.

aliment. **1:** that which nourishes; a food or nutriment. **1:** that which sustains or serves as a means of support.

alimentary. of or pertaining to food or nutrition.

alimentary canal (alimentary tract; digestive tract). the tube-like portion of most animals' digestive tract through which ingested materials pass. Ingesta enter via an orifice, the mouth, and undigested matter and waste (egesta) pass out through an anus in most taxa. The Platyhelminthes and Coelenterata have a single opening, and orifices are absent in those parasites and other animals that absorb liquid nutrients. Food moves through the canal from mouth to anus by the action of cilia or muscular contractions. Reverse movements may occur, as in the case of ruminant animals or animals that vomit or regurgitate food. The alimentary canal of vertebrates usually includes a mouth, pharynx, esophagus, stomach, and intestines.

alimentary tract. alimentary canal.

alimentation. feeding or the provision of nourishment. **forced alimentation**. forced feeding. **parenteral alimentation**. intravenous provision of nourishment; intravenous feeding. **rectal alimentation**. provision of nourishment by means of retention enemas.

aliphatic amino acid. *See* amino acid.

aliphatic hydrocarbon. a major class of hydrocarbons marked by arrangement of the constituent atoms into straight or branched-chain molecules. There are three subgroups: alkanes, alkenes, and alkynes, *q.v.* Alkanes are saturated and comparatively unreactive compounds; alkenes are unsaturated compounds that contain one or more double [C=C] bonds; and alkynes are unsaturated, containing one or more triple [C≡C] bonds.

alkalemia. excessive alkalinity of blood, as reflected in a pH of arterial blood that lies above the normal range (7.35 to 7.45 in humans).

alkali. **1:** any strong water-soluble base, especially a hydroxide of an alkali metal (group I) (e.g., potassium hydroxide, sodium hydroxide). They neutralize acids, react with acids to yield salts, react with fatty acids to yield soap, and turn red litmus paper blue. Because of their strong dissociation in water and their affinity for lipids, alkalis are highly corrosive and can cause rapid, deep erosion of tissue. *See* alkali poisoning, corrosion, corrosive poisoning. **2a:** alkali soil or corrosive alkali. **2b:** any strongly corrosive base, especially a metallic hydroxide (e.g., sodium, ammonium, and potassium hydroxides) or carbonate. Because of their high degree of ionization, solubility in water, and action on the fatty tissues, they cause rapid, deep, and very painful destruction of tissues. They tend to gelatinize tissue, giving it a grayish color and a slippery or soapy texture. *See also* alkali poisoning, corrosive, corrosion.

alkali chlorosis. *See* chlorosis.

alkali disease. **1:** (Western duck sickness; botulism of ducks) a form of botulism causing death of dabbling (puddle) ducks

(Family Anatidae) and, to a lesser extent, other ducks, loons, mergansers, geese, and gulls in North America. The disease is caused by ingestion of vegetation contaminated with botulinum toxin, type C produced by *Clostridium botulinum.* Tens of thousands of birds die in most years, but losses in the western United States may reach a million or more during great outbreaks. **2:** a chronic disease of livestock marked by emaciation, rigidity, and anemia due to grazing on plants having a high selenium content. Humans have also been affected occasionally by eating grain from plants grown on selenium-rich soils. **3:** blind staggers; an acute or subacute form of selenium poisoning.

alkali grass. a common name of *Zigadenus elegans* and *Z. venenosus*.

alkali metal. any of the generally univalent elements of group IA of the periodic table: lithium, sodium, potassium, rubidium, cesium, and francium. They are essential micronutrients that are easily absorbed and excreted by living organisms. They are the most strongly electropositive of metals, often reacting vigorously or even violently with water. The salts, especially in diseased organisms, are toxic.

alkali poisoning. a serious, often life-threatening type of corrosive poisoning due to ingestion of a strong water-soluble corrosive alkali such as one finds, for example, in the home or on the farm (e.g., ammonia, lye, and certain soap powders or other cleaning agents). Tissues of the gastrointestinal tract may be rapidly and deeply eroded. The victim may suffer excruciating pain and may quickly die due to edema of the throat and pharynx with closure of the respiratory tract, choking, and asphyxia. *See also* alkali disease, acid poisoning.

alkali reserve. **1:** the set of dissolved substances in natural water bodies that tends to maintain constant pH. **2:** the relative amount of buffers in circulating blood, especially sodium bicarbonate, that can neutralize acid.

alkali sacaton. *Sporobolus airoides*.

alkali soil. *See* alkali.

alkaline. **1:** pertaining to or having properties or reactions of an alkali. **2:** of or pertaining to water, an aqueous solution, or a body of water that has a pH above 7.

alkalinity. **1:** the state of being alkaline. **2a:** the degree of alkalinity. **2b**: the extent to which water or an aqueous solution, has a pH above 7.0. **3:** the number of milliequivalents of H^+ neutralized by 1 liter of seawater at 20°C.

alkalinization. alkalization.

alkaliphile (alkalophile). an alkaliphilic organism.

alkaliphilic (alkalophilic). occurring in, highly tolerant of, or flourishing in alkaline environments. *Cf.* alkaloduric.

alkaliphily (alkalophily). the condition or state of being highly tolerant of alkaline conditions in the environment.

alkaliphobe. an alkaliphobic organism.

alkaliphobic. avoiding or intolerant of alkaline habitats or conditions. *Cf.* alkaloduric.

alkaliphoby. the condition or state of avoiding or being highly intolerant of alkaline environments.

alkalization (alkalinization). the process of rendering a substance alkaline.

alkalizer. an agent that neutralizes acids or increases the alkalinity of a solution; one that brings about alkalization.

alkaloduric. exceptionally tolerant of alkaline conditions. *Cf.* alkaliphilic.

alkaloid. **1:** any of a large class of alkaline, bitter-tasting, biologically active, usually water-insoluble, nitrogenous organic compounds that are produced chiefly by plants. They are usually derivatives of the nitrogen ring compounds quinoline, isoquinoline, pyridine, and pyrrole. Most are crystalline solids; a few are liquids or gums. Many flowering (dicotyledonous) plants produce alkaloids, especially those of the families Solanaceae, Fabaceae, and Papaveraceae. Monocots rarely produce alkaloids, although many groups of lower plants (not algae) contain species that do. Most alkaloids are powerful, physiologically active agents. Most are toxic and many have therapeutic value. They typically exhibit pharmacologic specificity (e.g., as analgesics (morphine), tranquilizers (reserpine), respiratory stimulants (nicotine), vasoconstrictors (scopolamine), local anesthetics (cocaine), muscle relaxants (strychnine), and psychedelic agents (LSD, psilocybin), and effects are usually also strongly species-specific. **2:** the term alkaloid is reserved by some authorities for nitrogen-containing heterocyclic or aromatic compounds with pharmacologic activity and the general properties of amines. **3:** any of certain synthetic compounds (e.g., procaine) with structures similar to naturally occurring alkaloids. **amine alkaloid**. usually an ergot alkaloid such as ergonovine. **amino acid alkaloid**. usually an ergot alkaloid such as bromocriptine or ergotamine. **animal alkaloid**. **1:** a ptomaine. **2:** a leukomaine. **artificial alkaloid**. a synthetic alkaloid. clavine alkaloid. a type of ergot alkaloid (e.g., elymoclavine, ergoclavine). **ergot alkaloid**. any of a large group of alkaloids that are constituents of *Claviceps* (ergot); they contain lysergic acid in combination with various amine moieties. The four main types are clavine alkaloids, lysergic acids, lysergic acid amides, and ergot peptide alkaloids. They are very toxic and may produce nausea, vomiting, a weak rapid pulse, coma, and death. **piperidine alkaloid**. any alkaloid that contains the piperidine ring (e.g., nicotine). **plant alkaloid**. any alkaloid that occurs naturally in plants. **synthetic alkaloid**. any synthetic substance that is structurally similar to plant alkaloids. **tropane alkaloid**. any of a group of tropane-like alkaloids (e.g., hyoscyamine, atropine, cocaine alkaloids) derived from the amino acid ornithine. **true alkaloid**. any alkaloid (e.g., the isoquinoline alkaloids) that includes a nitrogen-containing heterocyclic nucleus linked by amide bridges.

alkalophile. alkaliphile.

alkalophilic. alkaliphilic.

alkalophily. alkaliphily.

alkalosis. increased pH of body fluids, usually including circulating arterial blood, above the normal range, by an excess of base or by the loss of CO_2 due to hyperventilation. **metabolic alkalosis**. that produced by an excess of base. **respiratory alkalosis**. an increase in the pH of blood above 7.45 due to excessive elimination of blood $C0_2$ due to hyperventilation.

alkamine (amino alcohol). an alcohol that contains an amine group, e.g., solanidine.

alkane (paraffin). any of a class of saturated, relatively unreactive, aliphatic hydrocarbons. The carbon "backbone" may have the form of a straight or branched chain or may be cyclic (e.g., hexane). Alkanes form saturated bonds whereby adjacent carbon atoms are joined by a single covalent (sigma) bond; they contain no unsaturated (double) carbon-carbon bonds. Alkanes are the primary constituents of petroleum products. Alkanes burn readily and sometimes explosively in air. Incomplete oxidation of alkanes or mixtures of alkanes (e.g., kerosene) in an oxygen-deficient atmosphere may produce dangerous quantities of carbon monoxide.

alkene (olefin). the systematic chemical name for a class of unsaturated aliphatic hydrocarbon. They are comprised of straight or branched chains of carbon atoms that contain at least one carbon-carbon double bond [C=C]; examples are ethene and propene. An alkene that contains two double bonds is termed a diene, those with three are trienes, etc. Alkenes are produced in large amounts during the cracking of crude oils and are common components of gasoline and other refined petroleum products. They are usually more toxic than alkanes of similar molecular weight and less toxic than aromatic hydrocarbons.

alkoxy. **1:** pertaining to, denoting, or containing an alkoxyl group. **2:** an alkylradical connected to a molecule by oxygen.

alkoxy-alkylated mercury compound. an organomercurial compound that contains an alkoxy radical; one of three types of organomercurials used as pesticides. *See also* organomercurial pesticides.

alkoxyl, alkoxyl group, alkoxyl radical. a univalent radical (e.g., methoxyl) composed of an alkyl group linked to oxygen.

alkyl. **1:** of, pertaining to, or containing an alkyl radical. **2a:** a monovalent aliphatic radical C_nH_{2n+1} (e.g., methyl, ethyl) which can be thought of as an alkane moiety that has lost one hydrogen atom. **2b:** a monovalent aliphatic, alicyclic, or aliphatic-aromatic hydrocarbon radical. **3:** a compound comprised of one or more alkyl radicals joined to a metal (e.g., sodium alkyl).

alkyl radical. *See* alkyl.

alkylated mercury compound (alkylmercury compound; alkylmercury; alkyl mercury). any mercury-containing alkane. *See also* organomercurial pesticides.

alkylating agent. any substance capable of forming covalent bonds and attaching alkyl groups (e.g., methyl, ethyl) to nitrogen and oxygen atoms of macromolecules such as the DNA bases. Such agents are highly reactive and often strongly carcinogenic and/or mutagenic. They can produce serious cytotoxic, immunotoxic, fetotoxic, teratogenic, and abortifacient effects, and they can cause widespread permanent damage to tissues. Effects to humans and domestic animals can be fatal. *See*, for example, *N*-dimethylnitrosamine, alkyl sulfonate.

alkylation. the addition of an alkyl group to a molecule. *See* alkylating agent.

alkyne. an unsaturated aliphatic hydrocarbon characterized by the presence of one or more triple [C≡C] bonds. The most important member of this group, and the simplest, is acetylene, HCCH.

all-aged. **1a:** the condition of a forest or a stand made up of trees of all or nearly all age classes. This is usually seen in a primary stand or forest that arose and has continued to develop naturally. Under such circumstances, new trees enter at various times and locations as space permits. **1b:** pertaining to a forest or stand as defined in 1a, above.

allamanda. *Cryptostegia*.

allantiasis (sausage poisoning). an obsolete term for botulism acquired from eating improperly preserved sausage.

allantoic. pertaining to the allantois.

allantoic membrane. allantois.

allantois (allantoic membrane). one of the three extraembryonic membranes that surround the mammalian fetus. It lies between the amnion and chorion and contributes to the formation of the placenta and umbilical cord; it is vestigial in humans. The allantois of birds and reptiles lies immediately beneath the porous shell, serving as a respiratory organ.

allaxis. metamorphosis.

allele (*See also* allelomorph). **1:** one of a pair or series of alternative forms of a gene or DNA sequence that occurs at a given locus on either of two homologous chromosomes and that separate during meiosis in a diploid organism. The possible number of alleles for a given gene is quite large. Only a few alleles, however, produce detectable effects on the phenotype. *See also* DNA polymorphism. Usually one of a pair of alleles is dominant to the other; in most cases, the wild type is dominant. *See also* multiple allele. **2:** the trait (phenotype) determined by an allele. **codominant allele**. either of two alleles that have an equal degree of dominance, both being fully expressed in the offspring. **dominant allele**. an allele that is expressed even when only one copy is present in the genotype. **multiple allele** (multiple allelomorph). any of a series of three or more alleles of a given gene, any two of which occupy the same locus on two homologous chromosomes in diploid cells. One allele may be dominant. **neutral allele**. either of two alleles that code for proteins that are slightly different structurally, but function identically in the phenotype. **recessive allele**. an allele that only expresses itself when two copies are present in the genotype.

allelochemic (allelochemical). **1:** relating to interspecific chemical interactions. **2a:** denoting a semiochemical agent of interspecies communication. **2b:** a waste product or metabolite produced by an organism that influences the growth, behavior, or population dynamics of another species.

allelochemical. allelochemic.

allelochemical interaction. any interaction between species that involves the release of biologically active chemicals. *See also* allelopathy, allomone, kairomone, signal.

allelochemics. **1:** plural of allelochemic. **2:** intraspecific chemical interactions that involve the release of biologically active substances.

allelomorph. **1:** allele. **2:** a trait produced by an allele. **multiple allelomorph**. **1:** multiple allele. **2:** a trait produced by a multiple allele.

allelomorphic. pertaining to an allele.

allelopathic. denoting or pertaining to allelopathy or to an allelopathic substance.

allelopathic chemical. allelopathic substance.

allelopathic substance (allelopathic chemical). **1:** an allelochemic; a waste product, excretory product, or metabolite that has a detrimental, usually inhibitory or regulatory, effect on organisms of species other than that producing the substance. **2:** any allelochemic used by humans as an herbicide. Included are any of a large number of naturally occurring herbicides produced by plants or microorganisms (mostly soil bacteria and fungi). Amygdalin, arbutin, caffeine, and gallic acid are considered allelopathic. *See* allelopathy.

allelopathy (biogenic toxicity). **1:** the production and release of substances that have deleterious interspecific effects (inhibition of germination, growth, or reproduction) at a distance by a plant or animal. Allelopathy plays an important role in nature and in agroecosystems (e.g., in weed control). *See* allelochemical. **2a:** the chemical inhibition of one organism by another regardless of species; antibiosis. **2b:** the release of a chemical by a plant that inhibits the growth of nearby plants (e.g., the release of juglone by black walnut leaves and roots.). **3:** an allelochemical effect. *See*, for example, 3-acetyl-6-methoxy benzaldehyde, arbutin, *Adenostoma fasciculatum*, *Arctostaphylos*, *Juglans nigra*. *See also* antibiont, amensalism. **indirect allelopathy**. the release of a chemical by one plant that kills or inhibits the growth of a second plant or microorganism that is essential to the growth or viability of a third plant.

allelotoxin. a toxic substance released by an allelopathic process.

allergen. an antigen (e.g., a foreign protein, microorganism, pollen) that induces a response in the immune system such that subsequent exposure to the same allergen produces an allergic reaction or specific hypersensitivity. The most common reactions are localized inflammation, watery eyes, and runny nose. Sometimes, a serious, often fatal systemic condition (anaphylaxis, anaphylactic shock) results. *Cf.* allergin.

allergenic. of, pertaining to, or acting as an allergen; able to induce an allergy; antigenic. *Cf.* allerginic, antigenic, immunogenic.

allergenicity. the ability of a substance to provoke an allergic reaction.

allergic. pertaining to an allergy or any response induced by an allergen.

allergic conjunctivitis (atopic conjunctivitis). **1:** hay fever. **2:** an allergic reaction of the conjunctiva of the eye to a specific antigen.

allergic response. allergic reaction.

allergic shock. anaphylactic shock; anaphylaxis.

allergie. allergy.

allergin. **1:** an antigen that triggers anaphylaxis. **2:** allergen.

allerginic. **1:** anaphylaxis-inducing. **2:** allergenic.

allerginicity. **1:** the ability of a substance to induce anaphylaxis. **2:** allergenicity.

allergize. to specifically sensitize or to make allergic by contact with or the introduction of a foreign substance (allergen) into the animal body.

allergized. specifically sensitized by contact with an allergen.

allergodermania. allergodermia.

allergodermia (allergodermania). an allergic dermatitis.

allergosis. any allergic disease.

allergy (allergie). **1a:** an immunologic state of a sensitive individual induced by an allergen. It is marked by the liberation of histamine and other molecules with histamine-like effects. **1b:** sensitivity to an allergen. **2:** sometimes used as a synonym for hypersensitivity, especially delayed hypersensitivity and class I (complement independent) hypersensitivity. **3:** that branch of medicine concerned with the study, diagnosis, and treatment of allergic states. *See also* allergic reaction. **atopic allergy**. atopy. **bacterial allergy**. hypersensitivity to a specific bacterial allergin. **cell-mediated allergy**. delayed hypersensitivity. **chemical allergy**. unusual sensitivity to a chemical due to prior exposure to that chemical or to one that is structurally similar. *Cf.* drug allergy. **contact allergy**. a cutaneous reaction to direct contact with an allergen. **delayed allergy**. delayed hypersensitivity. **drug allergy**. sensitivity to a drug (roughly synonymous with chemical allergy). **food allergy**. that in which the antigens include ingested foods, drugs, or other materials to which one has been sensitized. **immediate allergy**. immediate hypersensitivity. *See also* immune response, immune reaction.

allethrin (allyl cinerin). **1:** the generic name for 2-allyl-4-hydroxy-3-methyl-2-cyclopenten-1-one ester of chrysanthemummonocarboxylic acid, $C_{19}H_{26}O_3$. It is a clear, amber, viscous liquid insecticide or synergist. **2:** any of a class of synthetic analogs (pyrethroids) of naturally occurring pyrethrins such as cinerin, jasmolin, and pyrethrin. Toxicities are similar to those of pyrethrins. Allethrins are allethrolone esters and pyrethrolone esters of chrysanthemummonocarboxylic acids such as chrysanthemic acid, barthrin, and cyclethrin. They are viscous, water-insoluble, hepatotoxic and nephrotoxic liquid insecticides that are readily absorbed by the lungs, skin, and mucous membranes of humans and other vertebrates. **allethrin I** (2,2-dimethyl-3-(2-methyl-1-propenyl)cyclopropanecarboxylic acid 2-methyl-4-oxo-3-(2-propenyl)-2-cyclo-penten-1-yl ester; allethrone ester of chrysanthemummonocarboxylic acid). the commercial product. It is very toxic to laboratory mice. **allethrin II** (3-(3-methoxy-2-methyl-3-oxo-1-propenyl)-2,2-dimethylcyclopropane-carboxylic acid 2-methyl-4-oxo-3-(2-propenyl)-2-cyclopenten-1-yl ester; allethrolone ester of chrysanthemumdicarboxylic acid monomethyl ester). a toxic, oily, pale yellow insecticidal mixture. It is less toxic than allethrin I.

allethrolone ester of chrysanthemumdicarboxylic acid monomethyl ester. allethrin II.

allethrone ester of chrysanthemummonocarboxylic acid. allethrin.

alley. **1:** a narrow walkway or passage, especially a narrow street-like passageway between buildings for vehicles or sometimes for foot traffic. **2:** a walking space within a barn for cattle (e.g., a loafing alley, a feeding alley).

alli-. a prefix denoting or pertaining to the genus Allium which includes, e.g., onions, garlics, and leeks.

alliaceous. **1a:** of or pertaining to plants of the genus *Allium*. **1b:** belonging to the genus *Allium*. **2:** having the odor or taste of garlic, onion, and related plants.

allicin. an antibiotic isolated from garlic (*Allium sativum*).

allidochlor (2-chloro-*N,N*-di-2-propenylacetamide; α-chloro-*N,N*-diallylacetamide; *N,N*-diallyl-2-chloroacetamide; CDAA). an amide preemergence herbicide that persists in soil for as much as two months. It modifies RNA and protein biosynthesis and inhibits cell division in primary meristems. Effects vary, but usually include inhibition of photosynthesis. Effects on barley are similar to those induced by propachlor. Some grasses are especially sensitive.

Alliette. fosetyl aluminum.

alligator. common name of reptiles of the genus *Alligator*.

alligator farming. the rearing and management of alligators for the commercial production of leather and of meat for human consumption. Alligator leather is used in handbags, suitcases, and many articles of clothing such as coats, jackets, and shoes. Such farms must be licensed under state law. The meat is sold for human consumption.

Alligator. a genus of reptiles (Family Alligatoridae) represented today by two species. The entire head is broad with a rather blunt snout; the teeth do not protrude at the sides of the jaws as in crocodiles; they attack humans only in self defense. ***A. mississippiensis*** (American alligator). it occurs in drainage systems along the east coast of the United States from North Carolina to the Rio Grande; maximum recorded length is about 5 meters. ***A. sinensis*** (Chinese alligator). a species that occurs in the Yangkze-Kiang river system of China; maximum length is about 2 meters.

alligatoridae. the family of reptiles that includes the two living species of alligator. *See Alligator*.

alliometry. allometry.

Allium. a genus of some 300 species of bulbous herbs (Family Liliaceae), many of which have a strong, distinctive odor and flavor. Several species are native to North America, chiefly in the west. Some 70 "species" are cultivated today and have been under cultivation for thousands of years. Included among those grown for food are garlic, leeks, onions, and chives. *See also* scallion. ***A. ascalonium*** (shallot). a perennial plant with clusters of small onion-like bulbs. It is commonly used in salads, cooked vegetables, and stews. ***A. cepa*** (onion). a wild biennial herb with a firm white bulb. Numerous varieties are cultivated as table vegetables. Included are the red onion which is strongly flavored, the milder yellow onion, the bland white onion, the large, mild, Spanish and Bermuda onions, and pearl onions. The latter are small white onions that are usually pickled. ***A. fistulosum*** (Welsh onion). a leeklike plant popular in Asia. ***A. porrum*** (leek). a perennial herb, that grows wild in Europe and Asia, with a single, small, cylindrical bulb. It is also cultivated as a vegetable. It has a mild oniony flavor and is most often used in salads, soups, and stews. The leek was cultivated in ancient Egypt and was probably introduced to England by the Romans. It is the floral emblem of the Welsh, who wear the entire plant on their hats on St. David's Day. ***A. sativum*** (garlic). a perennial plant with a bulb comprised of small bulblets (cloves). Garlic cloves are used in cookery most often as a flavoring. They have also been used as a folk remedy for thousands of years. They are thought by many to act as a blood thinner, antioxidant, and cancer preventive. ***A. schoenoprasum*** (chive; cive). a hardy, perennial, bulbous herb with hollow leaves that is indigenous to Europe and Asia and found wild today in Italy and Greece. It is cultivated for use in salads and as an oniony flavoring in other foods. It is sometimes used as an ornamental border plant. *A vineale* (wild garlic). a wild plant often considered a weed because of the disagreeable odor and taste imparted to the milk of cows that feed upon them.

allo-. a prefix meaning different.

allobiosis. altered responses of an organism following physiological changes, especially those due to exposure to alterations in the environment.

allobiosis. altered responses of an organism following physiological changes or exposure to environmental changes.

allochemic. any secondary compound produced by plants that reduces or inhibits the use of the plant by herbivores.

allochoric. of or pertaining to a species that occurs in two or more similar communities within a region.

allochthonous (allocthonous). exogenous; of foreign origin, as in the case of living organisms, soils, rocks, or sediments that occur in a given ecosystem but originated elsewhere. *Cf.* xenogenous.

allochthonous organism. an organism that is foreign to the soil but that may enter it by some means as by precipitation, manure, or sewage. They may persist for some time but do not contribute in a significant way to ecologically significant transformations or interactions.

allocthonous. allochthonous.

allodiploid. a diploid produced by allopolyploidy.

allodiploidy. the state of being allodiploid.

allogamous. of or pertaining to allogamy.

allogamy. cross-fertilization of plants either by pollen from different flowers on the same plant or by pollen from genetically different plants of the same species. *Cf.* autogamy.

allogenic (allothogenic). **1:** pertaining to an ecologic succession. *See* succession (allogenic succession, autogenic succession). **2:** allogeneic.

allohexaploid. a hexaploid produced by allopolyploidy. *See* polyploidy. In *Triticum aestivum*, for example, two each of three different chromosome sets from three different parent species are present.

allohexaploidy. the state of being allohexaploid.

allolactose. a disaccharide isomer of lactose that occurs in milk. It is an effector molecule that permits activation of the *lac* operon by binding to the repressor molecule.

allometric. denoting or pertaining to allometry.

allometry (alliometry). **1:** growth of part of an organism relative to the growth of the whole organism or another designated part. **2:** the measurement or study of allometric growth.

allomone. **1a:** an allelochemical that affords an advantage to the organism that produces it through its effects on other species. Examples of allomones include repellents, poisons, venoms, attractants, counteractants, distractants, inductants, suppressants (e.g., antibiotics), and screens (e.g., the inks of cephalopods). **1b:** any substance secreted by an organism that confers an adaptive advantage on that organism. **2:** any coactone that serves to defend the producing organism. *See also* allelochemical, kairomone, signal.

allopentaploid. a pentaploid produced by allopolyploidy.

allopentaploidy. the state of being allopolyploid.

alloploid. **1:** of or pertaining to a hybrid organism or cell having two or more sets of chromosomes acquired from two different ancestral species. **2:** an alloploid organism or cell. Alloploids are designated as allodiploids, allotriploids, allotetraploids, allopentaploids, allohexaploids, etc., depending on the number multiples of such chromosomes present.

alloploidy. the state of being alloploid.

allopolyploid. an alloploid cell or organism that contains multiples of two homologous pairs of chromosomes derived from two or more species by hybridization. *See also* amphiploid.

allopolyploidy. the state of being allopolyploid.

allosome. a chromosome distinguished by structural or functional peculiarities.

allotetraploid (amphidiploid). a tetraploid produced by allopolyploidy. *See* polyploidy.

allotetraploidy (amphidiploidy). the state of being allotetraploid.

allothogenic. allogenic.

allotment. 1a: allocation, portion, share; apportionment; quota. 1b: quota, allowance. **2a:** on federal grazing lands an allotment is an area designated for use and managed for grazing of livestock. The number of livestock and season of use are prescribed for each allotment by the Bureau of Land Management or the Forest Service, depending on which agency has jurisdiction. **acreage allotment**. the number of acres of a specified crop that a farm is allowed to produce under provisions of permanent commodity price support law. Allotments are based on estimates of a farm's share of the national acreage needed to produce sufficient supplies of a particular crop.

allotriploid. a triploid produced by allopolyploidy. *See* polyploidy.

allowable cut. the amount of wood that can be removed from a landholder's property during a specified period without infringing on the predicted net growth of timber during that period.

allowable human daily intake. *See* acceptable daily intake.

allowable use. the amount of usage of rangeland considered to be appropriate until proper degree has been established. If the range is in good condition, 40 to 50% of the annual primary production is often acceptable.

alloy. 1: a solid or liquid mixture of two or more metals, or of a metal or metals with a nonmetal. Examples are brass, bronze, or carbon steel. **2:** a less costly metal mixed with a more valuable one. **3:** to produce an alloy (*See* def. 1). **4:** to reduce the value of a metal admixing it with a metal of lesser value (*See* def. 2). **5:** to adulterate; to debase or to reduce the value of a substance by admixture.

alluvial. denoting or pertaining to alluvium. *See also* soil (alluvial soil).

alluvial community. a community that develops and is sustained by periodic inundation and deposition of alluvial deposits (e.g., flood plain communities, river deltas).

alluvial cone. a delta shaped feature of a landscape, comprised of storm-washed soil and rock deposited at the mouth of a stream. Sometimes treated as a type of alluvial fan. The angle of slope, however, is greater than that of an alluvial fan.

alluvial deposit (alluvial sediment). silty or sandy sediments deposited on the bed of a stream or lake from slowing water currents. *Cf.* aeolian deposit.

alluvial fan. soil and rock detritus transported by swift-flowing water and deposited in the form of a fan at the mouth of a stream, most often as it flows onto a plain or flat, open valley at the base of a mountain slope. The angle of slope is less than that of an alluvial cone. Adjacent alluvial fans sometimes may merge to form terraces and piedmont plains.

alluvial pan. an alluvial cone or alluvial fan.

alluvial plain. a flat area (e.g., a flood plain) produced by deposits of alluvium.

alluvial sediment. alluvial deposit.

alluvio maris. 1: soil formed from deposits washed up from an ocean. **2:** sand, soil, or other deposited material along a seashore.

alluvion. land that has been built up by deposition of alluvium.

alluvium. sediment transported and deposited by streams. Particle size varies widely and may include both organic and mineral matter such as sand, small rocks, stones, and/or boulders. Alluvium, when composed of sediments bearing smaller, fine-textured materials, often produces very fertile soils. Thus, silt deposited by flooding rivers or by lakes is usually very fertile.

allyl aldehyde. acrolein.

allyl cinerin. allethrin.

allyl isosulfocyanate. allyl isothiocyanate.

allyl isothiocyanate (3-isothiocyanato-1-propene; isothiocyanic acid allyl ether; allyl isosulfocyanate; 2-propenyl isothiocyanate; volatile oil of mustard; mustard oil; Redskin). a common mustard oil,

$$CH_2CH{:}CH_2NCSouth$$

It is a vesicating, volatile, colorless to pale yellow, irritant substance with a strong, irritating odor. It is slightly soluble in water and alcohol. It occurs in *Brassica nigra* (black mustard seed), *q.v.*, and many other plants, it repels many insects but attracts butterflies of the genus *Pieris*. It is extremely toxic and is used as a fumigant, a war gas, and therapeutically as a counterirritant. The lethal oral dose for domestic cattle is about $10^{-3}\%$ body weight. *See also* sinigrin.

allyl sulfide (3,3′-thio-bis-1-propene; diallyl sulfide; thioallyl ether; garlic oil; "oil garlic"). a compound,

$$(CH_2{=}CHCH_2)_2S$$

used to dehydrate onions and in the manufacture of fruit and garlic flavorings. It is a colorless, nearly water-insoluble, irritant liquid with a garlicky odor. Acute exposures can cause unconsciousness, and long-term exposure can damage liver and kidneys.

2-allyl-4-hydroxy-3-methyl-2-cyclopenten-1-one ester of chrysanthemummonocarboxylic acid. allethrin (def. 1).

almond. *See Prunus amygdalus*.

Alnus (alder). a genus of deciduous trees and shrubs (Family Betulaceae). Some species are ornamental, while others are excellent sources of lumber, used especially for cabinet-making. ***A. glutinosa*** (black alder; European alder). a species indigenous to parts of Eurasia and North Africa. It is often grown as an ornamental.

aloe (bitter aloe). the thickened sap of various species of plants of the genus *Aloe* (Family Lilaceae). Formerly used as a cathartic, its use was largely discontinued because it can cause severe abdominal cramps. The active principle is aloe-emodin, *q.v.*

Aloe. a genus of perennial succulent herbs (Family Lilaceae) that are native chiefly to southern Africa, but also to tropical Africa and Madagascar. ***A. vera*** (Barbados aloe). a species, native to the Mediterranean region, that contains emodin. It is often grown in a pot or tub. The thickened sap is called aloe-emodin or bitter aloe (as are certain other species).

aloe-emodin (1,8-dihydroxy-3-(hydroxymethyl)-9,10-anthracenedione; 1,8-dihydroxy-3-(hydro-xymethyl)anthraquinone; 3-hydroxymethylchrysazin; bitter aloe; rhabarberone). a cathartic anthraquinoid that occurs free and as a glycoside in *Rheum* (rhubarb), *Cassia* (senna) leaves, and in various species of *Aloe* (Lilaceae). *Cf.* emodin.

alopecia. baldness; natural or abnormal loss or deficiency of hair.

Alopecurus alpinus (Arctic timothy). a grass (Family Gramineae) grown by Eskimos as feed for livestock.

Alopex lagopus (blue fox). *See* Ronnel.

alp. **1:** a high mountain pasture; one above timberline. **2:** a high mountain.

alpaca. **1:** *Lama pacos*. **2:** the soft, long, elastic wool of the alpaca.

alpestr-. prefix meaning alpine.

alpestrine. **1a:** living on mountains above the tree line. **1b:** alpine. **2:** of or pertaining to montane, subalpine plants.

alpha (symbols, α, A). **1:** first letter of the Greek alphabet, it has many applications in science. For terms beginning with alpha, see the root word. **2:** the first; first in order of magnitude. **3:** the beginning; the first in a series or classification. **4:** the closest to or first in position from a part of an organic molecule (e.g., the position adjacent to a phenyl group). **5:** an aromatic substituent linked to an aliphatic chain. **6:** the direction of a chemical bond opposite the plane of view.

alpin-. a prefix denoting or pertaining to alpine regions or situations.

alpine. **1:** pertaining to the Alps or other high mountains. **2:** pertaining to habitats, organisms, or conditions that occur between the tree line and the snow line in high mountains. **3:** alpestrine (in part).

alpine fir. *Abies lasiocarpa*.

alpine goat. a U.S. breed of dairy goat.

alsad (alsophyte). a grove plant.

alsike clover. *Trifolium hybridum*.

alsike poisoning. trifoliosis.

alsium. a grove community, one with trees and grasses.

alsophyte. alsad.

alt-. a prefix denoting or pertaining to height.

altai-. a prefix indicating Siberia; Siberian.

alter. **1:** change, modify, adjust, or convert. **2:** to neuter (sexually); to remove the gonads from an animal as by castrating or spaying. **3:** to remove the comb from a beehive. **4:** to prune, clip, or shear.

altern-. a prefix meaning alternate.

Alternaria solani. a fungus pathogenic to plants of the family Solanaceae, it produces alternaric acid, a withering agent. It also causes hemorrhagic syndrome in young domestic fowl.

alternaric acid. an acid isolated from the plant fungus, *Alternaria solani*. It is a withering agent that acts on the leaves of solanaceous plants at concentrations as low as 5 μg/ml.

alternate. **1a:** take turns, rotate, or interchange. **1b:** oscillate, fluctuate. **1c:** a backup, replacement, or standby. **1d:** sequential or successive. **2:** a surrogate, replacement, or representative. **3:** descriptive of leaves, petals, or twigs, for example, that are occur alternately on opposite sides of an axis, stem, or branch.

alternation of generations. a type of life cycle that occurs commonly in both plants and animals, especially parasites. Alternate generations usually differ not in appearance, but usually physiologically and behaviorally as well. In the most common type of alternation, one generation reproduces sexually and the next reproduces asexually. In bryophytes, for example, the dominant and most conspicuous generation is the sexual or gametophyte phase. In this phase, a zygote is produced that develops into the next generation of asexual (sporophyte) plants.

alternative test method. any method used to test products or other substances, especially for use or consumption by humans or livestock, that does not employ animals. Such methods include, for example, the use of cultures, bacteria, or living animal tissue *in vitro* (e.g., cloned skin cultures). Computer simulations alone or in conjunction with tests can often be used. The output from computer simulations that lack empirical substantiation should be treated as hypothetical.

alternifolius. having alternate leaves.

altherbosa. tall herbage.

altimeter. an aneroid barometer used to determine height above sea level.

altretamine (*N,N,N',N',N'',N''*-hexamethyl-1,3,5-triazine-2,4,6-triamine; 2,4,6-tris-(dimethylamino)-*s*-triazine; hexamethylmelamine; HMM). a very toxic insect sterilant and antineoplastic alkylating agent.

alumina. the native form of aluminum oxide, Al_2O_3. It occurs as corundum, a powder or crystalline material.

aluminium. aluminum.

aluminum (Al; aluminium (British, Canadian))[$Z = 13$; $A_r = 26.98$]. a soft, silvery white, moderately reactive, low-density metallic element with many commercial and industrial uses. It occurs naturally in shale, igneous rocks, clay, and most soils, usually as a silicate. Humans are rarely poisoned by aluminum except as indicated below. Aluminum is readily soluble in water of low pH. It is thus sometimes a major environmental poison in highly acidic water bodies where it leaches into the water from aluminum-containing bottom sediments. In such cases, it not only poisons fishes and other aquatic biota, but on rare occasions may threaten humans and even livestock by entering water supplies. It can damage the skin, eyes, and mucous membranes of the upper respiratory tract. While present in very small amounts in plants, it may be essential to the growth of some plants. It is abundant in all soils. Those having a pH greater than 5.5 and low amounts of organic matter may also contain aluminum in soluble or exchangeable form sufficient to kill certain plants.

aluminum phosphate (aluminum orthophosphate). a white, crystalline compound, $AlPO_4$, that is insoluble in water and ethanol and slightly soluble in hydrochloric and nitric acids. Its solutions are corrosive to tissues. It is used, for example, in paints and varnishes, pulp and paper, and pharmaceuticals. It occurs in large amounts in phosphate rocks in Florida, but because of its limited solubilities, it is only slightly available to plants in the soil.

aluminum sulfate. a water-soluble mineral that is usually produced by treating bauxite with sulfuric acid. It is used, for example, to purify water, to acidify or to reduce the alkalinity of soils, and in the tanning of leather.

aluminum *tris*(ethyl phosphite). fosetyl aluminum.

aluminum *tris*-(*O*-ethylphosphonate). fosetyl aluminum.

alumstone. alunite.

alunite (alumstone). a mineral that occurs commonly as fine-grained masses. It is a hydrous sulfate of potassium and aluminum, $A_{13}(SO_4)_2(OH)_6$.

alveolar gland. acinar gland.

alveoli. plural of alveolus.

alveolitis. inflammation of the alveoli. *See* alveolus.

alveolus (plural, alveoli). **1:** a small hollow, cavity, or blind sac. **2:** any of exceedingly numerous, thin-walled blind sacs (ca. 100 to 500 million in humans) at the terminus of bronchioles in the lungs of mammals and reptiles. They are about 250-350 μ in diameter, forming the respiratory surface of the lung. They have a surface area in humans of up to 100 m^2 at maximum inspiration, thereby providing not only an excellent respiratory exchange surface, serving the exchange of respiratory gases (carbon dioxide and oxygen), but also as an important portal of entry and/or elimination for volatile xenobiotics and for volatile products of toxicant metabolism. **3:** a terminal lobule or acinus of a racemose gland. **4:** any of the numerous tiny sacs of the milk secreting cells of the mammary gland, *q.v.* **5:** a cell in a honeycomb. **6:** a pore or pit in the perithecium of a fungus.

alyce clover. *Alysicarpus vaginalis*.

Alypia. *See* forester (def. 4).

Alysicarpus vaginalis (alyce clover). an annual herb, a forage plant (Family Fabaceae, formerly Leguminosae). It is a forage plant, grown also for hay and green manure. It is indigenous to central Asia and cultivated widely in warm regions.

amabil-. a prefix meaning lovely, loveliness.

amalgam (amalgamation). a liquid or solid alloy of mercury with one or more other metals.

amalgamate. to form or make an amalgam.

amalgamation. **1:** the process of making an amalgam. **2:** an amalgam.

Amanita (amanita). a genus of fairly large, gilled mushrooms (Family Amanitaceae); most species are toxic. They are common in America and Europe, where they most often occur on the ground in forested areas where they flourish in late fall and early winter. Amanitas in the United States occur chiefly in the mid-Atlantic states southward to Florida and west to Texas, where most species favor dry pine woods. *A. phalloides*, however, occurs on damp, sandy soils at mid to low elevations. Some species of *Amanita* are harmless, but most are poisonous. They probably account for at least 90% of fatal poisonings by mushrooms in the United States. The few edible species should be collected only by an expert on this genus. Two species are of special concern. ***A. muscaria*** (fly agaric; fly fungus; fly mushroom). a poisonous species with a broad, flat or convex, yellow to blood-red cap and white gills; the stalk is ringed and has a bulbous base. All parts are poisonous. It occurs under both evergreen and deciduous trees in the Rocky Mountains and Pacific Coast of North America; it occurs only rarely in the eastern United States. Symptoms of poisoning usually begin within 3 hours following ingestion and may include digestive distress with lacrimation, sweating, salivation, nausea, vomiting, diarrhea, abdominal cramps, excessive thirst, profuse sweating, vertigo, delirium, jaundice, convulsions, and possibly coma. Fatalities are rare in adult humans, although convulsions and coma may persist for as long as 12 hours and may require respiratory maintenance. ***A. phalloides*** (death cap; deadly amanita; deadly agaric; green death cap). the cap is smooth and greenish; the top of the stalk has a skirt-like ring and the base bears a sac-like cup. It grows in evergreen and deciduous forests, especially on damp, sandy soils and in urban areas at low elevations. It is abundant in Europe and North America and occurs in the United States from Massachusetts to Virginia and Ohio, the Pacific Northwest, and California. It is deadly and often mistaken for edible mushrooms if the ring on the stalk is ignored. All parts are poisonous. Toxic effects may not appear until 10 to 15 hours after ingestion. Symptoms include nausea, severe intermittent vomiting, profuse diarrhea, thirst, weakness, extreme abdominal pain, restlessness, delirium, hallucinations, coma, and late-developing jaundice. Remissions may alternate with these symptoms during a 6 to 8 day period, with rapidly developing renal failure and hepatic insufficiency. This may be followed by a slow recovery or death due to cardiac, renal, hepatic, or CNS lesions. Lacking proper treatment, death may occur in 5-10 days. One or two mushrooms, even though cooked can be fatal to an adult human; less is lethal to children. The case mortality rate among humans ranges from 50 to 90%. Signs of *A. phalloides* poisoning in feral animals and livestock are similar to those of humans. *See* amanitatoxin.

***Amanita* toxin**. amanitatoxin.

amanita. *Amanita*.

amanitatoxin (*Amanita* toxin; amanitotoxin). a toxic principle isolated from *Amanita phalloides*. It causes symptoms and lesions in mammals that are similar to those of humans caused by *A. phalloides*. It contains amanitine, α-amanitin, β-amanitin, and phalloidin.

amanitotoxin. amanitatoxin.

amar-. a prefix indicating bitterness.

Amaranthus (pigweed; carelessweed; redro ot). a genus of weedy, mostly annual herbs of North America (Family Amaranthaceae). *A. palmeri* and *A. retroflexus* accumulate amounts of nitrate sufficient to poison livestock and are believed to poison livestock at times.

amaryllid. a plant of the family Amaryllidaceae.

Amaryllidaceae (amaryllis family). a family of perennial plants with bulbs, corms or rhizomes. Many contain toxic, usually emetic, alkaloids (e.g., lycorine), and some are toxic to livestock. The bulbous members of the family (e.g., daffodils, narcissus, spider lilies, snowdrops, and snowflakes) can be especially dangerous to small children who may mistake them for food. Ingestion may induce severe and repeated episodes of vomiting with little or no diarrhea. Recovery is usually complete within 24 hours. *See Narcissus*.

amaryllis family. Amaryllidaceae.

Amaryllis (amaryllis). a genus of South African bulbous plants (Family Amaryllidaceae) that is toxic to livestock. There are many cultivated hybrids, but only one true species, *A. belladonna*. ***A. belladonna*** (amaryllis; belladonna lily; naked lady). a South African bulbous plant (Family Amaryllidaceae) with numerous cultivated hybrids that is toxic to livestock. In the United States, it grows well out of doors only in California.

amaryllis. common name of any of the numerous hybrids of *Amaryllis belladonna*.

amasesis. inability to chew (masticate).

aluminum orthophosphate. aluminum phosphate.

amasesis. the inability to masticate.

amaurosis. complete loss of vision, especially where no lesion or pathologic condition of the eye is evident. *See also* blindness. **intoxication amaurosis**. that caused by a systemic poison (e.g., ethanol, tobacco). **lead amaurosis**. that due to lead poisoning.

amaurotic. pertaining to amaurosis or to one suffering from amaurosis; afflicted with amaurosis.

amber. a fossil resin exuded from coniferous trees.

ambidextrous. neither right-handed nor left-handed, but able to use each hand well.

ambient. **1:** surrounding or enveloping. **2a:** pertaining to or denoting the environment (especially the background, local, or prevailing environment) in which an organism lives or components (as ambient environment, ambient air, ambient temperature). **2b:** pertaining to the preexisting or normal environment such as that obtaining prior to a forest fire. **3:** unconfined, outdoor, or out of doors.

ambient air. **1:** any unconfined portion of the atmosphere; open air. **2:** the air immediately surrounding an organism or other object of interest. **3:** outdoor air. This definition is the one most commonly encountered in air pollution science and in various government regulations.

ambient air level (AAL). any of a variety of state and local health-based guidelines for noncriteria air pollutants (toxics) in ambient air. They are based variously on risk assessments, NOAELs or LOAELs, and OELs (RELs, PELs, TLVs). AALs for a given chemical vary greatly among jurisdictions.

ambient air pollution. outdoor air pollution.

ambient air standard. the highest concentration of a given air pollutant that is permitted under law within a given jurisdiction. *See also* ambient air.

ambient temperature. the temperature of the surrounding medium (e.g., soil, air, water) in the immediate vicinity of a living system or object.

ambiguous. **1a:** vague, indefinite, unclear. **1b:** open to alternative interpretations or misinterpretation. **2:** of uncertain origin.

ambisexual (ambosexual; ambosexous; bisexual). in biology and agriculture: **1:** hermaphroditic; pertaining to a plant or animal that has both male and female reproductive organs. **2:** a plant or animal that has both male and female reproductive organs. **3:** sexually responsive to both sexes.

ambivorous. feeding on grasses and broadleaved plants.

ambivory. the selective feeding of an animal on grasses and broadleaved plants.

amble. **1a:** to walk at a slow, easy pace; to stroll. **1b:** a slow walk; a stroll. **2a:** of a horse: to walk at a slow pace in which the legs move in lateral pairs, usually with a four-beat rhythm. **2b:** the ambling gait of a horse.

ambler. a horse that ambles.

amblotic. likely to cause abortion.

Amblyomma (lone-star tick). a genus of showy, hard ticks (Family Ixodidae). Some North American species cause tick paralysis, *q.v. See also Dermacentor*. ***Amblyomma americanum*** (lone-star tick). a pest and vector of Rocky Mountain spotted fever that occurs chiefly in the southern United States and northern Mexico. Hosts include humans, dogs, livestock, and birds. The larvae, nymphs, and adults bite. ***A. cajennense*** (Cayenne tick). an important pest and vector of Rocky Mountain Spotted fever in humans, livestock, and wild animals; larvae, nymphs, and adults bite. It occurs from southern Texas southward into South America and the larger Caribbean islands. ***A. hebraeum*** (South African bont tick). a major vector of heartwater in southern Africa. ***A. maculatum*** (Gulf Coast tick). a major pest of livestock and humans in the southeastern United States. ***A. variegatum*** (tropical bont tick), an important pest of domestic livestock and vector of heartwater in Africa and the Caribbean.

ambosexous. *See* ambisexual.

ambivore. an animal that selectively feeds on grasses and broadleaved plants.

ambosexual. ambisexual.

ameba, amoeba. common name for *Amoeba spp.* and similar protozoans.

amebacidal. amebicidal.

amebacide. amebicide.

amebiasis (amoebiasis; amebiosis; amebism; amebic dysentery). a disease caused by the protozoan, *Entamoeba histolytica*, or other pathogenic amebas. *Cf.* entamebiasis. **canine amebiasis**. infection of dogs by *E. histolytica* acquired from humans. Dogs, however, rarely pass the cysts and do not serve as reservoir hosts. **hepatic amebiasis**. infection of the liver that may occur with or without prior dysentery.

amebic colitis. amebic dysentery. *See* dysentery.

amebic dysentery. amebiasis.

amebic. pertaining to, characteristic of, resembling, or caused by amebas.

amebicidal. destructive to amebas.

amebicide (amebacide). an agent that destroys amebas.

amebiosis. amebiasis.

amebism. amebiasis.

ameboid. amoeba-like.

amel-. a prefix denoting enamel.

amelia. the congenital, sometimes inherited, absence of one or more limbs. **autosomal recessive amelia** (porcine amelia). inherited amelia expressed in piglets. **porcine amelia**. autosomal recessive amelia of pigs.

ameliorate. to ease or improve (e.g., the condition of); to alleviate.

amelioration. moderation or improvement of an illness or condition.

ameliorative. able or tending to ameliorate.

amendment. **1a:** change or alteration. **1b:** revision. **1c:** addition. **1d:** (legislative) act, bill, reform, or rider. **soil amendment**. **1a:** any material added to soil to improve the chemical or physical characteristics of soil. **1b:** any material other than conventional fertilizers added to soil in order to improve its productivity. **1c:** material added to soil as a means of treating waste material.

amensalism. the production of a substance by one organism that inhibits the activity of one or more other organisms. *See also* antibiosis, antibiont, allelopathy.

American coral snakes. *See Micrurus*.

America white ipecac. *Euphorbia ipecacuanha*.

American alligator. *Alligator mississippiensis*.

American Association for Accreditation of Laboratory Animal Care (AAALAC). a nonprofit corporation, founded in 1965, that inspects by invitation and accredits laboratory animal care facilities and programs. The association encourages optimal care for laboratory animals through peer review, site visits, and specific recommendations to the responsible authority for each facility reviewed.

American bison. *See Bos bison*.

American Conference of Governmental and Industrial Hygienists (ACGIH). a professional organization in the United States that develops and publishes threshold limit values (TLV) for airborne toxicants in the workplace.

American copperhead. *Agkistrodon contortrix*.

American dog tick. *Dermacentor vatriabilis*.

American Farm Bureau Federation (AFBF). founded in 1920, it promotes the interests of farmers, especially by helping to secure state and national legislation.

American foulbrood. a disease of honey bee larvae, caused by *Bacillus larvae*, in which brood cappings are sunken, discolored, and perforated. In serious cases, an entire hive may be affected and production halted.

American hellebore. *Veratrum viride*.

American Horse Council, Inc. an organization formed in 1969 that represents and promotes the development of the entire horse industry in the United States.

American lance-headed vipers. snakes of the genus *Bothrops*.

American lanceheads. snakes of the genus *Bothrops*.

American pronghorn antelope. *Antilocapra americana*.

American red elder. *Sambucus pubens*.

American veratrum. *Veratrum viride*.

American water moccasin. *Agkistrodon piscivorus*.

American white hellebore. *Veratrum viride*.

American laurel. *See Kalmia*.

American wormseed. *Chenopodium ambrosioides*.

American yew. *Taxus canadensis*.

Amerikanische Lanzenottern. snakes of the genus *Bothrops*.

Ametabola. a group of several orders of primitive insects that undergo ametabolous development.

ametabolic. ametabolous.

ametabolic development. ametabolous development.

ametabolism. ametabolous development.

ametabolous (ametabolic). pertaining to ametabolous development.

ametabolous development (ametabolic de

velopment; ametabolism). the developmental pattern whereby offspring undergoing little or no metamorphosis but attain adulthood gradually through a series of molts with little or no change in external characteristics.

amethysti-. a prefix indicating violet color.

ametryn (2-ethamino-4-isopropylamino-6-methylmercapto-*s*-triazine; *N*-ethyl-*N'*-(1-methylethyl)-6-(methylthio)-1,3,5-triazine-2,4-diamine; ametryne). a methylthiotriazine similar in biological properties to atrazine and simazine but absorbed through foliage as well as the roots of green plants. Ametryne and related methylthiotriazines are used both as pre-emergence and postemergence herbicides.

ametryne. ametryn.

Amianthium muscaetoxicum. *Amianthium muscitoxicum*.

Amianthium muscitoxicum (= *A. muscaetoxicum*; fly-poison; staggergrass; cow-poison). a poisonous perennial herb (Family Liliaceae) that contains the steroid alkaloid, veratramine. It is highly toxic either by contact or ingestion. The bulb is especially toxic.

amiben. a substituted benzoic acid herbicide that inhibits germination and the development of amylase activity in intact barley seeds. Seeds of plants (e.g., those of squashes) that have low carbohydrate reserves are unaffected.

amide. **1a:** any organic compound that contains the $CONH_2$ radical; a primary amide. **1b:** any organic compound with the general formula $RCONH_2$ (primary amide), $(RCO)_2NH$ (secondary amide), or $(RCO)_3N$ (tertiary amide). All are basic, white, crystalline solids. **2:** any inorganic salt that contains NH_{2-}. They are formed by the reaction of ammonia with metals such as sodium and potassium. *See also* herbicide (amide herbicide).

Amidostomum anseris. a parasitic, bloodsucking nematode that invades the gizzard and, occasionally, the pro ventriculus and esophagus of domestic and wild ducks and geese (Order Anseriformes). Mortality among young birds is high.

aminacrine (9-acridinamine; 5-aminoacridine; 9-aminoacridine). a very toxic anti-infective agent and local antiseptic used in veterinary practice.

amine. a nitrogenous organic compound with the general formula R_3N where R can be hydrogen or a hydrocarbon group. Synthetic amines are prepared by reducing amides or nitro compounds. Naturally occurring toxic amines are uncommon but occur in some mistletoes (Family Loranthaceae) and some wild peas (*Lathyrus*); they also contribute to the toxicity of a number of mushrooms. *See also Acacia berlandieri* and *Claviceps*. **aliphatic amine**. any amine in which the substituted hydrocarbon groups are aliphatic hydrocarbon groups. All have strong odors. **aromatic amine** (arylamine). any amine derived from an aromatic hydrocarbon by the replacement of at least one hydrogen on the benzene ring by an amino group ($-NH_2$). Some bioaccumulate through food chains. The most common target organs among laboratory animals are the kidney, liver, gastrointestinal tract, and urinary bladder. Many are carcinogenic to laboratory animals and to humans. **biogenic amine**. any of a large group of naturally occurring, biologically active compounds that contain a primary amine group. Most are neurotransmitters, such as the catecholamines (adrenalin, noradrenalin, and dopamine) and the indoleamine, serotonin. *See also* biogenic. Other examples are histamine and serotonin. **higher aliphatic amine**. any amine in which one or more aliphatic hydrocarbon groups contain seven or more carbon atoms. Most are less poisonous lower aliphatic amines. **lower aliphatic amine**. an amine in which none of the aliphatic hydrocarbon groups contains more than 6 carbon atoms; all are highly flammable. They are used chiefly as intermediates in the manufacture of chemicals (e.g., polymers, agricultural chemicals, medicines). Lower aliphatic amines are among the more toxic substances used routinely on a large scale.

amino-. a prefix indicating the presence of an amino group, NH_2.

amino acid (aminoalkanoic acid; AA; aa). any of a class of organic acids (but see basic amino acid, below) that contains both the amino (NH_2) and carboxyl (COOH) groups. The amino group is usually bonded to the carbon atom adjacent to the carboxyl group (the α-carbon) (*See* α-amino acid). Free amino acids occur in both plant and animal tissue, where they take part in important cellular processes. Amino acids are produced chiefly by autotrophic plants and bacteria. The amino acids that occur commonly in proteins are alanine, arginine, asparagine, aspartic acid, cysteine, glutamine, glutamic acid, glycine, histidine, isoleucine, leucine, lysine, methionine, phenylalanine, proline, serine, threonine, tryptophan, tyrosine, and valine. **α-amino acid**. any amino acid in which the amino group is bonded to the carbon atom next to the carboxyl group; they have the general formula R-CH(NH_2)-COOH, where R is a side chain of varying length and structure. Except for glycine, α-amino acids occur as D- (dextro) and L- (levo) optical isomers. Only the L-isomers occur in proteins. The D-isomers occur chiefly in bacterial cell walls. The α-amino acids are fundamental components of living tissue as they combine to form peptides, polypeptides, proteins (some 24 amino acids occur in proteins), and many other important biological compounds. **acidic amino acid**. an amino acid that bears a net negative charge at neutral pH; these acids contain a second acid moiety and are hydrophilic (aspartic acid, cysteic acid, glutamic acid). **aliphatic amino acid**. an amino acid that has a side chain with no aromatic components (e.g., alanine, glycine, isoleucine, leucine, serine, threonine, valine). All but alanine and glycine are hydrophobic. **aromatic amino acid**. an amino acid that contains a benzene ring (phenylalanine, tyrosine, tryptophan). **basic amino acid** (dibasic amino acid). one that bears a net positive charge at neutral pH; they are hydrophilic and contain a second basic moiety (arginine, histidine, ornithine, lysine). Histidine is a weakly basic polar compound. **dibasic amino acid**. basic mino acid. **essential amino acid**. an α-amino acid essential to survival of a specific organism but not synthesized by the organism, either as free amino acids or in proteins. For humans the essential amino acids are isoleucine, leucine, lysine, methionine, phenylalanine, threonine, tryptophan, and valine. Arginine and histidine are also essential during periods of rapid growth. These amino acids are required also by albino laboratory rats. **hydrophilic amino acid**. any acid or basic amino acid. **hydrophobic amino acid**. one that is insoluble or nearly insoluble in water (isoleucine, leucine, methionine, phenylalanine, tryptophan, valine). tyrosine is somewhat hydrophobic at neutral pH. **nonessential amino acid**. one that can be synthesized by an organism in sufficient amounts to support life and health (alanine, glycine, and about 12 others are nonessential for humans; all are α-amino acids). **nonpolar amino acid**. one in which the group attached to the α-carbon is hydrophobic (α-aminobutyrate, leucine, valine). **polar amino acid**. a hydrophilic amino acid that occurs in aqueous solution as a dipolar ion (the acid and basic amino acids). **sulfur-containing amino acid**. These are cysteine and methionine. The latter is hydrophobic.

amino compound. an organic compound that contains an amino group.

amino group. NH_2, the essential component of an amino acid. Compounds with amino groups occur in numerous important or essential biological compounds. Numerous amino compounds are important toxicants (e.g., many polypeptides, proteins).

4-amino-PGA. aminopterin.

aminoacetic acid. glycine.

aminoacidemia. excessive amounts of specific amino acids in the blood.

5-aminoacridine. aminacrine.

9-aminoacridine. aminacrine.

aminoalkanoic acid. amino acid.

aminobenzene. aniline.

***p*-aminobenzenearsonic acid**. arsenilic acid.

2-aminobenzoic acid. anthranilic acid.

4-aminobenzoic acid. *p*-aminobenzoic acid.

4-aminobenzoic acid. *p*-aminobenzoic acid.

***o*-aminobenzoic acid**. anthranilic acid.

***p*-aminobenzoic acid** (4-aminobenzoic acid; vitamin B_x; bacterial vitamin H^1; chromotrichia factor; antichromotrichia factor; trichochromogenic factor; antigray-hair factor; anticanitic vitamin; PABA). a compound that occurs widely among biota, it is generally considered to be a growth factor in laboratory rats, the domestic fowl, and yeast and is the antigray-hair factor in laboratory rats. PABA is moderately toxic to laboratory mice and rabbits; less so to rats. It is used as a topical sunscreen, in research as a sulfonamide antagonist, and formerly as an antirickettsial.

aminocarb (4-(dimethylamino)-3-methylphenol-methylcarbamate ester; methylcarbamic acid 4-(dimethylamino)-*m*-tolyl ester; 4-dimethyl-amino-*m*-tolyl methylcarbamate). a tan, crystalline, slightly water-soluble, highly poisonous insecticide with a melting point of 93 to 94°C.

2-amino-5-chlorotoluene. 4-chloro-*o*-toluidine.

3-amino-2,5-dichlorobenzoicacid. chloramben.

α-amino-2,3-dihydro-3-oxo-5-isoxazoleacetic acid. ibotenic acid.

β-aminoethylglyoxaline. histamine.

4-amino-ethylglyoxaline. histamine.

3-(β-aminoethyl)-5-hydroxyindole. serotonin.

β-aminoethylimidazole. histamine.

3-(2-aminoethyl)-1*H*-indol-5-ol. serotonin.

4-aminofolic acid. aminopterin.

2-amino-5-guanidinovaleric acid. arginine.

α-amino-3-hydroxy-5-isoazoleacetic acid. ibotenic acid.

amino-(3-hydroxy-5-isoxazolyl)acetic acid. ibotenic acid.

***N*-[*p*-[[(2-amino-4-hydroxy-6-pteridinyl)methyl]-amino]benzoyl] glutamic acid**. folic acid.

3-(4-amino-2-methylpyrimidyl-5-methyl)-4-methyl-5,β-hydroxy-ethylthiazolium chloride. thiamine.

aminophen. aniline.

6,Δ,α-aminophenyl-acetamido penicillanic acid. ampicillin.

4-aminophenylarsonic acid. arsenilic acid.

aminopterin (*N*-[4-[[(2,4-diamino-6-pteridinyl)-methyl]amino]benzoyl]-L-glutamic acid; 4-aminofolic acid; 4-aminopteroylglutamic acid; 4-amino-PGA). a rodenticide, abortifacient, and antineoplastic agent, it is a folic acid antagonist that inhibits purine synthesis. It is teratogenic, causing nervous system defects, abnormalities of the extremities, micrognathia, hydrocephalus, and related abnormalities in the fetus.

4-aminopteroylglutamic acid. aminopterin.

6-aminopurine. adenine.

4-amino-*N*-2-quinoxalinyl-benzenesulfonamide. sulfaquinoxaline.

4-amino-*N*-2-thiazolylbenzenesulfonamide. sulfathiazole.

aminotriazole (amitrole; 3-amino-1,2,4-triazole (1*H*-1,2,4-triazol-3-amine; 3-amino-1*H*-1,2,4-triazole; 1,2,4-tri-azol-3-yl-amine; ATA). an amide herbicide that also exhibits antithyroid activity. It is a color less, crystalline solid, soluble in water and ethanol. It is a powerful plant growth suppressant and cotton defoliant that is readily absorbed by leaves and roots. It inhibits many aspects of growth and differentiation including cell division in the primary root meristems in some plants (e.g., *Pisum*, *Linum*). Aminotriazole causes albinism in developing leaves, which is irreversible if the application rate is high. It is extremely toxic to laboratory mice and rats, is carcinogenic to a number of laboratory animals, and is a suspected human carcinogen. All uses were banned in the United States in 1971. *Cf.* dichlobenil.

3-amino-1,2,4-triazole. aminotriazole.

3-amino-1*H*-1,2,4-triazole. aminotriazole.

4-amino-3,5,6-trichloropicolinic acid. picloram.

4-amino-3,5,6-trichloro-2-pyridine-carboxylic acid. picloram.

amiton (S-[2-(diethylamino)ethyl]phosphorothioic acid *O,O*-diethyl ester**;** *O,O*-diethyl *S*-(β-diethylamino)ethylphosphorothiolate). an extremely toxic contact insecticide and miticide. Amiton is a cholinesterase inhibitor.

amitosis. direct division of the nucleus and cell, without the splitting of the nucleus that occurs during mitosis. A new nucleus is formed that is probably not identical to the original nucleus. Amitosis occurs in some protozoans, in the embryonic membranes and cartilage of mammals, and the endosperm of flowering plants.

amitotic. of or pertaining to amitosis.

amitrole. aminotriazole.

ammeter. an instrument that measures the strength (intensity) of an electric current in amperes.

Ammi. a genus of branched annual herbs (Family Apiaceae) of the North Atlantic islands and the Mediterranean region. *Ammi majus* (bishop's weed) and *A. visnaga* seeds eaten by poultry produce severe photosensitization. Toxic concentrations of nitrate in the foliage have also been reported.

ammonia. **1:** true ammonia is a pungent, colorless, alkaline gas, NH_3. It is highly water soluble (a saturated solution at 0°C contains 36.9% ammonia) and is widely distributed in the environment, especially in surface waters. Ammonia is produced naturally when bacteria decompose nitrogenous compounds, such as those in manures. Ammonia may present a problem in enclosed livestock facilities. Natural waters are also frequently contaminated by ammonia from agricultural runoff, sewage effluents, and industrial wastes. Ammonia is quite poisonous to terrestrial and aquatic animals, especially in the unionized form. Aqueous solutions of ammonia such as ammonium hydroxide, NH_4OH, are alkaline and caustic. Common applications of ammonia in agriculture, and industry include use as a refrigerant, a pesticide, and in the manufacture of dyes and fertilizers (e.g., ammonium nitrate, ammonium sulfate, urea). Ammonia is important in intermediary metabolism and is the chief excretory product from protein metabolism in many bacteria, fungi, seeds, invertebrate animals, and aquatic vertebrates. In mammals, most endogenous ammonia is detoxified by conversion to urea which is temporarily stored in the urinary bladder. Ammonia is caustic and damaging to the skin, mucous membranes, and eyes. Inhalation of concentrated ammonia can cause severe respiratory distress and can be life threatening. Full recovery is usual in non-fatal exposures. **2:** ammonia water. **3:** often loosely applied to compounds such as ammonium chloride, ammonium hydroxide, benzalkonium chloride, and quaternary ammonium salts, all of which are poisonous and cause chemical burns on contact with tissue. Exposure to any of these substances can cause conjunctivitis, laryngitis, tracheitis, pulmonary edema, and pneumonitis. **anhydrous ammonia** (synthetic ammonia). It is used as a nitrogen fertilizer and is also the basic chemical feed stock for the industrial production of all other nitrogen fertilizers.Anhydrous ammonia is usually liquified under high pressure and sold to agriculturists in liquid form; often called aqua ammonia. When released from high pressure, it rapidly vaporizes. Consequently, the liquid ammonia must be injected into the soil and sealed. **aqua ammonia**. **1:** ammonia water. **2:** anhydrous ammonia. **aqueous ammonia**. ammonia water. **household ammonia**. ammonia water. **synthetic ammonia**. anhydrous ammonia.

ammonia fixation (ammonium fixation). adsorption of ammonium ions by clay minerals in the soil, thus rendering them insoluble and nonexchangeble.

ammonia intoxication. ammonia poisoning.

ammonia poisoning (ammonia intoxication; ammonia toxicity). poisoning from true ammonia, NH_3, of endogenous or exogenous origin. In many vertebrate animals, ammonia is detoxified in the liver; but in certain disorders, it may be shunted past the liver and may accumulate in blood with serious consequences that may compound the injurious effects of the precipitating condition. Aquatic plants appear to be less sensitive to ammonia than aquatic animals. **ammonia poisoning, acute**. a life-threatening condition caused by inhalation of concentrated ammonia in animals with lungs or by absorption of dissolved ammonia through gills in fish and other aquatic organisms. Concentrated ammonia gas is an asphyxiant causing spasms of the glottis and edema of the upper respiratory tract if inhaled. Death may swiftly intervene in the absence of prompt treatment. **ammonia poisoning, chronic, of fish**. signs may include lower rates of growth and development and impaired hatchability of eggs. Lesions of gills, liver, and kidney are also common. Percidae and Salmonidae appear to be the most sensitive families of fish.

ammonia solution. 1a: ammonia water. 1b: ammonium hydroxide.

ammonia toxicity. ammonia poisoning.

ammonia water (ammonia solution; aqueous ammonia; household ammonia; aqua ammonia; aqua ammoniae). a highly volatile, extremely irritant liquid; the fumes are extremely irritant to the eyes and lungs. The "ammonia" commonly used in dry cleaning, it is an aqueous solution of ammonium hydroxide. It is also used alone and in various formulations as a general or all-purpose cleaner in the home where it is a common source of poisoning. It can damage the skin, causing rashes, erythema, and chemical burns. The fumes are extremely irritating to the eyes and lungs and can be seriously harmful to individuals with respiratory problems such as colds, asthma, or bronchitis. Warning labels for household ammonia vary greatly; an example:

> "POISON: May cause burns. Call a physician. Keep out of reach of children."

Cf. ammonia, ammonium hydroxide.

ammoniac. ammoniacal.

ammoniacal (ammoniac). **1a:** pertaining to ammonia or its properties. **1b:** having the characteristics of ammonia. **2:** pertaining to a solution in aqueous ammonia.

ammoniated. containing or combined with ammonia; ammonia added.

ammonification. **1:** the act of infusing with ammonia, as in the manufacture of fertilizers. **2:** the condition of being impregnated or perfused with ammonia. **3:** the production and release of ammonia ions (ammonia) or derivative compounds during microbial decomposition of organic matter.

ammonify. **1:** to cause ammonification. **2:** to undergo ammonification.

ammonio-. a prefix indicating or pertaining to ammonia or ammonium.

ammonium. ammonium ion.

ammonium chloride (sal ammoniac; muriate of ammonia). a white, crystalline, slightly hygroscopic salt, NH_4Cl, of hydrochloric acid. It has a number of industrial and commercial uses and is important agriculturally as a fertilizer. Its value as a fertilizer is similar to that of ammonium sulfate, but it is more expensive and usually acidifies the soil more.

ammonium fixation. ammonia fixation.

ammonium fluoride (neutral ammonium fluoride). a granular powder, FH_4N, that yields ammonia and HF gas on heating. It is used to etch and frost glass, as a wood preservative, a moth-proofing agent, and an antiseptic in brewing beer. Symptoms of acute intoxication by ingestion are those of severe gastrointestinal distress with hemorrhagic gastroenteritis. These include nausea, vomiting, abdominal pain, diarrhea, muscular weakness, tremors, and convulsions. In severe cases, hyperpnea, depression, cardiovascular collapse, and death may ensue. Chronic poisoning mottles dental enamel and produces a generalized osteosclerosis with synostoses and calcification of ligaments and tendons.

ammonium fluosilicate. ammonium hexafluorosilicate.

ammonium group. a positively charged nitrogen atom joined by single bonds to four other atoms or groups, the simplest being NH_{4+}.

ammonium hexafluorosilicate(cryptohalite; diammonium hexafluorosilicate; ammonium fluosilicate; ammonium silicofluoride). a very toxic, odorless, crystalline powder used in pesticides. It is very toxic, chiefly causing severe gastroenteritis in humans. In fatal cases, symptoms may include dyspnea, tremors, convulsions, muscular weakness, and death from heart failure.

ammonium hydrate. ammonium hydroxide.

ammonium hydroxide (ammonia solution; aqua ammonia; ammonium hydrate). a colorless, toxic, corrosive, strongly alkaline liquid, NH_4OH, with a strong odor. It is an aqueous solution of up to about 30% ammonia. Both the liquid and vapor are highly irritant to the skin, eyes, and mucous membranes. When mixed with chlorine-containing products, it evolves deadly chloramine and, on heating, releases toxic and flammable compounds. Many cleaning products contain ammonium hydroxide but do not include it on the label. Ammonium hydroxide is the "ammonia" commonly used for dry cleaning and alone or in other products as a general purpose household cleaner; it is a frequent cause of poisoning in households. If swallowed, it causes gastrointestinal distress similar to that of alkali poisoning, including extreme pain, coughing, vomiting, a weak rapid pulse, and collapse. Later, the stomach and esophagus may perforate, with fever and intensification of abdominal pain with rigidity. Pulmonary edema, irritation, and even pneumonia may appear within 12 to 24 hours. Inhalation of highly concentrated solutions cause the lips and eyelids to swell, with temporary blindness; the victim may foam at the mouth and the skin appears reddish. *See also* ammonia (def. 3), ammonia water.

ammonium ion (ammonium). a positively charged univalent ion, NH_4^+, derived from ammonia. The herbicidal activity of ammonium salts (e.g., ammonium sulfate) is due largely to the presence of ammonium ions. The ammonia is very rapidly absorbed by the plant and quickly renders the normally acid and highly buffered sap alkaline, thereby quickly killing the plant.

ammonium nitrate. a colorless, highly water-soluble crystalline compound, NH_4NO_3, that is also soluble in alcohol and liquid ammonia. It is prepared commercially by reaction of nitric acid and ammonia. It is a popular fertilizer applied as a top dressing in the form of small, clay-covered pellets and in explosives. It is also used in solid-fuel rocket propellants, in pyrotechnics, and in the production of nitrous oxide.

ammonium silicofluoride. ammonium hexafluorosilicate.

ammonium sulfate (ammonium sulphate; sulfate of ammonia). a toxic, white to brownish-gray crystalline substance, $(NH_4)_2SO_4$, used as a rapid-acting herbicide. It is highly soluble in water, forming a corrosive, acidic solution; it is insoluble in ethanol and acetone. It is used also in fertilizers, in water treatment, fermentation, in fire-proofing formulations, and as a food additive. Toxicity is due chiefly to the usually rapid absorption and toxicity of the ammonium ion and to complexes formed with plasma proteins. Ammonia is very rapidly absorbed by plant cells and the cytoplasm, which is normally heavily buffered at an acid pH, may become alkaline. The increased alkalinity and the toxic action of ammonia on protoplasm causes rapid death of the cell.

ammonium sulphate. ammonium sulfate.

ammonium thiocyanate. a colorless, crystalline, water-soluble solid, NH_4SCN. It is an extremely toxic and rapid-acting plant poison. *See also* ammonia, ammonium ion.

Ammophila arenaria (European beachgrass). a tall, sturdy, perennial grass with long rhizomes that occurs naturally in coastal sands. The leaves have a ligule; the panicle tapers somewhat from the middle to the tip. Propagation is usually by transplantation of corms as most seeds are sterile and, in any case, few remain on coastal dunes long enough to germinate.

ammophilous. thriving or growing in sand or in sandy situations or habitats.

amnion (amniotic sac; indusium). the innermost of the extraembryonic membranes of many higher vertebrate animals (amniotes). It is a thin, strong, transparent membrane that contains the fluid (amniotic fluid) that bathes the fetus. Together with the placenta, it forms part of the afterbirth of placental mammals. *See also* Amniota.

Amniota. vertebrate animals (reptiles, birds, and mammals) whose embryos are enveloped by an amnion, *q.v.* Two additional extraembryonic membranes (lacking in anamniotes) are also present (the allantois and chorion) together with a yolk sac. These membranes are adaptations to terrestrial existence, allowing the embryo to develop within a fluid-filled cavity internally or in a shelled egg externally, rather than in an aquatic environment.

amniote. any animal belonging to the Amniota, *q.v.*

amniotic. of or pertaining to the amnion.

amniotic sac. *See* amnion.

Amo-1618. a plant growth retardant.

Amoeba. a genus of naked, lobular, protozoans (Class Sarcodina), usually of changing shape, that locomote by means of pseudopods. They are mostly aquatic but are also abundant in soils that are rich in organic debris. Included are both free-living and parasitic species, although many parasitic forms are now placed in other genera (e.g., *Entamoeba, Endolimax,* and *Iodamoeba*). Thus, in earlier classifications, *E. histolytica* was *Amoeba dysenteriae* and *Amoeba histolytica*.

amoebiasis. amebiasis.

Amoebotaenia. a genus of small intestinal tapeworms of birds, usually with less than 30 segments. ***A. cuneata*** (*A. sphenoides*). a species that commonly infests the domestic fowl, *Gallus domesticus*.

amoen-. a prefix indicating charm, loveliness; charming or lovely.

Amoracia lapathifolia (*A. rusticana*; *Cochlearia amoracia*; horseradish). a tall, coarse, white-flowered, perennial herb with a large taproot, the source of a condiment (horseradish). It is indigenous to Europe, but is widely cultivated elsewhere. *A. lapathifolia* escaped from cultivation in the United States and now occurs widely in moist situations throughout the eastern and midwestern states. Ingestion of tops or roots, which contain mustard oil, has produced acute gastroenteritis in cattle, sheep, and swine that can result in collapse and death. ***A. rusticana***. *A. lapathifolia*.

amorpha. lacking a definite form.

Amorphophallis (krubi). a genus of large Asiatic herbs (Family Araceae) with immense bulb-like roots. They are sometimes grown as greenhouse curiosities in temperate climates. All parts of the plant are toxic and, when chewed, produce symptoms similar to those of *Dieffenbachia*.

amorphophyte. a plant that bears asymmetric flowers.

amorphous. formless, shapeless, of indeterminant shape or structure, nebulous.

amotus. the elevated hind toe of certain birds.

AMP (adenosine monophosphate; adenosine 5′-dihydrogen monophosphate; 5′-AMP; adenylic acid). a mononucleotide ester (organic salt) of phosphoric acid and adenosine. Oxidative phosphorylation of AMP in mitochondria (catalyzed by adenylate kinase) can add more phosphate groups to yield the diphosphate (adenosine diphosphate) or the triphosphate (adenosine triphosphate). AMP is reconstituted by hydrolysis of ATP → ADP → AMP. *See* ADP, ATP.

5′-AMP. AMP.

amp. ampere.

ampel-. of or pertaining to a vine or vines; vinelike.

amperage. the strength or intensity of an electric current measured in amperes.

ampere (A; amp). the base SI unit of electrical current, it is equivalent to one coulomb per second. The ampere is formally defined, scientifically, as the constant current which, if maintained in two straight parallel conductors of infinite length, of negligible circular cross section, and placed one meter apart in a vacuum, would produce between these conductors a force equal to 2×10^{-7} N/m (newtons per meter) of length.

amperometric method. any electroanalytic technique based on the measurement of current or current difference. Depending on the technique used, such methods are termed either "amperometry" or "amperometric titration." Amperometric methods are widely used for air pollutant measurement and are the basis of most of the instruments that are used for the continuous monitoring of sulfur dioxide concentrations in the atmosphere; such instruments are often incorrectly termed "coulometric".

amperometric titration. *See* amperometric method.

amperometry. *See* amperometric method.

amph-, amphi-, ampho-. prefixes that indicate the presence of two sides; two or both (two-sided; double; around; surrounding); on both sides.

ampherotoky. reproduction parthenogenically of both males and females from an ovum, seed, or spore.

amphiarthrosis (plural, -ses). a joint permitting only slight motion, as between opposing vertebrae.

amphiaster. the achromatic spindle having two asters that forms during mitosis.

Amphibia. a class of anamniote tetrapods (Subphylum Vertebrata), comprised of some 2600 known species of toads and frogs (Anura), salamanders, and newts (Urodela). The skin of living amphibians is usually moist and lacks scales. They are typically pentadactyl (five toed), and the hind feet are often webbed; two nares communicate with the mouth cavity, and the mouth is often lined with fine teeth; the tongue is usually protrusible. The skeleton is bony, the skull bears two occipital condyles, and the pelvic girdle articulates with the sacrum. Ribs, when present, are not attached to the breast bone (sternum). All have a middle ear and ear drum (tympanum) that detect sound waves, a three-chambered heart, nucleated erythrocytes, respiration via the skin, the mucous membrane that lines the mouth, and gills or lungs. They are usually carnivorous, preying on insects and other small invertebrates, are typically oviparous, the eggs have a gelatinous covering and the larvae (tadpoles) are usually aquatic and breathe via gills. Habitats of adults are diverse, but most forms live in fresh-water or brackish-water, although arboreal and terrestrial forms occur.

amphibian. **1a:** amphibious; amphibiotic; able to live both on land and in water; having the properties of or behaving as an amphibian (defs. 2a, 2b). **1b:** pertaining to any life form defined in 2a or 2b. **2a:** any animal of the class Amphibia. **2b:** any life form adapted to or able to live in both aquatic and terrestrial environments.

amphicarpic. amphicarpous.

amphicarpogenous. of, or pertaining to fruit (e.g., peanut) that grows and matures below ground.

amphicarpous (amphicarpic; amphicarpus). bearing two types of fruit that may ripen at the same time.

amphicarpus. amphicarpous.

amphicribral. amphiphloic.

amphidiploid. allotetraploid.

amphidiploidy. the state of being allotetraploid (= amphidiploid).

amphigean. indigenous to both the New and Old Worlds.

Amphimerus. a genus of opisthorchid parasitic trematodes, probably transmitted by fish, that occur in the bile ducts of terrestrial vertebrates.

amphimictic. of or pertaining to amphimixis.

amphimixia. amphimixia.

amphimixis (amphimixia). true sexual reproduction whereby male and female gametes merge, forming a zygote. *Cf* apomixis.

amphioecious. denoting a population or species that has a wide but variable tolerance to habitat and environmental conditions as reflected in the presence of clines and subspecies. *Homo sapiens* can be regarded as amphioecious in many respects. This species, however, occupies diverse habitats and environments largely by virtue of an ability to modify the environment. Humans in the Antarctic, for example, bring a temperate "climate" with them in the form of heated buildings and protective clothing.

amphiphloic (periphloic; amphicribral). pertaining to or descriptive of plant vascular tissue in which the phloem surrounds the xylem. *Cf.* amphixylic.

amphiphyte. a plant with amphibious characteristics that grows at the margin of wetland and open water.

amphiploid. an allopolyploid cell or organism that has a diploid set of chromosomes from each parental species.

amphiploidy. the state of being amphiploid.

amphispermous. of or pertaining to plants (e.g., grasses) in which the seed is firmly enclosed by the pericarp.

amphivasal. amphixylic.

amphixenosis. a zoonosis that is maintained in nature by humans and lower animals.

amphixenosis. anthropozoonosis.

amphixylic (perixylic, amphivasal). pertaining to or descriptive of plant vascular tissue in which the xylem surrounds the phloem. *Cf.* amphiphloic.

amphor-. a prefix denoting or pertaining to an urn or pitcher; pitcher-like.

amphoteric. able to react either as an acid or a base (e.g., casein).

amphotericin B. a very toxic amphoteric polyene antibiotic and antifungal agent, $C_{47}H_{73}NO_{17}$, isolated from a strain of *Streptomyces nodosus*. It is available as the sodium deoxycholate complex which is used widely to treat systemic mycoses. It has nephrotoxic, abortifacient, fetotoxic, and teratogenic activities.

ampicillin (6,Δ,α-aminophenyl-acetamido penicillanic acid; beta lactamase). a semisynthetic, broad-spectrum antibiotic, $C_{16}H_{19}N_3O_4S$, derived from penicillin. It prevents bacterial reproduction by interfering with synthesis of the cell wall. It is used chiefly to treat enteric fevers, gonorrhea, and infections of the respiratory, urinary, and gastrointestinal tracts and of the ear, nose, and throat. Unwanted effects in humans may include nausea, vomiting, diarrhea, and a specific dermatitis. Anemia, hepatitis, and allergic reactions sometimes occur. A person allergic to penicillin will probably also react to ampicillin.

ampl-. a prefix meaning large, ample.

amplexi-. a prefix denoting clasping.

amplexicaul. stem-clasping.

amplissimus. very large.

amplitude. **1a:** extent, reach. **1b:** magnitude. **2:** the magnitude of displacement of a wave (or cycle) from its mean value. **ecological amplitude**. the range of tolerance of an organism, population, or species to an environmental factor or to environmental conditions more broadly considered.

ampul-. a prefix denoting or pertaining to a flask; flask-like in shape.

ampul. ampule.

ampule (ampul). a small, sealed glass container used to keep substances sterile.

ampulla. **1:** a flask-shaped structure. **2:** a saclike or funnel-shaped enlargement near the terminus of a duct or tube. **3:** a dilated or expanded part of a duct or canal

as the semicircular canals of the ear.

AMS. Agricultural Marketing Service.

amsinckia poisoning. poisoning of herbivores by *Amsinckia intermedia*. Signs of poisoning may appear in domestic cattle, horses, and swine following long-continued feeding on small amounts of seed. Affected animals are sometimes weak, moderately jaundiced, and usually fail to gain weight. Horses also exhibit neurologic effects (*See* walking disease). Effects in cattle are generally mild; swine fail to gain weight, sheep are relatively resistant, and domestic fowl are very resistant to amsinckia poisoning. Hepatic lesions may form that are similar to those caused by *Echium* and *Heliotropium*. *See* hard liver disease.

Amsinckia intermedia (tarweed; fiddleneck). an annual herb up to 1 m in height (Family Boraginaceae) that is native to California, Oregon, Washington, and Idaho. It is hepatotoxic and is especially dangerous when present as a weed in wheat fields. The hepatic lesions resemble those produced by *Echium* and *Heliotropium*. Symptoms in domestic cattle, horses, and swine may appear after long-continued feeding on small amounts of the seeds. Horses also exhibit neurologic effects. *See* "walking disease." Lethal concentrations of nitrate have also been reported in *Amsinckia*. *See* amsinckia poisoning.

amtofebrile. antipyretic.

amygdal-. a prefix meaning almond-like.

amygdalin (mandelonitrile-β-gentiobioside; amygdaloside). a bitter, white, crystalline substance,

$$C_6H_5CHCNOC_{12}H_{21}O_{10}$$

It is a cyanogenic glycoside that is soluble in water and ethanol and insoluble in ether. It occurs chiefly in seeds and other parts of various plants of the family Rosaceae, most notably in the seeds of species of *Prunus* (e.g., bitter almond, apricots, black cherries, peaches, plums) and certain other human food sources such as cassava. *See also* benzaldehyde, mandelonitrile.

amygdaloside. amygdalin.

mandelonitrile-β-gentiobioside. amygdalin.

amygdaline. pertaining to an almond; almond-like.

amygdaloid. **1:** resembling an almond or

tonsil. **2:** an extrusive or intrusive rock-containing gas cavity that contains secondary minerals.

amygdaloside. amygdalin.

Amygdalus communus. *Prunus amygdalus*.

Amygdalus communis amara. *Prunus amygdalus amara*.

amyl alcohol, primary active. 2-methyl-1-butanol.

amylaceous. starchy; of the nature of starch.

amylase (diastase). any of a class of enzymes that catalyze the hydrolysis of starch or glycogen to disaccharides and glucose. Amylase produced by the pancreas facilitates the digestion of starch in the small intestine. **alpha amylase**. an amylase present in wheat seeds. Flour made from wheat with excessive amounts of amylase yields bread that has a sticky texture.

amyloid. **1:** a waxy, translucent substance composed chiefly of protein that accumulates in various organs of animals that have certain diseases. **2:** a nonnitrogenous food consisting esp. of starch.

amyloid deposit. abnormal protein material deposited in tissues, usually as the result of a disease. Such deposits impair certain organs. *See* amyloidosis.

amyloidosis. **1:** an amyloid deposit in tissues or organs of an animal. **2:** any disease marked by amyloid deposits in organs or tissues.

amyloplast. **1a:** a plastid whose chief function is the storage of starch, normally as grains. They occur chiefly in the endosperm of roots, tubers, and seeds. **1b:** any plastid that stores and synthesizes starch (e.g., chloroplasts).

amyosthenia. muscular weakness.

amyosthenic. pertaining to or causing muscular weakness.

amyotrophia. amyotrophy.

amyotrophy (amyotrophia). muscular atrophy.

AN. acrylonitrile.

an-. a prefix meaning without, lacking. *Cf.* a-, A-.

ana-. a prefix meaning up, upward.

Anabaena. a toxic genus of cyanobacteria. *See also* Annie, Fannie, and Mike.

Anabaena flos-aquae. a poisonous species of Cyanobacteria. *See* Annie, Fannie, and Mike.

anabasine (3-(2-piperidinyl)pyridine; 2-(3-pyridyl)piperidine; neonicotine). a liquid, insecticidal alkaloid, $C_{10}H_{14}N_2$, produced by the herb *Anabasis aphylla* and *Tobaccum*. It is a colorless, nicotine-like liquid that darkens on exposure to air; it is miscible with water and soluble in ethanol and ether. Acute or subacute intoxication may cause confusion, vertigo, syncope, visual and auditory disturbances, nausea, excessive salivation, vomiting, photophobia, cold extremities, diarrhea, and clonic spasms.

Anabasis aphylla. a woody perennial herb (Family Chenopodiaceae) that secretes the insecticide anabasine. It is indigenous to the Caucasus and nearby areas about the Black and Caspian Seas.

anabasis. a period of increasing severity of a disease.

anabatic. **1:** pertaining to upward-moving air due to local heating of the land surface. *Cf.* katabatic. **2:** (obsolete) pertaining to any upward-moving air current. **3:** of or pertaining to anabasis.

anabatic wind. a wind or air current that moves uphill due to the effects of local heating of the land surface.

anabiosis. the reanimation or resuscitation of an organism following apparent death.

anabiotic. an agent that restores, reanimates, or resuscitates. *See* anabiosis.

anabolic. of, pertaining to, or promoting anabolism. *See also* catabolic.

anabolin. anabolite.

anabolism (constructive metabolism). the sum of those metabolic processes that produce complex cellular constituents (e.g., protein, fat, glycogen) from simpler precursors (e.g., the synthesis of proteins from amino acids). ATP, produced during catabolism, provides the necessary energy. Anabolic processes maintain an organism's structural and organizational complexity and integrity. During growth and at certain other times (e.g., periods of preparation prior to migration, reproductive activity, or the onset of winter), anabolic activities exceed catabolic activities. Anabolism is promoted, most notably by hormones such as the anabolic steroids. *Cf.* catabolism.

anabolite (anabolin). any product of anabolism.

anacanthus. a botanical term meaning spineless.

Anacardaceae (cashew family; sumac family). a family of some 60 genera of shrubs and trees, the cashew or sumac family, that are native to South America and southeastern Asia. Many are commercially important (e.g., cashew, mango); a number of genera produce poisons (e.g., *Anacardium occidentale*).

Anacardium occidentale (cashew). an evergreen tree (Family Anacardaceae), the nut of which, edible and commercially important. The shell of the nut contains an irritant oil that produces a contact dermatitis that appears to be identical to that of poison ivy. The active principles are mono- and polyhydric phenols. The sap is used to make an insecticidal varnish.

anacatharsis. severe, protracted vomiting.

anacathartic. **1a:** an emetic. **1b:** having the action of an emetic. **2:** an agent that induces vomiting.

Anacyclus pyrethrum. a prostrate perennial shrub (Family Asteraceae) endemic to Morocco, the flowers of which are a source of pyrethrin. *See also Chrysanthemum*.

anadipsia. extreme or excessive thirst. *Cf.* polydipsia.

anadromous. descriptive of a fish that leaves marine waters and ascends freshwater streams and spawn in the waters where they hatched. Such fish typically remain in marine waters during most of their adult lives, returning to freshwater streams just prior to spawning. Included are steelhead trout, most salmon, some cutthroat trout and Dolly Varden char, and lamprey eels. *Cf.* catadromous.

anaerbic compost. *See* compost.

anaerobe (anaerobic organism). an organism that can live and multiply in an environment that lacks molecular oxygen. The term is often applied only to anaerobic microorganisms. *Cf.* aerobe. **facultative anaerobe**. an organism that grows and reproduces in the presence or absence of molecular oxygen. **obligate anaerobe**. an organism that grows only under anaerobic conditions.

anaerobe. a microorganism that can live and multiply in the absence of molecular oxygen.

anaerobic. **1a:** pertaining to or denoting an anaerobe. **1b:** anoxybiotic, anoxybiontic; able to live in the absence of molecular oxygen. **1c:** pertaining to biological processes that occur in the absence of molecular oxygen (e.g., alcoholic fermentation, lactic acid fermentation, anaerobic respiration). **2:** pertaining to a process that occurs in the absence of molecular oxygen. **3:** of or pertaining to a milieu or environment that lacks molecular oxygen (e.g., excessively wet soil, certain bottom sediments of a water body). Anaerobic soils may result from excessive wetness. **strongly anaerobic**. pertaining to poorly-drained soil that remains wet or waterlogged for long periods of time. A mottled pattern of grays and browns results. **weakly anaerobic**. pertaining to a soil horizon that is anaerobic for short periods of time and moist for extended periods. A marbled or weakly mottled pattern usually results.

anaerobic organism. anaerobe.

anaerobiosis (anoxybiosis). **1:** the process of living in an environment free of molecular oxygen. **2:** The functioning of a tissue or organ in the absence of free oxygen.

anal. of or pertaining to the anus.

analgesia. absence of or insensitivity to pain, measured by the absence of or reduced response to painful stimuli.

analgesic (analgetic). **1:** pain-relieving; able to reduce or relieve pain. 1b: antalgic. **2:** a substance with analgesic properties, one used to relieve pain without loss of sensation or consciousness. The most widely-used analgesics are salicylates such as aspirin.

analgetic. analgesic.

analog (analogue). **1a:** that which is analogous to another being, object, property, or event. **1b:** a cell, tissue, organ, or other component in one species that is similar in function but not in structure, composition, or origin to that in another species. **2:** in chemistry, a compound that is structurally similar to another. **food analog**. a food product that resembles, and substitutes for, a conventional agricultural product such as meat loaf, sausages, or steaks made from soybean meal.

analogous. **1:** comparable; susceptible to comparison (in part or as a whole). **1b:** parallel. **1c:** similar, especially with respect to function. **2:** in evolutionary biology and allied fields: having a similar function but a dissimilar structure and evolutionary origin. *See* homologous. **3:** in commerce: having similar characteristics such as size, weight, cost, or value.

analogue. analog.

analysis of variance (ANOVA; variance analysis). a statistical technique used to define and isolate the sources of variability (variance) within a set of observations (variates). That is, to resolve the total variance of a set of variates into component variances associated with specified factors that affect the variates. It is especially useful for examining sources of uncontrolled variation in experimental conditions.

analysis unit. the basic inventory unit in forestry; such units are used to acquire basic data for use in forest planning models. An analysis unit may contain specified types or species of trees and may be further characterized by the site characteristics such as physiography, geographic location, or management practices.

anamnesis. **1:** recall, recollection, remembering; the ability to remember. **2:** that which is remembered. **3:** the past medical or veterinary history of a subject. **4:** immunologic memory; the condition of elevated immune response following exposures of immunocompetent cells subsequent to an initial exposure. Such responses are referred to as anamnestic or secondary immune responses.

anamnestic. **1:** aiding the memory. **2:** pertaining to a person's medical history or the veterinary history of an animal.

anamniote. any vertebrate animal (fish and amphibians), the embryos of which lack an amnion, allantois, and chorion. *Cf.* amniote.

anaphylactic. **1:** of or pertaining to anaphylaxis. **2:** exhibiting extreme sensitivity to a specific antigen.

anaphylactic intoxication. anaphylactic shock.

anaphylactic reaction. anaphylaxis.

anaphylactic shock (anaphylactic intoxication; allergic shock; anaphylaxis (in part)). a severe allergic reaction with sudden onset by a sensitized human or other animal due to prior exposure to a specific antigen (e.g., as by a bee sting, injection of a foreign serum, or blood transfusion). Anaphylactic shock in hu mans is usually marked by heightened irritability and violent cardiovascular activity. Skin eruptions, respiratory symptoms including difficult breathing, chest constriction, a violent cough, and cyanosis are also common. Convulsions, loss of consciousness, and death may result in severe cases. These reactions are due mainly to increased permeability of capillary endothelium and contraction of smooth muscle. *Cf.* anaphylaxis, allergic reaction, hypersensitivity.

anaphylactin. the specific antibody in a given case of anaphylaxis. It is formed following the initial encounter with the antigen and interacts with it during a subsequent challenge.

anaphylactogen. **1a:** a substance (antigen) that can sensitize or provoke susceptibility of a human or other animal to anaphylaxis such that subsequent challenge may induce anaphylaxis. **1b:** a substance (antigen) that can induce an anaphylactic reaction in a sensitized individual.

anaphylactogenesis. the sensitization, initiation, or development of anaphylaxis.

anaphylactogenic. able to induce anaphylaxis or to make a human or other animal susceptible to anaphylaxis. This term is usually applied to substances (anaphylactogens) that sensitize or provoke susceptibility of an animal to anaphylaxis.

anaphylactoid (pseudoanaphylactic). similar to or resembling anaphylaxis.

anaphylaxis (allergic shock; anaphylactic reaction). **1a:** an unusually severe allergic reaction, marked by immediate but transient hypersensitivity to a foreign protein (allergin, antigen) in a sensitized individual. **1b:** a severe reaction to horse serum or other antitoxic serum. *See* sickness (serum sickness). **1c:** induced sensitivity to any foreign substance. **2:** anaphylactic shock. **3:** originally used as a synonym for hypersensitivity in the sense of being a state of abnormally heightened sensitivity to an antigen due to prior exposure to the same antigen.

Anaplasma. a genus of pathogenic bacteria (Family Anaplasmataceae, Order Rikettsiales) that infect erythrocytes, appearing as spherical chromatic granules. Ruminants (Families Bovidae and Camelidae) are the hosts of these microbes which are transmitted by a number of species of ticks (most importantly by those of the genera *Boophilus* and *Dermacentor*) and by hepatophagous insects such as mosquitoes, horseflies *(Tabanus),* stable flies *(stomoxys),* and deerflies *(Chrysops). A. centrale* causes benign anaplasmosis in cattle; *A. marginale*, the type species. It causes malignant anaplasmosis in cattle; *A. ovis* causes anaplasmosis in sheep and goats. Reservoir hosts include various wild ungulates such as deer and antelope.

anaplasmosis (gallsickness). a type of infectious disease of ruminants of the families Bovidae and Camelidae, the causative agents of which are bacteria of the genus *Anaplasma*, *q.v.* **bovine anaplasmosis**. a benign disease of cattle caused by *A. centrale*. **malignant (bovine) anaplasmosis** (yellow teat disease). an acute disease of domestic cattle and related ruminants. Progressive jaundice, fever, anemia, hair loss, and cessation of milk flow are typical signs of this disease, caused by *A. marginale*, which is transmitted to the host by the tick, *Haemaphysalis cinnabarina punctata*. The disease is endemic in portions of the tropics and subtropics and has been reported from all of states of the continental United States. It is a disease of importance to the cattle industry. **ovine anaplasmosis**. a disease of sheep and goats caused by *A. ovis*. While usually benign, patent disease may develop under stressful conditions.

Anapsida. a subclass of the vertebrate class Reptilia. It contains modern turtles, tortoises, and certain of the earliest ancestral reptiles. There is one modern order (Chelonia).

anatol-. a prefix denoting or pertaining to Asia Minor.

anatomic, anatomical. **1:** of or pertaining to the structure (anatomy) of an organism. **2:** structural. **3:** pertaining to the anatomical sciences.

anatomy. **1:** the structure of an organism. **2:** the natural science that deals with the structure of living organisms. **3:** dissection.

ancestor. progenitor; an individual from whom a human or other animal is descended.

ancestor merit. an estimate of the production of future daughters by a bull based on the history of the bull's sire and maternal grandsire.

ancestral. of or pertaining to ancestry.

ancestry. **1a:** lineage; descent; heritage. **1b:** ancestors; progenitors; roots.

anchored dune. a sand dune that is stabilized by the presence of vegetation.

Ancistrodon. *Agkistrodon*.

ancona. a breed of domestic fowl that originated in Italy. Specimens have black feathers with white tips. The breed is productive, laying white eggs.

Ancylostoma (= *Ankylostoma*; Old World hookworms). a genus of nematodes that are parasitic in the duodenum where they attach to the villi of the mucosa. They nourishment themselves by sucking blood and may cause anemia in the host, especially one that is malnourished. The eggs pass out of the body via the feces; the larvae hatch in moist soil and develop into a worm-like (filariform) infective stage that enter the body of a new host via the skin and perhaps in drinking water. They are transported by way of the circulatory system to the lungs and upper respiratory tree; they are then swallowed and pass to the duodenum. ***A. braziliense***. a species with a single pair of ventral buccal teeth that parasitize dogs, cats, and occasionally humans. ***A. caninum***. a species that has three pairs of ventral teeth in the oral cavity. They commonly infest the intestines of dogs but also occur in human skin. ***A. ceylanicum***. a species that parasitizes the civet cat of Ceylon, and rarely, humans. ***A. duodenale***. a parasite of humans that is of widespread occurrence in the tropics and subtropics. It occurs rarely in the United States but is the most commonly encountered hookworm of humans in Europe, northern Africa, and eastern Asia. Humans usually acquire the disease by penetration of the filarial stage through the bare feet or sometimes other parts of the body that touch infected soil.

ancylostomatic. pertaining to hookworms of the genus *Ancylostoma*.

ancylostomiasis (ankylostomiasis; miner's disease; tunnel disease; intertropical or tropical hyphemia; hookworm disease). a disease of humans caused by *Ancylostoma duodenale*. Signs and symptoms include itching lesions at the points of entry of the parasite, anemia, eosinophilia, dyspepsia, wasting. Chronically, severely infected children may develop abnormal mental and physical traits, including a swollen abdomen.

Andalusian. a gray horse of medium size (15 hands or less) with large bones and well-shaped legs.

andesite. a dark, very finely crystalline extrusive rock of volcanic origin comprised largely of plagiocase feldspar with smaller amounts of other dark minerals (hornblende, biotite, or pyroxene).

andi-. a prefix referring to the Andes.

andisol. a black mineral soil formed from volcanic ejecta that contains 5% glass. It represents the eleventh soil order in the U.S. system of soil taxonomy.

andr-. a prefix denoting or pertaining to a male, maleness, or a male organ or part.

androdioecism, androdioecy. the process or condition whereby a species of plant produces male and female flowers on different plants.

androecium. the male reproductive component of a flower, comprised of the stamens collectively.

androgametophore. a plant that has male

sexual organs.

androgen (androgenic hormone; male hormone). any substance that stimulates the development of male characteristics in an animal. Androgens are secreted by the testis in all vertebrates and to some extent by the ovaries and adrenal cortex of mammals. Most are 19-carbon steroid hormones produced from acetate and cholesterol. They can also be synthesized chemically. Androgens promote the development and function of male sexual organs, regulate spermatogenesis, and promote the development and maintenance of male secondary sexual characteristics including sexual behavior. Androgens also facilitate the retention of nitrogen and synthesis of tissue-specific proteins. Many androgenic hormones are known; among the most common or important are testosterone, androsterone, androstenedione, and dihydrotestosterone. Androgens promote masculinization and are teratogenic. They can accelerate bone development in the fetus and masculinization of the female fetus. Disorders of androgen secretion or improper administration can cause serious, even life-threatening effects.

androgenesis. **1:** patrogenesis; male parthenogenesis; development of eggs that carry only paternal chromosomes; offspring with the paternal chromosomes only. **2:** the production of haploid plants from pollen grains that germinate within the anther.

androgenic hormone. androgen.

androgenic. **1a:** of, pertaining to, or having the properties of an androgen. **b:** having a masculinizing effect.

androgenous. tending only to produce male offspring. *Cf.* androgynous.

androgynal. androgynous.

androgynism. androgyny.

androgynoid. a male that resembles a female, or one that has hermaphroditic attributes.

androgynous (androgynal). **1:** having both male and female characteristics. **2:** sexually ambiguous; having neither clearly masculine nor clearly feminine appearance or behavior. **3a:** hermaphroditic; having both male and female reproductive organs. **3b:** the condition of having staminnate and pistillate flowers in the same inflorescence. *Cf.* androgenous.

androgyny (androgynism). the process or state of being androgynous.

Andromeda (andromeda; bog rosemary). a genus of attractive, widely distributed and cultivated shrubs (Family Ericaceae), these plants contain grayanotoxins, *q.v.* Other plants that contain grayanotoxins include *Kalmia latifolia* (mountain laurel), *Ledum* (Labrador tea), *Lyonia* (e.g., stagger bush), *Pieris* (pieris), and *Rhododendron* (rhododendrons, azaleas). *See* andromedotoxin.

andromeda. common name for shrubs of the genus *Andromeda*.

andromedotoxin. grayanotoxin.

andromonoecious. of or pertaining to a species of plant that produces male and bisexual flowers on the same plant.

androperianth. a fused perianth and stamen.

androphore. a stalk or column that supports the stamens of a flower.

androphyte. a male plant.

Andropogon furcatus (big bluestem). a tall, excellent forage grass of the prairies and open places in the central Great Plains and the southwestern United States It provides excellent forage for cattle.

Andropogon virginicus (broomsedge; broom grass; blue sedge; yellowsedge). a widely-distributed grass (Family Gramineae) along roadsides and poor soils in the southeastern United States It is often planted along embankments and freshly-backfilled sites to protect against erosion. It provides inferior livestock forage.

androstenedione. *See* androgen.

androsterone. *See* androgen.

anelectrode. anode.

anemia. **1a**: any condition characterized by an abnormally low number of red blood cells (erythrocytes) per cubic mm. in circulating blood; a low hemoglobin (hb) concentration (the amount of hb per 100 ml of blood), a low volume of packed red blood cells per 100 ml of blood (hematocrit), or any combination of these measurements. Below normal values in any or all of these measurements constitutes anemia. Of critical importance is the associated decrease in the oxygen-carrying capacity of the blood. **1b:** any disease or condition marked by anemia as defined above. **2:** any condition whereby the oxygen-carrying capacity of blood is reduced. Anemia can result from any of numerous causes including hemorrhage, a low rate of production or a higher than normal rate of destruction of erythrocytes, irregularities in the synthesis of hemoglobin, or hemolysis. Anemia is often marked by hypoxia, pallor of the mucous membranes and/or skin, shortness of breath, lethargy, fatigue, and palpitations of the heart. **acquired hemolytic anemia** ((nonhereditary) anemia (acute or chronic)). anemia associated with external or environmental factors such as, e.g., autoantibodies, infectious agents, and xenobiotic chemicals including medications, burns, certain phytotoxins, and venoms. **toxic anemia** (hemotoxic anemia). any anemia due to the destructive effects of a chemical, e.g., an herbicide or pesticide), a bacterial toxin, a metabolic poison, a venom. **aplastic anemia**. *See* hypoplastic anemia. **apparent anemia**. physiologic anemia. **deficiency anemia**. nutritional anemia. **deficiency anemia of suckling pigs**. a deficiency disease marked by a loss of vigor, low blood pressure, and paleness of the mouth. It is caused by failure of piglets that are housed indoors on wood or concrete floors to get enough iron and copper. Pigs maintained on dirt floors get ample iron and copper from the soil. **diphyllobothrium anemia** (fish tapeworm anemia). a rare type of pernicious anemia due to infection by *Diphyllobothrium latum* (the fish tapeworm). **equine infectious anemia** (horse malaria; swamp fever). a widespread contact (viral) disease of horses and other equids. Signs of the disease include progessive anemia associated with staggering gait, remittent fever, general debility, loss of muscle, and failure to thrive. The disease is transmitted by contact, by bloodsucking insects, and the use of unsterilized syringes and needles. **fish tapeworm anemia**. diphyllobothrium anemia. **genetic anemia**. any anemia that is due to or accompanies an inherited structural or functional defect of erythrocytes or erythropoietic tissue. **goat's milk anemia**. an iron deficiency anemia of infants maintained chiefly on goat's milk. **hemolytic anemia of the newborn**. a disease of foals, piglets, and puppies. **hemolytic anemia**. anemia due a higher than normal rate of destruction of mature erythrocytes in circulating blood and an inability of bone marrow to compensate. **hemorrhagic anemia**. that due to loss of blood through internal or external hemorrhage. **hemotoxic anemia**. toxic anemia. **hookworm anemia**. anemia that may accompany heavy infestations of hookworms, (e.g., *Ancylostoma duodenale* or *Necator americanus*). **hyperchromatic anemia**. hyperchromic anemia. **hyperchromic anemia** (hyperchromatic anemia). that in which the mean corpuscular hemoglobin concentration is greater than normal. **hypochromatic anemia**. hypochromic anemia. **hypochromic anemia** (hypochromatic anemia). that in which the mean corpuscular hemoglobin concentration is greater than normal. **hypoferric anemia.** iron deficiency anemia. **hypoplastic anemia** (aplastic anemia). a progressive nonregenerative group of anemias that can be life-threatening. This form of anemia is due to greatly depressed production of blood in bone marrow. This condition is often caused by ionizing radiation or chemicals. Aplastic anemia is the ultimate form of hypoplastic anemia. **infectious anemia**. that which results as a complication of infection. **iron deficiency anemia** (hypoferric anemia). a hypochromic microcytic anemia marked by low stores of iron in the marrow, low serum iron levels, and increased capacity of serum to bind iron. **isochromic anemia**. normochromic anemia. **lead anemia.**

anemia associated with lead poisoning and apparently due to defective hemoglobin synthesis. Iron is poorly incorporated into the porphyrin ring, if at all, during metabolic synthesis of hemoglobin. **local anemia**. anemia due to a reduced supply of blood to a body part, as in the occlusion of a vessel. **malignant anemia.** pernicious anemia. **megaloblastic anemia**. anemia due to premature release of erythroblasts from bone marrow due to folic acid antagonists (e.g., methotrexate) or antimalarials (e.g., pyrimethamine), or to vitamin B_{12} or folic acid deficiency. **milk anemia.** an iron deficiency anemia that sometimes occurring in infants maintained on a milk diet for an extended period of time. **nonhereditary hemolytic anemia**. acquired hemolytic anemia. **normochromic anemia** (isochromic anemia). anemia in which the mean concentration of hemoglobin in erythrocytes is within the normal range. **nutritional anemia** (deficiency anemia). that due to a deficiency in the diet of substances (e.g., copper, iron, protein) that are essential to the formation of red blood cells. **physiologic anemia** (apparent anemia). that due to overhydration, increased fluid volume of the blood.

anemic. pertaining to or characterized by anemia; exhibiting the characteristics of anemia.

anemic anoxia. *See* anoxia.

anemo-. pertaining to or referring to wind; thus anemone literally means wind flower.

anemochore. a plant (e.g., milkweed), the seeds of which are dispersed by wind.

anemochoric, anemochorous. of, pertaining to, or having the properties of an anemochore.

anemochory. wind dispersal.

anemogamy. the pollination of plants by windborne pollen.

anemohydrochory. dispersal by wind and water.

anemometer (wind gauge). an instrument used to measure the speed, velocity, or force of air flow.

anemophilous plant (anemophyte). a plant, such as a grass, that is pollinated by windborne pollen. In such plants, the petals of the flowers are absent or inconspicuous, produce no nectar, and have no scent. Stigmas and anthers are long and feathery, producing large quantities of dust-like pollen.

anemophilous. wind loving; pertaining to plants that are pollinated by wind.

anemophily. pollination by pollen conveyed from the anther to the stigma by wind.

anemophyte. an anemophilous plant, *q.v.*

anemoscope. an instrument used to indicate and/or record the direction of the wind.

anemotaxis. movement toward or away from the wind. *Cf.* aerotaxis.

aneroid. **1:** lacking fluid. **2:** of or pertaining to a type of barometer without mercury. A pointer indicates variations in air pressure, driven by the movement of the elastic wall of an evacuated chamber which therefore expands and contracts with changing pressure of the ambient air. **3:** of or pertaining to the mercury-free pressure gauge used with some sphygmomanometers.

anesthesia. **1:** the partial or complete loss of sensation by a body part or by the intact animal due to depression of nerve function by pharmacologic action, certain venoms, or neurologic disease. **2:** anesthesiology as a clinical specialty in medicine or veterinary medicine. **general anesthesia**. that which causes temporary loss of consciousness with complete inability to perceive normally painful stimuli. It is usually effected by inhalation or injection. **inhalation anesthesia**. general anesthesia produced by the inhalation of anesthetic vapors or gases. *See* anesthetic gas. **local anesthesia**. the temporary prevention or inhibition of local pain by use of a local anesthetic injected into the concerned tissue or site, by topical application, or by injection near the nerves that serve the field to be anesthetized (nerve block). **regional anesthesia**. the insensibility of a body part produced by interrupting afferent (sensory) nerve conduction from that part.

anesthetic gas. a vapor or gas such as ethyl ether, nitrous oxide, or methoxyflurane used to induce general anesthesia in breathing animals by inhalation.

anesthetics. **1:** plural of anesthetic (def. 2). **2:** the science or practice of anesthesia. **3:** those chemicals, collectively, that have an analgesic or anesthetic effect. *Cf.* anesthetic.

anesthetize. to treat with or bring under the influence of an anesthetic.

anesthetized. **1:** the absence of or diminished response to external stimuli accompanied by loss of the righting reflex; **2:** treated by or under the influence of an anesthetic.

anestrous. pertaining to anestrus.

anestrus (anoestrous, anoestrum). **1:** in mammals, a state of nonbreeding; a nonbreeding period. **2:** the normal period of reproductive quiescence between two breeding periods in mammals that reproduce periodically.

aneuploid (heteroploid). **1a:** having a chromosome number that is not an exact multiple of the basic haploid (monoploid) number. **1b:** an aneuploid organism.

aneuploidy (heteroploidy). a chromosomal aberration in which the number of chromosomes in a cell is not an exact multiple of the haploid number. One or more individual chromosomes may be represented more than twice (polysomy), only once (monosomy), or, rarely, not at all (nullisomy). Aneuploidy is normal in some organisms. *See also* aneuploid, chromosome aberration.

aneurysm. a localized, abnormal swelling or ballooning of a blood vessel.

aneurysmal, aneurysmatic. pertaining to an aneurysm.

anezeh. a breed of horse bred by a nomadic tribe in the Syrian desert.

angel's tears. *Narcissus triandrus*.

angelica (angelica tree). **1:** common name of any plant of the genus *Angelica*. **2:** *Aralia spinosa*.

angelica tree. *Aralia spinosa*.

Angelica. a genus of herbs (Family Apiaceae) that is native to temperate parts of North America, Eurasia, and New Zealand. It is a livestock forage in certain places, especially throughout the western United States Some species are poisonous, some therapeutic. Furthermore, some species of *Angelica* are sometimes mistaken for the extremely toxic water hemlock, *Cicuta maculata*. *A. archangelica* (garden angelica; wild celery). It was formerly blanched and eaten much like celery, and the stalks were also candied. Angelica is, however, moderately poisonous and possibly carcinogenic. Women have been poisoned by using it in futile attempts to induce abortion. ***A. venosa***. an extremely poisonous species.

angiocarp. a plant that bears angiocarpous fruit.

angiocarpic. angiocarpous.

angiocarpous (angiocarpic; gymnocarpous). **1:** of or pertaining to a fruit that is partially or entirely enclosed in a shell, involucre, or husk. **2:** pertaining to a fungus or lichen, the fruiting body of which is lies within the thallus.

angiogram (angiograph). a roentgenogram (x-ray record) that displays blood vessels containing a contrast medium.

angiograph. angiogram.

angiography. **1:** the preparation and analysis of angiograms. **2:** the study of blood vessels.

angiosperm. a flowering plant (Class Angiospermae), one in which the seeds are enclosed in an ovary . *Cf.* gymnosperm.

Angiospermae (Angiospermophyta; Anthophyta). a major, highly diverse class of vascular plants, the seeds of which, are produced within an ovary. It is a subdivision of the Spermatophyta, the seed plants, that includes also the gymnospermae. The angiospermae is comprised of two major subgroups: the Monocotyledonae and Dicotyledonae. *Cf.* Gymnospermae.

Angiospermophyta. Angiospermae.

angiostrongylosis (eosinophilic meningitis). an infection of humans and other mammals by nematodes of the genus *Angiostrongylus.*

Angiostrongylus. a genus of parasitic nematodes (Family Metastrongylidae), most of which invade respiratory or circulatory systems of rodents, carnivores, and marsupials. ***A. cantonensis***. a parasite of rodents caused by ingestion of infected mollusks. The larvae develop in the brain and migrate to the lungs, where they remain as adults. This species is although believed to cause eosinophilic encephalomeningitis in humans who live in the Pacific basin. Larvae have been found in the anterior chamber of the eye and in cerebrospinal fluid of persons living in Thailand who had ingested raw snails. ***A. costaricensis*** (= *Morerastrongylus costaricensis*). a parasite of rats and other rodents in Central America that occasionally infects humans. Following infection, the parasite invades and remains in the mesenteric arteries of the vertebrate host. ***A. Malaysiensis***. a parasite of rodents, it is endemic to Malaysia and may be responsible for an eosinophilic meningitis that occurs in that region. ***A. vasorum*** (= *Haemostrongylus vasorum*). a parasite of dogs and foxes that resides in the pulmonary artery and occasionally in the right ventricle of the heart. Common signs of infection are dyspnea and ascites. Thrombi may occur in the lungs, and affected animals occasionally die from cardiac insufficiency.

angiotensin II. a polypeptide with eight amino acid residues, it is a powerful vasoconstrictor and stimulator of aldosterone secretion from the adrenal cortex. It is formed from angiotensin I.

angle. an arithmetic, algebraic, or geometric function determined by the origin of two straight lines that emanate from a common point or two planes that emanate from a common line. *Cf.* slope.

angle of friction. angle of repose.

angle of repose (angle of friction). **1:** the angle between the horizontal plane and the plane of contact between two masses when the upper mass is just about to slide over the lower. **2:** sometimes used as a synonym of angle of rest, *q.v.*

angle of rest. the maximum slope at which a stack or accumulation of loose or fragmented solid material (e.g., crushed rock or loose soil) will stand without sliding or will come to rest when poured or dumped in a pile or on a slope. If this angle is exceeded, the material will slide downslope. *See* slope, angle, angle of repose (def. 2).

angle worm. *See* earthworm.

angleworm cast. the soil that accumulates above ground due to tunneling activity of earthworms.

angli-. English.

Ångstrom (symbol, Å; angstrom, Ångstrom unit; Å unit; tenthmeter). a unit of length, used chiefly to express wavelengths of optical spectra. It is defined by the wavelength of the red line of cadmium, 10^{-10} m.

angui-. a prefix referring to snakes or to snake-like or snake-shaped organisms or objects.

Anguina agrostis. a nematode that infests the seeds and hay of *Festuca rubra commutata*, *q.v.* It has caused fatalities in sheep, cattle, and other types of livestock. Common signs in sheep and cattle are tremors, ataxia, and collapse.

Anguina. a genus of telenchid nematodes that includes a number of important parasites of wheat and other grasses. A.

agrostis. a parasite that infests *Festuca rubra* (red fescue). It is responsible for the neurotoxic effects of this grass on livestock.

angul-. of or pertaining to an angle.

angus. Aberdeen Angus.

angust-. a prefix meaning narrow.

angustifoli-. a prefix meaning narrow-leaved.

Anguuina agrostis. a telenchid nematode, it is a plant parasite that infests *Festuca rubra*. It is responsible for the neurotoxic effects of this grass on livestock.

anhidrosis (anidrosis; adiaphoresis; isehidrosis). absence of sweating in any animal that normally has sweat glands. *See* anhidrotic.

anhidrotic (anidrotic; antihidrotic; antipers pirant; antisudorific). **1a:** of or pertaining to anhidrosis. **1b:** preventing or inhibiting perspiration. **2:** of or pertaining to the absence or reduction in number of sweat glands. **3:** an agent that inhibits perspiration.

anhydride. **1:** an oxide of a nonmetal (acid anhydride) or a metal (basic anhydride) that reacts with water to yield an acid or a base, respectively. An acid anhydride reacts with water to form an acid (e.g., sulfur trioxide, SO_3, reacts with water to yield sulfuric acid, H_2SO_4). A basic anhydride reacts with water to yield a base (e.g., calcium oxide, CaO, reacts with water to form calcium hydroxide, Ca(OH)2). Anhydrides of organic acids (e.g., acetic anhydride, maleic anhydride, phthalic anhydride) react with alcohols to form esters. Thus, acetic anhydride, $(CH_3\text{-}CO)_2O$, reacts with ethanol, C_2H_5OH, and yields ethyl acetate, $CH_3COOC_2H_5$, a useful solvent. Anhydrides also react with ammonia and primary or secondary amines to yield amides. **2:** a compound from which water has been extracted. It can react with water to yield the original compound.

anhydrite. anhydrous calcium sulfate.

anhydrous calcium sulfate (anhydrite). a compound, CaSO, used as a soil amendment.

anhydrous sodium borate. borax.

anhydrous sodium borax. borax.

anhydrous. lacking water; descriptive of an inorganic compound that lacks combined or adsorbed water. *Cf.* anhydride.

anidrosis. anhidrosis.

anidrotic. anhidrotic.

anilid. anilide.

anilide (anilid). a type of amide ($RNHC_6H_5$) in which the hydrogen of the amido group is replaced by a phenyl group. Anilides (e.g., alachlor, acetanilide) are *N*-acyl derivatives of aniline. *See also* herbicides (anilide herbicides).

aniline (benzeneamine; aniline oil; phenylamine, aminobenzene; aminophen; kyanol). **1:** a colorless or brownish, oily liquid, $C_6H_5NH_2$, with a characteristic odor and an acrid taste. It has numerous uses relevant to agriculture as in the manufacture of defoliants, dyes, fungicides, herbicides, insecticides, pharmaceuticals, polyurethanes and rubber, resins, and varnishes. Aniline is an ingredient of household products such as stove and shoe polishes, paints, varnishes, and inks. It is highly toxic to humans and is readily absorbed through the skin, digestive tract, or respiratory system. It commonly causes methemoglobinemia by oxidizing the iron in hemoglobin. Moderate exposure can produce anoxia with cyanosis. The case mortality rate in human infants who develop methemoglobinemia from exposure to aniline is 5-10%. Aniline is a known carcinogen. The fumes are poisonous if inhaled. *See also* acetanilid. **2:** any monocyclic aromatic amine that has a single amino group.

aniline oil. aniline.

aniline poisoning. *See* anilinism.

anilinism (anilism; chronic aniline poison

ing). chronic poisoning by aniline, marked by methemoglobinemia, aplastic anemia, vertigo, muscular depression, cardiac weakness, intermittent pulse, and cyanosis.

anilism. anilinism.

animal. **1:** any member of the Kingdom Animalia. **2:** any multicellular eukaryotic organism with holozoic nutrition, a capacity for spontaneous movement (locomotion), and usually rapid motor responses to stimulation. **3:** under law, any animal organism other than *Homo sapiens*. **4:** often improperly used in reference only to mammals. **cold-blooded animal**. *See* ectotherm. **domestic animal**. a tame animal that can live and breed normally in close association with and under the control of humans (e.g., horses, cattle, dogs, domestic cats). **experimental animal**. any animal used for experimental purposes. It may be a feral animal or a domestic animal held captive in the laboratory for some time prior to utilization to allow acclimatization to laboratory conditions. More often, it is a laboratory animal. **feral animal**. **1:** an animal that was formerly tamed but has since become wild. **2:** an animal that lives under natural conditions and has never been domesticated. **laboratory animal**. any animal used primarily for research and testing that is bred, born, reared, and dies in the laboratory or a supply house. Such animals typically represent strains or populations that have no genetic counterparts among wild animals of the same species. In some instances, there are no wild or nonlaboratory representatives of a species or strain. Many are highly inbred and few, if any, can be said to be representative of the species from which the original stock was drawn. In some cases, the actual species of such an animal is questionable. For purposes of scientific discourse, laboratory animals should generally be identified by an accepted or accredited common name, together with a breed or strain designation if known, and the immediate source if they come from a long-established population in a particular laboratory, institute, or supply house. In general, a scientific name should not be used, although the use of a genus and family name is desirable. Laboratory animals are typically housed, fed, and studied under conditions that wild populations of the same species (if such exist) do not experience and often do not tolerate. They are usually characterized in part by their small size and the ability to survive and reproduce under laboratory conditions. By contrast, the vast majority of animal species do not reproduce under laboratory conditions and many cannot survive. **poisonous animal**. any animal whose tissues, in whole or in part, contain poisons in sufficient quantity or concentration to injure or kill specified organisms that might encounter it under natural circumstances or might otherwise conceivably be exposed to it such that sufficient poison is absorbed to cause injury or death. Nearly all major classes of animals contain poisonous species or populations. *See also* poisonous, venomous. **sentinel animal**. an animal placed in a particular situation or environment to indicate or detect the presence of a hazard such as air or water pollutants, an infectious agent, or mine gas (methane). **venomous animal**. **1:** an animal that produces a venom. **2a:** an animal that can deliver a poison (called a venom, *q.v.*) to another animal by biting (usually by fangs), stinging, or in some cases by spraying or "spitting." **2b:** this definition extends def. 2a to include animals that deliver complex toxicants or venoms by contact (e.g., through breaks in the integument of the victim). *See also* poisonous, venomous. **warm-blooded animal** *See* endotherm. **wild animal**. a feral animal (def. 2).

animal black. bone black.

animal charcoal. bone black.

animal confinement facility. an area or shelter where animals are housed, fed, and reared or maintained.

animal developmental biologist. an embryologist who studies the developmental biology of animals.

animal drug. any drug developed and applied to the diagnosis, mitigation, treatment, cure, or prevention of disease in any animal. The safety and effectiveness of animal drugs and their use, under law in the United States, is assured by the Food and Drug Administration (FDA). To secure formal approval of an animal drug by the FDA, the sponsor or manufacturer of the drug must demonstrate in its pre-market approval application that the drug has proven to be "safe and effective" in scientific testing. They must also demonstrate a methodology to detect and measure any residue that may remain in edible animal products, and must show that edible animal products when ready-to-eat, are free from unsafe residues. When using animal drugs, both veterinarians and farmers must follow all restrictions placed on the use of the drug (withdrawal times) and must adhere to any warning labels or constraints stated on the drug label.

animal feeding operation. a feeding operation conducted in a confinement facility. **concentrated animal feeding operation**. a large animal feeding operation or facility; one that processes enough animals to be to be regulated as a potential point source of pollution under the Unified National Strategy for Animal Feeding Operations, *q.v.*

animal husbandry. the agricultural science and practice concerned with livestock production, especially for meat. Animal husbandry is concerned chiefly with management and production of beef cattle and horses.

animal industry (animal science industry). a broad field of applied science concerned with all aspects of livestock production including the preparation and processing of products from livestock production. *See* antibiotic feed.

animal month. one month's tenure on range of one animal.

animal protein. **1:** protein that occurs uniquely (or nearly so) in animals as opposed to plants. **2:** protein used in animal feeds that are derived from animal sources such as waste meat or meat products from meatpacking or rendering plants, surplus milk or milk products, and fish and other marine animals.

animal protein factor (APF). an obsolete term denoting undetermined growth factors in animal feed of animal origin (e.g., meat scraps, meat meal, fish meals, milk products).

animal science. animal science industry.

animal therapy. the use of pets and other domestic animals (as in a petting zoo) to provide therapy and, for example, to relieve loneliness or feelings of uselessness or rejection.

animal toxin. zootoxin.

animal type. a domestic animal that has a set of characteristics that is appropriate to or identifies it as bred for or suitable for a particular use (e.g, dairy type)

animal unit (AU). a standard unit of measurement, based on feed requirements, used to combine various classes of livestock according to size, weight, age, and use. **1:** in general, the animal unit for a given species or type of livestock is considered to be equivalent to a mature cow that weighs 454 kilograms (1000 pounds). It is applied to any feeding operation and is arrived at by a computation such as the following (*See* animal unit conversion factor): the number of slaughter and feeder cattle, plus the number of mature dairy cattle multiplied by 1.4, plus the number of swine weighing over 25 kilograms (52.5 lbs.) multiplied by 0.4, plus the number of sheep multiplied by 0.1, plus the number of horses multiplied by 2.0. **2:** on federal lands, an animal unit represents one mature cow, bull, steer, heifer, horse, or mule, five sheep, or five goats, all over six months of age.

animal unit conversion factor. in the United States, a numerical figure which allows conversion from one kind or class of animal to another. A conversion factor is satisfactory in respect to the amount of forage required to maintain an animal, but may have no application in determining stocking rates for range use for particular kinds or classes. The per-month conversion factors that are generally accepted in the U.S are: slaughter or feeder cow, 1.0; mature cow with calf, 1.0; mature dairy cow, 1.4; mature bull, 1.12; weaned calf, 0.6; yearling cow from 12 to 17 months, 0.7; yearling cow from 17 to 22 months, 0.75; two-year-old cow from 22 to 32 months, 0.9; horse, 2.0; swine weighing more than 25 kilograms (52.5 pounds), 0.4; sheep, 0.1; mule deer, 0.2; white-tail deer, 0.14; elk, 0.7; mature ewe with lamb, 0.2; weaned wether, 0.17; doe goat with kid, 0.17; buck, 0.17; weaned wether, 0.14; and grown horse, 1.25.

animal unit month (AUM). the amount of forage or grazing required to sustain one animal unit or equivalent for one month. Grazing fees for federal lands are charged by animal unit months or head-months.

Animal and Plant Health Inspection Service (APHIS). an agency of the USDA charged with protecting animal and plant health and safety. This is accomplished by inspections and the implementation of suitable regulatory and control programs. Thus, border inspections are conducted to prevent transboundary transmission of pests and disease, quarantine and eradication programs are developed and administered, and United States exports are certified to be in compliance with the plant and animal health standards of the importing country.

Animal Damage Control Program (ADC). an animal and plant health inspection service that was renamed the Wildlife Services (WS) Program, *q.v.*, in 1997.

Animal Welfare Act (P.L. 89-544) (August 24, 1966). a congressional act originally designed to curtail theft and mistreatment of dogs and cats for purposes of experimentation. Later amendments addressed additional concerns such as the regulation of shipping of pets by public transport and dog fighting; birds and additional mammals were brought under protection of the act. Farm animals are not covered by this law. The USDA, however, has the authority to "promulgate standards to govern the humane handling, care, treatment, and practices in experimental procedures to ensure that animal pain and distress are minimized." Legislation has been proposed, but not enacted, in recent years to regulate poultry and livestock confinement facilities, the diets of calves maintained for veal productiony methods of slaughtering poultryy and prohibiting the handling of nonambulatory (downer) livestock by dealers unless they are humanely killed.

anion. an ion that bears a net negative electrical charge. Examples are chloride, bicarbonate, and phosphate ions.

anion exchange. the process whereby an anion in solution exchanges with another anion previously bound to a solid, positively charged material (an anion exchanger). With respect to soil, anion exchange takes place between an anion in solution and any surface active material in the soil (e.g., clay, organic matter). *Cf.* cation exchange.

anion exchange capacity (AEC). defined with respect to soil, AEC is the total exchangeable anions that a soil can adsorb at a specific pH, expressed in SI units as centimoles of negative charge per kilogram of soil (cmole/kg). It may also be expressed in terms of milliequivalents per 100 grams of soil, usually at neutrality (pH 7.0), but may be referred to some other pH value.

anion exchanger. *See* anion exchange.

anionic. of or pertaining to a negatively charged ion (anion).

anis-, aniso-. prefixes meaning unequal.

aniseed. *See Pimpinella anisum.*

anisochromasia. **1:** any condition of nonuniformity of color or pigmentation. **2:** unequal distribution of hemoglobin in erythrocytes. Thus, under the compound microscope, the rim of the cell appears to be much more heavily pigmented than the center of the disk in reptiles and birds. This is because the mature (circulating) cell lacks a nucleus, and is thus biconcave and much thinner at the center. Anisochromasia is also pronounced in mammals that have certain types of iron deficiency anemia.

anisochromatic. not uniformly colored or pigmented.

anisomycin (1,4,5-trideoxy-1,4-imino-5-(4-methoxyphenyl)-D-xylo-pentitol3-acetate; [2R-(2α,3α,4β)]-2-[(4-methoxyphenyl)methyl]-3,4-pyrrolidinediol 3-acetate; 2-*p*-methoxyphenylmethyl-3-acetoxy-4-hydroxypyrrolidine). a crystalline compound, $C_{14}H_{19}NO_4$, with monoclinic needles. It is slightly soluble in water and most organic solvents, but is soluble in chloroform, ketones, and low molecular weight alcohols. It is a fungicide and antiprotozoal antibiotic produced by *Streptomyces griseolus* and *S. chromogenes* and is used chiefly as a fungus inhibitor and mildew preventive in vegetables.

ankyl-. a prefix that denotes a fusing or growing together of parts.

ankylose. to fuse in an immoveable articulation.

ankylosis. a stiffening or fixation of a joint by the formation of a fibrous or bony union across the joint due to disease.

ankylostomiasis. ancylostomiasis.

ankylotic. characterized by or pertaining to ankylosis.

annelid (annelid worm). common name of any of the segmented worms (Phylum Annelida) such as an earthworm or clamworm. Examples of genera with venomous species are *Eurythoe*, *Glycera*, *Hermodice*, and *Eunice*.

annelid worm. *See* annelid.

Annelida (segmented worms). a phylum of coelomate invertebrate worms with elongate, soft bodied, metamerically segmented bodies, often with paired setae ("bristles") on each segment. They have a soft, muscular, cylindrical body, a thin cuticle, a well-developed closed circulatory system, paired nephridia (usually metanephridia), a well-developed nervous system with a brain and ventral nerve cord, and a well-developed blood vascular system. The setae of some species form a stinging mechanism (e.g., *Eurythöe complanata*, *Hermodice carunculata*). Certain other species have hard, chitinous, biting jaws (e.g., *Glycera dibranchiata*, *Eunice aphroditois*).

Annie, Fannie, and Mike. the three genera of cyanobacteria that contain most of the species that are poisonous to animals are *Anabaena*, *Aphanizomenon* and *microcystis*. They are known collectively as Annie, Fannie, and Mike by water treatment professionals.

annu-. a prefix meaning annual.

annual. **1:** yearly; seasonal. **2:** annual plant (*See* plant). *Cf.* biennial, ephemeral, perennial. **hardy annual**. *See* plant.

annual bearing. **1a:** bearing fruit each year. **1b:** pertaining to a plant that produces fruit every year.

annual bluegrass. common name of *Poa annua*.

annual fescue. *Festuca megalura*.

annual meadow grass. common name of *Poa annua*.

annual poa. common name of *Poa annua*.

annual ring. **1:** the layer of growth of a perennial plant that occurs during one year as seen in cross section of the trunk or stem, branch, or root. **2:** the woody layer (xylem) produced by one year's growth of cambium. These layers appear as concentric rings when viewed in cross section.

annual sage. *Salvia reflexa*.

annul-. a prefix meaning ringed, annular.

annular. ring-shaped.

annulation. **1:** the formation of a ring or rings. **2:** a circular or ringlike formation, belt, segment, or part.

annulus. **1:** a ring. **2:** in biology, any ring-shaped structure such as an annual ring of a tree, the ring of tissue that remains on the stipe of the fruiting body of certain fungi, the annual zone on the scales of a bony fish, or one of the segments of an annelid worm.

anode (anelectrode; positive electrode). **1a:** a positive electrode, one where negative charges (electrons or ions) are withdrawn from the system. **1b:** an electrode to which negatively-charged particles are attracted and toward which they flow if unrestrained. **1c:** the positive pole of a galvanic battery or the electrode attached to it.

anoestrous. anestrus.

anoestrum. anestrus.

anomal-. a prefix meaning unusual, abnormal, or anomalous.

Anopheles. a cosmopolitan genus of dark brown, malaria-transmitting mosquitoes (Family Culicidae, Subfamily Anophelinae) distributed throughout much of Eurasia, Africa, and North, Central, and South America. The antennae appear feathery and the head bears a patch of pale "hair". In North America they commonly occur in deciduous or mixed forests and often about human habitation. The eggs are laid and the larvae develop in small, undisturbed puddles or pools of rainwater. Upon landing, these mosquitoes assume a posture in which the head points downward and the hind legs extend upward into the air. The females of many species transmit malaria to warm-blooded animals.

anophelicide. **1:** destructive to mosquitoes of the genus *Anopheles*. **2:** an agent that destroys anopheline mosquitoes.

anopheline. pertaining to mosquitos of the genus *Anopheles*.

Anoplura (Siphunculata). an order of wingless, eyeless, insect ectoparasites (sucking lice) that attack herbivorous mammals. Both nymphs and adults have sucking mouth parts, the body is compressed and transparent, and the legs are prehensile and bear sharp claws. Terrestrial mites are intermediate hosts. Humans may be infested by either of two varieties of *Pediculus*, the head louse or body louse. These lice are widely distributed, occurring under unhygienic conditions. They carry a number of diseases.

anorectic (anoretic; anorexic). **1:** denoting, pertaining to, or characteristic of anorexia, especially anorexia nervosa. **2:** an agent that causes anorexia.

anoretic. anorectic.

anorexia. loss of appetite; aversion to food.

anorexiant. a substance, event, or phenomenon that provokes anorexia.

anorexic. anorectic.

ANOVA. analysis of variance.

anoxemic. of, due to, or characterized by a serious lack of oxygen in circulating blood.

anoxia. **1a:** an absence or near absence of oxygen in the surrounding medium (e.g., soil, water, air), habitat (anoxicity), from inspired gases, arterial blood, interstitial fluid, or tissues of an animal. Anoxia may result, for example, from low atmospheric pressure, anemia, decreased or interrupted blood flow, or a decreased capacity of cells to utilize available oxygen. **1b:** a seriously reduced oxygen supply to cells or tissues of a living organism. **2:** formerly synonymous with hypoxia. **anemic anoxia**. severe anoxia in which oxygen is almost completely absent from circulating blood as, for example, carbon monoxide poisoning. **anoxic anoxia**. any extremely severe anoxic state of a living organism. **diffusion anoxia**. diffusion hypoxia (*See* hypoxia) in terrestrial or certain semi-terrestrial vertebrates that is severe enough to result in the absence of oxygen in alveolar gas. **histotoxic anoxia**. anoxia due to the failure of cells to utilize oxygen because of the poisoning of respiratory enzymes of the cells, as by amygdalin or other sources of cyanide, *q.v.* In higher animals, this condition is usually marked by higher than normal oxygen tension in arterial and capillary blood. *Cf.* hypoxia, oligoxia. **oxygen affinity anoxia**. anoxia due to the failure of hemoglobin to liberate oxygen. **stagnant anoxia**. severe stagnant hypoxia in which the free oxygen is absent from tissues.

anoxiate. to induce a state of anoxia, *q.v.*

anoxic anoxia. *See* anoxia.

anoxic. **1:** of, pertaining to, or characteristic of anoxia. This term may be used in reference to any type of living system. **2:** anoxia, anoxicity. **3:** said of a habitat that is devoid of molecular oxygen; anoxia, anoxicity. *See also* normoxic, oligoxic.

anoxicity. anoxia, *q.v.*, especially as applied to a habitat (e.g., a farm pond, a soil habitat).

anoxybiosis. anaerobiosis.

anoxybiotic. anaerobic (def. 1b).

ant. a common name of any insect of the family Formicidae. **army ant**. any of the chiefly tropical ants of the suborder Dorylinae. They travel in vast swarms, preying mainly on other arthropods, especially insects. **driver ant**. army ant. **legionary ant**. army ant.

Antabuse. disulfiram.

antagonism. **1a:** discord, conflict, or friction. **1b:** aversion, animosity, hostility. **2:** mutual resistance or opposition of force, tendency, or action between organisms, structures, agents, diseases, biologic processes, populations, or species. *Cf.* antipathy. **3a:** the interaction of two or more biologically active agents such that the effects of one or more of them are reduced or eliminated. **3b:** the condition whereby the combined effect of two or more chemicals to which a living system is simultaneously or sequentially exposed is less than additive. **4a:** the inhibition of one species or population by the action of another, especially inhibition of growth. Any of a number of mechanisms may be involved. Sometimes antagonism is manifested by the passive creation of unfavorable conditions by the presence and normal activities of one species or sometimes by specific behaviors (e.g., competition for limited resources such as food or nesting sites, or by the secretion of inhibitory substances such as antibiotics or other toxicants). **4b:** a relationship between two individual organisms, populations, or species such that both are adversely affected by the relationship. **5:** the opposing action of two muscles or sets of muscles such that the contraction of one is accompanied by relaxation of the other. **bacterial antagonism**. the destruction, injury, or inhibition of one species or strain of bacterium by the products of another. **biochemical antagonism**. functional antagonism in which two chemicals produce opposing biochemical effects. **chemical antagonism** (inactivation). in biology, antagonism that results from a chemical reaction between two compounds, resulting in less biologically effective products. **competitive antagonism** (receptor antagonism). antagonism resulting from the competition of two pharmacologically active agents for the same receptor site, thus producing a less than additive effect with simultaneous exposure. Carbon monoxide poisoning, for example, results from competitive antagonism with molecular oxygen for receptor sites on hemoglobin. **dispositional antagonism**. decreased activity of one substance due to the influence of another on its disposition (absorption, transformation, distribution, or excretion) in the body and consequent reduction in availability to the target organ. **functional antagonism**. that due to the interaction of two chemicals that produce opposing biochemical or physiological effects on the same system or process, thus effectively counteracting each other. **induction antagonism**. the reduction in biological or pharmacological action of one substance due to prior treatment with another substance that induces deactivating enzymes. **noncompetitive antagonism**. the phenomenon whereby an antagonist alters the receptor or otherwise prevents an agonist from exerting an effect. In this case, the efficacy of an agonist (e.g., a drug) is reduced as the concentration of a second agonist increases. **physiological antagonism**. func tional antagonism that affects a physiological process. **receptor antagonism**. competitive antagonism.

antagonist. **1a:** an adversary or assailant; **1b:** a competitor, challenger, or opponent. **2:** one who competes with or is opposed to another. **1:** an agent that opposes, resists, or neutralizes the action or effects of another. *Cf.* agonist. **2:** a substance that interacts with or occupies a receptor but does not initiate a response. **3:** a substance that reduces the toxicity of a toxicant. **4:** a muscle that opposes the action of another. Specifically, a muscle that acts in opposition to a "prime mover." **amino acid antagonist**. a chemical (e.g., an amino acid analog) that hinders the absorption, metabolism, or biological activity of an amino acid. When such amino acids are incorporated into proteins, they may produce significant, even disastrous, results (e.g., production of a teratology). Examples are ethionine, azaserine, and asparagine. **competitive antagonist** (receptor antagonist). a pharmacologically active agonist that competes with another for the same receptor sites in the same organism. A given dose of agonist will thus induce a less intense response when the competitive molecule is also present. **dispositional antagonist**. a substance that inhibits a biologically or pharmacologically active substance from reaching the target organ or that reduces the duration of its presence at the site of action. Ipecac is an example. **functional antagonist**. a substance that produces biochemical or physiological effects that reverse or oppose those produced by another. **noncompetitive antagonist**. a substance that acts against the action of a substance either by (1) blocking or altering the receptor or by binding irreversibly to a receptor site, thus preventing the agonist from exerting an effect, or by (2) acting at a distance from the receptor site without altering the affinity of the agonist for the receptor. **pharmacological antagonist**. a substance that inhibits the action of an agonist at its site of action. **physiological antagonist**. functional antagonist. **receptor antagonist**. competitive antagonist. **structural antagonist**. antimetabolite.

antagonistic (antipathic). having the action of an antagonist.

antagonistic effect. *See* antagonism.

antalgic. able to reduce responses to painful stimuli (i.e., to reduce signs associated with pain). *See also* analgesic.

ante-. a prefix meaning before.

antefebrile. preceding or prior to the onset of fever.

antelope. **1:** any of several ruminants of the family Bovidae that occur chiefly in Africa and Asia. All have hollow, unbranched horns that are not shed. **2:** the tanned hide or leather made from the hide of antelope. 3. a common, improper designation for *Antilocapra americana*, *q.v.*

antemortem. **1:** prior to death. *Cf.* postmortem. **2:** the examination that USDA meat inspectors are required to conduct of all live animals immediately prior to slaughter.

antenna (plural, antennae). **1:** a conductor, usually a wire or set of wires, that sends and/or receives electromagnetic waves; an aerial. **2:** one of a pair of jointed, movable, usually sensory appendages that project from the heads of insects and most other arthropods. Crustacea have two pairs of antennae; the anterior pair are small and are termed antennules.

antennae. plural of antenna.

antennule. a small antenna, especially either of the pair of antennae of crustaceans. *See* antenna.

anterior. **1a:** before, in front, or forward in time or space. **1b:** antecedent; foregoing. **2:** pertaining to or toward the head or forward end of the body of an animal when moving. **3a:** the front, as opposed to the back; the side or end of an organism furthest from the posterior (e.g., the tail end, rear, back, bottom, rump) side or end. **3b:** in quadrupeds, fish, most birds, embryos, and many other bilaterally symmetric animals, the head end is the anterior end of the body. **3c:** in humans and other primates, the forward plane or ventral surface of the body. **4:** in green plants, the side away from the main axis (e.g., the lower lip of a flower). **5:** antecedent in time; earlier; preceding. **6:** appearing early in a sequence.

anterior pituitary gland. anterior lobe of the pituitary gland. *See* pituitary gland.

anterior pituitary growth hormone. somatotrophin.

anterior presentation. the normal birth position of mammals in which the front feet and head emerge from the birth canal first.

anth-. a prefix denoting or referring to the term male as an adjective or modifier (e.g., male organisms, male body parts).

anthelminthic. anthelmintic.

anthelminthic drug. anthelmintic drug.

anthelmintic. (anthelminthic; antihelmintic; antihelminthic; helminthagogue; helminthic; helmintic; vermifuge). **1:** an agent that destroys or eliminates intestinal worms. **2:** able to destroy or eliminate intestinal worms (helmintic; vermifugal).

anthelmintic drug (anthelminthic drug). a medication used to treat infestations of intestinal worms. *See* anthelmintic.

anther. the terminal, sac-like, pollen-bearing part of a stamen in flowering plants. The pollen is contained in one or more (usually two) pollen sacs.

antheridium. a male reproductive structure (gametangium) of ferns, mosses, fungi, and algae that produces gametes.

antherozoid (spermatozoid). the motile male gamete generated in an antheridium.

anthesis. **1a:** flowering. **1b:** the period or process of expansion in flowers, from bud through full blossom to the setting of the fruit.

antho-, anth-. a prefix denoting or pertaining to flowers.

anthocarp. A composite fruit formed from several flowers.

anthocyan. anthocyanin.

anthocyanin (anthocyan). any of a class of water-soluble glucoside pigments of flowers and often other plant parts. Colors of anthocyanins range from red through violet to blue. They are the chief pigments that confer these colors on the flowers, fruits, and leaves. *Cf.* anthoxanthin.

anthoecology. the ecological discipline that treats the evolution and behavior of flowers in relation to their environment.

anthography. the scientific description of flowers; the discipline of describing flowers.

Anthonomus grandis (boll weevil). an American beetle that originated in Mexico and Central America. The larvae are major pests of cotton plants. They are the target of an Animal and Plant Health Inspection Service eradication program that is cooperatively funded and managed by cotton producers.

Anthony pig. the runt of a litter of pigs.

anthophagous. feeding on flowers. *Cf.* anthophilous.

anthophilous. attracted to or living in or among flowers. *Cf.* anthophagous.

anthophorous. pertaining to or descriptive of plants that bear flowers.

Anthophyta. Angiospermae.

Anthoranthum aristatum (annual vernal grass). an aromatic grass (Family Gramineae) sometimes considered a weed.

anthos. Greek, meaning flower.

anthotaxy. the arrangement of flower blooms.

anthoxanthin. a class of crystalline pigments that confer the pale yellow to ivory white colors of flowers and other parts of green plants.

anthrac-, anthraco-. prefixes meaning "carbon," "coal," "carbuncle," used in the formation of compound words such as anthracosis; anthracnose; anthracoid. anthraco- is used especially before a consonant.

anthracemia (anthrax septicemia). the presence of *Bacillus anthracis* in circulating blood, usually seen prior to infection of the skin or lungs.

anthracnose. **1:** any of a number of fungal diseases of plants characterized by dead spots on the leaves, twigs, or fruits. **2:** any one of several diseases, caused by parasitic fungi the Melanconiales. They attack the bean plants, grapes, melons, and cotton for example. *See* bird's eye rot.

anthranilic acid (2-aminobenzoic acid; *o*-aminobenzoic acid). a white or pale yellow, crystalline acid,

$$NH_2C_6H_4{\cdot}COOH$$

It is used as an intermediate in the synthesis of dyes, perfumes, and pharmaceuticals. The cadmium salt is used as an ascaricide in swine. *See also* cinnamyl anthranilate.

anthrax. a notifiable disease caused by *Bacillus anthracis*. It occurs most commonly among domestic cattle and sheep. Pigs and horses can also be affected. Humans are sometimes infected by contact infected meat products or by direct contact with diseased animals. The disease in humans is usually marked by hemorrhage and serous effusions in various organs and body cavities and subsequently, extreme prostration and death. *See also* charbon. **apoplectic anthrax**. a type of anthrax that affects livestock and other animals. Death may ensue within a few minutes to a few hours following infection. **cerebral anthrax**. a form of anthrax in which the bacilli have invaded the capillaries of the brain. It most often occurs in individuals with pulmonary or intestinal anthrax. In such cases, hemorrhagic meningitis complicates the effects of the pulmonary or intestinal anthrax. **cutaneous anthrax**. the form of the disease (eschar with edema; carbuncle) most common in humans, causing more than 95% of reported cases. It is acquired by handling infected material. Affected individuals develop a characteristic lesion. Initially a papule, it enlarges, forming a vesicle that splits open and discharges bloody serum. Within about 36 hours the vesicle turns into a bluish-black necrotic mass. This obvious condition is accompanied by severe systemic effects with vomiting, profuse sweating, high fever, eventual prostration, and often death. **intestinal anthrax**. a usually fatal form of anthrax in humans that is acquired by eating infected meat. The victim experiences chills, high fever, vomiting, headache, back pain, and pain in the extremities, bloody diarrhea and often bleeding from the mucous membranes and skin, and finally prostration, and usually death. **malignant anthrax**. a fatal form of the disease in cattle and sheep that also occurs in humans. It is marked by the development of hard ulcers at the point of inoculation and eventually extreme prostration and death. **pulmonary anthrax** (anthrax pneumonia; wool-sorters' pneumonia; wool sorters' disease). a form of anthrax in humans due to inhalation of spore-bearing dust. The presenting symptoms may be chills and fever followed by back pain and pain in the legs, dyspnea, coughing, rapid breathing, a rapid pulse, and extreme prostration.

anthrax septicemia. anthracemia.

anthrop-, anthropo-. prefixes that refer to human beings.

anthropeic. anthropic.

anthropic (anthropeic; anthropical). of or pertaining to humans, to their influence, or span of existence. *Cf.* anthropogenic.

anthropical. anthropic.

anthropobiology. the science that treats the biologic relationships among human races and among hominids generally.

anthropocentric. **1:** interpreting the world and especially the characteristics of non-human organisms in terms of human behavior and values. **2:** of or pertaining to the view that humans are the center of the universe.

anthropochory. dispersal by humans.

anthropogenesis. anthropogeny.

anthropogenic (anthropogenous). **1:** produced or caused directly or indirectly by humans (e.g., changes in soil due to plowing; increased death rate of aquatic organisms due to the pollution of water by humans). *Cf.* anthropic. **2:** pertaining to anthropogeny.

anthropogenic. produced by or derived from human activity.

anthropogenous. anthropogenic.

anthropogeny (anthropogenesis; anthropogony). the origin and development of hominids.

anthropogony. anthropogeny.

anthropoid. **1:** human-like in form and structure. **2:** one of the higher primates (Order Anthropoidea). Use of the term is often reserved for the great apes and monkeys.

Anthropoidea. an order of mammals that includes man, the great apes, and monkeys.

anthropology. the science that treats the origins and development of humans, including social and physical development and their cultural relationships, biological characteristics, and social customs and beliefs of humankind. *Cf.* anthroponomy.

anthropomorphic. of or pertaining to the view that natural phenomena and non-human entities, especially other living systems, have human characteristics.

anthropomorphism. **1:** the interpretation of natural phenomena, especially the behavior of living organisms in terms of human characteristics, especially human behavior. **2:** the sense or belief that nonhuman organisms and even nonliving entities and processes have human qualities, functional or structural. *Cf* theriomorphism.

anthroponomy. the science that treats the development of humankind, especially with respect to environmental relationships. *Cf.* anthropology.

anthropophilic. human-loving; applied usually to bloodsucking arthropods in which *Homo sapiens* is the sole or preferred host and to fungi that preferentially infest human skin.

anthropozoonosis (amphixenosis; zooanthroponosis). a zoonosis of animals in nature that is transmissible to humans (e.g, brucellosis, rabies).

Anthurium (tailflower). a genus of toxic, commonly cultivated tropical plants (Family Araceae) that have heart-shaped or strap-like leaves. ***A. scherzerianum*** (flamingo flower). a perennial herb indigenous to Central America that is grown in greenhouses for its ornamental value. It is valued for its bright scarlet spathe and coiled yellow spadix.

anti-. a prefix meaning against, opposite.

antiadrenergic (sympatholytic; sympathicolytic; sympathoparalytic). **1:** pertaining to the blocking of transmission by adrenergic nerves. **2:** antagonistic to, opposing, or inhibiting the action of adrenergic neurons. **3:** an antiadrenmergic agent. *Cf.* anticholinergic.

antiagglutinin. an antibody that acts specifially against the action of agglutinin.

antiallergic. of or pertaining to an agent or treatment that inhibits, prevents, or eases an allergic reaction.

antianaphylaxis. acting against or preventing anaphylaxis.

antianopheline. **1a:** detrimental to, destructive to, or used against mosquitoes of the genus *Anopheles* or their larvae.

1b: a chemical that destroys mosquitoes of the genus *Anopheles*.

antiantibody. an immunoglobin that acts as an immunogen when inoculated into another species, or sometimes as an autoantigen. Produced following the introduction into the body of an immunogen with which it then interacts.

antiauxin. a substance that inhibits the growth-regulating effects of an auxin.

antibacterial agent. antibacterial (def. 2).

antibacterial. **1:** destructive to or inhibiting the multiplication of bacteria. **2:** an antibacterial agent; a chemical or physical agent that kills (bacteriocidal) or reversibly suppresses (bacteriostatic) the multiplication of bacteria. Common antibacterials are antibiotics, antiseptics, and disinfectants. Some are selectively toxic, others are not. *Cf.* antimicrobial.

antibiont. a living organism involved in antibiosis.

antibiosis. **1:** an association of two species of organisms to the detriment of one, or between one species and a metabolic product of another (e.g., a toxin or antibiotic). **2**. the production of a metabolite (antibiotic) by a microorganism (e.g., a bacterium or fungus) that is toxic or inhibitory to certain other organisms. *See* allelopathy (def. 2), antibiotic, antibiont, probiosis.

antibiotic. **1:** pertaining to antibiosis. **2:** detrimental or destructive to living organisms. **3:** any of a variety of low molecular weight, soluble organic compounds with selective antimicrobial activity. Included are both natural and synthetic compounds. Many are used therapeutically in human and veterinary medicine. Antibiotics are generally effective at low concentrations against bacteria but are not effective against viruses. Many also produce toxic effects or allergic reactions in animals. Antibiotics that act chiefly against Gram-positive organisms (e.g., penicillin, lincomycin, erythromycin, tylosin) should not be used in guinea pigs and hamsters. Such antibiotics cause excessive growth of *Clostridium difficile* populations in the intestines which produce toxins that cause a rapidly-developing, fatal enterocolitis. Broad-spectrum antibiotics should not be used orally because of the direct effect on intestinal flora and must otherwise be used with caution. **4:** pertaining to an antibiotic (def. 4) or to the action of an antibiotic. **broad spectrum antibiotic**. an antibiotic that has a wide range of activity against both Gram-positive and Gram-negative bacteria.

antibiotic feed. any feed that contains antibiotics. The use of antibiotics in livestock and poultry feeds has caused much controversy as it is thought by many that the meat of animals given such feeds may pose a health risk to humans. The animal science industry, however, maintains that the use of such feeds offers no risk to humans.

antibiotic-resistant. pertaining to or characteristic of microorganisms that continue to multiply when exposed to antibiotic substances.

antibody (ab; Ab; antisubstance; immune body; sensitizer). **1:** a protein (immunoglobulin) elaborated into the blood or body tissues in response to the presence of a foreign substance (an antigen or a hapten associated with a carrier) with which it specifically combines, thereby inhibiting or neutralizing its action. **2:** a specific serum globulin, produced endogenously in response to the presence of an antigen. It binds with the specific antigen or sometimes certain structurally similar antigens if such are subsequently encountered. Antibodies are produced as part of the development of natural immunity or artificially by immunization. Some antibodies presumably exist naturally in the vertebrate body prior to or independent of the stimulus of an antigen. Most, however, are produced by B lymphocytes that have matured into plasma cells under the influence of a specific antigen. Antibodies confer humoral immunity for a period of time and in some cases provide lifetime immunity. *Cf.* sensitinogen, sensitizer. *See also* allergy, antigen, antigen-antibody reaction, immune response, immune reaction. **blood group antibody**. see blood group. **cell-bound antibody**. an antibody that forms on the surface of a sensitized lymphocyte, producing a cell-mediated (delayed) sensitivity to foreign proteins, cells, and tissues. **humoral antibody**. a free molecule, one of the two types of antibody (*See also* cell-bound antibody). Humoral antibodies are liberated into blood and tissue fluids where they may combine with bacterial toxins or coat bacteria, thereby promoting phagocytosis. **monoclonal antibody.** an antibody produced by a cell clone or a genetically homogeneous population of hybrid cells. The result in either case is a single line of cells that produce a specific antibody (immunoglobulin molecule).

antiboreal (austral; notalian). pertaining to cool or cold temperature regions of the southern hemisphere such as those that occur at high latitudes or high mountains. *Cf.* boreal.

anticancer drug. any medication that inhibits the growth of, disrupts, weakens, or destroys cancerous tissues with limited harm, if any, to healthy tissues.

anticanitic vitamin. *p*-aminobenzoic acid.

anticholinesterase (acetylcholinesterase inhibitor). any substance that inhibits or inactivates acetylcholinesterase, thereby preventing the hydrolysis of acetylcholine. Their action leads to an accumulation of endogenous acetylcholine with a resultant hyperactivity of cholinergic neurons, which can prove lethal. A number of organophosphorus or carbamate esters have anticholinesterase activity. Symptoms of acute poisoning can include irritability, tremors, and convulsions, with death usually resulting from respiratory failure. Acetylcholinesterase-inhibiting pesticides include aldicarb, carbaryl, ethion, fenthion, malathion, methomyl, and parathion. Further examples of anticholinesterases are physostigmine, tetraethyl pyrophosphate, snake venoms, and nerve gases. Their effects may be countered by the administration of atropine sulfate or pralidoxime iodide, which reactivate acetylchoiinesterase.

antichromotrichia factor. *p*-aminobenzoic acid.

anticoagulant. **1:** preventing or delaying coagulation. **2:** of or pertaining to the capacity of a substance to prevent coagulation, especially of blood. **3:** a substance that interferes with or prevents coagulation, especially the clotting of blood. Anticoagulants are the most widely used agents to control small mammal pests (especially rodents). Most applications are in cities, but their use in agricultural and forest environments are increasingly replacing the use of acute poisons such as compound 1080. Examples are dicoumarol, warfarin, heparin, phenindione, prothromadin, and many venoms. *See also* anticoagulant poison, snakebite, venom (snake venom).

anticoagulant poison. an anticoagulant used generally to control rodents and other small mammal pests. They are the most widely used agents for this purpose. Many small non-pest animals are unintended victims of such poisons.

anticodon. a triplet of sequential nucleotides within a loop of transfer RNA (tRNA) that pairs with a complementary triplet (codon) in messenger RNA. The matching of codon to anticodon provides the mechanism for translating genetic code into a spccific sequence of amino acids. *Cf.* codon.

anticonvulsant (anticonvulsive). **1:** an agent used to prevent, control, or to relieve convulsive states (e.g., epileptic seizures, seizures caused by toxic substances, nonspecific insults to the CNS). They can cause osteomalacia. **2:** pertaining to the capacity to prevent, control, or relieve convulsions. *See also* convulsant, convulsions, barbiturates.

anticonvulsive. anticonvulsant.

antidiuresis. decreased loss of water from the body via the kidneys.

antidiuretic. pertaining to the inhibition of water-loss via the kidneys (diuresis).

antidotal (antidotic). alexipharmic (def. 1a);

pertaining to or having the action of an antidote.

antidote (counterpoison; alexipharmic (def.2)). **1:** a substance that acts specifically to prevent, inhibit, inactivate, counteract, reverse, or relieve the action or poisonous effects of a toxicant. Many supposed antidotes are themselves toxic or otherwise potentially harmful and should not be used without special training or, with few exceptions, under appropriate medical supervision. **2:** in vernacular usage, any substance that prevents or relieves the effects of a poison is called an antidote. Definition 1 above is the only accepted professional usage. **chemical antidote**. a substance that unites with a poison to form a nontoxic compound. **mechanical antidote**. a substance that prevents the absorption of a toxicant. *See* true antidote. **physiologic antidote**. an agent that produces systemic effects that counter those of the specific toxicant under treatment. **true antidote**. a substance that acts by a specific mechanism related to that of the toxicant (e.g., *N*-methylpyridinium-2-aldoxime). A chemical that dilutes a toxicant (e.g., water, milk) or otherwise acts nonspecifically as a barrier to contact or absorption or one that acts to alleviate symptoms is not a true antidote. This term is sometimes used to emphasize that the vernacular usage of antidote (def. 2, above) is not intended; the term "specific antidote," though not recommended, is sometimes used. **universal antidote**. a hypothetical antidote that would be therapeutically effective against all poisons. No such antidote is known to exist. The use and effectiveness of any antidote depend on the type and amount of the toxicant encountered and the route of exposure or administration. Their use ordinarily depends also on the time delay between exposure and treatment, the condition (sometimes including age and sex) and history of the victim, and usually other factors.

antidotic. antidotal.

antiduretic hormone. *See* vasopressin.

antiemesis. relief or inhibition of nausea and vomiting.

antiemetic. **1:** preventing, suppressing, or relieving nausea and vomiting. **2:** an antiemetic drug; an agent that has antiemetic action (e.g., chlorpromazine, naloxone, perphenazine, prochlorperazine, promethazine). Chlorpromazine and other phenothiazines can block nausea and vomiting induced by a variety of causes. The nausea and vomiting induced by motion sickness can be controlled by a number of drugs (e.g., diphenhydramine, Dramamine, promethazine, and scopolamine).

antiemetic drug. antiemetic (def. 2).

antiestrogen. **1:** antiestrogenic (def. 1). **2:** any substance that chemically reduces or inhibits the biological action of an estrogen.

antiestrogenic (antiestrogen). **1a:** able to reduce or inhibit the biological action of estrogens. **1b:** antiestrogen (def.2)

antifebrile. antipyretic.

antifebrin. acetanilide.

antifeedant. an allelochemic produced by one organism that deters another organism from feeding upon it.

antiformin. alkaline sodium hypochlorite solution. *See* sodium hypochlorite solution.

antifungal (antimycotic). **1:** antagonistic to, destructive to, or able to suppress the growth of fungi. **2:** an antifungal agent.

antifungal agent. an antifungal; an antimycotic; a substance that is able to destroy or suppress the growth of fungi.

antifungal drug. a medication used to treat fungal diseases. *Cf.* antimycotic.

antigalactic. **1:** preventing or reducing the secretion of milk. **2:** an agent with antigalactic action.

antigen (Ag; immunogen). any substance, usually one of high molecular weight, such as a foreign protein or polysaccharide that, when introduced into the body of an animal, causes the formation of specific antibodies. Subsequent exposure to the antigen (following a latent period usually of 1-2 weeks) produces a reaction specifically with that antibody. The result may be a greatly increased specific resistance to infection by the antigen (immunity) or in some cases, structurally-related antigens. There are three classes of antigen: extrinsic, intrinsic, and occult. **blood group antigen** (blood group substance). any of a number of inherited antigens that occur on the surface of erythrocytes that determine a blood grouping reaction with specific antiserum. antigens of the ABO and Lewis blood groups also occur in saliva and other body fluids. **conjugated antigen**. conjugated hapten (*See* hapten). **complete antigen**. an antigen that stimulates the production of a specific antibody *in vivo* or *in vitro*. **extrinsic antigen**. an antigen that is not a cellular constituent. **Forssman antigen**. a heterogenetic antigen that occurs in dogs, cats, horses, sheep, turtles, the eggs of certain fish, and also in some varieties of corn and certain strains of bacteria (e.g., pneumococci). It does not occur in most rodents (exceptions are guinea pigs and hamsters, frogs, hogs, and most primates. **incomplete antigen**. hapten. **intrinsic antigen**. an antigen that is a cellular constituent. **occult antigen** (self antigen). an antigen that does not reach tissues that produce antibodies. **partial antigen**. hapten. **rhus toxicodendron antigen**. antigen extracted from the fresh leaves of *Rhus toxicodendron* (poison ivy). It is used to determine sensitivity to the contact dermatitis caused by its leaves. **rhus venenata antigen**. an extract of the fresh leaves of *Rhus venenata* (poison sumac). It is used to determine one's sensitivity to this plant or to alleviate the contact dermatitis caused by its leaves. **self antigen**. occult antigen. **sensitized antigen**. antigen-antibody complex. **species-specific antigen**. an antigen that occurs only in the fluids and tissues of individuals of a single species. Such antigens permit one to distinguish among individuals of closely-related or phenotypically similar species. **specific antigen**. an antigen that characterizes a single genus of microorganisms.

antigenic (immunogenic). able to act as an antigen; capable of inducing the production of an antibody. *Cf.* allergenic, immunogenic.

antigenicity (immunogenicity). **1a:** the state or condition of being antigenic. **1b:** the capacity of an antigen to induce a specific immune response. Antigenicity usually depends on the size and chemical configuration of a molecule. *See* antigen, hapten.

antiglobulin. a substance that blocks the action of globulin.

antigray-hair factor. *p*-aminobenzoic acid.

antihelminthic. anthelmintic.

antihelmintic. anthelmintic.

antihemolytic (hemosozic). **1:** preventing or inhibiting hemolysis; able to inhibit hemolysis. **2:** a substance that inhibits hemolysis.

antihemorrhagic. hemostatic; arresting hemorrhage; an agent that is able to prevent or arrest hemorrhage.

antiheterolytic. of, pertaining to, or having the properties of an antiheterolysin.

antihidrotic. anhidrotic.

antihistamine. any substance (most are complex, synthetic amines) that has an action such that it allays the effects of excess histamine produced by body tissues. Examples of antihistamines are diphenhydramine hydrochloride, chloropheniramine maleate, cimetidine, imidazole, tripelennamine hydrochloride, and promethazine. antihistamines are used to relieve the symptoms of allergies (e.g., hay fever), motion sickness, and stomach ulcers. They help to counteract some allergic manifestations of snakebite or horse serum, but play no role in the initial (acute) stages of snake venom poisoning. Even normal use of antihistamines can cause nausea, vomiting, fatigue, sedation, constipation, dryness of the mouth and upper respiratory tract, dysuria, and hypotension. One should avoid overuse.

antihistaminic. **1:** of or pertaining to antihistamines or to the action or effects of antihistamines. **2:** an agent with an action similar to that of an antihistamine.

antihormone. any substance that prevents or interferes with the action of a hormone.

antihydropic. **1a:** relieving or able to relieve generalized edema. **1b:** an agent that relieves generalized edema.

antiinflammatory. **1:** counteracting or having the capacity to counteract or reduce inflammation. **2:** an antiinflammatory drug or procedure that controls, counteracts, or reduces inflammation.

antiinflammatory drug. an antiinflammatory (def. 2) used to control inflammation not caused by infection.

antiknock agent. an organometallic compound added to fuels of internal combustion or spark-ignition engines in very small amounts to increase their resistance to knocking. In the past, tetraethyllead and/or tetramethyllead were most commonly used for this purpose. Halogenated compounds are also added to the fuel to facilitate the removal of lead after combustion.

Antilocapra americana (American pronghorn antelope; pronghorn; pronghorn antelope; prongbuck). a swift, antelope-like ruminant of the plains of western North America, including northern Mexico. Prior to settlement of western North America by Europeans, the pronghorn was represented by numbers as great as those of the American buffalo. While many populations have recovered well, this species is still endangered in parts of its range. Some are still taken by hunters, but its chief predators are wolves and coyotes. The pronghorn is the only living member of the family Antilocapridae. Not a true antelope, it is nevertheless closely related to the antelopes of Africa and Asia. The pronghorn is a social animal usually occurring in small bands or herds. It stands approximately 0.9 m at the shoulder and weighs approximately 45 kg. The body is light brown with white underparts and a white rump patch; the throat is marked by two white stripes. It has long, pointed ears and a short tail; both sexes bear horns with a bony core and a single branch or prong and a horny sheath which is lost and replaced annually. It browses mainly on shrubs, especially sagebrush, although it also eats grass. It is the swiftest mammal in North America, reaching speeds of about 100 km/hr (about 60 mph).

Antilocapridae. a family of the Ruminentia comprised of a single species, *Antilocapra americana*, the American pronghorn antelope.

antilysin. a substance that hinders the action of a lysin or prevents the process of lysis.

antilysis. the prevention, inhibition, or suppression of lysis.

antilytic. pertaining to, marked by, or capable of antilysis; inhibiting or counteracting lysis.

antimalarial. **1:** having the capacity to prevent or treat malaria; an antimalarial drug. **2:** an anti-malarial drug.

antimalarial drug. a substance used to prevent or treat malaria; an antimalarial (def. 2).

antimicrobial (antimicrobic). **1:** lethal to or inhibiting the multiplication of microorganisms, especially bacteria; bactericidal; inhibiting or interfering the pathogenicity of microbes; inhibiting mitosis. **2:** an antimicrobial agent, physical (e.g., extremes of temperature, ultraviolet radiation) or chemical (e.g., any of numerous antiseptics and disinfectants, sulfonamides), that destroys or reversibly suppresses the proliferation of microorganisms.

antimicrobial drug. an antibiotic; an antimicrobial (def. 2).

antimicrobic. antimicrobial.

antimony sodium oxide L(+)-tartrate. antimony sodium tartrate.

antimony sodium tartrate (antimony

sodium oxide L(+)-tartrate; sodium antimonyl tartrate). an anthelmintic used against schistosoma. An extremely toxic, transparent or whitish, hygroscopic powder or scales,

$$(SbO)NaC_4H_4O_6$$

It is used as an anthelmintic against *Schistosoma* infestations; used also as a leishmanicide in veterinary practice.

antimutagen. a substance that inhibits the activity of a mutagen. *See* mutagen.

antimycobacterial. antagonistic to or destructive of mycobacteria.

antimycotic. antifungal.

antiophidic. **1:** able to inhibit or relieve the effects of snakebite. **2:** any agent that opposes or relieves the effects of snakebite. *Cf*. antiophidica.

antiophidica. **1:** drugs that oppose or relieve the effects of snakebite. **2:** plural form of antiophidic (def. 2).

antioxidant (antoxidant; antioxidizer). **1:** any substance that inhibits or prevents oxidation. Some are important metabolically because they protect cells and tissues from attack by molecular oxygen, peroxides, or any active oxygen species (e.g., the superoxide radical). Included in this category are superoxide dismutases, reduced glutathione (GSH), alphatocopherol, ascorbic acid, beta-carotene, catalase, glutathione peroxidase, glutathione reductase, histidine, methionine, and polyunsaturated fatty acids. Antioxidants are used as preservatives to prevent or inhibit rancidity, spoilage, loss of potency of vitamins, or the development of other undesirable changes in foods, animal feeds, and other organic materials. A number of these applications are important in agriculture **2:** of or pertaining to an antioxidant.

antioxidation. the inhibition or prevention of oxidation.

antioxidizer. antioxidant.

antiparasitic. antagonistic to or destructive to parasites.

antipathic. antagonistic.

antipedicular. destructive to lice.

antipediculotic. **1a:** effective against pediculosis. **1b:** an agent that is effective in the treatment of pediculosis.

antiperspirant. anhidrotic.

antipneumococcic. **1:** antagonistic to or destructive of pneumococci. **2:** an agent that inhibits or destroys pneumococci.

antiprotozoal (antiprotozoan). **1:** antagonistic to or destructive to protozoa. **2:** an agent that destroys protozoa or inhibits their growth or reproduction.

antiprotozoan. antiprotozoal.

antipruritic. **1:** able to prevent or relieve itching. **2:** an agent that is able to prevent or relieve itching.

antipyresis. the use of antipyretics to reduce fever.

antipyretic (antifebrile; antithermetic). **1:** able to relieve or reduce fever. **2:** an agent (antifebrile, antithermetic, febrifuge), such as aspirin, that relieves or reduces fever.

antiqu-. a prefix meaning ancient.

antirobin. the antitoxin of robin.

antiscorbutic acid. ascorbic acid.

antisepsis. prevention of infection by the destruction or inhibition of the growth of infectious organisms.

antiseptic. **1:** pertaining to antisepsis. **2:** able to destroy microbes or inhibit their growth. **2:** any nonselective antimicrobial that is applied topically to kill or inhibit the growth of pathogenic microorganisms in living material; one that is able to effect antisepsis. Antiseptics are not usually harmful to human skin.

antisera. plural of antiserum.

antiserum (plural, antisera). a serum or body fluid that contains high levels of antibodies against a specific antigen. Antisera result naturally (through disease) or through immunization. Antisera are injected to give passive immunity against specific diseases. *Cf.* antitoxin, antivenin.

antisnakebite serum. snake antivenin. *See* antivenin.

antispasmodic. **1a:** preventing, easing, or relieving spasms or cramps. **1b:** an agent that relieves or prevents muscle spasms (e.g., of skeletal muscle, bronchi, gastrointestinal tract).

antispasmodic drug. an antispasmodic (def. 1b).

antistaphylococcic. **1:** antagonistic to or destructive of staphylococci or their toxins. **2:** a substance that destroys staphylococci.

antisubstance. antibody.

antisudorific. *See* anhidrotic.

antitetanic. **1:** preventing, relieving, or curing tetanus. **2:** an agent that prevents, relieves, or cures tetanus or tetanic spasms.

antithermetic. antipyretic.

antithyroid drug. an antithyrotoxic medication used to counter the effects of an overactive thyroid gland. *Cf.* antithyrotoxic.

antithyrotoxic. antagonistic to, inhibiting, preventing, or counteracting the toxic effects of thyroid gland secretions or extracts.

antitoxic. **1:** active against or able to neutralize a poison. **2:** of or pertaining to the nature of or action of an antitoxin. *Cf.* antidotal.

antitoxigen. antitoxinogen.

antitoxigen. antitoxinogen.

antitoxin (antitoxinum; immunotoxin; toxolysin). **1a:** an antibody induced by the presence of a toxin, including not only bacterial exotoxins, but also phytotoxins and zootoxins. **1b:** whole serum or the globulin fraction of serum from animals (e.g., horses, rabbits) that have been immunized by injections of a specific toxoid. Some examples are defined below. *See also* antiserum, antivenin. **diphtheria antitoxin**. antitoxin produced in the blood of horses that have been injected with diphtheria toxoid. The serum from the blood of such horses, when injected into humans, confers immunity to diphtheria. **bothropic antitoxin** (Bothrops antitoxin). an antitoxin specific for the venom of pit vipers of the genus *Bothrops* (Family Crotalidae). *Cf.* serum (antibothropic serum). **Bothrops antitoxin**. bothropic antitoxin. **botulinum antitoxin**. botulism antitoxin. **botulism antitoxin** (botulinum antitoxin). an antitoxin specific for a toxin of one of the strains of *Clostridium botulinum*. **crotalus antitoxin**. an antitoxin specific for rattlesnake venoms (i.e., of *Crotalus*).

antitoxinogen (antitoxigen). a substance that stimulates or induces the formation of antitoxin by an animal *in vivo*. *Cf.* antitoxin.

antitoxinum. antitoxin.

antitrichomonal. **1:** lethal to or resistant to trichomonads. **2:** a drug that is active against trichomonal infection.

antivenene. antivenin.

antivenin (antivenom; antivenene; antivenine). **1:** a specific antitoxin that acts against, inhibits, or reverses the effects of a venom. The antibodies in such immune serum preparations neutralize the venom of the taxon for which it was prepared. **2:** any serum, such as antiserum or antivenomous serum that contains a venom-specific (species-specific, taxon-specific) antitoxin used in the treatment of poisoning by envenomation, as in snakebite or bee stings. *Cf.* antidote. **antivenin (Crotalidae) polyvalent [USP]**. a sterile antitoxin that contains the specific venom-neutralizing globulins produced by horses that have been immunized against the venom of four crotalid snakes: *Crotalus adamanteus*, *C. atrox*, *C. durissus terrificus*, and *Bothrops atrox*. It is used in the passive immunization of envenomations by most crotalid snakes and against that of *Agkistrodon spp*. **antivenin (*Micrurus fulvius*) [USP]**. a sterile antivenin used in the treatment of bites by *Micrurus fulvius*, the Eastern coral snake. **black widow spider antivenin**. *Latrodectus mactans* antivenin. ***Latrodectus mactans* antivenin**. any antivenin used in the treatment of bites by *Latrodectus mactans*. **Lyovac® antivenin**. a *Latrodectus mactans* antivenin; normally used only for children. **polyvalent (Crotalidae) antivenin**. antivenin (Crotalidae) polyvalent [USP]. **polyvalent antivenin**. any antivenin used as an antidote or neutralizing agent against the venom of more than one species of animal. **polyvalent crotalid antivenin** (Wyeth, Inc., Philadelphia). an antivenin specific for the venoms of *Crotalus adamanteus* and *C. atrox*. It is to some degree effective against all crotalid venoms. **polyvalent crotaline antivenin**. antivenin (Crotalidae) polyvalent [USP]. **snake antivenin** (antisnakebite serum; antivenomous serum; Calmette's serum). any antiserum used against a snake venom. There are no other specific antidotes for snakebites. Antivenins are available for most snake venoms. Their use usually requires correct identification of the snake and their effectiveness is further affected by the site of the bite and the amount of time elapsed between bite and administration of the antivenin. Most fatalities in treated individuals are due to administration of the wrong antivenin or of too small a dose. Snake antivenins can also cause heart or respiratory problems, and their administration to persons who have not been bitten can be as dangerous as an actual envenomation.

antivenin therapy. the use of antivenins to treat snakebite and envenomations by other animals.

antivenine. antivenin.

antivenom. antivenin.

antivenomous. **1a:** antagonistic to or counteracting the action of a venom. **1b:** able to relieve or counteract the effects of a venom. **2:** pertaining to antivenin.

antivenomous serum. snake antivenin.

antiviral. **1:** destructive to or inhibiting the replication of a virus. **2a:** an agent that destroys or inhibits the replication of viruses. **2b:** an antiviral drug.

antiviral drug (antiviral, def. 2b). a medication used to treat infection caused by a virus.

antler. either of two bony horns from the head of cervids (deer and related taxa; Family Cervidae).

antoxidant. antioxidant.

antrum. a cavernous space as within a hollow organ.

ANTU (1-naphthalenylthiourea; 1-(1-naphthyl)-2-thiourea; 1-naphthylthiourea; α-naphthylthiourea; *N*-1-maphthylthiourea; α-naphthylcarbamide; krysid; chemical 109). a derivative of thiourea, ANTU is a nearly tasteless, odorless, gray powder,

$$C_{10}H_7NHCSNH_2$$

ANTU is insoluble in water and sparingly soluble in most organic solvents. ANTU is a highly selective rodenticide used to control *Rattus norvegicus* (Norway rat). It is an antithyroid agent that induces massive pulmonary edema with effusion into the pleural cavity and impaired breathing; fatty degeneration of the liver, and hypothermia with a high rate of mortality in the Norway rat. While highly toxic to *R. norvegicus*, it is less toxic to other species of *Rattus* and is relatively nontoxic to most domesticated animals with the exception of dogs, which are quite susceptible to ANTU poisoning. It may be moderately toxic to adult humans. The lethal dose to monkeys is about 4000 mg/kg. A suicidal adult male human ingested about 80 g of 30% ANTU rat poison together with a considerable amount of alcohol. He soon vomited, suffering no significant ill effects.

Anura (Salientia). the subclass of Amphibia (sometimes classified as an order of the subclass Lissamphibia) containing frogs and toads. All have a short vertebral column, lack tails as adults, and are highly adapted to jumping and usually swimming by virtue of powerful hind legs, a strong pectoral girdle that withstands the force of landing, and webbed feet. The larvae (tadpoles) are aquatic with gill slits and a tail; they undergo a usually rapid metamorphosis in which the tail is resorbed and the gill slits are replaced by lungs. Some, representatives of the families Bufonidae, Atelopodidae, Dendrobatidae, Discoglossidae, Hylidae, Phylomedusae, Pipidae, and Ranidae, are poisonous.

anuresis. the absence of urination. *Cf.* anuria.

anuretic. of or pertaining to anuresis.

anuria. an absence of or sharp decline in the secretion of urine. *Cf.* anuresis.

anus. **1:** the terminal (posterior or distal) portion and outlet of the alimentary canal of all metazoans except the Coelenterata and Platyhelminthes. It is the portal through which feces (i.e., unabsorbed food, food residues, and waste products) are eliminated. It is often controlled by a sphincter that relaxes, thus allowing the passage of feces. **2:** sometimes the muscular band (sphincter) at the terminus of the rectum that opens to allow the passage of feces is considered the anus.

anvil. a tool consisting of a block, base, or bed of wrought iron, sometimes overlaid with steel, which is used for shaping metal items by hammering (e.g., tools, horseshoes).

APA. Administrative Procedure Act.

apamin. a small (mol wt 2027.38) highly potent neurotoxic polypeptide,

$$C_{79}H_{131}N_{31}O_{24}S_4$$

It is a component of the venom of *Apis mellifera (= mellifica)*, the honey bee. Apamin is a CNS poison that is able to cross the blood-brain barrier.

APAP. acetaminophen.

Aparasin. a trade name for lindane.

apatite. a common mineral of varying color, calcium fluorophosphate,

$$Ca_5FP_3O_{12}$$

It occurs as individual crystals or in a mass. It was formerly used in the manufacture of phosphate fertilizers.

APC virus. adenovirus.

aperistalsis. the absence of peristalsis.

apert-. a prefix that denotes bareness, openness, or uncovered.

apetalous. lacking petals.

apetalus. **1a:** lacking petals. **1b:** sometimes designating a flower that lacks petals.

apex (plural, apices). the top, tip, or end of a pyramidal, conical, or pointed structure such as a twig or leaf.

APF. animal protein factor.

APH. actual production history.

APHA. American Public Health Association.

aphagia. complete failure to eat; inability to eat. **2:** dysphagia.

aphagous. of or pertaining to failure to feed. Applied especially to the adult stage of insects that do not feed.

aphagy. failure to feed.

Aphanizomenon. a toxic genus of cyanobacteria. *See also* Annie, Fannie, and Mike.

Aphasmida. Adenophorasida.

Aphasmidia. Adenophorasida.

APHIS. Animal and Plant Health Inspection

Service.

aphosphorosis. lacking or deficient in phosphorus.

Aphtiria. a trade name for lindane.

aphylaxis. the absence of immunity.

aphyllous. leafless.

aphyllus. leafless; without leaves.

Apiaceae (the carrot, celery, or parsnip family). a family of hollow-stemmed herbaceous plants that are widely distributed throughout temperate and subtropical regions. Some are weeds, others are important agricultural, ornamental, medicinal, garden, and household plants. Some are of great size; some are poisonous.

apiarist (beekeeper). a person who raises (keeps) bees.

apiary (bee yard). a place where bee colonies are kept and maintained, usually including tools. Included are hives and, in some countries, natural sites, often nests in trees or in the ground.

apical (apicalis). **1:** pertaining to an apex, pinnacle, summit, top, or tip. **1:** of, pertaining to, or at the apex or tip. *See*, for example, dominance (apical dominance), meristem (apical meristem). **2:** nearer the apex of a stem or other pointed structure than to a specific reference point. *Cf.* basal.

apical dominance. the inhibition of cell division, growth, and development of some or all lateral buds by the apical bud of a shoot.

apicalis. apical.

apices. plural of apex.

apiculat-. a prefix meaning pointed; having a point.

apiculture. the husbandry of bees for honey production.

apid. common name of any member of the family Apis (honeybees).

Apidae. a family of venomous and nonvenomous social bees (Order Hymenoptera). Certain solitary bees are included in some classifications. Apids characteristically build extensive combs of wax. A single colony may contain 10,000 to 40,000 individuals. There are three castes: a queen (the sole egg-laying female of a colony), drones (fertile males), and workers (sterile females). Bees are valuable to humans as pollinators and honey producers. *See also apis*, bee.

apiology. the natural science that treats the biology, natural history, rearing, and management of bees, especially those of the genus *Apis*.

apiphobia (melissophobia). a morbid fear of bees.

Apis (honeybees). a genus of venomous bees; the type genus of the family Apidae. Honeybees are distinguished from other genera of bees largely by wing venation. The venom is acidic and contains hemolytic enzymes. The sting causes local pain and a burning sensation with swelling, erythema, and whitishness at the wound site. Multiple stings may cause shock and respiratory distress and may require hospitalization in some cases. Fatalities are rare and are usually due to anaphylactic shock. ***A. cerana*** (Eastern honeybee; *A. indica*). the honey-producing honeybee of southern Asia. ***A. dorsata***. the large honeybee of Asia, the nest of which is suspended from a tree branch, rocky ledges, and other sheltered surfaces. It contains a single comb. ***A. florea***. the small honeybee of Asia. ***A. indica***. *A. cerana*. ***A. mellifera*** (*A. mellifica*; common honeybee; Western honeybee; European honeybee). a small, venomous, reddish brown and black bee with orange-yellow rings on the abdomen. It is worldwide in distribution. Feral colonies build nests in hollow trees, while domesticated colonies are kept in hives by beekeepers. Honeybees were introduced to North America by early settlers in the 17th century. Gathered honey is not ster ile and has been known to cause botulism when used in infant formula. ***A. mellifera adansonii*** (killer bee). **1:** a honeybee, native to Africa, that is extremely aggressive and attacks in swarms when disturbed, it was brought to Brazil and ac¶-cidentally released there in 1956. **2:** hybrids of the African honeybee and the common honeybee are sometimes given this subspecies designation. The genetic interrelations among these bees is much more complicated than indicated by subspecies designation. ***A. mellifica***. *A. mellifera*.

apisination. envenomation by the sting of a bee.

Apium graveolens (celery). a common table vegetable (Family Apiaceae). Toxic concentrations of nitrate that might be hazardous to foraging animals are occasionally reported.

aplasia. arrested or defective development of an organ or tissue.

aplastic. **1:** denoting, pertaining to, or characterized by aplasia. **2a:** displaying arrested or defective development of an organ or tissue. **2b:** anatomically undeveloped.

Aplopappus. synonym of *Haplopappus*.

apnea. **1:** absence or temporary cessation of breathing. **2:** asphyxia.

apneic. of, pertaining to, or affected by apnea.

apo-. a prefix meaning from.

apocrine gland. a sweat gland that functions in evaporative cooling and, in some cases, during periods of emotional stress.

Apocynaceae (dogbanes). a family of widely distributed trees, shrubs, and vines that are most abundant in the tropics. Many are grown as ornamentals and occasionally for pharmaceuticals or their edible fruit. They have a poorly developed corona, a well-developed latex system, granular pollen, and carpels that are often joined by the style and stigma. Some are poisonous. Well-known genera with poisonous species are *Apocynum* (dogbane, Indian hemp), *Nerium* (oleander), and *Vinca* (periwinkle).

Apocynum (dogbane; Indian hemp). a genus of poisonous, chiefly North American herbs (Family Apocynaceae) with opposite leaves and small white or pink flowers; some species are cultivated. They contain toxic digitalis-like cardioactive glycosides such as apocynin. ***A. androsaemifolium*** (dogbane; wild ipecac; milk ipecac; rheumatism weed). a dogbane found in dry thickets, and woodland borders throughout much of the United States and in Canada from Nova Scotia to Newfoundland. It contains apocynin. ***A. cannabinum*** (Canadian hemp; black Indian hemp). a plant of open ground, thickets, and woodland borders in southern Canada and the northern tier of states in the United States, extending from New England south to Florida in the eastern United States. It contains apocynin.

apogamous. arising by apogamy.

apogamy. apomixis in which the unfertilized female gamete or an associated cell of a plant produces an embryo. **parthenogenetic apogamy**. a type of apogamy that occurs in both pteridophytes and angiosperms. In pteridophytes, an unfertilized oospore gives rise to a haploid sporophyte that produces normal spores. In angiosperms, the unfertilized oosphere gives rise to the embryo. If the oosphere is haploid, the resulting plant is usually sterile and smaller than normal (e.g., *Solanum spp*.); if diploid, the plant will resulting be normal (e.g., *Taraxacum spp*.) **vegetative apogamy**. a type of apogamy that occurs in both pteridophytes and angiosperms. In pteridophytes, the haploid sporophyte develops from cells in the prothallus or from a cell in the archegonial wall. In angiosperms, a cell of the nucellus or one of antipodal cells develops into the embryo.

apogenic. sterile (def. 1b).

apogeny. sterility (def. 1a).

apomict. a plant produced by apomixis.

apomixis. any of several types of asexual reproduction that superficially resemble sexual reproduction even though fertilization does not occur. Examples are apogamy and parthenogenesis. In angiosperms, for example, an embryo may be produced from a diploid cell of the ovule and the remaining cells may develop into seeds. Such seeds produce plants (apomicts) that are genetically identical to the parent. In some plants, a pollen grain provides a necessary stimulus to apomixis, giving the appearance of sexual reproduction. *See also* apogamy, apospory, agamospermy, parthenogenesis.

aponeurosis. a broad, flat, fibrous, or membranous sheetlike tendon.

aponia. **1:** lack of pain. **2:** abstention from exertion.

aponic. **1:** pertaining to or characteristic of aponia. **2:** able to relieve pain or fatigue.

apophysis. an outgrowth or process.

apoplast. 1a: the complex of adjoining cell walls that are continuous throughout most plants. This is probably both a water reservoir and a significant pathway of water movement. **1b:** the complex of cell walls, xylem, phloem, and intercellular spaces. This constitutes the system through which most water and dissolved nutrients flow.

aposematic. having warning coloration that indicates to potential predators an animal (usually applied to insects) is unpalatable.

apospory. the development, without meiosis, of a gametophyte from a sporophyte.

appaloosa (buttermilk horse; Colorado ranger; dollar spot; leopard spot; rain drop). a western saddle horse, 14-15 hands high, with a short back, and a ground color of white, marked by dark spots, especially across the croup.

appeasement. **1a:** accommodation. **1b:** submission or conciliation. **2:** any type of agonistic behavior, other than flight, that serves to prevent an attack or prevent or stop a fight between individuals of the same species. **ritual appeasement**. appeasement based on signals, usually of submission, by an animal that is losing a conflict.

appendage. **1:** an attachment, extremity, limb (e.g., leg, arm, wing), or member that attaches to or arises from a larger structure (e.g., from an organ, from the body of an animal, from a larger appendage). **2:** an appendix.

appendic-. a prefix denoting or pertaining to an appendage.

appendix. a small, tubular, vestigial appendage that extends outward from the cecum of the large intestine.

appetite. **1:** hunger, thirst. **2:** craving or desire.

apple. **1:** fruit of the apple tree. *See Pyrus*. **2:** a suffix in the names of certain other plants, the fruits of which bear a superficial resemblance to the apple (def. 1). **cooking apple**. any variety of apple cultivated and marketed for cooking (e.g., stewing, baking) as opposed to eating raw. **eating apple**. any variety of apple cultivated and marketed for eating.

appraised stumpage price (appraised rate). the market price, estimated by the U.S. Forest Service, of timber to be cut and removed a national forest. This price is the advertised minimum for competitive bidding by potential buyers. It may not be lower than the base rates.

apple cider. the juice of pressed apples that has been allowed to ferment at least slightly. **hard cider**. fermented sweet cider. **sweet cider**. apple juice, usually lightly fermented.

apple cider vinegar (cider vinegar). vinegar produced from prolonged fermentation of apple cider.

apple cider yeast. *Saccharomyces malei*.

apple of Peru. *See Datura*.

apple tree. the common (domesticated) apple. *See Pyrus*.

appraised rate. *See* appraised stumpage price.

appraised value. an assessment of the present worth of real or personal property.

appressed. tightly flattened or compacted.

approach-withdrawal. a highly adaptive general phenomenon whereby a motile organism tends to approach weak stimuli (e.g., as those emanating from food) and to withdraw from strong stimuli (e.g., high temperature, loud noise, noxious odors). Underlying mechanisms are extremely diverse and this type of behavior in many forms of animals can be modified by experience. Numerous mechanisms have thus independently evolved whereby animals are attracted to and approach nutrient materials and avoid those which are irritant or toxic. Such mechanisms sometimes fail, as in the case of inexperienced animals or during unfavorable conditions when, for example, livestock on range or pasturage may eat alternative materials that they normally avoid and may thus be exposed to toxic or otherwise hazardous species.

appropriative right. a right to the use of irrigation water. This right is limited to beneficial use on the land according to the laws and regulations of individual states in the western United States *See* riparian doctrine.

approximate. **1a:** nearly correct; nearly accurate. **1b:** close to a desired result. **1c:** estimated. **1d:** relative. **2a:** to estimate; guess; approach. **2b:** to calculate or evaluate by hasty, crude, or inaccurate methods. **3:** approximation.

approximation. **1:** an estimate thought to be nearly correct. **2:** an estimate arrived at by the use of hurried, crude, or inaccurate methods. **3:** the process of estimating by the use of hurried, crude, or inaccurate methods.

appulsion. the breaking away and removal of a river bank by the force of wind and high waters.

apricot. *Prunus armeniaca*.

apricot kernel. apricot pit.

apricot pit (apricot kernel). the large, hard seed of the apricot fruit. *See also* amygdalin, plant poisoning, apricot-kernel oil.

apricot-kernel oil. either of two oils extracted from the pits (seeds) of *Prunus armenica* (apricot). **1a:** persic oil. A colorless or straw-colored oil expressed from the pit. **1b:** bitter almond oil. A toxic, colorless or yellow aromatic oil obtained from the pit by steam distillation.

aprocarb. propoxur.

apron. **1:** a garment that covers the front of the body, usually from the neck to below the knees or from the waist the knees and tied at the waist. It is designed to protect the wearer's clothing (e.g., while baking, cooking, shoeing horses). **2:** a metal plate or cover, usually vertical, placed on a machine in order to protect those who operate it. **3:** a continuous conveyer belt, usually made of metal links, rubber, or a rubber-like material used, for example, to carry commodities (e.g., fruits or vegetables) or other materials onto a "table" for sorting or into bins. **4:** fleshy folds on the neck of an animal or the fatty layer that covers an animal's belly. **5:** the fringe of long hair on the throat or chest of certain long-haired dogs such as the collie. **6:** a metal strip used to guide the sap of a tree into a cup that is suspended from it. **7:** a slab of concrete or stone that links adjoining areas. **8:** a broad paved area used for parked cars, as at the end of a driveway or on the side of a road. **9:** a structure or device designed to protect soil or the surface of the earth from the erosive action of moving water. **10:** a platform that receives water falling over a dam. It serves to dissipate the energy of the water before it enters the stream below the dam; it also serves to aerate the water. **11:** a deposit of sand, gravel, shale, or similar materials that extends from the base of a mountain or the edge of a glacier. **12:** to apply, put on, or furnish with an apron.

aptera. wingless; used especially with respect to wingless insects of a taxon that includes winged species.

apterism. winglessness; a condition whereby wings are absent. Used especially with respect to insects. *Cf.* apterous.

apterium. a featherless area, situated between feather tracts on the body of a bird.

apterous. wingless; lacking wings. Used largely with respect to wingless adult insects. *Cf.* apterism.

Apterygota. a subclass of primitive wingless insects that undergo little or no metamorphosis.

apterygote. **1:** apterygotous; of, pertaining to, or belonging to the Apterygota. **2:** common name of an insect that belongs to the Apterygota.

apterygotous. apterygote (def. 1).

apyretic. afebrile.

apyrexia. absence of fever.

apyrogenetic. apyrogenic.

apyrogenic (apyrogenetic). not producing fever.

AQI. Agricultural Quarantine Inspection.

aqu-. a prefix denoting or pertaining to water.

aqua ammonia. ammonia water; ammonium hydroxide.

aqua ammoniae. ammonia water.

aqua humus. a fertilizer comprised of water-dispersible humic and fulvic acid derivatives. It is used as a base for the preparation of various blends of liquid fertilizers containing nitrogen, phosphorus, and potassium.

Aquacide. diquat.

aquaculture (aquafarming; underwater agriculture). **1:** the cultivation of marine or fresh water aquatic plants or animals (most commonly fish, shellfish, shrimp and crawfish, and seaweed) in a controlled or managed environment. The "crops" may be grown in natural waters (sometimes using enclosures) or in controlled marine or freshwater environments. In the United States today, aquaculture is dominated by catfish production. *See also* fish farming, hydroponics, mariculture, pisciculture. **2:** The National Aquaculture Act of 1980 defines aquaculture as "the propagation and rearing of aquatic species in controlled or selected environments, including ocean ranching." Under this act responsibility for most aquacultural research and regulatory activities is divided among the Departments of Agriculture, Commerce, and Interior.

aquafarm. a tract of usually shallow water, often near the shore of a bay or inlet, that is used for aquaculture.

aquafarming. aquaculture.

aquagium. a natural or manmade watercourse that runs through marshy ground.

aqualin. acrolein.

aquatic **1:** of or pertaining to water, especially in the environment. **2:** occurring in, used in, conducted in, or practiced in or on water (e.g., aquatic sports, aquatic pollutants). **3a:** living chiefly or entirely in or on water or growing in water. **3b:** of or pertaining to an organism that lives chiefly or entirely in water. **4:** descriptive of plants that are adapted to partial or complete submergence.

aquatic rights. the right by ownership or permission to navigate or fish navigable waters.

aqueous. **1a:** of, pertaining to, or having the properties of water; watery. **1b:** in water (e.g., in aqueous solution); containing water.

aqueous ammonia. ammonia water.

aqueous humor. the transparent, watery fluid that occupies the space between the cornea and the lens (i.e., the anterior and posterior chambers) of the vertebrate eye.

aquiclude. any geologic formation that absorbs and retains water but does not allow its transport at a rate sufficient to supply aquifers, springs, wells, etc.

aquifer. a below ground geological formation, or group of formations, that contains and conveys groundwater in sufficient amounts to supply wells and springs that can be used to supply domestic, agricultural, and industrial needs. an aquifer can be comprised of saturated sands, gravel, fractures, and cavernous and vesicular rock. **confined aquifer**. an aquifer that contains confined groundwater is confined under pressures that are significantly greater than atmospheric pressure. **semi-confined Aquifer**. an aquifer that is partially confined by layer(s) of soil of low permeability through which recharge and discharge can occur.

aquifuge. a rock that has no interconnected pathways or interstices that connect to the outside and therefore neither absorbs nor transmits water.

aquil-. eagle-like.

$\mathbf{A_r}$. relative atomic mass.

arabian. one of the oldest known breeds of saddle horse, they are usually gray or chestnut in color and stand 14 to 15 hands high; body weight ranges from 350 - 1000 pounds. The breed is noted for endurance.

arable. **1a:** of or pertaining to tillable land. **1b:** (*Brit.*) tillage. **2:** able to be plowed; suitable for tillage and crop production.

arable land. land that is practical and economical for the cultivation of crops.

Araceae (arum family). a family of toxic plants, all of which produce symptoms similar to those of *Dieffenbachia*. Many are common house plants. Important genera are *Acorus*, *Amorphophallis, Anthurium, Arisaema, Caladium, Dieffenbachia*, and *Philodendron*. All parts of these plants are toxic and can cause severe irritation of mucous membranes, with swelling of the tongue and throat, salivation, nausea, vomiting, and diarrhea. Active principles vary among species but always include calcium oxalate.

arachic acid. arachidic acid.

arachidic acid (arachic acid; eicosanoic acid). a widely-distributed fatty acid,

$$CH_3(CH_2)_{18}COOH$$

It is a minor component of peanut oil, the oils of related plants, butter, and certain other fats.

arachidonic acid. an essential fatty acid derived from fats in food.

arachn-, aran-. prefixes that denotes or pertains to spiders.

arachnid. any arthropod of the class Arachnida (e.g., a spider, scorpion, tick, mite, harvestman).

Arachnida. **1:** a large class (sometimes ranked as a subclass of the class Arachnoidea) of invertebrates (Phylum arthropoda, Subphylum Chelicerata). Most arachnids are predaceous terrestrial forms such as scorpions, spiders, harvestmen, mites, ticks, and related forms. Arachnids are characterized by a cephalothorax (a fused head and thorax, the prosoma) that bears four pairs of walking appendages, a pair of usually sensory pedipalps, and a pair of prehensile buccal chelicerae that are often served by poison glands; the abdomen (opisthosoma) lacks appendages. The sexes are separate, development is direct (except for ticks and mites), compound eyes are lacking — the eyes are simple, most species breathe via tracheae, book lungs, or gill lungs, they eat only fluids, and excretion is by coxal glands and Malpighian tubules. **2:** formerly a subclass of the arthropod class Arachnoidea.

arachnidism (arachnoidism). **1:** systemic poisoning from the bite of a spider or other arachnid, especially that due to *Latrodectus mactans*. The condition is usually characterized by intense pain and muscular rigidity. **2:** any condition due to the bite of a spider. *Cf.* araneism.

arachnoid. **1:** of, pertaining to, or characterizing the Arachnida. **2:** any of various invertebrates that are related to or resemble arachnids. **3:** resembling a spider's web; membranous. **4:** arachnoid mater; the soft, weblike middle covering (meninges) of the central nervous system of tetrapods.

Arachnoidea. *See* Arachnida (def. 1).

arachnoidism. arachnidism.

Aralia spinosa (devils walking stick; angelica tree). a poisonous shrub or small tree with distinctive thick, club-like, spiny branches tipped with very large, doubly pinnately compound leaves. It occurs from Pennsylvania to Florida and west to Illinois and Iowa.

aran-. a prefix denoting or pertaining to spiders.

Araneae (Araneida). true spiders; the largest order (nearly 50,000 species are known) of arachnids. Nearly all are terrestrial and venomous, but only about 200 species are harmful to humans and other large animals. The body is comprised of an unsegmented cephalothorax and a short, usually unsegmented abdomen which are greatly constricted at their juncture; spinnerets that extrude silk are on the ventral surface of the abdomen in females of many species; the chelicerae are modified as short, pointed fangs with a venom duct in the terminal claw and used to kill or immobilize prey; the pedipalps are short and leg-like with crushing bases and are modified as accessory sex organs in the male; external respiration is by two or four book lungs and tracheae are sometimes present; digestion is external. Species that are dangerous to humans include the widow spiders, *Latrodectus mactans*, and related species, the brown or violin spider, *Loxosceles reclusa* (sometimes called the brown recluse), and related species, the jumping spiders, *Phidippus*, the tarantulas, *Aphonopelma* and *Pamphobeteus* species, the trapdoor spiders, *Bothriocyrtum* and *Ummidia* species, the so-called banana spiders, *Phoneutria* and *Cupiennius sallei, Lycosa* (wolf spider), and *Heteropoda*, the crab spider, *Misumenoides aleatorius*, the running spiders, *Liocranoides* and *Chiracanthium*, the orbweavers, *Neoscona vertebrata, Araneus* species, and *Argiope aurantia* (orange argiope), the running or gnaphosid spiders, *Drassodes*, the green lynx spider, *Peucetia viridans*, and the comb-footed or false black widow, *Steatoda grossa*.

araneid. **1a:** common name of any spider. **1b:** of or pertaining to spiders.

Araneida. Araneae.

Araneus (orbweavers). a genus of true spiders (Family Araneidae) that are dangerous to humans as allergic reactions are possible. ***A. marmoreus*** (marbled spider; false shamrock). a common spider of Europe and North America that builds in shrubs and trees. *See* Araneae.

aranose. resembling a spider web.

arbitrage. **1:** the simultaneous purchase and sale of the same securities, commodities, or foreign exchange in different markets to profit from unequal prices. **2:** arbitration (obsolete).

arbitrager (arbitrageur). a practitioner of arbitrage.

arbitrageur. arbitrager.

arbitration. the resolution of disputes by use of an impartial person (an arbitrator) selected by the parties. Decisions reached through arbitration are usually legally binding.

arbor. **1:** a tree. **2:** a small latticework structure used to support climbing plants and vines and to provide shade. **3:** a wooden platform used to sun-dry fibers. *See* arbor bolt.

arbor-. a prefix denoting or pertaining to trees.

arbor bolt. the main supporting rod or beam on a disk gang harrow.

arboraceous (arboreous). treelike; of or pertaining to trees.

arborator. a person who tends trees.

arboreal. **1:** of or pertaining to trees; treelike. **2:** arboreous; living in or among trees. **3:** adapted to living in or among trees.

arboreous. arboreal; arboraceous.

arborescent. having tree-like branches; branched or branching in the manner of a tree.

arboretum. **1a:** a collection of trees and shrubs grown for scientific study and exhibition. **1b:** a plot of ground where trees and shrubs are grown for scientific study and exhibition. **1c:** a botanical garden devoted chiefly to the cultivation, study, and exhibition of woody plants.

arboricide. a chemical that destroys trees.

arboriculture. the cultivation of trees and shrubs, especially of ornamental and shade trees.

arborist. a specialist in the cultivation and maintenance of trees and shrubs. Skills include the prevention, diagnosis, and treatment of tree diseases, and tree surgery.

arborvirus. arbovirus.

arborvitae (thuja; thuya). aromatic evergreen trees of the genus *Thuja* such as *T. occidentalis* and *T. plicata*, including numerous horticultural varieties.

arbovirus (arborvirus). any of a large, heterogenous group of RNA-containing viruses, most (if not all) of which are transmitted by blood-sucking arthropods (e.g., ticks, fleas, mosquitoes). Of the approximately 350 species, only about 45 species are pathogens.

arbre. a common name for *Blighia sapida* and its fruit in Haiti.

arbus-. referring to a tree.

arbutin (4-hydroxyphenyl-β-D-glucopyranoside; hydroquinone-β-D-glucopyranoside; arbutoside; ursin). a plant growth inhibitor, $C_{12}H_{16}O_7$, that is toxic to grasses (Family Gramineae). Arbutin occurs in the leaves of a variety of green plants, especially those of the family Ericaceae. Examples are *Adenostoma spp.*, *Arctostaphylos spp.*, *Bergenia crassifolia*, *Pyrus spp.*, and various species of *Vaccinium* (e.g., blueberries, cowberries, cranberries). *See* especially *Adenostoma* and *Arctostaphylos*. Arbutin is sometimes used therapeutically as a diuretic and urinary antiinfective agent. *See also* allelopathy.

arbutoside. arbutin.

arcadian. **1a:** simple, content. **1b:** rural, pastoral, provincial, or rustic. **1c:** suggestive of simple, innocent contentment. **1d:** descriptive of a place of simplicity and contentment. **2a:** of Arcadia. **2b:** a native of Arcadia. **3:** a Greek dialect spoken in Arcadia.

Arceuthobium pusillum (dwarf mistletoe). a very small mistletoe (Family Loranthaceae) with leaves reduced to tiny connate thin brown scales; the "berry" is dry and compressed. It is a parasitic epiphyte on coniferous trees, chiefly on the branches of *Picea* and rarely on *Larix* and *Pinus strobus* in North America from Newfoundland to Ontario, southward to the mid-Atlantic States, west to Michigan, Wisconsin, and Minnesota. *Cf. Phoradendron serotinum. See also Viscum album.*

arch. **1:** arc; curve. **2:** vault; span; dome. **3:** chief; major; main; primary. **4:** any bow-like part as the arch of a foot. **5:** a curved masonry construction (or one of steel, wood, etc.) that spans the top of an opening; it consists of a number of wedgelike stones, bricks, or other such pieces. **6:** a dam construction that has the form of a vertically-placed, barrel-like vault with the convex side toward the impounded water. **7:** to make or form into the shape of an arch, vault, or curve (e.g., the horse arched its neck; the trees formed an arch over road).

arch-. a prefix meaning first or primary; ancient.

arche-. **1:** a prefix meaning prior, original, or first as in archegonium, archetype. **2:** in scientific vernacular, a synonym of archi-.

archegonia. plural of archegonium.

archegonium (plural, archegonia). the female sex organ of mosses, liverworts, ferns, and related plants and of most gymnosperms. It is made up of a multicellular flask-like structure that contains a single ovum (oosphere).

archetype. **1a:** an original, early, or primitive model from which entities of the same type are based or copied. **1b:** a prototype. **2:** in biology, a progenitor or antecedant type of organism from which later species inherited basic or typical characteristics.

Archimedean screw. a hollow screw-like device or spiral tube used to lift water. The lower end of the screw is submerged or drilled into an aquifer or other water body. Water can then be lifted when the screw or tube is turned (e.g., by means of a handle) to a height above the water body and distributed, for example, into an irrigation ditch.

archin. emodin.

archont-. a prefix meaning noble, majestic.

arcifinious. descriptive of a land area that has natural boundaries such as mountains or rivers.

arct-. a prefix denoting or pertaining to bears.

Arctic Circle. an imaginary line on the surface of the earth that runs parallel to the equator. It lies at 66 1/2°N latitude, i.e., 23 1/2° south of the North Pole. It marks the northernmost point at which the sun can be seen at the winter solstice and the southernmost point at which the midnight sun is visible.

Arctic timothy. *Aloecurus alpinus*.

arctic. **1a:** pertaining to the region above the Arctic Circle. **1b:** pertaining to the region that lies north of the limit of tree growth or to the area above tree line on a mountain. **2:** of, pertaining to, or located at or near the North Pole. **3a:** pertaining to a climate or area that has a mean low temperature below 0° C during the coldest month of the year and below 10° C during the warmest month. **3b:** characteristic of the severe, extremely cold weather that occurs north of the Arctic Circle. **3c:** frigid. **4:** arising from the North Pole or the arctic region (e.g., an arctic wind). **5:** suitable for use in the arctic, as arctic boots. **6:** bleak.

Arctium (burdock). a genus of weedy plants (Family Compositae).

Arctostaphylos (bearberry). a genus of American and circumpolar woody evergreen shrubs (Family Aricaceae) characterized by alternate leaves, nodding flowers, and drupaceous fruit. Species such as *A. glauca, A. glandulosa*, and *A. uva-ursi*, together with other shrubs such as *Adenostoma fasciculatum*, comprise the hard chaparral community. These plants secrete phenolic compounds such as arbutin and *p*-hydroxy cinnamic acid that are washed to the ground by rainfall. These compounds inhibit the growth of nearby grasses and, as a result, substantial bare areas develop within and even beyond the edge of the community. Whenthe chaparral is destroyed by fire, grasses grow vigorously for a while but slow as the chaparral community reestablishes itself.

arcuat-. a prefix denoting or pertaining to a bow; bow-shaped.

Ardennes. a short, hearty breed of horse that originated in France.

are. an Asian land measure; 1 are equals 100 square meters.

area source. any small or large non-natural diffuse source (e.g., a feedlot, runoff from fertilized fields) of water or air pollution; one that cannot be classified as a point source (e.g., a smoke stack). *Cf.* point source.

arec-. a prefix denoting or pertaining to the

genus *Areca* or plants of the genus *Areca*.

areca. any palm of the genus Areca.

areca nut. betel nut.

areca nut palm. *Areca catechu*.

Areca (betel palm). a genus of palm trees of India and the Malay Archipelago. ***A. catechu*** (betel palm; betel nut; betel nut palm; areca nut; areca nut palm; betel nut seed). a species that produces betel nuts, **q.v.**

Arecaceae. Palmaceae.

arena-. a prefix indicating a relationship to sand.

arena. amphitheater; domain, realm, sphere of action (e.g., a sports stadium, livestock arena).

arenaceous. **1a:** sandy, sandlike. **1b:** containing or derived from sand. **2:** growing in sand.

arenarius. of or pertaining to sandy areas.

arene. aromatic hydrocarbon.

arenicolous. living in a sandy area.

areol-. a prefix meaning pitted; having an aereole.

arg-. a prefix denoting or pertaining to silver.

arg. abbreviation for arginine.

Argas. a genus of soft ticks (Family Argasidae), some species of which infest birds, including poultry, and sometimes humans. ***A. persinus*** (abode tick; fowl tick; Persian tick). a bloodsucking parasite of poultry that may transmit fowl spirochetosis. ***A. reflexus*** (pigeon tick). the cause of a cutaneous inflammatory lesion in humans.

argasid. **1:** any of the numerous ticks (soft ticks) of the family Argasidae. **2:** belonging to or pertaining to the family Argasidae.

Argasidae. a family of nocturnal, bloodsucking, soft ticks (Order Acarina). They remain concealed in crevices during the day.

Argentine bahia. *Paspalum notatum*.

argillaceous. of or pertaining to rocks, soils, or other material composed of clay or in which clay is a prominent component. *See also* clayey.

arginine (arg; 2-amino-5-guanidinovaleric acid). an essential amino acid in laboratory rats.

Argiope (garden spiders). a genus of large, predominantly black and yellow garden spiders (Family Araneidae). Two species are common throughout much of North America. ***A. aurantia*** (black and yellow garden spider; orange argiope). garden spiders with black, yellow, and silver markings. The pain from the bite is slight and localized. The venom is neurotoxic and usually produces no more than mild systemic symptoms in humans, although an allergic reaction is possible. ***A. trifasciata*** (banded garden spider). toxicologically similar to *A. aurantia*.

argut-. a prefix meaning sharp or pointed.

arid. **1:** extremely dry, lacking moisture, or parched as an arid land or an arid climate. **2a:** barren or unproductive due to a lack of moisture. **2b:** denoting a region or climate in which precipitation is too low to support crop production. *See also* arid land, arid climate.

arid climate. a climate in which precipitation is too low to support crop production, although sparse natural vegetation may be supported as in a desert or subdesert. In temperate areas, an arid climate is generally considered as one with less than ten inches or less of precipitation per year; in tropical areas, the value is taken to be as much as 20 inches per year. *See* arid land.

arid land. land in which precipitation is too low to support the production of crops without irrigation. This is usually taken to be less than ten inches in temperate climate and up to 20 inches in the tropics. Other factors such as farming practices, type of landscape, soil type and composition, and the seasonal pattern of precipitation are important determiners of precipitation effectiveness and amount required for crop production. *See* arid climate.

aridisol. a type of soil formed in very dry climates (deserts) where the soil is not moistened for more than a few weeks each year. Soil formation under these conditions involves the accumulation of salts and other minerals with very little organic material. A white layer (a calcite horizon) forms in the subsoil due to the accumulation of calcite.

aridity. **1:** dryness; lack of moisture; a condition of extreme dryness. **2:** barrenness.

aridity index. an index that indicates how much more precipitation, if available, could be lost to the environment by evapotranspiration at a specified location under existing conditions.

aril. **1:** a colored, pulpy or fleshy, outer covering of the seeds of certain plants. It derives from the funicle or base of the ovule. **2:** a fleshy outgrowth of certain seeds.

Arisaema (Indian turnip; dragon arum). a genus of about 60 species of tuberous, usually monoecious plants (Family Araceae) that are nearly worldwide in distribution; some thrive as house and garden plants. The flowers are often distinctive. Most species have an acrid, pungent sap, some hybridize, and all should be considered toxic, although the best known for its toxicity is the familiar *A. triphyllum*. An active principle in all cases is calcium oxalate. ***A. atrórubens*** (Jack-in-the-pulpit; Indian turnip; oignon sauvage; petit prêcheur). a species that occurs in rich woods and thickets from southern Canada to New England, South Carolina, Tennessee, Missouri, and eastern Kansas. The hood is purple or purple to bronze with pale longitudinal stripes and the leaflets are often pale or glaucous beneath. It is generally, but not always, taller than *A. triphyllum*. ***A. dracontium*** (green dragon; dragon root). a tall perennial herb with fleeting blossoms found on rich or alluvial woods, thickets, and swales from Florida to Texas and northward locally to southern New Hampshire, Vermont, southwestern Quebec, southern Ontario, Michigan, and Wisconsin. The toxic effects are similar to those of *A. triphyllum*. ***A. stewardsonii*** (Jack-in-the-pulpit; Indian turnip; bog onion). found in wet or swampy woods and thickets, from Nova Scotia to Minnesota, northern New Jersey, and Pennsylvania, and along the mountains to North Carolina. ***A. triphyllum*** (*A. atroubens*; Jack-in-the-pulpit; small Jack-in-the-pulpit; Indian turnip). a woodland herb with tripartite leaves and mottled flowers, it is one of the most familiar species of arum in eastern North America. It is usually smaller than *A. atrórubens* and is otherwise similar to it in appearance but occurs in wet woods, swamps, and peat bogs of the Coastal Plain and Piedmont of the eastern United States, from Georgia to Kentucky, northward to southeastern Massachusetts and southeastern New York. Plants range in height from about 0.3 to nearly 1 m; the flower is about 25 cm long, green, and often spotted with purple. The fruits are a brilliant red. The entire plant, but especially the rhizome, is poisonous. The sweet-tasting berries, sour leaves, flowers, and roots are most deadly. The poisons are aroin (related to coniine) and calcium oxalate. Ingestion can cause an intense burning sensation, irritation, blistering, and edema in the mouth and pharynx when plant material is chewed. This is probably due to the presence of calcium oxalate crystals. If material is actually swallowed, symptoms may also include severe gastroenteritis, hemorrhage, convulsions, mydriasis, coma, and death. *See also Arum maculatum*.

arist-. a prefix meaning hairy, bristly, or pertaining to a beard.

arista. a large bristle that extends from the third antennal segment of many flies.

arithmetic mean. the mean of a set of quantities obtained by adding the quantities in the set and dividing this sum by the number of quantities in the set.

Arizona coral snake. a common name of *Micruroides euryxanthus*

Arizona fir. *Abies lasiocarpa*.

arm (brachium). **1a:** the portion of the superior limb between the shoulder and the elbow. **1b:** commonly used to mean the whole superior limb in bipedal mammals.

arma-. a prefix meaning armed.

armadillo. any mammal of the family Dasypodidae. *See also Dasypus novemcinctus*.

Armillariella. a genus of mushrooms (Family Tricholomataceae). ***A. mellea*** (honey mushroom). a normally edible but sometimes toxic mushroom that causes gastrointestinal distress; complete recovery is usual within a few hours. ***A. tabescens*** (ringless honey mushroom). Effects are similar to those of *A. mellea*.

armyworm. the larvae of *Pseucaletia unipuncta*.

Aroclor. polychlorinated biphenyl.

arom-. a prefix pertaining to or indicating scent or odor.

aromatic. **1a:** fragrant. **1b:** scented; perfumed. **1c:** a fragrant substance; a fragrance. **2:** any of a class of organic compounds that have one or more benzene rings as part of their molecular structure. Benzene is an important compound of this class.

aromatic amino acid. *See* amino acid.

aromatic compound. aromatic hydrocarbon.

arousal. **1:** stirring to action; stimulating a strong response. **2:** awakening; waking up. **3a:** becoming aroused; awakening. **3b:** a strong behavioral responsive by an animal to certain stimuli such that it is stirred to action (flight, attack). **4:** sexual stimulation.

arpen. arpent.

arpent (arpen; arpenus). an old, variable French unit of land measure equivlent to about 0.845 acres. It is still used in the province of Quebec, Canada and in parts of Louisiana.

arpenus. arpent.

arrect-. a prefix meaning erect.

arrent. to give one license to use for a fee as cropland or farmland.

arrow crotalaria. *Crotalaria sagittalis*.

arrowgrass. *Triglochin*.

arroyo. a usually dry channel produced by periodic severe erosion due to sparse vegetation and heavy rainstorms. They occur in arid regions (as in the southwestern deserts of the United States), usually in alluvial slopes near the base of a mountain. *Cf.* wadi.

arsenate. any salt or ester of arsenic acid. They occur widely in nature and contaminate coal and metal ores. A number are used in pesticides, herbicides, fungicides, and algicides. They are carcinogenic and less toxic than arsenites. Arsenates uncouple oxidative phosphorylation by replacing inorganic phosphorus in the ATP complex. The breath and feces of poisoned animals or humans have a sweet garlicky odor. The individual suffers nausea, violent vomiting, acute gastroenteritis with a burning pain throughout the digestive tract, and may experience intense (sometimes bloody) diarrhea and dehydration. Severe poisoning may result in shock, convulsions, coma, paralysis, and death.

arsenic (As; arsenium; ratsbane)[$Z = 33$; $A_r = 74.9$]. a grayish-white, toxic, metalloid element. It occurs widely in soil, water, and a number of human foods, especially shellfish. Arsenic and many of its compounds are poisonous at low concentrations. It causes cardiac arrhythmias, toxic amaurosis, and cancer and is a well-known teratogen to a variety of vertebrate animals. It can be absorbed via intestinal and pulmonary routes and sometimes via skin. While used chiefly in herbicides and pesticides, arsenic and its compounds have a variety of applications, e.g., as pharmaceuticals, in catalysts, bactericides, fungicides, additives in animal feed, corrosion inhibitors, tanning agents, and wood preservatives. Human exposure to arsenic occurs chiefly in the manufacture of herbicides, insecticides, wood preservatives, and during the smelting of lead, copper, and zinc ores. Humans may also be exposed through exposure to contaminated drinking water, wood products impregnated with preservatives, and emissions from smelting, pesticide, and glass-manufacturing plants.

arsenic acid (orthoarsenic acid). **1:** the white, poisonous, water-soluble, crystalline hydrate, H_3AsO_4, of arsenic trioxide It is a defoliant and is used in the manufacture of insecticides, glass, and arsenates. **2:** the hydrate of arsenic pentoxide, *q.v.*

arsenic acid anhydride. arsenic pentoxide.

arsenic anhydride. arsenic pentoxide.

arsenic monosulfide. arsenic disulfide.

arsenic oxide. **1:** arsenic pentoxide. **2:** arsenic trioxide.

arsenic pentoxide (arsenic oxide; arsenic anhydride; arsenic acid anhydride; arsenic acid). a white amorphous, deliquescent solid, As_2O_5, that forms arsenic acid in aqueous solution and is soluble also in ethanol. It is used, for example, as an insecticide and herbicide. *Cf.* arsenic acid, arsenic trioxide. Arsenic trioxide and arsenic pentoxide are both often referred to as arsenic oxide.

arsenic sulfide. arsenic disulfide.

arsenic trioxide (crude arsenic; white arsenic; arsenious acid; arsenious oxide; arsenious anhydride). a white, odorless, tasteless, powder, As_2O_3, that is sparingly soluble in water but is highly soluble in most acids, alkalies, and in glycerol. It has many uses including the manufacture of pigments, ceramic enamels, and glass and as an insecticide, rodenticide, herbicide, sheep and cattle dip, wood preservative, and the preparation of other arsenic compounds. As_2O_3 is poisonous and is carcinogenic to humans. Both arsenic trioxide and arsenic pentoxide are sometimes referred to simply as arsenic oxide.

arsenic trisulfide (arsenic yellow; yellow arsenic sulfide; auripigment; orpiment; king's yellow; king's gold). a toxic compound, As_2S_3, that has many industrial uses (e.g., in the manufacture of glass, linoleum, oil cloth, photoconductors, and semiconductors) and is used as a pigment, in pyrotechnics, and in depilating animal hides.

arsenic yellow. arsenic trisulfide.

arsenical. **1:** a compound or formulation that contains arsenic. **2:** any biologically active substance that depends on the arsenic content for its action or effects. **3:** referring to or denoting arsenic or one of its compounds; containing arsenic. **4:** used colloquially in reference to any pesticide that contains arsenic. *Cf.* herbicide (organoarsenical herbicide); wood preservative.

arsenic disulfide (arsenic sulfide; arsenic monosulfide; red arsenic sulfide; realgar; red orpiment; red arsenic; ruby arsenic; red arsenic glass; C.I. Pigment Yellow 39; C.I. 77085). a toxic, water-insoluble, orange-red powder or deep red, orange, or black monoclinic crystals, As_2S_2 or AsS, that occurs naturally as the mineral realgar. It is used as a rodenticide, in fireworks to give an intense white flame, in pigments, and in depilating and tanning hides.

arsenilic acid (4-aminophenylarsonic acid; *p*-aminobenzenearsonic acid; atoxylic acid). a toxic substance synthesized from aniline and arsenic acid. It is used in the manufacture of arsenicals, as an additive to promote growth of poultry and swine and to control dysentery in swine. Swine are occasionally poisoned by rations containing excessive amounts of organic arsenical growth promoters. *See* arsenilic acid poisoning.

arsenilic acid poisoning. poisoning by arsenilic acid, mostly of swine due to rations that have excessive amounts of arsenilic acid and/or other arsenical growth promoters. The condition in swine may progress from partial paralysis (paresis) of the pelvic limbs and ataxia, to include also paresis of the forelimbs. Affected animals may assume a dog-like sitting position. Lesions are often those of myelin and axonal degeneration in peripheral nerves, in optic nerves, and optic tracts. Early withdrawal of arsenicals may result in complete spontaneous recovery.

arsenious acid. arsenic trioxide.

arsenious acid copper(2+) salt (1:1). cupric arsenite.

arsenious anhydride. arsenic trioxide.
arsenious oxide. arsenic trioxide.

arsenite. **1:** a trivalent negative ion, AsO_3^{3-}, in aqueous solutions of arsenic oxide, As_4O_6. **2:** any salt or ester of arsenous acid or any salt that contains trivalent arsenic. Arsenites occur widely in nature and are common constituents of pesticides. Many arsenites are antiparasitic or antiprotozoal agents. They are corrosive to tissues and are carcinogenic. Arsenites are generally more toxic than arsenates. Acute exposure of humans may cause intense nausea and retching, abdominal pain, severe diarrhea, and dehydration. The breath and feces have a garlicky odor. *See also* cupric arsenite, sodium arsenite.

arsenium. arsenic.

arseno-. a prefix indicating the inclusion of the element arsenic.

arson. the crime of deliberately setting fire to property (a building, automobile, forest). Legal definitions vary among governmental jurisdictions.

arsonic acid copper(2+) salt (1:1). cupric arsenite.

Artemisia (sage; sagebrush; mugwort). a genus of more than 200 species of plants (Family Asteraceae). They produce commercially important volatile oils (e.g., absinthe, wormseed, α-santonin) and are toxic if ingested in large amounts. Sages of the rangelands of the western United States are, however, generally considered good forage for livestock. *See* sage sickness. ***A. absinthium*** (wormwood; absinthium; absinth; absinthe). a coarse perennial herb formerly used as an ingredient of absinthe. ***A. californica***. a sagebrush which, together with certain others such as *Salvia leucophylla*, comprise the soft chaparral communities of southern California. It secretes volatile terpenes that adsorb to the soil and inhibit the growth of grasses, sometimes producing bare areas up to a few meters wide around the community.

arteriolar necrosis. arteriolonecrosis.

arteriole. any of the smallest branches of an artery; they terminate in capillaries.

arteriolonecrosis (arteriolar necrosis). necrosis of arterioles.

arteriosclerosis (hardening of the arteries). any of a number of diseases marked by degenerative changes in the arteries, including a thickening of the artery walls and an accumulation of calcium with loss of elasticity, a narrowing of the lumen, and reduced blood flow.

arteriospasm. a spasm of an artery.

artery. a blood vessel with characteristic elastic walls that conveys blood away from the heart to a capillary bed. It has an inner coat (tunica intima) of endothelial cells that rests on a thin elastic membrane (lamina elastica interna). The surrounding wall of the artery is comprised of strong elastic fibromuscular tissue. Except for the pulmonary and umbilical arteries, arterial blood is aerated and red in color.

artesian. denoting, characteristic or, or pertaining to ground water under sufficiently high hydrostatic pressure to cause it to well above the level of the aquifer that contains it.

artesian aquifer (confined aquifer). an aquifer in which the water is under sufficient hydrostatic pressure that the water can rise above the top of the aquifer.

artesian well. **1:** a well that has been drilled into an artesian aquifer. The hydrostatic pressure is usually strong enough to force the water to or near the surface. **2a:** a well that flows without being pumped. **2b:** a well in which the water rises to a point above the aquifer but not necessarily to the surface of the earth.

arthritis. a condition in which of the joints of mammals are inflamed, swollen, sore, and stiff.

arthro-. a prefix denoting a joint.

arthrogram. an x-ray of the joints of an animal taken with an arthroscope.

arthrogryposis. **1:** a persistent flexure or contraction of joint muscles in terrestrial vertebrates. **2:** a tetanoid spasm. **congenital arthrogryposis** (congenital articular rigidity, CAR). a myopathy characterized by congenital articular rigidity with other anomalies. It occurs in cattle, sheep, horses, and pigs. The condition can be lethal, but complete recovery may occur. Any of these plant teratogens may contribute to the disease: *Datura* (thorn apple), *Lupinus* (lupine), *Nicotiana* (tobacco), *Oxytropis* (locoweed), *Prunus serotina* (black cherry), *sorghum vulgare* (sorghum or broom corn), and *Tsuga* (hemlock).

arthrology. the scientific study of the structure and function of joints.

arthropod. any animal of the phylum Arthropoda.

Arthropoda. the largest animal phylum. Arthropods are distinguished by a chitinous exoskeleton, a segmented body with segmented appendages, a reduced coelom, an open circulatory system with an extensive hemocoel, a complete digestive system, and a nervous system comprised of a dorsal anterior brain and a ventral nerve cord with fused ganglia in each segment.

arthropodiasis. any condition of a vertebrate animal that is caused directly by an arthropod.

arthropodic. arthropodous.

arthropodous (arthropodic). pertaining to arthropods.

articul-. a prefix meaning jointed, articulate, or articulated.

articular. **1:** of or pertaining to the joints of a plant or animal. **2:** a bone of the lower jaw in bony vertebrates other than mammals. The maleus is the homolog of the articular in mammals.

articulate. hinged; joined.

articulation. **1:** an act or the process of articulating (e.g., speech) **2:** a speech sound, especially that of forming a consonant. **3a:** a surface such as a joint or hinge that apposes another surface so as to permit movement of one or both parts (e.g., the knee joint of a tetrapod). **3b:** the process of forming a joint. **4:** a joint, hinge, or point between two parts where separation may occur spontaneously (e.g., the point of attachment of a leaf to the main stalk of a plant). **5:** a node or the space between two nodes in the stem of a plant. **6:** a joint, as at the juncture of bones in higher vertebrates or of movable segments as in arthropods.

artificial flavor. food flavoring.

artificial insemination (AI). the mechanical placement of fresh or frozen semen from specially selected males into the vagina of a female mammal without sexual contact.

artificial parturition. induced abortion.

artificial pasture. pasture or grazing lands maintained under relatively intensive management such that some tillage and plantings may take place, foreign forage species may be introduced, and the pasturage further supported by irrigation, application of fertilizers, provision of additional feed in the field, and mowing.

artio-. a prefix that indicates an even number.

artiodactyl. having an even number of digits.

Artiodactyla. even-toed ungulates; the order of hoofed mammals (Subclass Theria) that have an even number of digits per limb (two or four). The axis of the limb lies between the third and fourth digit in swine and the hippopotamus, for example. The camel, deer, giraffe, antelope, and cow are two-toed.

Artocarpus communis (breadfruit tree). a tree (Family Moraceae) indigenous to Polynesia that is occasionally grown on the extreme tip of the Florida peninsula. This tree is characterized by thick foliage, thick oval leaves, small flowers, and the edible fruit (breadfruit) that it produces.

arum family. Araceae.

arund-. a prefix denoting or pertaining to reeds.

arundinaceous. pertaining to reeds or canes; reedlike; reedy.

arven-. a prefix denoting or pertaining to cultivation.

arviculture. the scientific study of field crops.

arvum. a grain field community.

aryl halide. *See* halide.

aryl-. a prefix indicating an aromatic radical.

arylamine. aromatic amine. *See* amine.

As. the symbol for arsenic.

ASA. acetylsalicylic acid.

Ascaridae. Ascarididae.

Ascaridata. Ascaridida.

Ascaridia. a genus of relatively large intestinal nematodes (Family Heterakidae) that cause ascaridiasis in birds. The life cycle is direct, with no intermediate host. They resemble ascarids in most respects. ***A. columbae***. a species that parasitizes domestic and wild rock doves, *Columba livia*. ***A. galli***. a common, nearly cosmopolitan parasite of the small intestine of the domestic fowl, turkey, geese, guinea fowl, and many other species of birds.

ascaridiasis. a disease caused by usually heavy infestations of the intestines of many wild and domestic birds by *Ascaridia spp*., *q.v.*

Ascaridida (Ascaridorida; Ascarididea; Ascaridata). an order of parasitic nematodes (Subclass Phasmidia, Class Rhabditida). of humans and domestic animals (e.g., species of *Ascaridia, Ascaris, Subuluris, Heterakis,* and *Anisakis*).

Ascarididae (Ascaridae). a family of large intestinal nematodes such as *Ascaris lumbricoides*. Hosts of various members of this family include humans, swine, sheep, cattle, horses, dogs, cats, and other mammals.

Ascarididea. Ascaridida.

ascaridole (1-methyl-4-(1-methylethyl)2,3-dioxabicyclo[2.2.2]oct-5-ene; 1,4-peroxido-*p*-menthene-2). an unstable, organic liquid peroxide. It is an anthelmintic terpene used against nematodes and is the active ingredient in wormseed oil. Ascaridole is toxic and can be lethal to humans and other animals. *See Chenopodium* (*C. ambrosioides*).

Ascaridorida. Ascaridida.

ascaris. any roundworm of the genus *Ascaris*.

Ascaris. a genus of large, heavy-bodied, nematode worms (Family Ascarididae) that parasitize the small intestine of mammals. The taxonomy of species may in some cases be in doubt. ***A. lumbricoides***. a parasite of the small intestine of humans, swine, and other domestic and wild mammals. One of the most common human parasites, it measures 8 to 12 inches in length. Victims may not develop specific symptoms, although signs and symptoms in humans may include restlessness, fever, and diarrhea. Humans usually become infested by ingesting food and water that contain embryonated eggs. The race (*A. lumbricoides suum*) or species (*A. suum*) found in swine is adapted to that host and is not normally transmitted to humans or conversely.

Ascarops strongylina. A small, geographically widespread, bloodsucking roundworm that parasitizes the stomach of domestic swine and wild boars in many parts of the world. It attaches to the gastric mucosa of the host, cause inflammation and ulceration in heavy infestations. The larvae develop in coprophagous beetles.

ascending. **1:** rising, moving upward, climbing; soaring. **2:** in botany, descriptive of rising or climbing growth of a plant or plant part. Such growth may be on a slant or with some lateral growth but tending overall to grow upward from the point of attachment. **3a:** in anatomy, descriptive of an organ or part of an organ that arises from or is attached to the superior side of a structure (e.g., the superior vena cava). **3b:** a structure or part of a structure, especially a tubular structure, that is oriented along the vertically axis of an animal's body (e.g., the ascending colon in humans).

Aschelminthes (Nemathelminthes). a group (formerly a phylum) of wormlike Metazoa that vary greatly in size. Included is the class Nematoda and other assorted pseudocoelomates (e.g., Gastrotricha, Nematomorpha, and Rotifera) which are now regarded as separate phyla. The group is comprised of unsegmented, bilaterally symmetric, usually cylindrical or filiform invertebrates with a complete alimentary canal but lacking a circulatory and respiratory system. Most are aquatic. The female is usually larger than the male.

ascidia. plural of ascidia.

ascidium (plural, ascidia). a baglike or pitcher-shaped part.

ascites (abdominal dropsy; hydroperitonia; hydrops abdominis; peritoneal dropsy). the accumulation of watery (serous) fluid in the abdominal cavity.

ascitic. of, pertaining to, characterized by, or denoting ascites.

ascitogenous. causing ascites.

Asclepiadaceae (milkweed family). a family of widely distributed herbs, shrubs, vines that are characterized by copious, milky sap and a pod that opens on one side when ripe and discharge small, flat seeds with a tuft of long silky down that favors dispersal by light winds. Many milkweeds have medicinal and/or toxic properties. *See*, for example, *Asclepias*, asclelpias poisoning.

Asclepiadora decumbens. *Asclepias asperula*.

asclepias poisoning. a sometimes fatal poisoning of herbivores by ingestion of plants of the genus *Asclepias*. Early signs among livestock are extreme depression, weakness, and staggering. Frequently repeated seizures are common in downed animals as are fever, miosis, and labored respiration. Gastrointestinal irritation may be mild to severe. Atonic crop accompanies depression in the domestic fowl.

Asclepias (milkweeds; silkweeds; swallow wort; mw). a genus of perennial herbs (Family Asclepiadaceae), the milkweeds. all have a poisonous, milky sap; some have medicinal value. Highly poisonous hybrids between species are common. Livestock in North America are occasionally poisoned, usually under conditions of drought or poor management. Symptoms produced by the various species differ only in intensity. The minimum lethal dose of the fresh green vegetation ranges from about 0.2 to 2000 g/kg body weight or more. *See* galitoxin, asclepias poisoning of livestock. *A. asperula* (= *Asclepiadora decumbens*). a species that occurs on open, dry soils from Kansas and Arkansas, south into Mexico. Toxicity is similar to that of *A. latifolia*. ***A. eriocarpa*** (wooly-pod milkweed). found on dry soils in California; it is moderately toxic. *See* milkweed poisoning of rabbits. ***A. galioides***. a species that is poisonous to livestock and fowl. ***A. incarnata*** (swamp milkweed). a species suspected of fatal poisonings of sheep. ***A. labriformis*** (labriform milkweed). a very toxic plant confined to old stream beds in sandy soils in Utah. ***A. latifolia*** (broadleaf milkweed). less toxic than *labriformis*, this species occurs in dry prairies and plains from Missouri, Kansas, and Colorado to Texas and Arizona. ***A. mexicana*** (Mexican whorled milkweed). a species found in scattered locations on the western slopes from southern Washington and eastern Idaho to California, western Nevada, and Mexico. It is among the least toxic species treated herein. ***A. pumila*** (low whorled milkweed; plains whorled milkweed). low to the ground (10 to 15 cm tall) with slender stems, it occurs in dry plains east of the Rocky Mountains from Montana and southwestern North Dakota to Texas and New Mexico. The toxicity is similar to that of *latifolia*. ***A. speciosa*** (showy milkweed). an herb that may attain a height of 2 m. It occurs in prairies and openings in Canada and the United States from Minnesota and British Columbia southward. Large amounts given experimentally have produced symptoms. ***A. subverticillata*** (= *A. verticillata*; whorled milkweed; eastern whorled milkweed). less toxic than *labriformis*. The western variety of this species occurs mainly on dry plains and foothills in the central and southern Great Plains into Mexico. The eastern whorled milkweed, less than 1 m tall, occurs in dry woods and open sterile soil in the southern to northeastern United States, Ontario, southern Manitoba, and southern Saskatchewan. ***A. syriaca*** (common milkweed). up to 2 m tall with coarse stems, it occurs in thickets, roadsides, and old fields in much of eastern North America. The young shoots and firm, ripe pods are cooked and eaten; they are suspected of some poisonings. ***A. tuberosa*** (butterfly weed; pleurisy root; chigger flower). a rough-haired plant with brilliant orange or yellow flowers, the stems, leaves, and roots of which are toxic. It occurs on dry sites in the eastern United States. It has been used as an emetic, diaphoretic, expectorant, and laxative. ***A. verticillata***. *A. subverticillata*.

Ascomyces. *Actinomyces*.

Ascomycetes. a large class of terrestrial, mycelial fungi (Eumycophyta) characterized by the formation (via sexual reproduction) of sac-like structures termed asci. Included are the yeasts, leaf-curl fungi, powdery mildews, black and green molds, cup fungi, morels, and truffles. Certain genera produce antibiotics and/or mycotoxins.

ascorbic acid (antiscorbutic acid; vitamin C). a white, crystalline, water-soluble vitamin, $C_6H_8O_6$, that occurs naturally in citrus fruits, green vegetables, etc., and is also produced synthetically. It is essential for normal metabolism, the prevention of scurvy, wound-healing, and tissue repair. Ascorbic acid is required in the diets of some mammals (e.g., primates, guinea pigs, the Indian fruit bat) but is synthesized by others.

ascospore. the haploid or sexual spore of ascomycete fungi. It forms within an ascus.

ASCS. Agricultural Stabilization and Conservation Service.

ascus. a large, sac-like structure, characteristic of the Ascomycetes, within which sexual spores (ascospores) are produced by meiosis.

-ase. a suffix denoting an enzyme. It is appended to the name of the substance or type of substance (substrate) upon which the enzyme acts (e.g., phosphatase, lipase, proteinase). Prior to the establishment of this convention, enzymes were usually denoted by the addition of the suffix -in; pepsin, ptyalin, and trypsin were so named.

asecretory. nonsecretory; without secretion.

Aselli's gland (Aselli's pancreas). a large lymph node that lies ventral to the abdominal aorta. It receives all the lymph from the intestines in many small mammals.

Aselli's pancreas. Aselli's gland.

asepsis. the state of being free of living pathogenic organisms; a state of sterility.

aseptic. of or pertaining to asepsis; free of living pathogenic organisms.

aseptic technique. the handling of sterile instruments or culture media in such a way that they remain free of contamination.

asexual. **1:** not sexual. **2a:** not involving sexual reproduction; descriptive of reproduction without the union of sperm and egg. **2b:** not involving sexual intercourse. *See also* reproduction (asexual).

ash. **1:** a tree of the genus *Fraxinus*. **2:** the solid noncombustible (mineral) residue that remains following complete combustion of a substance. **bone ash**. the residue that remains following the combustion of bone in air. It is rich in calcium and phosphates and has applications as a minor fertilizer. **fly ash**. ash of small particle size from the combustion of coal or other fuels. In the absence of dust separators, such ash is emitted from the smoke stack and may be transported great distances from the source. **total ash**. the mineral residue obtained by controlled incineration, e.g., of coal. **volcanic ash** (volcanic dust). fine particles of lava ejected during a volcanic eruption. Sometimes the particles are shot high into the atmosphere and carried long distances by the wind. **wood ash**. the powdery ash remaining from the complete combustion of wood. It is especially rich in potash and may be used as a fertilizer.

ashleaf maple. *Acer negundo*.

Asian ass. *Equus hemionus*.

Asian long-horned beetle. a serious pest (Family Cerambycidae) of hardwood trees in China, it has found its way to the U. S. where it is rapidly spreading and causing much damage because it has no known natural predators or parasites. As of 1998, it had been found in at least 14 states where it has already caused significant damage. The U.S. Animal and Plant Health Inspection Service (APHIS) is attempting to detect and destroy this beetle. Detection and control, however, are made extremely difficult as it bores deep within a tree to lay its eggs. They are thus difficult to find and, if found, probably impossible to eradicate by the use of pesticides or other established methods of control. Thus far, the only means of suppression is to locate and remove infected trees. This beetle has destroyed numerous trees in parts of the New York State, and has the potential to destroy millions of acres of hardwoods such as, horse chestnuts, poplars, willows, and elms if it becomes established in North America. Authorities believe that this insect has been entering the United States in solid wood packing materials from China. In late 1998, APHIS banned all shipments from China that contain such packing materials unless properlly treated to destroy this pest.

Asiatic class. domestic fowl that have heavily feathered legs and lay eggs with brown shells (e.g., Brahmas, Cochins, and Langshans).

Asiatic domestic cattle. *Bos indicus*.

Asimina triloba (papaw). a tree, the fruit of which is commonly eaten by humans, a small percentage of whom develop a contact dermatitis. Some individuals experience severe gastrointestinal symptoms.

asocial. not social, reclusive; solitary. Even the sexes may associate only when mating. Truly asocial animals may aggregate at times, e.g., about a common, usually temporary resource such as a fresh water pond.

asparagus broccoli. *B. oleracea var. botrytis*.

asparagus. **1:** any herb of the genus *Asparagus* (Family Lilaceae). **2:** the shoots, especially the young, edible shoots of *A. officinalis*. **garden asparagus**. *A. officinalis*.

Asparagus (asparagus). a genus of herbs (Family Liliaceae). ***A. asparagoides***. a delicate, twining plant (Family Liliaceae), cultivated by florists. The leaves are bright green and oval in shape. ***A. officinalis*** (asparagus; garden asparagus). a perennial herb indigenous to Europe and widely cultivated for its edible young shoots and for the rhizome and roots, which are used as a diuretic.

aspect. **1:** appearance; countenance; physical appearance; the manner of appearance. **2:** the seasonal appearance of vegetation, a species, or a type of plant, especially that associated with the changing seasons (e.g., winter aspect). **3a:** a facet or feature of a structure. **3b:** the side of an object viewed from a designated direction. Thus, the lateral aspect of a seated person. **3c:** the face or side elevation of any natural feature, slope face, or man-made structure. **4:** the compass direction toward which a slope faces; exposure. Aspect affects insolation and other physical factors of the environment (e.g. wind) that affect the microclimate experienced by local fauna, flora, and soils. **5:** the magnitude of exposure of an area or location to environmental variables.

aspection. periodic, usually seasonal, changes in aspect of the vegetation within a particular area, region, or community.

asper-. rough.

aspergillic acid (h2-hydroxy-3-isobutyl-6(methylpropyl)pyrazine-1-oxide). a yellow, crystalline, diketopiperazine-like mycotoxin, antifungal, and antibiotic,

$$C_{12}H_{20}N_2O_2$$

It is produced by certain strains of *Aspergillus flavus*. This compound is toxic to animals but only moderately toxic to bacteria.

aspergillomycosis. aspergillosis.

aspergillosis (aspergillomycosis). a respiratory disease caused by any of several species of *Aspergillus* and characterized by inflammatory granulomatous lesions of the skin and mucous membranes, including especially the ear, orbit of the eye, urethra, nasal cavities, and lungs. Bones, meninges, liver, kidneys, and other organs may be affected in severe cases. *A. fumigatus* is the usual cause of the disease in humans and fowl. In fowl, the air sac system is affected as well as the lungs.

aspergillotoxicosis. aspergillus toxicosis.

Aspergillus. a large genus of filamentous fungi (Family Moniliaceae, Class Ascomycetes), a number of which produce black, brown, or green spores. The conidiophores are elongated and bear many chains of basipetal conidia. Included are common molds and a few species that are pathogenic to humans, other animals, or plants. Some species, such as *A. flavus*, produce penicillin. *See also* aspergillustoxicosis. Only a few species are briefly described herein. ***A. clavatus***. a species isolated from soil and feces. It produces patulin, a carcinogenic mycotoxin, and the indole mycotoxins, tryptoquivaline and tryptoquivalone. ***A. flavus***. a species with yellow-green conidia that infests peanuts and grain. It produces aflatoxins B_1, B_2, G_1, G_2. *See also* aflatoxins M and other mycotoxins such as aspergillic acid, kojic acid, palmotoxin B_0, palmotoxin G_0, and β-nitropropionic acid. ***A. fumigatus*** (formerly *A. gliocladium*, *Eurotium malignum*). a thermotolerant species that grows in soil and manure. It produces various mycotoxins and antibiotics (e.g., aspergillin, fumagilin, fumigacin, fumigation, fumigatoxin, fumitremorgin A and B, and helvolic acid). It is the common cause of aspergillosis in humans and birds and is considered the primary pathogen in the case of birds. It sometimes occurs in infections of the ears, lungs, and upper respiratory tract of humans and other vertebrates. ***A. gliocladium***. a former name of *A. fumigatus*. ***A. nidulans***. a species that commonly occurs in soil; it produces nidultoxin. It sometimes causes aspergillosis in humans. ***A. niger***. a pathogenic species that occurs commonly in soil. It produces a mycotoxin, malformin A, which causes severe, persistent infections in humans and laboratory mice. The spores are black. ***A. ochraceus***. this species ferments the coffee berry, producing the characteristic aroma. It produces the mycotoxins ochratoxin A and ochratoxin B. ***A. parasiticus***. a species that produces aflatoxins. ***A. terreus***. a species that produces the epoxide mycotoxins, terreic acid and citrinin, and antibiotics. It may sometimes infect the upper respiratory tract and has been isolated from the external auditory meatus in patients with otomycosis (mostly in Japan and Taiwan). It occasionally causes aspergillosis in humans and animals. ***A. versicolor***. a species commonly found in soil and often in dried salted beef. It produces the bisfuranoid mycotoxin, sterigmatocystin.

aspergillus toxicosis (aspergillotoxicosis; aspergillustoxicosis). **1:** a mycotoxicosis caused by a member of the genus *Aspergillus*. **2:** an acute lethal disease of livestock caused by feed contaminated by *Aspergillus spp.* The disease is marked by widespread internal hemorrhage; death usually intervenes within a week. *Cf.* aflatoxicosus.

aspergillustoxicosis. aspergillus toxicosis.

asperity. roughness, as of a leaf.

asperous. rough to the touch.

Asphordylia websteri (alfalfa gall midge). an insect pest (Family Itonididae) of alfalfa in Arizona and New Mexico.

asphyctic. asphyxial.

asphyxia. **1a:** suffocation from insufficient intake of air. **1b:** a condition of imminent death due to lack of oxygen in respired air. **2:** severely impaired oxygen exchange (external or internal) by a living animal. **3:** a condition of extreme hypoxia or anoxia. *See also* asphyxiant.

asphyxial (asphyctic). pertaining to asphyxia.

asphyxiant. **1:** causing asphyxia; asphyxiating. **2:** an agent that is able to asphyxiate; one that deprives an organism, cell, tissue, or organ of oxygen. **chemical asphyxiant**. a chemical that blocks oxygen transport to cells (e.g., carbon monoxide) or oxygen utilization by cells or tissues (e.g., cyanide). **simple asphyxiant**. a gas or an aerosol that dilutes the air or oxygen that reaches a respiratory surface (e.g., nitrogen, hydrogen, helium). Respirable air containing high levels of simple asphyxiants lacks sufficient oxygen to support respiration.

asphyxiate. **1:** to cause or induce asphyxia. **2:** to suffocate, choke, smother, or expose to a noxious gas; to cause to lose consciousness or die by blocking or hindering normal breathing. **3:** to become asphyxiated.

asphyxiating. **1:** asphyxiant (def. 1). **2:** the induction of asphyxia. **3:** having the ability to suffocate or asphyxiate.

asphyxiation. **1:** the act of asphyxiating. **2:** the state of asphyxia.

aspidin. a toxic principle, $C_{25}H_{32}O_8$, isolated from aspidium.

aspidium. the dried rhizomes and stems of the fern *Dryoptens filix-mas* (European aspidium) or of *D. marginalis*. Although toxic, it is sometimes used to treat otherwise unmanageable tapeworm infestations.

Aspidosperma (quebracho; aspidosperma). a genus of tropical American trees and occasional shrubs (Family Apocynaceae) with small flowers, alternate leaves, compressed peltate seeds with a flat papery wing, and follicular fruit. ***A. quebrachoblanco*** (quebracho; white quebracho). a tree and important source of quebrachine, adiposamine, and aspidospermine. *See also* quebracho.

aspidosperma. a common name of plants of the genus *Aspidosperma*.

aspidospermine. a white to brownish, crystalline irritant, alkaloid, $C_22H_{30}N_2O_2$, from the dried bark of *Aspidosperma quebrachoblanco*.

aspirate. **1:** to draw or remove by suction. **2:** to remove (a fluid) from a body cavity by use of an aspirator or suction syringe. **3:** to inhale (fluid or a foreign body) into the bronchi and lungs, often after vomiting. **4:** the material that has been aspirated.

aspiration. **1:** removal of fluid from a cavity or space by suction (e.g, by use of a syringe or aspirator). **2:** accidental inhalation of fluids or objects into the bronchi and lungs.

aspirin. **1:** a derivative of salicylic acid, it is a white, crystalline substance, $C_9H_8O_4$, that is widely-used as an antiinflammatory agent and an analgesic to reduce fever and relieve the pain (e.g., of headache, rheumatism, gout, neuralgia). It dissolves and is absorbed only slowly in acid media (as in the stomach) and can irritate the stomachs of most mammals. The neutral form, acetylsalicylic acid, *q.v.*, is rapidly absorbed. **2:** loosely, an over-the-counter drug, an aspirin tablet.

ass. **1:** a common name for either of two living equines, *Equus hemonius*, the Asian ass, *E. asinus*, the African ass, and domestic breeds, *E. a. asinus*. *See Equus*. **2:** a common name of any equine of small size with a large head, long ears, and small hooves. **Asian ass**. *E. hemonius*. **African ass**. *E. asinus*. **domestic ass**. *See E. a. asinus*. **3:** burro, donkey, jackass.

assassin bug (cone nose; kissing bug; reduviid). any of numerous true bugs (Family Reduviidae) that feed chiefly on other insects. Some species are blood-sucking parasites of mammals.

assay. **1a:** analysis; test of purity; trial. **1b:** a procedure used to measure or identify a substance or a process. **1c:** the chemical analysis of a material to determine the proportion of constituents present. **2:** to subject to analysis; to perform an assay. **Ames assay**. Ames test. **autoimmune assay** (ANA test). a blood test to identify autoimmune disease. **biological assay** (bioassay). the establishment of the relative strength of a biologically active substance by comparing measured effects in test organisms to those of a reference standard. **competitive adsorption assay**. a competitive assay in which an antibody is adsorbed to a surface, and the test ligand is allowed to compete with a known amount of labeled ligand. **competitive binding assay** (displacement analysis; saturation analysis). a type of assay in which a binder competes for a labeled versus an unlabeled ligand. The free and bound ligands are separated, and the amount of ligand is determined by comparing bound and unbound ratios with reference standards. **four-point assay**. an assay that employs a mixture of two doses of test material and two doses of a standard. **immunochemical assay**. immunoassay. **immunoenzymatic adsorption assay**. an assay in which unlabeled ligand is adsorbed to a surface and exposed to labeled antibody and test ligand in solution. The amount of bound, labeled antibody is inversely proportional to the amount of free test ligand. **microbiological assay**. a biological assay that uses microorganisms as the test system. **sandwich adsorption assay**. an assay in which cold antibody is adsorbed to a surface to which the test ligand and cold ligand are added, followed by the addition of labeled antibody, thus the term "sandwich." The enzyme product is proportional to the concentration of bound ligand. **stem cell assay**. the determination of the effectiveness of a given drug against human cancer by the use of human tumor cell suspensions that are incubated with a number of drugs, suspended in agar, and plated over a layer of agar. The number of treated colonies that grow compared to the number of untreated (control) colonies that grow provides a measure of a drug's effectiveness. **toxicity assay**. toxicity test. *See also* test.

assessed value. the value of real estate, acreage, or sometimes personal property assessed by a governmental body for purposes of taxation.

assessment. a critical, often official, appraisal or evaluation of the nature, status, significance, or importance of an entity or process.

assimil-. a prefix meaning like unto.

assimilate. **1:** to absorb or incorporate. **2:** to absorb digested food or nutrients into the system (i.e., into protoplasm). **3:** to incorporate, interpret, and integrate new information into one's mind. This process may or may not involve the conscious mind. **4:** that which is assimilated. *Cf.* dissimilate.

assimilation. **1a:** the incorporation by a living organism of simple nutrients or many of the products of digestion into the complex organic compounds that make up the protoplasm living cells. **1b:** in heterotrophic organisms, such materials result from the digestion of food and subsequent absorption into the body. **1c:** in autotrophic organisms (e.g., green plants), elements and relatively simple inorganic molecules such as H_2O, CO_2, and N_2, absorbed from the environmental media (air, soil, water) through the roots provide the raw materials of assimilation. **2:** the incorporation, interpretation, and integration of new knowledge or experiences with the existing cognitive organization of the brain.

assimilation capacity (assimilative capacity). **1a:** the capacity of a biotic system (most commonly an ecosystem, especially a body of water) to receive a certain amount or to sustain certain concentrations of an environmental chemical or waste discharges without suffering significant or permanent harm or degradation. **1b:** the ability of a receiving system or ecosystem to cope with certain concentrations or levels of waste discharges without suffering any significant deleterious effects.

assimilative capacity. assimilation capacity.

assn. abbr. for association.

assoc. an abbreviation for association.

association (assn., assoc.). **1a:** the act of associating. **1b:** the state of being associated. **2:** an organized body of people (e.g., an organization, club, society, sorority) based on common interests, activities, or purpose. **3a:** a mental connection or relation among thoughts or perceptions. **3b:** a remembered or imagined thought or perception, linked to a person, object, or idea. **4a:** a bond, connection, or affiliation. **4b:** any of various types of weak chemical bonding (e.g., hydration, solvation, or the formation of complex ions). **5:** a large number of biotic species in a given geographic area that comprise a community.

Association of Official Analytical Chemists (AOAC). a North American professional scientific society with the major goal of fostering public health and safety in acquiring, developing, testing, and adopting standard, reliable chemical and biological methods for the testing and analysis of materials such as food, drugs, cosmetics, pesticides, feeds, fertilizers, air, water, or hazardous materials, that can affect public health and/or safety, environmental quality, or economic abuse of consumers.

aureole. a halo-like or ring-like feature; the pink aureole about the nipple of humans and certain other mammals.

Association of American Plant Food Control Officials (AAPFCO). an association, the officers of which are charged by law to enforce regulations regarding the sale of commercial fertilizer and fertilizer materials.

assurg-. a prefix meaning spreading out, sprawling, clambering.

assurgent. curving upward, e.g., as the stem of a plant.

aster. **1:** common name of plants of the genus *Aster*. **2:** sometimes used as a common name for specific plants or genera that are not true asters, such as *Callistephus chinensis* (China aster), *Xylorrhiza* (woody asters), and *Gutierrezia* (tansy asters).

Aster (asters, true asters, true hardy asters, starworts, Michaelmas daisies). a genus of perennial herbs (Family Asteraceae, formerly Compositae), many of which are indigenous to the temperate zone of North America. Some are facultative accumulators of selenium from soil. Examples are *A. ascendens*, *A. coerulescens*, *A. commutatus*, *A. ericoides*, *A. glaucoides*, and *A. laevis*. *See also Xylorrhiza*.

aster family. Asteraceae.

aster yellows. a dwarfing and yellowing of

asters and certain other plants, caused by a mycoplasma transmitted by a leafhopper.

Asteraceae. a family of herbaceous plants known variously as the aster family, the composite family, and the sunflower family. Formerly, Compositae. *See also Aster*.

asthenia. muscular weakness or debility; loss of muscular strength. *Cf.* abirritation, hyposthenia.

astheniant. **1:** weakening or debilitant. **2:** a debilitating agent. *Cf. See also* hypostheniant.

asthenic. weak or debilitated; pertaining to or characterized by asthenia; one who suffers from asthenia. *See also* hyposthenic.

astomous. lacking a mouth.

astr-. a prefix denoting or pertaining to stars.

Astragalus (milk vetches; poisonvetches; locoweeds). the largest genus of plants of the Family Fabaceae. Some species are harmless and afford good forage for livestock, and certain oriental species (e.g., *A. gummifer*) have medicinal value. A number of species, however, are poisonous. There are three distinct categories of poisonous species: (1) The poisonvetches are obligate selenophiles and the selenium content accounts for their chief toxic effects. Many of these species are distasteful to livestock and do not usually pose a hazard on the range if other forage is available. The species in this category that most often poison livestock are *A. bisulcatus*, *A. pattersonii*, *A. pectinatus*, and *A. racemosus*. (2) the true locoweeds (see loco) contain the mydriatic alkaloid, indolizidine-1,2,8-triol. Locoine was the name given to early toxic extracts from locoweeds. *See also Oxytropis*. (3) paralytic poisons. Certain species of *Astragalus* produce a distinctive type of paralytic poisoning of livestock with respiratory involvement. Symptoms in range cattle include poor condition, a dry or peeling muzzle, chronic cough, loss or huskiness of voice, anorexia, and incoordination, especially of the hind legs, which tend to knock together (hence the name "cracker-heel" disease, *q.v.*). Forced exercise causes a wheezing or even a roaring exhalation. Paralysis may develop, and sudden death from asphyxia is common in animals that are driven. Mortality is otherwise unpredictable. The syndrome is more severe in sheep and especially so in horses. Species responsible for this third type of poisoning include *A. emoryanus*, *A. michauxii*, *A. pubentissimus*, *A. sabulosus*, and *A. tetrapterus*.

Astrakhan. **1:** the fur of young lambs, with lustrous, closely curled wool, from Astrakhan; the lambskin of Karakul or Bokhara sheep. *See* Astrakhan cloth.

Astrakhan cloth. **1:** a fabric with curled pile that resembles Astrakhan fur. **2:** Astrakhan.

astringent. **1:** able to cause the contraction of tissues, arrest secretion, or control bleeding. **2:** an agent (e.g., alum, tannic acid, zinc oxide, zinc sulfate) that causes the effects in def. 1.

astrologer. a lyrical name for the cock of the domestic fowl, the herald of dawn.

asymmetric, asymmetrical (unsymmetrical). lacking symmetry.

at bay. cornered; treed; trapped.

at the end of the halter. denoting an animal sold "as is" with no guarantees or right of return.

ATA. **1:** alimentary toxic aleukia. *See* aleukia. **2:** aminotriazole.

atactic. pertaining to or characterized by incoordination or ataxia; uncoordinated.

atactostele. *See* stele.

atamasco lily. *Zephyranthes atamasco*.

atavism. reversion to, or reappearance of, an ancestral characteristic not observed in recent generations. The individual ex hibiting such a characteristic is often called a throwback.

ataxia (ataxy; dyssynergia; incoordination). **1:** incoordination or failure of muscular coordination of voluntary muscular action; often manifested by unsteady gait. **2:** a disorder of the neuromuscular system characterized by a loss of coordination and an unsteady gait.

ataxiadynamia. muscular weakness with lack of coordination. *See also* ataxia, incoordination.

ataxic. pertaining to, characterized by, or afflicted with ataxia.

ataxy. ataxia.

-ate. a suffix meaning having the condition or property of (thus, septate means having septa).

atheroma. fatty degeneration of the inner lining (tunica intima) of arteries.

atherosclerosis. a common type of arteriosclerosis in which fatty plaques develop on the inner lining (tunica intima) of medium and large arteries. The plaques form from smooth muscle cells, cholesterol, and other lipids. They may occlude the arteries, reducing the flow of blood; they also serve as sites where blood clots (thrombi) can form.

atlas. a book, map, or other format that presents information by means of illustrations rather than by description.

atmometer (evaporometer). an instrument used to measure the rate of evaporation of water from a moist surface.

atmosphere. **1:** air. **2:** any gas that surrounds a given body. **3:** milieu; surroundings, environment. **4:** the entire envelope; of air that surrounds the earth. **standard atmosphere**. a standard unit of pressure representing that exerted by a vertical column of mercury (760mm) at mean sea level at a latitude of 45^0. It is equivalent to 1000 gm/cm^2, 1,013,250 dynes/cm^2, 101,325 pascals (newtons/m^2) in the SI system. In the English system, this is equivalent to a force of 14.695916 pounds per square inch.

atmospheric dust. *See* dust.

atmospheric pressure. the pressure (force per unit area) of air at any location within the earth's atmosphere. *See* atmosphere (standard atmosphere).

atom. **1:** any extremely small item or particle; amote. **2a:** the smallest unit of an element that still retains the chemical characteristics of that element. **2b:** an elementary component of a molecule.

atomic number. *See* proton number.

atomize. **1:** to produce an atomic vapor. *See* atomization. **2:** to nebulize; to divide a liquid into minute particles by impact with a jet of steam or compressed air, or by passage through a porous mechanical device.

atonic. pertaining to atony.

atony. **1:** lack of tonus. **2a:** weakness; lack of energy. **2b:** debility. *See* abirritation.

atopic conjunctivitis. allergic conjunctivitis.

atopy (atopic allergy). a genetic predisposition to develop a particular type of allergy.

atoxylic acid. arsenilic acid.

ATP. (adenosine triphosphate; adenosine 5′-(tetrahydrogen triphosphate)). a nucleotide comprised of the purine base, adenosine, and D-ribose attached to three phosphate groups, two of which are high-energy groups. It is an essential energy-storage and energy-transfer component of all living cells. ATP is hydrolyzed to adenosine diphosphate (ADP) with the release of free energy and pyrophosphate (PP_i); ADP may then form adenosine monophosphate (AMP) with the release of additional free energy and PP_i. The energy so released, drives numerous endergonic metabolic processes such as active transport and muscle contraction. ATP is reconstituted chiefly from ADP during photosynthesis and oxidative phosphorylation. *See also* ADP, AMP.

atr-. a prefix meaning extreme; dark or black.

atrazine (2-chloro-4-ethylamino-6-isopropyl-amino-1,3,5-triazine; 6-chloro-*N*-ethyl-*N*′-(1-methylethyl)-1,3,5-triazine-2,4-diamine). a selective triazine derivative used as a pre- and post-emergence herbicide, characterized by high toxicity to a diversity of monocotyledonous and dicotyledonous plants when applied to the soil. It is slightly toxic to fish and slightly to moderately toxic to laboratory mice. Acute poisoning of domestic cattle and sheep can cause muscular spasms, fasciculations, unsteady gait, pulmonary congestion, and hyperpnea. Lesions may include degenerative changes in the adrenal glands, kidneys, and liver. Atrazine is a possible mutagen, carcinogen, and teratogen. It is degraded nonenzymically in corn to the inactive hydroxy derivative and is thus used extensively in controlling annual grasses and broadleaved weeds in corn. It has become a matter of public health and environmental concern because of runoff into surface waters. An EPA survey, conducted from 1988 to 1990, indicated that atrazine is the second most commonly detected pesticide residue in drinking water.

atriplex poisoning. atriplicism.

Atriplex (saltbushes). a genus of plants (Family Chenopodiaceae) that are facultative accumulators of selenium from soil. ***A. sentibaccata*** (Australian saltbush; creeping salt bush). a species that is tolerant of alkaline and saline soils. It is of some value as forage for livestock. While indigenous to Australia, it now occurs also along the California coast.

atriplicism (atriplex poisoning). poisoning due to ingestion of the spinach-like herb, *Atriplex littoralis* (Family Chenopodiaceae).

Atropa belladonna (deadly nightshade; belladonna; death's herb). a shrublike, perennial European herb (Family Solanaceae), the source of belladonna. It is the most notorious and probably the most dangerous of the solanaceous plants, which also include *Cestrum, Datura, Hyoscyamus, Mandragora, Physalis,* and *Solanum.* It contains atropine and various related alkaloids (e.g., 1-hyoscyamine, scopolamine, and tropane); the roots contain scopoletin and scopolin. All parts of the plant are toxic and toxicity is greatest at maturity. Validated poisonings are rare. Children may be attracted to the black berries, and consumption of as few as three of these can be fatal to a child. *See also* belladonna poisoning.

atrophia. atrophy (def. 1).

atrophic. **1:** pertaining to organisms, or stages in the life cycle of an organism, that do not feed. **2:** of or pertaining to atrophy.

atrophie. atrophy (def. 1).

atrophy (atrophia; atrophie). **1:** a wasting away or decrease in the size of a cell, tissue, organ, or part. Atrophy may result, for example, from loss of cells through death and resorption, reduced rates of cellular proliferation, ischemia, malnutrition, physical pressure, decreased activity, hormonal changes, or directly or indirectly from empoisonment. **2:** to undergo or to cause atrophy. **acute yellow atrophy** (acute parenchymatous hepatitis; Budd's jaundice; malignant jaundice; massive hepatic necrosis; Rokitansky's disease). atrophy of the liver marked by wasting, massive necrosis, and jaundice. It is a complication of fulminant hepatitis and is an infrequent and usually fatal complication of viral hepatitis and may result from toxic hepatitis due to poisoning, for example, by carbon tetrachloride, fluorinated hydrocarbons, or *Amanita phalloides*.

atropia. atropine.

atropine (*dl*-hyoscyamine; tropine tropate; tropic acid ester with tropine; *dl*-tropyl tropate; α-(hydroxymethyl)benzeneacetic acid 8-methyl-8-azabicyclo[3.2.1]oct-3-yl ester; $1\alpha H,5\alpha H$-tropan-3α-ol(+)-tropate; atropia). an alkaloid, $C_{17}H_{23}NO_3$, that occurs naturally in *Atropa belladonna* and related plants (*Datura*, *Hyoscyamus*); it is also prepared synthetically. It is used in medicine mainly as a muscle relaxant (e.g., to dilate the pupil of the eye or to inhibit movements of the intestine), as an antispasmodic, antisudorific, and anticholinergic. Atropine (as atropine sulfate) is an antidote for anticholinesterases and various other toxic substances such as the organophosphorous pesticides (e.g., parathion); artificial respiration may also be required. It is moderately toxic to adult humans. Infants and young children are, however, very vulnerable to atropine intoxication, which may result from its use in ophthalmologic treatments. *See also* atropinic effects, atropine poisoning.

attainment area. an area deemed to have air quality equal to or higher than the National Ambient Air Quality Standards for a given pollutant as defined in the Clean Air Act.

attaint. an injury (e.g., an abrasion or laceration) to a horse's leg inflicted by its own hooves.

attapulgite. a highly absorptive mineral clay from Florida and Georgia, it is the active ingredient in most fuller's earth. The chief component is a hydrous magnesium aluminum silicate. It is used, for example, in the manufacture of fertilizer, the conditioning of fertilizer products, as a suspending agent (e.g. of fertilizers), as a thickener of latex paint, and as an oil well drilling fluid.

attenu-. a prefix meaning attenuated; drawn-out to a point

attenuate. **1a:** diminish, lessen, reduce. **1b:** weaken, dilute. **2a:** weaken, enfeeble. **2b:** with respect to a microbe or other pathogen, to reduce its virulence.

attenuated strain. a strain of pathogenic microorganisms whose virulence has been lost or greatly reduced.

attractant. **1:** a substance that attracts. **2:** a substance produced by one organism or species that attracts another, usually conferring an advantage on the producing organism. Examples are sexual attractants and baits used by humans and other species to lure prey. *See* allomone.

attrition. **1:** reduction; decrease; decline. **2:** erosion by friction or corrosion.

atypical. **1a:** abnormal, unusual. **1b:** not typical; not corresponding to the normal form, type, pattern, color, condition, or location from that observed under similar circumstances. **2:** abberant.

Au. antigen.

AU. animal unit.

aubergine. the blackish-purple fruit of the eggplant (*Solanum melongena*) used as a vegetable.

auditory. pertaining to the structures of the ear that are concerned with hearing.

auditory canal. a tube in the outer ear of most vertebrates that leads to the tympanic membrane.

auger. **1:** a type of tool used to bore holes, it is typically a pointed, threaded screw with a handle used to bore holes, e.g., in wood or in soil. **2:** a conveyor or elevator in the form of a spiral or broad-threaded screw in an open tube. **3:** a large screw-shaped device on a combine harvester that carries the grain up into the grain tank. **soil auger**. one used to obtain samples of soil or to dig holes (e.g., for fence posts).

august-. a prefix meaning august, noble, majestic, or regal.

Aujesky's disease (infectious bulbar paralysis; pseudorabies; mad itch). a viral disease of animals, characterized by intense itching, convulsions, and paralysis. It is a serious, often fatal, disease of young pigs and cattle. Dogs and cats are also susceptible.

AUM, A.U.M. animal unit month.

aural. of or pertaining to the ear or to hearing.

aurant-. a prefix indicating or pertaining to the color orange.

aure-. a prefix denoting or pertaining to

gold; golden.

auri-. a prefix denoting or pertaining to an ear; having the shape of an ear; ear-like.

auripigment. arsenic trisulfide.

austr-. a prefix meaning southern or Australian.

austral. **1a:** southern; southerly. **1b:** pertaining to cool temperate regions of the southern hemisphere. **2:** antiboreal. *Cf.* boreal.

Australian nettle. *Dendrocnide moroides*.

Australian saltbush. *Atriplex sentibaccata*.

autarky. **1:** self-sufficiency; the practice of eating only foods produced locally. **2:** independence; economic independence as a national policy.

autecology. the ecology of the individual organism or species.

auto-. a prefix meaning self.

autoallergy. autoimmunity.

autoantigen (autoimmunogen). *See* autoimmunity.

autochory. self-dispersal of spores or seeds by an explosive mechanism.

autochthonal. autochthonous.

autochthonic. autochthonous.

autoclave. a vessel that can be sealed and the contents heated to a high temperature under high pressure. Such an instrument is used to sterilize surgical tools and other utensils and in the form of a household pressure cooker, to cook certain foods without loss of nutrients or flavor.

autocytolysis. autolysis.

autocytolytic. autolytic.

autoecious. denoting or pertaining to parasites that complete all stages of a life cycle on the same host as in the case of: (1) certain fungi, rusts, that parasitize plants; and (2) certain insects that remain on one host species throughout the year, for example, aphids that do not alternate between a primary woody and a secondary herbaceous host.

autogamous. self-fertilizing.

autogamy. **1:** self-fertilization; pollination of a flower by its own pollen. *Cf.* allogamy. **2:** conjugation in an individual organism by division of its nucleus into two parts that reunite, producing a zygote.

autogenic movement. autonomic movement.

autograft. a graft obtained from the tissues of the recipient.

autohemorrhea. reflex bleeding in insects.

autoimmune disease. **1a:** severe, clinical autoimmunity. **2:** any disease in which autoimmunity is a contributing factor.

autoimmune disorder. autoimmunity, *q.v.*; a condition in which the immune system produces antibodies that attack an individual's own tissues.

autoimmunity (autoallergy; autoimmune disorder; autoimmune disease). a condition or process whereby the immune system produces antibodies (auto antigens) that attack certain components of an individual's own cells or tissues. Effects (hypersensitivity) are usually harmful to the individual.

autoimmunogen. auto antigen. *See* autoimmunity.

autointoxication (autotoxicosis; autotoxemia; autotoxaemia; autotoxis). self empoisonment due to endogenous microorganisms or to intermediates and end products of metabolism. **intestinal autointoxication** (alimentary toxemia; intestinal intoxication). a disorder caused by the accumulation of intestinal poisons in blood.

autolysis (autocytolysis; self digestion). lytic disintegration of a cell or tissue by the action of cellular enzymes. This process takes place either within a lysosome or in the cytoplasm following the release of enzymes from lysosomes. It occurs in damaged tissues immediately prior to and after the death of a cell, or in developmental processes such as metamorphosis. Autolysis is responsible for the softening of fruits and vegetables after picking and of meat following the death of an animal. *Cf.* cytolysis.

autolytic (autocytolytic). pertaining to autolysis or the capacity to autolyze.

autolyze. to cause or to undergo autolysis.

automatic take-off. an apparatus that registers the end of milk flow in a milking machine, shuts off the vacuum, and retracts the milking unit from a cow's udder.

Automeris io (io moth). a reddish-brown moth (Family Saturniide), the hind wing of which has a characteristic black eyespot. The caterpillar is green with reddish-pink and white lateral stripes. *See* io-moth dermatitis.

autonomic. **1:** of, or pertaining to the autonomic nervous system (ANS) or to a part of the ANS. **2:** involuntary; reflex. **3:** pertaining to a chemical substance that affects tissues served by the ANS. **4:** the autonomic nervous system. **5:** due to endogenous influences (in botany). Thus, autonomic movement of a plant or plant part is that which results from internal growth changes.

autonomic movement (autogenic movement; spontaneous movement). Movements of plants induced by internal stimuli. They include growth itself, nutation, protoplasmic streaming, chromosome movement during nuclear division, locomotion of motile algae, fungi, spores, and0 gametes, and some movements mediated by pulvini. *Cf* paratonic movements.

autonomic nervous system (ANS). the division of the vertebrate peripheral nervous system that controls, but does not usually trigger, glandular secretion and the action of viscera and smooth (involuntary), cardiac, and some striated muscle. The ANS is controlled by the CNS and is thus thought by many scientists to respond only indirectly to external stimuli. There are two anatomically and functionally distinct divisions of this system, the sympathetic and parasympathetic nervous systems, *q.v.* Most autonomic effectors (e.g., muscles and glands) receive dual innervation from these divisions, although only sympathetic neurons serve the blood vessels of the skin and limbs. The visceral nervous system of arthropods is structurally and functionally analogous to the ANS.

autonomic poison. a poisonous substance that acts on the autonomic nervous system.

autoparasitism (secondary parasitism; superparasitism). parasitism of one parasite by another.

autopentaploid. having five sets of homologous chromosomes derived from a single parent species.

autopentaploidy. the state of being autopentaploid.

autopoisonous. poisonous to the producing organism. *See also* autotoxic.

autopollination. the spontaneous transfer of pollen from anthers to stigma during the opening of a flower.

autopolyploid. **1:** having more than two homologous sets of chromosomes derived from a single parent species. **2:** an autopolyploid cell or organism. *Cf.* allopolyploid.

autopolyploidy. the autopolyploid state.

autopsy. **1:** (postmortem examination; necropsy; necroscopy; thanatopsy; postmortem examination). the postmortem examination of an animal or human by gross dissection (accompanied as needed by histological, chemical, analytic, and other methods) to determine circumstances accompanying or causing the death, including the pathology of disease. This is usually required under law for an unexplained, unnatural, or violent human fatality. **2:** to perform a postmortem examination. **3:** a critical analysis of a past event or process, sometimes by reproducing the effect or the circumstances surrounding the event or process. For example, the duplication of an event, action, or circumstance that occurred during the course of a disease and observation of its relationship to the aggravation or amelioration of symptoms.

autosomal. of or pertaining to autosomes.

autosomal chromosome. autosome.

autosomal recessive amelia. porcine amelia *See* amelia.

autosome (autosomal chromosome). a chromosome other than a sex chromosome; a chromosome that does not contribute to the determination of sex.

autosterile. autoincompatible.

autotetraploid. having four sets of homologous chromosomes derived from a single parent species.

autotetraploidy. the state of being autotetraploid.

autotoxaemia. autointoxication.

autotoxemia. autointoxication.

autotoxic. **1a:** pertaining to autotoxicity. **1b:** pertaining to chemicals that are detrimental to individuals of the same species. **2:** autopoisonous; poisonous to the producing organism. *See also* autointoxication, autopharmacologic, autotoxicosis.

autotoxicity. **1:** the property or capacity of an endogenous chemical or excreta to destroy cells or tissues or interfere with metabolic processes of the producing organism or its progeny. **2:** self-destruction of a plant species by the production of chemicals that are released into the environment and directly inhibit the growth or otherwise harm plants of the same species. **3:** the production by an animal of chemicals that destroy or interfere with the normal activity of cells or tissues of the body. **4:** the state or condition of being autotoxic. **alfalfa autotoxicity**. production and release by alfalfa of a toxin that inhibits root growth and sometimes delays or reduces germination. Reseeding of alfalfa is not possible until the autotoxic chemical disintegrates or is dispersed. This may require a year or more.

autotoxicosis. autointoxication.

autotoxin. **1:** an autotoxic toxin. **2:** an autointoxicant.

autotoxis. autointoxication.

autotriploid. having three sets of homologous chromosomes derived from a single parent species.

autotriploidy. the state of being autotriploid.

autotroph. autotrophic organism.

autotrophic (holophytic; lithotrophic). **1:** capable of self-nourishment; said of all photosynthesizing plants. **2:** of or pertaining to organisms (*See* autotrophic organism) that can synthesize complex organic nutritive substances from simple inorganic nutrients, using CO_2 as the sole source of carbon. *Cf.* photoautotrophic, holophytic, holotrophic, holozoic, lithotrophic.

autotrophic nutrition (lithotrophic nutrition; holophytic nutrition). a mode of self nourishment whereby complex organic nutritive substances are manufactured from simple inorganic nutrients, using CO_2 as the sole source of carbon.

autotrophic organism (autotroph; holophytic organism; lithotrophic organism; producer). a producer (in ecology; see primary producer); an organism (e.g., a green plant) capable of growth and maintenance on purely inorganic media; one that produces food from inorganic substances. Autotrophs utilize carbon dioxide and nitrates or ammonium salts as sources of carbon and nitrogen, respectively. The energy source for these syntheses is either light via photosynthesis (photoautotrophic organisms; photoautotrophs) or inorganic chemical reactions independent of light energy via chemosynthesis (chemoautotrophic or chemosyn thetic organisms; chemolithotrophic organism; chemolithotroph; lithotroph). *See also* heterotrophic organism.

autumn crocus. *Colchicum*.

autumn galerina. *Galerina autumnalis*.

amudane. griseofulvin.

autumn lag. a period, usually from early autumn to early winter when rabbits, doves, pigeons, and many other birds in northern and north temperate latitudes slow down or cease reproducing.

auxin (growth substance; plant hormone). any organic substance produced in the apices of growing shoots and roots of plants that promotes or regulates growth of the producing plant when present in low concentrations. This is effected by the stimulation of cell elongation in the region just behind the apex. Auxins are very toxic to plants, inhibiting growth at higher concentrations. Consequently, both natural auxins and synthetic compounds that are closely related to auxins are used as herbicides (e.g., 2,4-D). Auxins also promote cell division in plant tissues and regulate phototropic and geotropic movements of plants. Auxins are antagonists of abscissic acid, retarding leaf abscission. The most widely distributed natural auxin is indoleacetic acid. Other natural auxins (e.g., indole-acetonitrile and indolepyruvic acid) also contain an indole ring. *See* herbicide (auxin herbicide).

auxin 1AA. 3-indolacetic acid.

auxo-. a prefix denoting growth or increase.

available. **1a:** accessible; obtainable. **1b:** handy; convenient. **1c:** on hand; on tap.

available element (available nutrient). an element (a macroelement or microelement, *q.v.*) in the soil that is readily taken up by plant roots.

available nutrient. available element.

available plant nutrient. a microelement or macroelement that occurs in the environment in a form that is readily absorbed and assimilated by a plant. Also, a chemical that is readily transformed into an available form in the soil during a single growing season.

available soil water. that water in a soil that is readily available to plants. *See* available water capacity.

available water capacity. the capacity of a soil to store water in a form that is readily available to plants. It is given as percentage by weight and equals the moisture content per unit weight at field capacity minus that at the wilting point.

avalanche. a large mass of snow and/or ice and debris that breaks free and moves rapidly down the slope of a mountain.

avalanche cone. debris (e.g., rocks, boulders, vegetation, ice, snow) deposited, usually in the shape of a fan or cone, at the base of the path of an avalanche.

Avena fatua (wild oats). a species of wild oats (Family Gramineae) that occurs as a weed among cereal crops. In most cases, it is controlled by used of selective herbicides.

Avena sativa (cultivated oats). **1:** a hardy cereal grain (Family Gramineae) cultivated in cool-temperate climates. This plant is variously involved in four types of livestock poisoning: nitrate poisoning, grass tetany, photosensitization, and smutty oat poisoning. **2:** the grain produced by this plant, which is used both for human consumption and in animal feeds.

average daily gain (ADG). an animal's average post-weaning daily gain in weight, determined by subtracting the weight at weaning from the weight to date and dividing the result by the number of days on feed.

avermectin (AVM). any of a group of broad spectrum antiparasitic antibiotics produced by the actinomycete, *Streptomyces avermitilis*. It is sprayed on wool as a protection against moths and beetles. The

oral toxicity of avermectins to poultry and livestock is moderate; fish and aquatic invertebrates are much more susceptible.
avermectin B_1. abamectin.

aversive. **1:** avoiding; characterized by avoidance (e.g., aversive behavior). **2:** noxious; promoting or stimulating avoidance. *Cf.* appetitive. *See also* approach-withdrawal.

Aves (birds). a class of some 9000 species of homeothermic (warm blooded), feathered, winged, egg-laying, vertebrates with a horny beak. The skeleton is made up of strong, lightweight, largely hollow bones that contain air sacs. Most are able to fly as adults. All birds except the flightless ratites (e.g., ostriches, casowary, emu) have a carinate (keeled) sternum (breast bone) to which the flight muscles attach.

avi-. a prefix denoting, pertaining to, or characteristic of birds.

aviadenovirus. a genus of viruses (family Adenoviridae) that includes types of viruses found in birds.

avian. of or pertaining to birds or to the class Aves.

avian encephalomyelitis. a viral disease of 1 to 3 week old chicks of the domestic fowl. It is a nervous disorder characterized by tremors of the head, incoordination, and sitting on the haunches.

avian tuberculosis. a disease of the domestic fowl and certain other birds, *Mycobacterium avium*. Affected birds typically become listless, lame, with diarrhea, impaired appetite, and loss of weight; the comb, wattle, and skin of domestic fowl become pale. Death may occur.

avian typhoid. a common infectious disease of fowl caused by *Salmonella gallinarum*. Signs of the disease include listlessness and postural effects with a slouching of the body with the head drawn in, drooping wings, ruffled feathers, diarrhea with pale yellow or greenish feces, and the comb and wattles become pale. Death may follow.

aviary. A large enclosure (e.g., a pen, flight cage, or building) that confines birds, usually with freedom to fly.

avicide. a substance with the intended use and ability to destroy birds. Avicides include organophosphates such as monocrotophos. *See* 3-chloro-*p*-toluidine.

aviculture. the breeding, care and rearing of birds in captivity either for research or commercial purposes.

avidin. a protein component of the uncooked egg white of birds and amphibians; it also occurs in the genital tract of many animals. Avidin binds to and inactivates biotin. High levels of avidin in the diet have been shown to cause biotin deficiency in laboratory rats and chicks of the domestic fowl. The white of cooked eggs does not produce biotin deficiency.

Avipoxvirus. a genus of viruses (Family Poxviridae) that includes the avian poxviruses such as the canarypox and fowlpox viruses.

avirulent. not virulent; not pathogenic; not productive of disease.

AVM. avermectin.

avocado. **1:** *Persea americana*. **2:** the fruit of *Persea americana*.

avoidance. **1:** aversiveness. **2:** the prolongation of dormancy of an organism, reducing vulnerability to environmental stressors.

avoirdupois. the English system of weights, used in most English-speaking countries except in science and medicine and the weighing of precious stones and metals. In this system, one ounce contains 16 drams; the pound is 16 ounces. The hundredweight, however, equals 100 pounds in the United States and 112 pounds in the United Kingdom.

avulsion. the abrupt transference of land held by one party to another without legal change of ownership. This may result, for example, by a sudden change in the channel of a river or as a result of flooding.

awl. A small hand tool with a sharp point used to punch holes in materials such as leather and heavy canvas.

awn (beard). a bristle-like projection from a glume or lemma of grasses (barley, oats), or from a fruit, or the tip of a leaf.

axenic-. a prefix denoting a system (e.g., a pure culture) in which all biological

populations are defined.

axenic. a well-defined biological population, usually a pure culture. Often used to indicate a cell culture that is free of microorganisms.

axial. **1:** the line about which a body turns. **2:** in the longitudinal axis.

axil. the upper angle between a leaf or other plant part and the stem from which it arises.

axill-. a prefix pertaining to or indicating axils (axillaris).

axillary. growing, developing, or situated in the axil of a leaf (e.g., a bud or flower).

axillary bud. a bud that develops in the axil of a leaf.

axillary flower. a flower situated in the axil of a plant.

axis. **1a:** the central line, stem, or cylinder of a structure; shaft; spindle. **1b:** the main stalk of a plant, compound leaf, or flower cluster. **2:** the second cervical vertebra.

axle. the rod, usually made of steel in motor vehicles and machinery, about which a wheel turns.

axon. the elongated, unbranched process of a neuron, the cytoplasm of which is continuous with that of the body or perikaryon. It conducts impulses away from the cell body of the neuron. Its volume is generally much greater than that of the cell body. *See* neuron.

axonal. of or pertaining to an axon or axons.

axunge. a thick, rich fat from the kidneys of geese and pigs.

Ayrshire. a hardy breed of dairy cattle that originated in southwest Scotland. It is deep and broad in the hips with narrow shoulders; the ground color is usually brown, but varies from red to mahogany, and is spotted with white. Bulls weigh about 1400 to 2000 pounds.

azeotrope (azeotropic mixture). a mixture of two or more compounds that has a constant boiling point. The vapor above the azeotrope has the same relative concentrations of the components as those of the boiling liquid. Azeotropes cannot be separated by fractional distillation.

azeotropic mixture. azeotrope.

azinphos ethyl. *See* azinphos methyl.

azinphos methyl (*o,o*-dimethyl-S-4-oxo-1,2,3-benzotriazin-3(4H)-yl methyl phosphorodithioate; "Guthion"). a toxic, brown, waxy solid, $_{12}N_3O_3PS_2$, which is soluble in water and in many organic solvents. It is a cholinesterase inhibitor that is readily absorbed by the skin of higher vertebrates; its toxicities are similar to those of azinphos ethyl. The maximum nontoxic oral dose for calves in one study was 0.44 mg/kg body weight; for cattle, 2.2 mg/kg; and for sheep, 4.8 mg/kg.

Azobacter. a genus of flagellated, rod-shaped, obligately aerobic, free-living nitrogen-fixing bacteria (Family Azobacteriaceae) present in soil.

azolla. a common name of ferns of the genus *Azolla*.

Azolla (azolla). a genus of tropical ferns that fix atmospheric nitrogen.

azotemia. azoturia.

azoturia (azotemia; blackwater; holiday disease; lumbago; Monday morning disease; Monday disease; myohaemoglobinuria; paralytic haemoglobinaemia). a common ailment of draft horses that have been kept on full feed even though idle. Affected animals typically sweat excessively, are reluctant to move and do so stiffly, limb muscles become stiff, and the animal sits or lies on one side. Animals may become lame and the urine may become red or brown.

azur-. of or pertaining to the color blue (azureum).

B

1206 Blue. FD&C Blue No. 1

1311 Blue. FD&C Blue No. 2

12070 Blue. FD&C Blue No. 2

***β*, B**. symbols for beta.

B. the symbol for boron.

b.p., **bp**. boiling point.

B.P. before present.

b.s., B.S. base saturation.

B.T.V. Boynton's tissue vaccine.

B-cell. B lymphocyte. *See* lymphocyte.

B-cell lymphocyte. B lymphocyte. *See* lymphocyte.

Ba. symbol for barium.

Baar's irrigator. a flexible hose used for irrigation that has sixteen folding sprinklers.

Babcock test. a traditional method of measuring the amount of butterfat in milk. It can be used to calibrate electronic testing devices used today.

babe. *See* baby.

Babesia (formerly *Piroplasma*). a genus of parasitic sporozoan bacteria (Family Babesiidae) that infect the red blood cells of the host where they multiply, causing babesiosis in most domestic animals. Those most often infected include dogs, cattle, and sheep. Several species also infect humans. Vectors include ixodid or argasid ticks. Other names for this genus are sometimes encountered in the literature (e.g., *Babesiella*, *Piroplasma*, *Nuttallia*). ***B. argentina***. *B bovis*. ***B. berbera***. *B. bovis*. ***B. bigemina***. the causative agent of bovine babesiosis; it is transmitted by ticks of the genus *Boophilus*, especially *B. annuiatus*. ***B. bovis*** (*B. argentina*; *B. berbera*) the cause of bovine babesiosis, transmitted by ticks of the genera *Boophilus*. ***B. caballi***. a cause of equine babesiosis throughout much of the world, including the southeastern United States Human infections are known. It is transmitted by ticks of the genera *Dermacentor, Hyalomma,* and *Rhipicephalus*. ***B. canis***. the cause of canine babesiosis in canids such as dogs, wolves, and jackals. The disease occurs in many tropical and subtropical areas in Central and South America, Europe, Asia, and Africa. In dogs, cases range from mild to severe. The tick *Rhipicephalus sanguineus* is the most common vector. ***B. divergens***. the cause of a disease of cattle that is similar to the disease caused by *B. bovis*. Most cases are known from western and central Europe. It also affects humans, reindeer, and perhaps other ruminants. The vector is *Ixodes ricinus*. ***B. equi***. a species that occurs in equids such as horses, mules, donkeys, asses, and zebras. Its distribution is similar to that of *B. caballi* but is more pathogenic. Humans are sometimes infected. ***B. felis***. the cause of feline babesiosis, infecting the domestic cat and other felids, mostly in Africa and India. ***B. gibsoni***. a species that infects canids such as dogs, wolves, and jackals, mainly in China, Sri Lanka, and India. The jackal suffers only mild effects, whereas the disease in dogs may be severe. ***B. microti***. a species that infects mice and voles (*Peromyscus* and *Microtus*) in North America; human cases are known from New England. The vector is *Ixodes dammini*. ***B. motasi***. the cause of causes acute and chronic babesiosis of ovids such as domestic sheep and goats in southern Europe, Africa, the Middle East, and parts of Asia. The vectors are ticks of the genera *Rhipicephalus, Haemaphysalis,* and *Dermacentor.* ***B. ovis***. a species that is widely-distributed in tropical and subtropical areas in the eastern hemisphere. It causes a disease (icterohematuria) of sheep and goats. ***B. trautmanni***. the cause of a sometimes fatal porcine babesiosis in southern Europe, central and western Asia, and Africa. It is transmitted by the tick, *Rhipicephalus sanguineus*.

Babesiidae. a family of protozoan parasites (Class Sporozoa, Order Piroplasmida) that invade the erythrocytes of a number of mammals. Ticks are the intermediate hosts. Species vary in shape, being round, oval, or piriform. Reproduction is by schizogony or binary fission.

babesiosis (piroplasmosis; Texas fever). a tick-borne protozoan disease infection of humans, livestock, and other animals caused by species of *Babesia*. In livestock and other animals, the disease is characterized by fever, weakness, listlessness, emaciation, severe anemia, diarrhea, and the presence of hemoglobin in the urine (hemoglobulinuria). The death rate is often highest among adult animals. **bovine babesiosis** (Texas fever; bovine hemoglobinuria; British redwater disease; redwater fever; tick fever). an infectious disease of cattle caused by *B. spp.* and transmitted by ticks. **canine babesiosis** (malignant fever). a disease of dogs caused by *B. spp.* **equine babesiosis** (equine biliary fever; biliary fever of horses). a disease of horses caused by *B. spp.*, marked by high fever, jaundice, and enlargement of the spleen and lymph nodes. **human babesiosis**. a rare disease of erythrocytes in humans. It is usually caused by *B. microti* in the United States and *B. divergens* in Europe. Affected individuals become pyretic, infirm, and develop a hemolytic anemia. The case mortality rate, however, is very low. In the United States, the disease is transmitted by the the northern deer tick and occurs most commonly on the coastal islands of the northeast.

babirussa. common name of swine of the genus *Babirussa*.

Babirussa (babirussa). a genus of wild swine (Family Suidae) of Celebes that are characterized by a pair of tusks that grow through the upper jaw, recurve and may eventually touch the forehead.

baby (babe, in part). **1:** an infant; a newborn (a neonate), usually a newborn child; sometimes a pet or infants of animals that have a sweet face or temperament. **2:** a modifier, as in baby beef. **3:** used in the common name of certain very small or infant-like mammals (e.g., bush baby). **4:** a coward or a crybaby. **5:** to pamper or spoil.

baccate. berrylike.

Baccharis. a genus of plant that invades the semi-arid regions of the United States and competes with forage grasses.

bacciferous. pertaining to a plant that produces berries.

Bacillaceae. a family of bacteria comprised of the genera *Bacillus* and *Clostridium*. The various species are identified on the basis of morphologic and biochemical criteria.

bacillary. pertaining to bacilli.

bacillary dysentery. shigellosis.

bacillary white diarrhea. an acute, infectious, highly fatal disease of chicks caused by *Salmonella pullorum*.

bacilli. plural of bacillus.

bacillomycin. an antibiotic produced by *Bacillus subtilis*.

bacillosis. a general infection by bacilli.

bacillus (plural, bacilli). any rod-shaped bacterium of the class Schizomycetes, especially those of the genus *Bacillus*. **grass bacillus**. *Bacillus subtilis*. **hay bacillus**. *B. subtilis*.

Bacillus. a genus of bacteria (Class Schizomycetes), various species of which synthesize one or more polypeptide antibiotics (e.g., gramicidins). Several species are pathogens of specific types of insects (*B. larvae*, *B. lentimorbus*, *B. popilliae*, *B. sphaericus*, and *B. thuringiensis*). ***B. alvei***. *B. larvae*. ***B. anthrax***. the only species of the genus that is an obligate pathogen of vertebrates, it is the cause of anthrax in humans, cattle, guinea pigs, mice, rabbits, sheep, and swine. Sporulation requires molecular oxygen as a matter of significance to the control of this pathogen. ***B. brevis***. a soil bacterium that is also found in air, dust, milk, and cheese. Some strains produce the antibiotic gramicidin. ***B. larvae*** (*B. alvei*). a pathogen that causes American foulbrood in bees. ***B. cereus***. a cause of two types of food poisoning, an emetic type and a diarrheal type, in humans and certain other mammals. ***B. mallei***. *Actinobacillus mallei*. ***B. sphaericus***. an insect pathogen associated with human and other mammalian disease, usually in already weakened or compromised hosts. ***B. subtilis*** (grass bacillus; hay bacillus). the type species of the genus, it occurs in soil and decomposing matter. Some strains produce antibiotics (bacitracin, subtilin, subtenolin, or bacillomycin). ***B. thuringiensis*** (Bt). a soil bacterium and insect pathogen that produces antibiotics that are effective against the army worm, cabbage worm, and the gypsy moth. It is used as a pesticide spray or dust. Certain genetically-engineered plants have received a gene from Bt that is responsible for the production of a protein that is lethal to a limited set of insect species by ingestion. Bt has been associated with infections of humans and other mammals. There are indications that certain beneficial insects, notably the monarch butterfly, are also harmed.

bacitracin. an antibiotic isolated from *Bacillus subtilis*.

back 40. back forty.

back cut. the final cut made in felling a tree. It is made on the side of the tree opposite that of the face cut (undercut).

back forty (back 40). **1:** an outlying, usually uncultivated acreage on a large farm or ranch. **2:** a slangy term used occasionally in friendly conversation referring to an unspecified, usually imaginary area.

back furrow. **1:** a synonym for esker. **2:** a ridge of soil thrown up by a turning plow onto a ridge previously formed by a plow working in the opposite direction.

back slope. *See* dip slope.

backboard. a board at the back of a cart or buckboard.

backburn. a fire set to burn against the wind in order to contain or control an advancing grass or forest fire.

backcross. **1a:** crossing an organism with one of its parent organisms. **1b:** a mating of an individual that is heterozygous for one or more pairs of alleles to a parental type that is homozygous for the same alleles. **2:** the offspring of a mating between a parent and its offspring.

backgrounding. the rearing of calves from weaning until they enter the feedlot.

backcrossing. in animal science, the breeding of a crossbred offspring back to a parent (usually a purebred).

backfat. the fat or amount of fat that covers the back (and ribs) of an animal or its carcass. It is a measure of yield grade.

backfat probing. measuring the thickness of the fat over the last rib of a pig using a probing knife.

backflow prevention device. a safety device that prevents water pollution or contamination by preventing the flow of water and/or chemicals in a direction opposite to that intended.

background concentration. background level.

background level (background concentration). the normal environmental concentration of a substance in the absence of inputs, directly or indirectly, by human activities. This term is often applied to ionizing radiation and to air pollutants that have both natural and anthropogenic sources. *See* background radiation.

background radiation. **1:** the sum of all naturally occurring radioactive (ionizing) radiation at a given location, including elementary particles from outer space (cosmic radiation) and terrestrial radiation from naturally occurring isotopes. **2:** the level of radioactive radiation (natural and anthropogenic) that would occur at a given location if the level were not influenced by a specified source.

backpack sprayer. a sprayer with plastic containers mounted on a backpack frame. It is used to apply herbicides or other chemicals.

backwash. **1:** backward flow of air or water, as from the propeller of a boat or aircraft. **2:** the result or aftermath of an incident. **3:** upward flow of water through a filter bed that cleans the filter after it is exhausted.

bacon. the back and sides of a pig that has been cured in brine or smoked. **green bacon**. unsmoked bacon.

bacon type hog (baconer). a hog bred and reared to produce a large proportion of lean meat as opposed to fat.

baconer. bacon type hog.

-bacter. a suffix meaning rod; often used in generic names of bacteria (e.g., *Aerobacter*).

bacteremia. **1a:** the presence of bacteria in circulating blood. **1b:** a disease caused by bacteria in circulating blood. **gram-negative bacteremia**. the presence of gram-negative bacteria in circulating blood. **gram-positive bacteremia**. the presence of gram-positive bacteria in circulating blood.

bacterial. of, pertaining to, or caused by bacteria.

bacterial decomposition (neutralization) of hazardous materials. the biodegradation, *q.v.*, of hazardous materials in the environment (e.g., hazardous wastes, herbicides and pesticides, petroleum) by the use of genetically engineered, oxidizing bacter in order to render them safe by conversion to simpler, and usually less toxic, molecules.

bacterial kidney disease (corynebacterium kidney disease; dee disease). a disease of bony fish including economically important cultured fish. The causative agent is a gram-positive bacterium *Renibacterium salmoninarum*. Infected fish become exophthalmic, blind, and emaciated; visceral lesions appear, especially in the kidney or body wall. Surviving fish are carriers of the disease.

bacterial leaf spot. *See* blackroot (def. 4).

bacterial neutralization of pesticides. *See* bacterial decomposition of hazardous materials.

bacterial vitamin H^1. *p*-aminobenzoic acid.

bacterial wetwood. a disease that affects poplar, elm (*Ulmus*), birch (*Betula*), maple (*Acer*), apple (*Malus*), and mountain ash (*Sorbus*). The disease is characterized by the oozing from cracks in the bark or from injuries (e.g., due to pruning, nail holes) of a brown slime flux (*See* slime flux). This disease is caused by infection of the heartwood by any of a number of common anaerobic soil bacteria. The bacteria absorb and metabolize nutrients from the wood, releasing fatty acids, methane, and carbon dioxide. The fatty acids become rancid, soaking the wood with smelly oils, while the gases exert pressure that forces the liquid out through the openings in the bark. The extruded liquid turns brown on exposure to air. The liquid absorbed by the heartwood raises the pH and increases resistance to decay.

bacterial wilt. any of a number of bacterial diseases of green plants characterized by the production of slime that obstructs the water-conducting tissue of the affected plant.

bacterial wilt of alfalfa. a disease of alfalfa, especially in damp areas, caused by *Corynebacterium insidiosum*. Affected plants yellow and bear dwarfed stems and leaves; the crown may rot.

bacterial wilt of corn (Stewart's disease). a bacterial disease especially of sweet corn in which the entire plant is affected. The leaves secrete a yellow exudate that results in long wavy streaks on the surface.

bacterial wilt of potato. the most serious

disease of potatoes in warm climates and a family of diseases that includes numerous crops, certain weeds, and other plants in some thirty families. The infective agents are various races of *Pseudomonas* (=*Ralstonia*) *solanacearum*. Plants that are highly susceptible to race 1 of *P. solanacearum* include the bell and chili peppers, eggplant, peanut, potato, tobacco, and tomato. Race 3 causes most cases of bacterial wilt of potato. It is insidiously transmitted chiefly as a latent infection of the seed tuber. It causes severe crop losses in tropical, subtropical, and warm temperate regions of Asia, Africa, and Central and South America. There is no practical chemical control of this disease. The only measures to date that are to some degree effective include integrated control components such as quarantine together with a restriction of the production of seed potatoes to prevent the spread of a disease outbreak, regulating the production of seed potatoes, and limiting the commercialization of ware potatoes.

bactericidal (bacteriocidal). **1:** destructive to, or inhibiting the growth or activity of bacteria. **2:** of or pertaining to bactericides.

bactericide (bacteriocide). **1:** any agent (e.g., an antibiotic) that destroys bacteria. *Cf.* germicide. Hypochlorite, various phenolics, mercuric chloride, and organic mercurials are commonly-used bactericides. **2:** bactericidal. **specific bactericide**. a bactericidal immune serum that acts on a single species of bacterium or a group of closely-related bacteria.

bactericidin. any substance in blood plasma that kills bacteria.

bacterin. A suspension of killed or weakened bacteria used as a vaccine to increase resistance to a specific disease.

bacteriochlorophyll. a light-absorbing pigment, closely related to chlorophyll, that occurs in photosynthesizing green sulfur and purple sulfur bacteria. The magnesium at the center of the tetrapyrrole ring is replaced by manganese. *See also* photosynthesis.

bacteriocidal. bactericidal.

bacteriocide. bactericide.

bacteriocin. any of a group of antibacterial proteins produced by various bacteria that selectively kill or inhibit, but do not lyse, other taxa of bacteria. They complex with specific sites on the cell wall, inducing a specific metabolic block. The different bacteriocins are often named for the group of bacteria that produce them. Thus, those produced by *Escherichia coli* are termed colicins, those by *Staphylococcus* are staphylococcins, etc.

bacteriocinogenic. producing bacteriocins.

bacteriolysant. **1a:** having the ability to lyse bacteria. **1b:** an agent that lyses bacteria. *See* lysin, bacteriolysis. *Cf.* bacteriolysin.

bacteriolysin. an antibacterial antibody that lyses bacteria. *See* bacteriolysis, lysin. *Cf.* bacteriolysant.

bacteriolysis. **1:** the dissolution or destruction of bacteria by any agent. *See* lysis. **2:** destruction of a bacterium by a specific bacteriolysin, with the release of the cell contents.

bacteriolytic. pertaining to, denoting, promoting, or able to destroy bacteria by lytic action.

bacteriolyze. to effect bacteriolysis.

bacteriophage (phage; phage particle). a virus that parasitizes bacteria, often destroying the host. Altered forms are used as vectors to clone DNA.

bacteriosis. a bacterial infection.

bacteriostat (bacteriostatic agent). any agent (e.g., low temperature, dyes, food preservatives, some antibiotics, weak antiseptics) that reversibly inhibits the multiplication of bacteria.

bacteriostatic. **1:** preventing the multiplication of bacteria without killing them. **2:** said of a substance having the ability to prevent multiplication of bacteria without killing them.

bacteriostatic agent. bacteriostat.

bacteriotoxin (bacterial toxin). an often ambiguous term; compare definitions 1 and 2. **1:** any toxin produced by a bacterium. There are a large number of such toxins. They are classed as bacterial endotoxins and exotoxins. Such toxins play important roles in a large number of disease processes and in food poisoning. **2:** sometimes used to denote a toxin that kills or inhibits the growth of bacteria.

bacterium (plural, bacteria). the common name of any of a large, extremely diverse assemblage of one-celled microorganisms that lack a distinct nucleus (prokaryotic) and usually have a rigid cell wall. Reproduction is by binary fission or asexual spores. The transfer of genetic material may be effected by conjugation (a sexual process) or mediated by virus or bacteriophage (transduction). Included are the actinomycetes, cyanobacteria (blue-green bacteria), eubacteria, iron bacteria, mycoplasmas, rickettsiae, spirochaetes, and sulfur bacteria. In terms of numbers and diversity of habitats, bacteria must be considered one of the most successful of life forms. Many are useful to humans, for example, as aids in food processing, pollution control, sewage treatment, waste treatment and genetic engineering). Many are, however, parasitic or otherwise harmful, and some secrete powerful toxins (e.g., botulinum toxin). *Cf.* microbe. *See also* exotoxin, endotoxin. **blue-green bacteria**. cyanobacteria. **denitrifying bacteria**. *See* denitrification.

Bacterium. a generic name of certain bacteria, *q.v.* It has been placed on the list of rejected names by the Judicial Commission and the International Committee on Systematic Bacteriology of the International Association of Microbiological Societies.

Bacterium albilineans. *See* leaf scald.

Bacterium coli. former name of *Escherischia coli*.

Bacteroidaceae. a family of non-sporeforming, motile and nonmotile, Gram-negative bacteria (Order Eubacteriales) that contain mostly obligately anaerobic, rod-shaped species that vary in size from minute forms to long, filamentous, branching filterable forms. These microbes occur chiefly in the intestinal tracts and mucous membranes of warm-blooded animals; some may be pathogenic. Body fluids are often essential to reproduction.

Bacteroides. a genus of Gram-negative, obligately anaerobic, motile or nonmotile, nonsporeforming bacteria (Family Bacteroidaceae). The motile species are peritrichous. ***B. nodosus***. the causative agent of foot rot in sheep and goats. **B. putredinis**. a species isolated from farm soil, humans feces, and foot rot of sheep.

bacteroidosis. a disease caused by bacteria of the genus *Bacteroides*, *q.v.*.

badlands, bad lands. **1:** eroded, deeply dissected land that is barren of vegetation and is unsuitable for agriculture. Badlands are found in North and South America and are characterized by rough ridges, crests and pinnacles, narrow ravines, and mesas. **2:** The Badlands; a heavily eroded, arid region in southwestern South Dakota and northwestern Nebraska.

BAF. bioaccumulation factor. *See* factor.

baffle. **1:** to perplex; to cloud, muddle, or obscure. **2:** any structure placed in a pathway or conduit that impedes or redirects flow or passage or that dissipates energy. **3:** plants situated and grown so as to make a wall or partition.

bag. **1:** sack; packet; pouch; purse; valise. **2:** to capture, take, or possess something. **3:** a unit of measurement in the coffee trade: 1 bag = 60 kg. **4:** an udder, especially that of a dairy cow or dairy goat. **5:** an external scrotum. **6:** to cut wheat with a sickle.

bag limit (hunter's bag). the total number of a game species that one is allowed to have in possession at one time according to law.

bagasscosis. bagassosis.

bagasse. **1:** the fibrous residue from sugarcane stalks that remains following crushing to extract the juice. Mixed with molasses it is used in livestock feeds. *See also* bagassosis. **2:** sometimes applied to sugar beet, sorghum, and sisal residues.

bagassosis (bagasscosis). a respiratory disorder resulting from inhalation of bagasse, the residue of sugarcane that remains following the extraction of sugar. The proximate cause of the disease is the inhalation of the spores of a fungus that infects sugarcane. An asthma-like attack usually ensues several hours after contact in an individual who has worked with cane for several months. With repeated attacks, the lungs become scarred with development of a pneumoconiosis.

bagot. a rare, long-haired breed of goat; both male and female have horns. The preferred coloration is a black head, neck, and shoulders and a white body.

bagpod. a plant of the genus *Sesbania*.

bagworm. common name for the larva moths of the family Psychidae.

bahama grass. *Cynodon dactylon*.

bahia grass. *Paspalum notatum*.

bail. **1:** to scoop out; empty; drain. **2a:** bail out; to give up or quit. **2b:** bail out; to help a person out of trouble or out of a difficult situation. **3:** a guarantee or security; a bond (as a bail bond). **4:** a bar that separates horses in a shed. **5:** a portable or mobile milking shed. **6:** the handle of a bucket.

Baileya multiradiata (desert baileya; cloth of gold). a small showy annual or perennial plant (Family Compositae) that is indigenous to the southwestern deserts of the United States. It occurs locally on dry sandy and graveled soils from Mexico north into Texas and California. Sheep are poisoned by ingestion of this plant and, in the absence of treatment, may die within several days or a few weeks following the appearance of symptoms. Signs of poisoning include sluggishness, anorexia, depression, and drooling of green saliva. There are occasional reports of red urine and a tendency to develop pneumonia among affected animals.

bait. material, often food, used to lure prey for the purpose of capture (e.g, by a trap or snare), management (e.g., placement of radio collars, health maintenance), or destruction. The bait itself may be poisoned or contain an immobilizing agent. Some baits are taxon specific (e.g., ant and roach baits); others are not. Nonspecific baits may kill valued, rare, or endangered species. A common example is the use of poisoned meats or dead animals to lure and kill a pest species (e.g., coyotes, snails) may also harm nonpest species (e.g., pets, golden eagles). Organisms other than humans that employ bait are fairly numerous (e.g., some spiders, angler fish, some fireflies). **snail bait**. any baits used to attract and kill snails. Such baits, until recently, contained ingredients that were attractive to many dogs. Newer formulations in the United States have dramatically decreased the number of dogs poisoned by metaldehyde, q.v.

bait shyness. the tendency for an animal to avoid a bait, the area around the bait, or the manner in which a bait is presented. Bait shyness may occur following a prior encounter with a bait.

baker. any large potato intended for baking.

balance. **1a:** equilibrium; symmetry. **1b:** the state or condition where two sides are equal with respect to some function or structural arrangement (e.g., in weight, length). **2:** stability; steadiness. **3:** remainder; remnant; residue. **3b:** surplus. **4:** a device used to measure weight (a measure or scale). **5:** the visual appeal of an animal due to a harmonious and symmetric relationship among body parts. **6:** to measure, fairly distribute, or apportion (e.g., to apportion the ingredients in a feed ration). **7:** the proportionate or relative amounts (e.g., of materials, commodities, resources) of land, labor, and capital under suitable management. bal ance in agriculture is the use resources that optimize sustainable productivity and profitability. One is thus concerned with the balance among resources such as land, equipment, capital, labor, and management.

balanced diet. *See* balanced ration.

balanced operation. a system of range management that provides sufficient feed and forage resources to livestock and wildlife to sustain life and satisfactory maintenance at all seasons of the year. To be successful, such an operation must secure, in an average year, a suitable margin of profit.

balanced ration (balanced diet). a daily allowance of feed that provides an animal with all essential nutrients in the correct proportions to maintain the animal and support growth, reproduction, and, as appropriate, functions such as lactation and gestation.

Balanites aegyptiaca. a genus of trees indigenous to the Near East, the berries of which contain a biologically active principle that is lethal to fish, tadpoles, mollusks, miracidia, cercariae, and tadpoles. This substance is used in drinking water as a prophylactic against schistosomiasis.

Balantidium coli. a parasitic bacterium of the intestinal tract of mammals, especially swine; infected humans can experience severe dysentery.

bald. **1a:** having little or no hair on the scalp. **1b:** to lose part or all of the hair of the scalp, especially in humans. **2:** lacking hair-like structures as in bald wheat (wheat that lacks the beard). **3:** bald; bald headed; bald faced; descriptive of an animal that has the appearance of baldness. For example, white on the face or head (e.g., of a bald eagle). A bald horse is one with a white stripe on the face from the forehead down to the nostrils. **4:** a small elevated treeless (usually grassy) site as on a dome or knob of an otherwise forested hill or mountain.

baldcypress. *Taxodium distichum*.

baldness. the condition of being bald.

Baldwin. a large, red, winter apple of good quality grown in New York and New England.

baldy. **1:** a polled or naturally hornless animal (usually said of bovines). **2:** a uncivil nickname for a person who is bald.

bale. **1:** a large, compressed bundle or pack of produce such as hay, straw, cotton, or wool that is usually bound with twine or wire. **2:** a standard pack for cotton. **3:** to gather, shape, and compress hay, straw, or grass into bales. **big bale**. a large bale of hay, straw, or silage that weighs 600-900 kg. **compressed bale**. **1:** a bale of cotton which has been compressed to form a small bale having a density of 22-1/2 lbs ft^3. **2:** a bale of hay reduced in size by compression. **gin bale**. a bale of compressed lint of approximately 500 lbs coming directly from the cotton gin.

bale accumulator. an attachment to a standard hay baler that gathers and automatically unloads about 8-12 bales.

bale chopper. a machine that chops up bales of hay, straw, or silage to provide feed or bedding of livestock.

bale ejector. an attachment to a standard hay baler that eliminates the need for hand loading by throwing bales into a trailing wagon.

bale mover. an apparatus that moves large round bales of hay.

baler. a machine or person who packages a material such as hay or straw into bales. **hay baler**. a person or machine that compresses hay into bales.

baling: 1: gathering, compressing, and shaping hay, grass, or straw into bales. **2:** compressing solid waste into blocks to reduce volume and simplify handling.

balk. **1:** a narrow strip of untilled land left to mark the boundary between fields. **2:** to shy; said of an animal that refuses to move.

ballistic separator. an apparatus that sorts organic from inorganic matter for composting.

balloon frame. a type of house or barn built on a wooden frame and lacking heavy timbers.

ballotin (ballottement; balloting). A method of palpating for a loose or floating object within the body cavity of an animal.

ballottement. ballotin.

ballotting. ballotin.

balsam. a resinous mixture of varying composition secreted by various species of evergreen trees. Many have a pleasant aroma; all are water-insoluble but are soluble in organic solvents. Their applications are diverse including use in fragrances, flavorings, cough syrups, the manufacture of chocolates, varnishes and lacquers, the manufacture of photographic paper, perfumery, medications, and microscopy.

balsam apple. wild balsam apple. *See Echinocystis lobata*.

balsam fir. *Abies balsamea*.

bamboo. **1:** any plant of the genus *Bambusa*. **2:** the stem of a bamboo plant.

Bambusa (bamboo). a genus of primitive grasses (Family Gramineae) with hollow, woody stems that are often thick enough for use in light construction such as platforms, pedestrian bridges, and abodes. The young shoots are succulent, and those of some species are used for food. ***B. vulgaris***. a species that grows mainly in temperate and subtropical regions of China, Japan, and North and South Korea. The shoots are cut, cooked, and eaten as vegetables.

banana. **1:** *Musa paradisiaca* var. *sapientum*. **2:** the soft, sweet, pulpy fruit which is high in food value and borne in clusters on this plant. Each fruit is 4 to 9 inches in length and is about 1 inch in diameter.

banco. a term used in the southwestern United States meaning a bank or emergent deposit of sand or silt produced by altered flow rates and shifting channels of a river.

band. **1:** a gathering or assembly. **2:** a belt, cord, or strap. **3:** a stripe, streak, or ribbon. **4a:** a group of animals, especially primates, as a band or troop of gorillas. **4b:** a flock of sheep.

band application. the spreading of herbicides, pesticides, fertilizers, or other chemicals over or next to each row of plants in a field.

band day. the amount of forage that is required by a band of sheep of a given size during one day on a designated plot of land at a given time or season.

banded cucumber beetle. *Diabrotica balteata*.

banded garden spider. *Argiope trifasciata*.

bane. **1a:** scourge, torment, or plague. **1b:** curse. **2:** a poisonous plant. **3:** a poison of plant origin, used especially in reference to those produced by specific plants such as those produced by *Aconitum* and *Arnica*. **4:** any poison (not a recommended usage). **5:** occasional vernacular usage: death, destruction. **6:** (obs.) to kill, especially by poisoning. **7:** (archaic) to afflict, injure or harm.

baneberry. **1:** the poisonous fruit of any plant of the genus *Actaea*. **2:** a plant of the genus *Actaea*.

Bang's disease. brucellosis of cattle.

bangtail. **1:** a racehorse, named for the usually banged or closely-docked tail of running horses in former times. **2:** a wild horse.

bank full height. that elevation of the bank of a stream that defines the cross-sectional area of an active stream channel.

banking. 1: preparation of a mound, ridge, embankment, or slope. **2:** the mounding of earth over the crown or about the base of a plant to retain moisture and protect from excessive heat.

bantam (banty). **1:** a small breed of domestic fowl; most are about half the size of the common domestic fowl. **2:** any small fowl. **3:** sometimes applied to a small person who is cocky (banty rooster).

banteng. a breed of cattle, the females of which are bright reddish brown and the bulls, black. They are used for meat and as draft animals in southeast Asia.

banty. bantam.

baobab. *Adansonia digitata*.

bapedi. a breed of black and white dairy cattle of Eastern Transvaal.

Baptisia (wild indigo; false indigo). a genus of leguminous herbs (Family Fabaceae, formerly leguminosae). ***B. leucantha*** (wild indigo). a species poison to grazing livestock. It contains quinolizidine alkaloids. ***B. tinctoria*** (wild indigo, rattleweed, horsefly weed). a hardy perennial herb with bright yellow flowers. The leaves and shoots are poisonous, causing severe diarrhea and anorexia in cattle. The active principles are cytisine and an alkaloid, quinolizidine.

baptitoxine. cytisine.

bar. **1:** a unit of pressure equal to 1 megadyne ($=10^6$ dyne)/cm^2 in the CGS system, 0.987 atmosphere, or 10^5 Pascals ($=10^5$Nm/m^2) in the SI system. **2:** a barlike section of bone or other tissue that connects two or more similar structures. **3a:** to block. **3b:** to ban or prohibit. **connector bar**. one of two converging ridges on the lower surface of a horse's hoof that are united by the frog and join the sole anteriorly.

barb. 1: any of numerous hairlike structures borne on the shaft (rachis) of a feather. **2:** in botany, a small, stiff, hairlike structure with hooks. **3:** any of the inner row of teeth in the peristome of certain mosses. **4:** the spines on the awns of grasses. **5:** any of the small sublingual projections of the mucous membrane near the openings of the submaxillary glands. **6:** a breed of horses that originated in Barbary and apparently was introduced into Spain by the moors. It is related to the Arab and probably to all modern breeds. **7:** a sharp projection on a fence wire.

barba amarillo. *Bothrops atrox*.

Barbados aloe. a common name of *Aloe vera*.

barban (4-chlorobut-2-ynyl-*N*-(3-chlorophenyl)-carbamate). a highly selective carbamate herbicide that retards germination in grasses. It is used on barley (*Hordeum*) and many varieties of winter wheat (*Triticum*) to control wild oats (*Avena fatua*). Applications up to 0.75 kg/ha to crops at the 2-3 leaf stage kill this grass with no significant damage to cereal grains and broadleaved crops. Barban is a skin irritant and is appears to be moderately toxic to humans.

barbed wire (bob wire; bobbed wire). single stranded wire, used largely in fencing, made of twisted wires to which wire barbs are connected at short intervals along its length. *See* barb.

barberry. **1a:** any plant of the genus *Berberis*. **1b:** the berry of such a barberry plant.

barbiturate. **1:** any of a class of organic compounds derived from barbituric acid (e.g., amobarbital, barbital, nembutal, pentobarbital, phenobarbital, secobarbital, secanol). All are CNS depressants used therapeutically as tranquilizers, hypnotics, anticonvulsants, and given as an aid to the induction of general anesthesia. **2:** denoting or having the properties of a barbiturate.

barbule. 1: one of the (usually interlocking) processes of the barbs of a feather.

bare root grafting. the joining of a scion to a plant by grafting it onto bare rootstock prior to onset of growth.

bare root transplantation. transplantation of a plant that has been removed from the ground with little soil remaining on the roots.

bareback. **1:** a fowl which is not yet fully feathered on the back. **2:** riding bareback; riding an unsaddled horse, ass, or mule.

bargaining association. a farm cooperative organized with the chief aim of favorably influencing farm prices and other aspects of commerce.

bargaining power. the ability, usually of an organized group, to favorably influence, for example, business transactions or market conditions.

barilla. common name of any plant of the genus *Halogeton*, especially *H. glomeratus*.

barium (Ba)[Z = 56; A_r = 137.36]. a toxic, reactive, divalent alkaline earth element; it is a silver-white powder or sometimes malleable metal. The powder is flammable at room temperature and should be stored under an inert gas. It is extremely reactive with ammonia, water, halogens, and most acids. The Ba^{2+} ion is myotoxic and is readily absorbed from the gastrointestinal tract. Thus, even partially soluble barium salts are highly toxic, causing severe hypokalemia (reduction of phosphorus in blood). Signs and symptoms in humans commonly include salivation, vertigo, nausea, vomiting, sweating, abdominal distress, violent and bloody diarrhea, bradycardia, cardiac arrhythmias, tinnitus, convulsions, and sometimes paralysis. Death from respiratory failure and cardiovascular collapse may occur. Signs in livestock include profuse salivation, sweating, violent peristalsis and convulsions, cardiac arrhythmias, palpitations, and sometimes paralysis. *See* barium.

barium carbonate. a solid, white, water-insoluble salt, $BaCO_3$, that occurs in nature as the mineral witherite. It is used as a rodenticide.

barium fluosilicate. a compound, BaSiF, used as an insecticide and rodenticide.

barium hydrate. barium hydroxide, octahydrate.

barium hydroxide (barium hydrate; caustic baryta). a white or colorless powder or crystalline compound, $Ba(OH)_2$. It readily dissolves in water to form a strong caustic base. All forms are poisonous. $Ba(OH)_2$ has many agricultural and commercial uses, e.g., in pesticides, in the saponification of fats, as a test for sulfides, and in the manufacture of alkali and glass. **anhydrous barium hydroxide**. the commercially available preparation; it usually contains 92-95% of the monohydrate. **barium hydroxide, monohydrate** (barium monohydrate). a white, slightly water-soluble, powder,

$$Ba(OH)_2 \cdot H_2O$$

It is soluble in dilute acids. **barium hydroxide, octahydrate** (barium hydrate; barium octahydrate; caustic baryta). a white powder,

$$Ba(OH)_2 \cdot 8H_2O$$

It is soluble in water, ethanol, and ether. It is used chiefly in organic preparations and analytical chemistry. **barium hydroxide, pentahydrate** (barium pentahydrate). toxic white flakes,

$$Ba(OH)_2 \cdot 5H_2O$$

Its applications are the same as those of the octahydrate.

barium monohydrate. barium hydroxide, monohydrate.

barium monosulfide. barium sulfide.

barium octahydrate. barium hydroxide, octahydrate.

barium pentahydrate. barium hydroxide, pentahydrate.

barium poisoning of livestock. poisoning of livestock by excessive use of products that contain barium. Affected animals sweat and salivate profusely and may convulse and experience a generalized paralysis.

barium sulfide (barium monosulfide; black ash). a poisonous, flammable, grayish-yellowish-green powder or lumps, BaS. It is used, for example, as a depilatory to remove hair from hides and as a flame retardant.

bark. **1:** the tissues, collectively, that lie outside the vascular cambium of the stems and roots of woody plants. The term is sometimes applied only to the tissues that arise from and lie external to the phellogen when such tissue is exposed by sloughing off of the epidermis. **2:** the outer covering of the trunk and branches of a tree or woody bush. It is tough or even hard and may be corky, papery, or leathery. **3:** to remove the bark from a tree or log. **4:** to surround with bark. **5:** to tan, cure, or dye using extracts from bark (e.g., tannin). **6:** a loud, sharp, vocalization characteristic of a dog and certain other animals.

barkhan. an isolated sand dune with a characteristic crescentic shape.

barley. *See Hordeum vulgare.*

barley yellow dwarf. a viral disease of cereals transmitted by certain species of aphids. It is characterized by a paling and yellowing of the leaves that starts at the tip.

barn. a farm building used chiefly to house or store hay, grain, livestock, farm machinery, and tools.

barn lot. barnyard.

barn raising. **1:** the process of constructing (raising) a barn. **2:** a gathering of neighbors convened to help erect a barn and to enjoy the accompanying socialization including food and drink.

barn rat. **1:** *Rattus norvegicus*. **2:** *See* agoraphobia.

barnevelder. a breed of domestic fowl.

barny. of or pertaining to an off-flavor of a dairy product marked by an odor like that of a cow barn and persistence on the palate.

barnyard (barn lot; farmyard). a usually fenced-in area of land about a barn.

barnyard fowl. any of the common domesticated species of fowl.

barnyard grass. *Echinochloa crusgalli*.

baro-. a prefix meaning pressure.

barograph (barometrograph). an instrument that measures and continuously records barometric (atmospheric) pressure.

barometer. an instrument that measures barometric (atmospheric) pressure.

barometrograph. barograph.

barophilic. of or pertaining to an organism that fails to grow or thrive in the absence of high atmospheric pressures.

barrack. a strong alcoholic liquor distilled mainly from dates, rice, and sap of the coconut palm.

barred. **1a:** blocked. **1b:** prohibited or banned. **2:** a pattern of alternating white and dark stripes on the plumage of the domestic fowl, *Gallus domesticus*. **3:** covered or protected by metal bars. **4:** to ban or prohibit.

barred Plymouth Rock (barred rock). a breed of domestic fowl valued for both its egg production and meat. Individuals grow to a weight of about 7.5 lbs (hens) or 9.5 lbs (cocks). The plumage is striped horizontally with black or nearly black alternating bars.

barrel (bbl.). **1:** a cylindrical wooden container with slightly bulging sides made of staves held together with flat, parallel, metal hoops; butt; cask; keg (a small barrel), **2:** the amount of material that such a container of a standard size can hold. **3:** any container, case, or part that is similar to a wooden barrel in form or function (hogshead, tub, vat, aluminum beer keg). **4:** the tube of a rifle or gun. **5:** any structure having the form of the vault of a barrel. **6:** the trunk or chest of a quadruped (e.g., of a horse, cow, primates, and certain other animals. **7:** to place or pack in a barrel or barrels. **8:** to polish or finish materials such as agates or metal items by tumbling in a barrel.

9: to speed, race, or zoom (e.g., he barreled his way through the crowd; the car barreled down the highway). **standard barrel**. a barrel that originated in the United States in association with flour milling; it was accepted as the common container for shipment. It is a wooden barrel that holds a net weight of 196 lbs of flour; 31 United States gallons (119 L) in the case of liquids; 42 U.S. gallons (159 L) for petroleum; and 105 U.S. dry quarts (115 L) for dry materials.

barren. **1a:** of or pertaining to a region or area of land that is open, bare, and/or empty of apparent life. **1b:** arid; desert-like; biologically unproductive. **2a:** pertaining to a sterile, infertile animal; a female unable to conceive or become pregnant; unable to produce offspring. **2b:** said of an female mammal that has undergone substantially more than one sexual or breeding cycle without bearing offspring. **2c:** childless. **2d:** not pregnant. **3:** denoting a plant that is unable to bear fruit. **4:** sterile; a nonbreeder. *See* sterility (def. 1b).

barren brome. *Bromus sterilis*.

barrenness. the condition of being barren in any sense, *q.v.*

barrens. an area of low plant diversity (e.g., a pine barrens, or sand barrens) relative to adjacent areas due, for example, to factors such as an unfavorable microclimate or poor soil conditions.

barrier. **1:** an obstruction or impediment (e.g., stratum corneum, blood-brain barrier). **2:** an obstacle (biological or nonbiological), within a living system, to the passage of substances. A number of examples are defined below. **blood-brain barrier** (BBB; blood-cerebral barrier; blood-cerebrospinal fluid barrier, hematoencephalic barrier). a barrier that separates the blood that circulates throughout the body from tissues of central nervous system (CNS). The BBB is impermeable to most ions and protein-bound chemicals, but allows passage of lipid-soluble molecules. It thus effectively protects tissues of the brain and spinal cord from the intrusion of most harmful molecules. **blood-cerebral barrier**. blood-brain barrier. **blood-cerebrospinal fluid barrier**. blood-brain barrier. **blood-testis barrier**. a permeability barrier to the passage of blood-borne materials into the seminiferous tubules of the testis. **filtration barrier**. a barrier that separates the blood in the glomerular capillaries of the kidney from that of the capsular space of the renal corpuscle. **gastric mucosal barrier**. a physiologic barrier that limits the diffusion of nonionic materials through the gastric mucosa. Certain agents (e.g., aspirin, bile salts, organic acids) impair the effectiveness of this barrier. **hematoencephalic barrier**. blood-brain barrier. **placental barrier** (placental membrane). the mammalian placenta physically separates fetal from maternal blood and limits the passage of large blood-borne molecules between the two. In at least some species, this barrier impedes the passage of many toxicants to some extent. Its effectiveness, however, probably varies considerably among placental mammals as a function of the number of cell layers (one to six).

barrier coating. a layer of a material (e.g., grout, caulk) applied to a surface to prevent passage of something through the surface. Barrier coatings are sometimes used, for example, to reduce or prevent corrosion or oxidation of metal surfaces or to protect against chemical impacts on various materials. *See also* barrier substance.

barrier substance. a substance applied to the skin as a barrier to irritants (often a cream or lotion). *See also* barrier coating.

barrosa. a breed of reddish-brown cattle with large, curved horns; used as a draught animal and for meat in northern Portugal.

barrow. a young, castrated male hog. Such animals are raised for pork.

barthrin (2,2-dimethyl-3-(2-methyl-1-propenyl)-cyclopropanecarboxylic acid (6-chloro-1,3-benzodioxol-5-yl)methyl ester; 2,2-dimethyl-3-(2-methylpropanyl)cyclopropanecarboxylic acid 6-chloropiperonyl ester;6-chloropiperonyl-2,2-dimethyl-3(2-

methylpropenyl)-cyclopropanecarboxylate; 6-chloropiperonyl chrysanthemumate; chrysanthemummonocarboxylic acid 6-chloropiperonyl ester). a synthetic analog of allethrin used as an a synthetic insecticide with low toxicity to humans and other mammals. It is an analog of allethrin, *q.v.*

barton. **1a:** the land of an estate. **1b:** a farm yard. **1c:** a poultry yard. **2:** farm buildings attached to a farm house.

Barzona. a breed of beef cattle developed in Arizona. The genetic background of the breed includes Africaner, Hereford, shorthorn, brahman, and angus stocks.

basal. **1a:** fundamental; basic; **1b:** elementary. **2:** of, pertaining to, at, attached to, or growing from the base of. *Cf.* apical.

basal application. the application of a pesticide or other chemical just above the soil line on plant stems or tree trunks.

basal area. **1:** the cross-sectional area of a tree near its base determined by the diameter at breast height (1.4 m above the ground). **2:** ground cover; the total area of ground covered by trees in a stand as the sum of the individual basal areas of trees occupying a particular area. This is usually given in square feet per acre. **3:** basal cover, ground cover; the surface area of soil occupied by a plant as measured near the ground.

basal feed. feed provided to livestock chiefly for its value as an energy source.

basalt. **1:** a crystalline igneous rock, volcanic in origin, that forms lava flows or minor intrusions. It is composed of feldspar and dark minerals such as pyroxine, augite, magnetite, and sometimes olivine. Native basaltic soils are dark and usually productive. **2:** basaltic rock; fine-grained igneous rock that contains basalt, diabase, and dolerite, and sometimes andesite. If the latter mineral is present, the rock is dark. **3:** in casual usage, any dark, fine-grained igneous rock.

basaltic rock. basalt (def. 2).

base. **1a:** basis. **1b:** core; essence. **2:** support; foundation; footing; the base or lower part; that part of an object which supports it; the part of a structure situated opposite the apex. **1c:** the part of an organ, extremity, or other body or plant part that lies nearest to its point of origin or attachment. **2a:** any substance with a pH greater than 7. **2b:** any substance that releases hydroxyl ions, OH^-, or other ions that combine with hydrogen ions in aqueous solution. Such solutions have a pH greater than 7. **2c:** any electropositive substance (a cation) that tends to unite with electronegative substances (an anion) to form a salt. **3:** in biochemistry, any nitrogenous organic compound such as an amine, an alkaloid, a ptomaine, or any of the purine and pyrimidine constituents (adenine, guanine, cytosine, thymine) of nucleotides (DNA, RNA) that act as a Lowry-Brønsted base. *See* base pair. **4:** in pharmacology, the main ingredient of a mixture. **Brønsted base**. Lowry-Brønsted base. **Lewis base**. any substance that tends to donate an electron pair (e.g., ammonia, the hydroxyl ion). **Lowry-Brønsted base** (Brønsted base). any substance that tends to accept a proton.

base acreage. average acreage supporting a particular crop on a farm. The planted acreage of the previous year is usually the accepted base acreage, but a 2-year average is sometimes used. **adjusted base acreage**. the base acreage adjusted for major environmental factors that affect acreage or productivity (e.g., fire, flood, drought, crop rotations). This value is used, for example, in acreage limitation programs.

base case. current socioeconomic conditions in relation to existing land or forest management strategies and the socioeconomic conditions to be expected if the existing strategy continues.

base level. the point at which erosion by running water stops (e.g., sea level, the margin of a lake).

base narrow. descriptive of a horse that stands with legs vertical but with front or rear legs close together.

base pair. a pair of bases (a purine and a pyrimidine) cross-linked by hydrogen bonds; adenine always pairs with thymine, and guanine with cytosine. Adenine and thymine are paired in the DNA molecule, guanine and cytosine in both the DNA and RNA molecule, and adenine and uracil in RNA. Such pairs hold the two helical chains of the Watson-Crick model together, conferring the helical configuration to the molecule. A base pair is the unit of measurement for DNA and RNA sequences. Since each chain in the DNA molecule serves as a template for the other, any change in a base pair produces a genetic mutation. *See also* base (def. 3).

base saturation (b.s.; B.S.). the extent to which the exchange sites of a material (e.g., soil) are occupied by exchangeable bases. It is expressed as a percent of cation exchange capacity.

base saturation percentage. the degree to which the adsorption complex of a soil is saturated with exchangeable cations other than hydrogen and aluminum.

base unit. any fundamental unit of the International System of Units, *q.v.*

base wide. descriptive of a horse that stands with legs vertical but with front or rear feet wide apart.

baseline data (baseline information). data that provide a standard (baseline) against which future measurements can be compared.

baseline information. baseline data.

bashi. a small, short-horned, East African breed of cattle that varies in color from red to fawn or black.

basi-. a prefix meaning most ventral; of or pertaining to a basal location.

basic amino acid. *See* amino acid.

basic anhydride. *See* anhydride.

basic commodity. *See* commodity.

basidia. plural of basidium.

basidiocarp. the fruiting body and most prominent feature of the higher basidiomycete fungi (Subclass Homobasidiomycetes). *See* Basidiomycetes.

basidioma. basidiocarp.

basidiomata. plural of basidioma. *See* basidiocarp.

basidiomycete. any representative of the fungal class, Basidiomycetes.

Basidiomycetes. a large class of mycelial fungi characterized by a spore-forming organ (basidium), which is the result of sexual reproduction. Included are smuts, rusts, jelly fungi, mushrooms, toadstools, puffballs, stinkhorns, bracket fungi, and bird's nest fungi. There are two subclasses: Heterobasidiomycetes (e.g., rusts and smuts), in which the basidium is deeply divided or septate, and Homobasidiomycetes (e.g., mushrooms and toadstools), in which the basidia are club-shaped and born on conspicuous fruiting bodies (basidiocarps, *q.v.*). Some species are edible, others are not; some are deadly poisonous.

basidiospore. a spore, usually formed on a basidium, that is produced by karyogamy and meiosis.

basidium (plural, basidia). the characteristic, club-like structure producing sexual spores in the Basidiomycetes, *q.v.* *See also* basidiocarp. In the higher members, basidia are produced by a basidiocarp, but in many lower basidiomycetes, they are borne individually on a hypha. A basidium produces a single haploid basidiospore.

basin. **1a:** a receptacle or vessel that contains or can contain fluids (e.g., a bowl, tub, or sink). **1b:** a pond, lagoon, lake, or reservoir. **1b:** a hollowed or cratered site used to hold water. **2:** a crater, depression, or hollow in the ground of any size, natural or man-made, that has no outlet. **3:** the drainage area or catchment area of a stream, lake, or inland sea. **earthen storage basin**. a pond or basin with earthen walls in which animal waste is temporarily stored without treatment. *See also* lagoon (anaerobic treatment lagoon).

basipetal. pertaining to the development of a plant from apex to base, such that the oldest structure is at the apex as in the case of leaves and flowers. *Cf.* acropetal, centripetal, centrifugal.

basipetally. **1a:** plant development that proceeds from apex to base. **1b:** the condition of a plant that develops in this manner.

basmati. a common variety of long-grain rice.

basophil (basophile; basophilic leukocyte). **1:** staining strongly with basic dyes. **2:** any cell in circulating blood that contains water-soluble granules that stain strongly with basic dyes. **3:** a type of granulocyte that releases histamine and serotonin from the granules in certain immune reactions. They also secrete anticoagulants. **4:** a type of secretory cell of the anterior pituitary.

basophile. basophil.

basophilic leukocyte. basophil.

bastard acacia. *Robinia pseudoacacia*.

bat guano. *See* guano.

bat. a mammal of the order Chiroptera. **vampire bat**. a member of the genus *Desmodus*. It is an important reservoir host of the rabies virus in Central and South America.

Batesian mimicry. mimicry in which the appearance of a prey species mimics that of an inedible species.

BATF. Bureau of Alcohol, Tobacco, and Firearms.

Batrachia. **1:** a synonym in some classifications for the amphibian subclass Anura (frogs and toads). *See also* Salientia. **2:** rarely encountered as a synonym for the vertebrate class Amphibia.

batrachian. **1a:** a frog or toad. **1b:** denoting or pertaining to frogs and/or toads.

batt. the long wooden handle of a scythe.

battery. **1:** group; set; series; array. **2:** a device that generates direct current electromotive force (voltage) by conversion of chemical, thermal, nuclear, or solar energy into electrical energy. **3:** cage battery, *q.v.* **cage battery** (battery). an array of battery cages that may house thousands of fowl. The use of such batteries is usually highly energy- and cost-efficient. Nevertheless, the risk of disease is high, pollution can be a serious problem, and conditions of such a system of housing are considered by many to be harsh or cruel. *See* battery cage, battery hen, freerange eggs.

battery acid. Sulfuric acid.

battery cage. a small wire cage, one of a large array of such cages (a cage battery). It is used to confine domestic fowl for any of several purposes (e.g., to raise chicks without the presence of a hen, to maximize egg production, to fatten fowl for market). *See also* battery, battery hen.

battery egg hen. a laying hen, usually one of four or five, confined in a battery cage with a view to maximizing egg production. The hens are crowded, with little room to move about or even to extend their wings; they are usually debeaked.

battery hen. battery egg hen.

batting. loosely matted, unspun cotton, flax, or wool.

bay leaf. a leaf of *Pimenta acris*, the bay-rum tree. These leaves are used as herbs in cooking, especially in soups. They are also a source for bay rum.

bay roan. **1a**: a coat color (brown with gray) of a horse. **1b:** denoting such a horse.

bay-rum tree. *See* bay leaf.

BBB. blood-brain barrier (*See* barrier).

bbl. abbreviation for barrel.

BCF. bioconcentration factor (*See* factor).

BCGF. B-cell replacing factor (*See* factor).

beachgrass. *See Ammophila arenaria.*

beak. **1:** a protruding, usually sharp, pointed mouth part or jaw that lies distal to other mouthparts. An example is the beak (nib) of a bird. **2:** the awn on the outer bracts of wheat. **3:** any part of a plant that resembles the beak of a bird.

beaker cell. goblet cell.

beam. **1a:** joist; rafter; building timber. **1b:** brace; crosspiece; girder; trestle. **2:** ray (e.g., a ray of light). **3:** to radiate, gleam, or shine. **4:** the central shaft or main frame of a plow; it supports all soil-engaging parts and is the part by which the plow may be drawn. **5:** The balance beam of a scale. **6:** an inclined board on which hides are dressed for tanning.

beamage. an estimate of the weight loss of a freshly-dressed carcass by evaporation during cooling.

bean. **1:** any of various leguminous plants (Family Fabaceae) that have edible seeds. *See*, for example, Fabaceae, *Phaseolus*. **2:** the highly nutritive seed of such a plant (def. 1). Beans are usually more or less kidney-shaped or flattened and born in a pod. Many can be eaten when fresh, dried, or ground into meal. **3:** certain seeds or bean-like structures that resemble beans, e.g., coffee bean (*See* coffee). **4:** a protuberance from the upper surface of the beak of certain wild and domestic water fowl.

bean family. Fabaceae.

bean group. those beans (e.g., navy, garden, and kidney beans) that can be inoculated by the same rhizobial bacteria.

bean pole. **1:** a tall stake used to support climbing beans. **2:** a tall, lanky person; a skinny person.

bean stem rot. a fungus disease of beans.

bean weevil. *Acanthoscelides obtectus.*

bean-hull poisoning. poisoning of livestock by ingestion of feed containing bean hulls infested by *Fusarium.* Signs are chiefly those of nervous dysfunction such as staggering, disturbances of behavior, impaired reflexes, decreased heart rate, hyperesthesia, convulsions, and respiratory failure. The toxins are mainly trichothecene mycotoxins.

beans. *See* Fabaceae.

bear corn. *Veratrum viride.*

bear's foot. *Helleborus foetidus.*

bear-grass. *Nolina texana.*

bearberry. a plant of the genus *Arctostaphylos.*

bearberry. *Rhamnus purshiana.*

beard. *See* awn.

beard tongue. *Penstemon.*

bearer. **1:** a porter; one who carries or supports. **2:** one who holds a check or draft for payment.

bearing. **1:** posture; carriage; the manner in which one bears oneself. **2a:** direction; orientation; location. **2b:** the direction of a line on the ground or on a map defined by the angle measured from some reference direction. This may be true north, magnetic north, or grid north. **3:** a replaceable machine part that bears the force and friction generated by another revolving part. Common types are ball bearings, roller bearings, and sleeve bearings. bearings are often lined with babbitt or a related metal. **magnetic bearing**. the horizontal angle between a map line or ground line and a magnetic meridian. A magnetic bearing differs from a true bearing by the angle of magnetic declination where the bearing is taken. **true bearing**. the horizontal angle between a map line or ground line and a geographic meridian.

bearing rein. neck rein.

bearwood. *Rhamnus purshiana*.

beast. **1a:** any living animal. **1b:** a lower animal, as distinguished from humans. **1c:** a four-footed animal; a terrestrial mammal other than a hominid. **2:** any fierce animal, especially a carnivore. **3:** a beast of burden, *q.v.* **4:** a human who behaves in a way that arouses contempt or abhorrence or fear in others.

beast of burden. beast (def. 3); an animal, usually a domestic animal such as an ass, camel, horse, mule, or ox, used for riding, hauling, or pulling.

beaver poison. *Cicutum maculata*.

bed. **1a:** a place to rest or sleep. **1b:** a bunk, cot, cradle, or crib. **2:** a layer or deposit; vein or lode (e.g., of ore). **3:** a small area of land especially prepared to grow particular plants (e.g, a flower bed, a bed of vegetables) **3:** an aggregation of clams, mussels, or oysters. **4:** the platform of a wagon or truck that supports material to be hauled.

bed load. the mass of particulates that are transported along the bed of a stream by rolling and bouncing.

bedding. **1:** absorbent litter (e.g., straw, leaves, shavings, sawdust, sand, or peat moss) on the floor of a stall or other animal housing for purposes of cleanliness and the comfort of resting animals. **2:** elevating the surface soil into a series of crowned parallel beds or lands (as by plowing) separated by shallow drains. **3:** the provision of surface drainage means of dead furrows that run parallel to a prevailing slope. The land between adjacent furrows is called a bed. **4:** the planting of small flowers into a flower bed. (5) Flowers that are suitable for planting in a bed.

bedding out. **1:** a method of managing sheep in which the flock is bedded on the same ground for a few nights; usually not more than three. **2:** the planting of a flower bed in a decorative way.

bedding plants. plants such as small annual flowers and table vegetables that are planted in beds.

bedload. sediment on or near the stream bed that moves, largely due to gravity, by sliding, rolling, or bouncing along the bottom, usually with relatively low velocity.

bedrock. the solid unweathered rock at the surface of the earth or at some depth beneath the soil, glacial drift, and other superficial deposits.

bee. **1:** any of numerous species of stout-bodied hymenopterous insects, chiefly of the families Apidae, Bombidae, and Xylocopidae. Most are venomous, flying, social (colonial) insects with both biting and sucking mouth parts and hairy bodies. Most are beneficial to humans. They collect pollen and produce honey and beeswax; some are predators or parasites of other insects. Some species are solitary. *See Apis*, *Bombus*, Hymenoptera, apamin, kinin peptide, mellitin, and venom (bee venom). **2:** a gathering of people within a community to socialize and carry out a common purpose or hold a friendly competition such as quilting, barn-raising, or corn-husking.

bee bread. pollen stored in the cells of a honeycomb and eaten by bees.

bee brush. a soft brush used to remove honey bees from combs.

bee bum. a beehive, especially one formed in a hollow log.

bee farming. the rearing of bees for the honey and beeswax they produce and for the pollination of agriculturally important plants such as alfalfa.

bee keeper. a person who rears bees, usually for honey.

bee louse. *Braula coeca*.

bee moth. any moth of the family Pyralidae. *See Achroia grisella* and *Galleria mellonella*.

bee pasturage. bee pasture.

bee pasture (bee pasturage). an area within flying distance from a beehive that contains plants attractive to the bees.

bee space. a passage within a beehive that is wide enough to allow a bee to pass but too narrow to permit comb construction.

bee sting. most species of bees have a stinger, *q.v.*, which may be used to sting organisms that disturb the hive. Such stings usually cause temporary local pain and edema with no systemic complications. Occasionally, a bee sting can induce anaphylaxis in sensitive individuals. *See also* bee, venom.

bee stinger. *See* stinger.

bee veil. a screen of cloth or wire worn over the head and neck to protect the wearer from bee stings.

bee wolf. *Philanthus triangulum*.

bee yard. apiary.

beechnut. *Fagus sylvatica*.

beef breed. any breed or line of cattle that has been developed for purposes of producing meat (e.g., angus, hereford). *Cf.* beef cattle.

beef bull. a bull reared to produce beef.

beef master. a breed of beef cattle that originated in Texas from a program of crosses of Brahman, Hereford, and Shorthorn cattle. Horned and polled lines occur. Color varies, but red and dun are most common.

beef middles. the middle portion of a cow's intestine used as sausage casing.

beef pritch. a rod inserted in a beef carcass to provide support while skinning. *See* gambrel.

beef shorthorn. *See* shorthorn.

Beef Carcass Data Service. a service that provides cattle producers with carcass evaluation data on their stock for a fee.

Beef Improvement Federation (BIF). a confederation of organizations and individuals concerned with the development and application of uniform procedures to the evaluation of beef cattle, the education of members and consumers, with confidence-building, and the promotion of their methods.

beefalo. an American breed of beef cattle, brown in color, derived from crosses of bison with charolais and hereford cattle.

beefiness. the quality of being beefy, *q.v.*

beefy. **1:** "well muscled"; a term denoting the desirable conformation of beef cattle as opposed to dairy cattle. The latter are not beefy; they are relatively thinner and more angular than the former. **2a:** denoting poultry that is fat and coarse. **2b:** pertaining to combs of poultry that are coarse and overgrown.

beehive. **1a:** a natural or man-made structure that houses a colony of bees. **1b:** the housing for a colony of honey bees including the frames in which the bees store honey. *See also* bee bum.

beekeeper. apiarist.

beer. an alcoholic beverage produced by fermenting malted barley with large quantities of water.

beet. any plant of the genus *Beta*.

beet flea beetle. *Chaetocnema coninna*.

beetle. any insect of the order Coleoptera. *See* H&D.

beetroot. *Beta vulgaris*.

beget. **1:** to father, sire, or procreate. **2:** bring about, cause, or engender.

beggar-tick. *Bidens frondosa*.

behavioral. pertaining to behavior.

behavioral response. **1:** behavior, *q.v.*, elicited by a specific stimulus or set of stimuli, usually external.

Belgian. **1:** *See* Belgian heavy draft. **2:** a breed of rabbits used for meat production.

Belgian blue. a breed of cattle obtained by crossing Friesians and shorthorns with native Belgian stock. **British Belgian blue**. a British variety that does not birth normally and must be delivered by caesarian section.

Belgian heavy draft (Belgian, Def. 1). a heavy, brown, powerful breed of draft horse with great traction that originated from the Flemish horse in Belgium. Stallions often weigh more than one ton.

belladonna lily. *Amaryllis belladonna*.

bellows. an instrument used for blowing air into a fire or across coals to increase the temperature or to maintain a good rate of combustion. This tool is used, for example, by blacksmiths.

bellwether. a wether, *q.v.*, with a bell hung around its neck, who leads a flock.

belly. **1:** the thickest circumference of an animal's body or of a body part (e.g., of an appendage, a muscle). **2:** the front surface of a thing as opposed to the back. **2:** a protruding or bulging surface. **3:** venter, paunch, abdomen; the front (e.g., of humans); the underpart (e.g., tetrapods) of the vertebrate body from the breastbone to the pelvis. This portion of the vertebrate body surrounds the abdominal viscera. *Cf.* dorsum. **4:** improperly, the stomach or womb. **5:** appetite; gluttony. **6:** the inside of a thing (e.g., the belly of a ship). **7:** to swell; swell out, fill out. **8:** belly crawl; to crawl on one's belly.

belted. **1:** any animal that has a band of color about the middle of the body that differs from that of the rest of the body. **2:** banded; descriptive of a fruit that has a band about the middle due to growth retardation that often results from frost injury to the fruit when young.

belted galloway. a medium-sized, polled, long-haired breed of beef cattle. The body is black with a white belt about the middle and a dense undercoat.

Beltsville. a breed of large, white turkeys.

Beltsville swine. either of two breeds of swine developed in Beltsville Maryland at the agricultural research center of the U.S. Department of Agriculture: **Beltsville No. 1**. a breed produced by crossing Danish Landrace and Poland China swine. The ground color is black with uniformly distributed white markings. **Beltsville No. 2**. a breed produced by a program of crosses using Danish Yorkshire, Duroc, Hampshire, and Landrace swine. The ground color is red with a white underline and sometimes black spots.

Bembix carolina (horseguard). a black-and-yellow sand wasp (Family Sphecidae) indigenous to the southern United States in areas that support horses or cattle. It kills and feeds horseflies and bottleflies to its larvae.

benazolin. an herbicide used to control broadleaved weeds.

bendiocarb (2,2-dimethyl-1,3-benzodioxol-4-ol-methylcarbamate; methylcarbamic acid 2,3-(isopropylidene-dioxy)phenyl ester). a slightly water-soluble contact insecticide, it is also poisonous to humans and other mammals by ingestion and absorption through the skin.

beneder. a breed of large, red or black spotted goats, bred in Somalia.

benefit/cost analysis. *See* cost-benefit analysis.

benign. harmless; mild (benign atmosphere); not malignant (benign tumor). *Cf.* malignant.

Benlate®. benomyl.

benomyl ([1-[(butylamino)carbonyl]-1*H*-benzimidazol-2-yl]carbamic acid methyl ester; 1-(butylcarbamoyl)-2-benzimidazolecarbamic acid methyl ester; methyl 1-(butylcarbamoyl)-2-benzimidazolecarbamate; Benlate). a colorless, crystalline solid, $C_{14}H_{18}N_4O_3$, used as an ascaricide and fungicide against eyespot in cereals. It is toxic to laboratory animals and humans; it is a mutagenic, teratogenic, and a reproductive poison.

bensulide (*N*-(2-mercaptoethyl)benzenesulfonamide). an herbicide that, like chlorthaldimethyl, inhibits division in the tips of root initials on *Cynodon dactylon* stolons. Like bromacil, it inhibits division in the meristem of oat roots with the appearance of binucleate cells and differentiation of tissues nearer the tip.

bentgrass. any grass of the genus *Agrostis*, especially *A. alba*.

benthic (benthonic). **1a:** denoting or pertaining to the benthos. **1b:** denoting or pertaining to life or existence on or near the bottom of a water body. **1c:** of or pertaining to life or existence at great depths in a very large water body (a major river, lake, sea, or ocean). **1d:** sometimes used in reference to any organism or biotic community found exclusively on the bottom of any body of water.

benthic organism. **1:** an aquatic organism that occurs on or near the bottom (e.g., of a stream or lake) or at great depths (e.g., in an ocean). **2:** benthos (def. 2).

benthic region. **1a:** denoting the benthos. **2:** the region of the floor or bottom layer of a body of surface water (e.g., a stream, pond, lake, or ocean), usually taken to include the surficial layers of bottom sediment and the water column immediately above.

benthiocarb. a thiocarbamate herbicide, it is an amber-colored, slightly water-soluble, liquid,

$$C_{12}H_{16}NOCl$$

Benthiocarb is moderately toxic to humans and other mammals by mouth. It is used to control sedges and grasses in rice paddies.

benthonic. benthic.

benthos. **1:** the floor of a sea or ocean; the depths; the deep. **2:** a collective term for benthic organisms; applied most frequently to marine organisms that live at great depths, but includes littoral and sublittoral organisms also. *Cf.* nekton, plankton.

Benzahex. a trade name for lindane.

benzene carbinol. benzyl alcohol.

benzamizole. isoxaben.

γ-benzene hexachloride (γ-BHC). lindane.

γ-benzene hexachloride. τ-hexachlorocyclohexane.

benzene hexachloride (BHC; hexachlorocyclohexane). **1:** a mixture of the isomers of 1,2,3,4,5,6-hexachlorocyclohexane, $C_6H_6Cl_6$. It is an insecticide formerly used to control cotton pests and also pea and bean weevil; it is used also as a seed dressing against wireworOctober 14,2001 m. BHC is no longer produced in the United States. It has been used as an ectoparasiticide for dogs and large mammals, but is highly toxic to domestic cats in concentrations that are effective against parasites. Lindane has largely replaced these usages. BHC is stored in body fat and excreted in milk. **2:** hexachlorocyclohexane.

gamma-benzene hexachloride. lindane.

benzene hexachloride. hexachlorocyclohexane.

benzene-methanol. benzyl alcohol.

benzeneamine. aniline.

benzenecarboxylic acid. benzoic acid.

1,2-benzenedicarboxylic acid dimethyl ester. dimethyl phthalate.

4-[2-benzimidazolyl]thiazole. thiabendazole.

benzoate (benzenecarboxylate). any salt or ester of benzoic acid.

benzoic acid (benzenecarboxylic acid; benzoyl hydrate; carboxybenzene; dra cylic acid; flowers of benzoin; phenylformic acid). a white, crystalline compound, $C_6H_4(OH)_2$, with an odor similar to that of benzoin or benzaldehyde. The carboxyl group binds directly to the benzene ring. Benzoic acid occurs naturally in some plants, as in benzoin resin (gum benzoin). It is used as a food preservative and fungistat. It is also used as an oral antiseptic, a diuretic, and expectorant. It is moderately toxic but has many highly toxic derivative compounds (e.g., benzidine, 3,3′-dichlorobenzidine). It is rapidly excreted as hippuric acid. Benzoylglucuronic acid is the detoxified form.

1-[5-(1,3-benzodioxol-5-yl)-1-oxo-2,4-pentadienyl]piperidine. piperine.

2*H*-1-benzopyran-2-one. coumarin.

1,2-benzopyrone. coumarin.

benzoyl hydrate. benzoic acid.

(E,E)-16-benzyl-6,7,8,9,10,12a,13,14,-15,15a,16,16-dodecahydro-5,13-dihydroxy-9,15-dimethyl-14-methylene-2*H*-oxycyclotetradec-[2,3-d]isoindole-2,18(5H)-dione. cytochalasin B.

Berberis (barberry). a genus of shrubs (Family Berberidaceae), the berries of which are a rich source of food for birds. Most are grown as ornamentals.

Bergenia crassifolia. a hardy, perennial garden herb (Family Saxifragaceae) with large, fleshy evergreen leaves that contain arbutin.

Berkshire, Berkshire swine. a breed of small, compact hogs that originated in Berkshire County, England. It arrived in the United States in large numbers in the first third of the 19 century and is now a common breed. It is long-bodied, with erect ears, a short head, and a black body, with a white snout, feet, and tail.

berm. **1:**a narrow shelf, mound, or horizontal area that breaks the continuity of a slope. Such a berm may be constructed as a means of diverting runoff, thereby reducing erosion. **2:** the edge of a drainage ditch, usually including the area to the edge of the spoil bank. **3:** a horizontal area parallel to the shoreline that marks the reach of a high tide. **4:** the shoulder of a large thoroughfare of highway.

bermat. chlordimeform.

Bermuda buttercup. *Oxalis pes-caprae*.

Bermuda onion. any variety of mild onion, *allium cepa*, introduced into the United States from Bermuda. Such onions are commonly cultivated on farms in the southwestern United States

Bermuda oxalis. *Oxalis pes-caprae*.

Bermudagrass. *Cynodon dactylon*.

Berrichon du Cher. a breed of French sheep; sometimes exported.

berry. **1:** any fleshy, succulent fruit with more than one seed that lacks a stony endocarp (e.g., banana, blackberry, strawberry, tomato). **2:** certain dry seeds such as the coffeeberry.

berry picker. a machine that harvests berries while straddling rows of vines and using flails as it progresses to shake the vines, thus causing the berries to detach and drop onto a conveyer belt.

berseem. *Trifolium alexandrinum*.

berseem clover. *Trifolium alexandrinum*.

Besnoitia. a genus of protozoan parasites (Family Besnoitiidae). ***B. besnoili***. the species that is the chief cause of besnoitiasis in cattle.

besnoitiasis (besnoitiosis). a disease of cattle primarily caused by *Besnoitia besnoili*. It causes the formation of cysts, mainly of the connective tissue of the skin, mucous membranes of the nose, and serous membranes. Signs initially include fever followed by hair loss and seborrheic changes in the skin.

besnoitiosis. besnoitiasis.

best management practices (BMP). a set of practices designed to maintain agricultural yields while reducing point- and nonpoint-sources of water pollution. The BMPs that are best fitted to local conditions are determined by the state water quality agencies or their designees.

beta (β, B). **1:** the second letter of the Greek alphabet. Its applications as a symbol or prefix in scientific terms are analogous to those of alpha. For terms beginning with beta, *See* the root word. **2:** the second. **3:** second in order of magnitude.

Beta (beet). a genus of some 15 species of widely cultivated, mostly biennial herbs and root crops (Family Chenopodiaceae) that are indigenous to Europe and the Middle East. ***B. vulgaris*** (beet; beetroot; sugarbeet; foliage beet; fodder beet; mangel; Mangelwurzel; mangold; chard; Swiss chard). a salad vegetable with a bright red root. It is the progenitor of several varieties that have been cultivated by humans for several centuries. Two botanical varieties, *B.v.* var. *crassa* (beet, beetroot) and *B.v.* var. *macrorhiza* (mangold), sometimes contain concentrations of nitrate that are toxic to livestock. ***B. vulgaris* var. *cicla*** (chard). a goitrogenic variety that also contains hazardous levels of soluble oxalates. Sugarbeet tops are sometimes used as forage by livestock. Nitrate poisoning of livestock has occurred following ingestion of sugarbeet tops. This has been associated with the use of 2,4-D as a weed killer. Death of livestock from consumption of the tops is also due to the presence of soluble oxalates in the tops. Beetroot crops grown for fodder are also toxic under certain conditions, producing hypocalcemia and occasionally acid indigestion.

betel nut (areca nut). **1:** *Areca catechu*. **2:** the seed of *Areca catechu*. Small pieces of the astringent kernel are often chewed in tropical areas, especially the East Indies. The nuts are sometimes mixed with other ingredients such as slaked lime and leaves of the plant. The nuts have an anthelmintic action.

betel nut seed. *Areca catechu*.

betel palm. any palm of the genus *Areca*, especially that of *A. catechu*.

bethanechol chloride (2-[(aminocarbon yl)-oxy]-*N,N,N*-trimethyl-1-propanaminium chloride; Duboid®; Urecholine®). a colorless or white crystalline substance or crystalline powder,

$$C_7H_{17}ClN_2O_2$$

It has a number of uses in human medicine and is also used as a cholinergic in veterinary practice.

Bettlach May disease. a deadly, paralytic disease of adult honey bees, *Apis mellifera*, that occurs chiefly in Switzerland. It is caused by ingestion of the pollen of certain types of buttercups, *Ranunculus*. The bee cannot fly and eventually dies.

Betula (birch). **1:** a genus of trees indigenous to the Northern Hemisphere. All have unisexual flowers in catkins; the bark often peels off in thin papery strips. The leaves are simple, serrated, and alternate. **1b:** the wood of the birch. It is close-grained, hard, and used chiefly in the manufacture of plywood, furniture, and interior finishes. **2:** a rod from a birch, used as a switch to administer a whipping. **3:** to whip with a birch rod.

Betulaceae (birch family). a family of deciduous trees and bushes that occur in both temperate and arctic regions. They often occur together with evergreens in northern coniferous forests. Genera that occur in North America include *Alnus* (alder), *Betula* (birch), *Corylus* (hazel), *Carpinus* (hornbeam; ironwood), and *Ostrya* (hop hornbeam; ironwood). A sixth genus, *Ostryopsis*, is limited to Mongolia. All have simple, toothed leaves and bear male and female flowers on separate plants. The female flowers occur in short clusters and the male flowers in drooping catkins. The nuts have a single seed. There are a few scrub species in this family, but a number of the trees have exceptional ornamental value and also provide valued hardwood, lumber, and edible fruits.

bevy. a covey (flock) of quail; a flock (wedge) of swans.

BHA. butylated hydroxyanisole.

BHC. benzene hexachloride.

BHT. butylated hydroxytoluene.

bi-. a prefix indicating two.

bial-, ptyal-, sialo-, siali-. prefixes denoting saliva.

bicollateral bundle. a vascular bundle that has two strands of phloem, one centripetal to and the other centrifugal to the xylem.

bicolor. having two colors; pertaining to an organism that has two colors.

bicornuate. having two horns.

bid. offer to buy something at a stated price. **opening bid**. the first bid for an item at an auction. **closing bid**. the successful bid for an item at an auction.

bidder. one who bids at an auction.

biddy. a popular name for a hen of the domestic fowl.

Bidens frondosa (beggar-tick). a weedy North American plant (Family Asteraceae) that sometimes contains amounts of nitrate that are toxic to livestock.

bident. a two-year old sheep.

Bidrin. dicrotophos.

biennial. 1: occurring every two years; of or pertaining to a cycle of two years. **2:** a biennial plant, *q.v.*

biennial plant. a green plant that completes its life cycle from germination to flowering, fruiting, seed production, and death in two years. The first year's growth is usually vegetative, but in the second year, flowers, fruits, and seeds are produced. Death of the plant follows.

BIF. Beef Improvement Federation.

bifoliate. having two leaves.

big bale silage. silage that is stored in big bales.

big bluegrass. *Poa ampla*.

big bluestem. *Andropogon furcatus*.

big head. *See* bighead.

big jaw. lumpy jaw.

bighead. 1: geeldikkop; photosensitization of sheep due to grazing on plants such as *Tetradymia*, *Tribulus terrestris*, and *Nolina texana* followed by exposure to sunlight. **2:** "big head" is sometimes used in reference to nutritional hyperparathyroidism in horses or goats.

bighorn. a species of wild sheep indigenous to the Rocky Mountain cordillera of North America.

bigleaf maple. *Acer macrophyllum*.

bikhroot. *Aconitum ferox*.

bilateral. having two sides, as in bilateral symmetry.

bile (gall). a dark yellow or brown-green alkaline digestive fluid secreted by the vertebrate liver. It is stored temporarily in the gall bladder (an organ lacking in many birds) and is conducted from the gall bladder to the duodenum of the small intestine via the bile duct under pressure by contractions of the gall bladder. Lecithin and the bile salts emulsify dietary lipids, thus facilitating digestion.

bile acid. any of the steroid acids (derivatives of cholesterol) produced by the vertebrate liver, such as taurocholic and glycocholic acids. They appear in bile as sodium salts that facilitate the emulsification of dietary fats. A large portion of the bile acids are recycled by enterohepatic circulation. Certain of the bile salts can cause cholestasis or hepatic necrosis and can be metabolized by the intestinal microflora into secondary bile acids, e.g., lithocholic acid, which are promoters of colonic carcinogenesis.

biliary fever of horses. equine babesiosis.

bilineurine. choline.

bilirubin. a yellowish-red bile pigment formed from hemoglobin and other heme pigments during the normal metabolic destruction of erythrocytes. It contributes to the yellowish color of urine and can cause jaundice if it accumulates in the blood. *See* bile.

biliverdin (biliverdinic acid; dehydrobilirubin). a green bile pigment,

$$C_{33}H_{34}O_6N_4$$

It is produced from hemoglobin during the normal metabolic destruction of erythrocytes. It is a precursor of bilirubin. *See* bile.

biliverdinic acid. biliverdin.

bill of lading. an agreement signed by a common carrier to deliver freight to a designated receiver.

billy goat. a male goat, especially a mature individual.

bin. a large storage container.

binapacryl (3-methyl-2-butenoic acid 2-(1-methylpropyl)-4,6-dinitrophenyl ester; 3-methylcrotonicacid 2-*sec*-butyl-4,6-dinitrophenyl ester; 2-*sec*-butyl-4,6-dinitrophenyl 3-methyl-2-butenoate; 2-*sec*-butyl-4-6-dinitrophenyB-methylcrotonate; 2-*sec*-butyl-4,6-dinitrophenyl senecioate; 3,3-dimethylacrylic acid 2-*sec*-butyl-4,6-dinitrophenyl ester; 4,6-dinitro-2-*sec*-butylphenyl β,β-dimethylacrylate; 3-methyl-2-butanoicacid 2-*sec*-butyl-4,6-dinitrophenyl ester; dinoseb methacrylate; senecoicacid 2-*sec*-butyl-4,6-dinitrophenylester). a very toxic compound,

$$C_{15}H_8O_6N_2$$

It is used as a fungicide and mitocide. *See also* dinoseb.

binary fission. the division of one cell into two cells by the formation of a septum as in most bacterial cell division.

binder. **1:** a binding agent; that which binds, attaches, fastens, or secures. **2:** a substance that binds a mixture of solid particles together (e.g., gums and waxes, starch, gelatin, sucrose).

bindweed. *Convolvulus*.

binomial nomenclature. an internationally agreed upon method of naming living organisms, devised by Linnaeus. Two names that are Latin in form are assigned to each species. The first desig-nates the genus, and the second designates the species.

bio-. **1:** a combining form indicating or denoting life. **2:** often used to replace the term biological; this usage should usually not to be encouraged, in part because it is often adopted without clear understanding of the original term. Such terms are also often adopted by the public as jargon that is claimed to be rooted in science (*See*, for example, biorhythm).

bioaccumulate. **1:** the process of accumulating an exogenous chemical within an organism to a concentration above that of the environment (medium) to which the organism is exposed. **2:** to progressively accumulate a chemical through a food chain. *See* bioaccumulation.

bioaccumulation. **1:** the tendency or process whereby an organism of accumulates or concentrates an exogenous substance from sources in its environment and by any or all routes of entry. This occurs when the intake of a chemical by an organism exceeds the rate of excretion. **2a:** the net accumulation of a substance from the environment of a living system by all routes of entry. **2b:** the state of having bioaccumulated a substance. **3:** operationally, bioaccumulation is the state whereby the concentration in an organism as a whole, or in selected tissues, is greater than the exposure concentration. Chemicals that bioaccumulate are often lipohilic and have a long biological half-life, usually because they are sequestered or accumulate in target organs within affected organisms and are not readily metabolized or excreted. Heavy metals accumulate, for example, in target organs and produce a toxic response when the concentration exceeds the threshold of response.

bioaccumulative. denoting or characterizing a substance in the environment that increases in concentration in exposed living organisms. Such substances are usually very slowly metabolized and excreted.

bioactivation. biological activation (*See* activation).

bioactive. jargon for biologically active (*See* active).

bioactivity. jargon for biological activity.

bioassay. biological assay (*See* assay).

bioaugmentation. a technique used in bioremediation where the rate of decomposition of pollutants (e.g., agricultural wastes) is accelerated by using suitable, highly active bacteria (often with bioenrichment). A technique used in bioremediation.

biocenosis (biocoenosis). **1:** a natural assemblage of organisms (such as a community), exclusive of components of the physical environment. *Cf.* biogeocenosis. **2:** sometimes used as a synonym of ecosystem; this is not recommended usage.

biochemical effect. *See* effect.

biochemical oxygen demand. biological oxygen demand.

biochemistry (molecular biology). the science that treats biochemical and molecular interactions within living cells and organisms. **agricultural biochemistry**. a major subdivision of agricultural chemistry, *q.v.*, that applies the theory and methods of biochemistry to the study of agricultural problems considered broadly. Included is the analysis and evaluation of life processes involved in the production of crops, poultry, and livestock and the assessment of biochemical properties and interactions of foods and a broad range of agriculturally important chemicals, including dyes, drugs, alcohols, fertilizers, soil conditioners, pesticides, herbicides, fungicides, and insecticides. Biotechnology and genetic engineering, especially applications to agriculture, have extensive biochemical components and receive much emphasis. *See also* agriculture, agricultural chemistry.

bioclimatology (ecoclimatology). the study of climate in relation the micro and macro environment of flora and fauna, their distributionpatterns, habitats, distribution, numbers, and types.

biocoenosis. biocenosis.

bioconcentration. biological concentration (*See* concentration).

bioconcentration factor (BCF). concentration of an environmental chemical in an aquatic organism divided by the concentration in the surrounding water. The BCF is a steady-state function and is valid only if tissue residues do not change with continued exposure. *See also* biomagnification factor.

biodegradability. the potential of a chemical to decompose under natural conditions due to biological processes (e.g, enzymatic activity, microbial oxidation). *See also* biodegradation.

biodegradable. **1a:** readily able to undergo biodegradation. **1b:** descriptive of a substance that readily decomposes under natural conditions as a consequence of biological processes (e.g., microbial oxidation, enzyme catalysis). **1b:** susceptible to biodegradation. **2:** pertaining to the tendency of a substance to rapidly decompose under natural conditions as a function of normal biological processes.

biodegradation (biological degradation). **1:** the partial or complete decomposition of a chemical into simpler substances by means of biological agents or processes (e.g., enzymatic action). **2:** any biological process such as microbial oxidation or enzyme catalysis that degrades substances. *See also* bacterial decomposition of hazardous materials.

biodiesel (methyl esters). a legal alternative fuel of commerce, biodiesel is registered with the Environmental Protection Agency as a pure fuel or fuel additive. It is usually produced (with glycerin as a by-product) by a catalyzed reaction of a vegetable oil or animal fat with methanol. Biodiesel can be used in diesel engines alone or blended with petroleum diesel. Biodiesel produces less carbon monoxide, particulate matter, and unburned hydrocarbons. Increased use of biodiesel would decrease the drain on petroleum reserves, decrease automotive pollution and bolster the market for soybean oil.

biodiversity. *See* diversity.

bioecology. **1:** ecology. **2:** the interdisciplinary ecological study that treats both plants and animals, their interrelationships, and interactions with the chemical and physical environment.

bioenergetics. **1:** the flow, transformations, and work of energy in and through living systems. **2:** the study of bioenergetics (def. 1).

bioenrichment. the addition of nutrients or oxygen to increase microbial breakdown of pollutants. *See* bioaugmentation.

biomass. **1a:** any living matter that can be converted to usable energy by means of biological or chemical processes. Examples are agricultural crops and crop residues, other plants and plant residues including grasses, wood and wood residues, animal wastes, feedstocks, and municipal wastes. Biomass is often expressed as the volume or weight of live or dry material, ash-free weight, standing crop, standing stock, or as energy in calories. Biomass is usually taken as a measure of biotic productivity. **1b:** the total mass (as weight) of specified living organisms or material of biological origin (e.g., corn, potatoes, total organic matter) in a given area or medium (e.g., soil). **1c:** any quantitative estimate of the total mass of specified organisms that comprise part or all of a population, area, habitat, or other defined unit. Frequently used in reference to vegetation. **1d:** the total weight of living matter in an area or in a specified population per unit area or unit volume as appropriate. **1e:** in forestry: the dry weight of all organic matter in a given ecosystem or of plant material that can be burned as fuel.

biogas. gas (methane plus carbon dioxide) produced by decaying waste such as animal refuse. *See* biomethanation.

biogenesis. 1: the theory that living organisms are always generated by living organisms rather than spontaneously from inanimate material. **2:** the production of a substance by a living organism.

biogenetic. biogenic.

biogenic (biogenetic). produced by living organisms (e.g., as coal, petroleum, biogenic amines).

biogenic toxicity. allelopathy.

biogeocenosis (biogeocoenosis). an ecosystem: the biocenosis, together with its biotope.

biogeochemical cycle. the repeated cycle through which a specific chemical element passes between biotic and abiotic components of the biosphere. Examples are the carbon cycle and nitrogen cycle.

biogeochemical prospecting. mineral exploration by means of trace element analysis of plants and parts of plants from known locations.

biogeochemistry. **1a:** the study of microbially mediated chemical transformations of geochemical interest, such as nitrogen or sulfur cycling. **1b:** The science that treats the chemical relationships between living organisms and abiotic components of the environment.

biogeoclimatic classification system. a hierarchical system of classifying ecosystems. It integrates regional, local, and chronological factors and combines climatic, vegetation, and site factors.

biogeoclimatic zone. a geographic area throughout which similar patterns of energy flow, vegetation, and soils occur due to a broadly homogenous macroclimate.

biogeocoenosis. biogeocenosis.

biogeography. the science of the geographic distribution of living organisms, populations, and communities in relation to their origins and the influence of past and present geographic and environmental factors. This science has important implications regarding the cultivation of crops and rearing of livestock.

bioindicator (biological indicator; indicator species). **1:** an organism, usually a plant, that is characteristic of a particular climate, soil, nutrient, habitat, or any special attribute of an environment. **2:** an organism (most commonly a plant or microorganism) that is used as a qualitative indicator of specific chemical activity in, or an aspect of the chemical composition of, a natural system, e.g., the presence of selenium in soil. *Cf.* biomonitor.

biolarvicide. an endotoxin of *Bacillus thuringiensis israelensis*. When the bacterium is ingested by mosquito larvae, the endotoxin is freed by stomach secretions and disrupts the ability of the gut to regulate electrolytes, thus killing the larvae.

biologic (biological). **1:** of or pertaining to living organisms or to life processes. **2:** immunization vaccines, bacterins, antigens, and antitoxins and other preparations made from living organisms and their products, intended for use in diagnosing, immunizing, or treating humans or animals, or in related research. The Animal and Plant Health Inspection Service is responsible for approving some animal biologics. **veterinary biologic**. any biologic used in veterinary practice. Veterinary biologics are regulated by the Animal and Plant Health Inspection Service, which licenses the production facilities and conducts a program to ensure that veterinary biologics are safe, pure, potent, and effective.

biologic agent. biologically active agent.

biological. **1:** of or pertaining to biology or to living organisms and life processes. **2:** biologic. **3:** any drug or pharmaceutically active chemical produced by, or made from, a living organism.

biological activity. *See* activity.

biological amplification. biomagnification.

biological concentration (bioconcentration). **1:** accumulation of an environmental chemical in biota. **2:** in the narrow sense, the accumulation of an environmental chemical in biota that results from or is associated with direct absorption or filtering as opposed to ingestion (see trophic concentration).

biological control. the control of a pest population by interfering with its life cycle by use of natural predators, parasites, pathogens, competitors, the introduction of abnormal (e.g., sterilized) organisms into the population, natural biological processes, or preparations of biological origin (e.g., plant or animal hormones, toxins). Such methods are used to complement, or as an alternative to, other control measures (e.g., chemical). The Animal and Plant Health Inspection Service program includes biological control methods in the control of several economically important pests of food and fiber crops.

biological degradation. biodegradation.

biological diversity. *See* diversity.

biological effect. *See* effect.

biological integrity. **1:** the ability of a living system to carry out all vital functions and to respond adaptively to external and internal stimuli. **2:** sustained maintenance of normal or characteristic structure and dynamics by a living system.

biological lag. the time required to produce

agricultural commodities and supply markets following a change in production. This lag is determined by the biological nature (e.g., time of planting of a crop; time to maturity) of agricultural products.

biological legacy. any feature that remains in an ecosystem, site, or landscape following a natural disturbance. Such a legacy may include a combination of microorganisms, dead and living trees and other plants, seeds, debris, humus, etc.

biological magnification. biomagnification.

biological marker (biomarker). any alteration in the structure or function of an organ, tissue, or cell of an organism that has induced exposure to a particular chemical or class of chemicals. In the broadest sense, any specific indicator of exposure to a specific biologically active agent could be considered a biomarker. In practice, the term is usually applied to distinct alterations (lesions) within cells, tissues, or organs of the affected organism that are specific to a given chemical. Ideally, a biological marker is any lesion that can be used in bioindication or biological monitoring for a given chemical. *Cf.* sign, symptom, indication, lesion, adduct.

biological mineralization. *See* mineralization.

biological oxygen demand (biochemical oxygen demand; BOD). a measure of the amount of oxygen consumed in the microbial oxidation of decomposable organic matter and the oxidation of reduced substances in water. Together with chemical oxygen demand, BOD is a measure of the pollution potential in water bodies (e.g., ponds, lakes and streams) and organic wastes. In practice, the BOD is the amount of oxygen (as mg O_2/liter water) taken up from a water sample that contains a known amount of dissolved oxygen kept at 20°C for (usually) 5 days (BOD5). The higher the BOD, the greater the organic pollution. Values above 20 for an effluent entering a clean river are generally considered polluting. BOD, together with the actual oxygen concentration of the water, can also be taken as a measure of the biomass that a body of water can support. BOD is also used to assess compost maturity. *See also* BOD5, CBOD5. *Cf.* chemical oxygen demand.

biological rhythm. any of numerous periodic (e.g., circadian (daily) rhythms, annual cycles, migratory cycles, molt cycles, hibernation, tidal rhythms) phenomena associated with living organisms. Such phenomena are partly endogenous, requiring only an external signal (zeitgeber) to bring the cycle into phase with environmental cycles (day, year, tidal cycles). The toxicity of a given xenobiotic chemical may vary substantially as a function of the phase or stage of a rhythm during which exposure occurs. Thus, the effect of a drug administered in the morning can differ from that given in the evening. *Cf.* biorhythm.

biological system. any living entity (e.g., tissue, organ, organism) or any interacting set of such entities (population, species, colony, community, ecosystem). Such systems contain nonliving parts.

biological treatment (biological waste treatment). a waste treatment technology that uses bacteria to decompose organic wastes.

biological waste treatment. biological treatment.

biologically active agent (biologic agent). any of numerous agents that can act upon a biologic target (heat, electricity, hormones, drugs, any source of mechanical injury).

biologically active. *See* active.

biologist. **1:** a scientist who studies living systems. **2:** one who studies or applies the findings of biologists (def. 1).

biology. **1:** the division of natural science that deals with living systems (e.g., cells, tissues, organs, organisms, populations, and biocenoses), life processes, interactions among living systems, and interactions of living systems with nonliving systems and environmental variables. **2:**

often loosely equated with the term "life sciences" (not recommended).

biolysis. death and/or subsequent decomposition of living (organic) matter by the chemical action of living organisms (usually microbes).

biolytic. **1:** pertaining to or denoting biolysis. **2:** destructive to life. **3:** denoting living systems that are capable of destroying living material.

biomagnification (biological magnification; biological amplification). **1a:** the phenomenon whereby the concentration of an environmental chemical increases from the original concentration at the point of entry into the environment (usually in the media — soil, sediment, water, or air — or at a relatively low trophic level) upward through one or more steps in a food chain. **1b:** operationally, biomagnification is the increasing concentration of an environmental chemical at higher trophic levels. **1c:** the tendency of a chemical to increasing concentrations at higher trophic levels through dietary accumulation. **2:** the increasing concentration of an environmental chemical (e.g., a pesticide, heavy metal, or other xenobiotics) within organisms as it progresses through a food chain. Substances that biomagnify (e.g., heavy metals, chlorinated hydrocarbons) are usually environmentally persistent and always have relatively long biological half-lives. *See also* factor (biomagnification factor).

biomarker. biological marker.

biome. **1:** a major biogeographic region or formation that has distinctive life forms and the characteristics of a climax community or complex of climax communities. Terrestrial biomes are characterized by the principal plants and plant formations, whereas the marine biomes are defined largely by the major fauna represented. The terrestrial biomes are tundra, northern coniferous forest, temperate deciduous and rainforest, temperate grassland, chaparral, desert, tropical rain forest, tropical deciduous forest, tropical grassland and savannah, and montane biomes. The marine biomes are not yet well established. Two examples are given here: **desert biome**. a biome characterized by less than 25 cm (10 in) of rainfall per year, a high rate of evaporation, and poor soil that supports little plant or animal life. **grassland biome**. a biome, usually of considerable geographical extent, where grasses form the dominant vegetation. Such a biome may be very stable largely due to temperature, sunlight, and soil conditions. Many grasslands have been transformed into land supporting livestock (as rangeland) or crops such as corn and wheat. Temperate and tropical grassland biomes differ in a number of significant ways.

biomedical (inclusive of veterinary science). usage and definitions of this term are varied and broadly overlapping because medicine, including veterinary medicine, as a science, is not a unified field of study but draws upon most of the natural sciences for its body of knowledge and applications. **1:** biological and medical; pertaining to the purely biological aspects of medical professions, especially with respect to research. **2:** pertaining to those biological sciences or components of a biological science that contribute to, support, or are directly involved in the study or practice of medicine and veterinary medicine. **3:** pertaining to those aspects of the science and practice of medicine that draw primarily on the biological knowledge. **4:** pertaining to or including medicine, veterinary medicine, and the biological and physical sciences. **5:** of or pertaining to the application of biological knowledge or disciplines to the study or practice of medicine and veterinary medicine.

Biomet TBTO. tributyl tin, tributyl tin compound.

biomethanation. anaerobic digestion that produces biogas, *q.v.* It is most often used in rural areas as it is the safest way of treating manure and human excreta. The biogas produced may be used as fuel or for light.

biometrician. a specialist in the field of biometry.

biometrics. biometry.

biometry (biometrics). the science that treats the statistical analysis of biological data, especially those based on the measurement of variation in living organisms. This is a field of some importance to agricultural scientists.

biomonitor (biological monitor). an organism used as a quantitative or semiquantitative indicator of specific chemical activity in, or an aspect of the chemical composition of, a natural system. If the effect registered by a bioindicator is sufficiently quantitative, the indicator may be termed a biomonitor. *Cf.* bioindicator.

bion. a living entity; the singular form of biota.

bionomics. **1:** bionomy. **2:** ecology. *See* bionomy.

bionomics. ecology.

bionomy (bionomics). the laws and principles that describe or regulate life processes; the science concerned with such laws. In application, bionomy is sometimes nearly synonymous with ecology.

biopesticide. a pesticide such as a virus, bacterium, pheromone, or a naturally produced plant compound that is biological in origin as opposed to synthetic pesticides.

biophylaxis. **1:** nonspecific protective reactions by an organism to pathogens, xenobiotics, or other entities that threaten its integrity. **2:** the protective reactions themselves (e.g., inflammatory processes, phagocytosis).

biophysics. application of the concepts, assumptions, theories, and methods of physics to the study of biological processes, including both life processes and physical processes (e.g., electrical phenomena, mechanics, luminescence) that occur in living systems.

biopsy. **1:** the removal of a small tissue sample or fluid for histological or other examination as an aid to diagnosis. **2:** a specimen obtained by biopsy.

bioremediation. the use of microorganisms, and sometimes other life forms, to remove, remedy, or detoxify polluted areas and environmental media (e.g., soils). *See* bioaugmentation, bioenrichment.

biotechnology. The scientific manipulation of living organisms, especially at the molecular genetic level, to produce useful products. Gene splicing and the use of recombinant DNA (rDNA) are prominent among techniques that are now in use.

center of origin. usually taken as the location or region where a particular crop first came under cultivation.

biorestoration. a term generally applied to the management of hazardous wastes in subsurface (e.g., soil) environments. Such methods are designed to convert organic wastes into microbial biomass and harmless by-products. Biorestoration involves stimulating either indigenous or introduced subsurface microflora to degrade xenobiotics in place. *See also* bioremediation.

biorhythm. a popular and commercial term, the definition of which varies with the commercial application or the state of ignorance of the user. Its use is science and agriculture is discouraged. *Cf.* biological rhythm.

biosecurity. the body of policies and methods used to protect a nation's food supply and agricultural resources from accidental contamination as well as acts of bioterrorism.

biosolid (sewage sludge). **1:** the dried residues of wastewater treatment. Formerly called sewage sludge. **2:** any organic solid resulting from wastewater treatment that can be usefully recycled.

biosphere. **1:** the part of the earth's crust, waters, and atmosphere that is habitable by living organisms. It extends from deep sediments below the oceans to an altitude of several thousand meters (geese have been observed flying at altitudes of at least 10,000 meters) in high mountains.

2: the global ecosystem comprising the entire earth and all the living organisms that inhabit it.

biostabilizer. an apparatus that converts solid waste into compost by grinding and aerating.

biostasis. the capacity of living organisms to maintain their status and integrity in the face of environmental changes. *Cf* homeostasis.

biostatistics. **1:** the science that treats relationships, variations, and differences among biological data. Biostatistics provides powerful tools for evaluating many types of agricultural information. **2:** vital statistics.

biosynthesis. **1:** the synthesis chemicals, especially cellular constituents, through metabolic processes. **2:** the synthesis of chemicals in the laboratory by means of reactions identical or nearly identical to those that occur naturally.

biosynthetic. of, pertaining to, or produced by biosynthesis.

biosystem. ecosystem.

biosystematics. experimental taxonomy (*See* taxonomy).

biota (singular, bion). **1:** living organisms in general. **2:** flora and fauna collectively, especially that of a particular region.

biotechnology. the adaptation and use of living organisms or their components to carry out defined biochemical and physiological processes that have agricultural, industrial, or other commercial or practical applications.

bioterrorism. a prospective if not emerging threat that might include, for example, the introduction of pests intended to kill food crops; the introduction of virulent disease at animal production facilities, poisoning food and water supplies for livestock and/or humans and contaminating blood supplies. In the United States, the federal government is increasing its efforts to improve biosecurity because of recognized vulnerabilities to citizens and to the military establishments. States are also developing strategies and teams of health professionals and others to anticipate, where possible, acts of bioterrorism and to respond as necessary. *See also* biosecurity.

biotic. **1a:** living. **1b:** denoting or pertaining to life or to any aspect or property of life; applied most often to characteristics of populations, living components of an ecosystem, or a region. **1c:** biological.

biotic community. a naturally occurring assemblage of plants and animals that live within the same environmental setting, are mutually interdependent, and more or less self-sustaining.

biotic potential. the inherent or theoretical capacity of a population to increase in numbers. This potential is never realized, partly or entirely due to unfavorable environmental conditions (e.g., competition, crowding). *Cf.* reproductive potential.

biotics. the natural science of life functions and vital activities.

biotin (W factor; *cis*-hexahydro-2-oxo-lH-thieno[3,4-d imidazoline-4-valeric acid). a member of the vitamin B_2 complex, it occurs in or is required by most organisms (e.g., in the synthesis of fatty acids).

biotope. a particular type of habitat or area that is defined by the particular association of organisms that typically occur in it. Examples are grassland, woodland, a south-facing slope of a hill. *Cf.* also ecosystem.

biotoxin. a term that is redundant with toxin (def. 1a). Any poison that is biogenic in origin.

biotrophic. pertaining to a relationship between two organisms in which one or both must associate with the other to obtain nutrients essential to life.

bioturbated. descriptive of bottom sediments that have experienced bioturbation, *q.v.*

bioturbation. the physical disturbance or mixing of bottom sediments by the activities of living, usually burrowing, organisms. Such activity can, for example, spread a pollutant to levels or areas of a sediment that have not been directly exposed or it can bring to the surface stable pollutants that have been sequestered for some time in lower (older) strata.

biparous. giving birth to two young.

bipartite. having two parts.

bipinnate (twice-pinnate; 2-pinnate; bipinnately compound). of or pertaining to a leaf that is pinnate with respect to both its primary and secondary leaflets.

bipinnately compound. bipinnate.

bipyridylium quaternary ammonium salt (bipyridilium compound; bipyridyl herbicide). a heterocyclic ammonium salt with a bipyridine nucleus. All are powerful, rapid acting herbicides. They quickly and strongly adsorb to soil and are rapidly absorbed by leaves. Thus, rain falling even within one-half hour of spraying does not reduce the effectiveness of these compounds. The best known (diquat, paraquat, and morfamquat) differ from typical quaternary ammonium compounds in phytotoxicity because they can form free radicals by reduction in photosystem 1 of the chloroplast. They are more toxic to animals than most herbicides.

birch. any of various deciduous trees or shrubs of the genus *Betula* and certain other members of the family Betulaceae.

bird. common name of any animal of the vertebrate class Aves, *q.v.* **cage bird**. any species or variety of bird that is adapted to confinement in a cage, usually small birds kept as pets or songsters in the home. **game bird**. **1:** sometimes applied to all gallinaceous or chicken-like birds (Galliformes) and certain other birds that are hunted for sport or food. **2:** any bird hunted for sport or for one's own use as food. Included are many galliform birds (e.g., pheasant, grouse, partridge, quail, wild turkey) and waterfowl (most ducks and geese), certain doves (e.g., mourning doves), and certain shorebirds (Charadriiformes) such as woodcock, snipe, and certain plovers. In the United States hunting is highly regulated. To hunt nonmigratory game birds, one usually requires a license issued by the state. In the case of migratory birds, especially migratory waterfowl, federal and international rules and regulations apply. Such laws are chiefly intended to protect the birds from excessive harvesting.

bird guano. guano.

bird's eye. a common name of herbs of the genus *Adonis*, especially *A vernalis*.

bird's-eye maple. the lumber and finished wood of the sugar maple, *Acer saccharium*, having natural markings in the grain that resemble eyes.

bird's-eye rot. anthracnose of grapes. Brown concave spots are formed on the stem and fruit.

birddog aircraft. an aircraft carrying the person who directs the fire bombing on a wildfire.

birefringence. the numerical difference in value between the highest and lowest refractive index of a mineral.

birth. passage of the offspring of mammals from the uterus through the birth canal to the outer world; the process of being born. **premature birth**. birth of a viable infant prior to the full term of gestation.

birth canal. the passageway through the cervix and the vagina through which the newborn mammal passes during birth.

birth defect. any inherited disorder or abnormality that affects newborn mammals. Some are chiefly genetic in origin, and others are primarily the effect of environmental factors.

birth rate. natality.

1,1-bis(*p*-chlorophenyl)ethanol. chlorfenethol.

1,1-bis-(*p*-chlorophenyl)methyl carbinol.

chlorfenethol.

bis(diethylthiocarbamyl) disulfide. disulfiram.

bis(diethylthiocarbamoyl). disulfiram.

***S*-1,2-bis(ethoxycarbonyl)ethyl *O,O*-dimethylphosphorodithioate**. malathion.

bis(tributyltin) oxide. tributyl tin; tributyl tin compound.

bis-(dimethylthiocarbamyl) disulfide. thiram.

bis(1-methylethyl)carbamothioic acid S-(2,3,3-trichloro-2-propenyl) ester. triallate.

3,6-bis-(carboxymethyl)-3,5-diazooctanedioic acid. edetic acid.

1,1-bis(*p*-chlorophenyl)-2-nitrobutane. Bulan.

β,β-bis(*p*-chlorophenyl)-β,β,β-trichlorethane. DDT.

1,-1-bis(*p*-chloro-phenyl)-2,2,2-trichloroethanol. dicofol. **3,7-bis(dimethylamino)phenazathionium chloride**. methylene blue.

3,7-bis(dimethylamino)phenothiazin-5-ium chloride. methylene blue.

bisethylxanthogen. dixanthogen.

bisexual (ambisexual; ambosexous). **1:** pertaining to an animal or plant that has active male and female organs. **2:** of, pertaining to, comprised of, or involving two sexes. **3:** hermaphroditic; having characteristics of both sexes, including behavioral characteristics. 4: expressing sexual interest in members of both sexes. 6: a bisexual plant, animal, or person.

2,2′-bis[8-formyl-1,6,7-trihydroxy-5-isopropyl-3-methylnaphthalene]. gossypol.

bish. *Aconitum ferox*.

bishma. *Aconitum ferox*.

bishop's weed. *Ammi majus*.

bishydroxycoumarin. dicumarol.

bis(3,5,6-trichloro-2-hydroxyphenyl)-methane. hexachlorophene.

1,4-bis(methanesulfonoxy)butane. busulfan.

2,2-bis(*p*-methoxyphenol)-1,1,1-trichloroethane. methoxychlor.

bis(1-methylethyl)carbamothioic acid S-(2,3-dichloro-2-propenyl) ester. diallate.

bis(2-methylpropyl)carbamothioic acid *S*-ethyl ester. butylate.

bison. *See Bos bison*, *Bos bonasus*.

bit. the metal part of a bridle that fits into the horse's mouth to enable the rider to control the horse by use of the rein, *q.v.*

bitch. a reproductively mature female canine (e.g., fox, wolf, jackal, coyote), especially the domestic dog, *Canis familiaris*.

bite. **1a:** to seize, wound, abrade, or puncture with the teeth or mouth. **1b:** to sting. **1c:** a puncture or injury produced by biting or stinging. **2:** a morsel of food; a small meal or snack.

biting ox louse. *Bovicola bovis*.

bitter. **1:** harsh; painful; distressing. **2a:** sharp, biting. **2b:** having a sharp, not sweet, usually unpleasant taste.

bitter almond. *Prunus amygdalus amara*.

bitter almond oil. *See* apricot-kernel oil.

bitter aloe. aloe; aloe-emodin.

bitter pit. a disease of apples.

bitter rubberweed. **1:** *Actinea odorata*. **2:** *Hymenoxys odorata*.

bitterbrush. *Purshia tridentata*.

bitternut hickory. *Carya cordiformis*.

bitterweed. **1:** *Actinea odorata*. **2:** *Hymenoxys odorata*. **3:** *Senecio glabellus*.

bitting. training a horse to yield to the bit prior to its use while riding.

biuret (allophanamide; carbamylurea). a crystalline compound,

$$NH_2CONHCONH_2 \cdot H_2O$$

It is formed from urea, that is soluble in water and ethanol. It is toxic to some crops and is a potential contaminant of some organic fertilizers.

black acacia. *Robinia pseudoacacia*.

black alder. *Alnus glutinosa*.

black and yellow garden spider. *Argiope aurantia*.

black Angus. Aberdeen Angus.

black ash. barium sulfide.

black baldy. a cow or calf, resulting from a hereford and angus mating, that has a black body with a white face and white markings.

black baneberry. *Actaea spicata*.

black blowfly. *Phormia regina*.

black carpenter ant. *Camponotus herculeanus pennsylvanicus*.

black cherry. *Prunus serotina*.

black choke. *Prunus serotina*.

black cloth. cloth used to cover greenhouse plants to provide short photoperiods to promote blooming of short-day plants.

black cohosh. *Actaea spicata*.

black cutworm. *Agrotis ypsilon*.

black disease (false blackleg). a highly contagious disease of cattle and sheep and occasionally of swine and horses. It is caused by *Clostridium oedematiens*. The signs are similar to those of blackleg and malignant edema. Vaccination can prevent development of the disease in animals at risk.

black earth. chernozen.

black elder. *Sambucus canadensis*.

black fly (buffalo gnat). any fly of the family Simuliidae (Order Diptera). The female frequently bites viciously; livestock is sometimes attacked by swarms of these insects. Larvae and pupae occur, sometimes in immense numbers, on rocks in swift streams. Several species are vectors of diseases affecting humans, livestock, and other animals.

black garget. ovine gangrenous mastitis.

black gum. a common name of *Nyssa sylvatica*, favored by foresters.

black hellebore. *Helleborus niger*.

black ill. *See* blackroot (def. 1).

Black Leaf 40. trademark for a formerly used insecticide consisting of a 40% solution of nicotine sulfate.

black locust. *Robinia pseudoacacia*.

black maple. *Acer nigrum*.

black medic. *Medicago lupulina*.

black mustard. *Brassica nigra*.

black nightshade. *Solanum nigrum*.

black pod. a severe fungal disease of *Theobroma cacao*.

black quarter. *See* blackroot (def. 1).

black rat. *Rattus rattus*.

black roan. **1a:** the color of a horse in which the coat is a relatively uniform mix of black and white such that the coat appears dark or blackish. **1b:** a horse of such a color. *Cf.* blue roan.

black sage. *Salvia mellifera*.

black sheep. **1:** any sheep with black fleece. **2:** a person who deviates from the accepted standards of behavior or who brings shame or embarrassment to his or her family or community.

black snake root. a common name of plants of the genus *Zigadenus*, especially *Z. venosus*.

black spot. **1:** a disease of plants caused by any of several fungi or bacteria. It is characterized by the presence of black spots on the fruit and leaves, lesions on twigs, defoliation, and rotting. **2:** any microorganism that causes the disease. **3a:** British: a dangerous area or place, especially a dangerous place in a road; one where accidents are frequent.

black spruce. *Picea mariana*.

black sugar maple. *Acer nigrum*.

black swallowtail. *Papilio palyxenes asterius*.

black tupelo. a common name of *Nyssa sylvatica*.

black Welsh mountain sheep. a small, black-faced, hardy breed of sheep with long, dark brown wool. A completely black strain also exists. They are a source of wool and mutton.

black turpentine beetle. *Dendroctonus terebrans*.

black-eyed Susan. **1:** a common name of a wildflower, *Rudbeckia hirta*. **2:** a casual name for *Abrus precatorius*.

black walnut. *Juglans nigra*.

blackface mountain sheep. a common Irish breed of sheep.

blackfaced sheep, blackface sheep. any breed of sheep with black faces.

blackjack pine. *Pinus ponderosa*.

blackleg. **1:** black quarter; quarter evil; black ill; an acute, usually rapidly fatal, infectious disease that affects the musculature of the upper legs of cattle and occasionally sheep. Animals between the ages of six months and two years are most commonly affected. The causative agent is a bacillus, *Clostridium chavoei*. Blackleg is a gas gangrenous disease with inflammation and swelling of muscles (especially of the shoulders, neck, and thighs) and associated pain and muscle damage. The disease responds partially to treatment with large doses of antibiotics. The disease is rapidly fatal, with death ensuing within 24 hours following the appearance of signs. In areas where the disease is endemic, vaccination of young animals can prevent the spread of the disease. **2:** a disease of potatoes caused by a complex of bacteria such as *Erwinia carotouora* and *E. atroseptica*. The bacteria cause a moist, black, soft rot from the seed to the stem in immature plants; the base of the stem turns black. **3:** a fungal disease of cabbage and other crucifers caused by *Phoma lingam* (Family Sphaeriodaceae). The fungi infect the seeds and produce lesions on the stem near the ground or on the petioles and leaves, which darken and become flaccid. **4:** bacterial leaf spot; a disease of geraniums caused by *Botrytis spp. and Rhizoctonia spp*. The disease is characterized by rotting stems that become black at the base.

blackseed. *Medicago lupulina*.

blacksided Trondheim. a small, white Norwegian breed of polled cattle with black spots.

blackstrap (blackstrap molasses). the liquid by-product of the crystallization of sugar from the liquid extracted from sugarcane or sugar beets. Blackstrap is used as a supplemental feed for cattle.

blackstrap molasses. blackstrap.

blacktongue. a disease of canids due to a deficiency of nicotinic acid (niacin). Signs include ulcers of the mouth, an inflammation of the alimentary tract, and extreme nervousness. **2:** a variant of anthrax in cattle and horses. The disease is marked by an ulceration of the tongue.

blackwater. **1:** azoturia. **2:** water that contains fecal material or food wastes.

bladder. a fluid-filled or inflatable sac-like organ. **swim bladder**. a two-chambered, gas-filled sac that serves as a hydrostatic organ in certain fish, allowing the maintainance of neutral buoyancy. It is located beneath the vertebral column and connects to the esophagus in certain fish. **urinary bladder**. a thin muscular sac that temporarily stores liquid or semiliquid wastes in certain invertebrate animals and in terrestrial vertebrates other than birds. *See also* gallbladder.

bladderpod. *See Sesbania.*

blade. **1:** a thin, flat leaf. **2:** the broad, flat, expanded part of a leaf. **3a:** a cutting tool such as a knife, axe, scalpel, or scythe. **3b:** the flat, sharp part or cutting edge of a cutting tool. **4:** a relatively thin, flat tool or part of a tool or structure (e.g., the blade of a bulldozer). **5:** the posterior part of a fowl's comb.

blanch. **1a:** to become pale, to whiten (as with bleach), or to fade. **1b:** to blush. **1c:** to make pale. **2:** to etiolate (*See* etiolation). **3:** to whiten a vegetable such as celery while it grows by etiolation, *q.v.*, by wrapping with paper or by mounding soil about the part to be whitened. **4:** to prepare vegetables or fruit for processing (e.g., canning) by heating them, usually in water or steam to inactivate enzymes.

bland. **1:** tasteless; insipid; Mild-flavored. **2:** dull; monotonous; uninteresting.

blast. **1:** a disease of sheep that involves flatulence. **2:** a plant disease in which a plant wilts presumably due to some unidentified component of the atmosphere.

-blast. a suffix meaning an embryonic precursor; a germ of something.

blausäure. hydrogen cyanide.

blaze. a white patch (e.g., a star) on the face of an animal. **2:** a mark or cut on a tree that marks a trail or timber to be felled.

bleach, chlorine. *See* bleach, household.

bleach, household. any product used in the home (also with applications on a farm) to bleach materials, usually fabrics. The chief active ingredients are usually hypochlorite salts, especially sodium hypochlorite which is produced by adding chlorine to a liquid solution of lye. Household bleaches contain other harmful ingredients such as chlorine, lye, artificial dyes, detergents, fluorescent brighteners, and synthetic fragrances. In the United States, household bleaches carry warning labels such as "CAUTION: keep out of reach of children. May be harmful if swallowed or may cause severe eye irritation. Never mix chlorine bleach with cleaning products containing ammonia or with vinegar. The resulting chloramine fumes are deadly. It also should not be used on silk, wool, mohair, leather, spandex, or on any natural fiber that is not colorfast as it can damage or discolor the fabric and cause colors to run." Note that such labels only warn against drinking liquid bleach. It is also a strong irritant, however. Contact with the skin and inhalation of fumes should be avoided. *See* hypochlorite.

bleaching clay. a clay in its natural state or after chemical activation that is able to adsorb dyes and other colorants.

bleaching powder. chlorinated lime.

bleat. a vocal sound ("baaa") made by a sheep or goat.

bleed. to lose blood, usually by rupturing, severance, or other damage to a blood vessel.

bleeding. the discharge of blood as from a punctured, incised, severed, or otherwise damaged blood vessel. **arterial bleeding**. bleeding from an artery, characterized by spurts of bright red blood. **capillary bleeding**. bleeding from capillaries. Capillary bleeding is diffuse, often persistent, but easily controlled, and occurs to some extent in all bleeding. **external bleeding**. the discharge of blood from the body's exterior. **gastrointestinal bleeding**. bleeding from the gastrointestinal tract. This should always be treated as a serious condition and the cause promptly determined. **internal**

bleeding. bleeding into internal body spaces. **occult bleeding**. inapparent bleeding. **venous bleeding**. the discharge of blood from a vein, characterized by continuous flow of dark red blood. *Cf.* hemorrhage.

bleeding disease. a disease of cattle caused by lying on moldy sweetclover hay. The effects are similar to that of bracken poisoning. Bleeding disease is characterized by uncontrollable internal bleeding, especially within the intestinal tract and under the skin from bruised muscles. The proximate poison is dicoumarin, a breakdown product of the glycoside coumarin.

bleeding heart. *See Dicentra*.

blepharism. spasm of the eyelid with continuous, involuntary blinking.

blepharitis. inflammation of the eyelid.

blepharospasm. twitching due to spasmodic contraction of the orbicularis occuli muscle.

Blighia sapida (some common names for the tree and the fruit: in Jamaica and the United States: akee; in Haiti: akee; aki; ackee; arbre; fricasse; in Cuba and Puerto Rico: vegetal). an evergreen tree indigenous to western Africa (Family Sapindaceae) and cultivated in southern Florida, the West Indies, and tropical America. The seeds, cotyledons, and unripe fruit contain highly toxic, deadly, hypoglycemic polypeptides. The aril of the unripe fruit contains hypoglycine A, the more toxic of the two. The whitish, ripe aril is cooked and eaten as a delicacy, but the water in which the fruit is cooked is toxic. Rancid fruit is also highly toxic. Ingestion of the unripe fruit by humans causes Jamaican vomiting sickness, which is characterized by severe emesis, drowsiness, stupor, coma and convulsions, with conspicuous hypoglycemia. Ingestion of the unripe fruit may cause defects in calves of domestic cattle similar to those of crooked calf disease. *See also* ackee poisoning.

blight. **1a:** any disease or infestation; pestilance. **1b:** to afflict; plague. **1c:** any disease, especially one that affects large numbers of individual organisms, such as a plague, pestilence, the effects of widespread drought. **2:** generic term for any sudden, severe plant disease that is characterized by withering and death without rotting of a plant or of the affected parts. The term is now applied chiefly to diseases caused by bacteria, viruses, and fungi. Other plant afflictions (caused by insects or unfavorable climatic conditions) that display similar symptoms may also be called blights. **3:** the causative agent of such a disease (def. 2, above). **4:** any cause of serious impairment or destruction. **5:** the condition or effects of deterioration or decay. **7:** to suffer blight. **8:** to destroy or cause to wither or decay.

blind. sightless; unseeing; unable to see; lacking useful sight.

blind area. in forestry, any location that cannot observed from a lookout post.

blind furrow. dead furrow.

blind nipple. blind quarter.

blind quarter (blind teat; blind nipple; inverted nipple). **1a:** an obstructed teat or a serious hereditary disorder of swine and first-calf heifers whereby a quarter of an udder fails to secrete milk. **1b:** a nonfunctional mammary gland.

blind seed. a disease of *Secale* (ryegrass) in which the seeds soften and shrivel. It is caused by the fungus *Phialea temulenta*.

blind staggers. alkali disease (def. 3); acute or subacute selenium poisoning. *See* selenium poisoning.

blind teat. blind quarter.

blindness. the state of being blind; loss of, lack of, or severe impairment of vision or the ability to perceive visual stimuli. *See also* amaurosis. **canine hereditary blindness**. an autosomal dominant condition of collies and certain other breeds of the domestic dog.

blissom. a ewe in heat.

blister. a small, liquid-containing sac or vesicle on the surface of the body or the lining or surface of an organ, usually less than 1 or 2 cm in diameter. **blood blister**. a blood-colored blister with some blood in the contents. **water blister**. a blister with nearly clear watery contents.

blister beetle poisoning. cantharidin poisoning of livestock, especially horses, although cattle and sheep are sometimes affected. Most reports of poisoning originate in the central and southwestern United States. The two-striped blister beetle (*Epicauta spp.*) sometimes swarms alfalfa hay while drying prior to baling. These beetles contain cantharidin, *q.v.* Ingestion of beetle-infested hay results in nephritis, cystitis, and hyperemia or ulceration of the oral, esophageal, and gastric mucosa. Clinical manifestations include colic, salivation, shock, frequent urination, and, occasionally, synchronous diaphragmatic flutter. Diagnosis is confirmed by detection of blister beetles in the hay and of cantharidin in the urine or stomach contents of affected animals. There is no certain treatment.

bloat. **1a:** to cause to swell (e.g., with water or air) or to make turgid (as a cell). **1b:** to cause or to result in the accumulation of gases in the digestive tract. **2a:** to fill to capacity or to overfill (stuff). **2b:** to swell; to become turgid. **3a:** the overextension of the first stomach (rumen) due to the accumulation of gas, or the condition that results from rupture of the rumen of domestic cattle. Fermentation of ingested forage and release of the gases produced take place in the rumen. These gases are normally expelled by belching. When this is not possible, the expanding gases may rupture the rumen. *See Delphinium.* **3b:** any flatulent digestive disturbance of livestock. **4:** bloat colic of horses due to distension of the gastrointestinal lumen. Causes vary.

blocky. having many sides with angular or rounded corners, as in the case of certain peds.

blocky foot. the condition of a horse's hoof where it has more vertical slope than normal.

blond d'Aquitaine. a large, horned, fawn-colored breed of beef cattle that originated in France.

blood. **1:** any fluid (often a complex tissue) that circulates in the vascular system of an animal. Blood typically distributes nutrients, oxygen, hormones, etc. to respiring tissues and absorbs waste products that are transported to the organs of elimination (e.g., lungs, gills, kidneys). Blood may also play specific roles in protection against infection and may scavenge dead cells. The blood often contains respiratory pigment, which is free in the plasma in most invertebrate animals but is contained in cells in vertebrates and certain invertebrates. The blood of vertebrate animals is considered a connective tissue. Systemic poisons are potentially transported from the site of entry via the blood to all parts of the animal's body, including target organs, detoxifying tissues, and organs of elimination. **2:** slang meaning full-blooded or descended from a single line or breed. **3:** descended from a registered breed. **4:** a market classification used to indicate the fineness of wool (e.g., quarter blood, low quarter blood, half blood). **5:** a market classification for a type of Spanish orange.

blood-brain barrier. *See* barrier.

blood flour. a feedstock of dried, powdered blood given chiefly to dairy calves on a minimum milk diet. *See* blood meal.

blood group substance. *See* antigen (blood group antigen).

blood meal. a feed made of animal blood that is heated to coagulation, drained, pressed to remove remaining water, ground to meal, and mixed with calf starter feed. *See* blood flour.

blood poisoning. a common nonmedical term that refers to the presence of infecting agents (e.g., bacteria) or their toxins in circulating blood of higher animals. *Cf.* septicemia, pyemia, toxemia.

blood pressure (BP). **1:** a measure of the pressure exerted on the wall of any blood vessel of an animal by the blood within.

In vertebrate animals blood pressure is generally measured for an artery (arterial tension), hence def. 2. **2:** the pressure (tension) exerted by blood circulating in the main arteries of a vertebrate. Numerous drugs and other chemical affect blood pressure. **high blood pressure**. hypertension. **low blood pressure**. hypotension.

blood spavin. a large varicose vein in a horse's leg inside of the hock, immediately above the site of bog spavin.

blood vessel. any of the closed tubular vessels that contain the circulating blood of an animal (e.g., artery, vein, venule, arteriole, and capillary).

bloom. **1:** to develop, flourish, or thrive. **2a:** to blossom or flower. **2b:** the state or condition of blossoming or flowering. **3:** an individual bud, blossom, or flower. **4:** freshness; a flush, glow, or radiance. **5:** descriptive of a dairy cow in early lactation when the coat is smooth, offering evidence of milking ability (dairy character). **3:** a waxy layer on the surface of certain fruits (e.g., apples, plums), leaves, stems, or other plant parts (e.g., apples). **6:** a proliferation of algae and/or higher aquatic plants in a body of water, especially algae or blue-green bacteria (Cyanobacteria, *q.v.*). **algal bloom**. a phytoplankton bloom marked by a sudden proliferation of algae that may form a surface scum and produce foul odors, such a bloom can adversely affect water quality and may indicate local, potentially hazardous changes in water chemistry. Formation of an algal bloom is favored by stagnation, warm temperatures, sunshine, and fertilization of a water body by entry of available phosphates and/or nitrates. Algal blooms have caused severe illness and extensive mortality among livestock, fish and wildlife, pets, and even humans in many areas of the world, including North America, experience life-threatening blooms. Decaying algal blooms can be especially toxic; they produce toxic compounds such as hydroxylamine and hydrogen sulfide. Algal blooms also deplete the oxygen in the water body at night through respiration and the process of decay, often killing aerobic aquatic organisms. Sustained algal blooms may lead to eutrophication of the water body. Poisonings attributed to algal blooms may sometimes be due to the presence of other organisms, most importantly the bacterium, *Claustridium botulinum*. *See also* alga, algal poisoning, *Gonyaulax catenella*, red tide. **aquatic bloom**. water bloom. **dinoflagellate bloom**. red tide. **gonyaulax bloom**. red tide. **gymnodinium bloom**. red tide. **phytoplankton bloom**. a type of water bloom in which a dense growth of algae and/or other phytoplankton, especially blue-green bacteria (*Cyanobacteria*), occurs in or on a water body (e.g., a lake or pond). An algal bloom is a phytoplankton bloom. **red dinoflagellate bloom.** red tide. **toxic bloom. 1a:** a toxic water bloom comprised of algae, the cyanobacteria, and/or other aquatic microorganisms. **1b:** algal bloom. **water bloom** (aquatic bloom). **1a:** the accumulation of microorganisms at or near the surface of a water body. **1b:** the accumulation, mainly of blue-green - bacteria, at the surface of a water body. Algal blooms, dinoflagellate blooms, and toxic blooms are types of water blooms. **2:** the microbiota (especially algae or the cyanobacteria) that make up a bloom. **3:** the scum formed by a water bloom as previously defined. **4:** the proliferation of higher aquatic plants.

blossom. **1a:** a bloom, bud, or flower. **1b:** to flower. **1c:** to grow, develop, thrive. **2:** the bloom or flower of an angiosperm. **3:** a bay or sorrel horse with an admixture of white in the coat.

blow-out. **1:** an excavation of loose soil or sand by wind. **2:** the blowout of a balloon tire. **3:** to lightly exercise or walk a horse to loosen it up for more intense exercise or to prevent chills or muscle stiffness following heavy work or running.

blowdown (windthrow). **1a:** uprooting by the wind. **1b:** a tree or trees uprooted and blown down by wind.

blowfly. a common name applied to a large number of often brightly-colored flies that infest wounds or lay eggs in rotting flesh.

blowing adder. *Heterodon contortrix*.

blowings. cotton waste from ginning.

Blue 1. FD&C Blue No. 1

Blue 2. FD&C Blue No. 2

blue bag. gangrenous mastitis marked by bluish discoloration of the udder, mainly of cattle, sheep, and goats.

blue cheese (blue-mold cheese). any cheese processed from cow's milk (or, in France, from sheep's milk) and ripened by a blue mold for two or three months. Such cheeses resemble Roquefort.

blue copper. the pentahydrate salt of cupric sulfate.

blue ear. a disease of swine in which sows may farrow prematurely.

blue eye. an distinct defect in the eye of a horse or mule put up for sale.

blue flag. a common name of plants of the genus *Iris*.

blue fox. *Alopex lagopus*.

blue gramagrass. *Bouteloua gracilis*.

blue gray. **1:** a calf produced by the mating of a white shorthorn bull with a Galloway or Aberdeen angus cow. **2:** a horse with a gray coat that contains scattered steel-blue hairs.

Blue John. vernacular for watered, skimmed, or somewhat sour milk that has a bluish tinge.

blue mold of tobacco. a disease of tobacco caused by the mold, *Peronospora tabicina* (Family Peronosporaceae). The leaves of affected plants are cupped, and one can observe a violet-colored, moldy growth on the undersurface of the leaf which becomes necrotic; young plants may die.

blue nose. **1:** a disease of horses. **2:** *See* acorn fruit.

blue roan. **1a:** the color of a horse with a coat of black hair interspersed with white such that the coat appears bluish or grayish in color. **1b:** a horse of such a color. *Cf.* black roan.

blue sedge. *Andropogon virginicus*.

blue skin. a condition marked by bluish spots on the skin that is considered a fault or disqualification for purebred sheep.

blue stain. blue discoloration of light colored wood due to secretions of fungi of the genera *Aureobasidium*, *Ceratocystis*, and *Lasiodiplodia*.

blue vitriol. the pentahydrate salt of cupric sulfate.

blue-faced Leicester. a medium-sized breed of sheep with long, dense, curly fleece. The ram is often used for crosses with smaller breeds.

blue-green algae. former common name of blue-green bacteria.

blue-green bacteria. a common name of microorganisms of the Cyanobacteria.

blue-mold cheese. blue cheese.

blueberry. *See Vaccinium*.

bluebonnet. a common name of herbs of the genus *Lupinus*.

bluebottle. a brightly colored blowfly whose maggots live in rotting flesh and sometimes in living sheep.

bluegrasses. grasses of the genus *Poa*. They are highly palatable range and pasture grasses and are valued highly as hay, as lawn grasses, and usefulness in erosion control.

bluejoint reedgrass. *Calamagrostis canadensis*.

bluejoint. *Calamagrostis canadensis*.

bluestone. the pentahydrate salt of cupric sulfate.

bluetop. *Calamagrostis canadensis*.

BMF. biomagnification factor (*See* factor).

BMP. best management practices.

boar. **1:** an adult male uncastrated swine. **2:** European wild boar. **European wild boar** (wild boar). *Sus scrofa*.

boar pig. a male pig of less than 1 year of age.

bob wire. barbed wire.

bobbed wire. barbed wire.

bobtail. an animal with a short tail.

BOD. biological oxygen demand.

BOD5. five-day biological oxygen demand: the amount of oxygen (mg O_2/liter of water) taken up from a water sample that contains a known amount of dissolved oxygen after storage at 20°C for 5 days. *See also* biological oxygen demand, CBOD5.

bodewash. dry buffalo or cow dung. Such dung (and that of other ruminants such as that of the Bactrian camel in the Gobi desert) has been and is used as fuel in treeless (e.g., desert and semidesert) areas. *See* buffalo chip.

Boer. a South African breed of white goats with red head markings that provide meat and skins.

bog. a type of wetland ecosystem in which organic matter accumulates as peat. Bogs are usually acidic, often with a green mat of moss, and are rich in plant residue which is slowly turned into peat. Bogs usually develop in a depression with precipitation serving as the major source of water. **blanket bog**. a thick, extensive area of bog.

bog asphodel. *Narthecium ossigragum*.

bog iron ore (bog iron). a spongy ferruginous deposit of hydrated oxide of iron (limonite) that forms in bogs and swamps by the oxidizing action of algae, bacteria, and/or the atmosphere on dissolved iron. It occurs as clumps and layers on level, sandy, acid soils that have underlain swamps or bogs.

bog iron. bog iron ore.

bog moss. peat moss.

bog onion. *Arisaema stewardsonii*.

bog rosemary. *See Andromeda*.

andromedotoxin (grayanotoxin I). **1:** a sometimes amorphous, sometimes crystalline, strongly emetic, cardiotoxic compound that occurs in the leaves of all toxic heaths (Family Ericaceae), including *Andromeda* (andromeda; e.g., bog rosemary), *Kalmia latifolia* (mountain laurel), *Ledum* (Labrador tea), *Lyonia* (e.g., stagger bush), *Pieris* (pieris), and *Rhododendron* (rhododendrons, azaleas). In humans and laboratory mammals, andromedotoxin paralyzes the motor nerve terminals in skeletal muscle. *See* grayanotoxin, *Rhododendron*. **2:** any grayanotoxin.

bog spavin (boggy in the hocks). a distention of the joint on the inner surface of the hock of a horse due to a buildup of fluid. A horse with this condition is definitely unsound.

bog spruce. *Picea mariana*.

boggy in the hocks. bog spavin.

bogland. a bog or boggy area.

Bohr effect. *See* effect.

boil down. to concentrate a liquid by boiling to remove volatile components (usually water) as in boiling down the sap from a sugar maple tree to produce a syrup or sugar.

boiler. a layer hen, sold for the table, because it is no longer laying.

boiling point. the temperature at which the vapor pressure of a liquid equals or slightly exceeds the vapor pressure of the atmosphere of its surroundings.

boiling-waterbath canner. a pot with a lid, rack, and cover that is of sufficient size that the jars of food to be canned can be covered with at least 1 to 2 in. of water.

bolaffinine. a very toxic protein isolated from the mushroom, *Boletus affinis* (Family Boletaceae).

Bolander water chestnut. *Cicutum californica*.

bold. **1a:** brave; daring. **1b:** confident; assured. **2:** descriptive of a horse with an assured gait. **3:** denoting or descriptive of the assured manner of an animal.

bole. **1a:** the base or widest part of a tree trunk. **2:** the trunk of a tree.

bolesatine. a toxic protein isolated from *Boletus satanas*. It inhibits *in vitro* protein synthesis.

bolete. a fleshy mushroom; any fungus of the genus *Boletus*. **red-mouth bolete**. *B. subvelutipes*.

Boletus (bolete). a genus of soft mushrooms (Family Boletaceae). Many species are edible, but a few are indigestible or strongly emetic if eaten raw, causing mild to severe gastroenteritis. A few (e.g., *B. affinis*) may be lethal to humans and to grazing animals (e.g., *Bos indicus*). ***B. affinis***. a possibly lethal species known from eastern North America, but more widely distributed. ***B. luridus***. a mushroom that produces disulfiram-like effects if alcohol is taken within 3 days after ingestion. ***B.satanas***. a European species that contains bolesatine (a toxic protein) and is considered by some to be toxic. ***B. subvelutipes*** (red-mouth bolete). a probably poisonous mushroom which, at some times in some places and should not be eaten. it occurs from Quebec to Virginia, west to Michigan in deciduous and mixed woods on the ground, usually near beech, balsam fir, eastern hemlock, or white spruce.

Bolivian coca. *Erythroxylum coca*.

boll. the seed pod of the cotton plant or flax plant.

boll number. the number of bolls per cotton plant.

boll period. the period of growth of cotton from the formation of bolls to their opening.

boll weevil. *Anthonomus grandis*.

bolling. a tree that has had its branches removed.

bollworm. *Heliothis zea*.

bolly. an unopened cotton boll that contains usable lint.

bolly cotton. the lint removed from bollies.

bologna bull. a medium-sized bull with little fat that is used to make bologna.

bolter. **1:** a plant that grows too fast, producing a low yield of seed. **1:** a sieve.

bolus. **1a:** any masticated bit of food that is ready to be swallowed. **1b:** a ball of regurgitated, partly digested, masticated food that is to be swallowed. **2:** a large "pill" or formed quantity of substance intended for dosing animals.

boma. a small enclosure, often of brush or brambles, used to enclose cattle, human campsites, or habitation. They offer protection from large predators. **thorn boma**. a boma made of thorn bushes.

bomb calorimeter. an apparatus used to measure the amount of heat released by a combustible substance, as in the determination of the caloric content, per gram, of livestock feeds.

Bombidae (bumble bee family). a family of medium to large social bees with long tongues. They sting viciously. The queen hibernates and initiates a new underground colony each spring. They are important flower pollinators, but many species have become rare because of sensitivity to pollutants. *See Bombus*.

Bombus (bumble bees). a genus of large bees (Bombidae). The venom is acidic and contains hemolytic enzymes. The sting causes local pain and a burning sensation, with swelling, erythema, and whitishness at the wound site. Common systemic effects are respiratory distress and shock. Multiple stings may require hospitalization. Fatalities are rare and are usually due to anaphylactic shock.

bone. a hard, rigid connective tissue that forms the bulk of the skeleton of bony fish and the higher classes of vertebrates. It is comprised of osteocytes, collagen fibers, and calcium salts (hydroxyapatite). Bones are usually hollow and in most vertebrates contains marrow, a blood-forming tissue. Bones form a framework of the body, make locomotion possible, and shield vital parts of the body (e.g., brain) from trauma. The calcium of bone is important in most species of bony vertebrates as a calcium reserve in the body. Calcium from bones forms the shell of bird eggs.

bone black (animal charcoal; animal black; bone charcoal; bone char). residue remaining from heating bones in a closed retort. Bone black is used, for example, in the manufacture of activated carbon, as a decolorizing agent, as a pigment in paint and shellac, and in water purification. Spent bone black is used in the manufacture of fertilizer.

bone char. bone black.

bone charcoal. bone black.

bone meal. animal bones or horns that have been steamed under pressure and ground into a fine powder. It is used as a fertilizer and as a mineral supplement for feeding farm animals, especially young animals. It contains 1.5 to 2.5% nitrogen, 12 to 15% phosphorus, and 20 to 34% calcium; significant amounts of magnesium are also present.

bone-crusher disease. dengue fever.

boner. a utility-grade beef carcass. The meat is low in fat, is not marbled, and is usually ground or processed.

boneset. common name of *Eupatorium perfoliatum*.

bonetail. *Bothrops atrox*.

bony fish. fish of the vertebrate class, Osteichthyes.

bood bud. bulblet.

BOR. Bureau of Reclamation.

borage. *Borago officinalis*.

Borago officinalis (borage). an herb with blue flowers and coarse, spiny leaves that is used to flavor lettuce salads.

boran. a usually white breed of East African beef cattle that is raised most commonly in Kenya.

borate ester. *See* borate.

borate salt. *See* borate.

borate. most commonly defined as any basic salt or ester of boric acid. **sodium borate** (sodium tetraborate; sodium pyroborate; borax decahydrate; tincal). a naturally occurring substance found in salt lakes, alkali lakes, and alkali soil. It is an odorless, white crystalline substance or powder,

$$Na_2B_4O_7 \cdot 10H_2O$$

It is used in cleaning solutions, herbicides, fertilizers, rust inhibitors, bleaches, as a detergent, as a flame-retardant fungicide for wood, and a laboratory reagent. It is toxic by inhalation or ingestion. *See also* borax, borism.

borax decahydrate. sodium borate (*See* borate.

borax. the commercial name for sodium borate. **anhydrous borax** (dehydrated Borax; anhydrous sodium borate; anhydrous sodium borax) a toxic, white, hygroscopic, nonflammable, free-flowing crystalline substance that forms a partial hydrate in humid air. It is used as an herbicide and in the manufacture of glass, enamels, and ceramics.

Bordeaux mixture. a mixture of cupric sul fate, calcium hydroxide, and water used as a fungicide on vines and fruit trees. It was originally applied to grape vineyards near Bordeaux, France. The mixture can injure certain crops. *See also* Burgundy mixture, cupric sulfate.

border disease. a viral disease of sheep in which ewes may abort or deliver weakened lambs, often with tremor. The disease is transmitted by the excreta of infected ewes.

border Leicester. a breed of longwooled sheep descended from the English Leicester. The head is bare and it has a pronounced Roman nose. The rams are used for breeding.

Bordetella. a genus of obligately aerobic, motile or nonmotile (motile species are peritricious) parasitic bacteria (Family Brucellaceae) that invade the mammalian respiratory tract. *B. pertussis* is the type species. ***B bronchiseptica***. the cause of atrophic rhinitis in swine, bronchopneumonia in rodents, and bronchopneumonia secondary to distemper in dogs.

boreal. northerly; pertaining to cool or cold temperature regions of the northern hemisphere such as those that occur at high latitudes or high mountains. Boreal regions or climates are more or less coincident with the northern coniferous zone, the taiga, with northern needle-leaf forest formations in general. *Cf.* antiboreal.

boreray. a rare breed of horned sheep that have cream-colored or gray fleece.

boric acid (boracic acid). a very toxic white, weakly acidic, crystalline substance, H_3BO_3. It is a nonirritating, weakly bacteriostatic substance and has been used widely in optic irrigants, as an anti-infective agent, and as a disinfectant. The effects of poisoning are usually reversible, but this substance can prove lethal. It can be absorbed through abrasions, burns, or breaks in the skin. In some countries, it is still a common cause of poisoning in infants. The oral MLD is about 2-3 g for human infants, about 5-6 g for children, and about 15-20 g for adults. Systemic poisoning may result from doses as low as 0.17-0.2g/kg body weight. Boric acid is a cumulative poison. Poisonings also arise from agricultural and industrial use and from careless usage of household products that contain boric acid. Unfortunately, it is not regarded by most people as poisonous; it resembles sugar in appearance and has been consumed by children and accidentally used on foods. Boric acid is toxic to all cells. Symptoms of chronic poisoning may appear following several weeks of exposure. *See* boric acid poisoning.

bornu white. a Nigerian breed of goat.

boroethane. diborane.

boron (B)[Z = 5; A_r = 10.81]. a trivalent metalloid element that occurs as a hard crystalline mass or brown powder; it forms borates and boric acid. The most common source of boron is naturally occurring disodium decahydrate tetraborate (borax). It is an essential trace element of plants. A deficiency of boron in soil causes, for example, heart rot in sugar beets and a necrosis of the inner bark of many deciduous trees. High levels of boron in soil can kill or stunt plants by inhibiting or suppressing germination and can induce premature ripening.

Borrelia. a genus of parasitic bacteria (family Treponemataceae) that occur as coarse, shallow, irregular spirals with finely tapered filamentous ends. These bacteria parasitise a wide variety of organisms and are usually found in blood or on mucous membranes. ***B. anserina***. the type species of the genus, it causes spirochetosis in fowl. It can be isolated from the blood of infected gallinaceous birds, waterfowl, and ticks. ***B. burgdorferi***. a species that causes Lyme disease in humans and borreliosis in dogs, cattle, and perhaps in horses. ***B. theileri***. the cause of borreliosis in cattle and a number of other mammals in Australia and South Africa. The vectors are the ticks, *Boophilus micropus* and *Rhipicephalus evertsi*.

borreliosis. any of a number of related diseases caused by bacteria of the genus *Borrelia*. **bovine borreliosis**. a disease of cattle caused by *B. burgdofferi*. Signs include a painful laminitis, arthritis, and synovitis. **canine borreliosis**. a disease of canines caused by *B. burgdofferi* and characterized by lameness due to an intermittent, migratory arthritis.

Bos. a genus (Family Bovidae) of cattle and related ruminants. ***B. bison*** (American bison; buffalo). a large animal with a massive head and shoulders with short, curly, brown hair. Large numbers of bison were slaughtered by "buffalo hunters" in the 19th century, and the numbers were reduced to a handful. Today they are protected, and most occurs in reserves and national parks in the United States and Canada. The bison can breed with domestic cattle, producing a cattalo or beefalo, *q.v. See also* charalo. ***B. bonasus*** (European bison). a species now found only in zoos and animal reserves. They are large animals, similar in appearance to the American bison, with massive heads and shoulders and short, curly, brown hair. ***B. bubalus*** (water buffalo). a common domestic bovine in the tropics. It is used for milk and as a draught animal. ***B. grunniens*** (yak). a large, stocky, brown or nearly black, shaggy ox, wild or domesticated, with large, curved horns. It is indigenous to Tibet and surrounding areas. ***B. indicus*** (Asiatic domestic cattle; brahman; zebu; Indian ox). a species of domestic cattle (Family Bovidae) of Asia and Africa. They have large ears, a large dewlap, short horns, and a prominent hump between the shoulders. They are used for milk and as a draught animal in warm climates. Several breeds of this species have been developed into the Brahman breed in the United States. ***B. taurus*** (European domestic cattle). cattle originally domesticated in Europe. Many breeds are now common throughout the temperate zones, and most cattle in the United States originated in Europe and are considered to be *Bos taurus*. Indeed, the first domestic cattle to arrive in the Western Hemisphere were brought by Columbus on his second voyage across the Atlantic. Cattle of this species are probably descended from the aurochs, a large European wild ox domesticated during the Stone Age that became extinct in the early 18th century. Another species, the smaller Celtic shorthorn, was the most important domestic ox of the Stone Age and may also have contributed to the ancestry of *B. taurus*.

boscage. a thicket formed of shrubs.

bosque. a wooded thicket, usually of tamarask, willow, and cottonwood, common to flood plains in the southwestern United States and portions of the central and northern Great Plains.

boss. **1:** a circumscribed, rounded or lumpy protuberance on an animal or plant. **2:** a metal stud on either side of the bit of a bridle. **3:** the dominant or even controlling individual among a group of animals.

bossy. a mild-mannered cow of which a person is fond.

bot. the larval stage of a bot fly.

bot fly. any of a number of species of robust, hairy flies (Order Diptera). They are often striking in appearance, being either black and yellow or gray. The larvae, or bots, are parasites of muscle tissue of humans and a number of domestic animals, especially herbivores such as sheep, goats, horses, camels, and deer. As an example, *See Gasterpophilus intestinalis*.

botanic, botanical. **1:** pertaining to, derived from, or characterizing plants or plant material. **2a:** any biologically active (usually therapeutic) substance produced by or derived from plants. **2b:** botanic insecticide; botanic pesticide; an insecticide or pesticide produced by or derived from plants (e.g., pyrethrum, rotenone, ryania, sabadilla). **3:** pertaining to or concerning the science of botany.

botanic insecticide. *See* botanic, def. 2b.

botanical insecticide. *See* botanic.

botanical pesticide. *See* botanic.

botanist. a scientist who studies plants or applies scientific knowledge to the care or use of plants.

botany. The natural science that treats plants.

bothropic. of, pertaining to, or produced by snakes of the genus Bothrops.

Bothrops (American lance-headed vipers; American lanceheads; lanceheads; hog-nosed vipers; Amerikanische Lanzenottern; palm vipers). a genus of small to large venomous snakes (Family Crotalidae) of tropical and southern South America and the West Indies. The head is broad, depressed, and distinct from the narrow neck; the body is relatively slender to stout and cylindrical or slightly compressed. Dorsal body scales are keeled, forming 19-35 non-oblique rows at the midpoint of the body. The large snakes of this genus, especially the bush and tree vipers, are very dangerous and are responsible for most snakebites in tropical Central America. *See also* antitoxin (bothropic antitoxin). ***B. asper***. former scientific name of *B. atrox*. ***B. atrox*** (formerly *B. asper*; Fer-de-lance; barba amarillo; bonetail; caicaca; cuatro narices; equis; manapare; tomigoff; torciopelo). the Fer-de-lance (French meaning lancehead) is called "barba amarillo" (yellow chin) in Spanish. It is a large venomous crotalid that occurs throughout much of the forested areas of Mexico, and Central and South America as far as Peru and northern Brazil, and is common on banana, cocoa, and coffee plantations and along streams. It occurs also in parts of the West Indies where it is common and greatly feared on sugar plantations. It is responsible for more human deaths in tropical Central America (and probably in the Americas altogether) than any other snake. The tail is tipped by a horny spine, and it is further distinguished by an olive-green, gray, or brownish ground color with a lateral pattern of dark, usually black-bordered triangles whose apices meet (or nearly meet) at the dorsal midline. The dark border may have a light edging. The belly is light cream to yellow with dark blotches. The fer-de-lance reaches a length of up to about 2.4 m. It often irritable, readily strikes (sometimes repeatedly), and is able to deliver a large amount of a powerful venom. Other species of tropical *Bothrops* and some species of *Trimeresurus* are sometimes called Fer-de-lance. *B. atrox* is most often confused with *B. jararaca*, *q.v.* Envenomation of humans causes local pain and bleeding from the bite, gums, nose, mouth, and rectum. Blood does not coagulate and seeps into muscles and the nervous system. Shock and respiratory distress precede death.

Botryomyces. an obsolete name of a number of presumed fungi that were associated with a disease of livestock and humans. The name of the disease was botryomycosis *q.v.*, a name which is still retained.

botryomycosis (actinophytosis) **1:** a chronic granulomatous condition of horses, cattle, swine, and humans that is usually caused by staphylococci; other bacteria may also be involved. The skin, and sometimes the viscera, is involved. Granules containing masses of bacteria appear in the pus.

botryomycotic. of, pertaining to, or affected by botryomycosis.

botryose. denoting or descriptive of plants with flowers that grow in clusters.

botrytis. a fungal disease of plants, especially young seedlings, caused by fungi of the genus *Botrytis*. Spots appear on flowers, leaves and stems of infested plants. The spots spread when wet, forming a gray, moldy growth.

Botrytis. a genus of parasitic fungi (Family Moniliaceae) the members of which cause botrytis.

bottlebrush. a common name of *Equisetum*.

bottle-feed. **1:** to nurse or feed infant mammals with milk or other nourishment from a nursing or "baby" bottle. **2:** to nurture, care for, or train with exaggerated concern and attention.

bottom. **1:** bottomland. **2:** the portion of a plow comprised of the moldboard, share, frame, and landside. **3:** endurance or stamina in a horse.

bottom board. the floor of a beehive.

bottom disease. crotalism.

bottomland (bottom; bottoms). an agricultural term for lowlands adjacent to a watercourse. Such land is subject to at least occasional flooding and usually has a rich, alluvial soil. *Cf.* alluvial plain, flood plain.

bottoms. *See* bottom land.

botulinic acid (allantotoxicon). a toxin in putrescent sausage.

botulinum poisoning. botulism.

botulism (botulinum poisoning; claustridium botulinum poisoning; Lamziekte). a rare but often rapidly fatal poisoning by ingesting botulinum toxin in food. The animals most frequently known to be affected are humans, domestic fowl, waterfowl, most livestock, and fish. Botulism is a serious and often fatal disease in most animals, but cats, dogs, and swine are fairly resistant to all types of botulinum toxin when administered by mouth. Species other than humans may acquire the toxin from carrion, contaminated vegetation, wounds, or directly from soil, alkaline muds, or bottom sediments where anaerobic conditions favor the proliferation of the causative bacterium, *Claustridium botulinum*. Paralysis is a feature of the disease in all species affected. Humans usually become intoxicated by consuming inadequately sterilized canned or bottled food that has become putrescent. Symptoms in humans usually appear within 12 to 36 hours following consumption of contaminated food. The first symptoms to present are disturbances of vision general weakness, and the victim is easily fatigued. The condition progresses to include nausea, vomiting, gastrointestinal distress, difficulty in swallowing, dysarthria, pronounced muscular weakness and paralysis, ptosis, loss of visual acuity, double vision, loss of pupillary reflexes, paralysis of the respiratory muscles, profuse sweating, a weak, rapid pulse, and sometimes diarrhea, with death from cardiovascular or pulmonary collapse. Congestion and hemorrhage may be noted in all organs, especially the CNS Suspected botulism requires prompt medical aid. Botulism in humans may be treated with type ABE botulinus antitoxin. *See also* alkali disease (Western duck sickness), allantiasis, limberneck.

botulism of ducks. alkali disease (def. 1).

boundary. **1:** border; margin; perimeter. **2:** the line that marks the edge of a property. **3:** a limit or barrier that confines an animal or any system to a definite place in space. such a boundary may be a natural barrier (e.g., a river, mountains, or even a change in soil type), man-made, even an imaginary boundary to which an animal is conditioned, or to which humans have agreed upon as a boudary.

boundary stone. a stone that marks an edge of a property.

bouquet. *See* nide.

Bourbon red. a breed of domestic turkey with white flight feathers, a white tail, and dark, reddish-brown body plumage.

Bouteloua (grama grass). a genus of grasses (Family Gramineae). a large number of species are native (or believed to be native) to prairies of western North America. They generally provide good to excellent forage for livestock. ***B. gracilis*** (blue gramagrass). a native North American grass used to reseed rangeland in the western United States ***B. eriopoda*** (black grama grass). an excellent forage grass of the arid southwestern rangeland of the United States

Bovicola. a genus of biting lice (Order Mallophaga); (considered by some to be a subgenus of *Damalinia*). ***B. bovis*** (*Trichodectes scalars*; biting ox louse; common red ox louse). a pest of cattle. ***B. caprae*** (*T. climax*). a pest of domestic sheep and goats. ***B. equi*** (*T. parumpilosus*; horse-biting louse). the common biting louse of horses, mules, and donkeys. ***B. ovis*** (*T. sphaerocephalus*), the common biting louse of sheep.

bovid. **1a:** any animal of the family Bovidae. **1b:** of or pertaining to the family Bovidae.

Bovidae. a family of the order Ruminentia comprising the hollow-horned ruminants such as true antelopes, cattle, oxen, goats, and sheep.

bovine. of or pertaining to cattle and related ungulates of the family Bovidae. It is sometimes restricted in usage to species of the genus *Bos*.

bovine acetonemia. bovine ketosis.

bovine anaplasmosis. *See* anaplasmosis.

bovine farcy. a slowly developing, chronic, fatal disease of cattle. The disease is marked by early development of swellings on the legs, especially above the hocks. The pathogen is *Actinomyces farcinicus*.

bovine genital trichomoniasis (tric; bovine venerial trichomoniasis). a sexually transmitted disease of cattle caused by *Trichomonas fetus*, a flagellate protozoan. Affected individuals are temporarily sterile. The disease is marked by difficulty in breeding, temporary sterility, and an accumulation of pus in the uterus in the abscence of fever. **tric**. bovine genital trichomoniasis.

bovine hemoglobinuria. bovine babesiosis.

bovine hemorrhagic syndrome (hemorrhagic syndrome of cattle). a disease of cattle incurred when on cornstalk pasturage or fed moldy corn, peanuts, or oats, and sometimes moldy commercial concentrates. The common cause of the disease is the fungus, *Aspergillus flavus*. Signs include excitement, incoordination, coma, and almost always death. Postmortem examination reveals tissue hemorrhages throughout the body. Young cattle are most susceptible, but swine, other livestock, and dogs are also susceptible.

bovine hypoglycemia. bovine ketosis.

bovine leucosis. a malignant disease of cattle.

bovine papular stomatitis. a widespread viral disease that affects young cattle. Papular and sometimes erosive lesions occur on the muzzle and mucous membranes of the mouth and pharynx.

bovine somatotropin (BST). a natural hormone of cattle that has been synthesized by genetic engineering. It is used in certain countries to increase milk production, presumably by as much as 20%.

bovine spongiform encephalopathy (BSE; mad cow disease). a slowly progressive, incurable, fatal brain disease of cattle that was first reported in Great Britain in 1986. The disease is normally transmitted by the use of feed additives that contain protein from parts of infected ruminants. Affected individuals initially become agitated and aggressive. BSE has not yet been found in United States cattle, although other BSE-like animal diseases are found in that country, including scrapie in sheep and goats. BSE-infected meat is also believed to be connected to a new strain of Creuzfeldt-Jacob disease in humans.

bovine tuberculosis. a highly contagious disease of cattle that causes severe economic losses, especially to owners of dairy herds. Infected herds must be quarantined. The Animal and Plant Health Inspection Service works with producers to eradicate the disease. This has not yet been accomplished.

bovine venereal trichomoniasis. bovine genital trichomoniasis.

bovine viral diarrhea (mucosal disease). an infectious viral disease of cattle, characterized by acute erosive stomatitis, gastroenteritis, and diarrhea.

bowel edema. a bacterial infection of swine due to sudden changes in diet and management. It occurs most often in pigs 8-14 weeks old. Affected individuals may have seizures, stagger, and become paralyzed. Some cases run a rapid fatal course.

box. **1:** carton; crate; container. **2:** a compartment, a stall. **3:** to pack or bundle. **4:** to fight, hit, or spar. **5:** common name of *Buxus sempervirens* and *Pachysandra procumbens*. **6:** to corral animals. **7:** the lowest commercial grade of softwood. **8:** a device that divides water into two or more ditches in an irrigation system. **9:** a box canyon; one that has only a single point of access. **10:** a method used to brace the branches of trees. **11:** to chip back on the underside of a cut in the trunk of a tree to prevent it from splitting when the tree falls. **12:** a hole made in the base of a pine tree from which to collect sap for the purpose of extracting turpentine and resin.

box elder. *Acer negundo*.

box scraper. an implement mounted on a plow that is used to level and grade the soil.

box stall. a stall in which an animal can be confined without being tied.

box the heart. to cut boards from the sides of a log, leaving the heartwood as a single piece.

boxwood family. Buxaceae.

box. *Pachysandra procumbens*.

boxwood. *Buxus sempervirens*.

Boynton's tissue vaccine (B.T.V.). a vaccine used to prevent hog cholera.

BP. blood pressure.

brace. **1:** that which strengthens or supports. (e.g., a prop, support, strut, beam, post, stay). **2:** a rod connected to the beam of a plow. **3:** to prop, buttress, bolster, strengthen, or fortify. **3:** a curved tool used to hold and rotate bits. **3a:** a set of two birds (a team), especially waterfowl and other game birds as a brace of ducks, partridges, or pheasants. **3b:** a pair of dead game birds.

brace comb. a section of honeycomb that is attached to and lies between other combs.

brachial gland. one of the lymph nodes in the arm.

Brachiaria brizantha. an annual grass (Family gramineae) that sometimes causes hepatogenic photosensitization in livestock. *See also* phylloerythrin.

brachiate. **1:** to move through the canopy of woods or forest by use of the arms as in the case of monkeys and smaller primates. **2:** branched in opposite pairs as when branches grow from opposite sides of a stem, as in the case of ash and maple trees.

brachium. arm.

brachycardia. Bradycardia.

brachyism. a dwarfing of plants in which the internodes are shortened.

bracken. the preferred common name for any fern of the genus *Pteridium*, especially *Pteridium acquilinum*.

bracken fern. *Pteridium*.

bracken poisoning. an acute fatal hemorrhagic disease of cattle and horses caused by eating large amounts of bracken (*Pteridium*) or hay that contains 20% or more of bracken; the toxic principle is thiaminase. The condition is marked by uncontrollable internal bleeding; in horses, the disease is manifested by necrologic signs; in cattle, by pancytopenia. *Cf.* bleeding disease.

bracken staggers. a condition of horses resulting from the ingestion of *Pteridium* (bracken). Signs and symptoms include incoordination due to thiaminase-induced thiamine deficiency. Affected animals may die.

brackish. **1:** a term applied to natural waters, indicating a saline concentration that is intermediate between that of fresh and salt water **2:** of low salinity.

bract. a small, specialized, often scalelike leaf or leaflike part that usually occurs at the base of a flower, flower cluster, fruit, or fruits. *Cf.* spathe.

bracteole (bractolet). a small or secondary bract, usually situated on a flower stalk.

bractolet. bracteole.

bradyarrhythmia. any disturbance of the

heart's rhythm resulting in bradycardia, *q.v.*

bradycardia (bradyrhythmia; brachycardia; bradysystole). an unusually slow heartbeat, usually defined as a rate below 60 beats per minute in resting humans. *Cf.* tachycardia.

bradycardiac (bradycardic). pertaining to or characterized by bradycardia.

bradycardic. bradiacardiac.

bradypnea. abnormally slows rate of breathing. *Cf.* dyspnea.

bradyrhythmia. Bradycardia.

bradysystole. bradycardia.

brae. a hillside; a steep bank or slope.

Braford. a breed of domestic cattle that is a cross between Herefords and Zebus.

Brahma. domestic fowl of the Asiatic class. Color patterns vary among light (Columbian pattern), dark (silver-penciled), and buff. The breed is characterized by the pea comb, yellow skin, and heavily feathered shanks and toes. The eggs are brown.

brahman. *Bos indicus*.

Brahman. a hump-backed breed of beef cattle, one of several breeds of Asiatic domestic cattle, *Bos indicus*, that were developed in the United States. Brahmans tolerate high temperatures and are resistant to many insects and tropical diseases.

braid. the coarsest grade of wool.

braining. **1:** a method of killing an animal by crushing the skull with a heavy blunt instrument. **2:** fowl slaughter; a method of slaughtering fowl by inserting a knife through the roof of the mouth into the rear of the brain; the medulla oblongata is severed and the fowl dies quickly. **3:** removing the brain of a carcass.

brake fern. *Pteridium aquilinum*.

brake. **1:** *See Pteridium*. **2:** a cluster or thicket of large ferns; an area overgrown with ferns and shrubs.

bramble fruit. a fruit of a plant of the genus *Rubus*.

bramble. **1:** any plant of the genus *Rubus*, such as the blackberry, rapberry or dewberry. **2:** any spiny clinging woody bush or shrub.

bran. the seed coat of the wheat seed and other cereal grains which is typically separated from flour and used as roughage, as a source of some vitamin B, and as feedstuff for animals, especially cattle and poultry.

branch wilt of walnut. a fungal disease of *Juglans regia*, the English or Persian walnut (Family Juglandaceae),due to infestation by *Hendersonula toruloidea* (Family Sphaeropsidaceae). The fungus initially infests twigs and smaller branches, but eventually invades larger branches also. The leaves of affected trees wilt suddenly in summer.

branched. having branches; branching;diverging; divided.

branchi-. prefix meaning gill.

branchia (plural, branchiae). a gill or gill-like structure.

branchiae. plural of branchia.

branchial. of, pertaining to, or resembling gills.

branching broccoli. *B. oleracea* var. *botrytis*.

brand. **1:** a figure burned with a hot iron on an animal's hide, especially of a horse or cow, to identify the owner. **2:** to mark an animal's hide with a hot iron to indicate ownership. **3:** the iron or steel die used in branding animals. **4:** a German cheese made from milk curd and butter and moistened with beer during aging. **5:** the official name of a product which may include an acronym or a logo. *See* brand name.

brand artist. a person who brands cattle or one who forges brands.

Brand Book. an official record of brands used to mark the cattle of different owners.

Brand Inspector. a government official who authenticates ownership of cattle being sold at licensed auctions and other outlets.

brand name. the name under which a product, class of products, or a service is produced and/or marketed by a particular firm. Of importance to agriculturalists is the rapidity with which new brand names and revised brand names of drugs appear. Because of this, physicians and veterinarians sometimes run a serious risk of failing to identify potential drug interactions that may affect individuals treated with more than one drug and in some cases may have difficulty in selecting the current drug of choice in a given treatment program.

branding bee. an annual, partly social community event during which livestock (cattle or horses) born within the previous year are branded.

branding iron. an iron rod to the tip of which a registered brand is welded. This devised is used to apply the brand to the owner's livestock. The tip of the iron bearing the brand is heated (historically in an outdoor wood fire) and applied to the hide of a cow or horse burning through the hair and burning the brand permanently into the skin. Conventional branding irons are now being replaced, to some extent, by electric irons and by colored plastic ear tags.

branding paint. a paint (some are indelible) used to place identifying marks on livestock, especially sheep.

brangus. a hardy breed of beef cattle developed from Brahman and Angus breeds in the United States

brasada. land marked by brush or thickets in the southwestern United States

Brassica. a genus of plants (Family Brassicaceae, formerly Cruciferae) that includes a wide variety of crop species, forage plants, and weeds (e.g., mustards, rape, kale, cabbage, brussels sprouts, broccoli, cauliflower, rutabaga, turnip). The classification of taxa within *Brassica* is complicated, with numerous synonyms and much confusion in the literature pertaining to some species or varieties. A number of *Brassica* are goitrogenic and thyrotoxic. The active principles in such cases are often components of mustard oils. Some also accumulate concentrations of nitrate under certain growing conditions that are toxic to livestock, and some have been implicated in trefoil dermatitis and rape scald. Several varieties are known to cause pulmonary emphysema of cattle (*See* emphysema). *See also erodium, Medicago, Trifolium*. ***B. arvensis***. *B. kaber*. ***B. caulorapa***. *B. oleracea* var. *gongylodes* ***B. hirta*** (white mustard). a rather hairy annual herb with ovate leaves that grows to a height of about 1.3 m. The seeds are a commercial source of mustard and mustard oil and the foliage is used as greens. The seeds are toxic and goitrogenic. *See*ds or seed meal fed to livestock in small amounts cause no problems, but cattle and sheep fed routinely on mustard seed meal, pod-bearing plants, or stubble are sometimes poisoned. The animals may suffer a sometimes fatal gastroenteritis. ***B. juncea*** (Indian mustard; mustard greens). an annual plant with narrow lobed leaves that reaches a height of about 1.3 m. It is used for spring greens. It can cause severe gastroenteritis in livestock. ***B. kaber*** (*B. arvensis*; charlock; wild mustard). Introduced into North America from Europe, charlock is now a widely distributed common annual weed of cereal crops. *See*ds or pod-bearing plants can cause a severe gastroenteritis when ingested by cattle, swine, or sheep. ***B. napobrassica***. *B. napus* var. *napobrassica* ***B. napus*** (rape; cultivated rape). a widely cultivated annual plant that is toxic to livestock. It is goitrogenic and thyrotoxic. The toxic material, L-5-vinyl-2-thiooxazolidone, can be extracted from the seed meal with hot water. The main effects of rape seed meal can be counteracted by the addition of iodinated protein to the diet. The seed also contains at least three glycosidic iso thiocyanates (mustard oils). Fresh rape can cause hepatogenic photosensitivity in grazing animals and is known also to cause pulmonary emphysema of cattle (*See* emphysema). Toxic concentrations of nitrate have also been reported. ***B. napus* var. *napobrassica*** (*B. napobrassica*; *B. oleracea* var. *napobrassica*; rutabaga; yellow turnip; swedish turnip). a hardy, highly goitrogenic biennial plant. Concentrations of nitrate toxic to herbivores have also been reported. ***B. nigra*** (black mustard). a branching annual herb with feathery leaves that grows to a height of 2 m. It is a source of mustard oils and condiments. Where it has escaped cultivation, it is considered a weedy pest. The seeds are toxic and goitrogenic. ***B. oleracea*** (broccoli, kale, wild cabbage, etc.). an annual or perennial species that includes the largest number of varieties within the genus. Some may contain concentrations of nitrates that are toxic to grazing animals, and many varieties have thyrotoxic or vesicant activity. ***B. oleracea* var. *acephala*** (common kale). grown for its foliage, used by humans as greens. It is goitrogenic, producing symptoms of hypothyroidism in livestock. Fresh kale is sometimes a cause of pulmonary emphysema of cattle (*See* emphy sema). ***B. oleracea* var. *botrytis*** (synonym *B. o.* var. *italica*; broccoli; asparagus broccoli; sprouting broccoli; branching broccoli; Italian broccoli; calabrese). a popular vegetable, it is toxic to livestock. It is goitrogenic, producing symptoms of hypothyroidism. ***B. oleracea* var. *capitata*** (cabbage). a variety that is goitrogenic, producing symptoms of hypothyroidism. ***B. oleracea* var. *gemmifera*** (brussels sprouts). a vegetable, it is an annual or biennial herb developed from wild cabbage for its small edible heads. It is toxic to livestock, producing symptoms of hypothyroidism. ***B. oleracea* var. *gongylodes*** (synonym, *B.caulorapa*)(kohlrabi, stem turnip). toxic to livestock, it is goitrogenic, producing symptoms of hypothyroidism. ***B. oleracea* var. *italica*.** *B. oleracea* var. *botrytis*. ***B. oleracea* var. *napobrassica*.** *B. napus var. napobrassica*. ***B. rapa*** (field mustard). an annual or biennial mustard with long, lobed leaves and flat or globular roots. ***B. rapa var. pekinensis*** (Chinese cabbage, pak-choi). it is goitrogenic, producing symptoms of hypothyroidism. ***B. rapa* var. *rapifera*** (turnip). a goitrogenic root vegetable that sometimes accumulates concentrations of nitrate that are toxic to livestock. Fresh turnip sometimes causes pulmonary emphysema in cattle (*See* emphysema).

Brassicaceae (mustard family). a large family (formerly Cruciferae) of at least 3000 species of dicotyledonous plants. Wild forms occur chiefly in the north temperate latitudes. Most are herbs with alternate leaves and flowers. The flowers have four petals in the shape of a cross; they are inserted into a raceme or corymb. Many mustards are important as crops. Some are toxic to grazing herbivores. *See Brassica*.

Braula coeca (bee louse). a usually harmless ectoparasite of honey bees, the larvae of which attack the honeycomb.

brawn. **1:** strength; might; muscle. **2:** hardened skin. **3:** a pig fattened for meat production. **4:** the boiled, pickled flesh of a boar.

brawny. big; strong; muscular.

braxy. an acute hemorrhagic, bacterial inflammation of the abomasum of sheep. It afflicts Norwegian, Scottish, and Faroe Island sheep. Diseased individuals usually die within a few hours. Signs of the disease are rarely observed, but include dullness, loss of appetite, and difficult breathing. The causative organism appears to be *Clostridium septicum*.

bray. The voice of a donkey or mule.

bread. a food staple made from flour or meal, usually from cereal grains. Commercially prepared breads often contain potentially harmful ingredients or additives, *q.v.*, that may not be on the label. Examples are butylated hydroxyanisole (BHA), butylated hydroxytoluene (BHT), colors, flavors, hydrogenated shortening, mineral oil, and pesticide residues. French breads are usually free of additives. *See also* bread poisoning.

bread poisoning. a colloquial term for poisoning by seeds of plants of the genus *Senecio* (groundsels). This plant has caused many human deaths in parts of Africa where the seeds sometimes contaminate grains from cultivated plants. The alkaloids present in the seed cause necrosis and enlargement of livers and rupturing of veins.

breadfruit tree. *Artocarpus communis*.

breadfruit. The starchy fruit of *Artocarpus communis*, the breadfruit tree. It is used as a vegetable in Polynesia.

break ground. to plow land for the first time.

break. **1:** to part, separate, or detach. **2:** to snap, split, crack, or shatter. **3a:** to tame, gentle, or domesticate a wild animal. **3b:** to break in or train a horse to wear a saddle and carry a rider. **4:** to mill grain by passing it through break rolls.

breakbone fever. dengue fever.

breaking ground ("breaking"). **1:** plowing native soil. **2:** land that has never been plowed before.

breast. **1:** in bipedal animals, the outer, front part of the trunk of the body or the front part of the body from the neck to the abdomen; the chest. **2a:** part of the trunk of quadrupeds that corresponds to that of the breast of bipeds. **2b:** the brisket. **3:** either of the pair of mammae occurring on the chest in humans, especially after puberty. Those of the female become enlarged and secrete milk following the birth of a child; those of the male usually remain rudimentary.

breast height. in forestry, the height (1.3m or 4.2 ft) above ground level at which the breast high diameter of a standing tree is measured.

breast high diameter. the diameter of a tree at breast height, *q.v.*

breath. a single inhalation or a single breathing cycle.

breathe. **1a:** to inhale; to take gaseous mixtures, especially air, into the respiratory system. **1b:** to inhale and exhale.

breathing (pneusis). the entrance and exit of gaseous mixtures (usually air) into and out of the respiratory system by inhalation (inspiration) and exhalation (expiration). *See also* respiration (def. 2). **irregular breathing**. aperiodic, variable, or erratic breathing. *See also* dyspnea. **labored breathing**. dyspnea. **shallow breathing**. breathing with abnormally low intake (low tidal volume).

bred heifer. a young cow during her first pregnancy.

breeches flower. *Daicentra cucullaria*.

breeching. a strap, fastened about the hind quarters of a horse, that permits it to push backwards to stop or to back up a cart or similar conveyance.

breed. **1:** to reproduce; to produce offspring. **2:** to cause the birth of young, as in raising stock. **3:** a group of animals, developed and maintained by humans, that have common origin and characteristics that distinguish them from other such groups within the same species. **4:** sort; kind; group; (as in a breed apart). **5:** a genetic line; a lineage; stock; strain; variety. **6a:** to improve or develop a breed of livestock, as by selective mating. **6b:** to produce a desired type of plant or animal by selection and crossing one variety with another to produce a new variety with the features desired by the breeder. **7:** to alter the genetic pattern of plants or animals to increase their value (e.g., increased yield) or marketability.

breed average milk production. the average milk production of cows of a particular dairy breed.

breed character. a distinguishing characteristic of an animal such as the shape and size of horns, ear, quality and type of pelage, body color, and markings.

breed true. the characteristic ability to

uniformly transmit a selected phenotype to an offspring.

breed type. the set of features that distinguish one breed from another.

breeder. **1a:** an animal, plant, or person that reproduces (breeds). **1b:** an animal or plant used for breeding. **2a:** a specialist in breeding animals, especially one who breeds new forms of plants or animals. **2b:** a person who raises plants or animals primarily for breeding purposes or for the production of breeding stock. The owner of the dam of any species or breed of livestock at the time of breeding.

breeder tom. a tom turkey used for breeding, which is then marketed between April and August.

breeding stock. quality animals used for breeding in order to maintain and improve the quality of the flock or herd.

brewers' grain. a by-product of brewing, it is the residue that remains from barley after malting. It is a valuable source of fiber and protein for cattle.

briar (brier). **1a:** any woody-stemmed, thorny plant. **1b:** a thorny stem or twig. **2:** plants of the genus *Smilax* (greenbriar). **3:** *Erica arborea* (European white heath). a shrub (Family Aricaceae) native to southern Europe (France and Corsica). The roots are used in making pipe bowls. **4:** a pipe made of brierroot.

brier. briar.

bright bay. a horse with a light, glossy, tan coat.

brindle. an animal (e.g., a brindle boxer) having a composite hair coat color having a uniform mixture of gray, brown, or tawny hairs with irregular streaks or sometimes spots of black or white.

brindled. gray, brownish, or fawn-colored, with dark or light streaks or spots.

brine. a solution of sodium chloride used to preserve certain foods.

brine poisoning. salt poisoning.

brisket. **1:** the breast, *q.v.*, or chest of a quadruped; it lies between the forelegs. **2:** the meat of the breast as here defined.

bristle. a stiff, sharp hairlike structure, or barb of a plant or animal.

British alpine. a large, black, Swiss type goat with white markings on the face and legs that produces excellent milk. The males have a large beard.

British cattle breed. any breed of cattle such as Angus, Hereford, and shorthorn that was developed in Great Britain.

British dane. a breed of dairy cattle.

British holstein. a breed of dairy cattle developed in Great Britain from the holstein.

British lop. a rare breed of swine with very large, droopy ears. They are raised chiefly in the southwest of England.

British milksheep. a medium-sized polled dairy sheep with a white face and legs.

British Oldenburg. a British sheep that yields high quality wool in good quantity.

British redwater disease. bovine babesiosis.

British romagnola. a breed of beef cattle developed in Great Britain from stock originating in northern Italy.

British Saanen. a breed of white dairy goat that produces a high yield of milk. It originated in Switzerland.

British saddleback. a breed of swine developed from Wessex and Essex breeds. it is black with a white band around the front of the trunk. It is used chiefly to breed hardy crosses.

British Thermal Unit (BTU). the amount of heat (ca. 252 calories) that will raise the temperature of one pound of water one degree Fahrenheit under standard conditions of pressure and temperature.

British Toggenburg. a dark brown breed of goat developed from the Swiss Toggenburg.

British white. a rare, medium-sized, white breed of hornless, short-legged cattle with colored points. They are used as suckling cattle for beef production.

broad bean. *Vicia faba*.

broadcast. **1:** to spread or scatter seeds, chemicals, or other materials uniformly over a large area. **2:** to plow, throwing the soil in one direction, allowing the field to remain level.

broadcast application. the spreading of chemicals, especially pesticides and fertilizers, over a large area. *Cf.* band application.

broadcast burning (slash burning). controlled burning over a designated area within well-defined boundaries as a precautionary treatment to reduce fire hazard following logging and to prepare the site for replanting.

broadcast fertilizer distributor. a machine that scatters fertilizer. Mounted machines usually have a capacity of up to 600 kg (ca. 1300 lb), while self-contained units may have a capacity of 5 tons. There is usually a hopper above and a spinning disk below that scatters the fertilizer.

broadcast planting. broadcast seeding.

broadcast seeding (broadcast planting). the scattering of seeds freely over the ground by hand, with the use of a machine (a broadcaster), or from aircraft, rather than sowing them in rows or in the ground (as by drill seeding). This may be followed by harrowing.

broadcaster. a machine that broadcasts seeds.

broadleaf milkweed. *Asclepias latifolia*.

broadleaved. descriptive of plants (e.g., trees, weeds) that have large leaves. *See* broadleaf, evergreen.

broadleaved evergreen. an evergreen tree that has large leaves (e.g., rhododendron).

broadleaved tree. a deciduous tree that has wide leaves rather than needles.

broccoli. *B. oleracea* var. *botrytis*.

brodifacoum (3-[3-(4′-bromo[1,1′-biphenyl]-4-yl)-1,2,3,4-tetrahydro-1-maphthalenyl]-4-hydroxy-2H-1-benzopyran-2-one). a whitish, poisonous, water-insoluble powder that is exceptionally toxic to rats. It is used as a rodenticide.

broiler. a young domestic fowl, usually 6 to 8 weeks old, weighing 3 to 5 pounds; raised chiefly for meat.

broken coat. a condition of rabbits in which the undercoat is exposed in places due to broken or missing guard hairs.

broken ear. a broken cartilage in the ear of a rabbit that prevents the ear from being carried erect.

broken flat. descriptive of land that has been subject to broadcast plowing.

broken knees. descriptive of a horse that has scarred knees, a sign that it has fallen. Such a horse might be one that stumbles readily.

broken mouth. a condition of sheep and goats, usually at an age of 5 or 6 years, in which some permanent teeth are missing.

broken terrain (broken land). terrain characterized by the presence of numerous sharply cut ravines or valleys.

broken wind (heaves). a disease of old horses in which the alveoli are dilated or ruptured. The most obvious sign of the disease is forced, double exhaling.

broken winded. said of a horse that is disabled by dialted or ruptured alveoli in the lungs.

bromacil (5-bromo-3-*sec*-butyl-6-methyluracil**;** 5-bromo-6-methyl-3-(1-methylpropyl)2,4-(1*H*,3*H*)-pyrimidinedione; 5-bromo-6-methyl-3-(1-methyl-propyl)uracil). a nonselective uracil herbicide,

$$C_9H_{13}BrN_2O_2$$

bromacil is used largely as a weedkiller on non-crops. It strongly inhibits root growth of oats by blocking division in the meristem; it also affects the integrity of cell walls and membranes; chloroplast development in oat leaves is partially inhibited. Bromacil is moderately toxic by ingestion to humans and laboratory animals.

bromadiolone (3-[3-(4′-bromo[1,1′-biphenyl]-4-yl)-3-hydroxy-1-phenylpropyl]-4-hydroxy-2*H*-1-benzopyran-2-one; 3-[α-[p-(p-bromophenyl)-β-hydroxyphenethyl]-benzyl]-4-hydroxycoumarin). a whitish, slightly water-soluble powder used as an anticoagulant rodenticide.

bromatoxism. See food poisoning.

brome. *See Bromus*

bromegrass. *See Bromus*.

bromethalin (*N*-methyl-2,4-dinitro-*N*-(2,4, 6-tribromophenyl)-6-(trifluoromethyl)benzenamine). a pale yellow water-insoluble crystalline substance used as a rodenticide. It is an uncoupler of oxidative phosphorylation. It is highly poisonous not only to rats and mice, but to cats, dogs, and presumably other mammals. Acute exposure of humans causes headache, confusion, personality changes, seizures, coma, and death. Warfarin-resistant strains of laboratory mice and rodents succumb readily to bromethalin.

***O*-(4-bromo-2,5-dichlorophenyl)-*O*-methyl phenylthiophosphonate**. leptophos.

bromomethane. methyl bromide.

3-(*p*-bromophenyl)-1-methoxy-1-methylurea. metobromuron.

bromoxynil (3,5-dibromo-4-hydroxybenzonitrile; 3,5-dibromo-4-hydroxyphenyl cyanide; 2,6-dibromo-4-cyanophenol; broxynil). a postemergence herbicide that is less active than ioxynil. The first visual effects are the appearance within 24 hours of necrotic spots on leaves; tissue damage spreads and the plant eventually dies. It is very toxic to laboratory mice.

Bromus (brome grass). a genus of annual, biennial, and perennial temperate zone grasses (Family Graminae) with few, many-flowered spikelets. ***B. arvensis*** (field bromegrass); a winter annual grass that develops a fibrous mass of roots that is valuable in controlling erosion and improve soil tilth. *See*dlings are winter-hardy and should be planted in the fall. ***B. catharticus*** (rescue grass). concentrations of nitrates toxic to livestock have been reported. ***B. inermis*** (awnless, smooth, or Hungarian brome grass). Sometimes becomes heavily ergotized and therefore hazardous. *See Claviceps*. ***B. sterilis*** (barren brome) a widespread weed that affects winter cereals.

bronchi. plural of bronchus.

bronchial tube. bronchus.

bronchial. of or pertaining to bronchi.

bronchitis.

bronchitis. an acute or chronic inflammation of the mucus membrane of the bronchi. It is often caused by viruses but can be due to an irritant. **infectious avian bronchitis** (gasping disease). a specific, highly transmissible viral disease of young birds in which respiratory passages are blocked by exudate. It can spread rapidly within a flock of poultry, often causing heavy losses of chicks and greatly impaired production by laying birds. **verminous bronchitis** (hoose). bronchitis complicated by bronchopneumonia. It is caused by lungworm infestation of the bronchi. It commonly occurs in cattle, swine, and sheep and rarely in other species of livestock.

bronchus (plural, bronchi; bronchial tube). either of two tubes in air-breathing vertebrates that conduct air between the trachea and the lungs. They branch from the lower end of the trachea, each communicating with a lung. They subdivide further into lobar, segmental, and subsegmental bronchi.

bronco. **1a:** originally an untamed mustang but now commonly used for any horse used in a rodeo. **1b:** any horse that bucks, cannot be gentled, and is difficult or impossible to ride. **bucking bronco**. an unbroken horse ridden in rodeos.

bronopol. a milk preservative.

Brønsted acid. *See* acid.

bronze turkey. the largest breed of the domestic turkey, *Meleagris gallopavo*, the plumage of which is a brilliant reddish-bronze color. It is the heaviest. An adult tom may weigh more than 36 lbs; an adult hen, up to 20 lbs.

bronze. **1:** the ground color of the bronze turkey. **2:** the metallic bronze-colored hue that one sees on certain birds, as it sometimes occurs on plumage of black or partially black domestic fowl (considered a defect).

brood. **1a:** group of offspring (siblings), especially birds, produced at the same time. **1b:** a litter. **1c:** a single generation of insects. **1d:** the eggs, larvae, and pupae of bees. **2:** to incubate eggs or raise and protect chicks prior to fledging or independence. **3a:** designation of an adult animal used for breeding purposes, e.g., a brood mare, brood cow, or brood sow. **3b:** designation of a hen that is inclined to incubate eggs. **sealed brood**. the pupae of bees. **unsealed brood**. the eggs of bees.

brood cow. a cow used for breeding purposes.

brood hen. a domestic fowl that is inclined to brood eggs.

brood hen. **1:** a hen of the domestic fowl that is inclined to brood eggs. **2:** a brood hen (def. 1) that is used to hatch eggs and tend young chicks.

brood mare (broodmare). a mare used for breeding purposes.

brood sow. a sow used for breeding purposes.

brooding. the incubation of eggs and rearing, usually by the hen, of chicks or nestlings prior to fledging or independence.

broodmare. brood mare.

broody. **1:** descriptive of a hen that is inclined to incubate eggs or brood chicks. **2:** in judging livestock, descriptive of an animal, especially a gilt or sow, that displays characteristics indicating that she will prove to be reproductively efficient.

broom. common name of two closely related genera of shrubs, *Sparteum* (*Cytisus*) and *Genista*, although it is sometimes applied to other leguminous shrubs with long, slender branches. *See Cytisus*

broom corn. a cultivated grain sorghum. *See Sorghum vulgare*.

broom grass. *Andropogon virginicus*.

broom groundsel. *Senecio spartioides*.

broom poisoning. poisoning caused by eating *Sparteum junceum* (=*Cytisus scoparius*), *q.v.* Poisoning is due to the presence of sparteine and cytisine within the plant.

broom-snakeroot. *See Gutierrezia*.

broomsedge. *Andropogon virginicus*.

broomtail. **1:** a wild horse or one raised on the open range. **2:** a "cow pony" with a short bushy tail.

broomweed. ***Gutierrezia microcephala*.**

brown alga. *See* alga.

brown algae. *See* Phaeophyta.

brown dog. *Rhipicephalus sanguineus*.

brown fat. adipose tissue in which oxidative phosphorylation can be uncoupled from phosphorylation and heat generation.

brown foot rot. a disease of wheat seedlings.

brown heart. a common boron-disease of beets and turnips.

brown leaf spot of tobacco. a fungal disease caused by *Alternaria tenuis*. Plants may be infected during periods of high temperature and humidity, is marked by large, brown, circular spots on the mature leaves that appear following harvesting.

brown lung. byssinosis.

brown lung disease. byssinosis.

brown patch. a fungal disease of grasses that presents during periods of high temperature and humidity. The turf develops scattered brown patches which eventually coalesce; a large area may eventually become brown and die. The disease is caused by *Pellicularia filamentosa* (Family Thelephoraceae).

brown rat. *Rattus norvegicus*.

brown rice. rice from which the husk has been removed but has not been polished; the bran layers thus remain.

brown rot fungus. a fungus that attacks cellulose and hemicellulose in wood, leaving dark-colored lignin and phenolic residues.

brown spot. a fungal disease that affects many plants. The fruit and foliage of infested plants exhibit irregular, brownish lesions; cankers develop on the stem.

brown spot of rice (rotten neck of rice). a fungal disease of rice seedlings in which the leaves discolor and develop reddish-brown spots and the stalk rots at the base of the head. Yield may be greatly reduced. The pathogen is *Helminthosporium oryzae* (Family Dematiaceae).

brown spot of soybeans. a widespread defoliating disease of *Glycine max* caused by the fungus, *Septoria glycines* (Family Sphaeropsidaceae).

brown stem rot. a disease of *Glycine max* (soybeans), caused by a fungus, *Cephalosporium gregatum* (Family Moniliceae). The disease is marked by the sudden appearance of blighted, withering leaves.

brown Swiss. a medium-sized breed of gray-brown cattle developed in Switzerland for milk, meat production, and use as draft animals. In the U.S, these animals are used only as dairy cows and veal.

Brown Spot of Smooth Bromegrass. a common disease of *Bromus inermis* caused by the fungus, *Helminthosporium bromi* (Family Dematiaceae). Signs include the development of small, dark brown, oblong leaf spots; these merge, forming large yellow areas. The leaves eventually turn brown and die. In the United States, plants become diseased during wet spring weather, usually becoming evident in June.

browning root rot of cereals. a fungal disease of cereals in which the root rots. It occurs in marshy areas in Central and North America. The pathogen is *Pythium arrhenomanes* (Family Pythiaceae).

brown-tail moth. *Euproctis chrysorrhea*.

brown-tail rash. brown-tail moth dermatitis (*See* dermatitis).

brown-ticked gray. **1a:** a horse that has very small, scattered, brown spots on a coat of gray. **1b:** the color of such a horse.

browse. **1:** to feed on the above-ground parts of plants (e.g., foliage, twigs, and tender shoots of woody plants). **2:** the edible above-ground parts of woody plants (shrubs, trees, vines) that are within reach of herbivores.

browser. an animal that browses.

brucella. common name of bacteria of the genus *Brucella*.

Brucella. a genus of rod-shaped bacteria. It is the type genus of the family Brucellaceae. *See* brucellosis, undulant fever.

brucellae. plural of brucella.

brucellosis (contagious abortion; contagious bovine abortion; infectious abortion; Bang's disease; epizootic abortion). an infectious disease of cattle that results in reduced milk yields, infertility, spontaneous abortion, inflamed genitalia in males, and often death. The disease also affects goats, sheep, and swine and can be transmitted to humans (*See* undulant fever). The pathogenic organisms responsible for the disease are the bacteria *Brucella abortus, B. melitensis,* and *B. suis*. brucellosis in animals is usually transmitted by contact, contaminated feed, water, and excrement. Prevention and control of brucellosis include vaccination of calves, blood tests of adults, and slaughter of infected animals. The Animal and Plant Health Inspection Service (APHIS) conducts a program intended to eradicating this disease from United States cattle herds. *See also* undulant fever.

brushing. habit of a horse.

brussels sprouts. *B. oleracea* var. *gemmifera*

bryology. the botanic science that treats mosses.

BSE. bovine spongiform encephalopathy.

BSF. busulfan.

BST. bovine somatotropin.

Bt. abbreviation and common name of *Bacillus thuringiensis*.

bu. bushel.

buccal cavity. the mouth or oral cavity of vertebrate animals and certain advanced invertebrates. In vertebrates it communicates with the pharynx.

buccal gland. *See* salivary gland (minor salivary gland).

bucco-. a prefix meaning cheek.

Buchloe dactyloides (buffalo grass; early mesquite). a short, valuable, native forage grass (Family Gramineae) of the Great Plains of western North America where it is sometimes used as a lawn grass. It is a codominant of the midgrass prairie.

buck. a sexually mature male; a term applied to any of a number of animals (e.g., deer, goat, rabbit, sheep).

buckaroo. an American cowboy.

buckbrush. *Ceanothus cuneatus*.

bucked knees. descriptive of a horse or cow with knees that bend outward.

bucket tree. a sugar maple that produces sufficient sap to require the hanging of a bucket on it.

buckeye. *See Aesculus*.

buckling. a male goat, 1 to 2 years of age.

buckskin. **1:** a disease of citrus fruit, usually grapefruit and oranges, caused by a mite, *Phyllocoptes oleivorous*. The rind of affected plants becomes leathery and encrusted. **2:** a horse, the light gray-brown ground color of which is similar to that of a deer. **3:** the hide of a buck deer. **6:** a viral disease of the sweet cherry in which the fruit drops early or fails to ripen.

buckthorn. **1:** a common name of plants of the genera *Ceanothus Rhamnus* (Family Rhamnaceae). **2:** sometimes erroneously applied to *Karwinskia humboldtiana*. *See* buckthorn toxins.

buckthorn polyneuropathy. *See* neuropathy.

buckthorn toxins (Coytillo; Tullidora). a mixture of anthracenones in the fruit of *Karwinskia humboldtiana* that produce an often fatal demyelinating neuropathy of peripheral nerves in humans and all forms of livestock. *See Karwinskia humboltiana*.

buckwheat. *Fagopyrum sagittatum*.

buckwheat family. Polygonaceae.

buckwheat poisoning (fagopyrism). **1:** an allergic reaction (*See* reaction) of grazing animals due to ingestion of *Fagopyrum sagittatum*. **2:** fagopyrism; a rare primary photosensitivity in animals that consume *F. sagittatum*.

buckwheater. a novice or incompetent farmer, one who "can't even grow buckwheat."

bud. **1a:** a primordium having the potential to grow and develop into a fully differentiated structure or organism. **1b:** an organism or part of an organism that has not yet matured. **2:** a small lateral or terminal protuberance that contains an embryonic shoot including the growing stem tip surrounded by leaf and/or flower primordia and often covered tightly by specialized bud scales or by a resinous and/or hair-like layer. **3:** an incompletely opened flower **4:** an outgrowth from an organism that eventually detaches and differentiates into a new individual. **5a:** to set buds (said of a green plant). **5b:** to begin growing from a bud. **6:** to produce or develop from buds.

budburst. the onset of flower or leaf bud growth.

budding. asexual reproduction, as by yeast, in which the "bud" forms as a protuberance of the parent cell. It eventually detaches, forming a smaller daughter cell.

budworm. the larva of *Acleris variana* (black-headed budworm) and *Choristoneura fumiferana* (spruce budworm). Both are serious forest pests in Canada and the United States

bufa-, bufo-. prefixes that indicate a relationship to or origin from toads (Family Bufonidae).

bufencarb (3-(1-ethylpropyl)phenol methylcarbamate mixture with 3-(1-methylbutyl)-phenyl methylcarbamate (1:3); methylcarbamic acid *m*-(1-ethylpropyl)phenyl ester mixture with *m*-(1-methylbutyl)phenyl ester; metalkamate). a very toxic, very dangerous cholinesterase inhibitor used as an insecticide; it is readily absorbed through the human skin.

buff Orpington. a breed of large domestic fowl raised for meat and eggs. They are brown in color with a red face, comb, and wattles. They are raised for eggs and meat; the eggs are brown.

buffalo. *See Bos bison*.

buffalo chip. dry buffalo dung used, as in the case of the dried dung of other ruminants, in otherwise fuelless environments; was used as fuel. Cow dung is used as fuel today in many developing countries.

buffalo gnat. black fly.

buffalo grass. *Buchloe dactyloides*.

buffalo wallow. any of numerous, shallow depressions in the Great Plains of North America, thought by many to be bathing (wallowing) holes of bison.

buffer. **1:** a substance or mixture that, in solution, resists large changes in pH when acidic or alkaline material is added. A buffer molecule prevents large changes in pH either by combining with H+ or releasing H+ into solution. A buffer is usually a mixture of a weak acid and its conjugate base (e.g., acetic acid and sodium acetate), or a weak base and its conjugate acid (e.g., ammonium hydroxide and ammonium chloride). Bicarbonates and proteins, for example, act as buffers in biological materials. **2:** to treat with a buffer. **3:** to insulate or protect a system from change due to external factors. **4:** to lessen the effect or impact of something. **5a:** in agronomy, a substance (e.g., carbonates, clay, humus) that prevents a rapid change in pH when acids or alkalis are added to the soil. **5b:** something that separates two entities. **5c:** anything that insulates or protects a system from external factors that may cause an impact. **riparian buffer**. *See* buffer strip (def. 1).

buffer strip (riparian buffer). **1:** a narrow strip of permanent, erosion-resisting vegetation that is often planted along contours between cultivated strips on sloping land or below a cultivated field near the bank of a body of water. Such strips can be effective in slowing and reducing the flow of water or the velocity of the wind and thus retain soil and sediment on the farm. This measure helps to prevent off-site pollution. **2:** a strip of land adjacent to a road, trail, water body, or recreation area to reduce disturbance by outside sources. In particular, such a strip screens disturbed (e.g., logged) areas from public view and preserves the esthetics of the surrounding area.

bug kill. loss of timber to insect pests.

bugle weed. *Ajuga remota*.

Bulan (1,1′-(2-nitrobutylidene)bis[4-chlorobenzene];1,1-bis(*p*-chlorophenyl)-2-nitrobutane; 2-nitro-1,1-bis(*p*-chlorophenyl)-butane). an insecticide that is very toxic, *by mouth*, to laboratory rats.

bulb. **1:** any rounded, enlarged, especially terminal part of a structure (e.g., the bulb of a thermometer). **2:** any bulb-shaped structure or protuberances of a living organism (e.g., olfactory bulb). **3a:** a usually subterranean, more or less globular bud, from which leaves emerge above and a stem, merely a flat disk, from which the roots extend appears below (bulbous plants include onions, lilies, and narcissus). Bulbs of many plants are poisonous. Because they are onionlike in appearance, they are sometimes eaten. **3b:** a fleshy underground organ of perennation and usually of vegetative propagation in many plants of the Liliaceae (e.g., *Convallaria*, *Gloriosa*, *Hyacinthus*, *Ornithogalum*, *Veratrum*) and Amaryllidaceae (e.g., *Galanthus*, *Hymenocallis*, *Leucojum*, *Narcissus*), spider lilies, snowdrops, and snowflakes. It consists of a highly modified shoot with a very short stem surmounted by swollen scale leaves or thickened leaf bases that store food. Growth begins with the elongation of the adventitious roots, followed by sprouting of the apical bud, usually followed by the growth of one or more additional buds. Growth is initially supported by nutrients stored in the bulb. The bulbs of many plants (as in the genera named above) are poisonous and may resemble edible bulbs such as onion. *Cf.* corm. **4:** a plant that grows from a bulb (def. 3).

bulbel. bulblet.

bulbil. *See* bulblet.

bulblet (bulbel; bulbil; bood bud). a small bulb or bulblike structure that develops above ground, typically in the axil of a leaf, but sometimes elsewhere (e.g., among or in the place of flowers. Bulblets are produced, for example, by onions, pineapples, and tiger lilies. They are able to produce a new plant. **aerial bulblet** (bulbil; bulbel; bood bud). a bulblet, capable of producing a plant, that grows in the axils of leaves (e.g., of the tiger lily) or that replaces flowers as in the onion.

bulbourethral gland. Cowper's gland.

bulbous bluegrass. *Poa bulbosa*.

bulbous water hemlock. *Cicutum bulbifera*.

Bulgarian milk. Bulgarian buttermilk.

Bulgarian sour milk. Bulgarian buttermilk.

Bulgaricus milk. Bulgarian buttermilk.

bulk bin. a large bin used to store feed.

bulk density. *See* density.

bulk density of soil. the mass of dry soil per unit bulk volume. This value is a measure of the combined volume of soil solids and pore space.

bulk milk tank. a refrigerated stainless steel vessel in which milk is rapidly cooled at a temperature of 2 to 4°C (35 to 39°F) and stored until picked up by a bulk tank truck.

bull. any of a number of sexually mature male animals such as elk, moose, or cattle. **2:** the central bar of a harrow.

bull holder. bulldog (def. 1).

bull pen. **1:** a small enclosure for the confinement of bulls. **2:** a gathering place, staging area, or living quarters for a group of men. **3:** The sale ring at a live stock auction.

bull thistle. *Cirsium lanceolatum.*

bullbriar. a common name of *Smilax rotundifolia*.

bulldog (bull holder). **1:** a steel ring fastened into the nostrils as an aid in controlling a bull. **2:** in a rodeo, to grasp the horns of a steer or cow, twist to the side, and throw the animal to the ground.

bulldog calf. bovine achondroplasia.

buller. a cow that is difficult to breed; one that has been in heat (estrus) for a long time.

bulling. a heifer or cow in heat (estrus) and showing signs of readiness to be served.

bullock. a castrated bull. **2:** a young male ox. *See* steer.

bumble bee. a common name applied to any of a wide variety of large bees, especially *Bombus* (Family Bombidae), that nest underground in small colonies. They are locally valuable for their honey and pollination of flowers.

bumblefoot. a condition of poultry and other fowl in which the foot or foot pads are hardened, swollen, and bear disabling sores. The infection, thought to be bacterial, may result from injury to the foot.

bunch. **1:** a cluster or clump, as of plants or grapes. **2:** to assemble, congregate, or cluster. **3:** a wound of a pig that fails to heal following castration. **4:** to skid logs together prior to loading for transport. **5:** to stack harvested crops in a field.

bunch grass. a grass that grows in bunches or in tussocks. This growth habit results from the absence of rhizomes and stolons.

Bunostomum. a genus of hookworms (Family Ancylostomatidae; Subfamily Necatorinae), members of which parasitize cattle, sheep, goats, and other herbivores.

bunt (high smut; low smut; pepper brand; smut ball; stinking smut). a fungal disease of wheat due to infestation of the germinating plant by the smuts, *Tilletia caries* or *T. foetida*. The head of the mature plant is undersized and bluish-green in color, the glumes are typically loose, and the grains are filled with smut spores that appear as an oily black powder with an offensive odor. Bearded varieties of wheat may drop their awns.

bunt order. the order of a dominance hierarchy, or peck order, in groups of cattle and pigs. It is expressed as the order in which the animals feed.

Bur artichoke. the flower head of an artichoke. Together with the oval bracts, it used as a vegetable.

bur clover. *Medicago hispida*.

burden. **1a:** hardship; ordeal. **1b:** strain or stress. **2:** load; weight. **3:** the total activity of a radiopharmaceutical retained by an organism at a designated time following administration. **body burden**. the total amount of a xenobiotic chemical, especially if toxic or hazardous, that occurs in a living organism at a specified time. **chemical burden**. the total amount of a chemical above background or baseline that occurs in any component of a living organism or ecological system (e.g., soil burden). **environmental burden**. the total amount of a chemical above background or baseline that occurs in a specified region or environment.

burdening. **1:** establishing, imposing, or sustaining a burden. **2:** the imposition or addition of a xenobiotic chemical to any natural entity, biotic or abiotic, over and above that which is normal.

burdock. a plant of the genus *Arctium*.

Bureau of Alcohol, Tobacco, and Firearms (BATF). a bureau of the U.S. Department of Treasury. Among other activities, it is charged with the issuance of permits for the production of ethanol.

Bureau of Customs. an agency within the U.S. Treasury Department. It collects duties and taxes due on imported merchandise and baggage. Customs agents are also charged with agricultural pests, certain products, and diseases out of the United States

Bureau of Land Management (BLM). a bureau within the Department of Interior that has exclusive jurisdiction over 268 million acres of federally owned lands, mostly in the western United States and Alaska.

Bureau of Reclamation (BOR). a bureau within the Department of the Interior that is charged to manage, develop, and protect water and related resources. In particular, it is responsible for building dams and canals and providing water to local water districts. The districts then sell water to agricultural producers.

burglar (robber). a horse of good condition and appearance, with the exception of a slight defect that can be temporarily corrected or masked during a sale.

Burgundy mixture. a copper-containing fungicide that is more potent than Bordeaux mixture and is used primarily against rusts. It contains 60 g sodium carbonate and 50 g cupric sulfate in 51 g water. *See* Bordeaux mixture, cupric sulfate.

burl. **1:** a bump or knot-like growth on the trunk of a tree. **2:** distorted grain in wood surrounding the pits of undeveloped buds. Such wood is considered valuable if it is undamaged and of sufficient size.

Burma bean. **1:** *Phaseolus limensis*. **2:** the seed of *P. limensis*.

burning. *See* combustion.

burr. glanders

burro (ass (in part); donkey; jack ass). *See Equus a. asinus*.

burrow. a hole dug into the earth by an animal such as a mole, badger, prairie dog, or fox.

burrow weed. common name of any plant of the genus *Haplopappus*.

burry. covered with burrs (e.g., as clothing); descriptive of wool that contains excessive amounts of burrs and/or other fibrous inclusions (e.g., dross, seeds).

bursal. pertaining to a bursa.

bush. **1:** a low shrub or small, bushy tree. **2:** the bush; rough, wild land dominated by bushes and small trees, usually covering a large area as in South Africa and Australia. **3:** the brushy tail of an animal such as the horse. **4:** to train a plant to take on a desired shape by the use of stakes, tie wires, and other devices. **5:** to oblige a seller to accept a lower price for a horse than was bid at auction. **sugar bush**. a grove of sugar maple trees.

bush bean. a green bean that grows on a bush rather than a vine or climbing runners. *See Phaseolus*.

bush hog. **1:** a device, pulled by a tractor, that is used to cut vegetation and to clear land. It has one or more blades that rotate in a plane parallel to the ground. **2:** to use a bush hog.

bushel (bu.). a dry volume measure of varying weight for grain and produce that equals four pecks or eight gallons (2150.42 in^3). A bushel of wheat, soybeans, or white potatoes each weighs 60 pounds; that of corn, rye, grain sorghum, or flaxseed weighs 56 pounds; a bushel of barley, buckwheat, or apples weighs 48 pounds.

bushwhacker. a hook used to cut brush.

busulfan (1,4-butanediol dimethanesulfonate esters; 1,4-bis(methanesulfonoxy)butane; 1,4-di(methanesulfonyloxy)butane;1,4-di-(methylsulfonoxy)butane; methanesulfonic acid tetramethylene ester; tetramethylene bis(methanesulfonate); busulfan; BSF). a crystalline, acetone-soluble alkylsulfonate. It is an extremely toxic alkylating agent used in agriculture as an insect sterilant. It is carcinogenic, teratogenic, and also depresses the immune system.

butachlor ((*N*-(butoxymethyl)-2-chloro-*N*-(2,6-diethylphenyl)acetamide; *N*-(butoxymethyl)-2-chloro-2′,6′-diethylacet anilide; 2-chloro-2,6-diethyl-*N*-(butoxymethyl)acetanilide). a preemergence selective anilide herbicide that is absorbed through the roots. Its herbicidal properties and applications are similar to other anilide herbicides, but it is also used to control grass and broadleaved weeds in seeded and transplanted rice paddies. It is moderately toxic to laboratory rats.

***n*-butane**. *See* butane.

butane. an extremely stable, but highly flammable, colorless gas, with a slight odor, at room temperature under standard atmospheric pressure. Butane is a hydrocarbon (an alkane),

$$CH_3CH_2CH_2CH_3$$

used as a fuel gas. It is a constituent of liquefied petroleum gas (LP gas). There are two isomers (*n*-butane and isobutane). One gallon of liquid butane weighs 4.87 pounds and produces 31 cubic feet of gas. One cubic foot of butane gas produces 3266 BTUs.

butanedioic acid mono(2,2-dimethylhydrazide). daminozide.

1,4-butanediol dimethanesulfonate esters. busulfan.

butanoicacid2,2,2-trichloro-1-(dimethoxyphosphinyl)ethyl ester. butonate.

1-butanol (butyl alcohol; 1-butyl alcohol; *n*-butyl alcohol; butyric alcohol). a colorless, water-soluble, flammable, irritant, toxic liquid:

$$CH_3(CH_2)_2CH_2OH$$

2-butanol (*sec*-butanol; *sec*-butyl alcohol; 2-butyl alcohol; butylene hydrate; secondary butyl alcohol; ethyl methyl carbinol; methyl ethyl carbinol; 2-hydroxybutane; *dl-sec*-amyl alcohol). a colorless, flammable liquid that readily decomposes in the environment, possibly contributing to oxygen depletion of aquatic habitats. 2-butanol is used to produce fish protein concentrate from fish meal. 2-butanol is also used as a flavoring agent and occurs naturally in food and beverages. It appears to be nontoxic to aquatic organisms.

***tert*-butanol** (*tert*-butyl alcohol; 2-methyl-2-propanol; tertiary butanol; *tert*-butyl hydroxide; trimethyl carbonol; trimethyl methanol). a colorless liquid or white crystalline solid, $(CH_3)_3COH$, that is soluble in water and ethanol. It is used as a solvent, an octane booster, a moderately toxic denaturant of ethanol, in flavors and perfumes, and in several manufacturing processes. At normal ambient levels it is not toxic to fish, amphibia, crustacea, algae, or bacteria. In laboratory animals, it causes physical dependence and postnatal effects from exposure *in utero*. In humans, it is also a skin irritant.

butcher. **1:** one who slaughters, dresses out animals, or sells the meat. **2:** livestock that was reared for or is destined for slaughter. **3:** to perform the functions of a butcher (def.1).

Buteo jamaicensis. red-tailed hawk (*See* 3-chloro-*p*-toluidine).

butilate. *See* butylate, def. 2.

butinol-*N*-(3-chlorophenyl)carbamate. an aryl carbamate herbicide used to control *Avena fatua* (wild oats) in cereal crops.

Butinox. tributyl tin, tributyl tin compound.

butonate (butanoic acid 2,2,2-trichloro-1-(dimethoxyphosphinyl)ethyl ester; butyric acid ester with dimethyl (2,2,2-trichloro-1-hydroxyethylphosphonate; *O,O*-dimethyl-2,2,2-trichloro-1-*n*-butyryloxyethylphosphonate; *O,O*-dimethyl2,2,2-trichloro-1-phosphonoethyl butyrate). a colorless, rather oily liquid. It is a cholinesterase inhibitor used as an insecticide and as an anthelmintic in horses.

butopyronoxyl (3,4-dihydro-2,2-dimethyl-4-oxo-2*H*-pyran-6-carboxylic acid butyl ester; butyl3,4-dihydro-2,2-dimethyl-4-oxo-2*H*-pyran-6-carboxylate; butyl mesityl oxide oxalate; α,α-dimethyl-α'-carboxydihydro-γ-pyrone butyl ester). a yellow or amber to pale reddish-brown liquid used as an insect repellent. It is toxic if in gested and may cause liver damage.

(*N*-(butoxymethyl)-2-chloro-*N*-(2,6-diethylphenyl)acetamide. butachlor.

***N*-(butoxymethyl)-2-chloro-2′,6′-diethylacetanilide**. butachlor.

butt. **1a:** the end, usually the broad end, of a structure; the stub or stump. **1b:** the thick basal end of a tree trunk or plant stem. **1c:** the above-ground remainder of the trunk of felled or fallen tree; the stump. **2:** to strike with the forehead or horns, usually from a running start, as by a ram or billy goat. **3:** in butchering, the broad (upper) half of a ham or shoulder. **4:** in Australia, a bundle (in Australia) that holds 196 lbs. of greasy wool or 112 lbs. of scoured wool. **5:** the target of a practical joke. *See also* barrel.

butt cut. **1:** the first log to be cut above the stump in harvesting trees. **2:** butt log; tanbark; the bark removed from the butt of a tree prior to felling.

butt log. *See* butt cut.

butter. a fatty substance, high in saturated fats and a solid at room temperature, that is made from milk or cream by churning. Salt and artificial coloring may be added. Commercially prepared butter may contain no more than 20% of nonfat constituents by weight. It has many uses in cooking and as a spread for bread and other baked goods. **sweet butter**. unsalted butter.

butter bean. **1:** any of the several types of lima bean grown for the table in the southern United States **2a:** *Phaseolus lunatus*. **2b:** the seed of *P. lunatus*.

butter oil (gee). liquid, clarified butterfat that is used to make process butter. It can be shipped unrefrigerated.

buttercup. *See Ranunculus*.

butterfat. milk fat.

butterfly. **1:** a common name of certain insects of the order Lepidoptera, *q.v.* **2:** a breed of English rabbit.

butterfly weed. *Asclepias tuberosa*.

buttermilk. a liquid dairy product with a distinctive acidic taste. **natural buttermilk**. the by-product of churning cream into butter. **Bulgarian buttermilk** (Bulgarian milk, Bulgarian sour milk, Bulgaricus milk). a thick, acid-tasting yogurt produced by the action of *Lactobacillus bulgaricus* on sterile milk at a relatively high temperature. **cultured buttermilk**. that which is usually made from skim milk by inoculating skim milk with a starter culture of *Lactobacillus* and incubating until the lactic acid content reaches about 0.8 percent. Salt is usually added to enhance the flavor.

buttermilk horse. appaloosa.

button clover. *Medicago orbicularis*.

button medic. *Medicago orbicularis*.

buttress. **1:** a brace; prop; foundation. **2:** backing; defense. **3:** a swelling on a tree near its base, commonly seen on trees on wet soils in the humid tropics and subtropics.

(5-butyl-1*H*-benzimidazol-2-yl)carbamic acid methyl ester. parabendazole.

***n*-butyl carbinol**. *n*-amyl alcohol, primary.

5-butyl-2-(carbomethoxyamino)benzimidizole). parabendazole.

***N*-butyl-*N*′-(3,4-dichlorophenyl)-*N*-methylurea**. neburon.

butyl 3,4-dihydro-2,2-dimethyl-4-oxo-2*H*-pyran-6-carboxylate. butopyronoxyl.

2-*sec*-butyl-4,6-dinitrophenol. dinoseb.

2-*sec*-butyl-4,6-dinitrophenyl 3-methyl-2-butenoate. binapacryl.

2-*sec*-butyl-4-6-dinitrophenyl 3-methylcrotonate. binapacryl.

2-*sec*-butyl-4,6-dinitrophenyl senecioate.

binapacryl.

***tert*-butyl hydroxide**. *tert*-butanol.

butyl mesityl oxide oxalate. butopyronoxyl.

butylate. 1: to introduce a butyl group into a substance. **2:** a selective, preplanting thiocarbamate herbicide that is moderately toxic to test mammals (bis(2-methylpropyl)carbamothioic acid *S*-ethyl ester; diisobutylthiocarbamic acid *S*-ethyl ester; *S*-ethyl-*N,N*-diisobutylthiocarbamate; butilate; diisocarb).

butylated hydroxyanisole (BHA). a toxic mixture of 2- and 3-*tert*-4-methoxyphenol, it is an antioxidant used for fats and oils and is sometimes used as a preservative in drug formulations. Its use in foods is restricted in the United States, but it may be used in breads. BHA can cause depression of growth rate, weight loss, liver damage, baldness, and fetal abnormalities. It has been linked to hyperactivity and behavioral disturbances in children. *See also* bread, butylated hydroxytoluene.

butylated hydroxytoluene(2,6-di-*tert*-butyl-*p*-cresol; BHT). a toxic, flammable, white, crystalline solid,

$$[C(CH_3)_3]_2CH_3C_6H_2OH$$

BHT dissolves in organic solvents. It is used as an antioxidant in various products, including food products and animal feeds. Its use in foods in the United States is restricted, but it may be used in breads. BHT causes depression of growth rate, loss of weight, damage to the liver, baldness, and fetal abnormalities in laboratory animals. It is carcinogenic to mice and also appears to promote existing tumors. It is a suspected human carcinogen. BHT has been linked to hyperactivity and behavioral disturbances in children. *See also* bread, butylated hydroxyanisole.

1-(butylcarbamoyl)-2-benzimidazolecarbamic acid methyl ester. benomyl.

2,6-di-*tert*-butyl-*p*-cresol. butylated hydroxytoluene.

butylene hydrate. 2-butanol.

5-butylpicolinic acid. fusaric acid.

5-butyl-2-pyridinecarboxylic acid. fusaric acid.

butyric acid ester with dimethyl (2,2,2-trichloro-1-hydroxyethylphosphonate. butonate.

butyric anhydride. a combustible, water-white liquid,

$$(CH_3CH_3CH_2CO)_2O$$

It is used to drive bees from honeycombs. The odor is similar to that of rancid butter and stale sweat.

Buxaceae (the boxwood family). a small family of chiefly tropical or subtropical trees, shrubs, and perennial herbs comprised of about six genera. *See Buxus*.

Buxus sempervirens (box; common box; boxwood). **1:** a slow-growing, broadleaved evergreen hedge plant or tree (Family Buxaceae) of southern Europe, northern Africa, western Asia, and widely cultivated elsewhere. The yellowish flowers grow in small clusters and produce seed capsules containing three seeds. All parts of the plant are poisonous. Despite a presumably repellant odor and bitter taste, grazing animals sometimes die from eating clippings from box hedges. Cattle, horses, swine, and sheep are highly susceptible. Early signs are mainly those of gastrointestinal distress and sometimes convulsions.

by. in animal husbandry: sired; begotten by a male.

byre (cow house). *See* cow barn.

byssinosis (brown lung; brown lung disease; Monday fever; Monday dyspnea; cotton-dust asthma; stripper's asthma; cotton-mill fever). a respiratory disease due to inhalation of cotton dust by textile workers and workers who process soft hemp and flax. Long-continued exposure usually causes irreversible damage to the lungs and eventual death. The disease is reversible if detected early.

byssinotic. **1:** of or pertaining to byssinosis. **2:** a victim of byssinosis.

byssophthisis. obsolete term for byssinosis.

C

c. abbreviation for centi- in the International System of Units.

C. the symbol for carbon.

C.I. 11380. FD&C Yellow No. 3.

C.I. 16035. FD&C Red No. 40.

C.I. 16155. FD&C Red No. 1.

C.I. 42053. FD&C Green No. 3.

C.I. 42095. FD&C Green No. 2.

C.I. 42640. FD&C Violet No. 1.

C.I. 52015. methylene blue.

C.I. 73015. FD&C Blue No. 2

C.I. 77085. arsenic disulfide.

C.I. 77947. zinc oxide.

C.I. acid blue 1. FD&C Blue No. 2

C.I. acid green. FD&C Green No. 2.

C.I. basic blue 9. methylene blue.

C.I. Colour Index.

C.I. food blue. FD&C Blue No. 1

C.I. food blue 1. FD&C Blue No. 2

C.I. food green 2. FD&C Green No. 2.

C.I. food green 3. FD&C Green No. 3.

C.I. food red. FD&C Red No. 1.

C.I. food red 6. FD&C Red No. 1.

C.I. food yellow 10. FD&C Yellow No. 3.

C.I. food violet 2. FD&C Violet No. 1.

C.I. pigment white 4. zinc oxide.

C.I. Pigment Yellow 39. arsenic disulfide.

C.I. solvent yellow 5. FD&C Yellow No. 3.

C3 proactivator. properdin factor B.

cabbage. **1:** *Brassica oleracea* var. *capitata*. **2:**

cachectic. pertaining to or exhibiting cachexia.

cachectin. tumor necrosis factor α.

cachexia (cachexy). any generalized, intense, constitutional illness marked by wasting. Cachexia may result from metabolic disorders, chronic disease, poisoning, or emotional distress. **cachexia mercurialis**. that due to chronic mercurial poisoning. **saturnine cachexia.** that due to chronic lead poisoning.

cachexy. cachexia.

cacodylic acid (dimethylarsinic acid; hydroxydimethylarsine oxide; Phytar). a moderately to very toxic organic arsenical,

$$(CH_3)_2As(O)OH$$

It is used as a desiccant and nonselective postemergence herbicide to control weeds in non-crop areas. Toxicity is greatest to vertebrates when ingested because the acidic gastric juices transform cacodylic acid to trivalent arsenic. Cacodylic acid is the major meta

bolite that is excreted in urine of humans and dogs following ingestion of trivalent or pentavalent inorganic arsenic. *See also* sodium cacodylate, biomethylated arsenic.

cacogenesis. the inability to to hybridize.

Cactaceae (cactus family). a family of perennial, succulant plants native to the

Western Hemisphere. They have reduced leaves, showy flowers, and often spiny stems. *See also Lophophora williamsii.*

cactinomycin (actinomycin C). an extremely toxic mixture of actinomycins C_2, C_3, and actinomycin D (dactinomycin) produced by *Streptomyces chrysomallus*. It was formerly used as an antineoplastic agent.

caeca. plural of caecum.

caecum, plural, caeca. *See* cecum.

3-caffeoylquinic acid. chlorogenic acid.

caicaca. *Bothrops atrox*.

calabrese. *B. oleracea* var. *botrytis*.

caladium. common name of any herb of the genus *Caladium*.

Caladium (caladium; elephant's ear). a genus of attractive, large-leaved herbs (Family Araceae). All parts contain calcium oxalate and other toxic principles. When plant material is chewed, the needle-sharp crystals of calcium oxalate are released by ruptured plant cells and penetrate the tissues of the mouth, tongue, and throat, quickly producing a severe burning sensation and irritation of the mucous membranes with swelling of the affected tissues; nausea; vomiting; diarrhea; salivation. Occasionally direct systemic effects are seen. On occasion, the swollen tongue may become immobile, interfering with speech, swallowing, and breathing; obstruction of the airway is a rare complication.

Calamagrostis canadensis (bluejoint reedgrass; bluejoint; bluetop.). a perennial forage grass (Family Gramineae) of medium quality. It occurs widely throughout the United States at elevations up to 4000 m, usually in marshy or poorly-drained soil. Also called bluejoint, bluetop.

calciferol. vitamin D_2.

calciphilic. "calcium loving"; thriving in soils that are rich in calcium salts.

calico bush. *Kalmia latifolia.*

calciferous. containing or producing calcium carbonate.

calcinogenic. **1:** calcium-forming. **2:** causing the deposition of calcium salts in tissues. *See* calcinosis.

calcinosis (exudative calcifying fasciitis; calcium gout; hypercalcification). a condition in which abnormal amounts of calcium salts are deposited in various tissues of the animal body. **enzootic calcinosis** (hypercalcification). a disease of herbivores (known especially in ruminants and horses) characterized by extensive calcification and degeneration of soft tissues (especially the heart, lungs, and tendons) with weight loss, emaciation, and attendant muscular, postural, locomotor, respiratory, and cardiovascular problems. It is a chronic, progressive disease with a variable, but often high mortality rate among livestock (apparently more than 50% in some areas). The victim becomes weak and listless, breathing becomes shallow and diaphragmatic, and the pulse rate increases with mild exercise. The etiology is not always clear, but can be the result of poisoning by certain plants that contain 1α,25-dihydroxycholecalciferol(1α,25 [OH],-D_3) glycoside, a derivative of calcitriol, or other biologically active substances that imitate the calcinogenic activity of calcitriol. Such plants include *Cestrum diurnum*, *Trisetum flavescens*, Nierembergia vietehii, *Solanum esuriale*, *S. torvum*, and *S. malacoxylon*. All of these plants contain calcitriol glycoside or a substance that imitates its calcinogenic activity. This disease sometimes results from mineral imbalances in the soil.

calcite. the most common form of naturally occurring calcium carbonate, CaCO, calcite is a crystalline mineral. It is an ingredient in limestone, marble, and chalk and sometimes used as a diluent for insecticidal dusts.

calcitic dolomite. a carbonate rock that contains from 10-50% calcite and 50-90% dolomite.

calcium (Ca) [Z = 20; A_r = 40.08]. a moderately soft, low-melting reactive metal. It is an essential component of and the most abundant mineral in living systems. Calcium is the fifth most abundant element, occurring naturally only in combined form (e.g., as calcium carbonate, calcium phosphate), mostly in large deposits of chalk, limestone, or marble and in seawater at a concentration of about 400 g/ton. Because of its relatively large atomic radius and low ionization potential, calcium is a very electropositive element with a valence of 2.

calcium arsenate (tricalcium arsenate). a very toxic, slightly water-soluble white powder,

$$Ca_3(AsO_4)_2$$

It is used as a molluscicide and as an insecticide to control cotton pests.

calcium arsenite. a white, granular powder, $CaAsO_3H$, used as a germicide and insecticide. It is a mixture of variable composition, prepared by passing steam over a dry mixture of calcium oxide and arsenic trioxide. It is toxic by ingestion and inhalation.

calcium borogluconate. a compound administered by injection to cattle affected by milk fevers due to calcium deficiency.

calcium carbimide. calcium cyanamide.

calcium carbonate (chalk). a widely distributed mineral, formed from animal remains.

calcium caseinate. a heterogeneous complex, it is the form of casein found in milk. *See* casein.

calcium chlorate. a white, water-soluble, heat-labile, crystalline compound,

$$Ca(ClO_3)_2 \cdot 2H_2O$$

Calcium chlorate is used as an herbicide, insecticide, and seed disinfectant.

calcium cyanamid. calcium cyanamide.

calcium cyanamide (calcium carbimide; calcium cyanamid; nitrolime; CCC). a colorless, crystalline, phytotoxic substance, $CaNC \equiv NC$, used as a fertilizer, defoliant, herbicide, pesticide, and as an anthelmintic in veterinary practice. It is poisonous by all routes of exposure and is moderately toxic to humans by ingestion.

calcium fluorophosphate. apatite.

calcium gout. calcinosis.

calcium hydroxide (slaked lime). a mixture of calcium oxide and water, used to improve soil quality. *See* Bordeaux mixture.

calcium oxalate (ethanedioic acid calcium salt). a white crystalline salt, CaC_2O_4, of oxalic acid. It is insoluble in water or body fluids. The crystals occur in large quantities in the tissues of certain plants, especially in those of the Family Araceae (arums such as *Caladium*, *Colocasia*, *Dieffenbachia*, *Dracunculus*, *Philodendron*). The crystals probably cause the intense burning sensation and oral edema that results from chewing the plant. This is partly effected by penetration of the oral mucosa by the crystals and partly by the chemical properties of the salt. See also oxalic acid.

calcium oxide (quicklime). a compound used in agriculture to reduce soil acidity.

calcium phosphate. any of several naturally occurring phosphates of calcium found in some rocks and in animal bones. They are used, for example, as fertilizers and food additives.

calcium sulfate (gypsum). sometimes used as a soil amendment to improve the quality of soil affected by salt.

calcium supplement. the addition of calcium to the diet of livestock in feeds, or by injection to improve electrolyte balance.

calf. a cow less than one year of age. **bull calf**. a male calf. **heifer calf**. female calf.

calf scours. *See* scours.

calices. plural of calyx.

caliciform cell. goblet cell.

California agaricus. *Agaricus californicus*.

California bur clover. *Medicago hispida*.

California Mastitis Test (CMT). a test used to determine the somatic cell content in milk. This is accomplished by use of a reagent that reacts with the chromatin material of cells in the milk to form a gel.

California water hemlock. *Cicutum californica*.

calix. calyx.

calomel. mercurous chloride.

caltrop. *Tribulus terrestris*.

Calvin cycle. the biochemical pathway of carbon dioxide fixation in many autotrophic organisms.

calyx (plural, calices; calix). **1:** the sepals of a flower taken as a whole. **2:** a cup-shaped portion of the renal pelvis that encircles one or more renal papillae in vertebrate animals. The calyx may consist of separate or fused sepals. **3:** any of a number of roughly cup-shaped zoological structures.

cambium. meristematic tissue of roots and shoots that produces both phloem and xylem. *See also* phloem, xylem, alburnum, annual ring, wood.

camel dung. the dung of dromedaries or Bactrian camels; the dried dung is used as fuel in the desert or semidesert where these animals are employed as beasts of burden.

campfire. **1:** a small outdoor wood fire as at a campsite, used for cooking, warmth, and/or light. Such fires are, by convention, no more than 1 m in height and 1 m in diameter. **2:** a gathering held around such a fire.

Camponotus herculeanus pennsylvanicus (black carpenter ant). a large ant of the eastern United States It attacks wooden structures such as tree trunks and stumps, telephone poles, and window sills.

Canada balsam. *Abies balsamea*.

Canada thistle. *Cirsium arvense*.

Canadian yew. *Taxus canadensis*.

cane sorghums. forage sorghums. *See Sorghum vulgare* .

canicola fever. human leptospirosis (*See* leptospirosis).

canid. common name of any animal the family Canidae.

canning. the preservation by use of air-tight rigid containers, usually of glass or metal. Methods of canning are prescribed for the particular type of food to be preserved in order to prevent spoilage. **aseptic canning**. the heating of food prior to canning to destroy pathogenic and food spoilage microorganisms; placement of the food in cans that have been sterilized and sealed to prevent reentry of microbes.

canopy. the uppermost layer of vegetation in a forest, woodland, or other woody community, comprised of the crowns of trees or sometimes shrubs. **closed canopy**. descriptive of vegetation, especially of trees, when the crowns of upper-story trees form a canopy in which the leaves of adjacent trees are touching and intermingled so that direct light cannot reach the forest floor.

canopy closure. the progressive reduction of space between crowns of trees as they grow and the crowns spread laterally, increasing canopy cover.

cantharides camphor. cantharidin.

cantharidin (hexahydro-3a,7a-dimethyl-4,7-epoxyisobenzofuran-1,3-dione; 2,3-dimethyl-7-oxabicyclo-[2.2.1]heptane-2,3-dicarboxylic anhydride; exo-1,2-*cis*-dimethyl-3,6-epoxy-hexahydrophthalic anhydride; cantharides camphor). a white powder or colorless, crystalline compound, $C_{10}H_{12}O_4$, that is slightly soluble in water, acetone, chloroform, and ethanol. It is secreted by certain beetles of the family Meloidae of which the Spanish fly, *Lytta vesicatoria*, is an important source. Cantharidin is used also in veterinary medicine as a vesicant, rubefacient, and counterirritant. It is a powerful irritant to all cells and tissues, and (as Spanish fly) has an unwarranted reputation as an aphrodisiac.

cantharidin poisoning (cantharidism; blister beetle poisoning). **1:** poisoning of humans by the misuse of cantharides. Symptoms appear immediately following ingestion and may include severe skin irritation, blistering of mucous membranes, abdominal pain, nausea, diarrhea, vomiting of blood, severe fall in blood pressure, hematuria, coma, and death due to respiratory failure. Lesions commonly encountered may be severe and include necrosis of esophageal and gastric mucous membranes, intense congestion of blood in the genitourinary tract, renal damage, and hemorrhagic changes in the ovaries. **2:** blister beetle poisoning of livestock.

cantharidism. cantharidin poisoning.

canthus. **1:** the angle where the eyelids meet on either side of the eye. **2:** canthus rostralis.

canthus rostralis. the angle between the flat crown of the head of a snake and the side, between the snout and head. *Cf.* canthus.

cantil. *Agkistrodon bilineatus*.

cap. **1a:** a lid or covering. **1b:** an upper limit or ceiling; a roof. **2:** a layer of highly impermeable material such as clay positioned over the top of a closed landfill to prevent intrusion by rainwater and to minimize production of leachate.

caper spurge. *Euphorbia lathyris*.

capping. a hard crust on the soil's surface that may result from the use of heavy machinery or heavy rains.

capriculture. the study and practice of goat keeping and breeding.

capsicum. the dried fruit of any of various species of *Capsicum*. It is used as an irritant and carminative.

Capsicum. a rather diverse, mainly tropical genus (Family Solanaceae) of shrubby, perennial plants that includes the cayanne (red) pepper. Many are grown for their fruit. *See* capsaicin, capsicin, capsicum.

capsular. of or pertaining to a capsule.

carbamate (carbamoate). a salt or ester of carbamic acid, some of which occur naturally in the blood and urine of mammals; others are manufactured xenobiotics with pesticidal activity. Those used as pesticides (e.g., carbaryl, carbofuran, pirimicarb, propoxur) generally have lower dermal toxicities than most organophosphate pesticides and are less persistent environmentally than the organochlorine insecticides. Carbamates vary greatly in toxicity from essentially nontoxic to lethal in trace amounts. Some are immunotoxic. The insecticidal carbamates kill animals mainly through direct, relatively reversible inhibition of anticholinesterase, although they also affect other enzymes. *See* carbamate poisoning.

carbamate poisoning. poisoning by carbamates is often by carbamate insecticides (*See* insecticide). The insecticidal carbamates (e.g., aldicarb) kill animals chiefly by direct, relatively reversible inhibition of anticholinesterase, although other enzymes are also affected. Some are immunotoxic. Carbamates may cause convulsions and even death through contact or ingestion in humans and other animals; some are lethal in trace amounts. Symptoms in mammals include excessive salivation, lacrimation, miosis, and bradycardia. Death is usually due usually due to bronchoconstriction with consequent respiratory failure. If death does not intervene, recovery is completely. Atropine is often a useful antidote.

carbamic acid ethyl ester. urethane.

carbamoate. carbamate.

carbanilic acid isopropyl ester. propham.

carbaryl (1-naphthalenol methylcarbamate; methyl carbamic acid 1-naphthyl ester; 1-naphthyl *N*-methylcarbamate). a white powder used mainly as a broad-spectrum contact insecticide and pediculicide that acts by reversibly inhibiting cholinesterase. It is a skin irritant and is generally regarded as moderately toxic to mammals. Symptoms of poisoning in humans may include dizziness, nausea, excessive salivation, vomiting, abdominal cramps, diarrhea, blurred vision, headaches, bronchoconstriction and dyspnea, muscle twitching, incoordination, slurred speech, general weakness. In severe cases, the symptoms can progress to convulsions, cyanosis, coma, respiratory failure, and death. Carbaryl is carcinogenic to humans and may also cause birth defects. *See also* anticholinesterase.

Carbicron. dicrotophos.

carbofos. malathion.

carbofuran (2,3-dihydro-2,2-dimethyl-7-benzofuranol methylcarbamate; methyl carbamic acid 2,3-dihydro-2,2-dimethyl-7-benzofuranyl ester; 2,2-dimethyl-2,3-dihydro-7-benzofuranyl-*N*-methylcarbamate; 2,2-dimethyl-7-coumaranyl *N*-methylcarbamate). a white, highly toxic, water-soluble, solid carbamate. It is a plant systemic insecticide, acaricide, and nematocide. It is also used in some household products to eradicate pests and should be considered highly dangerous, especially in households with children. Carbofuran reversibly inhibits cholinesterase. It is absorbed by the roots and leaves of plants and by insects feeding on the plant material. The minimum toxic dose in cattle and sheep is 4.5 mg/kg body weight; the lethal oral dose is 18 mg/kg in the laboratory rat and 9 mg/kg in sheep and cattle. Swine have been poisoned after drinking water contaminated by this compound. The oral LD_{50} in the dog is 19 mg/kg body weight. *See* insecticide (carbamate insecticide).

carbohydrate. a compound with the approximate general formula, $(CH_2O)_n$. Included are a wide range of substances from simple sugars (e.g., glucose) to polymers of high molecular mass (e.g., starch, cellulose, glycogen). Most carbohydrates play a role in storing and transferring energy. Some have structural or additional physiologic functions.

α-2-carbomethoxy-1-methylvinyl dimethyl phosphate. mevinphos.

2-carbomethoxy-1-methylvinyl dimethyl phosphate. mevinphos.

2-carbomethoxy-1-propen-2-yl dimethyl phosphate. mevinphos.

carbon (C). [$Z = 6$; $A_r = 12.01$]. a nonmetallic tetravalent element. The chief sources of carbon are biological, e.g., coal, oil, gas, carbonates, and atmospheric CO_2. Carbon combines to form a vast array of compounds and is a universal component of living systems.

carbon balance. the concentration of carbon released into the atmosphere compared to the amounts stored in the oceans, soil, vegetation, and fossil fuel (e.g., coal, petroleum).

carbon bisulfide. carbon disulfide.

carbon cycle. the sequence of reactions by which carbon dioxide is converted to organic forms (e.g., by photosynthesis or chemosynthesis) and cycled through the biosphere, usually with partial incorporation into sinks (e.g., sediments and soils), returning to its original gaseous form by respiration or combustion.

carbon dioxide (carbonic anhydride; carbonic acid gas). a colorless, odorless, tasteless gas, CO_2, with density about 1.5 times that of air. The concentration of carbon dioxide in the atmosphere is 0.034% by volume. Enormous amounts enter the atmosphere from the combustion of fossil fuels such that the concentration of carbon dioxide is increasing by about 0.27% annually.

carbon disulfide (carbon bisulfide). a clear, colorless, or slightly yellow highly flammable liquid that usually has a strong, disagreeable odor due the presence of impurities. It is a serious fire and explosion hazard and is toxic to humans by skin absorption. It is used, for example, as a solvent, to degrease seed meals (e.g., of castor and cottonseed), and to fumigate seeds infested by insects.

carbon fixation. the conversion of carbon dioxide and other single-carbon inorganic compounds to organic forms (e.g., sugars).

carbon-nitrogen ratio. the ratio of organic carbon to nitrogen (C/N) by mass in soil or organic material.

carbonic acid gas. carbon dioxide.

carbonic anhydride. carbon dioxide.

carbophenothion (phosphorodithioic acid *S*-[[(4-chlorophenyl)thio]methyl]*O,o*-diethyl ester; *S*-[[(*p*-chlorophenyl)thio]methyl] *O,o*-diethyl phosphorodithioate; *O,o*-diethyl *S*-(*p*-chloro-phenylthio)methyl phosphorodithioate). a light amber, nearly water-insoluble liquid pesticide used on fruits, nuts, vegetables, and fiber crops. It is also used in certain household products to eliminate pests. It is extremely toxic to laboratory rats, dairy cattle, sheep, and goats and should be considered dangerous to humans.

carboxybenzene. benzoic acid.

carboxyl group. a group, -COOH, bound to a carbon skeleton as in the carboxylic acids and fatty acids.

20β-carboxy-11-oxo-30-norolean-12-en-3β-yl-2-*o*-β-D-glucopyranuronosyl-α-D-glucopyranosiduronic acid. glycyrrhizic acid.

Carduus (nodding thistle; plumeless thistle). a genus of hardy, robust, spiny-leaved annual or perennial herbs with purple, tubular flowers (Family Compositae) that are indigenous to southeast Asia. Toxic concentrations of nitrate are occasionally reported.

carelessweed. *Amaranthus*.

Carolina buckthorn. *Rhamnus caroliniana*.

Carolina ipecac. *Euphorbia ipecacuanha*.

Carolina jessamine. *Gelsemium sempervirens*.

Carolina yellow jessamine. *Gelsemium sempervirens*.

carp. *See Cyprinus carpio*.

carrot. *Daucus carota* var. *sativus*.

carry. to keep or stock animals on a given piece of land, as farmland.

carrying capacity. **1:** the maximum number of organisms of a particular species that can be maintained and sustained indefinitely in a given ecosystem. Usually denoted by K, it is the upper asymptote of the logistic equation. **2:** in forestry, the average number of livestock and/or wildlife that can be sustained on a management unit, given the management objectives for the unit. **3:** in recreation management, the amount of use a recreation area can sustain without a loss of environmental quality. **4:** in wildlife management, the maximum number of animals of a specified species that a specified area or ecosystem can sustain during a given period of the year. **5:** in recreation management, the amount of use a recreation area can sustain indefinitely without a decline in quality.

cartel. an alliance or arrangement among commercial or industrial ventures or nations with the aim of limiting competition or establishing a monopoly in a market.

Carya cordiformis (bitternut hickory). a deciduous shade tree (Family Juglandaceae) that is grown also for its timber, which is used, for example, to make furniture, ladders, wooden implements and tool handles. The wood is also used as a household fuel and for smoking meat.

caryopsis. the seed of cereals and grasses.

casava. cassava.

cascara buckthoorn. *Rhamnus purshiana*.

casein. **1:** caseinogen, a colloidal aggregate of related proteins together with phosphorus and calcium that occurs in cow's milk and cheese. It is a white, amorphous, tasteless, solid. The components are water-insoluble but dissolve in alkaline and salt solutions. Casein (in the form of calcium caseinate) is precipitated from milk by the action of rennin. **2:** paracasein; rennet casein; rennet curd, a phosphoprotein that binds to calcium and precipitates as a cream-colored curd when milk is curdled by rennet, either by direct bacterial action or when added in solution. It is the chief protein of cheese and is used also in the manufacture of casein plastics. **3:** the preserved stomach of a suckling calf, lamb, or pig.

caseinogen. a synonym for casein (def. 1).

cash crop. a crop grown for sale rather than for use on the farm (e.g., as food or feed).

cashew. *Anacardium occidentale*.

cashew family. Anacardaceae.

cask. a barrel.

cassava (casava; manioc; tapioca plant). **1:** any of several widely cultivated plants of the genus *Manihot* (e.g., *M. esculenta*, *M. utilissima*). The tubers are often a local staple food source and are the source of tapioca. The raw tubers often contain lethal concentrations of cyanide, but are generally rendered harmless either by heating the peeled tubers or by repeated mashing and soaking in water. **2:** the rootstock of *M. esculenta*. *See also Manihot*.

caste. any of several types of morphologically and physiologically distinct adult individuals that occur in the colonies of certain social insects, especially the Hymenoptera and Isoptera. Thus, for example, honeybee colonies are made up of workers (sterile females), a queen (a fertile female), and drones (males).

castration. the removal of testicles. In the case of livestock, castration reduces sexual activity and facilitates the fattening of meat animals.

cat's claw. *Acacia greggii*.

catabolite repression. inhibition of any of a variety of inducible enzymes at the level of transcription by glucose or other compounds that serve as ready source of carbon.

catbrier. a common name of *Smilax rotundifolia*.

catclaw. *Acacia greggii*.

catclaw acacia. *Acacia greggii*.

caterpillar rash. caterpillar dermatitis (*See* dermatitis).

catharsis (purgation). **1:** the loosening of the bowels or the induction of defecation. Some poisons have a violent cathartic action. The primary danger of such an effect is that large amounts of fluid are lost from the body. This may result not only in dehydration, but also in a life-threatening loss of electrolytes. Catharsis is sometimes used to remove unabsorbed poisons from the digestive tract. **2:** a freudian term (psychocatharsis) relating to the method of recalling past (especially repressed) experiences with a view to releasing emotional tension or anxiety.

cathartic (purgative; eccoprotic). **1:** pertaining to catharsis. **2:** an agent that stimulates bowel movements; a powerful laxative. Care should be taken in their use and all should be considered poisonous, some violently so.

cation. an ion that carries a positive electrical charge, one that moves toward the cathode (e.g., potassium, sodium, calcium, magnesium) Cations that commonly occur in soils are ammonium, calcium, hydrogen, magnesium, potassium, and sodium.

cation exchange. the process of interchange between a cation in solution and one bound to a solid negatively-charged material (a cation exchanger). With respect to soil, cation exchange takes place between a cation in solution and any surface-active material in the soil such as clay or organic matter. *Cf.* anion exchange.

cation exchange capacity (CEC). defined with respect to soil, CEC is the total exchangeable cations that a soil can adsorb at a specific pH, expressed in SI units as centimoles of positive charge per kilogram of soil (cmolc/kg). It may also be expressed in terms of milliequivalents per 100 grams of soil, usually at neutrality (pH 7.0), but may be referred to some other pH value. As applied to soils, this term is synonymous with base-exchange capacity but is more precise.

cation exchanger. *See* cation exchange.

cationic. of or pertaining to a positively charged ion (cation).

catkin. a type of spike or raceme with a flexible rachis and thus often with a drooping inflorescence. The flowers in the cluster are small, stalkless, and usually unisexual. *Cf.* spike.

cattle. **1a:** plural of cow. **1b:** a collective name for ruminant mammals of the genus *Bos*, especially those of the domesticated species, *Bos taurus* and *B. indica*. Cattle generally have unbranched horns comprised of a horny layer surrounding a bone extension of the skull. Unlike the antlers of deer, the horns of cattle are not shed. Some cattle are naturally hornless. **beef cattle**. any set, mix, or breed of cattle reared to produce meat. *Cf.* beef breed. **dairy cattle**. any set, mix, or breed of cattle reared to produce milk for human consumption.

cattle dung. cow dung.

caustic baryta. barium hydroxide, especially the octahydrate.

Cavia. a genus of rodents (Family Caviidae) that includes the guinea pig, a domesticated form of unknown origin. The guinea pig is a small, stout-bodied, short-eared, very short-tailed (the tail cannot be discerned externally) hystricomorph rodent. It is raised for food in South America, as a pet, and as a laboratory animal in biological research. There are numerous variants; the parent species was probably agouti in coloration. While it is sometimes considered to be a distinct species, *C. cobaya*, it is more probably a variety of a feral South American species (possibly *C. porcellanus* or *C. cutleri*). It is especially unwise to assign a scientific name to domestic variants and lines used in laboratories and sometimes in agriculture as eqwell. *See also* cavy.

cavies. plural of cavy.

Caviidae. a family of rodents that includes the domesticated guinea pig, *Cavia*.

cavy (plural, cavies). a common name of any of several South American rodents (Family Caviidae), especially *Cavia porcellus*. All are short-tailed, rough-coated herbivores. The domesticated guinea pig, *Cavia*, is a cavy.

cayenne tick. *Amblyomma cajennense*.

cc. cubic centimeter.

CCC. **1:** calcium cyanamide. **2:** chlorcholine chloride.

CD. chlordecone.

CDAA. allidochlor.

CDEC. sulfallate.

CDM. chlordimeform.

Ceanothus cuneatus (buckbrush; wedgeleaf ceanothus). a shrub (Family Rhamnaceae) of the rangelands of the western United States that serves as forage for livestock.

Ceanothus sordiatus (buckthorn; jimbrush). a thorny shrub. It is the alternate host for the fungus that causes the crown rust of oats.

CEC. cation exchange capacity.

ceca. plural of cecum.

cecal. **1:** pertaining to a cecum. **2:** terminating blindly as a pouch or in a cul-de-sac.

ceco-, cec-. affixes denoting a cecum. *See also* typhlo-.

cecum (plural, ceca; caecum). any hollow outpocketing or pouch-like structure, large or small, from a digestive or other organ. In amniotes, a cecum arises at or near the junction of the large and small intestines. In animals such as the horse, the cecum is large and contains cellulose-digesting flora. A number caeca with of secretory and absorptive functions occur in the midgut of insects.

cedar. common name for a number of mostly coniferous evergreen trees. *See*, for example, *Thuja*. **true cedar**. common name of trees that belong to the genus *Cedrus*, q.v.

Celastraceae (the stafftree family). a widely distributed family of trees and often climbing shrubs. *See Euonymus.*

celery. *Apium graveolens*.

celeryworm. a common name of the larva of *Papilio palyxenes asterius*.

celiac sprue. gluten-induced enteropathy.

cell wall. the cellulose layer or structure that lies outside the cytoplasmic membrane of a plant cell. It supports and protects the membrane and maintains the shape of the cell.

cellulose. a glucose polysaccharide (with beta-1,4-linkage) that is the main component of the plant cell wall. It is the most abundant polysaccharide in nature.

Celsius temperature scale (formerly centigrade temperature scale). a temperature scale in which 0 degrees is the freezing point of water at standard atmospheric pressure, almost exactly zero degrees, and the boiling point is almost 100 degrees. Temperature, Θ_c in degrees Celsius (^{o}C), is related to the Kelvin scale by the equation:

$$\Theta_c = T_k - 273.15.$$

central nervous system (CNS). the chief neurobehavioral coordinating mechanism of vertebrates and higher invertebrates, it is characterized by a high concentration of nerve cell bodies and synapses and is often closely associated with primary or special sense organs (olfactory, optic, auditory). The cerebral region of the CNS (the cerebral ganglia of invertebrates and the brain of vertebrates) coordinates and integrates the activities of the whole nervous system. *See also* peripheral nervous system.

cenobiology. biocenology.

centi- (c) a prefix to units of measurement that indicates a quantity of 0.01 (10^{-2}) times the value indicated by the root. *See* International System of Units.

centigrade temperature scale. former name of the Celsius temperature scale.

-centric. a suffix **1:** indicating the presence of a center of a specified kind or number. **2:** having a specific object or entity as a focal point or matter of interest.

cephaeline methylether. emetine.

cephalopathy. encephalopathy.

Cercocarpus (mountain mahogany). a genus of small semideciduous shrubs and small evergreen trees of western North America (Family Rosaceae) that are often browsed by animals. They are cyanogenic.

cereal (cereal grains). any of a number of grasses that are cultivated for the large seeds or grains that they produce. **2:** the gains themselves, the most common uses being for the preparation of flour or meal or feeds for cattle.

cereal grains. *See* cereal.

cereal leaf beetle. *Oulema melanopus*.

cerebropathia. encephalopathy.

cerebropathy. encephalopathy.

cerumen (earwax). a soft waxy secretion of the ceruminous glands of the external auditory meatus.

ceruminal. pertaining to cerumen.

ceruminiferous gland. any gland that

secretes a waxy substance (e.g., earwax). *See*, for example, beeswax, cerumin.

ceruminous. pertaining to cerumen.

CF. chronicity factor (*See* factor).

Chaetocnema concinna (beet flea beetle). sugar beet pest.

chaff. the seed coverings (e.g., glumes, husks) or small pieces of stems or leaves separated from the seed during threshing or processing. **2:** finely cut hay or straw used as animal feed.

chalice cell. goblet cell.

chalk. calcium carbonate.

charbon. an infectious disease of animals, especially herbivores, due to the presence of *Bacillus anthracis* in the blood.

chard. *Beta vulgaris*.

charlock. *Brassica kaber*.

charm. a flock of finches.

Chemhex. a trade name for lindane.

chemical 109. ANTU.

chemical agent. any substance or active principle that can produce an effect on a living organism by interacting with endogenous substances. Included are an extremely diverse array of drugs, poisons, and substances with other types of action.

chemical analysis. analysis, by chemical means, to determine the chemicals that are present or the amounts, concentrations, or proportions of constituents that are present in a substance. *See also* qualitative analysis, quantitative analysis.

chemical ecology. the science of chemical relationships within and among biotic populations.

chemical lure. a chemical substance secreted by one organism that attracts another. *See* kairomone.

chemical oxygen demand (COD). the amount of oxygen depleted in a water sample due to chemical oxidation by sulfuric acid-potassium dichromate solution. The result is a measure of total oxidizable compounds in a sample. *See* biological oxygen demand.

chemical. 1: any substance; an element, compound, or a mixture of two or more substances. **2:** pertaining to a substance or process that changes the structure of one or more substances.

chemiculture. hydroponics.

chemigated water. irrigation water that has been subjected to chemigation, *q.v.*

chemigation. the addition of one or more chemicals to irrigation water.

chemoautotroph. chemoautotrophic organism.

chemoautotrophic organism (chemoautotroph; chemolithotrophic organism; chemolithotroph; chemosynthetic organism; chemotrophic organism). includes certain bacteria that derive the energy that drives carbon dioxide assimilation and biosynthesis from the oxidation of reduced inorganic compounds. As examples, nitrifying bacteria oxidize ammonia to nitrate or nitrite, iron bacteria convert ferrous compounds to ferric, and colorless sulfur bacteria convert hydrogen sulphide to sulfur. *See* autotrophic organism, photoautotrophic organism.

chemoheterotroph. a chemoheterotrophic organism.

chemoheterotrophic (chemoorganotrophic; chemozootrophic). descriptive of organisms that obtain energy from oxidation/reduction reactions using organic electron donors. *Cf.* chemolithotrophic.

chemoheterotrophic organism. an organism characterized by chemoheterotrophic nutrition (*See* chemoheterotrophic).

chemolithotroph. chemoautotrophic

organism.

chemolithotrophic organism. chemoautotrophic organism.

chemoorganotroph. a chemoorganotrophic organism.

chemoorganotrophic. chemoheterotrophic.

chemoorganotrophic organism. a chemoheterotrophic organism. *See* chemoheterotrophic.

chemosterilant. a substance that causes reproductive sterility. Such chemicals are sometimes used to control a pest population by sterilizing a portion of the males. This is often accomplished by releasing large numbers of laboratory-reared sterile males (e.g., as with *Drosophila*, fruit flies) into a pest population at appropriate times. To the extent that the population is limited by reproductive rate and the sterile males compete successfully with normal males, the number of offspring in the next generation will be reduced.

chemosynthesis. the synthesis of organic compounds using chemical energy from the oxidation of simple organic substrates. *Cf.* photosynthesis.

chemosynthetic organism. chemoautotrophic organism.

chemotaxis. Oriented movement of a motile organism with reference to a chemical agent. **negative chemotaxis**. a chemotaxis in which the organism moves away from a particular chemical stimulus or that moves down a chemical gradient. **positive chemotaxis**. a chemotaxis in which the organism moves toward a particular chemical stimulus or that moves up a chemical gradient.

chemotrophic organism. chemoautotrophic organism.

chemotrophic. pertaining to a chemotrophic (chemoautotrophic) organism or to the chemoautotrophic mode of nutrition.

chemozoophobe. a chemozoophobous plant. A plant that produces noxious chemicals (allelochemics) that inhibit consumption by herbivorous animals.

chemozoophobous. pertaining to or denoting plants that inhibit consumption by herbivorous animals by producing noxious chemical substances (allelochemics).

chemozootroph. a chemozootrophic organism.

chemozootrophic. chemoheterotrophic.

chemozootrophic organism. a chemoheterotrophic organism. *See* chemoheterotrophic.

Chenopodiaceae (goosefoot; pigweed). a cosmopolitan family of some 75 genera of weedy herbs and shrubs, many members of which contain high concentrations of oxalate salts. *See Chenopodium*.

Chenopodium (goosefoot; pigweed). a nearly cosmopolitan genus of toxic or potentially toxic, mainly weedy plants (Family Chenopodiaceae). Concentrations of nitrate toxic to livestock have also been reported. ***C. ambrosioides*** (American wormseed; Mexican tea; Spanish tea). a strong-smelling annual herb of tropical America, now widely distributed throughout most of eastern North America. The raw leaves are toxic but have been used in medicinal teas. It is also cultivated for wormseed oil, an irritant used as an anthelmintic. The active principle is the terpene ascaridole. Concentrations of nitrate toxic to livestock have been reported. ***C. botrys*** (Jerusalem oak; feather geranium). a strong-smelling annual herb with toxicities similar to those of *C. ambrosioides*. The leaves are toxic. Amounts of nitrate toxic to livestock have also been reported.

chernozem (black earth). dark, fertile soil with a high organic content. It occurs extensively in the Great Plains of North America and the steppes of Russia.

cherry. a common name of various species of *Prunus* (e.g., black cherry, cherry laurel) and a number of unrelated plants.

cherry, black. a common name any of these

species: *Prunus serotina*, *P. demissa*, *P. melanocarpa*, and certain others.

cherry laurel. *Prunus caroliniana*.

chewing fescue. *Festuca rubra commutata*.

chick. **1:** a newborn or young bird of any flying species prior to the onset of flight. **2:** a nestling. **3:** the still downy young of precocious species (e.g., waterfowl and gallinaceous birds). **4:** in the medical literature the term is often loosely applied specifically to the chick of *Gallus domesticus*, the domestic fowl.

chick embryo. **1:** the embryo of any bird, but most commonly applied to gallinaceous birds and waterfowl. **2:** in the biomedical sciences, this term often refers only to the embryo of *Gallus domesticus*.

chicken. **1:** *Gallus domesticus*. **2:** the flesh of *G. domesticus*, especially used as a food.

chicken Stomach worm. *Yetrameres americana*.

chickweed. *Stellaria media*.

Chiendent. *Agropyron repens*.

chigetai. *Equus h. hemionus*.

chigger flower. *Asclepias tuberosa*.

chill starvation. a condition of very young lambs that is due to severe reduction in body heat from severe weather.

chinawood oil. tung oil.

Chinese cabbage. the common name of oriental varieties of *Brassica rapa* (field mustard), in particular *B. r. pekinensis* (*q.v.*) and *B. r. Chinensis* (pak choi).

Chinese gooseberry. *Actinidia chinensis*.

Chinese layering. air layering.

Chinese white. zinc oxide.

Chiricanthium (running spiders). a genus of tiny, greenish-white or pale yellow spiders with neurotoxic venom that cling tenaciously when biting, causing an intensely painful wound. Only four of the approximately 160 species are known to bite humans. Systemic effects in humans and other large animals are usually mild, although an allergic reaction is possible. ***C. diversum***. the species that most often bites humans, it ranges in length from 7-16 mm. The abdomen is ovoid and the color varies from yellow, green, or greenish white to red-brown. The chelicerae are strong and the legs are long and hairy. They often must be forcefully removed from the site of the wound. Persistent pain about the wound, restlessness, lymphadenitis, and lymphadenopathy are characteristic.

chi-square test. X-square test. *See* test.

chitinase (chitodextrinase; polyp glucosaminidase). any of a class of enzymes secreted by chitinivores that catalyzes digestion of chitin to *N*-acetylglucosamine; some exhibit lysozyme activity.

chitinous of or pertaining to chitin; composed of chitin.

chitobiose. the disaccharide repeating unit in chitin. It is similar to cellobiose, differing only by the replacement of the hydroxyl group on carbon-2 by an *N*-acetylamino group on carbon-2.

chitodextrinase. chitinase.

chive. *Allium schoenoprasum*.

chlamydial bovine abortion. epidemic bovine abortion.

chlamydial ovine abortion. endemic abortion of ewes.

chloramben (3-amino-2,5-dichlorobenzoic acid). a white, solid substituted benzoic acid used as a selective preemergence herbicide. It is carcinogenic to some laboratory animals.

chlorazine (2-chloro-4,6-bis(diethylamino)-1,3,5-triazine). a solid triazine herbicide that is insoluble in water but soluble in alcohols, hydrocarbons, and ketones.

chlorcholine chloride (CCC). a retardent of plant. It may act by inhibiting gibberellin synthesis. *See also* mevalonic acid.

chlordane (1,2,4,5,6,7,8,8-octachloro-2,3,-3a,4,7,7a-hexahydro-4,7-methano-1H-indene; 1,2,4,5,6,7,8,8-octachloro-3a,4,7,7a-tetrahydro-4,7-methanoindan; 1,2,4,5,6,7,8,8-octachloro-4,7-methane-3a,4,7,7a-tetrahydroindane). a colorless or pale yellow, viscous, flammable, liquid chlorinated diene hydrocarbon (a cyclodiene), $C_{10}H_6Cl_8$. It is a contact and stomach poison used as an insecticide and fumigant. Its use was discontinued in the United States in 1987. Chlordane does not readily degrade in the environment and is thus a long-term risk to many life forms. It is lipophilic and accumulates in body fat and bioaccumulates in food chains. Poisoning of mammals and birds can occur via ingestion, inhalation, or contact. Chlordane is mildly to very toxic to laboratory animals and is slightly more toxic to humans than DDT. Large doses are carcinogenic to some laboratory animals and probably to humans. Repeated or chronic exposure may damage the liver, kidneys, and spleen. Symptoms of acute intoxication in vertebrate animals include irritability, hyperexcitability, tremors, and convulsions punctuated by deep depression.

chlordecone (1,1a,3,3a,4,5,5a,5b,6-decachloro-octahydro-1,3,4-metheno-2*H*-cyclobuta[*cd*]-pentalen-2-one; Kepone; CD). a hepatotoxicant, nephrotoxicant, and neurotoxicant, chlordecone is used as an insecticide to control ants and roaches and as a fungicide. Its use was largely terminated following contamination (first noted in bald eagle and osprey eggs) of the James River in the United States. Chlordecone is very toxic to laboratory rats and causes hepatocellular carcinoma in laboratory rats and mice. Its neurotoxic effects are due to blocking of Na^+/K^+-ATPase. Cholestyramine, a nonabsorbed anion exchange resin, increases the rate of elimination of chlordecone. *See also* poisoning (chlordecone poisoning).

chlordecone poisoning (Kepone poisoning). poisoning by chlordecone; intoxication of animals is characterized by neurological signs and symptoms such as tremors and exaggerated startle response. Factory workers exposed to this substance experienced chest pain, weight loss, skin rash, muscular weakness, incoordination, arthralgia, slurred speech, nervousness, visual and auditory hallucinations, anxiety, irritability, and short-term memory loss. *See also* chlordecone.

chlordiazepoxide (7-chloro-*N*-methyl-5-phenyl-3*H*-1,4-benzodiazepin-2-amine 4-oxide; 7-chloro-2-methylamino-5-phenyl-3*H*-1,4-benzodiazepine 4-oxide; metaminodiazepoxide;methaminodiazepoxide; clopoxide). a benzodiazepine, *q.v.*, used as an anxiolytic and a veterinary tranquilizer. It is a controlled substance (a depressant) listed in the U.S. Code of Federal Regulations (*Cf*R).

chlordimeform (*N'*-(4-chloro-2-methylphenyl)-N,*N*-dimethylmethanimidamide; *N'*-(4-chloro-*o*-tolyl)-N,*N*-dimethylformamidine; chlorophen-amidine; chlorphenamidine; chlorophedine; bermat; spanon; CDM). an insecticide and acaricide that is effective against a wide range of pests that are resistant to carbamates and organophosphates. It is an effective ovicide against a variety of lepidopterous insects. Chlordimeform is very toxic to male laboratory rats, somewhat less toxic to female rats, and is carcinogenic to some laboratory animals. Symptoms of acute intoxication in humans include dermatitis, hemorrhagic cystitis, abdominal pain, dysuria, nocturia, hematuria, and proteinuria. Death from respiratory collapse may follow.

chlorfenac (2,3,6-trichlorobenzeneacetic acid; 2,3,6-trichlorophenylacetic acid; fenac). an herbicide used to control annual weeds, *Agropyron repens*, and seedling perennials in maize and *Beta vulgaris*. It is moderately toxic to laboratory rats.

chlorfenethol (4-chloro-α-(4-chlorophenyl)-α-methylbenzenemethanol; 4,4'-dichloro-α-methylbenzhydrol; *p,p'*-dichlorodiphenylmethyl carbinol; 1,1-bis(*p*-chlorophenyl)-ethanol; 1,1-bis-(*p*-chlorophenyl)meth-yl carbinol; DMC). a colorless, crystalline, water-insoluble compound used as a miticide on ornamental plants with effects similar to those of DDT. It is very toxic to laboratory rats.

chlorfenvinphos (phosphoric acid 2-chloro-1-(2,4-dichlorophenyl diethyl ester; *O,O*-diethyl-*O*-[2-chloro-1-(2,4-dichlorophenyl)vinyl] phosphate; 2,4-dichloro-α-(chloromethylene)benzyl alcohol diethyl phosphate; CVP). an amber, liquid insecticide and acaricide used mainly to control flies, lice, mites, and ticks on cattle. The minimum oral toxic dose for cattle is about 22 mg/kg body weight. The acute oral LD_{50} for laboratory rats is 10-39 mg/kg body weight. It is also toxic to fish.

chlorhydrin. chlorohydrin.

α-chlorhydrin. α-chlorohydrin. *See* chlorohydrin.

chlorinated camphene. toxaphene.

chlorinated cyclodiene insecticide. cyclodiene insecticide.

chlorinated diphenyl. polychlorinated biphenyl.

chlorinated diphenylene. polychlorinated biphenyl.

chlorinated organic insecticide. organochlorine insecticide.

chlorinity. **1a:** a measure of the amount or concentration of dissolved chlorine in water, especially salt water. **1b:** total weight of dissolved chlorine per unit volume of solvent. *Cf.* chlorosity, salinity.

chlormequat chloride (2-chloro-*N,N,N*-trimethylethanaminium chloride; (2-chloroethyl) trimethylammonium chloride; chlorocholine chloride; choline dichloride). a phytocidal bipyridilium quaternary ammonium salt used as a plant growth regulator. It is a white, water-soluble, very hygroscopic, crystalline solid, with a fish like odor,

$$[ClCH_2CH_2N^+\text{-}CH_3)_3]Cl.$$

It is extremely toxic to humans and to laboratory mice and is corrosive in aqueous solution.

chloro-IPC. chlorpropham.

chloroacetic acid (chloroethanoic acid; monochloroacetic acid; MCA). an herbicide used also in the manufacture of various organic chemicals. Chloroacetic acid and its salts (chloroacetates) are very toxic to laboratory animals and presumably to humans. It is irritating to the skin and mucous membranes. It has been used chiefly as a contact pre-emergent herbicide and defoliant to control weeds in *Brassica oleracea* (kale), *B. oleracea* var. *capitata* (cabbage), *Allium ampeloprasum* var. *porrum* (leeks), and sprouts.

γ-chloroallyl chloride. 1,3-dichloropropene.

2-chloroallyl diethyldithiocarbamate. sulfallate.

chloroalonil. chlorothalonil.

chloroazotic acid. *aqua regia*.

7-chlorobicyclo-[3.2.0]hepta-2,6-dien-6-yl dimethylphosphate. heptenophos.

chlorobiphenyl. polychlorinated biphenyl.

2-chloro-4,6-bis(diethylamino)-1,3,5-triazine. chlorazine.

4-chlorobut-2-ynyl-*N*-(3-chlorophenyl)-carbamate. barban.

***m*-chlorocarbanilic acid isopropyl ester**. chlorpropham.

4-chloro-α-(4-chlorophenyl)-α-trichloromethyl)benzenemethanol. dicofol.

4-chloro-α-(4-chlorophenyl)-α-methylbenzenemethanol. chlorfenethol.

chlorocholine chloride. chlormequat chloride.

2-chloro-2,6-diethyl-*N*-(butoxymethyl)-acetanilide. butachlor.

α-chloro-*N,N*-diallylacetamide. allidochlor.

2-chloro-2-diethylcarbamoyl-1-methylvinyldimethylphosphate. phosphamidon.

6-chloro-3-(*O*,*o*-diethyldithiophosphorylmethyl)benzoxazolone. phosalolne.

2-chloro-2′,6′-diethyl-*N*-(methoxymethyl)-acetanilide. alachlor.

2-chloro-*N*-(2,6-diethylphenyl)-*N*-(methoxymethyl)acetamide. alachlor.

3-chloro-1,2-dihydroxypropane. α-chlorohydrin. *See* chlorohydrin.

7-chloro-4,6-dimethoxycoumaran-3-one-2-spiro-1′-(2′-methoxy-6′-methylcyclohex-2′-en-4′-one). griseofulvin.

2-chloro-*N,N*-di-2-propenylacetamide. allidochlor.

chloroethanoic acid. chloroacetic acid.

2-chloroethanol. ethylene chlorohydrin. *See* chlorohydrin.

2-chloro-4-ethylamino-6-isopropyl-amino-1,3,5-triazine. atrazine.

2-[[4-chloro-6-(ethylamino)-1,3,5-triazin-2-yl]amino]-2-methylpropanenitrile. cyanazine.

2-[[4-chloro-6-(ethylamino)-*s*-triazin-2-yl]amino]-2-methylpropionitrile. cyanazine.

6-chloro-*N*-ethyl-*N*′-(1-methylethyl)-1,3,5-triazine-2,4-diamine. atrazine.

(2-chloroethyl)trimethylammonium chloride. chlormequat chloride.

chlorogenic acid (3-[[3-(3,4-dihydroxyphenyl)-1-oxo-2-propanyl]oxy]-1,4,5-trihydroxycyclohexanecarboxylic acid; 1,3,4,5-tetrahydroxycyclohexanecarboxylic acid; 3-(3,4-dihydroxycinnamate; 3-caffeoylquinic acid; 3-(3,4-dihydroxycinnamoyl)quinic acid). a plant growth inhibitor that occurs in green coffee beans and in the leaves, fruit, and other tissues of dicotyledenous plants. *See also Helianthus annus*.

4-chlorobut-2-ynyl-*N*-(3-chlorophenyl)-carbamate. barban.

chlorohydrin (chlorhydrin). any of a number of compounds obtained by substitution of chlorine for some hydroxyl groups of a group of glycols or polyhydroxy alcohols. **α-chlorohydrin** (3-chloro-1,2-propanediol; 3-chloro-1,2-dihydroxypropane; α-monochlorohydrin; β,β'-di-hydroxyisopropyl chloride; glycerol α-monochlorohydrin; 3-chloropropylene glycol; α-chlorhydrin). a clear to straw-colored, sweet liquid, systemic poison,

$$CH_2ClCHOHCH_2OH$$

It is a very toxic rodent sterilant. *Cf.* epichlorohydrin. **ethylene chlorohydrin** (2-chloroethanol; 2-chloroethyl alcohol; glycol chlorohydrin). a colorless, highly toxic liquid,

$$ClCH_2CH_2OH$$

It has a faint ethereal odor, boils at 129°C and is miscible with water and ethanol. It is used as a solvent and in organic synthesis, including the synthesis of insecticides. It is toxic by ingestion, skin contact, and other routes. The nervous system, spleen, liver, and lungs may be affected. Inhalation of the vapor can be fatal. It is also a mild skin and severe eye irritant and is probably mutagenic. Ethylene chlorohydrin may react violently with oxidizing agents and reacts violently with water and steam to yield toxic and corrosive compounds. *See also* chlorohydrin, epichlorohydrin.

τ-chloroisobutylene. 3-chloro-2-methylpropene.

2-chloro-*N*-isopropylacetanilide. propachlor.

4-chloro-2-methylalanine. 4-chloro-*o*-toluidine.

4-chloro-6-methylalanine. 4-chloro-*o*-toluidine.

7-chloro-2-methylamino-5-phenyl-3*H*-1,4-benzodiazepine 4-oxide.

chlordiazepoxide.

4-chloro-2-methylbenzeneamine. 4-chloro-*o*-toluidine.

4-chloro-α-(1-methyl-ethyl)benzeneacetic acid cyano(3-phen-oxyphenyl)methyl ester. fenvalerate.

(O,O-diethyl-O-(3-chloro-4-methyl-2-oxo-2*H*-1-benzopyran-7-yl)-phosphorothioate). coumaphos.

(4-chloro-2-methylphenoxy)acetic acid. MCPA.

7-chloro-*N*-methyl-5-phenyl-3*H*-1,4-benzodiazepin-2-amine 4-oxide. chlordiazepoxide.

***N*'-(4-chloro-2-methylphenyl)-N,*N*-dimethylmethanimidamide**. chlordimeform.

3-chloro-2-methyl-1-propene. 3-chloro-2-methylpropene.

3-chloro-2-methylpropene (CMP; 3-chloro-2-methyl-1-propene; τ-chloroisobutylene; methallylchloride; β-methallylchloride; β-methallyl chloride; 2-methylallyl chloride; isobutanyl chloride). a colorless, volatile liquid with a disagreeable odor, C_4H_7Cl, used as an insecticide, fumigant, and reagent. CMP is an irritant to the eyes and respiratory tract and is mildly toxic by inhalation. It is carcinogenic to some laboratory animals. CMP is a serious fire hazard.

chloronitrous acid. *aqua regia*.

***S*-(6-chloro-2-oxobenzoxazolin-3-yl)methyl diethyl phosphorothiolothionate**. phosalolne.

chlorophedine. chlordimeform.

chlorophen-amidine. chlordimeform.

***N*-[[(4-chlorophenol)amino]carbonyl]-2,6-difluorobenzamide**. diflubenzuron.

chlorophenothane. DDT.

chlorophenoxy compound. any of a number of noninsecticidal organohalide pesticides and growth regulators that profoundly affect the growth and structure of plants. They are widely employed in herbicide formulations. Examples are 2,4-D, 2,4,5-T (*See also* Agent Orange), chlorophenoxyacetic acid, and chlorophenoxypropionic acid. They produce intermediates that cause downward (epinastic) bending of the plant, cessation of growth, tumor formation, and secondary root induction. Meristematic cells fail to divide, and cells which normally elongate enlarge only radially. Parenchyma cells of mature plants swell, divide, and produce callus tissue (i.e., tissue that usually develops only at an injured surface) and expanding root primordia. Roots no longer elongate and root tips swell. Young leaves fail to expand further; roots fail to absorb water and salts; photosynthesis is inhibited and phloem becomes plugged. These compounds modify nucleic acid metabolism in plants and alter numerous enzyme systems. They are generally moderately or very toxic to animals. All should be considered hazardous to humans, some extremely so.

chlorophenoxyacetic acid (*p*-chlorophenoxyacetic acid; 4-chlorophenoxyacetic acid; 4-CP; CPA; PCPA). as with other chlorophenoxy compounds, *q.v.*, chlorophenoxyacetic acid, $C_8H_7O_3Cl$, has profound effects on the growth and structure of plants. It is used as an herbicide and as a growth regulator for tomatoes and peaches.

4-chlorophenoxyacetic acid. chlorophenoxyacetic acid.

***p*-chlorophenoxyacetic acid**. chlorophenoxyacetic acid.

(3-chlorophenyl)carbamic acid 1-methylethyl ester. chlorpropham.

3-(*p*-chlorophenyl)-1,1-dimethylurea. nonuron.

***N'*-(4-chlorophenyl)-*N*,*N*-dimethylurea**. nonuron.

***S*-[[(*p*-chlorophenyl)thio]methyl]*O,o*-diethyl phosphorodithioate**. carbophenothion.

***S*-(2-chloro-1-phthalimidoethyl)-*O,O*-diethylphosphorothionate**. dialifor.

chlorophyll. the chief photosynthetic pigment of higher plants and algae. It is located in the grana of chloroplasts and consists of a porphyrin ring with a magnesium atom at the center that is esterified to a phytol (a type of long-chain aliphatic alcohol). Chlorophyll is a pigment that absorbs light chiefly in the red and blue-violet portions of the light spectrum. Different species of chlorophyll have different absorption maxima. **chlorophyll a** and **chlorophyll b**. species of chlorophyll that occur in higher plants. **chlorophyll c** and **chlorophyll d**. species of chlorophyll that occur in algae. *See also* photosynthesis, bacteriochlorphyll.

Chlorophyllum molybdites (green-spored lepiota). a very common poisonous mushroom (Family Agaricaceae) that grows in lawns, meadows, and pastures from New York to Florida west to Michigan, Colorado, and California. It is a potent purgative, causing one to several days of violent purging. Symptoms appear within one-half to 3 hours following ingestion and include severe nausea, vomiting, diarrhea, and abdominal pain. Hospitalization may be necessary.

6-chloropiperonyl chrysanthemumate. barthrin.

6-chloropiperonyl-2,2-dimethyl-3(2-methylpropenyl)-cyclopropanecarboxylate. barthrin.

6-chloropiperonyl-2,2-dimethyl-3(2-methylpropenyl)-cyclopropanecarboxylate; 6-chloropiperonyl chrysanthemumate. barthrin.

chloroplast. a chlorophyll-containing organelle of photosynthetic eukaryotes.

3-chloro-1,2-propanediol. α-chloroydrin. *See* chlorohydrin.

chloropropham. chlorpropham.

3-chloropropylene glycol. α-chloroydrin. *See* chlorohydrin.

chlorosis. **1:** a condition of green plants in which normally green parts become pale yellow or even white. It may result from disease (esp. certain virus infections), senescence, exposure to various air pollutants, mineral deficiencies (esp. of iron, magnesium, or copper), or poor lighting. **2:** a disorder mainly of adolescent women of the 19th century marked by hypochromic erythrocytes and a greenish-yellow skin discoloration. **alkali chlorosis**. the yellowing of foliage by soluble salts in the soil when present in excessive amounts.

chlorosity. total chlorine per liter of seawater. *See also* chlorinity.

chlorothalonil (2,4,5,6-tetrachloro-1,3-benzenedicarbonitrile; 2,4,5,6-tetrachloro-3-cyanobenzonitrile; tetrachloroisphthalonitrile; *m*-tetrachlorophthalodinitrile; *m*-TCPN; 2,4,5,6-tetrachloro-1,3-dicyanobenzene; 1,3-(4-dicyano)-2,4,5,6-tetrachlorobenzene; 1,3-dicyanotetrachlorobenzene; chloroalonil; chlorthalonil). a nearly water-insoluble crystalline substance used as a bactericide, nematocide, and as an agricultural horticultural fungicide. It is slightly toxic orally to laboratory rats, and moderately toxic, *i.p.*, but is carcinogenic to some laboratory animals. Very toxic fumes, NO_X, Cl^-, and CN, are released when it is heated to decomposition.

chlorotic. Pertaining to or having the characteristics of chlorosis, especially the yellowing of leaves.

(4-chloro-*o*-toloxy)acetic acid. MCPA.

3-chloro-*p*-toluidine (CPT). a selective nephrotoxic (sometimes hepatotoxic) avicide. It is extremely toxic to most birds but is only slightly toxic to the laboratory rat. Acute exposure produces renal necrosis in all sensitive species and, in quail, a focal hepatic necrosis also occurs. Death in some species appears to result from a complex of causes and is occasionally due to cardiovascular and respiratory effects. Renal deacetylase activity is high in the very sensitive starling and domestic fowl but is undetectable in the nonsusceptible

Buteo jamaicensis (red-tailed hawk) and laboratory rat.

4-chloro-*o*-toluidine (2-amino-5-chlorotoluene; 4-chloro-2-methylalanine; 4-chloro-6-methyl-alanine; 4-chloro-2-methylbenzeneamine; 4-chloro-2-toluidine). a moderately toxic substance, C_7H_8ClN, if ingested. It is carcinogenic to some laboratory animals and is a confirmed human carcinogen and probable mutagen. *Cf.* 4-chloro-2-toluidine hydrochloride.

4-chloro-2-toluidine. 4-chloro-*o*-toluidine.

***N'*-(4-chloro-*o*-tolyl)-N,*N*-dimethylformamidine**. chlordimeform.

2-chloro-1-(2,4,5-trichlorophenyl)ethenyl phosphoric acid dimethyl ester. tetrachlorvinphos.

2-chloro-1(2,4,5-trichlorophenyl)vinyl dimethyl phosphate. tetrachlorvinphos.

7-chloro-2′4,6-trimethoxy-6′-methylspiro[benzofuran-2(3*H*),1′-[2]cyclohexene]-3,4′-dione. griseofulvin.

2-chloro-*N,N,N*-trimethylethanaminium chloride. chlormequat chloride.

chlorphenamidine. chlordimeform.

chlorpropham ((3-chlorophenyl)carbamic acid 1-methylethyl ester; *m*-chlorocarbanilic acid isopropyl ester; isopropyl-*m*-chlorocarbanilate; isopropyl *N*-(3-chlorophenyl)carbamate; chloro-IPC; chloropropham; CIPC). a carbamate herbicide and plant growth regulator, it is the 3-chloro derivative of propham. It inhibits gibberellin-induced α-amylase synthesis in barley endosperm and is thus thought to inhibit gene derepression (i.e., the onset of metabolic changes) but not processes that are already in progress. Chlorpropham is generally effective against monocotyledons, but not dicotyledons, and is used to control *Digitaria* (crabgrass). Chlorpropham is moderately toxic to laboratory rats. *See also* propham.

chlorpyrifos (phosphorothioic acid *O,O*-diethyl *O*-(3,5,6-trichloro-2-pyridinyl) ester; *O,O*-diethyl *O*-3,5,6-trichloro-2-pyridyl phosphorothioate; chlorpyrifos-ethyl). chlorpyrifos and its *O,O*-dimethyl analog, chlorpyrifos methyl, are used as insecticides and acaricides; it is a popular termiticide. The LD_{50} of chlorpyrifos in goats has been reported as 500 mg/kg body weight orally and 97 mg/kg in the laboratory rat. Bulls, in comparison to calves, steers, and cows, are highly susceptible to a single dose. This agent in pour-on preparations has poisoned bulls, particularly of exotic breeds. Chlorpyrifos, is extremely toxic to fish, birds, and aquatic invertebrates and is very toxic to laboratory rats. *See also* termiticide, Dursban *[Dow]*®.

chlorpyrifosethyl. chlorpyrifos.

chlorthal methyl. chlorthal-dimethyl.

chlorthal-dimethyl (DCPA; 2,3,5,6-tetrachloro-1,4-benzenedicarboxylic acid dimethyl ester; 2,3,5,6-tetrachloroterephthalic acid dimethylester; dimethyl-2,3,5,6-tetrachloroterephthalate; chlorthal methyl). a toxic aromatic compound,

$$C_{10}H_6Cl_4O_4$$

It is used as a preemergent herbicide. It inhibits the rooting of stolons in *Cynodon dactylon* by completely inhibiting cell division in the root primordia. Cell enlargement may continue for a while, however. Chlorthal-dimethyl blocks mitosis in maize and onion roots at metaphase, and the meristematic tissues become disorganized with the development of many hypertrophied cells.

chlorthalonil. chlorothalonil.

chlorthiepin. endosulfan.

Choanichthyes. a subclass of Osteichthyes that includes lungfishes and coelacanths. *Cf.* Actinopterygii.

choker. a noose of wire rope used to skid or lumber yard logs.

choking staggers. moldy corn poisoning.

cholera. **1:** any condition characterized by profuse vomiting and diarrhea (seldom used). **2:** Asiatic cholera. **Asiatic cholera** (cholera). an acute epidemic infectious disease of humans caused by *Vibrio cholerae* (*V. comma*), a comma-shaped bacterium. Symptoms appear suddenly one to five days following infection. These include vomiting, profuse, watery diarrhea, and dehydration. The loss of fluids and electrolytes may be intense and untreated patients may suffer extreme dehydration, shock, rapid collapse, and death with no lesions of the intestinal mucosa. The fluid loss results from the action of a toxin, choleragen, that promotes the passage of fluid from circulating blood into the large and small intestines. Infection is always from consumption of food or water that is contaminated by the vibrio. *See* choleragen, the cholera toxin. Originally confined to Asia, cases are now seen worldwide. A few cases occur annually in the United States. *See also* choleragen, *Vibrio cholerae*. **buffalo cholera**. a disease of bison that become epidemic among domestic animals.

cholera vibriao. cholera vibrio.

cholera vibrio (cholera vibriao). the bacterium, *Vibrio cholerae* (*V. comma*) that causes cholera. *Cf.* vibrio.

cholest-5-en-3β-ol. cholesterol.

cholesterin. cholesterol.

cholesterol (cholest-5-en-3β-ol; cholesterin). a monohydric secondary alcohol, of the cyclopentenophenanthrene system, with the formula, $C_{27}H_{45}OH$. It is comprised of four fused rings with one double bond. It occurs widely among animals and is the principal sterol of higher vertebrates, found in all body tissues, associated mainly with lipids. In vertebrate animals, including humans, it occurs especially in the brain and spinal cord, fats and oils, and plasmalemmas. It is the main constituent of gallstones. In humans and other higher animals, some occurs as the sterol and some as a lipid (i.e., esterified with higher fatty acids) in blood serum. Most cholesterol in higher animals is synthesized in the liver; some is dietary. It is the parent compound of bile acids and many steroids (e.g., the steroid hormones), and its presence in excess in the blood of humans and certain other animals contributes importantly to cardiovascular disease.

choline (2-hydroxy-*N,N,N*-trimethylethanaminium; (β-hydroxyethyl)trimethylammonium; bilineurine; hydroxyethyl trimethyl ammonium hydroxide; lipotropic factor; transmethylation factor; trimethylammonium ion). one of the B complex vitamins, it is essential to normal fat and carbohydrate metabolism and is a precursor of acetylcholine. It occurs in most animal and many plant tissues. Choline is an essential nutrient for some animals and microorganisms; humans, however, synthesize adequate amounts.

choline dichloride. chlormequat chloride.

cholinergic nervous system. parasympathetic nervous system.

cholinergic. **1a:** stimulated, activated, or conveyed by choline (acetylcholine). **1b:** pertaining to or denoting a nerve cell or fiber that releases acetylcholine (ACh) as a neurotransmitter. The voluntary motor neurons of vertebrates, parasympathetic and preganglionic sympathetic neurones, and some invertebrate neurones are cholinergic. **2:** a drug or other agent that produces effects similar to those of the parasympathetic nervous system; a parasympathomimetic agent.

cholinesterase. **1:** any enzyme such as acetylcholinesterase that catalyzes the hydrolysis of choline esters. They are inhibited by physostigmine (eserine). *See also* pseudocholinesterase. **2:** sometimes used in text as a synonym for acetylcholinesterase.

cholinesterase-inhibiting pesticide. any of a group of anticholinesterases used as pesticides (e.g., carbaryl, ethion, fenthion, malathion, methomyl, parathion). *See* anticholinesterase.

Chondrostei. a subdivision of the Actinoptgerygii comprised of primitive fishes that include sturgeons.

chorion. **1:** One of the three extraembryonic membranes that surround the mammalian fetus. It lies external to and covers the amnion and yolk sac. **2:** in botany, the outer covering of a seed.

Choristoneura fumiferana. *See* budworm.

Christmas rose. *Helleborus niger*.

chromatid. either of two daughter strands of a duplicated chromosome that are joined at the centromere during mitosis and meiosis.

chromatin. the diffuse, thread-like material seen in the cell nucleus during interphase. It condenses into chromosomes during mitosis. The main constituents are DNA and proteins.

3-chloro-1,2-dibromopropane. dibromochloropropane.

chromosomal aberration (chromosomal mutation). any change (mutation) in the number, structure, or sequence of genes reflected in gross chromosomal structure or number. That is, any modification (detectable by light microscopy) of the normal chromosome complement as a result of deletion (gaps, breaks), duplication, inversions, or translocations of chromosomal material. Many toxicants as well as ionizing radiation can cause chromosomal aberrations. Alkylating agents can cause chromosome breakage by cross-linking with DNA. A chromosomal mutation.

chromosomal mutation. *See* chromosomal aberration.

chromosome (karyomite). **1a:** any of a number of deeply staining rod-like or thread-like bodies that occur in the nucleus of eukaryotic cells. These are composed largely of DNA in association with RNA and protein (mostly histones) and comprise a linear sequence of hundreds of genes. The number of chromosomes is usually constant in any given species. **1b:** the condensed form of chromatin (DNA plus nucleoproteins). It is visible during mitosis. **2:** sometimes used in reference to the genetic material of prokaryotes and viruses that consist exclusively of DNA (RNA in certain viruses) with no associated protein. It is never visible as discrete thread-like or rod-like structures, as in eukaryotes.

chromosome aberration. *See* aberration.

chromotrichia factor. *p*-aminobenzoic acid.

chronic effect. *See* effect.

chronic obstructive pulmonary disease. pulmonary emphysema (*See* emphysema).

chronic obstructrive pulmonary disease. pulmonary emphysema (*See* emphysema.

Chrysanthemum (chrysanthemum). a large genus of widely cultivated annual and perennial Old World herbs (Family Asteraceae). Many are toxic, producing a contact dermatitis with erythema and vesiculation in sensitive individuals. The active principle is a resin. ***C. (Pyrethrum) cinariifolium***. a perennial herb indigenous to Dalmatia and Montenegro. It grows to 30-40 cm, with many stems and white flower heads. The dried, ground flowers are the major source of pyrethrum insecticides. Cultivated pyrethrums are derived form *C. coccineum*. *See* pyrethrum flowers. ***C. coccineum*** (= *C. roseum*). a hardy perennial herb, indigenous to western Asia, which grows to about 1 m in height; the flowers, which bloom in late spring, are various shades of red, pink, and white. It is a source of pyrethrum flowers, *q.v.* ***C. marschalli***. a hardy perennial indigenous to western Asia. It is also a source of pyrethrum flowers. ***C. roseum***. *C. coccineum*.

chrysanthemum. a common name of many of the plants of the genus *Chrysanthemum*.

chrysanthemummonocarboxylic acid 6-chloropiperonyl ester. barthrin.

1,3-dichloropropylene. 1,3-dichloropropene.

Cicuta (water hemlock). a genus of native

North American herbs (Family Apiaceae) that inhabit moist situations. Included are a number of species, of which the most important toxicologically is *C. maculata*. All contain convulsant toxins, most notably cicutoxin (found in *C. maculata* and *C. virosa*), which is the most violent poison produced by plants in North America. Numerous poisonings have occurred because these plants were mistaken for parsnips, artichokes, or other rots. The odor is similar to that of wild parsnip. Most poisonings occur during early growth in the spring. Children have been poisoned by making peashooters and whistles from the hollow stems. ***C. bolanderi***. *C. californica*. ***C. bulbifera*** (bulbous water hemlock). an herb of the northern United States. It is known to poison livestock. ***C. californica*** (= *C. bolanderi*; California water hemlock; Bolander water hemlock). a perennial species that occurs in middle western California. It is known to poison livestock. ***C. douglassi*** (Douglas water hemlock). a species found along the Pacific Coast states and British Columbia. ***C. maculata*** (water hemlock; cowbane; musquash root; beaver poison). a perennial herb of bogs, marshes, and swampy areas in eastern North America and west to the Great Plains. It has small white flowers and may grow to 2 m. All parts can poison fatally, especially the root. Symptoms may include those of gastroenteritis, convulsions (may be persistent), and respiratory depression. The active principle is the resin, cicutoxin. ***C. occidentalis*** (western hemlock). a plant of the Rocky Mountain states west to the Pacific Coast. ***C. vagans*** (tuber water hemlock; Oregon water hemlock). a water hemlock of the Pacific Northwest region of the United States.

cider vinegar. apple cider vinegar.

cilia. plural of cilium.

ciliate. **1:** having cilia. **2:** a protozoan that locomotes by means of cilia on the surface of the cell.

cilium (plural, cilia). **1:** an eyelash. **2:** short, threadlike appendages that extend from the surface of some protozoa. The cilia beat rhythmically, propelling the organism through the medium.

Cimex lectularius (bedbug). an insect (Family Cimicidae) that feeds by sucking the blood of animals and humans.

cinerin. either of two naturally occurring insecticidal pyrethrins, cinerin I and II, isolated from pyrethrum flowers. Both substances are water-insoluble, viscous liquids that oxidize in air, rapidly becoming inactive; they are hepatotoxic and nephrotoxic. Symptoms of acute intoxication in laboratory mammals include diarrhea, convulsions, prostration, and death due to respiratory paralysis. *See also* pyrethrin (def. 2).

cinerin I. *See* cinerin.

cinerin II. *See* cinerin.

cinmethylin (*exo*-(±)-methyl-4-(1-methylethyl)-2-[(2-methylphenyl)methoxy]-7-oxabicyclo-[2.2.1]heptane; (±)-2-*exo*-(2-methylbenzyl-oxy)-1-methyl-4-isopropyl-7-oxabicyclo[2.2.1]-heptane). a preemergence grass herbicide. It is moderately toxic orally to laboratory rats.

CIP. cleaned in place.

CIPC. chlorpropham.

Cirsium arvense (Canada thistle). a spiny-leafed plant (Family Asteraceae) introduced to North America from Eurasia. Concentrations of nitrate toxic to livestock have been reported.

Cirsium lanceolatum (bull thistle). a spiny-leafed weed found in pastures and waste places. When numerous, this plant, because of the prickly leaves can disrupt grazing.

citric acid (1,2,3-tricarboxylic acid). a slightly (laboratory rats) to moderately (laboratory mice) toxic, moderately strong, tribasic carboxylic acid, $C_6H_8O_7$, with colorless, translucent prisms or powder. It is odorless, with a strongly acidic taste, soluble in water and ethanol, and decomposes before boiling. It occurs in the juice of many fruits, especially citrus fruits. Lethal intoxication of laboratory rats is marked by signs of metabolic acidosis and calcium deficiency. It plays an important role in cellular respiration in both plant and animals cells as a component of the tricarboxylic acid cycle (citric acid cycle; Krebs cycle) wherein it is converted to isocitrate by aconitase, *q.v.* *See* citrate.

citric acid cycle. tricarboxylic acid cycle.

Citrullis vulgaris. a species of plants (Family Cucurbitaceae) comprised of watermelons and citron. All other melons are varieties of *C. melo*.

cive. *Allium schoenoprasum*.

Cladosporium. *See* aleukia, fagicladosporic acid.

clams. actinomycosis.

class. **1:** a taxonomic group within the hierarchy of taxonomic classification that ranks between phylum or division and order. A class consists of a number of closely related orders; it is usually a large and easily recognized group of organisms. Several related classes form a phylum or subphylum. A class of plants takes the ending -opsida or -phyceae, whereas animal classes typically end with -a (some exceptions are the vertebrate classes Aves, Osteichthyes, Chrondrichthyes). **2:** in statistics, any category or group of like observations; a set.

Claustridium botulinum. the bacterium that causes botulsim, *q.v.*

claustridium botulinum poisoning. botulism.

clavus. ergot.

clay. **1:** a soil component comprised of particles less than 0.002 mm in diameter. **2:** a composite of weathered minerals that exhibit surface and cation exchange attributes.

clayey. **1:** argillaceous. **2:** clay-like; having the properties of clay; comprised largely of clay.

clean. **1a:** neat, orderly, or tidy; spotless. **1b:** sanitary. **1c:** unadulterated; uncontaminated, unpolluted. **2:** said of a cow and certain other livestock that have shed their afterbirth. **2:** descriptive of the natural, wholesome flavor of dairy products.

cleaned in place (CIP). referring to milk-handling equipment that is cleaned without disassembly by circulating washing solutions through it.

cleaning. a treatment made in a stand of trees that are not past the sapling stage for the purpose of freeing desirable species from less desirable species of the same age that overtop them or are likely to do so.

clear cut (clearcutting). a forest management technique in which all the trees in one area or stand are harvested at the same time. Under certain soil and slope conditions, it can contribute sediment to water pollution. *Cf.* clearcut.

clear length. the portion of the bole of a tree that lies below the first branches or stubs of branches.

clearcut. a forested area from which all merchantable trees were recently been harvested. *Cf.* clear cut.

clearcutting. clear cut.

clearcutting silvicultural system. a silvicultural system in which an entire stand of trees from an area of one hectare or more is removed in a single harvest. The stand is replaced by a new even-aged stand.

clearcutting with reserves. a variation of the clearcutting silvicultural system in which some trees are retained, for purposes other than regeneration.

climatology. the science that treats climate, climatic conditions and phenomena, and the interactions between climate and other environmental phenomena. **agricultural climatology** (agroclimatology). the science that treats climate and various climatic factors with respect to their effects on agricultural land, plants and animals (including effects on growth rates and yield), and the gathering and use of climatological data to establish optimum, marginal, and limiting climates for any given crop or type of livestock. *See also* meteorology (agricultural meteorology).

climax. the most advanced successional stage of a plant community. It is considered to be a state of dynamic equilibrium with the prevailing environment and is thus considered the final stage of succession.

climbing nightshade. *Solanum dulcamara*.

clinical effect. *See* effect.

clinometer. a simple instrument used to measure vertical angles or slopes. It is used in forestry, for example, to measure distance and tree heights.

cloak fern. *Notholaena sinuata*.

clod. a large mass of soil particles, often produced by disturbances such as tillage or logging, that may be difficult to break down into fine soil (tilth) for cultivation. *See also* soil aggregation.

clofenotane. DDT.

clonal. **1:** pertaining to a clone. **2:** in agronomy, propagation from a bud.

clonal deletion. in immunology, the inactivation or destruction of lymphocyte clones that would be expected to produce immune reactions against the animal's own body.

clonal selection. the mechanism whereby exposure to an antigen results in the activation of specific T-cell or B-cell clones (*See* clone), yielding an immune response.

clone. **1:** a cell or organism that is genetically identical to a parent cell or organism. Plant clones are produced asexually, for example, from cuttings or suckers; clones of animals are usually produced asexually by transferring the nucleus of a somatic cell to an ovum and allowing normal embrygenesis to occur. **2a:** any of a population of cells produced by mitotic divisions of a single parent cell. **2b:** an organism descended from a single cell considered identical to that of the parent organism. **3:** a copy of a DNA fragment that is replicated by a phage or plasmid. **4:** to produce a clone. **DNA clone**. a DNA fragment or gene from an external source that has been inserted mechanically into plasmids, introduced into a host bacterium (e.g., *Escherichia coli*) and reproduced by the host. **lymphocyte clone**. in immunology, any of a set of lymphocytes that share identical antibodies.

cloning. **1:** the natural production of a clone by the mitotic division of a progenitor cell with the production of genetically identical daughter cells or clones. **2:** obtaining a group of genetically identical cells from a single cell or making copies of a gene by use of genetic engineering techniques.

cloning vector. a DNA molecule that can bring about the replication of foreign DNA fragments.

clonogenic. arising from or denoting a clone.

Clophen. polychlorinated biphenyl.

clopoxide. chlordiazepoxide.

2-chloro-*N*-(1-methylethyl)-*N*-phenylacetamide. propachlor.

close breeding. inbreeding that involves, for example, brother-sister, sire-daughter, or son-dam matings.

closewool. designating or referring to a breed of sheep that has thick, dense wool.

clostridium (plural, clostridia). common name of any member of the genus *Clostridium*.

Clostridium. a large genus of anaerobic (some are aerotolerant), mostly motile, spore-forming soil bacteria (Family Bacillaceae) that contain Gram-positive rods. Some species produce extremely potent exotoxins. ***C. botulinum***. the causal agent of botulism. This bacterium occurs naturally under anaerobic or slightly anaerobic conditions in soil, bottom sediments, and organic materials, especially vegetables. It is a common cause of food poisoning, especially from improperly preserved foods. There are six main types of *C. botulinum* that are differentiated by their secretion of antigenically distinct but toxicologically similar neurotoxins. Human botulism is caused mainly from types A, B, E, and F; type Cα causes botulism in birds; types Cβ and D cause botulism in cattle. Botulinum toxin binds irreversibly to nerve terminals, preventing the release of acetylcholine. The toxin is inactivated by heating at 80-100°C for a sufficient time. The toxin is one of the most potent known and is chemically similar to that of *Abrus precatorius*, the precatory bean. The toxin has been extensively investigated as a chemical warfare agent. ***C. histolyticum***. the causal agent of gas gangrene. ***C. perfringens***. the most widely distributed disease-causing organism known, occurring in soil, dust, on food, and in the intestinal tracts of humans and other warm-blooded animals. Ingestion of contaminated food results in diarrhea and abdominal pain within about 12 hours following ingestion. ***C. tetani***. a common anaerobic soil bacterium, it is the causal agent of tetanus (def. 3). This bacterium forms spores that can remain alive in soil for many years. It most commonly enters the animal body via puncture wounds, growing in dead or damaged tissue that has an inadequate blood supply and thus, low oxygen tension. The toxin (tetanospasmin) secreted by *B. tetani* is a potent neurotropic agent that interferes with eurotransmitters, such as acetylcholine, causing tetanus (lockjaw).

clot (coagulum). a semisolid mass, as of blood or lymph. **blood clot**. a jelly-like clot formed within the body or externally by the agglutination of erythrocytes. **external clot**. a blood clot formed outside a blood vessel. **internal clot**. a blood clot formed within a blood vessel.

clot formation. *See* coagulation (blood coagulation).

clotbur. a plant of the genus *Xanthium*.

cloth of gold. *Baileya multiradiata*.

clotting. the activity of the blood and blood vessels that causes blood to form a jelly-like clot, usually near an injury that helps to slow or stop bleeding. The body's clotting mechanism is slowed by the use of anticoagulants in treating certain diseases or anticoagulant poisons or venoms.

clotting agent. coagulant.

cloud grass. *Agrostis nebulosa*.

clover. a common name of herbs of the genus *Trifolium*. *See also Medicago*.

clover disease. trifoliosis.

clowder. clutter (def. 2).

cluster pine. *Pinus pinaster*.

clutch. the full set of eggs laid by a bird during a single nesting cycle.

clutter. **1a:** to litter, scatter, or strew. **1b:** disarray or disorder. **2:** confusion. **3:** a group of domestic cats (also called clowder).

Clydesdale. a breed of heavy draft horse developed in Scotland. The ground color is brown or black with white feathers at the feet.

clyers. *See* actinomycosis.

C_m. clearance factor (*See* factor).

CMDP. mevinphos.

CMP. 3-chloro-2-methylpropene.

CMT. California mastitis test.

CMU. nonuron.

-CN the cyanide radical (*See* cyanide.)

CNS. central nervous system.

CoA. coenzyme A.

coagulant (clotting agent). **1:** an agent that causes a liquid such as blood or lymph to coagulate. **2:** able to cause coagulation. **blood coagulant**. an agent that causes blood to clot following, for example, damage to a blood vessel. *See also* coagulation, snakebite, venom (snake venoms).

coagulate. **1:** to clot; to form a clot (e.g., a blood clot). **2:** to congeal, curdle, solidify, or thicken. *See also* coagulation. **3:** to form a coagulum (*See* coagulation, def. 2).

coagulated. clotted or curdled.

coagulation. **1a:** the process of clotting. **1b:** the alteration of a dispersed system into a liquid phase and an insoluble mass (the clot or curd). **2:** the disruption of tissue with the formation of amorphous mass (coagulum) as by photocoagulation or electrocoagulation. **blood coagulation** (clot formation). the process of blood clot formation by conversion of whole blood or plasma into a gel. Blood coagulation is a complicated process involving the sequential activation of a number of soluble proteins (clotting factors) in blood plasma with consequent production of a fibrin network. Coagulation can be initiated by various blood disturbances, by injury to the blood vessel walls, or the introduction of coagulants in the cardiovascular system. If the blood coagulation system is deficient, uncontrolled bleeding may result from even slight trauma, external or internal. Anticoagulants (e.g., warfarin, hydroxycoumarins, many snake venoms) that impair the blood coagulation system can produce severe, even fatal hemorrhage. *See*, for example, coumarin, cytochalasins, fibrin.

coagulum. clot. *See also* coagulation.

coarse woody debris (CWD). logs and stumps of trees, intact or rotting, that provide habitat for plants and animals and serve as a source of plant nutrients.

coast goldenbush. *Haplopappus venetus*.

cobra venom cofactor. properdin factor B.

coca. *See Erythroxylum*.

coccus. a spherical bacterium.

Cochlearia moracia. *Amoracia lapathifolia*.

cock. **1:** the adult male of many kinds of birds. **2:** in poultry, a male that is more than 18 months old.

cockerel. a male domestic fowl less than 18 months old.

cockerell. *Hymenoxys lemmoni*.

cocklebur. a plant of the genus *Xanthium*.

COD. chemical oxygen demand.

codon. a triplet of sequential nucleotides in messenger RNA that directs the placement of a particular amino acid into a polypeptide chain. This is accomplished by complementary base pairing with an anticodon in transfer RNA. *Cf.* anticodon.

coelom. the body cavity of vertebrate and certain higher invertebrate animals. It is lined with epithelium of mesodermal origin. *See also* abdominal cavity.

coelomate. **1a:** any animal that possesses a true coelom. Included are the Chordata, which includes the vertebrates, and certain invertebrate phyla: Annelida Arthropoda, Bryozoa, Chaetognatha, Brachiopoda, Echinodermata, Hemichordata, and Mollusca. **1b:** of, or pertaining to, animals that have a true coelom.

coenobiology. biocenology.

coenzyme (cofactor). an organic molecule that binds to an enzyme and acts as a cofactor in the sense that it is necessary if the enzyme is to be catalytically active. (examples are coenzyme A, NAD, FAD, pyridoxal phosphate. Some coenzymes form a tightly bound prosthetic group (e.g., NAD and NADP) that acts as an acceptor of electrons or functional groups. Such coenzymes are not easily removed.

coenzyme A (CoA). a coenzyme comprised of adenine, ribose, three phosphate groups, pantothenic acid (a vitamin of the B complex), and mercaptoethylamine. It plays a key role in acetylation and is involved in the "initial" step of the tricarboxylic acid cycle during which the acetyl group is introduced into the cycle.

cofactor. *See* coenzyme.

coffee. **1:** the seed of the coffee tree, genus *Coffea*. These beans when roasted are pulverized and brewed with hot water to form a widely used beverage. Coffee is grown chiefly in countries where highly toxic pesticides are still used. **2:** the beverage prepared from coffee beans.

coffeebean. a plant of the genus *Sesbania*.

coffeeberry. *Rhamnus purshiana*.

coffeeweed. a plant of the genus *Sesbania*.

cohesion. **1:** the attraction between identical molecules. **1b:** cohesive forces that hold solids or liquids together.

cohosh. a plant of the genus *Actaea*.

coil gland (convoluted gland). a gland in which the secretory portion is coiled, tangled, or convoluted.

Colchicum (autumn crocus; meadow saffron). a genus of poisonous cormaceous plants (Family Lilaceae). They are unrelated to the true crocus (*Crocus spp.*), although the flowers are very similar. Poisoning by these colchicine-containing plants occurs after a partially dose-dependent latent period of several hours. Such poisoning may result from ingestion of leaves, seeds, bulbs, or flowers. They are believed to cause defects in calves similar to crooked calf disease. ***C. autumnale*** (autumn crocus; fall crocus; meadow saffron; naked ladies). a small (ca. 1/3 m tall), frequently cultivated, poisonous perennial herb (Family Liliaceae). This extremely dangerous plant is native to the United Kingdom and mainland Europe, occurring also throughout much of Eurasia. It frequents damp meadows and woodsy areas of England and Wales and some parts of Scotland. Colchicine, the chief active principle, is concentrated in the corms (solid bulbs at the base of the stem). *See*ds, leaves, and flowers are also toxic, causing a burning pain in the mouth and stomach, with symptoms of gastroenteritis, anuria, hematurea, and prostration. The corm has sometimes been fatally mistaken for an onion. Livestock are usually poisoned by consuming the plant, but humans and suckling animals may be poisoned by drinking the milk of an affected animal. Symptoms in most animals are mainly those of gastrointestinal irritation, including a burning pain in the mouth and stomach, prostration, hematuria, and anuria.

colbred. a white-faced, medium-sized breed of sheep. Cross-bred offspring are noted for high fertility and high milk production.

cold-blooded. *See* ectotherm.

Coleoptera (beetles). an order of insects, it is the largest order in the animal kingdom, containing more than 300,000 known species worldwide. The forewings (elytra) are tough and armor-like and, when folded, cover the membranous hindwings.

coleoptile. a cylindrical sheath that surrounds and protects young shoots of grasses and cereals while still below the soil surface.

colic **1:** pertaining to the colon. **2:** acute, usually intermittent, painful spasms of any visceral organ. **3:** in young infants, spasmodic gastrointestinal pain evidenced by crying and irritability.

colicin. any bacteriocin produced by *Escherichia coli*.

colicky. denoting, resembling, or affected by colic or the symptoms of colic.

coliform bacteria. Gram-negative, non-spore-forming, facultative, rod-shaped, eubacteria. They occur in the intestinal tract of higher vertebrates and obtain energy by fermentation (e.g., of lactose) or by aerobic respiration. Coliform bacteria may contaminate water via fecal pollution and are a potential source of poisoning or disease. They often serve as an indicator of fecal contamination of water supplies. *Escherichia coli* and *Enterobacter* are important examples.

coliform index. a rating of the purity of water based on a count of fecal bacteria in a sample.

collapse. **1a:** breakdown, disintegration, or failure, *q.v.* **2a:** attack; seizure. **3a:** a sudden weakness or cessation of normal activity as in exhaustion, prostration, or cardiovascular or respiratory failure or insufficiency. **3b:** a state of extreme physical depression or prostration with failure of circulation. **4:** catastrophe, destruction, disintegration, or ruin. **4:** an abnormal caving in of the walls of any part of a hollow organ. **5:** to fail or break down completely. **5a:** to fall. **5b:** to fall into ruin. **6:** to founder; fail. **7:** to faint or swoon. **cardiovascular collapse**. complete failure of the cardiovascular system. **circulatory collapse**. shock; severe insufficiency or complete failure of the peripheral or cardiac circulation other than that due to congestive heart failure. **respiratory collapse**. extreme insufficiency of external respiration.

collapsing. **1:** falling into extreme and sudden prostration that resembles shock; **2:** failing; breaking down. **3:** experiencing a seizure. *See also* collapse.

collective farm. a state-controlled agricultural production unit comprised of workers from a number of farm households or villages.

collective group. species aggregate (*See* aggregate).

colloid fraction. **1:** with respect to soils, this function is the relative amount of colloids in a soil. **2:** organic and inorganic matter having very small particle size and a correspondingly large surface area per unit mass.

colluvium. material deposited on a slope as a result of erosion or mass transport.

colon bacillus. *Escherichia coli*.

colonial bentgrass. *Agrostis tenuis*.

colonic. pertaining to the colon.

color. **1:** the characteristic sensations produced by light waves that strike the retina of the eye of vertebrates such as birds and a few mammals such as humans and other primates. **2a:** a hue, pigment, or tint. **2b:** a dye, stain, or pigment. **3:** to dye, paint, stain, or tint. **food color** (food coloring). any additive used to color a food product to confer a specific, deirable color that appears natural or normal. There are seven food colors or food dyes now generally in use. All are made from coal tar and are identified by the color name and a number. Several million pounds of these additives are consumed annually in the United States alone. See, for example, erythrosine, FD&C Blue 1, FD&C Blue 2, FD&C Red 40. *Cf.* pigment. **artificial color**. any synthetic or manufactured color that does not also occur naturally. Such colors are almost always coal-tar derivatives. Almost all are carcinogenic.

color index. colour index.

Colorado ranger. appaloosa.

Colorado rubberweed. *Actinea odorata*.

colour index (C.I.; color index). a British system of color classification. Colors are indexed according to hue, brightness, and saturation as an aid to identification or reproduction.

colt. a male foal.

Colubridae. the largest and most cosmopolitan family of snakes, comprising more than three-fourths of the known species of snakes. The family is morphologically, behaviorally, and ecologically diverse. Both jaws hold solid or grooved teeth but no enlarged hollow fangs. In most cases, the head is wider than the neck. Most species are nonvenomous. Africa is the only continent that has dangerously venomous colubrid snakes. *See* *Lampropeltis*, adder.

Columba livia (rock dove; rock pigeon). a common, widely-distributed pigeon

(Family Columbidae) indigenous to Europe, western Asia, India, and North Africa, but in historic times it successfully invaded the New World. Many breeds of domestic pigeons and homing pigeons derive from this dove. The wild form is typically bluish-grey, with black wing bars, an irridescent neck, and a white rump; variants from this color pattern are fairly common, but all have a white rump. They originally occurred largely on rocky coasts, but now occupy cities and towns in such numbers that they are often considered pests.

Columbia sheep. a relatively large, hornless, white-faced breed of medium-wool that was developed in the United States largely for use on western range, from crosses of the Lincoln and Rambouillet breeds. These sheep are also successful on farms and are excellent breeders.

Columbus grass. a common name of grasses of the genus *Sorghum*.

combine (harvester; combine harvester). **1:** a machine, usually self-propelled with cabin for the driver, that is used to harvest a wide variety of crops. It cuts the crop, conveys it to a threshing mechanism, and sorts the grain or seed from the chaff or straw. The chaff is blown out the back of the combine and the straw remains strewn on the ground. The grain or seed is lifted into a hopper from which it is loaded onto wagons or trucks. **2:** to harvest using a combine.

combine harvester. *See* combine.

combustible. **1a:** capable of combustion. **1b:** a substance or material that is capable of combustion.

combustion. **1:** burning; the oxidation of any substance at such a rapid rate that light and sensible heat are released. **spontaneous combustion**. ignition of a mass of material by chemical oxidation within the mass and without external ignition.

comfrey. any herb of the genus *Symphytum*.

commensal. an organism that participates in commensalism.

commensalism. a symbiotic relationship in which one organism benefits without affecting the other(s). *See* mutualism, parasitism, symbiosis.

commercial thinning. a silvicultural treatment whereby an overstocked stand is thinned out by removing trees that are large enough to have some commercial value. This treatment is intended to improve the health and growth rate of the remaining crop of trees.

commodity. **1:** an asset or posession. **2: 3:** goods or merchandise. **4:** a manufactured product. **basic (farm) commodity**. **1:** any of six agricultural crops, corn, cotton, peanuts, rice, tobacco, and wheat, that have been declared by permanent law to require federal price support. **2:** any commodity considered essential or very important to an agricultural economy (e.g., certain crops, eggs, milk, hides).

common almond. *Prunus amygdalus*.

common box. *Buxus sempervirens*.

common broom. *Sparteum junceum*.

common buckthorn. *Rhamnus cathartica*.

common carp. *Cyprinus carpio*.

common elder. *Sambucus canadensis*.

common flax. *Linum usitatissimum*.

common groundsel. *Senecio vulgaris*.

common honeybee. *Apis mellifera*.

common horse bot fly. *Gasterophilus intestinalis*.

common kale. *B. oleracea* var. *acephala*.

common milkweed. *Asclepias syriaca*.

common nightshade. *Solanum nigrum*.

common red ox louse. *Bovicola bovis*.

common salt. sodium chloride, especially

that found in nature.

common sunflower. *Helianthus annus*.

communicable. *See* contagious.

community. **1:** a group of interacting individuals that reside in a defined area or that have characteristics in common that are not shared by the general population. **3:** society as a whole. **4:** a biocenosis. **5:** population. This usage is not recommended. **biotic community**. **1:** biocenosis. **2:** in paleontology, a group of species that frequently occur together.

compaction. closely packed feed in the gastrointestinal tract, causing digestive disturbances such as constipation.

compartment. **1a:** any convenient portion of an organism, an ecosystem, or any entity that can be defined and quantified for purposes of study or one that is homogeneous with respect to certain properties (e.g., water, air) especially one that is semiautonomous in some respects (e.g., soil).

complete metamorphosis. *See* metamorphosis.

component. **1:** a constituent or ingredient. **2:** fundmental; integral; constituent. **2:** The nutritional solids in milk, notably fat,protein, lactose, and minerals.

Compositae. *See* Asteraceae.

composite family. Asteraceae.

compost. 1: a mixture of organic residues (e.g., garbage, manure, trimmings from plant material, peat, and other degradable waste) and soil such that certain bacteria in the soil break down the biodegradable materials into organic fertilizer. Lime and chemical fertilizers may also be added. **2:** to bring about the decomposition of organic material (as in def. 1) by soil bacteria in the presence of air to form a humus-like organic fertilizer. *See also* composting. **3:** the stabilized product of composting. *See also* composting. **aerobic compost**. a composting environment inhabited by aerobic bacteria. Aerobic composting generates more heat proceeds faster than anaerobic composting. Temperatures may reach 140°F or more. At such a temperature pathogens, weed seeds, and fly ova are destroyed. There are little or no unpleasant odors. Enclosed aerobic composting is the most rapid composting method. **anaerobic compost**. a composting environment that is free of molecular oxygen. It is inhabited by anaerobic bacteria that obtain oxygen from the organic wastes. Temperatures may reach 130°F; digestion requires more time than in aerobic composting, offensive odors are generated and pathogens may survive. **mature compost**. compost that has been brought to a stable state. It is rich in readily available plant nutrients, poor in phytotoxic acids and phenols, and low in readily available organic compounds. **mushroom compost**. Cellulose-rich compost that is high in organic matter (e.g., manure and straw) that has undergone the initial decomposition stage of a controlled process. It is used by mushroom growers. **semi-mature compost**. compost in which the organic matter has been reduced by 40-50% and is in what is termed the mesophilic stage, having passed through a thermophilic stage.

compost activator. *See* activator.

composting. the decomposition of organic material naturally by soil bacteria in the presence of air with the formation of a humus-like organic fertilizer. Composting can be facilitated by mechanical mixing and aeration, by dropping the materials through a vertical series of aerated chambers, or by forming compost heaps (sometimes in pits) in the open air and mixing or turning it occasionally. *See* compost (def. 2).

compound 1080. sodium fluoroacetate.

compound 118. aldrin.

compound 42. warfarin.

compound gland. a gland with repeatedly branching excretory ducts, the smallest of which drain the secretory portion of the gland.

Compound 269. endrin.

Compound 497. dieldrin.

Compound 711. isodrin.

Compound F-2. zearalenone.

concentrator. accumulator.

conception. formation of the zygote by fertilization or penetration of the ovum by a spermatozoon.

conception rate. in livestock, the total number of resulting conceptions divided by the total number of services.

cone nose. assassin bug.

cone rake. a tool used to collect cones from the crown of a standing tree. It is usually lowered from a helicopter over the crown of a tree. Cones or cone-bearing branches are then removed from the tree by the rake.

confined aquifer. artesian aquifer.

confinement. **1a:** isolation. **1b:** restraint; restriction. **2:** the housing of livestock in, for example, corrals, pens, or cages in order to achieve year-round production.

conformation. in agriculture, the general shape of an animal.

congenital leukoderma. albinism.

congenital leukopathia. albinism.

conidia. plural of conidium.

conidium (plural, conidia). a nonmotile, asexual spore produced by mitosis and formed from the ends or sides of a hypha. Conidia are produced in large numbers by the asexual phase of basidiomycete soil fungi.

conifer. **1:** a vernacular descriptive term for cone-bearing trees and shrubs (gymnosperms) such as those of the family Pinaceae. Any gymnosperm of the family Coniferae.

conifer release. the release of established coniferous trees from suppression by undesirable trees and shrubs which have overtopped them. The latter are thinned out to improve growth of the affected conifers.

coniferous. **1:** having the charasteristics of a conifer. **2:** descriptive of a tree that has cones.

Conium maculatum (poison hemlock; deadly hemlock; water hemlock; spotted cow bane; spotted parsley; poison fool's parsley; California fern; Nebraska fern; winter fern). an extremely poisonous neurotoxic herb that grows to a height of about 1 m before flowering, it resembles Queen Anne's lace (wild carrot). Poison hemlock is easily distinguishable from Queen Anne's lace if the leaves are directly compared; also, while the main leaf stalks of Queen Anne's lace are "hairy," those of poison hemlock are smooth. While indigenous to Europe (Family Apiaceae), it has become widely distributed in North and South America. *C. maculatum* is fetotoxic in domestic cattle, causing deformities in the calves that are similar to those of crooked calf disease.

conk. a hard, spore-containing fruiting body of a wood-decaying fungus that projects beyond the bark of an affected tree. It contains spores of a wood-decaying fungus.

constructive metabolism. anabolism.

consumer. **1a:** one who consumes or uses. **1b:** one who purchases, uses, maintains, or disposes of a product or service as opposed to one who manufactures or sells a product. **1c:** the end user of a product or commodity. **2:** a heterotrophic organism.

consumer product. any tangible goods or personal property that enters into commerce and that is normally used for personal, family, or household purposes. Included is any such product intended to be attached to or installed in or on any real property.

contagious (communicable; infectious; transmissible). of or pertaining to a disease that is readly spread within or among species by direct (e.g., physical contact) or indirect contact, as in the case of a vector-borne disease.

contagious abortion. brucellosis. *See also* abortion.

contagious bovine abortion. brucellosis.

contaminant. **1:** any substance that enters a system where it does not normally occur or in amounts substantially greater than are normally present. Man is an important agent of contamination.

contemporary. **1:** coexistent; concurrent. **2:** descriptive, for example, of cattle of the same breed that are born, raised, and begin lactating at about the same time.

contour. profile, outline, shape.

contour cropping. a form of contour farming in which different crops are planted across a slope in order to retard erosion.

contour farming. a general method of cultivating crops on sloping ground in which plowing and planting are done across the slope (i.e., following the contours of the land) rather than down slope.

contour interval. the difference in elevation between adjacent contour lines on a map.

contour line. a line on a map that connects points of equal elevation.

contour map. a topographic map which portrays relief by means of lines (contour lines) that connect points of equal elevation. Such maps have a uniform contour interval.

contour ridging. a method of contour farming in which the land is plowed on the level across a slope, raising a ridge that helps to retain water.

contract production. the commitment of a firm to purchase a commodity from a producer at a price formula agreed upon in advance.

contractual framework. the situation where forest management practices are chiefly regulated by contracts.

contralateral. of, pertaining to, affecting, or on the opposite side.

control point. any of a system of points selected on the basis of location, elevation, or both, used as fixed references in locating or positioning map features.

convoluted gland. coil gland.

Convolvulus (bindweed; glorybind; wild morning glory). a genus of cultivated annual or perennial herbs or shrubby twining or horizontal plants (Family Convolvulaceae) in which toxic concentrations of nitrate have been reported.

convulsant (convulsivant). **1:** of or pertaining to the ability to induce convulsions. **2a:** any agent that causes or precipitates convulsions. **2b:** a convulsant drug or poison such as strychnine, brucine, and other alkaloids from nux vomica.

convulsion (seizure). an involuntary, usually violent, generalized (whole-body) spasm or series of spasms of voluntary muscle. Convulsions are activated by the brain and are usually accompanied by a period of unconsciousness. Convulsions are symptomatic of a variety of disorders and disease states. **clonic convulsion**. a convulsion with intermittent muscular contractions alternating with periods of relaxation. **coordinate convulsion**. a convulsion with clonic contractions that are similar to normal voluntary movements. **crowing convulsion**. a convulsion of the larynx. **epileptiform**. any convulsion with loss of consciousness. **local convulsion**. any localized spasm (e.g., of a single muscle or body part). **tetanic convulsion**. tonic convulsion. **tonic convulsion** (tetanic convulsion). a convulsion with sustained muscular contraction and rigidity of the body. The limbs are fixed in position. **toxic convulsion**. that caused by the action of a toxin on the nervous system. **uremic convulsion**. that due to failure of the kidneys to excrete products of protein digestion (e.g., urea) normally.

convulsivant. convulsant.

convulsive. of, pertaining to, producing, or characterized by convulsions.

cookfire. any fire used for cooking.

coonntie. a common name of *Zamia integrifolia*.

cooper. **1:** to make or repair wooden barrels. **2:** a person who makes or repairs casks, barrels, etc. **3:** to do the work of a cooper. **4:** to furnish or fix; fix up.

cooperage. **1:** the business of a cooper. **2:** the place where one coopers. **3:** items made by a cooper. 4. the fee or price paid for the work of a cooper.

cooperative. a type of business owned and controlled by the people who use its services.

Cooperative Extension Service. an adult education and development organization supported by public funds in the United States. It was established in 1914 by the Smith-Lever Act. Its broad goal is to plan, execute, and evaluate learning experiences designed to help people learn and understand the skills necessary for solving farm, home, and community problems. Organized at the federal, state, and county levels, it is one of the largest public education programs in the world. Services include educational programs based in large measure on the results of research, especially that of the Agricultural Research Service of the U.S. Dept. of Agriculture and the state land-grant colleges and universities. Four-H clubs for youth are also sponsored by the service.

copper (cuprum; Cu)[Z = 29; Ar = 63.55]. a golden-red metallic transition element that occurs in nature mainly as the sulfide. Copper compounds are white, except for the oxide which is red, and form blue solutions. Elemental copper is an excellent conductor of heat and electricity, with low toxicity. Copper is a constituent of cytochrome *c* oxidase, which catalyzes the transfer of electrons from cytochrome a_3 to oxygen in respiration. Copper is also a component of the protein, plastocyanin thus also playing a role in photosynthetic electron transport. Soluble copper salts are strong irritants and hepatotoxicants. As a component of phylloerythrin, copper causes hepatogenic photosensitization in livestock.

copper acetate arsenite. cupric acetoarsenite.

copper ethanoacetoarsenate. cupric acetoarsenite.

copper fungicide. any inorganic fungicide that contains copper (e.g., Bordeaux mixture, Burgundy mixture). Copper fungicides are often based on copper oxychlorides or copper oxide; they may also be blended with other types of fungicide.

copper monosulfate. cupric sulfate.

copper poisoning (toxemic jaundice). poisoning, acute or chronic, due to soluble copper salts, all of which are strong irritants and hepatotoxicants. Among livestock, sheep are affected most often. Acute poisoning is usually observed after accidental administration of excessive amounts of soluble copper salts, which may be present in anthelmintic drenches, mineral mixes, or improperly formulated rations. **chronic hepatogenous copper poisoning**. a form of chronic phytogenous copper poisoning due to the repeated ingestion of plants such as *Heliotropium europaeum* or *Senecio spp*. for a period of several months. These plants contain hepatotoxic alkaloids, which cause excessive copper retention by the liver. **chronic phytogenous copper poisoning**. a form of copper poisoning that results from long-term repeated ingestion of plants such as subterranean clover (*Trifolium subterraneum*). In this case, a mineral imbalance is incurred that results in excessive copper retention. Plants that are not hepatotoxic contain normal amounts of copper and low levels of molybdenum. Grazing animals are most commonly affected.

copper spot. a fungal disease of grasses characterized by coppery or orange spore masses that the cover the leaf blades. The causative agent is *Gloeocercospora sorghi*.

copper sulfate. cupric sulfate.

copperas. ferrous sulfate.

copperhead. *Agkistrodon contortrix*.

copse. a small dense woodland or stand of trees.

coral snake. *See Micruroides euryxanthus*, *Micrurus*.

coralberry. a plant of the genus *Actaea*.

cord. a 128 cubic foot stack of roundwood, split or whole, with or without bark, and neatly stacked with all pieces of the same length and orientation. A 4'x4'x8' stack of firewood equals one cord.

corduroy, **corduroy road**. a roadway over which logs, saplings, or poles have been placed transversely. Fill material is usually added. Such roads today are usually temporary and used in wet areas or areas with soft subsoils to allow traffic to pass safely and without interruption.

core (decore). **1:** the center, heart, or middle. **2:** the crux, essence, or heart of the matter. **3:** to remove the core of a fruit.

coreid bug. leaf-footed bug.

corium. dermis.

cork. the layer of dead tissue that lies immediately beneath the bark of a tree, especially the thick layer that develops in the cork oak.

corm. a short, fleshy, erect, underground stem that is usually more broad than high and covered by membranous scale-like leaves. It closely resembles a bulb but is more stemlike. *Caladium* and *Colchicum* grow from corms. Corms are often poisonous as in the above-named genera. *Cf.* bulb.

corn. *Zea mays*

corn belt. *See* feed grains and livestock belt.

corn borer (European corn borer). **1:** the larva of *Pyrausta nubilalis*, the European moth. It is a serious pest that feeds on and destroys corn and certain other plants, including beans and potatoes. **2:** any of a number of other insect larvae that feed on corn.

corn borer moth. *See Pyrausta nubilalis*

corn cockle. *See Agrostemma*.

corn earworm. *Heliothis zea*.

corn poppy. *Papaver rhoeas*.

corn root-rot organism. *Giberella zeae*.

corn-lily. *Veratrum californicum*.

cornstalk disease. **1:** a disease syndrome in cattle in which nitrosohemoglobin in blood is formed due to the ingestion of corn of high nitrogen content. **2:** moldy corn poisoning, *q.v.*, a rare disease of livestock (primarily cattle) caused by the ingestion of moldy corn forage.

Cornus (dogwood). a genus of ornamental shrubs, trees, and occasional herbs (Family Cornaceae) that are native to north temperate regions. The fruit is slightly toxic.

corolla. **1a:** the collective structure formed by the arrangement of the entire set of petals of a flower. **1b:** the inner circle of petals is usually the conspicuously colored component of a flower head.

corrective (corrigent). **1:** of or pertaining to an action or process of adjusting, altering, changing, modifying, repairing, amending, rectifying, remedying, or setting straight that which is injurious, malfunctioning, or that does not conform to a norm. **2a:** improving precision or accuracy. **2b:** improving suitability. **3:** alleviating or counteracting that which is undesirable, harmful, damaging, or injurious. **4:** a drug or medication that modi fies, alleviates, or remedies, or corrects harmful or undesirable effects of another drug.

corrigent. corrective.

corrosion. **1:** the dissolving or wearing away of metal caused by a chemical reaction (e.g., corrosion of pipes by water). **2:** erosion or disintegration of tissue as by a strong acid or strong base.

corrosive. **1:** causing or productive of corrosion. **2:** a corrosive chemical that reacts with a surface (e.g., living tissue), eroding it, causing it to deteriorate, or destroying it.

corrosive alkali. *See* alkali.

corticotrop(h)in. ACTH.

Corydalis (corydalis). a genus of toxic, usually biennial, leafy-stemmed, herbs (Family Fumaraceae; alternatively Family Paparivaceae, Subfamily Fumarioideae) that are indigenous to North America. The flowers, which are an irregular blue, purple, yellow, or rose, are similar to those of *Dicentra*. A number of species can poison grazing livestock. The alkaloids isoquinoline, aporphine, protoberberine, protopine, and others have been isolated from various species. *C. caseana* (synonym, *Capnoides caseana*) has caused the most livestock fatalities (sheep and cattle). Symptoms may appear within a few minutes to a few hours following ingestion. The symptoms of sheep and cattle are similar to those due to ingestion of *Dicentra*. They include restlessness and patent depression, followed by twitches especially of the facial muscles, convulsive clonic spasms of 5 to 15 minutes in duration. During the spasms, muscular rigidity is readily induced by any strong stimulus such as a loud sound. Biting or snapping at objects is common as is scouring, bleating, or bawling. In severe poisoning, the animal soon weakens, becomes quiescent, and dies following several seizures. Otherwise, recovery is rapid and uneventful. *Cf.* corydalis.

corydalis. **1:** common name of any plant of the genus *Corydalis*. **2:** the dried tuber of *Dicentra (Bicuculla) cucullaria* or *D. canadensis* (Family Fumariaceae). It contains corydaline, bulbocapnine, isocorydine, corytuberine, corycavine, corybulbine, corydine, fumaric acid, acrid resin, and protopine. *Cf.* Corydalis.

corynebacterium kidney disease. bacterial kidney disease.

Corynebacterium diphtheriae. the bacterium that causes diphtheria via secretion diphtheria toxin (*See* toxin).

Corynebacterium insidiosum. the bacterium that causes bacterial wilt of alfalfa, *q.v.*

cosmetic appearance. the external appear-appearance of a fruit or other agricultural commodities independent of their yield, taste, and nutritional value. Included are changes in appearance due to superficial damage or alterations that do not significantly alter these values. Grades and standards set by the Agricultural Marketing Service for many agricultural commodities are said by consumer and environmental groups to cause harm because they invite excessive pesticide use just to make fruits and vegetables attractive. Agricultural interests generally believe, however, that consumers prefer blemish-free produce.

Cosmetic Blue Lake. FD&C Blue No. 1

cosmopolitan (cosmopolite). worldwide in extent, distribution, or influence; pandemic.

cosmopolite. cosmopolitan.

cosset lamb. a lamb that is reared by hand.

cost-benefit analysis (benefit/cost analysis). a method for comparing alternate courses of action by an assessment, usually in both economic and social terms, of their direct and indirect benefits and costs.

Cotinus nitida (green June beetle). a hardy, elongate, somewhat flattened pytophagous insect (Family Scarabaeidae). It inhabits gardens, orchards, open woods, and crops, especially above sandy soils. The adults feed on the foliage and ripening fruits (especially peaches) of many trees, shrubs, and certain other plants (e.g., hollyhocks). The larvae destroys root vegetables and especially tobacco seedlings.

Cotswold sheep. a large, white-faced, hornless breed with a broad, flat back, and moderately deep body. The wool is coarse, heavy, and hangs down in ring lets. It was developed in the Cotswold Hills in England; in the United States it most commonly occurs in the Northwest. This breed is frequently crossed with the Merino and Rambouillet sheep.

cotton. **1:** soft, white fibrous material that surrounds the seeds of a number of tropical plants. **2:** cotton fabric. a soft fabric produced from cotton. **American Egyptian cotton**. an important variety of cotton (Family Malvaceae) grown in several western states. **upland cotton**. The predominant variety of cotton (Family Malvaceae) grown in the United States.

cotton bagging. **1:** material such as jute used to cover bales of cotton. **2:** any coarse cotton cloth used to make bags.

cotton rat. a rat of the genus *Sigmodon*.

cotton-dust asthma. byssinosis.

cotton-mill fever. byssinosis.

cottonmouth. *Agkistrodon piscivorus*.

cottonmouth moccasin. *Agkistrodon piscivorus*.

cottonmouth water moccasin. *Agkistrodon piscivorus*.

cottonseed. seed of the cultivated cotton plant, *Gossypium spp*. It contains about 20% protein and 20% oil. The oil contains gossypol, *q.v.*, and a number of other toxic pigments, some of which are more toxic than gossypol. Meal produced from cottonseed may contain about 40% protein and a very low percentage of oil if solvent extraction is employed in its manufacture. It is thus usually a good food supplement for livestock. Nevertheless, occasional intoxications of livestock may occur, either from poorly controlled manufacture of the meal or when the meal comprises too high a proportion of the diet. The gossypol content of the swine diet must be less than 100 mg/kg body weight to be considered safe. *See* gossypol poisoning.

cottonseed meal. *See* cottonseed.

cottonseed oil. *See* cottonseed.

coturnism. food poisoning caused by the common European quail, *Coturnix*, and characterized by impaired speech, dyspnea, weakness, nausea, paresthesia of the lower legs, partial paralysis, and occasionally death.

cotyle. a *cup-shaped* structure.

cotyledon (seed leaf; ear leaf). an embryonic organ of angiosperms and gymnosperms that stores and digests reserve materials. In some plants, the cotyledons are flat and leaflike, in others (where they function in food storage) the cotyledons are fleshy. The cotyledons may expand when the seed germinates. Angiosperms usually have either one cotyledon (monocots) or two (dicots). *See*ds of gymnosperms, however, may have numerous cotyledons. In most plants, the cotyledons emerge above the soil with the seedling as it grows. They are not true leaves.

cotyledonary. cup shaped; resembling half a bean seed.

couch grass. *Agropyron repens*.

coumafuryl (3-(α-acetonylfurfuryl)-4-hydroxycoumarin). a white, nearly water-insoluble powder, $C_{17}H_{14}O_5$. It is a coumarin anticoagulant. This compound or its sodium salt is used as a rodenticide. It is highly toxic by ingestion and inhalation.

coumaphos (O,O-diethyl-O-(3-chloro-4-methyl-2-oxo-2*H*-1-benzopyran-7-yl)-phosphorothioate). a crystalline substance, $C_9H_3O_2(CH_3)ClOPS(OC_2H_5)_2$, that is insoluble in water. It is an anticholinesterase used as an insecticide and anthelmintic. It is used, in particular, against cattle grubs and a number of other ectoparasites and for treatment of areas holding cattle. The maximum concentration that may be safely used on adult cattle, horses, and hogs is 0.5%.

***p*-coumaric acid** (3-(4-hydroxyphenyl)-2-propenoic acid; *p*-hydroxycinnamic acid; β-[4-hydroxyphenyl]-acrylic acid). a crystalline substance that is only slightly soluble in cold water, but soluble in hot water, ethanol, and ethyl ether; nearly insoluble in benzene and ligroin. It is a plant growth inhibitor structurally related to ferulic acid. It is the aglycon of coumarin glycoside. Coumarin is the anhydride of *o*-coumaric acid.

coumarin (2*H*-1-benzopyran-2-one; 1,2-benzopyrone; *cis-o*-coumarinic acid lactone; cumarin; coumarinic anhydride; tonka bean camphor). a very toxic, white crystalline substance, $C_9H_6O_2$, with an odor of new-mown hay. It occurs naturally in many plants (e.g., clovers), especially in the fruit or test where it inhibits germination. The seeds of such plants germinate only when the coumarin is removed or destroyed, e.g., by photolysis. It is used, for example, in the production of the anticoagulant, dicoumarin (dicumarol).

coumarin anticoagulant. any anticoagulant (e.g., warfarin, coumafuryl) that contains coumarin as the active component. These tend to be less toxic than indandione anticoagulants, *q.v.* Effects of acute exposures include hemorrhage and sudden death with no warning. In subacute cases, symptoms include anemia, weakness, pale mucous membranes, dyspnea, moist rales, bloody feces, and scleral, conjunctival, and intraocular hemorrhage. Staggering, ataxia, blood-tinged froth around mouth and nose, and CNS signs may appear when hemorrhage occurs in the brain or spinal cord.

***cis-o*-coumarinic acid lactone**. coumarin.

coumarinic anhydride. coumarin.

counterpoison. antidote.

county agent (extension agent; farm and home advisor; agricultural agent; home demonstration agent). an agricultural professional who acts jointly as an agent of the county, the State Cooperative Extension Service, and the U.S. Department of Agriculture. Chief responsibilities are to act as a source of information, education, and assistance to the local population on matters relating to problems and needs associated with the home, homemaking, the farm, and community problems.

covalent. denoting or pertaining to nonionic chemical bonds, those in which electrons are shared by two atoms.

cover. **1a:** lid; top; **1b:** refuge; shelter. **2:** the area of ground covered by vegetation of a given species at a given location. **basal cover**. *See* basal area. **ground cover**. *See* basal area, crop (cover crop). **litter cover**. dead material (litter) on the soil surface. **soil cover**. a layer of uncontaminated soil that overlies contaminated soil. **vegetative cover**. plant material on or near the soil surface.

covering. *See* service.

covey. a small flock of quail or partridge. *See also* bevy.

cow. **1a:** singular of cattle. **1b:** a mature female bovid. **2:** a mature animal of the species *Bos indicus* or *B. taurus*. **3:** a mature female of certain species of mammal other than bovids (e.g., elephant, whale). **4:** sometimes in the vernacular, a domestic bovine of either sex of any age. **beef cow**. a cow reared to produce beef. **dairy cow**. a domestic cow bred and used for the production of milk for human consumption or one that is used to raise replacement dairy heifers.

cow barn. a barn that houses cattle. The equivalent terms in the United Kingdom are byre or cow house.

cow chip. the dried dung of cattle.

cow dung (cattle dung). the dung of cattle. *See also* bodewash, camel dung, buffalo chip.

cow heifer. a cow that has calved only once.

cow house. byre (*See* cow barn)

cow killer. a large whitish, yellow, and red wasp (Family Mutillidae) covered with long hairs. The female is wingless and delivers a powerful sting.

cow pony. cow horse; a horse used by ranchers to herd, round-up, or otherwise work livestock. Many cow ponies are descended from mustangs, but have also been crossed with other breeds of horse to the extent that nearly all working horses of the western United States are of mixed blood.

cow pony. cow horse (*See* horse).

cow-poison. *Amianthium muscitoxicum.*

cowbane. *Cicutum maculata.*

cowberry. *See Vaccinium.*

cowitch. *Rhus radicans.*

Cowper's gland (bulbourethral gland). an accessory gland of the male reproductive system that secretes a fluid, in advance of the seminal fluid, that bathes and neutralizes the wall of the urethra.

cowperitis. inflammation of Cowper's gland.

cowshed. a small barn or shed that protects cattle from wind and precipitation.

coxal gland. *See* Arachnida.

coyotillo. a common name of *Karwinskia humboldtiana.*

4-CP. chlorophenoxyacetic acid.

CP 4742. sulfallate.

CPA. chlorophenoxyacetic acid.

CPT. 3-chloro-*p*-toluidine.

crab. **1:** a common term that usually refers to any of the numerous species of crustaceans comprising the order Decapoda, especially those with a broad cephalothorax and a small abdomen that lies against the ventral surface of the cephalothorax. *Birgus latro* (coconut crab) and several species of tropical reef crabs (e.g., *Atergatus floridus*, *Carpilius maculatus*, and *Demania toxica*) are poisonous. The antennae are small and the walking legs are well-developed. **2:** sometimes used in reference to crayfish or any large decapod crustacean.

crabseye. *Abrus precatorius.*

cracker-heel disease. a disease of range cattle, sheep, and horses caused by certain species of *Astragalus*, *q.v.* Cattle exhibit poor condition, a dry or peeling muzzle, chronic cough, loss or huskiness of voice, anorexia, and incoordination, especially of the hind legs which tend to knock together (hence, "cracker-heel" disease). Forced exercise causes a wheezing or even a roaring exhalation. Paralysis may develop, and sudden death from asphyxia is common in animals that are driven. Mortality is otherwise unpredictable. The syndrome is more severe in sheep and especially in horses. Species responsible for this type of poisoning include *A. emoryanus*, *A. michauxii*, *A. pubentissimus*, *A. sabulosus*, and *A. tetrapterus*.

crampiness. a painful condition of confined cattle, especially of bulls, characterized by involuntary contractions of muscles of the hind legs. Affected animals shift from foot to foot.

cranberry. *See Vaccinium.*

Craniata. Vertebrata.

craw. **1:** the crop of a bird or an insect. **2:** the stomach of certain invertebrates.

crazy chick disease. a disease of chicks that is apparently due to a diet that is deficient in vitamin E or too rich in lipids. Affected birds suffer paralysis and cannot remain standing.

cream-of-tattar tree. *Adansonia digitata.*

creamery. a milk processing plant, or sometimes a butter-producing factory.

creep-feed. to feed animals in a creep feeder.

creeping bentgrass. *Agrostis palustris*.

creeping fescue. *Festuca rubra* var. *commutata*.

creeping indigo. *Indigofera endecaphylla*.

creeping lady's sorrel. *Oxalis corniculata*.

creeping red fescue. *Festuca rubra commutata*.

creeping saltbush. *Atriplex sentibaccata*.

creeping turfgrass. *Agrostis tenuis*.

Crenosoma vulpis. a species of lungworm (Family Metastrongylidae) that parasitizes small carnivores in Europe, Asia, and North America such as the dog, fox, raccoon, and wolf. It invades the bronchi, causing bronchitis.

creosote. an oily, colorless or yellowish to dark green-brown toxic and flammable mixture of phenols. It is oily with a characteristic, pungent odor and is obtained by distilling coal tar or wood tar. Creosote is used as a wood preservative and disinfectant, has been used externally by humans as an antiseptic, and can be taken internally as an expectorant in chronic bronchitis. It is a skin irritant and carcinogen. Exposed workers may develop pulmonary cancer, or cancer of the skin, larynx, or nasal cavity after 12-30 years of exposure.

crepuscular. of or pertaining to twilight (dusk or dawn), a term used especially with respect to animals that are most active at these periods of the day.

cresol (tricresol; hydroxymethylbenzene; methyl phenol). a mixture of the three isomeric cresols, *m*-, *o*-, and *p*-cresol, obtained from coal tar. It is a common component of disinfectants and is used in some dandruff shampoos. It is readily absorbed through the skin and the membranes of the pulmonary system. Intoxication may be due to ingestion, inhalation, or absorption through the skin. Symptoms of cresol intoxication may include dyspnea, irritability, depression, hyperactivity, and unconsciousness. Nausea, vomiting, and a burning sensation of the mouth and throat are also common. Cresols can damage the liver, kidney, spleen, pancreas, and central nervous system. The three isomers have properties similar to those of phenol, but are less toxic. ***m*-cresol** (3-methylphenol; metacresol). a moderately toxic phenolic compound used as a local antiseptic, in disinfectants, fumigants, and explosives. It often occurs mixed with *ortho*- and *para*- analogs as cresol or cresylic acid, light yellow liquid, with a melting point (m.p.) of 11°C and a boiling point (b.p.) of 203°C. ***o*-cresol** (2-methylphenol; orthocresol). a moderately toxic solid (m.p.) 31°C, b.p. 191°C) phenolic compound used as a disinfectant. ***p*-cresol** (4-methylphenol; paracresol). a crystalline solid (m.p. 36°C, b.p. 202°C) with a phenolic odor that is used as a disinfectant.

crib. a usually slatted bin or box that holds fodder.

crinum lily. *Crinum*.

Crinum (crinum lily). a genus of toxic bulbous plants (Family Amaryllidaceae). ***C. americum*** (swamp lily). the only species native to the United States. They are toxic to livestock. ***C. asiaticum*** (St.-John's-lily; poison bulb). a commonly cultured potted plant.

Crioceris (asparagus beetle). either of two leaf beetles, both the larvae and adults of which feed on the asparagus plant.

crisis. **1:** a turning point in an acute disease generally marked by an intensification of symptoms followed by recovery, stabilization, or worsening of the condition. **2:** an emergency. **abdominal crisis**. severe abdominal pain. **cholinergic crisis**. a condition characterized by hyperactivity of cholinergic nerve pathways. This may result, for example, from poisoning by organophosphate or carbamate pesticides or from other anticholinesterases. **salt-losing crisis**. acute loss of sodium with dehydration, hypotension, and sudden death due to acute vomiting. **thyroid crisis**

(thyroid storm). a sudden increase in severity of the symptoms of thyrotoxicosis, marked by extreme tachycardia and fever; sometimes fatal. **true crisis**. a fall in temperature and pulse rate.

critical. **1a:** grave; serious; dangerous. **1b:** crucial; decisive; pivotal. pertaining to or denoting a state of crisis. **2:** in danger of death. **3:** crucial, decisive. **4a:** of or pertaining to a turning point. **4b:** of sufficient size, quantity, or intensity to effect a major change or turning point (e.g., a critical concentration). **4a:** of doubtful or uncertain outcome; characterized or attended by uncertainty. **5:** derogatory; disparaging.

critical effect. *See* effect.

crop. **1:** a batch or lot; any set of items formed or produced together or at the same time. **2:** harvest; yield. **3:** cultivated plants. **4:** the plants of a given kind cultivated or harvested at the same time. **5a:** plants grown for food, fiber, forage, or commerce. **5b:** a group of plants or animals grown (usually in one place) in a controlled, organized fashion and harvested for profit or subsistence. **6:** a plant or animal product or commodity produced for commercial reasons (e.g. a maple syrup crop, a sugar crop, a honey crop). **7:** The young animals produced on a farm usually in a given year (e.g., a crop of yearlings, a crop of calves). **8:** the total yearly production from a specified area (e.g., the Ohio corn crop). **9a:** to prune, clip, shear, or trim. **9b:** to gather or harvest (e.g., flowers, wheat), especially by cutting. **10:** to plant, grow, or produce a crop; to yield a crop. **11:** the yield of a cultivated plant during a designated time frame (e.g., seasonal, annual). **11:** to farm as a sharecropper. **12a:** to cut or lop off (e.g., the wattles of a fowl, the branch of a tree). **12b:** to cut off the upper parts of a plant. **12c:** to clip the ears or tail of an animal (e.g., a show dog) or to cut off part of the ear of a mammal (usually as a means of identification). **12d:** to clip, trim, prune, shear, or otherwise cut close (e.g., wool or grass). **13a:** something that has been cut or trimmed or results from cutting or trimming. **13b:** very short or closely cut hair. **14:** earmark; an identification notch or mark in the ear of an animal, especially a cut across the upper part of an ear. **15:** to bite or nip off. **16:** to graze or browse, especially by nipping off tender buds or shoots or by nipping of grass or other vegetation close to the ground. **17:** a thin-walled distensible pouch (gullet) in the lower end of the esophagus of many birds. A single opening communicates with the lumen of the esophagus. This organ is highly developed in many grain-eating birds such as the domestic fowl (*Gallus*), but is rudimentary or even absent in many birds. This organ receives, stores, and softens food before passage into the stomach. In many species, the food undergoes some digestion in the crop prior to their offering it to the chicks. The food in the crop of certain birds is enriched by secretion of a proteinaceous material by cells that line the crop (as in pigeons and doves) or by regurgitation from the proventriculus (as in the budgeriger). **18:** a sac-like enlargement of the alimentary canal of earthworms, some insects, and certain other invertebrates that serves the functions of food storage and partial digestion. **19:** the honey sac of a honey bee, *Apis millifica*. **20a:** the part of the chine of a quadruped (e.g., a domestic cow) that lies immediately behind the withers. **20b:** a cut of meat from this region: short ribs or spare ribs. *See also* chine. **21:** a working unit in forestry and related fields which is usually equal to 10,000 boxes (e.g., a crop of turpentine). In general, this unit represents material coming from a tract of some 250 acres, comprising about 5000 trees. **22:** that part of a tanned hide produced by cutting it in half along the backbone and then trimming off the belly. **23:** a riding whip that has a short, straight shaft and loop. **24:** the shaft or handle of a whip. **25a:** an outcrop. to appear at the surface; to outcrop. **25b:** an outcrop. **banner crop**. a crop of high quality and/ or yield. **basic crop**. any crop that is considered essential or very important to an agricultural economy. *Cf.* commodity (basic (farm) commodity). **cover crop**. **1:** a crop planted to cover the soil and thus to prevent drying of the soil and erosion. *See also* cover. **2:** a crop grown to give protection to another crop. **3:** a crop grown to provide cover for small animals, especially game birds.

crop dusting. aerial dusting.

crop glands. glands in the crop of male and female pigeons and doves (Family Columbidae) that secrete crop milk, *q.v.*, in response to stimulation by prolactin.

crop milk. a proteinaceous, milk-like secretion of the crop (or crop glands, *q.v.*) in pigeons and doves that is fed to the young. It contains mostly sloughed off epithelial cells. The process is stimulated by prolactin during the nesting season.

crop tree. a tree in a young stand or plantation intended to be retained until mature for an interim or final harvest.

cropland. land used primarily to produce row crops, close-growing crops, and fruit and nuts. **fallow cropland** (summer fallow). cropland that remains idle during the growing season. It may be managed for future use (e.g., tilled or sprayed to control weeds). **idle cropland**. that used only as pasture, for plants grown to improve soil quality, or for orchards. **summer fallow cropland**. cropland in semiarid areas that is taken out of production during a cropping season chiefly to conserve moisture for the next growing season. A common practice in dry-land wheat farming is to rotate half of the cropland to summer fallow each year.

cross ditch. in forestry, a suitably reinforced ditch dug across a road at an angle and to a sufficient depth to divert both road surface water and ditch water away from the road.

cross-drain culvert. a culvert that conveys ditch water from one side of a road to the other.

crossed eyes. strabismus.

cross-eye. esotropia, a form of strabismus.

cross-eyed. strabismic. *See* strabismus.

cross-fertilization. the pollination of one individual plant by another. *See also* allogamy.

cross-pollination. pollination of a plant by one that hs a different genetic constitution.

cross-resistance (cross-tolerance). the condition whereby an organism is resistant (or tolerant) to a substance other than the one that originally induced resistance in the population. Insect populations, for example, often become resistant to a given insecticide. These organisms may then show resistance to certain other insecticides to which they have not before been exposed.

cross-tolerance. cross-resistance.

crotalaria poisoning. crotalism.

Crotalaria (rattlebox; crotalaria). a genus of small herbs (Family Fabaceae, formerly Leguminosae) found in warm areas, often on sandy soils of fields and along roadsides. Several species contain pyrrolizidine alkaloids and are known to be toxic; some can prove lethal. The raw seeds are especially toxic. They cause liver cell necrosis, enlargement of certain hepatic cells, and rupture of veins. Crotalaria are involved mostly in the poisoning of livestock. ***C. retusa***. all parts contain monocrotaline. Consumption of seeds has caused major losses of domestic fowl with effects similar to those of *C. spectabilis* but less acute. Cattle rarely consume *C. retusa*. ***C. sagittalis*** (arrow crotalaria; rattlebox; rattleweed; wild pea). an annual herb (Family Fabaceae) of the eastern and southern United States It is poisonous; horses can be killed by ingesting crotolaria fresh of in hay. ***C. spectabilis***. a coarse annual herb of fields and roadsides of the southern United States. The toxic principle is monocrotaline. This plant has caused severe losses among domestic cattle, horses, swine, and domestic fowl and lesser losses among goats, sheep, and mules. In chronic poisoning of the domestic fowl (80-100 seeds), signs and lesions may not appear until a month or more following consumption. In acute poisoning (ca. 320 seeds or more), fowl die within 1-10 days. Symptoms include darkening, scaling, and congestion of the comb, depression, severe diarrhea with green feces, and retention of seeds in the crop. Lesions include petechial hemorrhages of the abdominal and thoracic serosa, a smaller than normal, often ruptured, marbled liver, and petachiae and suffusions of visceral fat, heart, and skeletal muscle. In less acute cases, massive hemorrhage and ascites may present. In chronic poisoning, development and progression of the above symptoms are slowed and the birds may become severely anemic with a paling of the comb. Lesions of chronic poisoning include necrotic enteritis, ascites, partial splenic atrophy, and cirrhosis of the liver. Symptoms of chronic poisoning of cattle may include nervousness or excitement, bloody feces, prostration, and death. Lesions commonly include profuse hemorrhage of the thoracic and abdominal serosa. In long-lasting cases, ascites and jaundice may be observed and hemorrhage is less severe. Symptoms of cattle sometimes include weakness, emaciation, incoordination, and constipation or diarrhea. *See* crotalism.

crotalaria. a common name of herbs of the genus *Crotalaria*.

crotalism (bottom disease; crotalaria poisoning). a venoocclusive disease of grazing animals (sometimes humans) caused by ingestion of plants of the genus *Crotalaria* (either by grazing or by ingesting hay that contains *Crotalaria*) and sometimes *Senecio* or *Heliotropum*). The condition is marked by weakness, emaciation, stupor, congestion, and hemorrhage of the liver and spleen. **2:** crotalaria poisoning (*See* poisoning).

Crotalus (rattlesnakes; klapperschlangen). a genus of New World, live-bearing, venomous snakes (Family Crotalidae). The eyes are small with vertically elliptical pupils, the head is broad, and the neck is narrow. They have large recurved fangs that are periodically replaced. The body is moderately slender to stout and more or less cylindrical. A jointed rattle comprised of cornified, button-like structures forms the tip of the tail in nearly all species. The larger species prey mainly on small mammals and the smaller species prey chiefly on lizards. Effects of envenomation may include pain, swelling, and edema about the bite, and thirst, nausea, vomiting, weakness, a weak rapid pulse, hypotension or shock, tingling or numbness (especially about the tongue and mouth), respiratory distress, and blurred vision. In serious cases, coma and death may ensue. *C. adamanteus* and *C. atrox* cause most of the snakebite fatalities in the United States. *See also* crotalid venom poisoning, rattlesnake. ***C. adamanteus*** (eastern diamondback rattlesnake). the only large rattlesnake in the southeastern United States, it reaches a maximum length of about 2.4 m. It occurs in old fields, dry pine flatwoods, sandhill or longleaf pine and turkey oak country, and palmetto thickets in coastal lowlands from North Carolina through Florida and eastward to Louisiana, from sea level to about 150 m. The side of the head is characterized by the presence of distinct, diagonal, whitish stripes. The fangs are long and the venom produced is considerable. These snakes feed chiefly on rabbits, squirrels, and birds. They are extremely dangerous, often irritable, and sometimes strike without warning, and the venom is highly destructive to tissue and the bite is sometimes fatal. *See* L-amino acid oxidase. ***C. atrox*** (western diamondback rattlesnake). a snake of arid and semiarid open or lightly wooded terrain, including rocky hills, deserts, and coastal sand dunes. It often occurs also in cultivated areas and in or near farm buildings. It reaches a length of about 2.15 m. The ground color is reddish, buff, brown, reddish, or gray, and there are characteristic lateral diagonal stripes on the head; the posterior-most stripe extends to the angle of the mouth; the tail is ringed with broad black and grey or white stripes. The range extends from central Arkansas to southeastern California southward through most of Texas into northern Vera Cruz and southern Sonora with a relict population in Oaxaca, Mexico. The fangs are long and the venom produced is considerable. This species is extremely dangerous, often irritable, and sometimes strikes without warning; the bite can seriously damage tissue and is sometimes fatal.

croup. the portion of the back of a horse

near the root of the tail.

crow poison. *Zigadenus densus*.

crowfoot. *See Ranunculus*.

crown density. *See* density.

Crown land. land that is owned by the Crown, as in Canada and England. *Cf.* Federal land.

crown-of thorns. *Euphorbia milli*.

crucifer. **1:** any plant of the mustard family, Brassicaceae (formerly Cruciferae). *See* especially *Brassica*. **2:** pertaining to a crucifer (def. 1).

Cruciferae. *See* Brassicaceae.

crude. **1a:** unprocessed; unrefined. **1b:** raw; in a natural state. **2:** crude oil (plural, crudes). *See* petroleum.

crude arsenic. arsenic trioxide.

crude oil. petroleum.

cruels. actinobacillosis.

crumbles. crushed feed pellets.

crustecdysone. β-ecdysone. *See* ecdysone.

cry. **1:** to weep or sob. **2a:** to call, yell, shout, or scream. **2b:** to bellow, roar. **3:** a pack of hounds.

cryptohalite. ammonium hexafluorosilicate.

Cryptostegia (allamanda). a genus of tropical vining shrubs (Family Asclepiadceae) that are often grown as ornamental plants. ***C. grandiflora*** (rubber vine, purple allamanda, pink allamanda). a thick-stemmed toxic woody vine with thick glossy leaves. It is widely cultivated in India for rubber from the sap. It has become naturalized on roadsides and waste areas in southern Florida. It has caused fatalities in humans from severe gastroenteritis.

cuatro narices. *Bothrops atrox*.

cube. *See Lonchocarpus*.

cubé. common name for shrubs of the genus *Lonchocarpus*.

cucumber. *Cucumis sativus* and certain other plants of the family Cucurbitaceae.

cucumber family. Cucurbitaceae.

Cucumis sativus (cucumber). the common garden cucumber (Family Cucurbitaceae). Concentrations of nitrate in the foliage toxic to livestock have been reported.

cucurbit. any plant of the family Cucurbitaceae.

Cucurbita maxima (squash). a vine-like herb with several varieties; they are common garden vegetables (Family Cucurbitaceae). Concentrations of nitrate in the foliage that are toxic to cattle have been reported.

Cucurbitaceae (cucurbits; cucumbers; gourds). a family that includes many agriculturally and commercially important plants (e.g., cucumbers, watermelon, squashes, gourds, and various ornamentals). The sap of most species contains cucurbitacins, a class of highly poisonous triterpenes. *See* cucurbitacin.

culm. the stem of a grass or sedge. **malt culm**. the roots and shoots of a partly germinated barley plant. Malt culms are used for livestock feeds.

cultivar. a variety of a particular species of plant that was developed under cultivation. A cultivar does not occur in nature, but is nevertheless often thought of as subspecies.

cultivate. **1a:** to till and otherwise prepare the soil to grow crops. **1b:** to grow a crop.

cultivated aconite. *Aconitum napellus*.

cultivated barley. *Hordeum vulgare*.

cultivated carrot. *Daucus carota* **var.** *sativus*

cultivated grain sorghum. *See Sorghum vulgare*.

cultivated oats. *Avena sativa.*

cultivated rape. *Brassica napus*.

cultivated rye. *Secale cereale*.

cultivated soybean. *Glycine max*.

cumarin. coumarin.

cumulative action. cumulative effect (*See* effect).

cumulative effect. *See* effect.

cupric. pertaining to copper, especially the divalent ion.

cupric acetoarsenite. ((acetato)trimetaarsenitodicopper; copper ethanoacetoarsenate; Paris green; copper acetate arsenite; emerald green; French green; imperial green; mineral green; Mitis green; parrot green; Schweinfurt green; Vienna green). a bright blue-green, crystalline powder,

$$Cu(C_2H_3O_2)_2 \cdot 3Cu(AsO_2)_2$$

It is used as an insecticide and in antifouling compositions and wood preservatives. It is very toxic to laboratory rats. Toxicity is due mainly to the arsenic content. Symptoms of intoxication in humans include gastrointestinal disturbances, muscular cramps, tremors, and peripheral neuritis.

cupric arsenite (arsonic acid copper(2+) salt (**1**:1); arsenious acid copper(2+) salt (**1**:1); Scheele's green). a poisonous, yellowish-green, highly water-insoluble, crystalline powder,

$$(CuHAsO_3 \text{ or } Cu(AsO_2)_2)$$

of varying composition. Formerly used in wallpapers, it is now used chiefly as an insecticide, fungicide, and wood preservative.

cupric sulfate (copper sulfate; copper monosulfate; hydrous copper sulfate; sulfuric acid, copper(2+) salt (**1**:1)). a water-soluble, hygroscopic salt, $CuSO_4$, which occurs as grayish-white to greenish-white crystals or an amorphous powder; hydrous copper sulfate is a blue crystalline substance. Cupric sulfate is used in copper-plating baths and has various other commercial applications. It is used therapeutically as an antidote to phosphorus poisoning and as a topical antifungal. It is also used in veterinary practice as an emetic, anthelmintic, and as a feed additive in copper deficiency of ruminants. Copper sulfate is a powerful algicide that acts by inhibiting photosynthesis. Toxic fumes of SO_x are released on decomposition by heating. **anhydrous cupric sulfate**. a toxic, highly irritant, hygroscopic, grayish-white substance, $CuSO_4$, consisting of rhombic crystals or powder; it is water soluble but nearly insoluble in ethanol. It is used as a fungicide and to detect and remove water from organic compounds, especially alcohols. The monohydrate salt (dried cupric sulfate),

$$CuSO_4 \cdot H_2O$$

The monohydrate is hygroscopic, water-soluble, off-white powder; it is nearly insoluble in ethanol. The pentahydrate salt (blue copper; blue vitriol; bluestone; Roman vitriol; Salzburg vitriol),

$$CuSO_4 \cdot 5H_20$$

The pentahydrate is moderately toxic by ingestion and consists of large blue triclinic crystals, blue granules, or a light blue powder, all of which are soluble in water, methanol, and glycerol and are slightly soluble in ethanol. The pentahydrate has numerous uses, e.g., as an agricultural algicide, bactericide, fungicide (in Bordeaux mixture), and herbicide; it is also used in insecticidal mixtures, as a wood preservative, food and fertilizer additive, in paints and dyes, as a precursor of other copper salts, and in electroplating solutions. It is a strong irritant and systemic poison in animals. Ingestion by humans leaves a metallic taste and can cause pain in the mouth, esophagus, and stomach, vomiting, diarrhea, shock convulsions, paralysis, coma, and death. Lesions may include extensive capillary damage and kidney and liver damage. Toxic fumes of SO_x are released on decomposition by heating. *See also* poisoning (cupric sulfate poisoning).

cuprophyte. a plant adapted to, or having a high tolerance to, copper in the soil. Such plants are often used as an indicator of copper-containing soils.

cure. **1:** to heal; restore. **2:** to preserve by natural or artificial means (e.g., by drying, salting, smoking, pickling. **3:** a remedy, treatment, medication, or relief.

curing. **1a:** healing, restoring. **1b:** preserving, drying, pickling, smoking, or salting. **barn curing**. the curing of hay by use of natural or heated air. This frees the fields of hay and reduces the risk of damage to hay by inclement weather. **field curing**. the drying (curing) of hay in the field due by sunlight and warm, dry breezes and convection currents.

currant. **1:** currant, gooseberry; a plant of the genus *Ribes*. **2:** the fruit (currant, gooseberry) of a plant of the genus *Ribes*. **3:** dried, seedless raisons prepared and used chiefly in Greece from various varieties of *Vitis vinifera*. They are used, for example, in pastries and breads.

cut period. in forestry, the time interval between major harvests in the same stand.

cutis. skin.

cutis vera. dermis.

cutleaf nightshade. *Solanum triflorum*.

cutting cycle. in forestry, the planned, recurring time interval between successive harvests in a crop or stand of trees.

CVF. cobra venom factor (*See* factor).

CVP. chlorfenvinphos.

CWD. coarse woody debris.

cwt. hundredweight.

cyanazine (2-[[4-chloro-6-(ethylamino)-1,3,5-triazin-2-yl]amino]-2-methylpropanenitrile;2-[[4-chloro-6-(ethylamino)-*s*-triazin-2-yl]amino]-2-methylpropionitrile). a white solid used as a selective pre- and postemergence herbicide to control weeds in crops such as alfalfa, corn, cotton, sorghum, soybeans, and wheat. It is very toxic orally to laboratory rats and mice.

cyanic acid sodium salt. sodium cyanate.

cyanide. **1:** any of numerous, highly toxic, compounds (e.g., hydrogen cyanide (HCN), potassium cyanide (KCN), and sodium cyanide (NaCN)) derived from hydrogen cyanide that contains the CN group and forms hydrocyanic acid in water. Cyanide, in its various forms, is toxic by all routes of exposure. They are used in fumigants, soil sterilizers, fertilizers, and rodenticides. Livestock losses sometimes result from improper or malicious use of these compounds. **2:** the -CN radical or the CN^- ion. It is a cellular asphyxiant, inhibiting cytochrome oxidase, thus blocking cellular respiration. Cyanides occur naturally in many plants, usually in the form of glycosides. Cyanide is odorless but reputedly has the odor of bitter almonds. Benzaldehyde, a breakdown product of amygdalin, is the source of this odor which may be present briefly when amygdalin is the source of the cyanide. All cyanide antidotes (e.g., amyl nitrite) are themselves poisonous.

cyanide poisoning. poisoning by a cyanide compound. The most important cause of cyanide poisoning among domestic animals is ingestion of such plants as arrow grass (*Triglochin spp.*), Johnson grass (*Sorghum halepense*), Sudan grass (*S. sudanense*), common sorghum (*S. bicolor [vulgare]*), wild black cherry (*Prunus serotina*), chokecherry (*P. virginiana*), pincherry (*P. pensylvanica*), and flax (*Linum usitatissimum*). These plants contain cyanogenetic glycosides that release hydrocyanic (prussic) acid when attacked by acids such as the hydrochloric acid found in the stomach of many animals. The cyanogenetic glycoside content of the plant is increased by heavy nitrate fertilization, wilting, trampling, and plant disease. Very young, rapidly growing plants contain greater quantities of the glycoside. Spraying of cyanogenetic plants with herbicides may enhance the toxic hazard.

Ruminants are more susceptible than monogastric animals, and cattle more so than sheep. All antidotes for cyanide poisoning are themselves poisonous.

cyanide poisoning of herbivores. a fairly common disease of herbivorous animals, caused by ingestion of cyanogenic plants or plant parts. The cyanide is usually present in glucosides which are hydrolyzed in the alimentary canal, yielding hydrocyanic acid. Some farm chemicals such as fungicides or insecticides may cause cyanide poisoning. Hydrogen cyanide and its salts are extremely poisonous to *Homo sapiens* and a wide array of terrestrial and aquatic animals.

cyanide poisoning of livestock. cyanides are used in fumigants, soil sterilizers, fertilizers, and rodenticides. Livestock losses sometimes result from improper or malicious use of these products. The most important cause of cyanide poisoning among domestic animals is ingestion of such plants as arrow grass (*Triglochin spp.*), Johnson grass (*Sorghum halepense*), Sudan grass (*S. sudanense*), common sorghum (*S. bicolor [vulgare]*), wild black cherry (*Prunus serotina*), chokecherry (*P. virginiana*), pincherry (*P. pensylvanica*), and flax (*Linum usitatissimum*). These plants contain cyanogenetic glycosides that yield hydrocyanic acid when hydrolyzed by enzymes during digestion. The cyanogenetic glycoside content of these plants is increased by heavy nitrate fertilization, wilting, trampling, and plant disease. Very young, rapidly growing plants contain greater quantities of the glycoside. Spraying of cyanogenetic plants with herbicide may increase the toxic hazard. Cattle are more susceptible than sheep, and ruminants are generally more susceptible than monogastric animals.

cyanihydric acid. hydrogen cyanide.

cyanobacteria (singular, cyanobacterium). a common collective name for members of the phylum Cyanobacteria.

Cyanobacteria (blue-green bacteria; formerly blue-green algae; Cyanophyceae; Cyanophyta; Myxophyta). a homogeneous group of closely related unicellular, prokaryotic, photosynthetic, colonial or filamentous nonmotile microorganisms. Formerly considered algae, they are now usually considered a phylum of the Kingdom Monera. Most occur in surface water bodies or in moist terrestrial situations as free-living or symbiotic associates of lichens or fungi. Some freshwater forms such as *Anabaena flos-aquae*, *Aphanizomenon flos-aquae*, and *Microcystis aeruginosa* ("Annie, Fannie and Mike") are extremely poisonous. These species occur worldwide in warm, calm, nutrient-rich waters where they sometimes form toxic blooms, *q.v.* Human intoxication is rare, but dogs, cats, domestic fowl, livestock, fish, and various wildlife species have sometimes been killed in large numbers by drinking water heavily populated with toxic blue-green bacteria. Dangerously toxic waters usually look like a thin blue or grayish paint. Livestock have been photosensitized from drinking water containing high concentrations of cyanobacteria. The toxic principles of this phylum have not been identified in natural populations (but *See Microcystis aeruginosa*). Toxic concentrations of nitrate have been reported in certain forms.

cyanobacterium. any microorganism of the phylum Cyanobacterium.

cyanoethylene. acrylonitrile.

cyanogenesis. the production of or elaboration of cyanogen or hydrocyanic acid as in the synthesis of cyanogen glycosides by plants such as *Prunus lawrocerasus* (cherry laurel), *Lotus corniculatus* (birdsfoot trefoil), and *Trifolium repens* (white clover).

cyanogenetic (cyanogenic; cyanogenous). **1a:** pertaining to cyanogenesis. **1b:** capable of producing cyanogen or hydrocyanic acid. **2:** pertaining to or denoting living organisms that produce hydrocyanic acid or its immediate precursors.

cyanogenic. cyanogenetic.

cyanogenous. cyanogenetic.

α-cyano-3-phenoxybenzyl-2-(4-chlorophenyl)-3-methylbutyrate. fenvalerate.

(±)-α-cyano-3-phenoxybenzyl-(±)-*cis,trans*-3-(2,2-dichloro-vinyl)-2,2-dimethylcyclopropane carboxylate. cypermethrin.

(*S*)-α-cyano-3-phenoxybenzyl-(1R)-*cis*-3-(2,2-dibromovinyl)-2,2-dimethylcyclopropane carboxylate. deltamethrin.

(*R,S*)-cyano(3-phenoxyphenyl)methyl(*R,S*)-*cis,trans*-3-(2,2-dichloroethenyl)-2,2-cyano-(3-dimethyl-cyclopropane carboxylate. cypermethrin.

cyano-(3-phenoxyphenyl)methyl 4-chloro-α-(1-methylethyl)benzeneacetate. fenvalerate.

cyano(3-phenoxyphenyl)methyl ester. deltamethrin.

Cyanophyceae. Cyanobacteria.

Cyanophyta. Cyanobacteria.

cycad. any plant of the family Cycadaceae, especially those of the genus *Cycas*.

Cycadaceae (cycads; the cycad family). a family of palm-like, but nonflowering plants, that bear cones. *See* Cycadales, *Cycas*.

Cycadales. an order of gymnosperms that includes the Cycadaceae. Plants of this order are thought to cause defects in calves of domestic cattle that are similar to those of crooked calf disease. *See also* *Cycas*.

Cycas (cycad; sago palm). a genus of palm-like tropical plants (Family Cycadaceae) the trunks of which may reach a height of more than 6 m and a diameter of 75 cm. The seeds (nuts) of the following two species contain a toxic glucoside, cycasin, *q.v.* ***C. circinalis*** (queen-sago; fern palm; false sago palm; crosier cycas). a sago palm with leaves up to 2.6 m. ***C. revoluta*** (king-sago). a sago palm with leaves that are usually less than 1.3 m long. Flour made from the nuts of these plants are associated with hepatotoxicity, carcinogenicity, teratogenesis, and neurotoxicity. *See also* cycasin.

cyclane. alicyclic hydrocarbon.

cyclical lactone. cytochalasin.

Cyclo-Werrol. hypericin.

cyclodiene insecticide (chlorinated cyclodiene insecticide). any of a class of chlorinated insecticides based on a cyclodiene ring structure. Included are aldrin, dieldrin, endrin, chlordane, heptachlor, endosulfan, and isodrin. All are highly toxic with similar effects. The United States has banned the use of aldrin and dieldrin.

***N*-cyclohexylcyclohexanamine**. dicyclohexylamine

3-cyclohexyl-6,7-dihydro-1*H*-cyclopentapyrimidine-2,4-(3*H*,5*H*)-dione. lenacil.

3-cyclohexyl-1,5,6,7-tetrahydro-2*H*-cyclopentapyrimidine-2,4(3*H*)-dione. lenacil.

cyclohexyl-5,6-trimethyleneuracil. lenacil.

3-cyclohexyl-5,6-trimethyleneuracil. lenacil.

cyclopiazonic acid. an indole mycotoxin produced by *Penicillium cyclopium* and related species and by several species of *Aspergillus*. It is a common contaminant of cheese, corn, and peanuts and occurs widely in mammalian tissues (notably voluntary muscle). It is used as an antifungal agent. It is extremely toxic to laboratory mice and rats, ducklings, and calves of domestic cattle. Effects vary among mammalian species. It is myotoxic, hepatotoxic, nephrotoxic, and cytotoxic to laboratory mammals. Symptoms of intoxication can include, for example, catalepsy, hypothermia, hypokinesia, incoordination, ptosis, sedation without loss of righting reflex, tremor, and convulsions.

Cyclosan. hypericin.

Cyclostomata. a subclass of the vertebrate

class Agnatha, the lamprey eel and hagfishes.

cyclostome. a fish of the Cyclostomata.

cyme (cymose inflorescence). a type of inflorescence in which the tip of the main axis forms a flower and growth is continued by a lateral bud, which also forms a flower, growth continuing by a bud on its axis. The oldest flowers of the inflorescence are thus at the top of the plant and the youngest are at the bottom. **dichasial cyme** (dichasium). a cyme in which in which two lateral buds continue the growth of each branch. **monochasial cyme** (monochasium). a cyme in which only one lateral bud continues the growth of each branch. *Cf.* inflorescence (racemose inflorescence).

cymose inflorescence. cyme.

Cynodon dactylon (Bermudagrass; scutch grass; bahama grass; devil grass; doob). a perennial forage grass (Family Gramineae) of the southern United States, it is considered a weed in cultivated fields. It is also used to seed lawns and has produced several hybrids.

cypermethrin (3-(2,2-dichloroethenyl)-2,2-dimethylcyclopropanecarboxylic acid cyano(3-phenoxyphenyl)methyl ester; (±)-α-cyano-3-phenoxybenzyl-(±)-*cis,trans*-3-(2,2-dichloro-vinyl)-2,2-dimethylcyclopropane carboxylate; (*R,S*)-cyano(3-phenoxyphenyl)methyl(*R,S*)-*cis,trans*-3-(2,2-dichloroethenyl)-2,2-dimethyl-cyclopropane carboxylate). a viscous, semisolid, synthetic type II pyrethroid, used effectively as a stomach and contact insecticide on a wide variety of insects and agricultural pests. It is insoluble in water, but soluble in acetone, methanol, methylene dichloride, and xylene. It is used in veterinary medicine as an ectoparasiticide. Cypermethrin is neurotoxic, depolarizing the neurone by increasing the permeability of the neuronal membrane to sodium. It is not environmentally persistent, however, and is no more than moderately toxic to most vertebrate animals tested. Effects of intoxication on humans include gastroenteritis, nausea, vomiting, diarrhea, paresthesia of the tongue and lips, hyperexcitability, incoordination, syncope, convulsions, muscular paralysis, collapse, and death due to respiratory paralysis.

cypress spurge. *Euphorbia cyparissias*.

Cyprinus carpio (common carp). a medium to large, robust, moderately depressed, bony fish (Family Cyprinidae) native to Asia, but now introduced and even cultured in warm and temperate waters of the world. The milt of this fish is the source of a toxic substance, cyprinin.

Cypripedium hirsutum (lady's slipper; moccasin flower). a genus of plants (Family Orchidaceae) indigenous to temperate portions of Asia, Europe, and North America. The hairs of stems and leaves are toxic, causing a contact dermatitis. The active principles are fatty acids.

cypromid. an amide herbicide. Applied to the foliage, amide damage takes the form of localized or general necrosis, depending on the dose applied. Cypromid also inhibits growth, probably as a secondary effect.

Cythion. malathion.

cytisine (1,2,3,4,5,6-hexahydro-1,5-methano-8H-pyrido-[1,2-a][1,5] ciazocin-9-one; baptitoxine; sophorine; ulexine). a bitter, highly toxic crystalline alkaloid found in *Laburnum anagyroides* [= *Cytisus laburnum*] and many other members of the family Fabaceae. Ingestion may produce nausea, vomiting, convulsions, frothing at the mouth, extreme drowsiness, and coma. Cytisine was formerly used as an antiemetic and antitussive agent. *See also* cyticism and broom poisoning. *See also* broom poisoning, cyticism, sparteine.

cytisism. poisoning caused by *Laburnum anagyroides* [= *Cytisus laburnum*] or cytisine. The condition is marked by a burning in the mouth and pharynx, thirst, nausea, vomiting, diarrhea, prostration, and an irregular pulse. Symptoms may also include visual disturbances, aphasia, delirium, and unconsciousness. Death from respiratory paralysis may occur.

Cytisus (broom; cytisus). a genus of stiff or spiny deciduous and evergreen shrubs (Family Fabaceae, formerly Leguminosae) that are native to Europe, western Asia, and northern Africa, but some are widely cultivated elsewhere. They have showy racemose flowers, a two-lipped calyx, and grow to a height of 3m. ***C. laburnum*** (laburnum, golden chain). the leaves and seeds are toxic, producing effects similar to nicotine: nausea, intense vomiting, dysphagia, severe vomiting, diarrhea, prostration, irregular pulse and respiration, delirium, twitching and coma; renal damage may occur. The active principle is cytisine. ***C. scoparius***. *Sparteum junceum*.

cytisus. any plant of the genus *Cytisus*.

cyto-. a prefix indicating relationship to or denoting a cell.

cytochalasin (cyclical lactone). any of a large class of cytotoxic macrolide mycotoxins with a highly substituted hydrogenated isoindole ring fused to a macrocyclic ring (a carbocyclic or a lactone) comprised of 11-14 atoms and is a carbocycle or a lactone. Cytochalasins were originally isolated from *Helminthosporium dematioideum*, *Metarrhizium anisopliae*, and *Rosellina necatrix*, but are now known from a wide variety of *Phoma*-type fungi. Cytochalasins have various unusual effects on animal cells. They interfere with cytokinesis, for example, by inhibiting the formation of contractile microfilaments. The resultant cell is multinucleate. They also induce extrusion of cell nuclei and reversibly inhibit cell movement. They reportedly inhibit such processes as glucose transport, phagocytosis, aggregation of platelets and clot contraction, release of growth hormone, and thyroid secretion. *See also Phoma*. Cytochalasins are used in medical research to elucidate cellular mechanisms, especially that of microfilaments.The toxicology of most of the more than 20 known cytochalasins is poorly known, and most information is based on *in vitro* studies. *See also* macrolide mycotoxins. **cytochalasin B** (7,20-dihydroxy-16-methyl-10-phenyl-24-oxo-[14]cytochalasa-6(12),13,21-triene-1,23-dione; (E,E)-16-benzyl-6,7,8,9,10,12a,-13,14,15,15a,16,16-dodecahydro-5,13-dihydroxy-9,15-dimethyl-14-methylene-2*H*-oxycyclotetradec-[2,3-d]isoindole-2,18(5H)-dione; phomin). initially obtained from *Phoma spp*. as phomin, it accounts for the phytotoxicity of *P. exigua* and others. *See also Phoma*. It has been extensively studied biologically and is used mainly as a tool in cytological research. It affects a number of cellular processes *in vitro*, including cytokinesis, single-cell movement, morphogenesis of epithelia, and contractile processes such as that of mammalian smooth muscle and cardiac cells.

cytochalasin D. a mycotoxin from *Chaetomium globusum*. It is extremely toxic (*s.c.*) to laboratory mice. **cytochalasin E**. extremely toxic to laboratory rats and guinea pigs. Acute intoxication (*i.p* or *p.o.*) of laboratory rats may cause ataxia, drowsiness, cyanosis, and coma. Death occurs within a few hours from circulatory collapse due to rapid, massive extravasation of plasma. Brain edema, pulmonary hemorrhage, injury to the vascular wall, and congestive degenerative changes and necrosis of liver, kidney, spleen, and small intestines are sometimes reported. **kodo-cytochalasin-1**. a species that is extremely toxic to laboratory mice.

cytochrome. any of a class of heme-containing conjugated protein pigments involved in electron and/or hydrogen transport in cells due to a reversible change in valency of heme iron. They are widely distributed in plant and animal tissues. The four main types are cytochromes *a* (*See* cytochrome c oxidase), *b*, *c*, and *d*. They transfer electrons from flavoproteins to molecular oxygen. This is effected by the porphyrin component that contains a chelated iron ion which reversibly changes state between the oxidized (ferric, +3) and reduced (ferrous, +2) oxidation states.

cytochrome *c* oxidase (cytochrome *a*). a cytochrome that catalyzes the transfer of electrons from cytochrome oxidase (cytochrome a_3) to oxygen in respiration. *See* cytochrome oxidase.

cytocide. an agent that destroys cells.

Cytodites nudus (air sac mite; pulmonary mite). a mite (Order Acarina) that infests the external respiratory system birds including the air sac system.

cytokinin (phytokinin, kinin). **1:** any of a naturally occurring class of plant growth promoters. They are derivatives of the base adenine. They stimulate cell division in the presence of auxin, induce flowering in some plants, delay senescence, and induce the breaking of dormancy in axillary buds and some seeds. Unusually high endogenous concentrations of cytokinins occur in plant diseases such as crown gall and witches broom. **2:** synthetic cytokinins, some of which are substituted phenylureas. *See also* growth inhibitor, mevalonic acid, tumor-inducing principle.

cytologist. a scientist whose field of study is cytology.

cytology. the study of the structure, physiology, development, reproduction, and life history of cells, especially by means of light or electron microscopy.

cytosol. cytoplasm exclusive of the mitochondria and endoplasmic reticulum.

cytosolic. of, pertaining to, or contained within the cytosol.

D

δ. delta.

Δ. delta.

D&C Blue No. 4. FD&C Blue No. 1

deltamethrin (3-(2,2-dibromoethenyl)-2,2-dimethylcyclopropanecarboxylic acid cyano(3-phenoxyphenyl)methyl ester; (*S*)-α-cyano-3-phenoxybenzyl-(1R)-*cis*-3-(2,2-dibromovinyl)-2,2-dimethylcyclopropane carboxylate; decamethrin; esbecythrin). an extremely toxic, synthetic, broad-spectrum contact and stomach pyrethroid insecticide and agricultural pesticide. Acute oral toxicity to vertebrates appears to be generally low. Effects on humans are those of acute gastrointestinal irritation, with numbness of the lips and tongue, hyperexcitability, syncope, incoordination, muscular paralysis, convulsions, and death from respiratory collapse.

2,4-D ((2,4-dichlorophenoxy)acetic acid). a synthetic auxin used as an herbicide. It selectively poisons broad-leaved plants, upsetting their growth and metabolism, causing them to overgrow and die. It also interferes with biochemical pathways that process nitrates in some plants with consequent accumulation of nitrate. 2,4-D is a constituent of Agent Orange. It is very toxic and an irritant to the eye and gastrointestinal tract. Symptoms of intoxication in humans typically include weakness, lethargy, diarrhea, spastic myotonia, ventricular fibrillation, and cardiac arrest; and sometimes hypermetabolism, hyperpyrexia, convulsions, and coma. Large doses or long-continued exposure can produce severe, protracted peripheral neuropathy and sometimes brain damage in humans. *See also* chlorophenoxy compounds, nitrate poisoning.

D plus rule. a general rule for thinning of crop trees of various types. In each case, a specified number is added to the diameter of the tree at breast height. A D+4 rule, for example, indicates that a 16 inch tree needs 16+4 or 20 feet of growing space for optimal growth.

D ring. a piece of metal in the shape of a D or 0 that links two separate straps or other parts of a harness.

d. deci-. Do not confuse with da.

da. deca-. Do not confuse with d.

dacry-, dacryo-, decry-. prefixes indicating or pertaining to tears or to the lacrimal sac or duct.

dactinomycin (actinomycin D). a cytotoxic polypeptide isolated from the bacterium *Streptomyces parvullus*, used in the treatment of certain cancers.

daffodil. common name of a number of hardy species *Narc Narcissus* that bear a single, large, trumpet-shaped flower. This term is often used incorrectly to refer to any species of hardy narcissus.

dairy. **1a:** a building on a farm used to cool milk prior to release from the farm. **1b:** dairy farm; a farm or section of a farm devoted to the production and marketing of milk. **3:** a company that receives milk from farms, packages it, and delivers it to customers within the region. **3b:** a company that processes milk to produce commodities such as cheese and butter. **2:** a store that sells milk and milk products.

dairy character. the aspect of a cow that presents evidence of milk-producing ability. *See also* bloom (def. 5).

dairy cow. a domestic cow, *Bovis*, bred and used for the production of milk for human consumption or one that is used to raise replacement dairy heifers.

dairy farm. *See* dairy.

dairy goat breed. any group of dairy goats that have a common genetic origin and identifiable (breed specific) traits (e.g., color pattern). United States breeds include Alpine, LaMancha, Nubian, Oberhasli, Saanen and Toggenberg.

dairy herd improvement (DHI). the improvement of dairy herds especially with respect to milk production and profitability. This is accomplished by use of an official DHI testing plan that requires supervision and compliance with all official DHI rules. *See* Animal Improvement Programs Laboratory; Dairy Herd Improvement Association; Dairy Herd Improvement Registry.

dalapon (2,2-dichloropropionic acid; α,α-dichloropropionic acid). a chlorinated aliphatic acid resulting from the substitution of one methyl group by a chlorine atom of TCA. The sodium salt is used as an herbicide to selectively control annual and perennial weed grasses on arable land. It is absorbed through roots and leaves. Dalapon is slightly toxic to laboratory mice. Dalapon and other halogenated aliphatic acids inhibit plant growth, induce leaf chlorosis, and at higher concentrations cause leaf necrosis. They affect many plant processes.

dallis grass. *Paspalum dilatatum*.

Dalmatian insect powder. pyrethrin.

dam. **1:** a barrier, barricade, or obstruction, as a dam across a river or stream that acts as a barrier. **2:** to block, check, barricade, or restrain. **3:** a female parent, e.g., of a horse and certain other mammals. **4:** the female parent of any of a number of domestic and wild mammals including horses, asses, dairy goats, and cows. **donor dam**. a genetically superior dam from which embryos are removed and transferred to recipients who enter gestation and bear the young. This permits superior dams to produce more offspring than normal.

damage. **1:** harm or injury with impairment; the harm, detriment or loss sustained as the result of an injury. **2:** the loss or temporary or permanent impairment of an organ or bodily part. *Cf.* harm, injury, trauma.

Damalinia. *See Bovicla*.

daminozide (ALAR; butanedioic acid mono-(2,2-dimethylhydrazide); *N*-(dimethylamino)succinamic acid; succinic acid 2,2-dimethylhydrazide). a pesticide used to delay ripening of apples, Concord grapes, other fruit, and sometimes peanuts. It reddens and firms apples, making them less likely to drop before growers are ready to pick them. Daminozide is carcinogenic to laboratory mice and rats and perhaps to humans. Residues of this compound remain on the treated produce until purchase.

damping-off. any of a number of fungal diseases of seedlings that is contracted either before or immediately after emerging from the soil. It is characterized by rotting of the stem at the soil level and eventual collapse of the entire plant.

danger tree. hazardous tree.

Dasypodidae. a family of 21 species of New World armored, burrowing, chiefly nocturnal mammals. While insects form the bulk of the diet, armadillos are omnivorous. They have strong claws and the head and body are covered by a carapace that consists of a horny material and bony plates. This body armor hangs down about the body and is divided into flexible bands across the back. The teeth are peglike teeth and lack roots and enamel. Armadillos are found from Argentina to Panama; one species reaches the southern United States. In some regions where they occur, they are used as food.

Dasyproctidea. the agouti family. *See Agouti*.

Dasypus novemcinctus (nine-banded armadillo; peba; Texas armadillo). the only species of armadillo (Family Dasypodidae) found in the United States, it occurs from Texas and Louisiana to Argentina. It is characterized by the presence of nine hinged bands of bony plates. It weighs up to about 6.4kg (15 lb) and is about 76 cm (30 in) long and 15 cm (6 in) tall at the shoulder. It is used as food in some areas where it occurs. Females usually give birth to identical quadruplets.

datura. **1a:** a plant of the genus *Datura*. **1b:** jimsonweed seed. It is a toxic anticholinergic. *See* poisoning (datura poisoning).

datura poisoning (daturism). poisoning due to ingestion of plants of the genus *Datura*. All classes of livestock have been poisoned, and human poisoning is not uncommon. Poisonings of humans are usually due to accidental ingestion of seeds. about 4-5 g of seeds or leaves may prove fatal to a child. Symptoms vary greatly among victims. Initial symptoms include a dry mouth, intense thirst, disturbances of vision, flushed skin, and hyperexcitability. Fever, tachycardia, incoherence, delirium, and even violent behavior may follow. In severe cases, effects may include CNS depression, respiratory depression, and circulatory failure. Fatal poisonings are marked by convulsions, coma, and death; early stages of poisoning are sometimes bypassed.

Datura (datura; thornapple; jimsonweed; apple of Peru; Jamestown weed; tolguacha). a genus of narcotic, poisonous, warm-climate annual or perennial plants (Family Solanaceae) with large, showy flowers. They contain several tropane alkaloids in high concentration (ca. 0.25-0.7%), the most important being atropine, hyoscyamine, and hyoscine (scopolamine). Toxic concentrations of nitrate have also been reported. All classes of livestock have been poisoned, and human poisoning is not uncommon. Most accidental poisonings are due to ingestion of seeds. Approximately 4-5 g of seeds or leaves may fatal to a child. Effects vary greatly. Initial symptoms include a dry mouth and intense thirst, disturbed vision, flushed skin, and hyperexcitability. Pyrexia, tachycardia, incoherence, delirium, and even violent behavior may follow. In fatal poisoning, convulsions, coma, and death ensue, and early stages are sometimes bypassed. ***D. metel***. a source of scopolamine. ***D. stramonium*** (thornapple; jimsonweed; stramonium; Jamestown weed; stinkweed; devil's apple; devil's trumpet; mad apple). an extremely dangerous, subcosmopolitan annual weed with a height of about 2 m. It is distributed widely throughout the United States on gravelly or rich alluvial soils. This plant probably causes more poisonings of humans than any other plant. The fruit is extremely poisonous and hallucinogenic when ingested. Some individuals suffer an allergic contact dermatitis. It is used by drug cultists, sometimes with lethal results. Children have been poisoned by the seeds, nectar, or the fruits that contain the seeds. Animals sometimes eat the foliage. Early symptoms of intoxication in humans are a dry throat, gaiety, laughter, and mild delirium. Pupillary enlargement may appear early and persist for several days in recovering individuals. Further symptoms may include dimness of vision, pupillary dilation, giddiness, delusions, delirium and sometimes mania, unconsciousness, and even death. This plant is teratogenic and a cause of arthrogryposis in livestock. ***D. suaveolens***. Effects similar to those of *D. stramonium*, *q.v.* Used by drug cultists, sometimes with lethal results.

daturism. datura poisoning.

Daucus carota* var. *sativus (cultivated carrot). a biennial herb and common garden vegetable and food crop (Family Apiaceae). The root is normally edible but may contain toxic concentrations of nitrates when growth has been especially rapid. This sometimes occurs also in the above-ground portions of the plant. Livestock have been poisoned on occasion, apparently by nitrates, when feeding on the leaves.

DBCP. dibromochloropropane.

1,2-DCE. ethylene dichloride.

DCPA. chlorthal-dimethyl.

***p,p'*-DDD**. 1,1-dichloro-2,2-bis(*p*-chlorophenyl)ethane.

DDE. dieldrin.

DDT (1,1′-(2,2,2-trichloroethylidene)bis[4-chlorobenzene]; 1,1,1-trichloro-2,2-bis(*p*-chlorophenyl)ethane; β,β-bis(*p*-chlorophenyl)-β,β,β-trichlorethane; chlorophenothane; clofenotane; dichlorodiphenyltrichloroethane; dicophane; gesarol; pentachlorin; *p,p'*-DDT, etc.). a white, colorless, odorless, water-insoluble, crystalline solid, DDT is a polychlorinated, lipophilic, environmentally persistent contact pesticide, larvicide, ectoparasiticide, and topical pediculicide that is toxic to many vertebrate animals. The *p,p'*-isomer is weakly estrogenic and has the greatest toxicity in insects; the *o,p'*-isomer is less toxic and more estrogenic. DDT is neurotoxic and can cause serious damage to the CNS It has been a serious cause of reproductive failure in birds and fish. It is not generally considered to be highly toxic to mammals, but note that DDT is very toxic to adult laboratory rats, while only moderately toxic to newborns. Humans can be exposed by ingestion or by absorption through the skin or respiratory tract. Acute toxicity is due mainly to the direct action of DDT on axonal membranes to increase sodium conductance with a consequent increase in excitability that can result in multiple impulses, tremors, and tetanus. Acute effects in humans include vomiting (which may be delayed), numbness about the lips, tongue and face, headache, sore throat, weakness, tremors of the head and neck muscles, ataxia, tonic and clonic convulsions, coma, and death from respiratory or cardiac collapse within 2-24 hours. There are no lesions. Chronic exposure can cause muscular weakness, dermatitis, liver damage, CNS degeneration, agranulocytosis, convulsions, coma, and eventual death. It is carcinogenic in a number of animals; it produces liver tumors in laboratory mice, but apparently not in humans. The acute dermal toxicity to humans is low. Largely because of its persistence in the environment, accumulation in fat, and induction of eggshell thinning in predatory birds, the United States Environmental Protection Agency cancelled all registrations and interstate sales of DDT in 1973. *See also* DDD, DDT dehydrochlorinase, dieldrin, eggshell thinning.

***p,p'*-DDT**. DDT.

DDVP. dichlorvos.

de-. a prefix meaning away from, cessation, or without. It sometimes has an intensive force and, often, a privative or negative connotation.

de minimus. the legal doctrine that states that the law does not deal with insignificant things. It is applied, for example, to the use of chemicals (e.g., some fertilizers, some herbicides) that are of very low risk.

deactivate. *See* inactivate.

deactivation. *See* inactivation.

dead. **1:** deceased, numb, lifeless, inert, or inanimate. **2:** pertaining to an organism or part of an organism having ended a state of life; no longer living. **3:** having the appearance of death. **4:** devoid of life. **5:** the dead; that which is dead (def. 1).

dead furrow (blind furrow). a double furrow made by a plow by throwing the two furrow slices to the side in opposite directions.

dead head. **1:** a sunken or partially sunken log in a body of water. **2:** blighted or dwarfed seed stalks that produce low yields of poor quality seed. **3:** deadhead; to move to a known or specified location without the aid of maps or direction-finding tools.

dead lake. an aged lake, one that has high amounts of bottom sediment and is usually overgrown with aquatic and emergent plants.

dead man. a heavy weight such as a log or stone that is partly or entirely buried in the ground to serve as an anchor.

dead mouth. **1:** a horse whose mouth is not sensitive to the bit or reins. **1a:** descriptive of a horse that no longer responds to direction by use of the rein and bit.

dead nettle. *Lamium amplexicaule*.

dead ripe. descriptive of fruit or grain, for example, that is completely ripe.

dead water. **1:** motionless surface water. **2:** the eddy immediately behind the stern of a boat.

dead weight. **1:** the sensible weight of a

limp body such as a sack of flour or a dead animal. **2:** a slaughtered animal with all of the viscera removed.

deadly agaric. *Amanita phalloides*.

deadly amanita. *Amanita phalloides*.

deadly cort. a common name of the deadly mushroom, *Cortinarius gentilis*

deadly galerina. *Galerina autumnalis*.

deadly hemlock. *Conium maculatum*.

deadly lawn galerina. *Galerina veneneta*.

deadly nightshade. *Solanum nigrum*.

deaf ear. **1:** a scab of cereal grains. **2:** the fold in the skin immediately below the ear of a domestic fowl.

death. **1:** demise; extinction; the cessation of life of a living system or a part thereof (e.g., an organism, appendage, population, community, species). Death may be a gradual process, not all parts of a system dying at the same time. The death of an organism is thus usually taken either as a point where the integrity of the individual fails or at a point where the process of dying is thought to be irreversible. The death of a cell or microorganism is usually followed by rapid dissolution. **2:** the permanent loss of the normal state of dynamic equilibrium typical of a live organism. **3:** the permanent cessation of all vital functions of a living organism or part of a living organism. **4:** that point or phase in the process of dying where the integrity of an individual fails or at a point where the process of dying is thought to be irreversible. **5:** the cause of dying.

death camas. a common name of plants of the genus *Zigadenus*.

death camass. variant of camas.

death cap. *Amanita phalloides*.

death loss. the loss of livestock through death from whatever cause as opposed to theft, straying, sales or other such causes.

dB. symbol for decibel.

debauchure. deboucher.

debeaker. a device used to cut off the tip of the beak of young domestic fowl, turkeys, and pheasants to limit injuries from pecking other fowl, cannibalism, and the breaking and eating of eggs.

debilitant. **1:** able to weaken or debilitate. **2:** a medicine used to reduce excitement. **3:** that which weakens or debilitates (e.g, a pathogen).

debilitating. **1a:** weakening, enfeebling, devitalizing, or crippling. **1a:** said of a trauma or disease that is enfeebling, crippling, or damaging. A degree of irreversibility is often implied. **1b:** denoting an agent such as a poison or pathogen that causes enfeeblement, or a serious decrement in health or performance of an individual, often to the point of helplessness.

debilitation. debility, enfeeblement, weakness, or severe loss of condition.

debility. **1:** a disability, frailty, impairment, or infirmity. **1b:** illness or sickliness. **2:** weakness or loss of tonicity of organs or of bodily functions.

deblossom. to remove blossoms from a plant to increase vegetative growth.

deboucher (debauchure). **1:** the mouth of a river or stream. **2:** the site where an underground stream or a spring issues forth above ground. **3:** the juncture between a tubular passages and a larger passage or chambers.

debris. trash; rubble; waste. **land clearing debris**. a forestry term meaning yard waste, prunings, and/or stumps six inches or greater in diameter that result from land-clearing operations.

debris cone. a cone-shaped or fan-shaped deposit of soil, sand, gravel, and boulders deposited by a mountain stream where it slows as when entering the floor of the mountain.

debris guard (debris screen). a grating at the intake of a stream or channel to protect the downstream side from debris.

debris dam. a barricade built across the channel of a stream in order to halt the flow of debris.

debris screen. debris guard.

deca-, **deka-**. Prefix meaning ten in the International System of Units and the metric system.

1,1a,3,3a,4,5,5a,5b,6-decachloro-octahydro-1,3,4-metheno-2*H*-cyclobuta[*cd*]-pentalen-2-one. chlordecone.

decaliter. ten liters.

decamethrin. deltamethrin.

decapitate. **1:** to cut off the head of; to behead. **2:** to severely prune back the branches of a tree or bush as an aid to rejuvenation.

decay (organic decay). **1:** to decompose; molder, putefry. **2:** (moldering, putrefaction) the decomposition of organic matter, with nearly total oxidation. **3:** a term in atmospheric chemistry that denotes the rate or course of disappearance of a pollutant (e.g., as the result of absorption or chemical reaction at the earth's surface, removal by precipitation, or by chemical transformation into another substance). *See also* decompose.

deci- (d). a prefix to units of measurement that indicates a quantity of one-tenth (10^{-1}) the value indicated by the root term in the metric system and the International System of Units.

decibel (dB): a unit of sound measurement. In general, an increase of ten decibels indicates a doubling in loudness.

decidua. the endometrium during pregnancy, including the ectopic tissue that surrounds the fertilized mammalian ovum. It is a thick, highly modified tissue. It is shed and eliminated from the uterus together with the placenta, following parturition. *See also* endometrium.

deciduous. **1:** subject to shedding or falling off seasonally or during a specific stage or interval of growth as in the case, for example, of antlers, hair, teeth, deciduous placenta, and leaves. **2:** describing, designating, or pertaining to perennial plants that shed leaves seasonally and are normally leafless for some time during the year. **3:** pertaining to the leaves shed by a deciduous plant. *Cf.* evergreen. **4:** ephemeral; transitory; impermanent; not lasting.

deciduous forest. a forest composed chiefly or exclusively of deciduous trees. Deciduous forests usually have a more complex structure than evergreen forests. They produce valuable hardwood timber such as teak and mahogany from the tropics and oak, maple, and beech from cooler areas.

deciduous plant. a plant that loses its leaves at certain stages of growth, under certain environmental conditions, and/or periodically, often annually.

deciduous tree. a tree that sheds its leaves at a certain season and remain bare for some time. In tropical areas the leaves drop during the hot season with consequent conservation of moisture. Deciduous trees of cool climates shed leaves during the autumn; this offers protection against the cold and frost of winter.

decistere. 0.1 cubic meter (3.5315 cubic feet).

declivity. **1:** a descending slope from a point of reference (e.g., from the peak of a hill). **2:** a downward deviation from the horizontal. *Cf.* acclivity.

decompose. **1:** to separate or resolve a compound into simpler compounds and/or elements. **2:** to bring about chemical disintegration of organic matter. **3:** to disintegrate or decay.

decomposer. a heterotrophic organism (e.g., a bacterium, fungus) that metabolically decomposes nonliving organic matter into simpler organic and inorganic compounds and/or elements. *See* decomposition.

decomposition. **1:** the process or condition of being decomposed. **2:** the separation or resolution of a compound (organic or inorganic) into simpler compounds and/or elements. **3a:** decay, disintegration, or putrefaction of material of biologic origin. **3b:** the metabolic (enzymatic) conversion of organic matter into simple organic and inorganic compounds, usually by the agency of bacteria or fungi and with the liberation of energy. **aerobic decomposition**. decay and decomposition of organic matter by microorganisms in the presence of molecular oxygen, usually by aerobic bacteria and fungi. **anaerobic decomposition**. decomposition of organic matter due to the activity of anaerobic microorganisms in an anaerobic environment.

decore. to core a fruit.

decorticate. **1:** lacking the cortex or outer layer. **2:** to remove the cortex or outer layer (e.g., as cortex of an organ; the bark of a tree; the peel of a fruit).

decoy hive. a bee hive sited with a view to attracting roving swarms of honey bees.

decreaser. decreaser plant.

decreaser plant (decreaser) a species of plant that occurred in original vegetation in area, that decreases in density or cover as a result of overuse.

decrudescence. a decline or subsidence of a process or condition such as the decrudescence of the gonads in certain animals (e.g., most birds) following a breeding season.

decumbent. of, pertaining to, or descriptive of plant stems that lie along the ground; only the apices rise up.

decurrent. extending downward as in the case of leaves that extend down the stem of a plant.

decussate. descriptive of a leaf pair that grows at a right angle the pairs growing immediately above and below.

dee disease. bacterial kidney disease.

deed restriction. a restriction or covenant placed on property described in a deed.

deep-litter system. absorbent materials such as straw or sawdust that are mixed with hydrated lime on the floor of a hen house to about a depth of about 6 in. (15 cm). This material is periodically raked over but the birds' droppings are not removed. When all of the fowl are sold the litter may be changed.

deep percolation loss. water lost by percolation through the soil to a depth below the reach of plant roots.

deep-rooted. denoting or descriptive of a plant that extends some roots deeply into the soil.

deep soil. a thick soil, often one that is more than 91 cm (36 in) in depth.

deep well. **1:** a well, the water level of which, lies below the level from which water can be raised by a suction pump. **2:** a well that maintains a water level more than 25 ft. below ground at sea level,

deering velvetbean. *Stizolobium deeringianum*.

deeryard. an area where deer congregate in winter. One that affords feed and protection from the inclement weather.

DEET. a pesticide and widely used insect repellent. It is neurotoxic and can cause brain disorders, slurred speech, difficulty in walking, tremors, and even death. There are documented cases of acute nervous system effects in children exposed to DEET; fatalities occurred. DEET is moderately toxic to laboratory rats and rabbits.

defecation. the voiding of feces.

defect. **1a:** a fault, flaw, blemish, imperfection, or weakness. **1b:** in agriculture, a defect (def. 1) that impairs the yield, value, marketability, or usefulness of crops, livestock, or of other farm commodities. **2:** departure from the standards or specifications set down for a breed of livestock. **heritable defect**. one that is grounds for disqualification for registration as a purebred.

deferred grazing. the practice of allowing pasturage to produce enough vegetation to support livestock before animals are permitted on the land. This is common practice, for example, in the rangelands of the western United States

deferred pricing. the practice of determining the price of a commodity after the product has been transferred to the buyer.

deficiency. **1:** inadequacy, lack, shortage, or insufficiency. **2:** defect; flaw. **3:** with respect to plants or animals, an inadequate amount of essential or important nutrients. **4:** the absence or deficiency of a chromosome segment.

deficiency disease. any condition of a plant or animal that is due to deficiencies in nutrients, foods, or other materials required for health.

deficit forest. a forest in which existing stands cannot provide enough harvest volume to provide long-term sustained yield.

deflation. **1:** a decline in consumer prices. **2:** the removal of fine soil particles from the surface soil by wind. *See* desert pavement.

deflea. to remove fleas from an animal, usually by use of a ectoparsiticide of insecticide.

deflocculate. to break down soil aggregates of clay into small particles by use of a deflocculating agent.

deflocculating agent. a chemical such as sodium carbonate, used bring about deflocculation or dispersion of particles.

deflorate. to remove the blossoms or flowers from a plant.

defoliant. an herbicide such as 2,4-D and 2,4,5-T that causes leaves to fall from trees and other growing plants.

defoliation. the loss or removal of leaves from plants, often by chemical means. *See*, for example, Agent Orange.

defoliator. an agent that damages plants, especially trees, by destroying leaves or needles.

deformity. a physical abnormality or deviation from the norm, impairment, or imperfection in a plant or animal. a deformity may be congenital, heritable, or due to disease or injury.

deforestation. clearing a forested area of trees on a more or less permanent basis (e.g. clearing for industrial, residential, or agricultural use).

deglutinate. to remove gluten from cereal grains.

deglutition. the process of swallowing.

degossypolized Cottonseed Meal. cottonseed meal from which the gossypol has been reduced to a level of not more than 0.04% free gossypol or has been converted to an inactive form. Such meal must also meet specifications for use as livestock feed.

degradable. referring to materials, usually waste materials, that degrade in the environmental media (air, soil, or water) due to chemical, physical, and/or biological phenomena. *See* biodegradable.

degradable waste. waste material that is degradable according to the above definition into products that are less noxious or offensive than the original waste. Ultimately, such waste will presumably yield simple inorganic compounds and nutrients.

degradation. **1:** deterioration; degeneration. **2:** the reduction of a chemical to less complex forms. *Cf.* decomposition. **3:** the diminution of biological productivity or diversity.

degras. crude wool fat.

degree of grazing. the degree to which an area is cropped by grazing. Classes are: ungrazed, lightly grazed, moderately grazed, closely grazed, and severely grazed.

deguelia root. derris root.

degumming agent. a chemical agent used to remove mucilaginous materials (e.g., gums, resins, proteins) from fats.

dehorning clipper. sharp-bladed shears with long handles used to remove the horns of cattle.

dehull. to remove hulls as from grains or seeds. *See also* dehusk.

dehusk. to husk; to remove husks as from ears of corn.

dehydrated alfalfa stem meal. a meal used as animal feed and comprised mainly of dried alfalfa parts, chiefly stems, that remain after the leaves have been removed. The fresh material must have been dried at a temperature of at least 100°C for no longer than 40 minutes.

dehydrated borax. anhydrous borax. *Cf.* Borax.

dehydrated food. food from which most of the water is removed during processing. Such foods are of much value where weight (or storage space) is an important quality (as when backpacking or in the case of soldiers under conditions of protracted combat). Dehydration also extends shelf life, sometimes dramatically, and reduces the need for refrigeration of some foods.

dehydrated soybean hay meal. a meal used as animal feed comprised of entire ground soybean plants excusive of stems, straw, and foreign materials. The plant material is artificially dried when fresh. The meal must be reasonably free of other plant materials and may not contain more than 33% crude fiber. It must have been artificially dried when freshly cut.

dehydration. **1:** the loss of water, as from a substance, cells, or tissues. **2:** the state of having a lower than normal water content. **3:** a procedure used to reduce the water content, e.g., of a chemical compound, food, tissues, an organism.

dehydrobilirubin. billiverdin.

dehydrogenase. any enzyme that catalyzes the removal (transfer) of a hydrogen atom from a molecule. *See also* oxidoreductase.

deka-. deca-.

delactation. **1:** cessation of milk production by a dairy animal. **2:** weaning; termination of suckling.

Delaney amendment. *See* Delaney clause.

Delaney clause. a clause of the Food Additive Amendment of the U.S. Food, Drugs, and Cosmetic Act, which stipulates that no substance known to induce cancer in any animal may be a component of any food.

delayed contact sensitivity. allergic contact dermatitis (*See* dermatitis).

delayed effect. *See* effect.

delayed response. delayed effect (*See* effect).

deleterious. damaging, injurious; unwholesome.

deleterious effect. *See* effect.

deletion. **1:** an omission or cancellation. **2:** absence of a chromosome segment that includes one or more genes.

delignification. the removal of lignin from wood or materials made of cellulose.

deliming. the removal of lime used to dehair an animal hide by use of a sulfuric acid bath.

delimitation. to establish or set the boundary or limits of; preventing the spread of a disease or intoxicating process (as in the case of snakebite) within an organism or a community.

deliquesce. **1a:** to dissolve gradually and liquefy by absorbing moisture from the air. **1b:** to become soft or to liquefy with time as in the case of some mushrooms. **2:** to disappear as if by melting. **3:** to divide repeatedly, resulting in finely divided parts, as in the venation of certain leaves.

deliquescence. the process of deliquescing or the condition of readily absorbing moisture from air, thereby liquefying or becoming moist.

deliquescent. pertaining to or exhibiting deliquescence; tending to liquefy or to form an aqueous solution by absorbing moisture from air.

dell. a small wooded valley.

Delnav®. dioxathion.

delousing. the destruction and removal of lice by the use of a chemical agent, often an insecticide.

delphinium poisoning. larkspur poisoning.

Delphinium (larkspur; delphinium; staggerweed). a genus of hardy toxic annual and perennial erect herbs (Family Ranunculaceae) commonly known as larkspur or delphinium. They are native to Asia Minor, Europe, and the United States. The genus can be easily identified by the distinctive flowers, which have a spur that projects to the rear. The crude extract of many species contains a number of complex neurotoxic diterpenoid alkaloids that are similar to those found in *Aconitum*. All parts of the plant are toxic, but the various species and hybrids are not uniformly toxic. Young leaves and seeds are most deadly. Accurate species identification is not always possible, especially because poisoning of livestock usually occurs before flowering. Within species, toxicity varies with season, the part eaten, the species of animal affected, individual susceptibility, and other factors. Larkspur is a great killer of livestock. New growth of certain species is lethal to cattle when it makes up less than 1% of the forage. ***D. barbeyi*** (staggerweed). The seeds are highly toxic. Most livestock losses to this plant occur among cattle throughout the western United States. Horses will ingest larkspur, but rarely if ever consume a lethal dose; sheep are highly resistant, and losses to larkspur poisoning are rare. *See also* larkspur poisoning.

delphinium. a common name of most perennial forms of the plant genus *Delphinium*, *q.v.*

delta (δ, Δ). fourth letter of the Greek alphabet. It has a number of applications in mathematics and science. **1a:** Δ signifies any of a number of things that resemble it (e.g., a river delta). **1b:** Δ indicates change. **2:** δ is used somewhat arbitrarily to denote or refer to one or more related chemical substances. **3:** denoting the fourth position from a particular atom or group in an organic molecule as in δ-diketones.

deltoid. triangular in shape as in the Greek letter delta (upper case).

deltoides. triangular.

Demaree. in honey-bee keeping, a method of controlling swarming in which the queen is separated from most of the brood.

demersal. fish or other aquatic animals that feed on benthic organisms.

demersed. living or growing under water (applied especially to plants).

Demeter. the Greek Goddess of agriculture. *See* Ceres.

demeton (demeton-*O* + demeton-*S*; phosphorothioic acid *O,O*-diethyl *O*-[2-(ethylthio)ethyl] ester mixture with *O,O*-diethyl *S*-[2-(ethyl-thio)ethyl]phosphorothioate; diethoxy thiophosphoric acid ester of 2-ethylmercaptoethanol; *O,O*-diethyl 2-ethylmercaptoethyl thiophosphate; *O,O*-diethyl *O*(+*S*)-2-(ethylthio)ethyl phosphorothioate mixture; mercaptophos; Systox). a pale yellow to light brown liquid systemic insecticide,

$$_8H_{19}O_3PS_2{\cdot}C_8\text{-}H_{19}O_3PS_2$$

with a faint odor of sulfur. It is a mixture, of demeton-*O* and demeton-*S*. Demeton is slightly soluble in water, but soluble in most organic solvents. It is a powerful anticholinesterase that is extremely toxic to a wide variety of animals and a deadly poison to humans by all routes of exposure; it readily penetrates the skin. It is a cumulative poison; only acute intoxication is known. It is easily taken up by plants, rendering the leaves and fluids toxic to insects and probably other herbivores. Signs and symptoms in humans include blurred vision, giddiness, headache, nausea, diarrhea, weakness and discomfort in the chest. It is teratogenic and a probable human mutagen. This highly dangerous pesticide, which is readily absorbed through the skin, is used in some household products to eradicate pests.

demeton-*S* (diethyl-*S*-(2-ethioethyl)thiophosphate; *O,O*-diethyl-*S*-(2-ethioethyl)phosphorothioate; *O,O*-diethyl-*S*-ethyl-2-ethylmercaptophosphorothiolate; *O,O*-diethyl-*S*-2-(ethylthio)ethyl phosphorothiolate; 2-(ethylthio)-ethanethiol *S*-ester + *O,O*-diethyl phosphorothioate). an insecticide,

$$C_8H_{19}O_3PS_2$$

Demeton-*S* is toxic to humans and to a variety of animals by ingestion, intraperitoneal, and subcutaneous routes. *See* demeton.

demeton-*O* + demeton-*S*. demeton.

demeton-methyl sulphoxide. demeton-*O*-methyl sulfoxide.

demeton-methyl. demeton-*S*-methyl.

demeton-*O*-methyl (*O,O*-dimethyl-*O*-ethylmercaptoethyl thiophosphate; *O,O*-dimethyl-2-ethylmercaptoethyl thiophosphate, thiono isomer; *O,O*-dimethyl-*O*-2-(ethylthio)ethyl phosphorothioate). an insecticide,

$$C_6H_{15}O_3PS_2$$

It is toxic to humans, many higher animals, and invertebrates probably by all routes of exposure. It is also teratogenic and has reproductive effects. Very toxic fumes, PO_x and SO_x, are released on decomposition by heating. *See* demeton.

demeton-*O*-methyl sulfoxide (demeton-*S*-methyl sulfoxide; demeton-methyl sulphoxide; *O,O*-dimethyl-*S*-(2-ethionylethyl)phosphorothioate; dimethyl-*S*-(2-ethionylethyl) thiophosphate; *O,O*-dimethyl-*S*-[2-(ethylsulfinyl)ethyl] phosphorothioate; *O,O*-dimethyl-*S*-(2-ethylsulfinyl)-ethyl thiophosphate; *O,O*-dimethyl-*S*-ethylsulfinylethyl phosphorothiolate; *S*-[2-(ethylsulfinyl)ethyl]-*O,O*-dimethyl phosphorothioate). an insecticide. It is toxic to humans, many other higher animals and invertebrates by most or all routes of exposure. It is a probable human mutagen. *See* demeton.

demeton-*S*-methyl (methyl demeton; demeton-methyl; phosphorothioic acid *O*-[2-(ethylthio)ethyl] *O,O*-dimethylester mixed with *S*-[2-(ethylthio)ethyl]; *O,O*-dimethylphosphorothioate; demetonmethyl; methylmercaptophos; methyl systox). an oily liquid,

$$C_6H_{15}O_3PS_2$$

demeton-*S*-methyl is slightly soluble in water. It is used as an insecticide to control aphids. It is toxic to humans and many higher animals and invertebrates probably by all routes of exposure and is probably mutagenic. *See* demeton.

demeton-*S*-methyl sulfoxide. demeton-*O*-methyl sulfoxide.

demetonmethyl. demeton-*S*-methyl.

demi-. a prefix denoting one-half or lesser; synonymous with hemi-, semi-.

demonstration plot. a plot of ground used to demonstrate proven agricultural practices, procedures, and techniques of production.

den. a chamber in which acid-rock slurry produced during the manufacture of su perphosphate is allowed to degas, solidify, and cure.

denaturation. **1:** an irreversible process in which the three-dimensional (tertiary) structure of a protein is altered (e.g., by heat, strong acid). **2:** the unwinding of double-stranded DNA and dissociation into two single strands.

dendritic. **1:** denoting a branching habit as DNA-DNA hybridization. that of a bush or tree; a term often applied to plants, the venation of the leaves of many higher plants, and to riverine systems having many irregular-branching tributaries.

dendrium. an orchard community.

dendrochronology. the determination of the age of trees by the examination of tree rings.

Dendrocnide moroides (Australian nettle). plants (Family Urticaceae) that induce a nonimmunologic contact urticaria by means of hollow stinging hairs that penetrate the skin and inject irritants that produce an almost immediate burning sensation and pruritis.

Dendroctonus borealis (Alaska spruce beetle). a wood-boring beetle (Family Scolytidae) that infests white and Englemann spruce in Alaska and northwestern Canada. Adults bore into the inner bark of a tree, and the larvae bore into the wood.

Dendroctonus terebrans (black turpentine beetle). an insect pest, the larvae of which feed on the inner bark of the trees, causing them to die.

dendrograph. a recording dendrometer that continuously records the growth of a tree in terms of changing diameter.

dendrology. the branch of botanical science that treats trees. It includes the work of both basic and applied scientists.

dendrometer. an instrument used to measure the outside diameter of trees, including the bark.

dengue fever (breakbone fever; bone-crusher disease). an acute infectious disease caused by an arbovirus and transmitted by the bite of a mosquito (*See Aedes*). This disease occurs in warm climates (e.g., the southern United States, South America, the Mediterranean countries, India, the Philippines, and Indonesia). There is both an epidemic and sporadic form. Symptoms, which appear following an incubation period of five to eight days, include chills, fever, severe headache, sweating, pain in the joints, and prostration. These symptoms temporarily subside in 2-4 days only to return within a few hours to two days. The fever reappears accompanied by a generalized rash. Recovery, with frailty and low blood pressure, may be prolonged. *See also* dengue hemorrhagic fever.

dengue hemorrhagic fever. a severe form of dengue fever. It is a leading cause of death of children in Southeast Asia and has become more frequently encountered in tropical South America in recent years. The disease may be complicated by hemorrhage, shock, and encephalitis. Good supportive care is required, but no specific treatment is available.

denitrification. the anaerobic biological reduction of nitrate nitrogen to gaseous nitrogen (dinitrogen), nitrous oxide, or nitrites and ammonia; also the conversion of ammonia to molecular nitrogen. In terrestrial environments, this is usually accomplished by facultative aerobic soil bacteria (denitrifying bacteria) under conditions of low environmental oxygen tension. However, denitrification is generally considered to take place on a global basis chiefly in oxygen deficient marine environments. **chemical denitrification**. nitrate reduction as a result of chemical reactions. **dissimilatory denitrification**. nitrate reduction accomplished by microbial activity or by chemical reactions involving nitrite. *Cf.* dissimilatory nitrate reduction to ammonia.

denitrify. to effect denitrification.

denitrifying. pertaining to conditions favorable to, or organisms that can accomp lish, denitrification.

density. **1:** mass per unit volume. **2a:** the number of entities or observations within a given area or volume. **2a:** The number of individual organisms a given population (e.g., elk in a national forest; insects on a tree) per unit area or, in some cases, volume. **2b:** the number of trees, basal area, volume, or other measure of density per acre or hectare at a given location. **3:** the closeness or degree of packing of certain materials such as wool fibers on a wool producer such as a sheep. **bulk density**. mass per unit volume (g/cc) of undisturbed soil, dried to constant weight at 105°C (221°F). *See also* bulk density of soil. **crown density**. the density, depth, or amount of foliage of a tree crown.

density dependence. the relationship between an environmental factor that affects population growth or survival as a function of population density, e.g., predation, parasitism.

density independence. the relationship between an environmental factor that affects population growth or survival independent, e.g., of a climatic variable such as rainfall, flooding, or prolonged drought)

dent corn. *Zea mays intentata*.

dent-, denti-, dento-. combining forms meaning teeth or relating to the teeth.

dental cup (dental star). a small, dark depression rimmed by hard, shiny-white enamel in the wearing surface of the incisors of horses. It disappears with wear.

dental star. dental cup.

dentate. **1:** of, pertaining to, or descriptive of a leaf, the margins of which have regular serrations. **2:** having sharp, relatively large, coarse teeth that are perpendicular to the margin of the jaw.

denticulate. having small, dentate teeth.

dentition. **1:** the total complement, type, or arrangement of teeth of a particular animal or type of animal. **2:** the growth and development of teeth.

denude. **1:** to strip the soil of vegetation. **2:** to lay bare.

denutrition. **1a:** to lack nutrition. **1b:** to withhold nutrition. **2:** failure to digest or metabolize food materials nutritionally useable forms.

25-deoxyecdysterone. ponasterone A.

deoxygenated. pertaining to a substance or an environment that is depleted of free oxygen; anaerobic.

deoxygenation. the removal of free oxygen from a substance or an environment.

3-[(6-deoxy-α-L-mannopyranosyl)-oxy]-14,16-dihydroxy-19-oxocard-20(22)-enolide. adonitoxin.

depauperatum. stunted.

dependent plant. plant that does not produce its own food from sunlight and available nutrients but lives on other organisms, living or dead.

depergelation. the process of thawing permafrost.

depilate. to remove hair from the skin or from pelts.

depilatory. **1:** able to remove hair from a mammal. **2:** a substance used to remove hair from a live mammal. **3:** a substance applied to the inner side of a pelt to loosen fibers from the skin.

depletion. **1a:** consumption; utilization. **2:** overuse; exhaustion; dissipation. **3:** the harvesting or result of harvesting natural renewable resource at a rate greater than the rate of renewal. **4:** in forestry, the harvesting of growing stock. **5:** an income tax allowance that reflects the purchase price of marketable timber, usually on fee simple land.

depletion curve. **1a:** in hydraulics, a graphical representation of water depletion from storage-stream channels, surface soil, and ground water. A depletion curve can be drawn for base flow, direct runoff, or total flow. **1b:** the depletion of water from storage-stream channels, surface soil, and ground water. A water depletion curve can be drawn for base flow, direct runoff, or total flow.

deposit. **1:** the process of transferring material (e.g., dust, sediment, pollutants) from one location or medium to a new position or surface by natural processes such as wind, water, ice, gravity, or human activity. **2:** material so deposited.

deposition. **1:** the amount of matter (e.g., a pollutant) that settles out of the air onto a surface (usually taken as the amount per unit area during a specified period of time). **2:** the amount of an aerosol or particulate matter that remains in the lung following exhalation.

dequeen. to remove the queen bee from a colony.

Dermacentor. a genus of wood ticks (Family Ixodidae) that transmit Rocky Mountain spotted fever; some species cause tick paralysis. They are parasitic on numerous mammals, including humans. *See* tick, Ixodidae, Ixodoidea, *Amblyomma*. ***D. andersoni*** (Rocky Mountain wood tick). a tick that introduces a venomous saliva on biting which can produce a flaccid ascending motor paralysis that progresses from the lower extremities. If the tick is promptly removed, rapid and complete recovery is usual. ***D. vatriabilis*** (American dog tick). a vector of bovine anaplasmosis, Rocky Mountain spotted fever, and tularemia.

dermatitis. inflammation of the skin. **actinic dermatitis**. any reaction of the skin to any photochemically active type of radiation (e.g., sunlight, roentgen rays, ultraviolet light). **allergic contact dermatitis** (delayed contact sensitivity). a delayed type of induced sensitivity (allergy) arising from a subsequent exposure of the skin to a specific sensitizing substance (allergen). The condition is marked by varying degrees of erythema, edema, and vesiculation. *See Datura*. **allergic dermatitis**. any dermatitis due to allergy (e.g., allergic contact dermatitis, contact-type dermatitis, cosmetic dermatitis, systemic lupus erythematosus). **brown-tail moth dermatitis** (brown-tail rash). a caterpillar dermatitis resulting from contact with setae or cocoon of the brown-tail moth, *Euproctis chrysorrhoea*. **caterpillar dermatitis** (caterpillar rash). a transient, sometimes painful allergic contact dermatitis from contact with caterpillar setae that penetrate the skin. **chemical dermatitis**. an allergic contact dermatitis or a primary irritation dermatitis due to contact with chemicals. It is usually characterized by erythema, edema, and vesiculation of the exposed area. **contact dermatitis**. an acute dermatitis induced by contact with any of a variety of substances. The proximate cause may be primary chemical irritation, mechanical injury, toxicity, phototoxicity, allergy, photoallergy, or a combination of these. **contact-type dermatitis**. any dermatitis that resembles contact dermatitis but is caused by an ingested or injected allergen, usually a medication. It is characterized by widespread cutaneous distribution. **cosmetic dermatitis**. any contact dermatitis caused by the application of a cosmetic preparation. It may be due to a primary irritation or to allergic sensitization. **exfoliative dermatitis** (dermatitis exfoliativa; pityriasis rubra; Wilson's disease). an erythematous dermatitis with generalized exfoliation and scaling of the skin. It is sometimes due to a severe drug reaction or in other cases associated with lymphomas or any of various benign dermatoses. **grass dermatitis**. meadow-grass dermatitis. **industrial dermatitis**. occupational dermatitis. **insect dermatitis**. dermatitis caused by contact with insects, usually moths or caterpillars. **io-moth dermatitis**. dermatitis resulting from contact with irritant setae of the caterpillar of *Automeris io*, the io moth. **meadow-grass dermatitis** (meadow dermatitis, grass dermatitis, dermatitis striata pratensis bullosa; phytophlyctodermatitis). a phototoxic dermatitis caused by contact with *Agrimonia eupatoria* and subsequent exposure to sunlight. There is an eruption of vesicles and bullae that occur in strips and unusual groupings reflecting the pattern of plant contact. It is often contracted during sunbathing. **nickel dermatitis**. an allergic dermatitis due to nickel or metals such as stainless steel that contain nickel. **occupational dermatitis** (industrial dermatitis). any contact dermatitis contracted as a result of one's employment. **paraffin dermatitis**. a chronic dermatitis caused by exposure to paraffin. With high exposure rates there is risk of scrotal cancer. **photoallergic dermatitis**. *See* photoallergic photodermatitis. **photocontactdermatitis**. an allergic contact dermatitis caused by the action of sunlight on skin sensitized by contact with an allergen such as sandalwood oil, hexachlorophene, or a halogenated salicylanide. *Cf.* phototoxic contact dermatitis. **phototoxic contact dermatitis**. a rapidly developing sunburn-like dermatitis of skin areas exposed to sun following cutaneous contact with a photosensitizing substance. *Cf.* photocontact dermatitis. **phototoxic dermatitis**. *See* phototoxic photodermatitis, phototoxic contact dermatitis. **poison ivy dermatitis**. *Rhus* dermatitis. **poison oak dermatitis**. *Rhus* dermatitis. **poison sumac dermatitis**. *Rhus* dermatitis. **primary contact dermatitis** (primary irritant dermatitis; primary chemical dermatitis). a chemical dermatitis resembling that from contact with an irritant or corrosive agent. It results from contact with substances that are toxic to cutaneous tissues. The degree of injury varies with the potency of the irritant, its concentration, and duration of contact. The lesions are usually erythematous and papular but, in more severe cases, may be purulent and necrotic. The eye is especially susceptible to injury, and keratoconjunctivitis and temporary blindness may result if the eye is involved. Primary contact dermatitis is frequently caused are spurges of the family Euphorbaceae, especially *Ranunculus spp*. (buttercups), *Daphne mezereum* (daphne), and *Capsicum frutescens* (wild pepper). **primary dermatitis**. a dermatitis that is induced directly by some agent; not an allergic response. **primary irritant dermatitis**. primary contact dermatitis. **rhus dermatitis**. *Rhus* dermatitis (Toxicodendron dermatitis). ***Rhus* dermatitis** (*Toxicodendron dermatitis*). an allergic contact dermatitis due to cutaneous exposure to urushiol from certain plants of the genus *Rhus* (synonym, *Toxicodendron*), e.g., poison ivy, poison oak, and poison sumac. **schistosome dermatitis** (cercarial dermatitis; marine dermatitis; seabathers eruption; swimmer's itch; swimmer's dermatitis). an itching dermatitis caused by infestation of the skin by blood flukes of the genus *Schistosoma* that are not normally parasites of *Homo sapiens*. The condition occurs in persons who wade or swim in saltwater. **swimmer's dermatitis**. schistosome dermatitis. ***Toxicodendron* dermatitis**. *Rhus* dermatitis. **traumatic dermatitis**. any dermatitis caused by an irritant or a physical agent. **trefoil dermatitis**. trifoliosis. *See also* phytodermatitis, phytophotodermatitis, photodermatitis, photoallergic photodermatitis, radiodermatitis.

dermatitis exfoliativa. exfoliative dermatitis.

dermatitis simplex. erythema.

dermatitis striata pratensis bullosa. meadow-grass dermatitis.

dermatomycosis. a dermatosis caused by a fungus.

dermatopolyneuritis. acrodynia (def. 2).

dermis (cutis vera; corium; skin; true skin). the thick, inner, living layer of the skin of vertebrate animals. It lies beneath the epidermis and sometimes also beneath an outer cuticle or horny layer and is bordered below by a thick, dense layer of connective tissue. The dermis typically consists of a dense, flexible, connective tissue matrix; an insulating layer of fat, a rich blood supply essential to temperature control, nerves, sensory organs, and glands (usually). In most mammals, it contains hair follicles, and in birds, feather papillae are included. The dermis of amphibians and fishes includes sensory nerve endings, chromatophores, and a rich blood supply that serves external respiration. Fish scales and certain bones of the skull (dermal bone) are dermal in origin.

dermotropic. affecting especially the skin.

Derrin. rotenone.

derris. **1:** common name of any plant of the genus *Derris*. **2:** a preparation of derris roots and stems that contains insecticidal rotenoids such as rotenone.

Derris (derris). a genus of more than 80 species of Old World tropical woody vines and shrubs with long climbing branches (Family Fabaceae, formerly Leguminosae). The roots and stems of some species contain rotenone and other insecticidal rotenoids. Resin from the roots of a few species are used by natives as arrow poisons or to temporarily paralyze fish in streams to ease the harvest. They are commercial sources of rotenone and other insecticidal rotenoids. ***D. elliptica***. cultivated in the Dutch East Indies and British Malaya. The roots are dried and powdered or the resin is extracted, and marketed under various brand names as contact insecticides. ***D. malaccensis***. cultivated in the Dutch East Indies and British Malaya for use as noted for *D. elliptica*.

derris root (tuba root; deguelia root). the dried resinous root of plants of the genus *Derris*, especially of *D. elliptica*. and *D. mallaccensis*. It contains insecticidal rotenoids such as rotenone. The dried powdered root is very toxic by mouth.

DES. diethylstilbestrol.

Descurainia pinnata (tansy mustard). a mustard (Family Brassicaceae) that is widely distributed throughout the southern United States. Range cattle have been poisoned from long-continued consumption of large quantities of this plant. *q.v.* Partial or complete blindness is the first symptom to present. Subsequently, the affected animal loses the use of its tongue and cannot swallow. The affliction is thus referred to as "paralyzed tongue," *q.v.*

desert. an arid region having scant vegetation, often partly covered by sand, and able to support a limited and specially adapted animal population. **cold desert**. a region of extreme cold that may be covered with perpetual snow or ice. Such deserts cover about one-sixth of the earth's surface, whereas warm deserts cover about one-fifth of the land surface of the earth.

desert baileya. *Baileya multiradiata*.

desert crust. a hard surface layer (hardpan) in a desert that contains calcium carbonate, gypsum (calcium sulfate), or other cementing substances.

desert pavement. a layer of gravel or stones remaining on the surface of the ground in deserts after the removal of fine material by wind. *See* deflation.

desiccant. a chemical agent that absorbs moisture (e.g., activated alumina, calcium chloride, silica gel, zinc chloride); some are able to fatally desiccate plants or invertebrate animals.

desmethoxyviridiol. a mycotoxin from *Nodulisporium hinnuleum*. It is extremely toxic orally to the cockerel of the domestic fowl.

desmetryne. an atrazine herbicide (a methylthiotriazine) that is similar in biological activity to atrazine and simazine but is absorbed through foliage as well as the roots. It is used both as a preemergence and postemergence herbicide.

Desmodus. *See* bat.

detassel. to remove the tassel.s from corn.

detergent. a product used in conjunction with water to clean dirt from surfaces. All are surface active agents that reduce the surface tension of water and emulsify fats and oils in contact with the water. Modern detergents are synthetic and are classified as anionic, cationic, or nonionic, depending upon the ionization of the surface active moiety of the compound and, hence, their mode of action. Within each of these classes, there is a wide range of chemical structures and toxicities. **anionic detergent**. the surface active moiety of the detergent bears a negative charge at neutral pH. **cationic detergent**. the surface active moiety bears a positive charge at neutral Ph. This type of detergent occurs chiefly in dishwashing detergents and fabric softeners. **dishwashing**

detergent. any detergent used to wash dishes and other kitchen utensils in a dishwasher. Most contain chlorine in a dry form that becomes active when wet, with the release of chlorine fumes which may seep out of the dishwasher into the kitchen. Accidental ingestion and, to a lesser degree, inhalation can be harmful. With reasonable care, the use of dishwashing detergent is considered safe for most people. Many users, however, are affected by headache, fatigue, burning eyes, and difficult breathing when exposed to the small amounts of chlorine released during normal dishwashing. Dishwashing detergent may contain other toxic materials. *See* dicyclohexylamine and ethylene oxide. **nonionic detergent**. the surface active moiety is nonionic at neutral pH. This type of detergent acts by hydrogen-bonding.

development. **1:** progressive change in structure or behavior from conception to death. **2:** ontogeny; regulated growth and differentiation of an individual, including cellular differentiation, histogenesis and transformations of cells and tissues into organs (organogenesis).

developmental biologist. embryologist.

developmental biology. embryology.

devil grass. *Cynodon dactylon*.

devil's apple. *Datura stramonium*.

devil's bit. *Veratrum viride*.

devil's hopvine. a common name of *Smilax rotundifolia*.

devil's tobacco. *Veratrum viride*.

devil's trumpet. *Datura stramonium*.

devils walking stick. *Aralia spinosa*.

dew poisoning. trifoliosis.

DFP. diisopropyl fluorophosphate.

DHI. dairy herd improvement.

dhurrin (*S*-α-(β-D-glucopyranosyloxy)-4-hydroxybenzeneacetonitrile; *p*-hydroxymandelo-nitrile-β-D-glucoside; β-D-glucopyranosyloxy-L-*p*-hydroxymandelonitrile). a cyanogenetic nitrile glycoside,

$$C_{14}H_{17}NO_7$$

Dhurrin has been isolated from various sorghums such as *Sorghum almum*, *S. halepense*, and *S. vulgare* (var. *sudanense*). It accumulates in the plant especially under conditions of drought, high soil nitrogen, and low phosphorous. The highest concentrations occur in young growth. When the plant is ingested (e.g., by livestock), dhurrin undergoes hydrolysis, yielding parahydroxy benzaldehyde, glucose, and hydrocyanic acid.

Di-Syston. disulfoton.

diabatic. occurring with a transfer of heat, as a diabatic process. *Cf.* adiabatic.

Diabrotica balteata (banded cucumber beetle). a pest (Family Chrysomelidae) of beans, other legumes, and cucurbits.

diacetylmethane. acetylacetone.

dialifor (phosphorodithioic acid *S*-[2-chloro-1-(1,3-dihydro-1,3-dioxo-2*H*-isoindol-2-yl)ethyl] *O,O*-diethyl ester; phosphorodithioic acid *O,O*-diethyl ester *S*-ester with *N*-(2-chloro-1-mercaptoethyl)phthalimide; *O,O*-diethyl *S*-(2-chloro-1-phthalimidoethyl)phosphorodithioate; dialifos). an extremely toxic, water-insoluble, white crystalline compound. It is an insecticide and acaricide used to control pests of citrus fruits, grapes, and pecans.

dialifos. dialifor.

diallate (bis(1-methylethyl)carbamothioic acid S(2,3-dichloro-2-propenyl) ester; diisopropyl-carbamic acid S-2,3-dichloroallyl ester; S-2,3-dichloroallyl-diisopropylthiocarbamate). a selective thiocarbamate herbicide used mainly to control *Avena fatua* (wild oats) in cereal crops. It inhibits shoot growth largely by interfering with cell elongation. It is also a mitotic poison at higher concentrations. It is incorporated into the top 10 cm of soil just prior to sowing the crop. Diallate is very toxic to laboratory rats.

diallyl sulfide. allyl sulfide.

***N,N*-diallyl-2-chloroacetamide**. allidochlor.

diameter inside bark. DIB.

2,5-diamino-4,6-diketopyrimidine-3-β-D-glucoside. vicine.

1,2-diaminoethane. ethylenediamine.

***N*-[4-[[(2,4-diamino-6-pteridinyl)-methyl]amino]benzoyl]-L-glutamic acid**. aminopterin.

diammonium hexafluorosilicate. ammonium hexafluorosilicate.

diammonium phosphate. a white, crystalline, water-soluble compound,

$$(NH_4)_2HPO_4$$

It is used, for example, as a fertilizer and in fire extinguishers.

dianat. dicamba.

diarrhea. **1a:** abnormally frequent passage of soft or watery fecal discharges. In some cases, the stool may contain pus, blood, or mucus. It is a common symptom of gastrointestinal disturbances including that caused by any of numerous poisons. Diarrhea can be acute or chronic; it can limit the uptake of water and salts by the body with resultant dehydration and/or electrolyte imbalances. **1b:** the condition characterized by def. 1a. **bacillary white diarrhea**. an acute, infectious and highly fatal disease of chicks, caused by *Salmonella pullorum*. **cachectic diarrhea**. that associated with cachexia; **choleraic diarrhea**. acute diarrhea that resembles that of cholera. The stools are serous and attended by circulatory collapse. **critical diarrhea**. that which occurs at the crisis of a disease or which causes a crisis. **dysenteric diarrhea**. that in which the stools are bloody with mucous. **irritative diarrhea.** that due to irritation of the intestine by poisonous substances, unsuitable food, purgatives, various drugs, or other ingested substances. **osmotic diarrhea**. that due to the presence of osmotically active nonabsorbable solutes in the intestine. **serous diarrhea** (watery diarrhea). that which produces soft feces with large amounts of serous fluid. **traveler's diarrhea**. that which occurs among travelers, especially those who visit tropical or subtropical areas that have poor sanitation. It probably most often results from infection with enteropathogenic *Escherichia coli*. **watery diarrhea**. serous diarrhea.

diastase. amylase.

diatomaceous earth (diatomite). a geologic deposit of fine, grayish, chalk-like, sileceous powder or a porous, rigid material composed chiefly or entirely of fossilized diatoms. It is used to filter out solid waste in wastewater treatment plants and is also used as an active ingredient in some powdered pesticides.

diatomite. diatomaceous earth.

diatoms. microscopic unicellular or colonial algae (Class Bacillariophyceae). All have siliceous cell walls that persist following death of the cell.

diazinon [Chevron]® (*O,O*-diethyl-*O*-(2-isopropyl-4-methyl-6-pyrimidinal)phosphorothioate). a colorless, slightly water-soluble liquid insecticide and parasiticide,

$$[C(CH_3)CHC_4N_2H(CH_3)O]PS(OC_2H_5)_2$$

It is used on open areas. In 1986, the U.S. Environmental Protection Agency banned its use in open areas such as sod farms because it endangered those migratory birds that gather in large numbers on such sites. The ban did not apply generally to its use in agriculture, or on lawns of homes or commercial properties. Its use against fire ants is permitted. Diazinon is toxic by inhalation, ingestion, and skin absorption in higher vertebraes. Young calves apparently tolerate diazinon sprayed at a concentration of 0.05% but are poisoned at 0.1%. Adult cattle may safely be sprayed weekly for an indefinite period. Young calves are poisoned by ingestion at a concentration of 0.88 mg/kg body weight but seem to tolerate 0.44 mg/kg. Adult cattle tolerate 8.8 mg/kg body weight orally but are poisoned by 22 mg/kg. Sheep tolerate 17.6 mg/kg body weight, but are poisoned at 26 mg/kg.

diazotroph. an organism that can use molecular nitrogen as its sole nitrogen source and thus fixes nitrogen. *See* nitrogen fixation.

DIB (diameter inside bark). the diameter of a tree or log excluding the bark.

dibasic amino acid. basic amino acid (*See* amino acid).

dibber. dibble.

dibble (dibber). **1:** a pointed, hand-held tool used to make holes in soil to plant seeds, bulbs, etc. and to transplant small plants, especially tree seedlings. **2:** to set (e.g., a plant) in holes made with a dibble **3:** to use a dibble.

dibenzothiazine. phenothiazine (def.1).

diborane (diborane hexahydride; boroethane). the simplest stable borane at normal ambient temprature and pressure, it is a highly reactive, toxic, colorless gas, B_2H_6, with an offensive, sickly-sweet odor. Diborane reacts violently with oxidizing materials; it is highly flammable and a serious fire hazard. It is soluble in carbon disulfide, but decomposes in water. It is a strong irritant and inhalation may result in pulmonary edema. Diborane has many uses, including that of an insect repellent.

diborane hexahydride. diborane.

1,2-dibromo-3-chloropropane. dibromochloropropane.

dibromochloropropane (1,2-dibromo-3-chloropropane; 3-chloro-1,2-dibromopropane; DBCP). a dense, light yellow or brown, environmentally persistent liquid with a pungent odor,

$$ClCH_2CHBrCH_2Br$$

It is used as a soil fumigant and agricultural nematocide. It is an irritant and CNS depressant.

1,2-dibromo-3-chloropropane (DBCP). a potent nematocide formerly used extensively as a soil fumigant.

1,2-dibromoethane. ethylene dibromide.

l,l-dibromoethane. a heavy liquid that was formerly used widely as a grain and soil fumigant for insect control and as a component of leaded gasoline for scavenging lead from engine cylinders.

3-(2,2-dibromoethenyl)-2,2-dimethylcyclopropane carboxylic acid. cyprmethrin.

dicamba (3,6-dichloro-2-methoxybenzoic acid; 3,6-dichloro-*o*-anisic acid; 2-methoxy-3,6-dichlorobenzoic acid; dianat). a translocated postemergence herbicide,

$$C_8H_6Cl_2O_3$$

It is used to control dock (Genus *Rumex.*, Subgenus *Lapathum*; Family Polygonaceae) in established grassland and to control bracken, *Pteridium spp*. Dicamba is poisonous to cattle and sheep and is moderately toxic to laboratory rats.

***S*-(1,2-dicarbethoxyethyl) *O,O*-dimethyldithiophosphate**. malathion.

Dicentra (dicentra; bleeding heart; Dutchman's breeches; squirrel corn; turkey corn). a genus of low, often stemless North American and Asian perennial herbs (Family Papaveraceae). All parts of the plant are toxic. ***D. eximia*** (turkey corn; staggerweed). potentially toxic to livestock. ***D. cucullaria***. (Dutchman's breeches, breeches flower). toxic to cattle. Symptoms are similar to those caused by ingestion of *Corydalis*. The animals run back and forth, however, with the head held unusually high and the animals fall in a characteristic position of emprosthotonos. ***D. canadensis*** (squirrel corn). a North American woodland plant, it is potentially toxic to livestock.

dicentra. common name of plants of the genus *Dicentra*.

dicentric. said of a chromosome or chromatid) that has two centromeres.

dichasial cyme. *See* cyme.

dichasium. *See* cyme.

dichlobenil (2,6-dichlorobenzonitrile). a selective pre- and postemergence, amide herbicide applied to the soil that disrupts many aspects of growth and differentiation. It destroys phloem, cambium, and parenchyma tissue in alligator weed *(Alternanthera philoxeroides)*. It prevents cell nuclei in oat (*Avena*) root tips, for example, from entering mitosis. Such blocking of cell division results in a blackening and death of the growing points in many plant species; this resembles the effects of boron deficiency. Its chief preemergence effect is the disruption of cell division in growing primary meristems.

dichloricide. *p*-dichlorobenzene.

dichloricide aerosol. Strobane®

dichloride mothproofer. Strobane®

S-2,3-dichloroallyldiisopropylthiocarbamate. diallate.

3,6-dichloro-*o*-anisic acid. dicamba.

2,6-dichlorobenzonitrile. dichlobenil.

***p*-dichlorobenzene** (1,4 dichlorobenzene; dichloricide; PDB). a white, crystalline, aryl halide insecticide, $C_6H_4Cl_2$, with a penetrating odor. It is moderately toxic to humans by all routes of exposure. Long-term chronic or intermittent exposure can cause liver and kidney damage. It is moderately toxic to laboratory animals by most portals of entry, but mildly toxic by the subcutaneous route. It is an eye irritant, a probable teratogen, a reproductive poison, a suspected carcinogen, and depressant. Severe (acute) clinical poisonings are unkown. Sources of exposure include mothballs and household deodorizers.

1,4-dichlorobenzene. *p*-dichlorobenzene.

1,1-dichloro-2,2-bis(*p*-chlorophenyl)ethane (1,1′-(2,2-dichloroethylidene)bis[4-chlorobenzene]; tetrachlorodiphenylethane; TDE; dichlorodiphenyldichloroethane; *p,p*′-DDD; *p,p*′-TDE). a nonbiodegradable pesticide and a metabolite of DDT in mammals. It was once used on many agricultural crops, but all applications were halted in the United States in 1972, largely due to its probable role in eggshell thinning among raptorial birds.

***O*-(2,5-dichloro-4-bromophenyl)-*O*-methyl phenylthiophosphonate**. leptophos.

2,4-dichloro-α-(chloromethylene)benzyl alcohol diethyl phosphate. chlorfenvinphos.

dichlorodiphenyldichloroethane. 1,1-dichloro-2,2-bis(*p*-chlorophenyl)ethane.

***p,p*′-dichlorodiphenylmethyl carbinol**. chlorfenethol.

dichlorodiphenyltrichloroethane. DDT.

α,β-dichloroethane. ethylene dichloride.

1,2-dichloroethane. ethylene dichloride.

***sym*-dichloroethane**. ethylene dichloride.

3-(2,2-dichloroethenyl)-2,2-dimethylcyclopropanecarboxylic acid cyano(3-phenoxyphenyl)methyl ester. cypermethrin.

dichloroethylene. ethylene dichloride.

1,1′-(2,2-dichloroethylidene)bis[4-chlorobenzene]. 1,1-dichloro-2,2-bis(*p*-chlorophenyl)ethane.

3,6-dichloro-2-methoxybenzoic acid. dicamba.

3′,4′-dichloro-2-methylacrylanilide. dicryl.

4,4′-dichloro-α-methylbenzhydrol. chlorfenethol.

2,4-dichloro-1-(4-nitrophenoxy)benzene. nitrofen.

(2,4-dichlorophenoxy)acetic acid. 2,4-D.

2,4-dichlorophenyl *p*-nitrophenyl ether.

nitrofen.

***N′*-(3,4-dichlorophenyl)-*N,N*-dimethyl-urea**. diuron.

***N′*-(3,4-dichlorophenyl)-*N*-methoxy-*N*-methylurea**. linuron.

3-(3,4-dichlorophenyl)-1-methyl-1-*n*-butylurea. neburon.

***o*-(2,4-dichlorophenyl)-*o*-methyl isopropylphosphoramidothioate**. DMPA.

***N*-(3,4-dichlorophenyl)-2-methyl-2-propenamide**. dicryl.

***N*-(3,4-dichlorophenyl)methylacrylamide**. dicryl.

di(*p*-chlorophenyl)trichloromethylcarbinol. dicofol.

(*N*-(3,4-dichlorophenyl)propanamide. propanil.

dichlorophos. dichlorvos.

1,3-dichloropropene (1,3-dichloropropylene; α,-γ-dichloropropylene; γ-chloroallyl chloride). a toxic, flammable liquid,

$$ClCH_2{=}CHCl$$

with a chloroform-like odor. The technical grade is a mixture of *cis-* and *trans-* isomers. It is used as a soil fumigant, nematocide, a solvent for oils and fats, in drycleaning, and in metal degreasing. It is moderately toxic orally to laboratory rodents; dermal toxicity to laboratory rabbits is moderate; it may be a human mutagen.

3′,4′-dichloropropionanilide. propanil.

α,α-dichloropropionic acid. dalapon.

2,2-dichloropropionic acid. dalapon.

α,-γ-dichloropropylene. 1,3-dichloropropene.

4,4′-dichloro-α-(trichloromethyl)-benzhydrol. dicofol.

dichlorovos. dichlorvos.

dichlorvos (phosphoric acid 2,2-dichloroethenyl dimethyl ester; phosphoric acid 2,2-dichlorovinyl dimethyl ester; O,O-dimethyl O-(2,2-dichlorovinyl) phosphate; dichlorophos; dichlorovos; DDVP).

$$(CH_3O)_2P(O)CH{:}CCl_2$$

An organophosphate, it is an amber, liquid, anticholinesterase that is used as a broad-spectrum insecticide and miticide used especially to control house flies and mosquitoes and to control both ecto- and endoparasites of domestic animals. Dichlorvos is moderately toxic to humans and extremely toxic to some animals. The minimum toxic dose is 10 mg/kg body weight in young calves and 25 mg/kg in horses and sheep. The oral LD_{50} in laboratory rats has been reported as 25-80 mg/kg. This compound is rapidly metabolized and excreted.

dicofol (4-chloro-α-(4-chlorophenyl)-α-trichloromethyl)benzenemethanol; 4,4′-dichloro-α-(trichloromethyl)benzhydrol; 1,-1-bis(*p*-chloro-phenyl)-2,2,2-trichloroethanol; di(*p*-chlorophenyl)trichloromethylcarbinol; DTMC). a neurotoxic organochlorine pesticide used mainly as a nonsystemic miticide on cotton and citrus fruits. The chemical name is 4,4′-dichloro-α-(trichloromethyl)benzhydrol. It interferes with the $Ca_2{}^+$, K^+, Na^+, and ATPases in insects and mammals. It is moderately toxic to laboratory rats and is carcinogenic to some laboratory animals. Symptoms of intoxication in humans may include nausea, vomiting, restlessness, fearfulness, tremor, convulsions, coma, respiratory collapse, and death.

dicophane. DDT.

dicotyledonous. **1a:** having dicotyledons. **1b**. pertaining to flowering plants.

dicoumarin. dicumarol.

dicoumarol. dicumarol.

dicrotophos (phosphoric acid 3-(dimethylamino)-methyl-3-oxo-1-propenyl dimethyl ester; phosphoric acid dimethyl ester, ester with *cis*-3-hydroxy-*N,N*-dimethylcrotonamide; 3-(dimethoxyphosphinyloxy)-*N,N*-dimethyl-*cis*-crotonamide; dimethyl 2-dimethylcarbamoyl-1-methylvinyl phosphate; dimethyl 1-dimethylcarbamoyl-1-propen-2-yl phosphate; Bidrin; Carbicron; Ektafos). the dimethyl phosphate of 3-hydroxy-*N,N*-dimethylcrotonamide, dicrotophos is an extremely toxic and hazardous liquid pesticide; it is fully miscible with water, ethanol, and xylene and somewhat soluble in kerosene. The commercial grade is brown. It is an anticholinesterase used as an insecticide and in some household products to control pests.

dicryl (*N*-(3,4-dichlorophenyl)-2-methyl-2-propenamide; 3′,4′-dichloro-2-methylacrylanilide; *N*-(3,4-dichlorophenyl)-methylacrylamide). a postemergent amide herbicide which, when applied to foliage, produces a localized or general necrosis that is dose dependent. It also inhibits growth. Dicryl is slightly toxic to laboratory rats.

Dictyocaulus. a genus of thin, elongate nematode lungworms (Family Metastrongylidae) that invade the respiratory tree of herbivorous mammals. There is no intermediate host, the life cycle is direct. The host is infected by ingested larvae. ***D. arnfieldi***. a species that infests the bronchi of equids, especially donkeys, horses, and mules. Most hosts are asymptomatic unless the infestation is heavy. ***D. filario*** (large lungworm; thread lungworm). the common lungworm of sheep, camels, and many feral ruminants as well. Infested animals may suffer serious consequences. Common signs include coughing, dyspnea, anemia, and emaciation. ***D. viviparus*** (hoose; husk). the common lungworm of many ruminants including cattle and deer. They commonly occur in the upper respiratory tract including the trachea, bronchi, and bronchioles. Affected animals have a characteristic chronic cough.

dicumarol (3,3′-methylenebis[4-hydroxy-2H-1-benzopyran-2-one]; bishydroxycoumarin; Dicoumarol; Dicoumarin; Dicumol; Dufalone; Melitoxin). a vitamin K antagonist that decreases the activity of vitamin K-dependent clotting factors in blood plasma. It inhibits the production of prothrombin in the liver, and increases prothrombin time. Dicumarol is used therapeutically as a long-acting, oral anticoagulant with a half-life in humans of four days. Dicumarol occurs in *Melilotus* (sweet clover) hay and is also prepared synthetically. It is the proximate cause of bleeding disease in cattle and can affect stock feeding on fresh clover or on hay.

Dicumol. dicumarol.

1,3-(4-dicyano)-2,4,5,6-tetrachlorobenzene. chlorothalonil.

1,3-dicyanotetrachlorobenzene. chlorothalonil.

dodecahydrodiphenylamine. dicyclohexylamine.

dog button. strychnine.

dicyclohexylamine (*N*-cyclohexylcyclohexanamine; dodecahydrodiphenylamine). a clear, colorless, liquid dicyclic aliphatic secondary amine used as a plasticizer, catalyst, and in insecticides, as corrosion inhibitors, antioxidants, and detergents. It produces effects similar to those of cyclohexylamine. Dicyclohexylamine is, however, less polar, more lipid soluble, more toxic, and much more likely to be absorbed in toxic amounts through the skin.

dieldrin (DDE; ((1aα,2β,2aα,3β,6β,6aα,-7β,7a)-3,4,5,6,9,9-hexachloro-1a,2,2a,3,-6,6a,7,7a-octahydro-2,7:3,6-dimethanonaphth[2,3-b]-oxirene; 1,2,3,4,10,10-hexachloro-6,7-epoxy-1,4, 4a,5,6,7,8,8a-octahydro-endo,exo-1,4:5,8-dimethanonaphthalene; Compound 497; Insecticide No. 497; HEOD). a light tan, nonflammable, environmentally persistent, chlorinated diene insecticide which is no longer manufactured or used in the United States. It is a stereoisomer of endrin. Dieldrin is toxic to an extremely wide spectrum of organisms and accumulates especially in aquatic organisms and is associated with reproductive failure in birds and fish. No other compound is known to produce such a degree of egg-shell thinning in birds as DDE.

diene. alkene.

diestrus. *See* estrus.

diethoxy thiophosphoric acid ester of 2-ethylmercaptoethanol. demeton.

***O,o*-diethyl *S*-(*p*-chloro-phenylthio)methyl phosphorodithioate**. carbophenothion.

***O,O*-diethyl *S*-(2-chloro-1-phthalimido-ethyl)phosphorodithioate**. dialifor.

***O,O*-diethyl *S*-(β-diethylamino)ethy-2-phosphorothiolate**. amiton.

***O,O*-diethyl dithiobis[thioformate]**. dixanthogen.

***O,O*-diethyl 2-ethylmercaptoethyl thiophosphate**. demeton.

***O,o*-diethyl *S*-ethylmercaptomethyl dithiophosphate**. phorate.

***O,O*-diethyl *O*(+*S*)-2-(ethylthio)ethyl phosphorothioate mixture**. demeton.

***O,o*-diethyl *S*-(ethylthio)methyl phosphorodithioate**. phorate.

***N,N*-diethyl-2-(1-naphthalenyloxy)propanamide**. napropamide.

***N,N*-diethyl-2-(1-naphthyloxy)propionamide**. napropamide.

(*0,O*-diethyl-*O*-(*p*-nitrophenyl) phosphate. paraoxon.

diethyl *p*-nitrophenyl phosphate. paraoxon.

***O,O*-diethyl *O*-*p*-nitrophenyl phosphorothioate**. parathion.

***O,O*-diethyl *O*-3,5,6-trichloro-2-pyridyl phosphorothioate**. chlorpyrifos.

***S*-[2-(diethylamino)ethyl]phosphorothioic acid *O,O*-diethyl ester**. amiton.

diethylcarbamodithioic acid 2-chloro-2-propenyl ester. sulfallate.

***O,O*-diethyl-*O*-[2-chloro-1-(2,4-dichlorophenyl)vinyl] phosphate**. chlorfenvinphos.

***O,O*-diethyl *S*-(2-chloro-1-phthalimido-ethyl)phosphorodithioate**. dialifor.

***O,O*-diethyl *S*-(β-diethylamino)ethyl-phosphorothiolate**. amiton.

diethyldithiocarbamic acid 2-chloroallyl ester. sulfallate.

3-(*O,o*-diethyldithiophosphorylmethyl)-6-chlorobenzoxazolinone. phosalolne.

***O,O*-diethyl-*S*-(2-ethioethyl)phosphorothioate**. demeton-*S*.

***O,O*-diethyl-*S*-ethyl-2-ethylmercaptophosphorothiolate**. demeton-*S*.

diethyl-*S*-(2-ethioethyl)thiophosphate. demeton-*S*.

***O,O*-diethyl-*S*-ethylmercaptoethyl dithiophosphate**. disulfoton.

***O,O*-diethyl-*S*-2-(ethylthio)ethyl phosphorothiolate**. demeton-*S*.

***O,O*-diethyl-*O*-(2-isopropyl-4-methyl-6-pyrimidinal)phosphorothioate**. diazinon [Chevron]®.

***O,O*-diethyl-*O*-[p-(methylsulfinyl)-phenyl]phosphorothioate**. fensulfothion.

diethyl-*p*-nitrophenyl monothiophosphate. parathion.

diethylstilbestrol (DES). a synthetic estrogen that was formerly used in the United States to promote growth of cattle and sheep and in the treatment of estrogen-deficiency disorders in veterinary medicine. It is carcinogenic, however, and produces other adverse effects, and its use in food-producing animals is no longer permited under regulations enforced by the Food and Drug Administration.

diethylxanthogenate. dixanthogen.

diffusion. **1:** a scattering or dispersion, e.g.,

of matter, energy, or ideas. **2:** a net movement of very small particles from a region of high concentration to a region of low concentration within a solution as a result of diffusion (defs. 3 and 4). In a closed system, a net movement will continue until the diffusible components are evenly dispersed. Diffusive movement within solids is extremely slow. **3:** the process whereby ions, molecules, and other very small particles within a gas, liquid, or solid disperse or intermingle as a result of random (thermal) molecular motion. **4:** the process whereby molecules move down a gradient of chemical potential until the gradient is abolished. **5:** the spontaneous mixing of one substance with another when in contact or separated only by a permeable barrier (e.g. a biological membrane). *See* osmosis. **facilitated diffusion** (facilitated transport). a type of carrier-mediated transport of molecules across a biological membrane down a concentration gradient. It proceeds faster than passive diffusion, but does not work against a concentration gradient. **passive diffusion**. movement across a cell membrane in which the rate of diffusion of the solute is directly proportional to its concentration gradient across the membrane. The gradient is the driving force for such movement. Unless qualified, the term diffusion usually refers to passive diffusion.

diffusion anoxia. *See* anoxia.

diflubenzuron (*N*-[[(4-chlorophenol)amino]-carbonyl]-2,6-difluorobenzamide). a stomach and contact insecticide, larvicide, and ovicide that acts as a growth regulator, blocks DNA synthesis, and inhibits the synthesis and incorporation of chitin into the insect cuticle. It is slightly or moderately toxic by ingestion to laboratory mice and rats, moderately toxic to laboratory mice, *i.p.*, and moderately toxic to laboratory rabbits via the percutaneous route of administration. The no-effect dietary dose for the laboratory rat has been reported as 40 mg/kg body weight. Diflubenzuron is, at most, moderately toxic to *Anas platyrhynchos* (mallard duck) and *Colinus virginianus* (bobwhite). *See* dose (median lethal dose), no-effect level.

DIFP. diisopropyl fluorophosphate.

digester. **1:** a cylindrical metal container used in papermaking to chemically separate lignin from cellulose. **2:** an enclosed composting system in which waste materials are mixed and aerated.

digestion. **1:** in biology, digestion is the breakdown by a living organism of complex molecules (typically food materials) into smaller, usually less complex molecules by mechanical and/or chemical processes. The process may be external as in the digestion of organic wastes by bacteria, or internal as in the digestion of food by most animals. In the latter case, digested material can then usually be absorbed, metabolized, and often assimilated by the organism's cells. Many poisons that are toxic by other routes of exposure may be detoxified through digestion if ingested. **chemical digestion**. that accomplished by digestive enzymes. **extracellular digestion**. **1:** digestion in which the digestive enzymes are released from cells, usually specialized cells, onto food that is either in an alimentary canal or cavity (e.g., as in vertebrate animals). **2a:** the biochemical decomposition of organic matter, resulting in partial gasification, liquefaction, and mineralization of pollutants. **2b:** the liquefaction of organic wastes by microbial action as in activated sludge. This is, of course, an application of biological digestion to engineering problems.

digestive tract. alimentary canal.

2,3-dihydro-2,2-dimethyl-7-benzofuranol methylcarbamate. carbofuran.

3,4-dihydro-2,2-dimethyl-4-oxo-2*H*-pyran-6-carboxylic acid butyl ester. butopyronoxyl.

6,7-dihydrodipyrido[1,2-a:2′,1′-c]-pyrazinediium dibromide. diquat.

dihydrotestosterone. *See* androgen.

1α,25-dihydroxycholecalciferol(1α,25-[OH],-D_3) glycoside. a derivative of calcitriol that may be responsible for some cases of enzootic calcinosis (*See* calci nosis) in herbivores.

3-(3,4-dihydroxycinnamate. chlorogenic acid.

3-(3,4-dihydroxycinnamoyl)quinic acid. chlorogenic acid.

2,2′-dihydroxy-3,3′,5, 5′,6, 6′-hexachlorodiphenylmethane. hexachlorophene.

1,8-dihydroxy-3-(hydroxymethyl)anthraquinone. aloe-emodin.

1,8-dihydroxy-3-(hydroxymethyl)-9,10-anthracenedione. aloe-emodin.

β,β'-di-hydroxyisopropyl chloride. α-chloroydrin. *See* chlorohydrin.

7,20-dihydroxy-16-methyl-10-phenyl-24-oxo[14]cytochalasa-6(12),13,21-triene-1,23-dione. cytochalasin B.

3-[[3-(3,4-dihydroxyphenyl)-1-oxo-2-propanyl]oxy]-1,4,5-trihydroxycyclohexanecarboxylic acid. chlorogenic acid.

2,6-diiodo-4-cyanophenol. ioxynil.

3,5-diiodo-4-hydroxybenzonitrile. ioxynil.

3,5-diiodo-4-hydroxyphenyl cyanide. ioxynil.

diisobutylthiocarbamic acid *S* ethyl ester. *S*-ethyl-*N,N*-diisobutylthiocarbamate. *See* butylate, def. 2.

diisocarb. *See* butylate, def. 2.

diisopropyl fluorophosphate (DFP; DIFP; diisopropyl phosphorofluoridate; diisopropylfluorophosphonate; diisopropylfluorophosphate; fluostigmine; isofluorphate; isoflurophate; isoflurophosphate; isopropylfluophosphate; phosphorofluoridic acid bis(1-methylethyl) ester; phosphorofluoridic acid diisopropyl ester; diisopropylphosphofluoridate). an extremely toxic fluorophosphate alkyl ester,

$$[(CH_3)_2CHO]_2P(O)F$$

It is an oily, liquid, cholinergic that irreversibly inhibits cholinesterase and proteolytic enzymes by reacting with serine residues. It is used in insecticides and nerve gas and as a miotic. DFP has pronounced miotic effects even at concentrations that are chemically undetectable. It forms hydrogen fluoride in the presence of moisture.

diisopropyl phosphorofluoridate. diisopropyl fluorophosphate.

diisopropyl-carbamic acid S-2,3-dichloroallyl ester. diallate.

diisopropylfluorophosphonate. diisopropyl fluorophosphate.

diisopropylphosphofluoridate. diisopropyl fluorophosphate.

diisopropylthiocarbamic acid S-(2,3,3-trichloroallyl) ester. triallate.

dilation of the pupil of the eye. *See* mydriasis.

dimetan. Dimetan®.

Dimetan® (dimethylcarbamic acid 5,5-dimethyl-3-oxo-1-cyclohexen-1-yl ester; 5,5-dimethyldihydroresorcinol dimethylcarbamate; dimetan). a synthetic neurotoxic carbamate insecticide. It is a cholinesterase inhibitor with an action similar to that of physostigmine.

1,4-di(methanesulfonyloxy)butane. busulfan.

dimethoate (phosphorodithioic acid *O,O*-dimethyl *S*-[2-(methylamino)-2-oxoethyl] ester; phosphorodithioic acid *O,O*-dimethyl ester, ester with 2-mercapto-*N*-methylacetamide). a systemic and contact insecticide and acaricide. A cholinesterase inhibitor, it is very toxic to laboratory rats.

[(dimethoxyphosphinothioyl)thio]butanedioic acid diethyl ester. malathion.

3-(dimethoxyphosphinyloxy)-*N,N*-dimethyl-*cis*-crotonamide. dicrotophos.

3-[(dimethoxyphosphinyl)oxy]-2-butenoic acid methyl ester. mevinphos.

dimethyl 1,2-benzenedicarboxylate. dimethyl phthalate.

4-dimethylamino-*m*-tolyl methylcarbamate. aminocarb.

[*O,O*-dimethyl *S*-1,2-bis(ethoxycarbonyl)-ethylphosphorothionate]. malaoxon.

***O,O*-dimethyl *O*-(2,2-dichlorovinyl) phosphate**. dichlorvos.

dimethyl 2-dimethylcarbamoyl-1-methylvinyl phosphate. dicrotophos.

dimethyl 1-dimethylcarbamoyl-1-propen-2-yl phosphate. dicrotophos.

***O,O*-dimethyl dithiophosphate of diethyl mercaptosuccinate**. malathion.

***O,o*-dimethyl *o*-(2-ethyl-4-ethoxypyrimidinyl)-6-thionophosphate**. etrimfos.

***O,O*-dimethyl *O*-(4-methylmercapto-3-methylphenyl) thionophosphate**. fenthion.

***O,O*-dimethyl *O*-(4-methylthio-3-methylphenyl) thiophosphate**. fenthion.

***O,O*-dimethyl *O*-[4-(methylthio-*m*-tolyl] phosphorothiolate**. fenthion.

***O,S*-dimethyl phosphoramidothioate**. methamidophos.

***O,o*-dimethyl *S*-phthalimidomethyl phosphorothionate**. phosmet.

O,O-dimethyl O-(2,4,5-trichlorophenyl)ester. Ronnel.

***O,O*-dimethyl-*S*-(*N*-methyl-*N*-formylcarbamoylmethyl)phosphorodithiolate**. formothion.

dimethyl 2,3,5,6-tetrachloroterephthalate. chlorthal-dimethyl.

3,3-dimethylacrylicacid 2-*sec*-butyl-4,6-dinitrophenyl ester. binapacryl.

***N*-(dimethylamino)succinamic acid**. daminozide.

4-(dimethylamino)-3-methylphenol-m-ethylcarbamate ester. aminocarb.

2-(dimethylamino)-*N*-[[(methylamino)-carbonyl]oxy]-2-oxoethanimidothioic acid methyl ester. oxamyl.

4-(dimethylamino)-3-methylphenol-methylcarbamate ester. aminocarb.

4-dimethyl-amino-*m*-tolyl methylcarbamate. aminocarb.

dimethylarsinic acid. cacodylic acid.

[(dimethylarsino)oxy]sodium As-oxide. sodium cacodylate.

2,2-dimethyl-1,3-benzodioxol-4-ol-methylcarbamate. bendiocarb.

1,1′-dimethyl-4,4′-bipyridinium. paraquat.

dimethylcarbamic acid 3-methyl-1-phenyl-1*H*-pyrazol-5-yl ester. pyrolan®.

dimethylcarbamic acid 1-[(dimethylamino)carbonyl]-5-methyl-1*H*-pyrazol-3-yl ester. dimetilan.

dimethylcarbamic acid 5,5-dimethyl-3-oxo-1-cyclohexen-1-yl ester. Dimetan®.

dimethylcarbamic acid ester with 3-hydroxy-*N,N*,5-trimethylpyrazole-1-carboxamide. dimetilan.

dimethylcarbamic acid 3-methyl-1-phenyl-1*H*-pyrazol-5-yl ester. pyrolan®

2-dimethylcarbamoyl-3-methyl-5-pyrazolyl dimethylcarbamate. dimetilan.

***O,O*-dimethyl-1-carbomethoxy-1-propen-2-yl phosphate**. mevinphos.

α,α-dimethyl-α′-carboxydihydro-γ-pyrone butyl ester. butopyronoxyl.

1,1-dimethyl-3-(*p*-chlorophenyl)urea. nonuron.

***O, O*-dimethyl-*S*-1,2-di-(ethoxycarbonyl)ethyl phosphorothiolo thionate**. malathion.

1,1-dimethyl-3-(3,4-dichlorophenyl-urea).

diuron.

2,2-dimethyl-2,3-dihydro-7-benzofuranyl-*N*-methylcarbamate. carbofuran.

5,5-dimethyldihydroresorcinol dimethylcarbamate. Dimetan®.

***O,O*-dimethyl-*O*-[*p*-(dimethylsulfamoyl)-phenyl]phosphorothioate (generic)**. famphur.

***N,N*-dimethyl-2,2-diphenylacetamide**. diphenamid.

***N,N′*-dimethyl-γ,γ′-dipyridylium**. paraquat.

dimethyl-*S*-(2-ethionylethyl) thiophosphate. demeton-*O*-methyl sulfoxide.

***O,O*-dimethyl-*S*-(2-ethionylethyl)phosphorothioate**. demeton-*O*-methyl sulfoxide.

***O,o*-dimethyl *o*-(2-ethyl-4-ethoxypyrimidinyl)-6-thionophosphate**. etrimfos.

***O,O*-dimethyl-*O*-ethylmercaptoethyl thiophosphate**. demeton-*O*-methyl.

***O,O*-dimethyl-2-ethylmercaptoethyl thiophosphate, thiono isomer**. demeton-*O*-methyl.

***O,O*-dimethyl-*S*-[2-(ethylsulfinyl)ethyl] phosphorothioate**. demeton-*O*-methyl sulfoxide.

***O,O*-dimethyl-*S*-ethylsulfinylethyl phosphorothiolate**. demeton-*O*-methyl sulfoxide.

***O,O*-dimethyl-*S*-(2-ethylsulfinyl)-ethyl thiophosphate**. demeton-*O*-methyl sulfoxide.

***O,O*-dimethyl-*O*-2-(ethylthio)ethyl phosphorothioate**. demeton-*O*-methyl.

1,1-dimethylhydrazine (*unsym*-dimethylhydrazine; *asym*-dimethylhydrazine; *N,N*-dimethylhydrazine; UDMH). a colorless, highly flammable, highly toxic, hygroscopic, water-miscible liquid, and a powerful reducing agent.

$$(CH_3)_2NNH_2$$

It is a plant growth regulator, a constituent of rocket and jet fuels, and a stabilizer for organic peroxide fuel additives. It is very toxic, orally, to laboratory rats and mice and moderately toxic by inhalation and skin contact. It is an irritant to the eyes, skin, and mucous membranes and a CNS stimulant, sometimes causing convulsions. UDMH is teratogenic and a confirmed carcinogen.

***asym*-dimethylhydrazine**. 1,1-dimethylhydrazine.

***N,N*-dimethylhydrazine**. 1,1-dimethylhydrazine.

***unsym*-dimethylhydrazine**. 1,1-dimethylhydrazine.

dimethylketal. acetone.

dimethylketone. acetone.

***N',N'*-dimethyl-*N*-[(methylcarbamoyl)oxy]-1-thiooxamimidic acid methyl ester**. oxamyl.

***N,N*-dimethyl-α-methylcarbamoyloxyimino-α-(methylthio)acetamide**. oxamyl.

***O,O*-dimethyl-*S*-(*N*-methyl-*N*-formylcarbamoylmethyl)phosphorodithiolate**. formothion.

4-(dimethylamino)-3-methylphenol-methylcarbamate ester. aminocarb.

2,2-dimethyl-3-(2-methylpropanyl)cyclopropanecarboxylic acid 6-chloropiperonyl ester. barthrin.

2,2-dimethyl-3-(2-methyl-1-propenyl)cyclopropanecarboxylic acid (6-chloro-1,3-benzodioxol-5-yl)methyl ester. barthrin.

(2,2-dimethyl-3-(2-methyl-1-propenyl)cyclopropanecarboxylic acid 2-methyl-4-oxo-3-(2-propenyl)-2-cyclo-penten-1-yl

ester. allethrin I.

2,2-dimethyl-3-(2-methyl-1-propenyl)cyclopropanecarboxylic acid 2-methyl-4-oxo-3-(2-propenyl)-2-cyclo-penten-1-yl ester allethrin I.

2,2-dimethyl-3-(2-methyl-1-propenyl)-cyclopropanecarboxylic acid (6-chloro-1,3-benzodioxol-5-yl)methyl ester. barthrin.

***O,O*-dimethyl-*O-p*-nitrophenylphosphorothioate**. methyl parathion.

dimethylnitrosamine. *N*-nitrosodimethylamine.

2,3-dimethyl-7-oxabicyclo-[2.2.1]heptane-2,3-dicarboxylic anhydride. cantharidin.

***o,o*-dimethyl-S-4-oxo-1,2,3-benzotriazin-3(4H)-yl methyl phosphorodithioate**. azinphos methyl.

***o,o*-dimethyl-S-4-oxo-1,2,3-benzotriazin-3(4H)-yl methyl phosphorodithioate**. azinphos methyl.

***N,N*-dimethyl-α-phenylbenzenacetamide**. diphenamid.

***N,N*-dimethyl-*N*-phenylurea**. fenuron.

***O,O*-dimethylphosphorothioate**. demeton-*S*-methyl.

dimethyl phthalate (1,2-benzenedicarboxylic acid dimethyl ester; phthalic acid dimethyl ester; methyl phthalate; dimethyl 1,2-benzenedicarboxylate; DMP). a highly toxic, colorless, slightly aromatic, synthetic ester that is slightly water soluble but is normally soluble in organic solvents. It is used as a solvent and plasticizer for cellulose acetate and cellulose acetatebutyrate compositions, in lacquers, resins, perfumes, and as an insect repellant. It is a persistent and widespread environmental pollutant. Dimethyl phthalate is an irritant to the eyes and mucous membranes but not to the skin. It is a CNS depressant.

1,4-di(methylsulfonoxy)butane. busulfan.

***O,O*-dimethyl-2,2,2-trichloro-1-*n*-butyryloxyethylphosphonate**. butonate.

***O,O*-dimethyl 2,2,2-trichloro-1-phosphonoethyl butyrate**. butonate.

dimethyltrichlorophenyl thiophosphate. Ronnel.

***N,N*-dimethyl-*N*'[3-(trifluoromethyl)-phenyl]urea**. fluometron.

1,1-dimethyl-3-(α,α,α-trifluoro-*m*-tolyl)urea. fluometron.

dimetilan (dimethylcarbamic acid 1-[(dimethyl-amino)carbonyl]-5-methyl-1*H*-pyrazol-3-yl ester; dimethylcarbamic acid ester with 3-hydroxy-*N,N*,5-trimethylpyrazole-1-carboxamide; 2-dimethylcarbamoyl-3-methyl-5-pyrazolyl dimethylcarbamate). a carbamate insecticide. It is extremely toxic orally to laboratory rats and moderately toxic dermally.

dimexan. a preemergent contact thiocarbonate herbicide used to control seedling dicotyledonous weeds before the crops emerge. It is also used as a preharvest desiccant for onions and peas. It is nonpersistent in soil.

dinitramine. a yellow solid used as a preemergence herbicide for annual grass and broad-leaved weeds in cotton and soybeans. It is carcinogenic to some laboratory animals.

2,4-dinitro-6-*sec*-butylphenol. dinoseb.

4,6-dinitro-2-*sec*-butylphenyl β,β-dimethylacrylate. binapacryl.

dinitrocresol (DN; DNC; DNOC; 2-methyl-4,6-dinitrophenol; 4,6-dinitro-*o*-cresol; 3,5-dinitro-2-hydroxytoluene; 3,5-dinitro-*o*-cresol; dinitro-*o*-cresol; 4,6-dinitro-2-methylphenol; Sinox). a solid, yellow compound,

$$CH_3C_6H_2(NO_2)_2OH$$

dinitrocresol is slightly water soluble and soluble in ethanol, ether, and acetone. It exists in nine isomeric forms, of which 3,5-dinitro-*o*-cresol is the most important commercially. It is toxic to humans and other animals. It is highly phytotoxic and cannot be used safely on or near actively growing plants. DNC is used chiefly as an herbicide to control weeds in cereal crops, an insecticide, and a dormant ovicidal pesticide spray for fruit trees. It is also used in preharvest desiccation of potatoes. It is an extremely toxic, cumulative poison.

3,5-dinitro-*o*-cresol. dinitrocresol.

4,6-dinitro-*o*-cresol. dinitrocresol.

dinitro-*o*-cresol. dinitrocresol.

2,6-dinitro-*N,N*-dipropyl-4-(trifluoromethyl)benzenamine. trifluralin.

3,5-dinitro-N^4,N^4-dipropylsulfanilamide. oryzalin.

2,6-dinitro-*N,N*-dipropyl-α,α,α-trifluoro-*p*-toluidine. trifluralin.

2,6-dinitro-*N,N*-dipropyl-4-trifluoromethylaniline. trifluralin.

dinitrogen. molecular nitrogen N_2; atmospheric nitrogen.

dinitrogen fixation. *See* nitrogen fixation.

dinitrogen monoxide. nitrous oxide.

3,5-dinitro-2-hydroxytoluene. dinitrocresol.

4,6-dinitro-2-methylphenol. dinitrocresol.

α-dinitrophenol. 2,4-dinitrophenol.

2,4-dinitrophenol (α-dinitrophenol; 2,4-DNP). a substance that is extremely toxic to plants and animals. Effects are similar to those of dinitrocresol, but less toxic. It is an intermediate in numerous organic syntheses, a wood preservative, insecticide, an indicator of pH, and a reagent used to detect ammonium and potassium ions. 2,4-DNP is extremely toxic to mammals, including laboratory rats, cattle, and humans.

dinocap (DNOCP). an acaricide and fungicide used mainly to control powdery mildew and other summer diseases of apples. The U.S. Environmental Protection Agency determined that dinocap is teratogenic to laboratory rabbits and recommended restricted usage in 1986.

dinoseb (2,4-dinitro-6-*sec*-butylphenol; 2-(1-methylpropyl)-4,6-dinitrophenol; 2-*sec*-butyl-4,6-dinitrophenol; DNBP). a nitrophenolic formerly widely employed as a selective herbicide against some weed species in cereal crops and also as an insecticide and miticide. It is extremely toxic to laboratory rats. Dinoseb is also a suspected teratogen in women who are exposed to it in early pregnancy. It may also cause sterility in human males.

dinoseb methacrylate. binapacryl.

dinoterb. a selective nitrophenol herbicide used chiefly for postemergence weed control in *Medicago sativa* (alfalfa, lucerne) and preemergence weed control in peas and beans.

2,3-*p*-dioxanedithiol *S,S*-bis(*O,o*-diethyl) phosphorodithioate. dioxathion.

dioxane-2,3-diylethylphosphorodithioate. dioxathion.

dioxathion (*p*-dioxane-2,3-diylethylphosphorodithioate; phosphorodithioic acid *S,S'*-1,4-di-oxane-2,3-diyl *O,O,O',O'*-tetraethyl ester; 2,3-p-dioxanedithiol *S,S*-bis(*O,O*-diethyl phosphorodithioate; "Delnav"). a viscous, nonvolatile, nonflammable, brown, nearly water-insoluble liquid,

$$C_4H_6O_2[SPS(OC_2H_5)_2]_2$$

It is a dangerous organophosphate cholinesterase inhibitor used as an insecticide and miticide. It is toxic to humans and laboratory animals by all routes of exposure. Studies indicate that dioxathion is extremely toxic, orally, to laboratory rats and extremely toxic, dermally, to female rats; it is significantly less toxic to male rats. It is hydrolyzed by heating and by alkali.

dioxin. an imprecisely defined family of more than 70 chlorinated compounds of which TCDD, *q.v.*, is the most widely known and often called "dioxin."

dip. **1a:** immerse; submerse. **1b:** to ladle, bail, or scoop. **2:** an aqueous solution of a chemical used for dipping animals, mainly sheep to remove ectoparasites. Most dips today contain organophosphates.

dipper. a deep trench used for dipping.

dipping. the process of immersing animals in a dip to remove ectoparasites.

dip slope (back slope; outface). a slope of the land's surface that conforms roughly to and is determined by the underlying rocks.

diphenadione. 2-diphenylacetyl-1,3-indanedione.

diphenamid (*N,N*-dimethyl-α-phenylbenzenacetamide; *N,N*-dimethyl-2,2-diphenylacetamide; *N,N*-dimethyl-α,α-diphenyl acetamide; 2,2-diphenyl-*N,N*-dimethylacetamide). a whitish crystalline slightly water-soluble organic compound used as a soil-applied, preemergence selective anilide herbicide for food crops, fruits and ornamentals. It is absorbed through the roots and inhibits root elongation. It is moderately toxic by oral, intraperitoneal and subcutaneous routes; it is probably mutagenic. Other soil-applied amides with similar action include alachlor, diphenamid, naptalam, and propachlor.

2-diphenylacetyl-1,3-indanedione (diphacinone; diphenadione). an indandione anticoagulant, a yellow, odorless powder or crystalline substance,

$$C_{23}H_{16}O_3$$

that is insoluble in water. It is an indandione anticoagulant used as a rodenticide.

2,2-diphenyl-*N,N*-dimethylacetamide. diphenamid.

diploid. **1a:** having two homologous sets of chromosomes. **1b:** of or pertaining to a cell of a eukaryote that has two complements of chromosomes, each derived from a different haploid gamete.

diploid cell. a eukaryote cell that contains two chromosome complements. *See* haploid cell.

diploidy. the diploid state.

4-(dipropylamino)-3,5-dinitrobenzenesulfonamide. oryzalin.

dipropylcarbamothioic acid *S*-ethyl ester. EPTC.

***N,N*-dipropyl-2,6-dinitro-4-trifluoromethylaniline**. trifluralin.

dipropylthiocarbamic acid *S*-ethyl ester. EPTC.

dipsogenic. inducing thirst; having the capacity to induce thirst.

Diptera. an order of insects, the true flies (e.g., fruit fly, horse fly, house fly, tsetse fly, gnat, midge, mosquito). All undergo complete metamorphosis; they have sucking or piercing-and-sucking mouthparts. A few species are wingless, all others have a single pair of functional wings. The hind wings are reduced to knobbed balancing organs (halteres). A number of forms are transvectors of disease to humans, livestock, and/or other animals.

diquat (6,7-dihydrodipyrido[1,2-a:2′,1′-c]-pyrazinediium dibromide; 1,1′-ethylene-2,2′-di-pyridylium dibromide; FB/2; diquat dibromide; Aquacide; Reglone). a bipyridylium, quaternary, ammonium salt used as a rapid-acting contact herbicide, desiccant, and defoliant. It is a convulsant and, at high doses, a gastrointestinal irritant. It is very toxic to laboratory rats, mice, and guinea pigs. Symptoms of acute diquat poisoning include acute hyperexcitability, convulsions, and gastrointestinal irritation and distension. *See also* paraquat.

diquat dibromide. diquat.

direct effect. *See* effect.

direct metamorphosis. *See* metamorphosis.

disaster payment. a direct payment to crop producers by the federal government when planting is prevented or crop yields are exceptionally low because of adverse weather and related natural conditions.

disc-, disco-. combining forms indicating a disk, relation to a disk, or discoidal shape of properties.

discharge. **1:** liberate; release. **2:** flow; issue forth. **3:** to shoot or fire (e.g., a rifle). **4:** dismiss; fire. **5:** execute; perform; fulfill. **6:** secrete, emit. **7:** excrete. **8:** release; emission; flow. **9:** a detonation or explosion; firing (as of a rifle). **10:** dismissal; release. **10:** execution, performance. **11:** secretion; emission. **12**. excretion.

disease. **1a:** morbus; ailment; illness; sickness; affliction; disorder; abnormal condition; infirmity; malady; complaint. **1b:** deviation of a living organism or part of an organism from a healthy or normal condition. **1c:** a physical disorder with a specific physiological basis and recognizable signs and symptoms. **1d:** any condition in which life functions are impaired, interrupted, or fail. *See also* disorder.

disease vector. an organism that harbors disease organisms of plants or animals and conveys or transfers them to host organisms. Ticks and mosquitos are well-known vectors of various diseases.

disinfect. to destroy pathogenic or otherwise harmful microorganisms, or to otherwise remove them from an object, by chemical or physical means.

disinfectant. **1:** having the capacity to disinfect. **2a:** any agent or process, chemical or physical, that disinfects. A given disinfectant may not destroy all types of microorgisms and they are usually ineffective against spores. **2b:** any of a wide variety of nonselective antimicrobial agents or processes that kill or inhibit the growth or activity of pathogenic microorganisms in or on nonliving material. Disinfectants (often used as aerosols sprays) contain any of a variety of toxic volatile chemicals that are dangerous when inhaled. Cresol, *q.v.*, is a frequent component of disinfectants. Other toxic chemicals frequently found in disinfectants are phenol, ethanol, formaldehyde, ammonium hydroxide, and chlorine.

disinfection. the process of destroying most living organisms (usually microbial pathogens) in water or on surfaces. *Cf.* sterilization.

disintegration. **1a:** deteriorating; crumbling; falling apart. **1b:** decaying; decomposing. **1c:** decay; decomposition. **2:** the process and result of the weathering and breakdown of rocks or minerals.

disk hiller. an attachment to a cultivator that throws up a hill of soil around the plants in the row.

disk mower. a mower that cuts hay by means of blades on rotating disks mounted on an arm.

dismount. to descend, alight, or get off as from a horse or motorcycle.

disodium ethylenebis[dithiocarbamate]. nabam.

disodium 3-hydroxy-4-((2,4,5-trimethylphenyl)azo)-2,7-naphthalenedi-sulfonate. FD&C Red No. 1.

disodium 3-hydroxy-4-((2,4,5-tri-methylphenyl)azo)-2,7-naphthalenedisulfonic acid. FD&C Red No. 1.

disodium 3-hydroxy-4-((2,4,5-trimethylphenyl)azo)-2,7-naphthalenedisulphonate. FD&C Red No. 1.

disodium indigo-5,5-disulfonate. FD&C Blue No. 2

disodium monomethanearsonate (DSMA). the disodium salt of methanearsonic acid,

$$CH_3AsO(ONa)_2$$

It is a selective postemergence herbicide used to control grass and weeds in cotton and rubber plantations. It is moderately toxic to laboratory rats.

disodium or dipotassium salt of 2,4,5,7-tetraiodofluoroscein. erythrosine.

disorder. an illness, disease, or any structural or functional disturbance or malfunction of an organism generally, or of a body part, regardless of cause. This term is roughly synonymous with disease, *q.v.*

disseminated. **1a:** dispersed; distributed; widely scattered. **1b:** sown; spread; strewn, scattered. Thus, the seed was disseminated by the wind.

dissimilatory nitrate reduction to ammonium (DNRA). the use of nitrate as an alternate electron acceptor by organisms in the absence of molecular oxygen, accompanied by the reduction of nitrate to ammonium. *Cf.* denitrification.

dissolved oxygen. *See* oxygen.

distress. **1a:** anguish; misery; pain; suffering (mental or physical). **1b:** anxiety; tension. **2a:** mental or physical suffering. **2b:** grief. **3:** any detrimental condition due to stress.

disulfide. disulfiram.

disulfiram (tetraethylthioperoxydicarbonic diamide; bis(diethylthiocarbamoyl); bis(diethylthiocarbamyl) disulfide; disulfide; tetraethylthiuram disulfide; teturamin; TETD; TTD; Antabuse). a light gray, slightly odorous, water-insoluble powder,

$$[(C_2H_5)_2NCS]_2S_2$$

It is soluble in carbon disulfide, benzene, and chloroform. It has many industrial applications and is used also as a fungicide, seed disinfectant, and as a chelator in copper and nickel poisoning. Under the name Antabuse, it is used in the treatment of alcohol abuse.

disulfoton ("Di-Syston"; phosphorodithioic acid *O,O*-diethyl *S*-[2-(ethylthio)ethyl] ester; *O,O*-diethyl-*S*-ethylmercaptoethyl dithiophosphate; thiodemeton; dithiodemeton). an organophosphate,

$$(C_2H_5O)_2P(S)SCH_2CH_2SCH_2CH_3$$

It is a yellow liquid when pure and is soluble in water; the technical grade is brown. Disulfoton is an anticholinesterase used as a systemic insecticide and acaricide; it is also a component of some household products used to eradicate pests. It is highly poisonous to vertebrates.

dithiobis[thioformic acid] *O,O*-diethyl ester. dixanthogen.

2,4-dithiobiuret (thioimidodicarbonic diamide). a crystalline compound,

$$C_2H_5N_3S_2$$

It is used in the manufacture of insecticides and rodenticides and to prevent wilting of plants. It is highly toxic by all routes of exposure and is extremely hazardous.

dithiocarbamate fungicide. any of a number of fungicides based on dithiocarbamic acid. They are chelates that consist of metal salts of dimethylthiocarbamate and ethylenebisdithiocarbamate anions and are named according to the metal ion present. Thus, maneb is the manganese salt of dimethyldithiocarbamate anion and zineb and nabam are the zinc and sodium salts, respectively. Ferbam is the iron salt of the ethylenebisdithiocarbamate anion, and ziram is the zinc salt of this ion. These fungicides are used commonly in agriculture because they are effective against fungi, and though they are immunotoxic, their toxicities to animals are relatively low. Certain products of the environmental degradation of these fungicides (e.g., ethylenethiourea) are a matter of concern.

dithiodemeton. disulfoton.

1,2-dithiolane-3-pentanoic acid. thioctic acid.

1,2-dithiolane-3-valeric acid. thioctic acid.

5-(1,2-dithiolan-3-yl)pentanoic acid. thioctic acid.

5-(1,2-dithiolan-3-yl)valeric acid. thioctic acid.

6,8-dithiooctanoic acid. thioctic acid.

diuresis. the free or excessive discharge of urine.

diuresis. the secretion and passage of large volumes of urine. Diuresis occurs in diabetes mellitus, diabetes insipidus, in chronic interstitial nephritis, as a result of anxiety or fear, and when stimulated by diuretics (*q.v.*).

diuretic. **1a:** promoting the production of urine. **1b:** a substance that promotes the production of urine.

diuron (*N*′-(3,4-dichlorophenyl)-*N,N*-dimethyl-urea; 1,1-dimethyl-3-(3,4-dichlorophenyl-urea)). an extremely persistent preemergence herbicide with low water solubility. It is very toxic to laboratory rats. A cumulative poison, repeated doses can cause anemia and methemoglobinemia in laboratory rats if hydrolyzed endogenously.

diversity. variety; heterogeneity. **biological diversity** (biotic diversity; biodiversity). the diversity of living organisms, now and in the past; not only the diversity of species and other taxa, but of biological processes, genes, ecosystems, and every level of organization and function. Scientists generally use this term in reference to diversity of life on Earth or at designated locations, habitats, communities, or regions as represented by species diversity, intraspecies (genetic) diversity, and ecosystem diversity. Most biologists, many agriculturalists, and many government officials believe that maintaining biodiversity and the retention of the greatest possible number of existing species and habitats is exceedingly important, because it is the only way to retain a diverse genetic pool that can be tapped for useful, beneficial, and perhaps essential characteristics and commodities in the future. Biodiversity is the framework of nature that makes life as we know it possible (*See*, for example, genetic diversity). In agriculture, biodiversity can be considered a production system, one that is characterized by the presence of multiple plant and/or animal species, as contrasted with the genetic specialization of monoculture. **biodiversity**. biological diversity. **biotic diversity**. biological diversity. **ecosystem diversity**. the diversity of ecosystems throughout the world from various microcosms, to tide pools, to the tundra, and the vast, poorly known ecosystems of the oceans. As in the case of species diversity and genetic diversity, ecosystem diversity is being eroded. As one example, large areas of the semiarid regions of Africa and, indeed, even in the United States, are undergoing desertification at the hands of humans. **genetic diversity**. intraspecies diversity with respect to genetic structure. **intraspecies diversity** (intraspecific diversity; genetic diversity). diversity of genomes and phenotypes with species. Often ignored because of the importance of species diversity and the current rapid rate of extinctions, intraspecific diversity may be just as important ecologically and to agriculture. As one example, intraspecies diversity provides resources for genetic resistance to pests and diseases. **intraspecific diversity**. intraspecies diversity. **phenotypic diversity**. intraspecies diversity with respect to the variety of characteristics of a certain type that occur (e.g., of color phases or patterns). **species diversity**. **1:** the absolute number of species in an area, community, assemblage, or sample of living organisms. Species diversity is lower, generally, in agroecosystems than in natural ecosystems. Fewer kinds of plants and animals are usually present. Considered globally, while new breeds and cultivars do appear, farmers tend to favor fewer and fewer of these and extinctions or near extinctions of old breeds are frequent. Agriculturalists are also contributing their share to the Earth's growing pollution problem, with sometimes serious effects on ecosystem viability. **2:** a standard measure that combines the number of species in a community such that a low value (low diversity) indicates either that few species are present or that abundance of those present are unequal; the converse of this indicates high diversity. Diversity is generally regarded as a stabilizing feature of a community.

dixanthogen (thioperoxydicarbonic acid diethyl ester; dithiobis[thioformic acid] *O,O*-diethyl ester; *O,O*-diethyl dithiobis-[thioformate]; bisethylxanthogen; diethylxanthogenate; ethylxanthic disulfide; preparation K; EXD). a nearly water-insoluble, non-persistent, preemergent contact thiocarbonate herbicide with an onion-like odor. It is used to control seedling dicotyledonous weeds before the emergence of crops. It is also used in insecticide formulations and as an ectoparasiticide. Dixanthogen is very toxic to laboratory rats, *by mouth*.

djenkol bean. *Pithecolobium lobatum*.

DMC. chlorfenethol.

DMDT. methoxychlor.

DMN. *N*-nitrosodimethylamine.

DMNA. *N*-nitrosodimethylamine.

DMP. dimethyl phthalate.

DMPA ((1-methylethyl)phosphoramidothioic acid *o*-(2,4-dichlorophenyl) *o*-methyl ester; isopropylphosphoramidothioic acid *o*-2,4-(dichlorophenyl) *o*-methyl ester; *o*-(2,4-dichlorophenyl)-*o*-methyl isopropylphosphoramidothioate). **1:** an herbicide and plant growth regulator, it is very toxic orally to laboratory rats; it is neurotoxic to chickens. It is moderately toxic der- mally. **2:** DMPA (or DMPA®) is also an acronym for dimethylolpropionic acid.

DN. dinitrocresol.

DNBP. dinoseb.

DNC. dinitrocresol.

DNOC. dinitrocresol.

DNOCP. dinocap.

2,4-DNP. 2,4-dinitrophenol.

DNPT. parathion.

DNRA. dissimilatory nitrate reduction to ammonium.

DO. dissolved oxygen (*See* oxygen).

dock. common name of herbs and shrubs of the genus *Rumex*.

doddie. Aberdeen Angus.

1,1a,2,2,3,3a,4,5,5,5a,5b,6-dodecachlorooctahydro-1,3,4-metheno-1*H*-cyclobuta(c,d)pentalene. mirex.

dodecachloropentacyclo(3,2,2,0^{2,6},-0^{3,9},0^{5,10})decane. mirex.

dog's mercury. an herb of the genus *Mercurialis*.

dogbane. a plant of the family Apocynaceae, especially one of the genus *Apocynum*.

dogbanes. *See* Apocynaceae.

dogwood. *Cornus*.

doll's eyes. *Actaea alba*.

dollar spot. appaloosa.

domestic ass. *See E. a. asinus*.

domestic fowl. *Gallus domesticus*.

domestic market. a market within the boundaries of the country where a commodity is produced.

domestic rat. *Rattus norvegicus*.

dominant. descriptive of an allele whose phenotype is expressed even in the presence of a recessive allele. *See also* allele, phenotype, recessive.

dominant gene. a gene whose phenotype is expressed when present in a single copy.

donkey. an ass or burro. *See* ass.

doob. *Cynodon dactylon.*

dormancy. **1:** a state of inactivity. **2:** a period in which a seed does not germinate or in which a plant does not grow. The natural breaking of dormancy may require particular environmental information such as winter weather (cold treatment) and day length or changes in photoperiod (photostimulation), and may require particular conditions of temperature, moisture, and nutrient availability.

dormant season. in agronomy, that part of the year when plants are dormant. In mid and high latitudes, this roughly coincides with the period when frosts occur; day length may also play a role in the induction and breaking dormancy, as may dry periods in some climates.

dormin. obsolete name for abscisic acid

dorsal. **1a:** pertaining to or nearer the dorsum, *q.v.* **2a:** pertaining to the upper surface of a leaf or other plant structure. **2b:** nearer to the surface of a petal that faces away from the axis of a flower.

Dorset sheep. a medium-sized breed in which both rams and ewes are horned. It was developed in England, and, while repeatedly introduced into the United States, they are not widely popular.

dorsum. **1a:** the part of an animal's body opposite the venter or belly. **1b:** the surface of a chordate that is nearest to the notochord. **2:** the surface of a prostrate organism or structure that is furthest from the substrate (e.g., as in flatworms and fern prothalli).

double cropping. the consecutive production of two crops of similar or dissimilar commodities on the same land within the same year.

double spruce. *Picea mariana.*

doubling time. the time required under specified conditions for a population to double in number or biomass.

douglas fir. *Pseudotsuga menziesii.*

douglas pine. *Pseudotsuga menziesii.*

douglas spruce. *Pseudotsuga menziesii.*

Douglas water hemlock. *Cicutum douglassi.*

downer. nonambulatory livestock.

dracylic acid. benzoic acid.

draft. **1:** a current of air. **2a:** a school of fish. **2b:** the fish taken in one drawing of a net. **3:** the taking of a single drink; the amount thus drunk. **4a:** pulling. **4b:** traction. **5:** the effort required to pull an implement through soil. **6:** a draft animal.

draft animal. an animal used to pull or bear heavy loads.

draft horse. a horse used to pull or bear heavy loads.

dragon arum. *See Arisaema.*

dragon root. *Arisaema dracontium.*

drainage. in agriculture, the process of removing excess water from the soil, either by ditches or by underground conduits.

draught. British, meaning draft.

dredging. the removal of bottom sediments ("muds") from a water body (e.g., river, stream, lake, reservoir) using a machine with a large scoop (a dredge). The process and loss of sediments disturbs the ecosystem and causes silting that can kill aquatic life. Dredging of contaminated sediments can expose aquatic life to otherwise sequestered toxicants such as heavy metals and other persistent toxicants. In the United States, dredging is regulated under Section 404 of the Clean Water Act.

dried cupric sulfate. the monohydrate salt of copper sulfate.

dried egg white. albumen.

dropping, fecal dropping. **1a:** a scat. **1b:** more generally, feces.

dropsy. edema.

drove. a group of certain kinds of animals that are being moved or herded, especially over a long distance (e.g., cattle, sheep. *See also* flock, herd.

dry. **1a:** arid; sere. **1b:** thirsty. **2:** to dehydrate, dessicate, or parch. **3:** to preserve salt, or cure. **4:** of or pertaining to a temporarily or permanently nonlactating female mammal.

dry adiabatic lapse rate. *See* lapse rate.

dry deposition. *See* acid deposition.

dry gangrene. *See* ergotism (gangrenous ergotism); *See also* fescue foot.

Drymaria (drymary). a genus of small branched annual or perennial herbs (Family Caryophyllaceae). ***D. arenarioides*** (Alfombrilla). a small, luxuriant, short-lived perennial herb that occurs on acid soils of northern and central Mexico. It is very toxic to livestock, although horses are not known to be affected. Toxicity, symptoms, and lesions in cattle are similar to those of *D. pachyphylla*. ***Drymaria pachyphylla*** (inkweed, drymary). a small, branching, prostrate summer annual that occurs on alkaline clay soils of west Texas, southern New Mexico to southeastern Arizona, and to central Mexico. Serious cattle losses have been recorded, mostly in drought years when other forage is unavailable. In experimental animals, symptoms usually appear within 18-24 hours following ingestion of a toxic dose. Death usually ensues within two hours following the appearance of symptoms. Symptoms include restlessness, anorexia, diarrhea, arching of the back, depression, and often coma and death. Survivors recover in about two days.

drymary. *See Drymaria.*

DSMA. disodium monomethanearsonate.

DTMC. dicofol.

duck ratten. *Veratrum viride.*

ductless gland. endocrine gland.

Dufalone. dicumarol.

Dugaldia hoopesii. synonym of *Helenium hoopesii*.

dule. a flock of doves.

dune (sand dune). a usually unstable hill or ridge of sand that has been formed by the action of winds. Dunes occur most commonly on sandy deserts such as the Sahara and on sea coasts. **anchored dune**. a dune that has been stabilized by usually well-rooted vegetation. Such dunes can be destroyed by wave action, by persistent winds or winds of unusual force, and by destruction due to human presence.

dune grass. *Elymus arenarius*.

Duroc. a large, rugged, solid red breed of swine developed in the northeastern United States during the 1800s. It is the result of crosses among a number of red strains. It is a breed noted for hardiness and productivity.

Dursban *[Dow]*®. a brand name for insecticides that contain chlorpyrifos. They are used chiefly to control chinch bugs around the Gulf of Mexico and to control ticks on cattle and sheep in Australia.

dust. **1:** fine dry particles or powder that are light enough to remain suspended in air for some time. Dust particles of a respirable size, which are toxic or to which toxicants adhere, can be quite hazardous. **2:** in air pollution science, dust is defined as the largest particulate matter contributing to air pollution, or those particles with a diameter of about 10μ. Exposure of air-breathing animals to dusts can induce any of a large variety of focal respiratory diseases, including pneumoconioses with pulmonary fibrosis and obstructive lung disease. Certain toxic dusts, when inhaled, may produce systemic poisoning, allergic reactions, dermatoses, and cancer. **atmospheric dust**. minute particles found at all elevations in the atmosphere that remain airborne due to slight air currents or settle slowly.

Sources are numerous and include smoke and various products of combustion, dust raised by winds from bare soil, especially very dry or disturbed soils (as in deserts, plowed fields, and roadways), ash from volcanic eruptions, pollen, or salt spray from the oceans. Except in the case of dust storms (usually originating from dry, or dry plowed, land), the highest levels of atmospheric dust usually occurs at low levels over cities; the lowest levels occur high above major oceans. **cotton dust**. a fine powder generated during the processing of raw cotton. Inhalation can cause byssinosis (brown lung disease). Symptoms include tightness of the chest and dyspnea. If detected early, the condition is reversible, but long-continued exposure to cotton dust may be permanently crippling and ultimately fatal. *See* byssinosis. **non-proliferative dust**. dusts other than proliferative dusts. They do not induce fibrotic changes in the lung, and their effects are generally reversible; they are cleared from the lungs by phagocytosis. **proliferative dust**. any type of dust that is not readily removed from the lungs by phagocytosis or other clearance mechanisms. Such dusts accumulate in the lung and may cause irreversible lung damage via fibrotic hardening known as a pneumoconiosis (examples are asbestosis, byssinosis, silicosis). **respirable dust** (respirable particles). suspended particulates that can be deposited to a significant extent in the lungs. There is little agreement about the type and size of particles involved, and the term is used variously by different authors. The size range of respirable particles varies with the depth and rate of breathing and probably with any of a number of taxon-specific factors. **volcanic dust**. volcanic ash (*See* ash).

Dutch liquid. ethylene dichloride.

Dutch oil. ethylene dichloride.

Dutch rush. *Equisetum hyemale*.

Dutchman's breeches. *See Dicentra*.

dwarf mistletoe. *Arceuthobium pusillum*.

dysentery. inflammation of the intestines marked by the frequent passage of feces, usually with blood and mucus. The two most common causes are infection with a bacillus (see bacteria) of the genus *Shigella*, and infestation by an ameba, *Entamoeba histolytica*. Both bacillary and amebic dysentery are spread by fecal contamination of food and water and are most common under conditions of poor sanitation. Dysentery, while chiefly occurring in the tropics, may occur in any climate. **amebic dysentery** (amebiasis; amebic colitis). a type of dysentery caused by the protozoan Entamoeba histolytica, characterized especially by ulceration of the large intestine. *Cf.* amebiasis. **bacillary dysentery** (shigellosis). that caused by the presence of bacilli of the genus *Shigella* in contaminated food.

dyspnea (shortness of breath; difficult or labored breathing, respiratory distress). difficult or forced and often irregular breathing. Dyspnea can arise from numerous conditions (e.g., extreme fear, asthma, cardiac insufficiency) or environmental factors. It can be induced directly or indirectly by any of numerous toxic substances including asphyxiants, pulmonary carcinogens, irritant gases, cardiotoxicants, or histamine. **cardiac dyspnea**. difficult or distressful breathing due to heart disease.

dyssynergia. ataxia.

dysuria. painful of difficult urination.

dziggetai. *Equus h. hemionus*.

E

ϵ. symbol for epsilon.

E. symbol for epsilon.

E.P. edible portion.

e$^-$. an electron.

E127. erythrosine.

E-600. paraoxon.

EAE. endemic abortion of ewes.

eagle fern. *Pteridium aquillinum.*

ear. **1a:** the organ of sound collection and hearing in vertebrate animals. It is also an organ of equilibrium. **1b:** any of diverse organs that sense vibrations in the medium surrounding the organism (e.g., air, water) at frequencies greater than several cycles per second. **1c:** often used to designate only the external ear or ear lobe (e.g., of mammals). **2:**the grainbearing spike of a cereal plant such as corn, rye, or wheat. **3:** to produce an ear (def. 2).

ear corn. unshelled corn.

ear leaf. a cotyledon.

ear marking (ear notching). the notching or distinctive removal of parts of the external ears livestock, for purposes of identification of the individual animal or its owner.

ear notcher. a punch used to cut a set of identifying notches in the ears of an animal.

ear notching. *See* ear marking.

ear tag. an identifying tag, fastened to the ear of an animal.

eared. **1:** having ears. **2:** an animal restrained by its ears. **3:** horned; indicating the presence of earlike tufts of feathers on the heads of some birds (e.g., owls).

eared cattle. cattle of Brahman extraction.

early. **1a:** first; initial; beforehand; in advance; premature. **1b:** occurring before the usual, normal, expected, or designated time (as an early spring). **2:** descriptive of a variety of plants that normally produce buds, leaves, mature fruit, etc. earlier in the year or season than normal. **3:** denoting a period of planting or harvesting that predates the expected or usual time.

early meadow grass. common name of *Poa annua.*

early mesquite. *Buchloe dactyloides.*

early speed. the ability of a horse to gather maximum speed in one or two strides.

earth. **1:** a planet in the solar system. **2a:** soil; dirt. **2b:** the ground; land.

earth gall. *Veratrum viride.*

earth up. to cultivate by throwing earth up about a plant as in the hilling of potatoes.

earth tide. the tide-like motion of the surface of Earth due to the same forces that produce tides in the oceans and other large masses of surface water. Earth tides occur twice each day, fluctuating from 7-15 cm (3-6 in.).

earthen dam. a dam across a waterway of various composition but including soil and soil-like mineral matrices such as clay, gravel, rocks, sand, and silt.

earthing. the mounding or banking up of soil about a plant.

earthworm (angle worm; fishing worm). any burrowing terrestrial oligochaete such as those of the genus *Lumbricus*. They are eaten by many kinds of wildlife, and some chemical elements can reach con centrations in earthworms that are potentially hazardous to wildlife at higher trophic levels. Earthworms bioconcentrate heavy metals to levels about one order of magnitude above that of the surrounding soil. Even in the absence of appreciable environmental contamination, some species of earthworms often contain high concentrations of Pb, Zn, and Se. Those of the genus *Eisenoides*, for example, have a remarkable ability to concentrate lead. The average bioconcentration factor for lead in *E. loennbergi* in one study was 52, while in other earthworms, the BCF for lead was less than one.

earthworm farm. a farm that produces earthworms to sell, for example, as fish bait.

earwax. cerumen.

east coast fever. an often lethal theileriasis of bovids, including domestic cattle and buffalo. It is a serious problem in East and Central Africa. The causative agent is the protozoon, *Theileria parva*. The sporozoites are transmitted to cattle by the tick, *Rhipicephalus appendiculatus*. Effects range from mild to fatal and may include persistent fever (107°F (42°C) or more), tearing and nasal discharge, swelling of lymph nodes may be pronounced, and anorexia with loss of condition and enfeeblement. Diarrhea and dyspnea are common features in fatal cases. Proper treatment can often prevent fatalities, and recovered animals are immune to infection by most strains of *T. parva*, but they may be carriers.

easter. a high wind or a rain storm with high winds that comes from an easterly direction.

eastern bracken. *Pteridium acquilinum*.

eastern diamondback rattlesnake. *Crotalus adamanteus*.

eastern fir. *Abies balsamea*.

eastern spruce. *Picea mariana*.

eastern whorled milkweed. *Asclepias subverticillata*.

Eastern honeybee. *Apis cerana*.

Eastern poison oak. *Rhus toxicodendron*.

easy. steady; a command to a horse to move slower. **2:** said of an animal that has a weak back or pasterns.

easy gait. any of a horse that is conveniently slow and easy on the rider.

easy keeper. a domestic animal that requires a minimum of feed and care to maintain.

eating. **1:** ingesting; devouring; consuming; chewing. **2:** eating away; eroding; corroding; rotting. **3:** descriptive of a food that is suitable for eating without processing, as an eating apple.

eating ripe. descriptive of fruit that is fully ripe and suitable for eating as-is.

EBA. epidemic bovine abortion.

EBDC. ethylene bis(dithiocarbamate).

ecad. a plant or a population of a species of plant that differs morphologically from the norm due to environmental influences.

eccentric. **1:** peculiar; idiosyncratic; unconventional; bizarre. **2:** descriptor of an a-symmetric tree trunk.

eccoprotic. cathartic.

eccrine gland. a coiled tubular gland (notably a sweat gland) other than an apocrine gland that is present and widely distributed in the skin. *See also* exocytosis.

eccrine secretion. exocytosis.

ecdemic. of or pertaining to a disease that is occurs far from the area of its origin.

ecdysial gland. any of several ectodermal glands (e.g., prothoracic) of insects.

ecdysone (molting hormone). **1a:** any of a class of steroid hormones of insects and crustaceans that regulate pupation; they stimulate growth and molting. More than 30 are known. **1b:** any of a set of hor mones with a similar action, including some compounds from certain plants or invertebrate animals and some synthetics. **α-ecdysone** (2,3,14,22,25-pentahydroxy-cholest-7-en-6-one). the main ecdysone of insects and crustaceans and the first to be isolated (in 1954). **β-ecdysone** (2,3,14,-20,22,25-hexahydroxycholest-7-en-6-one; 20-hydroxyecdysone; ecdysterone; crust-ecdysone; isoinokosterone). the main ecdysone of plants.

ecdysterone. β-ecdysone.

ece. habitat.

ecesis. The establishment of an invading plant in a community, by virtue of its ability to germinate, grow, and reproduce in the new environment.

Echinochloa crusgalli (barnyard grass). a grass (Family gramineae) in which concentrations of nitrate toxic to livestock have been reported.

Echinocystis lobata (wild balsam apple; prickly cucumber). a species of green plant (Family Cucurbitaceae) indigenous to the United States. The fruit typically explodes when ripe, shooting out seeds.

eclosion. the process of hatching from an egg.

eco-. a prefix indicating a relationship to environment.

ecocide. willful destruction of a portion of the environment.

ecoclimate. the climate within a particular habitat.

ecoclimatology. bioclimatology.

ecologic, **ecological**. of or pertaining to ecology.

ecological effect. *See* effect.

ecological system. ecosystem.

ecological unit. in forestry, a land area characterized throughout by the presence of similar biotic and physical (geological, climatic) environments.

ecology (bionomics, in part). **1:** bioecology, *q.v.*; the science of interrelationships within and among populations of living organisms and of populations or individual living organisms with their environment. Ecology embraces all aspects of the natural sciences that help to understand or modify ecological relationships. **2:** the relationships of living systems with each other and with their environment (e.g., the ecology of the bald eagle). **3:** frequently used erroneously as a synonym for environment ("protect the ecology"). *See also* autecology, ecopedology, ecophysiology, synecology. **agricultural ecology** (agroecology). a rapidly developing field of study and application that aims, in part, to provide frameworks whereby sustainable agriculture may be achieved while preserving or conserving existing local or regional environments. This science considers humans as integral components of agroecosystems. *See also* agriculture (contemporary agriculture). **landscape ecology**. the science that treats the distribution patterns of communities and ecosystems, the ecological processes that affect those patterns, and changes in patterns and environment through time. **forest ecology**. the study of ecological relationships among forest biota and their interactions with abiotic components of the environment. Today, a major concern of this science is to acquire knowledge to improve forest management practices, especially with a view to facilitating sustained yield.

economic form. any type of living system considered to be of economic importance to humans. Included are insects that prey upon pests and insects that pollinate agricultural crops or produce honey. Many pests fall into this category.

economic poison. any chemical used to control pests or to defoliate cash crops.

Economic Research Service (ERS). an agency of the USDA that conducts research in and analysis of agricultural economics.

ecopedology. the science of the ecological

relationships and potential of soils, including the ecological interrelations of its living (e.g., microbiota, plant nematodes, plant roots) and nonliving components.

ecophysiology. the science of the physiological adaptations of organisms with each other and with their environment generally.

ecotoxic. of or pertaining to the capacity of a substance to damage, destroy, or disrupt plant or animal populations, biotic communities or other ecological associations or processes.

ecotoxicity. the capacity of a substance to injure or damage a population, community, ecosystem, or a significant level of ecological organization above that of an individual organism. The ecotoxicity of a substance is a complex function of the range of its toxicities to living organisms in the environment or ecosystem of concern, the types of organisms that are most affected, its persistence in the environment, and the potential for bioaccumulation or biomagnification. Numerous other factors can affect the ecotoxicity of a substance in a particular environment. In practice, the ecotoxicity of a substance is usually based on the toxicity of a substance to one or more test species, persistence in the environment, and its tendency to bioaccumulate. The ecotoxicity of most substances is now largely a matter of conjecture.

ecotoxicology. the science that treats naturally occurring and anthropogenic toxic chemicals in the environment and their interactions with ecological systems. The focus of ecotoxicology is on populations and communities. Ecotoxicology also considers harm to species, populations, or biotic communities that may not be reflected in obvious harm to individuals (e.g., reproductive impairment, changes in competitive relationships, population decline). In practice, this discipline draws on scientists whose advanced education has been primarily in ecology, biogeography, various related fields of biology, chemistry, or toxicology. It includes the study of direct and indirect harmful effects on humans (e.g., food-chain transfer). Ecotoxicology subsumes a portion of biotoxicology.

ectad. outward; toward the exterior.

ectal. external; outer.

ecto antigen (exo antigen). **1a:** an antigen (usually a toxin) that is easily separated from its source (usually a bacterium), e.g., by shaking. **1b:** any antigen formed in the ectoplasm of a bacterium.

ectocrine. **1a:** an ectohormone; a hormone that is secreted into the surroundings and affects the metabolism of other organisms. **1b:** pertaining to substances synthesized by a plant or products of their decomposition that stimulate or inhibit the growth of nearby plants. **1c:** a substance with ectocrine (*See* defs. 1a, 1b) properties. **ecological ectocrine**. any substance synthesized by one species which, when released into the environment, influences metabolic processes of another species (e.g., vitamins produced by ruminants which are ingested by other animals). *Cf.* endocrine, exocrine.

ectohormone. *See* ectocrine.

ectoparasite. an external parasite, one that lives on the surface of the host's body.

ectoparasiticidal. lethal to ectoparasites.

ectoparasiticide. an agent that, when applied directly to the exterior of the body, destroys ectoparasites.

ectopic. out of place; foreign.

ectopic pregnancy. a pregnancy in which the fetus is implanted outside the uterus (e.g., in the uterine tubes, or in a body cavity).

ectotherm (poikilotherm). an animal whose body temperature varies with that of the environment. Sometimes improperly referred to as "cold-blooded."

ectothermic. **1:** of or pertaining to the heat economy of and ectotherm. **2:** pertaining to an organism that does not regulate body temperature metabolically, one that has a body temperature that depends on heat derived from the environment.

eczema. a cutaneous inflammatory, noninfectious condition (acute or chronic) of the skin of humans and animals that is marked by some combination of noninfectious vesicles, papules, pustules, scales, crusts, and/or scabs. Eczema most often results from contact with external agents (e.g., certain toxic or irritant chemicals or allergens). Constitutional or predisposing, possibly genetic, factors may contribute. Whereas chronic eczema usually benefits from treatment, it often recurs. **facial eczema** (pithomycotoxicosis). a mycotoxic disease of grazing livestock in which toxic liver injury commonly results in photosensitization. It occurs most commonly in New Zealand, Australia, several South American countries, occasionally in Europe, and probably also in North America. Sheep, cattle, and farmed deer of all ages may acquire the disease, but it is most severe in young animals. The mycotoxin, sporidesmin, a secondary metabolite of *Pithomyces chartarum*, a saprophyte that grows on dead litter in pastures. Photosensitization and jaundice appear about 10-14 days following ingestion. At that time, affected animals frantically seek shade. The disease is typified by erythema and edema of photodermatitis in unpigmented skin that rapidly results from even brief exposures to sunlight. Stress is severe, with death intervening one to several weeks after appearance of the photodermatitis. Characteristic liver and bile duct lesions are seen in affected animals. In acute cases showing photodermatitis, livers are initially enlarged, icteric, and have a clear lobular pattern. There is subsequent atrophy with marked fibrosis. The shape is distorted and large modules of regenerated tissue appear on the visceral aspect. The urinary bladder usually exhibits hemorrhagic or ulcerative erosions of the mucosa stained with bile pigment and localized edema.

EDA. ethylenediamine.

Edam. a round, solid, usually well salted, Dutch cheese made from cow's milk that contains about 2.5% fat. It is dipped in red wax prior to marketing.

edaphic. **1:** of or pertaining to soil. **2:** due to, arising from, or influenced by factors inherent in soil or sediments rather than climatic or other environmental factors.

edaphology. the science that treats the relationships among soils, their formation, interactions between soils and plants, and use of the land by humans.

edathamil. edetic acid.

edatope. refers to a specific type or combination of soil moisture and soil nutrient regimes.

EDB. ethylene dibromide.

EDC. ethylene dichloride.

eddy-chamber. an atomizing chamber in a nozzle.

edema (dropsy). **1a:** an excessive accumulation of serous fluid in body cavities or intercellular spaces (often under the skin). The increase in interstitial fluid volume may result from any of a number of factors such as increased venous pressure, decreased capillary permeability, decreased plasma protein, a lowered gradient in osmotic pressure between the capillary and the tissue served, or insufficient lymphatic drainage. *Cf.* ascites, dropsy. **1b:** the associated swel-ling of tissues. *See also* ascites, dropsy. **conjunctival edema**. swelling of conjunctival tissue, a common response to noxious stimuli. **pulmonary edema**. excessive accumulation of fluid in the lungs. It hinders alveolar absorption of oxygen, is often progressive, and may be fatal. Pulmonary edema can result from any of a number of diseases, from excess bronchial secretion as that induced by cholinesterase inhibitors or from damage to the pulmonary epithelium by chemical irritants. Other proximate causes include elevated venous pressure, decreased capillary permeability, decreased plasma protein, or insufficient lymphatic drainage. Pulmonary edema interferes with oxygen uptake and can be fatal. **wattle edema** (edema of the wattles). enlargement of the wattles of a fowl due to the presence of an inflammatory liquid. The condition of unknown origin.

edema of the wattles. *See* edema (wattle edema).

edemagen. an irritant that injures capillaries with resultant edema; not a true inflammagen, *q.v.*

edematigenous. edematogenic.

edematization. any process that causes edema.

edematogenic (edematigenous). able to produce or to cause edema.

edematous. swollen; characterized by or affected by edema.

edetic acid (*N,N'*-1,2-ethanediylbis[*N*-(carboxymethyl)glycine]; 3,6-bis-(carboxymethyl)-3,5-diazooctanedioic acid; ethylenebisaminodi-acetic acid; ethylenediamine-*N,N,N',N'*-tetraacetic acid; (ethylenedinitrilo)tetraacetic acid; ethylenediaminetetraacetic acid; ethylenediaminetetraacetate; ethylenedinitrilotetraacetic acid; ethylenediamine tartrate; edathamil; EDTA; EDTA acid). a caustic, toxic, colorless, flammable, crystalline substance,

$$(HOOCCH_2)_2\text{-}NCH_2CH_2N(CH_2COOH)_2$$

It has an ammoniacal odor, and is slightly soluble in water but insoluble in common organic solvents. Edetic acid is a powerful chelator of +2 and +3 cations. The sodium salt binds heavy metals most strongly and can form as many as six bonds. It is used as a chelator of metals in inorganic chemistry; it is used in the treatment of heavy metal poisoning (e.g., lead poisoning, especially of livestock), and in calcinosis therapy. EDTA has many uses in the food-processing industry (e.g., the capture and removal of miniscule metal particles that enter food from the rollers, scrapers, and blenders that are used to process foods). While edetic acid can complex with inorganic lead, it also causes a rapid and potentially injurious depletion of calcium. It is thus the calcium chelate of edetic acid that is injected in the treatment of lead poisoning. Otherwise, a net loss of calcium would accompany solubilization and excretion lead. If injected into and about the bite wound, edetic acid can help to curb the effects of certain viper venom enzymes on tissues of a victim. In such cases, concentrations of about 0.025 to 0.05 M edetic acid seems to retard the development of tissue damage and necrosis. Edetate is teratogenic and probably mutagenic; it is also a blood coagulant. Both liquid edetic acid and the vapor are extremely caustic to the eyes, skin and mucous membranes, and other body tissues, as are concentrated solutions of its salts.

edge. *See* ecotone.

edge bone. aitchbone.

edge effect. the tendency toward greater diversity and density of plant and animal populations in an ecotone, *q.v.* Sometimes taken to include environmental conditions (e.g., microclimatic factors) that obtain within the ecotone between two ecosystems.

edge firing. a controlled burn in which fires are set at the perimeter of the target area and permitted to spread inward.

edible. **1a:** comestible; food. **1b:** descriptive of items (food) that are suitable for ingestion by humans; fit to eat; comestible.

edible portion (E.P.). that part of a food item that is actually edible.

EDTA. edetic acid.

EDTA acid. edetic acid.

effect. consequence; outcome; result; any change, result of, or response to a stimulus, agent, action, force, or antecedent event. **additive effect**. the result of the action of two or more agents (e.g., substances), actions, or processes which, when used in combination (simultaneously or sequentially), produce a total effect that is equivalent to the arithmetic sum of the individual effects of each agent experienced alone. **adverse effect** (adverse reaction; deleterious effect). **1:** adverse reaction; a harmful or injurious effect as by a medication, poison, or ionizing radiation. **2:** any effect (immediate or delayed) that is injurious or otherwise harmful to a living system, including progeny. *See also* unintended effect, critical effect. **adverse drug effect**. **1:** an unintended adverse effect of a drug administered in normal amounts. *See also* unintended effect. **2:** an adverse effect of a drug regardless of cause (e.g., abuse, overdosage, error). **antagonistic effect**. any change brought about through antagonism or antagonistic actions, interactions, or relationships. *See* antagonism. **Baldwin effect**. a sequential process in which characters acquired due to environmental factors are replaced by heritable characteristics. **biochemical effect**. an effect observed at the biochemical level. Effects of a stimulus or other agent (e.g., a drug or poison) observed at higher levels of organization (e.g., physiological, whole organism, population, ecosystem) usually involve biochemical effects and mechanisms. **biological effect**. any change in an organism induced by a stimulus or a biologically active agent. **Bohr effect**. the effect of carbon dioxide on the oxygen-carrying capacity of the respiratory pigment of the blood. **negative Bohr effect**. reduction in the volume of oxygen that the respiratory pigments can hold when dissolved carbon dioxide acidifies the blood. This response maximizes the exchange of oxygen between the respiratory surfaces (gills or lungs) and body tissues, because CO_2 tension is low at the respiratory surface where oxygen is absorbed and high in the body tissues where oxygen is released. **clinical effect**. any clinically observable effect. **chronic effect**. a permanent, or semipermanent, effect (e.g., of a drug or disease), usually of an adverse nature, that develops over a long period of time. **critical effect**. the first adverse effect observed in a particular organ in association with a known concentration of a particular substance under specified conditions. The critical effect defines the critical concentration. *See* concentration (critical concentration), critical organ. **cumulative effect** (cumulative action). any result of repeated equivalent exposures to a biologically active agent or stimulus in which the effect of any subsequent exposure is more pronounced than that of the initial exposure. **delayed effect** (delayed response). an effect that occurs or develops following a substantial interval of time following exposure to a stimulus or substance. Carcinogenic effects of chemicals may be delayed by as much as 20-30 years. **deleterious effect**. adverse effect. **direct effect**. any effect that is induced directly by the action of a substance or a stimulus. **ecological effect**. any effect expressed at an ecological level of organization as an effect on a population, biotic community, or ecosystem (e.g., eutrophication; a change in diversity, primary or secondary productivity, soil pH, or sedimentation rates). **environmental effect**. any effect on elements or aspects of the environment. Ecological effects are a subset of environmental effects. **excitotoxic effect**. injury or death due to the induction of sustained, repetitive discharge. **founder effect**. the exhibition of genetic drift, *q.v.*, by a population that was established by only a few founding individuals. Thus, colonies of individuals on isolated oceanic islands are frequently the descendants of very few initial colonzing members, and such populations rapidly become genetically distinct from the parent population from which the colonizing individuals derive. **functional effect**. a change in the function or behavior of a target tissue or organ. **genetic effect**. any effect on the genetic material (e.g., a gene mutation, formation of a DNA adduct). **graded effect**. an effect whose magnitude or intensity is related to the level of exposure or dose. **greenhouse effect**. the increase in atmospheric temperature due to an increase in concentration of carbon dioxide. This is analogous to the elevated temperature of a greenhouse due to the properties of the glass roof and sides such that ultraviolet radiation enters relatively freely, and is absorbed by and heats surfaces in the greenhouse which releases long-wave (heat) radiation that does not readily pass through the glass but interacts with the air, elevating the temperature. Similarly, short-wave radiation from the sun passes through the atmosphere and is absorbed by the earth's surface, which releases long-wave radiation. Much of this escapes through the atmosphere, but some is absorbed by atmospheric water vapor, ozone, and most importantly CO_2. As the concentration of atmospheric CO_2 increases, more heat radiation is absorbed by the earth's atmosphere, causing a rise in temperature. Evidence in recent decades indicates a slow, persistent increase in the concentration of atmospheric carbon dioxide. This is thought to be due to combustion of fossil fuels (e.g., coal and petroleum products). The associated elevation in atmospheric temperature will have both direct and indirect effects on agriculture. **health effect**. any effect on the health of a living system, usually an individual or a specified population. **idiosyncratic effect**. any abnormal or aberrant response of a living system. Such an effect may be genetically determined or the result of an unusual and unknown history. **immediate effect** (immediate response). any effect that occurs during exposure to a substance or a stimulus or that develops rapidly and appears following a single exposure to the causative agent. **indirect effect**. any effect that is not directly induced by the action of the causative agent or stimulus but results from internal processes initiated by the causative agent or as the result of modified responses to the environment. Thus, a partially paralyzed (direct effect) animal can no longer compete adequately and consequently dies (indirect effect). **indirect systemic effect**. any systemic effect produced by a pronounced local effect such that the damage may itself produce systemic effects. The kidneys, for example, may be damaged by a severe acid burn on the skin in which none of the acid is distributed to the kidneys. **irreversible toxic effect**. **1a:** a permanent effect; one that persists indefinitely after cessation of exposure and elimination of the toxicant from the system. **1b:** a toxic effect that does not respond to treatment. **latent effect**. an effect that does not present initially following exposure to a biologically active agent or stimulus. Thus, signs of lung cancer induced by chronic exposure to cigarette smoke may not appear for many years following initial exposure. **lethal effect**. any effect to which the death of a living system can be directly attributed. **local effect** (local injury). any effect that occurs at the site of first contact between a chemical and a biological system. Local effects are usually produced by highly reactive (e.g., caustic, oxidant, irritant) substances or by radiation. **morphologic effect**. any change in the gross or microscopic structure of organs and tissues (e.g., neoplasia, necrosis, atrophy) in response to a biologically active agent. **nephrotoxic effect**. any toxic effect (direct or indirect) on the kidney. **neurotoxic effect**. any toxic effect (direct or indirect) on the nervous system. **no effect**. the failure of an active agent (e.g., a drug) or stimulus to elicit an observable effect. In the case of exposure to a drug or infective agent, the individual is said to be asymptomatic. **nonlethal effect**. a biological effect that does not impair viability or cause death. **oxygen effect**. enhancement of radiosensitivity of cells by a high ambient concentration of oxygen. **quantal effect**. an effect whose magnitude or intensity bears no relationship to the intensity of exposure or the dose. The incidence of occurrence of such an effect may, however, be dose/exposure dependent. **residual effect**. an effect that remains following withdrawal of the causative agent; one that remains after other effects are no longer observed. **reversible effect**. a temporary effect; one that ameliorates and ultimately dissipates following termination of exposure to a stimulus or biologically active agent. **reversible toxic effect**. a toxic effect in which target tissues repair or regenerate following injury. The degree of reversibility depends upon the type and extent of injury and the capacity of the tissues to regenerate. **side effect**. unintended effect. **sublethal effect**. any effect that is deleterious to an organism, but which does not impair its viability in an otherwise normal environment. The distinction between sublethal and lethal is not absolute nor always clear. **synergistic effect**. a response in which the combined effect of two or more stimuli or biologically active agents (e.g., poisons) to which a biological system has been concurrently exposed is much greater than the sum of the effects of each agent acting alone. **systemic effect**. an effect that occurs at remote sites in the target system following absorption and transport of a biologically active agent, usually by the blood vascular system, from the point of entry. **therapeutic effect**. any biological effect of a chemical substance or stimulus that is associated with healing. **toxic effect**. any of a great variety of signs, symptoms, or complaints observed in a victim of poisoning or envenomation, including radiation poisoning. All such effects are the result of biochemical or biophysical interactions between a toxicant and/or its metabolites and the affected organism. Toxic effects have been variously classified according to the location of the effect, level of organization (e.g., biochemical level, ecosystem level), organs affected, time of onset, processes affected, destructiveness, degree of reversibility, or the mechanisms involved. **toxic systemic effect**. any toxic effect that follows absorption and distribution of the toxicant to sites within the organism that are remote from the point of entry. **unintended effect** (side effect; unintentional effect). any effect of a substance or process other than those intended on the basis of design, selection, or application. Nearly any action, product, or technology has effects other than those for which it was developed and applied; only rarely are such effects beneficial. A drug or medication, for example, typically has multiple effects on the dosed organism. The spectrum of effects usually varies with dose, other drugs administered, the individual, the species, and various environmental factors. A drug, however, is selected for medical use on the basis of one or a limited number of effects that are desired in a given instance. Other effects that are unintended or unwanted for a given application are usually referred to as side effects. For many chemicals, so-called side effects are not well documented; sometimes these unintended (and often unpredicted) effects are significant clinically and may be more pronounced than the intended effect. Technologies, industrial, and commercial processes are designed to produce certain products and effects intended to benefit humans in some way. All such activities also produce unintended effects, some of which may be harmful. Common illustrations of unintended effects are releases (e.g, into the work space or into the general environment) of toxic substances and other hazardous materials during manufacture, transport, usage, and/or disposal. **unintentional effect**. unintended effect.

effete. barren, sterile, or unreproductive, especially with respect to animals that do not produce young or plants that fail to bear fruit.

efficacy. **1a:** competence; efficiency. **1b:** mastery; effectiveness.

effluent. **1:** any fluid discharged from a given source into the external environment. **2:** in the environmental sciences and environmental protection, this term usually refers to wastes discharged into surface waters or to wastewater (treated or untreated) that flows from a treatment plant, sewer, or industrial outfall into a lake or waterway.

efosite Al. fosetyl aluminum.

egesta. undigested material discharged (egested) from the body (e.g., feces, vomitus, regurgitated pellets). *See also* egestion, excreta.

egestion. elimination of undigested materials from the body by any of a number processes (e.g., defecation, vomiting, regurgitation).

egg. **1a:** the female gamete or mature sex cell, especially if fertilized. **2a:** ovum. **2b:** in birds, reptiles, and certain other vertebrates, a fertilized ovum together with nutritive tissues and surrounded by protective tissues (e.g., membranes, a shell) that undergoes development without attachment to the female. It may pass from the body and hatch externally (as in birds, most reptiles, amphibians, and insects), or remain within the body of the female and hatch internally (as in certain snakes). **3:** the mammalian ovum. **addled egg**. an inedible egg of a fowl. One that is rotten or in which the yolk and white have become intermixed. **clear egg**. an infertile egg. **isolecithal egg**. an egg in which the yolk is distributed evenly throughout.

egg albumin. albumen.

egg white. *See* albumen.

eggplant. *Solanum melongena esculentum*.

eggplant blight. a disease of *Solanum melongena* caused by the fungus *Phomopsis vexans* (Family Sphaerioidaceae). Growth of affected seedlings is inhibited, leaves are spotted with brown, and the skin becomes discolored with depressed circular lesions as the first sign of rot. Eventually the entire fruit may decay.

eggshell, of birds. the calcified envelope of a normal avian (bird) egg. It lies immediately external to the two membranes (composed of keratin fibers) that enclose the white and yolk of the egg. There are numerous channels through the shell (pores) that permit the exchange of respiratory and other gases between the egg and the surrounding environment. The eggshell by virtue of its composition and structure affords mechanical protection to the egg within. If the shell is too thin or otherwise weakened, the egg may be easily crushed as by the weight of the incubating adult. *See* eggshell thinning.

eggshell thinning. the production of a thinner than normal eggshell in birds that have been exposed to organochlorine compounds (especially DDE) and have accumulated residues. This has brought a number of large birds (e.g., peregrine falcons, pelicans) to the verge of extinction. When such large birds incubate affected eggs, their body weight may be sufficient to break such eggs, injuring or destroying the developing chicks. Following the banning of DDT in the United States and many other nations, most of the endangered bird populations have rebounded.

Egyptian millet. *Sorghum halepense*.

eicosanoic acid. arachidic acid.

EIS. environmental impact statement.

Ektafos. dicrotophos.

Elaeagnus umbellata (autumn olive). a deciduous, open grown shrub, often found on acidic soils of low fertility, that may attain a height of 3-5 m (10-15 feet). It fixes atmospheric nitrogen and produces large number of red berries eaten by various wildlife species.

Elaeis guineensis (African oil palm). an ornamental palm tree (Family Palmaceae) grown in the southern United States It is important to the oil trade in Africa.

elephant's ear. a common name of herbs of the genus *Caladium*.

elder. *See Sambucus*.

elderberry. *Sambucus canadensis*.

electrocortin. aldosterone.

electrolyte acid. sulfuric acid.

electron acceptor. an oxidant; a substance that accepts electrons during an oxidation-reduction reaction.

electron donor. a reductant; a substance that donates electrons in an oxidation-reduction reaction.

electron transfer chain. *See* electron transport system.

electron transport system (electron transport chain; electron transfer chain; respiratory chain). a series of electron carriers within a mitochondrion or chloroplast. Electrons progress along this chain (sometimes together with hydrogen ions) by a series of redox reactions. The energy of oxidation is thus captured in a graded way and ATP is generated. The electron carriers include cytochromes, quinones, and ferredoxin. *See* especially cytochrome, cytochrome c, cytochrome oxidase, cytochrome *c* oxidase, cytochrome P-450. *See also* copper, mitochondria, NAD. podophyllotoxin,

element. an identifiable component, part, constituent, rudiment, factor, process, or condition. **chemical element**. a substance that cannot be chemically decomposed into structurally simpler components with the exception of radioactive elements. They combine variously to form all molecules. All are poisonous under some circumstances and all can form naturally occurring or synthetic toxic compounds. There are 92 naturally occurring elements and 17 transuranic elements (those of atomic mass greater than uranium). Ele ments heavier than lead are radioactive and unstable. **available element**. any chemical element in solution in the soil that is readily absorbed by plants; in higher plants, absorption takes place via roots or similar structures. *Cf.* nutrient (available nutrient). **essential element**. macronutrient. **essential trace element**. a trace element essential to critical physiological processes; one that is an essential component of nutrition. Included are Zn, Mg, Cu, Ni, Mn, Co, Cr, I, Me, and Se. Their essential roles are performed as part of the structure of metabolically important molecules (e.g., enzymes, pigments) or are important cofactors or catalysts of metabolic processes. Excessive amounts are toxic. **trace element**. **1:** an element that occurs in such minute quantities in a specified type of system (e.g., soil, an organism) that they may not be quantified by conventional chemical assays. Most are essential to life in small quantities. **2:** essential trace element.

elemental mercury. the element mercury.

Eleusine indica. (goose grass). a grass (Family gramineae) that sometimes contains concentrations of nitrate toxic to livestock.

eluviation. the loss or removal of material, as a suspension, from a soil horizon.

Elymus. (wild rye; lyme grass). a genus of wild, tall, annual or perennial Northern Hemisphere grasses (Family Gramineae) that may become heavily ergotized and therefore hazardous. A few species are planted on embankments or along garden borders. *See Claviceps*. ***E. arenarius*** (sea lyme grass; dune grass). a stout grass of Eurasia, used to bind shifting sand. Also called dune grass.

emaciation. an extremely lean, wasted, enfeebled condition of the body. It is sometimes seen in chronic disease.

embryo. **1:** the early multicellular stages in the growth and development of an organism. **2a:** the prenatal stage of a mammal (between the fertilized egg (zygote) stage and the fetal stage) during which organogenesis occurs. The embryo is more sensitive to many poisons than at later stages in development. **2b:** in *Homo sapiens*, the conceptus is designated as an embryo from the 2nd through the 8th week following fertilization, at which time organogenesis is essentially complete. *See* fetus. **adventive embryo**. A nonzygotic plant embryo that develops from the cells of the nucellus. This is common in *Citrus spp*.

embryo recipient. as applied to livestock, any female that carries the embryo of another individual, gives birth, and serves as a surrogate mother. *See also* dam (donor dam).

embryo sac. **1:** a sac that contains the embryo in animals. **2:** the mature female gametophyte in angiosperms. It is situated within the ovule and consists of eight or fewer cells, bounded by a membrane, with no cellulose walls between them.

embryocidal. **1:** of or pertaining to anything able to kill an embryo. **2:** a substance that destroys embryos.

embryofetoxic. poisonous to the embryo and fetus.

embryogenesis. the process of embryo formation and development.

embryologist (developmental biologist). a scientest who treats the development of embryos. *See* embryology.

embryology (developmental biology). **1:** the science or study of the development of embryos. **2a:** animal embryology; animal development biology; the study of development from fertilization of the egg to birth or hatching. **2b:** the study of prenatal development. **comparative embryology**. the science that treats relationships among embryos of different taxa.

embryonic. pertaining to or resembling an embryo.

embryopathy. any morbid or pathological condition in an embryo or fetus.

embryotoxic. toxic to an embryo.

embryotoxicity. **1:** the quality of being em

bryotoxic. **2:** any toxic effect on an animal embryo due to prenatal or prehatching exposure.

emerald green. cupric acetoarsenite.

emergency. **1:** a sudden, usually unanticipated event or situation that is potentially harmful, requiring a rapid response; an accident. *Cf.* crisis. **chemical emergency**. an emergency that arises from an incident or sudden situation whereby the safety of workers, residents, the environment, or property is seriously and immediately threatened by the presence of toxic or otherwise hazardous chemicals. Such is usually due to an accidental release or spill of hazardous chemicals and requires immediate action to contain the chemicals or otherwise minimize the hazard. **medical emergency**. **1a:** an emergency that seriously and immediately threatens the health or life of one or more humans. **1b:** a condition that requires urgent medical treatment (e.g., cardiac arrest, inhalation of certain pesticides). **1c:** a procedure that must be performed immediately (e.g., cardiopulmonary resuscitation). **toxicological emergency**. **1:** any exposure to a poison where the poison or conditions of exposure are unknown. **2:** acute exposure to a potentially lethal or damaging poison, whether intentional, accidental, or natural (e.g. snake bite, possibly poisonous mushrooms), is a medical emergency, even in the absence of signs or symptoms. **veterinary emergency**. **1:** trauma, disease, suffering, or any condition that threatens the health or life of one or more domestic animals, especially if circumstances indicate imminent threat to additional animals or to humans. **2:** any serious condition that requires immediate diagnosis and/or treatment. **3:** a suspected communicable disease that might affect humans or could spread to other animals in a herd or to a larger area or region.

emesis. vomition, vomiting, the act of vomiting.

emetic (vomitive; vomitory). **1:** pertaining to or capable of causing vomiting (emesis). Note that some animals cannot vomit. **2:** any chemical that causes vomiting (e.g., apomorphine, syrup of ipecac, strong soapy water, or a strong solution of sodium chloride or mustard).

emetine (6′,7′10,11-tetramethoxymetan; cephaeline methylether). the principal alkaloid, $C_{29}H_{40}N_2O_4$, of ipecac. Emetine and its dihydrochloride are extremely toxic to laboratory rodents. It is used therapeutically as an antiamebic and is used in veterinary medicine in lung worm infections. *See also* ipecac.

emetism. poisoning by an overdose of ipecac. The victim usually suffers acute inflammation of the pylorus, intense vomiting, diarrhea, and sometimes coughing and suffocation.

emetocathartic. **1:** able to produce both emesis and catharsis. **2:** an substance that produces both emesis and catharsis.

emigration. the deliberate, usually oriented departure of an organism from the habitat, community, ecosystem, location, region, or country in which it has been living. *See* immigration.

eminent domain: the power and authority of the government, subject to constitutional and statutory restrictions, to acquire private land for public use, with just compensation paid to the landowner.

emiocytosis. exocytosis.

emission. **1:** emanation; discharge; a release. **2:** natural or anthropogenic pollution discharged into the atmosphere, as from surfaces, vents, residential chimneys, smokestacks, and automobile, locomotive, or aircraft exhaust or commercial or industrial facilities. **3:** a measure of the extent to which a given emission source discharges a pollutant, usually expressed as the amount per unit time (rate) or as the amount per unit volume of gas emitted.

emission standard. the maximum amount of air pollutants, including toxic emissions, that are legally allowed to be discharged from a single mobile or stationary source.

emodin (1,3,8-trihydroxy-6-methyl-9-10-anthracenedione; 1,3,8-trihydroxy-6-methyl-anthraquinone;4,5,7-trihydroxy-2-methyl-anthraquinone; archin; frangulic acid; frangula emodin; rheum emodin). a cathartic anthraquinoid that occurs mostly as the rhamnoside in roots of *Rheum rhabarbarum* (rhubarb), *Rhamnus fran-gula* (alder buckthorn), and *Rhamnus purshiana* (cascara buckthorn). It also occurs in *Rumex* and related genera. Emodin is a constituent of a number of irritant laxative drugs. Emodin laxatives act chiefly on the small intestine and are not effective if passage to the small intestine is delayed. Repeat dosing of large animals, especially horses, is to be avoided. Emodin is extremely toxic to certain protozoans such as *tetrahymena pyriformis* and *Escherichia coli* mutants.

emphysema. **1:** a pathologic distension of tissue due to gas or air in the interstices. **2:** pulmonary emphysema. **pulmonary emphysema** (chronic obstructive pulmonary disease). a chronic destructive disease of the lungs. They are congested and characterized by an abnormal increase in the size of the air spaces distal to the terminal bronchioles accompanied by destructive changes in their walls; alveoli may be ruptured the remaining alveoli become enlarged. **toxic pulmonary emphysema of cattle** (respiratory syndrome of cattle). a conspicuous disease that occurs most often in animals pastured on fresh *Brassica napus* (rape), and sometimes on fresh *B. oleracea* var. *acephala* (kale), *B. rapa*. var. *rapifera* (turnip), and rarely *Medicago sativa* (alfalfaa). Its occurrence is unpredictable and the mortality varies from about 5-35%. Signs of the disease usually appear in some animals within a week or ten days following pasturage on succulent rape. Affected animals stand apart from the herd, head extended and breathing is labored typically with an expiratory grunt. Examination reveals an elevated heart rate, paresis of stomach and intestines, and slight jaundice. The severity of dyspnea increases with exercise. subcutaneous cervical and lumbar emphysema may develop within a few days. The liver is variably necrotic, and varying portions of the gastrointestinal tract are inflamed. The pulmonary lesions are unusual. The lung is congested, distended and edematous with numerous air pockets. Histologically, most of the alveoli are ruptured.

empirical. pertaining to information, inference, and theory gained from or based, on experience and observation. *See also* experiment, extrapolation.

emptysis (hemoptysis). expectoration of blood or blood-stained mucus arising from bleeding in the lungs, respiratory tree, or larynx.

emulsifiable concentrate. a concentrated pesticide that can be dispersed in water as fine suspension.

emulsification. the process of producing an emulsion. Fat globules, for example are emulsified in the intestines by the detergent action of bile.

emulsifying agent. a substance that effects emulsification such as bile, agar, gums, lecithin, and soaps.

emulsion. a suspension of tiny drops of one liquid within another, as of oil in water.

encephalitogen. a substance (e.g., lead) that can cause encephalitis.

encephalitogenic. causing or tending to cause encephalitis.

encephalopathia. encephalopathy.

endangered. jeopardized; threatened; imperiled.

endangered species. any species of living organism faced with the likelihood of extinction, especially those that have been so declared by legal authority (e.g., the U.S. Department of Interior and/or designated state agencies). In the United States, the Endangered Species Act contains the requirements for declaring a species endangered. A number of species in this category have been threatened by toxic environmental chemicals either directly or indirectly (e.g., by effects on prey or other species with which they normally interact).

endangerment. hazard; risk; peril. **1a:** the

act of placing in danger. **1b:** the state of being in danger or placed in danger.

endemic abortion of ewes (EAE; endemic ovine abortion; chlamydial ovine abortion; enzootic abortion of ewes). a disease of sheep marked by abortion, stillbirth, or premature lambing; placental retention sometimes occurs. The disease usually has little effect on the health of the ewes, although some deaths occur. The disease is known from the United States, several European countries, and South Africa. The infectious agent is a strain of *Chlamydia psittaci* similar to that which causes epidemic bovine abortion, *q.v.*

endemic ovine abortion. endemic abortion of ewes.

endemic. pertaining to a disease (endemic disease) that is continually present in a particular population or region but with low mortality.

endo. isodrin.

endo isomer. isodrin.

endocolitis. inflammation of the colon.

endocrine gland (ductless gland; gland of internal secretion). a ductless gland or group of secretory cells that secrete hormones directly into the blood, lymph, or other body fluids. Such glands occur in all vertebrates and many invertebrates.

endoenteritis. inflammation of the intestinal mucoso.

endogastritis. inflammation of the gastric (stomach) mucosa.

endogenetic. *See* endogenous.

endogenous (endogenetic). **1:** growing from within. **2:** of or pertaining to any property, substance, or part that occurs or arises naturally within a system. Examples are endogenous rhythms and internal secretions. *Cf.* exogenous, extrinsic.

endometrium. the mucous membrane that lines the inner surface of the uterus. Followiing fertilization of the ovum, the endometriumm serves as nutrient and protective tissue in which the ovum is implanted. *See* decidua.

endophyte. a symbiotic or parasitic organism that grows within a plant.

endosperm. a specialized, usually triploid, nutritive tissue of the embryo sac of some angiosperms. It is the product of double fertilization.

endospore. **1:** any spore formed within a sex organ or sporangium. **2:** a thick-walled sexual spore that develops within certain bacteria (e.g., *Bacillus*, *Clostridium*) under unfavorable conditions. These spores are heat- and drought-resistant and can remain alive for many years.

endosulfan (6,7,8,9,10,10-hexachloro-1,5,5a,6,9, 9a-hexahydro-6,9-methano-2,4,3-benzodioxa-thiepin-3-oxide; 1,4,-5,6,7,7-hexachloro-5-nor-bornene-2,3-dimethanol cyclic sulfite; 1,2,3,4, 7,7-hexachlorobicyclo[2.2.1]-2-heptene-5,6-bisoxymethylene sulfite; chlorthiepin). the commercial product is a nephrotoxic, extremely hazardous, brown crystalline or liquid chlorinated diene insecticide and miticide, $C_6H_6Cl_6O_3S$, used on ornamental flowers, vegetables, forage crops, and to control tsetse flies and termites.

endothall (7-oxabicyclo-(2,2,1)-heptane-2,3-dicarboxylic acid). a soil and contact herbicide, $C_8H_8Na_2O_5$. It is a preemergence and postemergence herbicide and defoliant used to control weeds in *Beta vulgaris* (sugarbeet) and spinach acreage. It has defoliant and desiccant properties and can thus be used as a preharvest treatment in a number of crops. Endothall is a strong irritant to the eyes and skin. It also affects amylase activity, proteolytic activity, and mitosis.

endotherm (homeotherm). an animal, such as a mammal or bird, that maintains a nearly constant body temperature, despite fluctuations in the external environment.

This is accomplished by physiological and behavioral heating and cooling mechanisms. Endothermic animals are sometimes referred to as "warm-blooded." *See* ectotherm.

endothermic. **1:** of or referring to the maintenance of a nearly constant body temperature by physiological and behavioral means. **2:** pertaining to an endotherm. *See also* ectothermic.

endrin (3,4,5,6,9,9-hexachloro-1a,2,2a,-3,6,6a,7, 7a-octahydro-2,7:3,6-dimethanonaphth[2,3-b]-oxirene; 1,2,3,-4,10,10-hexachloro-6,7-epoxy-1,4,4a,5,-6,7,8,8a-octahydro-*endo,endo*-1,4:5,8-dimethanonaphthalene; mendrin; nendrin; hexadrin; Compound 269; experimental insecticide No. 269). a white, flammable, highly poisonous, water-insoluble, crystalline, chlorinated, cyclodiene insecticide. It is a stereoisomer of dieldrin with effects similar to those of dieldrin and aldrin. Endrin is teratogenic, hepatotoxic, and neurotoxic, attacking the CNS It is highly toxic (much more toxic than DDT) to humans and other vertebrates; approximately 7 g is lethal to a human. Symptoms of intoxication include nausea, mental disorientation, convulsions and in severe cases, respiratory collapse and death. Endrin is carcinogenic to some laboratory animals and is probably mutagenic. It is no longer manufactured or used in the United States.

energy. **1:** the capacity to do work. Energy is manifested in many forms such as light and other electromagnetic waves, motion (kinetic energy), potential (energy of position; stored energy), electrical, heat, chemical, nuclear, and sound waves. **2:** the ability to engage in vigorous activity.

English pea. *Pisum sativum*.

English yew. *Taxus baccata*.

ENT 7796. a trade name for lindane.

entamebiasis (entamoebiasis). a disease caused by *Entamoeha histolytica*. *Cf.* amebiasis. *See also* amebic *dysentery*.

entamoebiasis. entamebiasis.

enteramine. serotonin.

enteric bacteria. bacteria that inhabit the intestinal tract, some of which are pathogenic (e.g., *Salmonella*, *Shigella*).

enteritis. inflammation of the intestine.

enteropathogen. a pathogenic organism (e.g., *Entamoeba histolytica*, *Escherichia coli*) that can cause disease of the intestinal tract.

enteropathogenic. of or pertaining to the production of disease in the intestinal tract by a pathogenic organism. *See* enteropathogen.

entomologist. a basic or applied scientist who studies the biology of insects; a student or practitioner of entomology.

entomology. the branch of zoology concerned with the study of insects, including insect taxonomy, anatomy, physiology, behavior, ecology, biogeography, economics, etc. **economic entomology**. the science that treats economic forms of insects (*See* economic form), including not only pests, but beneficial insects such as honey bees. The goal of economic entomology is to increase the benefits of insects to humans and to reduce the disbenefits. **forest entomology**. the branch of entomology concerned with the study of forest insects, especially with respect to their impacts (favorable and unfavorable) on forests and forestry strategies and practice. Included also, for example, are aspects of insect behavior, physiology, biogeography, life stages, and economics, especially to the extent that such knowledge is necessary for the control of economically important forest insects.

entomopathogenic. **1:** pathogenic to insects. **2:** of or pertaining to any agent (e.g., a substance or an organism) that is pathogenic to or causes disease in insects.

entomophagous. insectivorous.

entomophilous. **1a:** literally, insect loving; **1b:** descriptive of flowers (plants) that are pollinated by insects. Such flowers attract insects often by their bright colors or at tractive scent or by the presence of available nectar or excess pollen.

entomophily. pollination of plants by insects; pollination in which insects transport pollen from the anther to the stigma. The pollen of such plants is sticky, thus readily adhering to insects. The anthers and stigmas are often situated so that a visiting insect will touch both.

entomotoxic. insecticidal; toxic to insects.

entomotoxicity. the capacity of a substance to poison insects.

entozoa. plural of entozoon.

entozoon (plural, entozoa). an animal parasite that infests the internal organs or tissues of an animal.

entrails. the abdominal viscera and or intestines of an animal body.

enuresis. involuntary urination or loss of urine.

envenomation. **1:** a complex, multiple empoisonment in which a mixture of toxic materials or venom is introduced into an animal by means of a bite, strike, impalement, or sting. The seriousness of any envenomation usually depends upon numerous factors such as the amount and kind of venom delivered, the nature, number, location, and depth of the bites or stings, the size and age of the victim, the sensitivity of the victim to the venom, and the pathogens present in or on the sting, mouth, and spine. The bite or sting of a venomous animal does not always result in envenomation; even so, there is risk of tetanus or of a communicable disease. *See also* snake venom poisoning. **2:** the injection of venom into a living organism by a venomous animal by means of a bite, sting, spine, or other structure. *See also* snake venom poisoning.

environment. **1a:** the complex of all external entities, conditions, and circumstances that affect life generally. **1b:** the entire complex of chemical, physical, and biotic factors (materials, processes, circumstances, and conditions) that may act upon, be perceived by, or otherwise affect or potentially affect a living organism, a type of organism, population, community, or ecosystem. **1c:** locale milieu; the immediate surroundings; the aggregate, at a given moment, of all external conditions and influences to which a system or organism may be subjected. Among epidemiologists, including veterinary epidemiologists, environment is taken to include everything that humans or a target animal population might expected to encounter (e.g., food, air, water, soil, feces, drugs), although the focus of an investigation might focus on a few environmental factors. **2:** a major habitat type (e.g., marine, terrestrial, rain forest, desert, lake). **abiotic environment**. **1:** the nonliving components, collectively, of an ecosystem; the physical and chemical factors in the environment that do not arise from biological activity. **2:** that part of the environment, in reference to an individual organism, taxon, or living system, that consists of nonbiological factors. *Cf.* biotic environment, abiotic, factor (abiotic). **biotic environment**. **1:** the complex of living components of an environment considered broadly. **2:** the sum of living organisms, their products, and activities that may affect a living system (i.e., an individual organism, taxon, population, community, or ecosystem). **external environment**. the term environment, unless qualified, is usually taken to refer only to the external environment; that is, those entities and factors (physical, chemical, and biotic) that lie outside the body of an organism. **human environment**. collectively, everything that humans encounter (e.g., food, air, water, drugs, heat); individually, everything outside the body including other humans. **internal environment**. those components and processes within an organism that result from or contribute to metabolic processes, including extracellular body fluid. **physical environment**. the abiotic component of the environment, including all physical and chemical environmental factors. **toxicologic environment**. those aspects of environment that are directly or indirectly toxic to a living system or that affect toxicity or the sensitivity of target and nontarget organisms to one or more toxic chemicals, or that may affect rates of exposure to toxic chemicals. *See also* factor (modifying factor).

environmental assessment. a written analysis prepared pursuant to the U.S. National Environmental Policy Act to determine whether a federal action would significantly affect the environment and thus require preparation of a more detailed environmental impact statement.

environmental audit. **1:** an independent assessment of the current status of an individual's or an organization's compliance with applicable environmental requirements pursuant to an ongoing activity. **2:** an independent evaluation of an individual's or organization's environmental compliance policies, practices, and controls.

environmental effect. *See* effect.

environmental protection agency (EPA). any of many governmental agencies in the United States and elsewhere. **U.S. Environmental Protection Agency**. an independent agency of the U.S. federal government that was established by Presidential Executive Order in 1970. This agency is directed by congress to protect the land, air, and water resources of the United States. It bears the chief responsibility, nationally, for regulation of pollution. This is accomplished by assessing impacts, developing and enforcing standards and regulations to control pollutants and their impacts on human health and the environment, and by establishing and conducting control programs that help to manage pollutants of air, water, and other environmental media, radiation, solid and toxic wastes, pesticides, and noise. The agency also sponsors research in support of its mission.

environmental resistance. limitations on the reproductive potential of a population due to environmental variables such as crowding, unfavorable climate, and the action of predators, parasites, or competitors.

environmental runoff. *See* runoff.

Environmental Impact Statement (EIS). a document, required under the U.S. National Environmental Policy Act, that is to be prepared by federal agencies prior to undertaking a major project or making legislative proposals that might significantly affect the environment. It is a decision-making tool that describes the positive and negative effects on the environment that might result from the undertaking and lists alternative actions.

environs. **1a:** surroundings; environment. **1b:** neighborhood; vicinity.

enzootic abortion of ewes. endemic abortion of ewes.

enzyme activity. *See* activity.

enzyme. a biological catalyst that enables metabolic processes to proceed rapidly at physiological temperatures and pH. Enzymes are denatured and become ineffective at temperatures only somewhat above their optimum and are also destroyed by a pH beyond a fairly narrow range. An enzyme usually catalyzes a specific reaction or a specific type of reaction in a living organism. Each enzyme is a highly specialized protein that catalyzes biochemical reactions by dramatically accelerating the conversion of one molecule (the substrate) into another (the product) by combining with the substrate and forming an intermediate enzyme-substrate complex. Upon completion of the reaction, the product and the enzyme separate and the chemically unaltered enzyme is able to complex with another substrate molecule. Enzyme catalysis is essential to nearly all cellular metabolism. Furthermore, enzymes typically act in highly integrated and regulated sequences or pathways, as in the electron transport system and pathways of protein synthesis.

eolation. aeolation.

eolian. aeolian.

eolian deposit. aeolian deposit.

eosinophil (eosinophile). **1:** readily staining

with the acid stain, eosin. **2:** a granulocyte that stains with eosin.

eosinophile. eosinophil.

eosinophilic meningitis. angiostrongylosis.

EPA. environmental protection agency; in the United States, this acronym is usually taken to be that of the U.S. Environmental Protection Agency.

Epicaerus imbricatus (imbricated snout beetle). a hard-bodied, plant-infesting insect pest (Family Curculionidae). The larvae feed on a wide variety of plants. Among agriculturally important plants, they are especially destructive to strawberries.

epichroic. discoloring.

epidemic bovine abortion (EBA; chlamydial bovine abortion; foothill abortion of cattle). an infectious disease of cattle that is endemic in the foothills of southern Oregon, California, and Nevada, but occurs also in other parts of the United States The disease is marked by abortion or delivery of infirm calves, although the affected cows show no signs of disease and generally recover rapidly following delivery. The infectious agent of endemic EBA is generally considered to be a strain of *Chlamydia psittaci* which is transmitted by a soft-shelled tick, *Ornithodoros coriaceus*. A spirochaete similar to *Borrelia* may also be involved. *See also* endemic abortion of ewes.

epidemic cellulitis. equine viral arteritis.

epidermides. plural of epidermis.

epidermis (plural, epidermides). **1:** the nonvascular outermost layer of cells covering the body of an animal or plant. It may consist of a single layer of cells as in invertebrate animals and may secrete a protective cuticle. In most vertebrate animals, the epidermis makes up the outer layer of the skin; it consists of several stratified layers and continually undergoes renewal. The epidermis of plants usually consists of a single layer of rectangular cells that fit closely together. The intact epidermis (together with the cuticle, if present) affords one degree or another of protection as a barrier to mechanical and chemical damage of living tissue as well as attack by pathogenic organisms. *See also* skin.

epinasty. a striking and rapid response of the shoots of many dicotyledonous plants exposed to auxins. There is a pronounced downward bending of leaves and petioles often with random bending of the younger internodes. All auxin-type herbicides promote epinasty. It is thus a characteristic symptom of the action of auxin herbicides on susceptible plants.

epiphyte (aerophyte; air plant). **1:** a plant that grows on and is anchored by another plant (the phorophyte). The relationship is not parasitic; the epiphyte does not depend on the phorophyte for water or nutrients, but derives these from the atmosphere. **2:** any organism that grows on the surface of a plant.

epiphytotic. **1:** of or pertaining to a disease epidemic in plants. **2:** a disease epidemic in plants. *Cf.* epidemic, epizootic, epidemiological.

epirheology. the science of the effects of exogenous factors on plants.

episcleritis. inflammation of the sclera (or 'white') of the eye.

epizoa. plural of *epizoon*.

epizoicide. an agent that destroys epizoa (*See* epizoon).

epizoon (plural, epizoa). an animal parasite that lives on the outer surface of the body of the host.

epizootic. **1:** of or pertaining to an epidemic disease among non-human animals. **2a:** a disease epidemic among animals other than humans. **2b:** any disease that affects a large number of animals (other than humans) in a given area or population. *Cf.* epidemic, epiphytotic.

epizootic abortion. *See* brucellosis.

EPN (O-ethyl O-*p*-nitrophenyl phenylphos

phonothioate (phenylphosphonothioic acid O-ethyl O-(4-nitrophenyl) ester; ethyl *p*-nitrophenyl benzenethiophosphonate). a neurotoxic, liquid or light yellow, water-insoluble, crystalline compound, with an aromatic odor. It has a melting point of 36°C and decomposes in alkaline solution.

$$C_6H_5P(C_2H_5O)(S)OC_6H_4NO_2$$

EPN is used as an insecticide against cotton pests, a miticide (sometimes together with methyl parathion) on fruit crops, and sometimes as a household pesticide.

Eptam. a thiocarbamate herbicide that is particularly useful for controlling *Agropyron repens* and *Cyperus spp.*

EPTC (dipropylcarbamothioic acid *S*-ethyl ester; dipropylthiocarbamic acid *S*-ethyl ester; *S*-ethyl dipropylthiocarbamate). a preplanting thiocarbamate herbicide that is soluble in water and miscible with benzene, ethanol, toluene, and xylene. It selectively inhibits shoot growth, while root growth remains unaffected or even enhanced. It is particularly useful in controlling *Agropyron repens* and perennial *Cyperus spp.* EPTC is moderately toxic orally to laboratory rats.

Equidae. a family of cursorial, herbivorous mammals (Order Perissodactyla). They are characterized by the presence of a single hoofed digit on each leg. They are adapted to plains, grassland, and semidesert environments and feed largely on grasses. *Equus*, *q.v.*, is the only modern genus.

equine. **1:** pertaining to a member of the family Equidae, especially one of the genus *Equus*, *q.v.* **2:** any mammal of the genus *Equus*.

equine biliary fever. equine babesiosis.

equine nervous syndrome. *See* moldy corn poisoning.

equine viral arteritis (epidemic cellulitis; pinkeye). an acute, contagious disease of horses. Signs exhibited by the dam vary but usually include fever, leukopenia, lacrimation, conjunctivitis, catarrh, edema, and abortion. Outbreaks are sporadic, mortality of dams is low, and the disease is often undiagnosed. The infectious agent is a virus.

equis. *Bothrops atrox*.

equisetosis (equisetum poisoning). poisoning of livestock by ingestion of *Equisetum*. In North America, equisetosis is most common in horses and is due solely to *E. arvense*. Hay containing 20% or more of *E. arvense* produces symptoms of a potentially fatal equisetosis in horses within 2-5 weeks. Poisoning of cattle is uncommon and rarely fatal because the ruminal bacteria produce an adequate internal supply of thiamine. Symptoms are similar to bracken poisoning (see *Pteridium*). Treatment with massive doses of thiamine can reverse the disease except in the terminal stages. *See Equisetum*, equisitine, *Pteridium*, thiaminase.

equisetum poisoning. equisitosis.

Equisetum (horsetail; mare's tail; horsetail fern; foxtail; meadow pine; rush; bottlebrush; snake pipes). a genus of small, perennial flowerless, nearly cosmopolitan herbaceous wetland plants (Family Equisetaceae) with leaves reduced to scales. The chief toxic principle is thiaminase, but these plants also contain several alkaloids: equisitine, palustrine, 3-methoxypyridine, and nicotine. Ingestion by nonruminant livestock (most notably horses) may result in thiamine deficiency and eventual death. Cattle and other ruminants are not susceptible to equisetosis since the ruminal bacteria produce substantial amounts of thiamine. While generally considered only moderately toxic, horsetails sometimes accumulate toxic amounts of nitrates and sometimes selenium during periods of rapid growth; cattle and other livestock may be susceptible to poisoning at such times. Children, using the hollow stems of horsetails as blowguns, have occasionally been fatally poisoned. ***E. arvense***. a weedy species common to sandy soils in moist fields, and along roadsides and railroad embankments throughout the United States and Canada. In North America, equisetum poisoning of horses is not uncommon and is due almost entirely to *E. arvense*. Hay that includes 20% or more of *E. arvense* produces symptoms of equisetosis in horses within 2-5 weeks. Poisoning of cattle by *E. arvense* is rarely fatal. ***E. hyemale*** (Dutch rush; scouring rush). a Californian species poisonous to livestock. The common names derive from its use for scouring. ***E. laevigatum***. a Californian species known to poison livestock. ***E. limosum***. a European species known to poison livestock. ***E. palustre***. a species that is locally abundant in wet meadows and along the margins of streams and lakes from Newfoundland and the northern tier of the United States into Alaska. This plant sometimes poisons cattle with loss of condition, hyperexcitability, diarrhea, muscular weakness, and decreased milk production. horses, sheep, and cattle are rarely poisoned. ***E. ramosissimum***. a South African species, it is known to poison livestock, especially sheep.

Equus. a genus of hoofed, fast-running, herbivorous mammals adapted to plains or grassland environments (Family Equidae). The teeth are adapted to grinding; the foot bones are elongate and terminate in a single toe covered by a hoof. The presence of intestinal protozoa makes it possible to digest cellulose. All species have a mane on the neck and a long tuft of hair on the tail that is used like a whisk to repel insects. Most species can interbreed, but the offspring are usually sterile. ***Equus asinus*** (African ass). the common wild ass of Northern Africa. There are several domestic breeds. ***E. a. asinus*** (donkey; domestic ass; burro). a domestic variety of the ass, commonly known as the donkey. It is used chiefly as a beast of burden and the breeding of mules by matings with a horse, *E. caballus*. ***E. burchelli*** (plains zebra). a species found throughout Africa south of the Sahara. It stands about 120 cm (4 ft) tall at the shoulder and has small ears. The stripes are very broad; striping varies greatly in pattern among the several races of this species, as well as among individuals of the same race. It can run swiftly, up to 40 miles per hour. These zebras usually occur in close association with other grazing animals such as wildebeest and antelopes. ***E. caballus*** (modern horse; wild horse). an equine with large hoofs, a short-haired coat, a long mane, and a long tail. Individuals are sexually mature at about the age of two years; full size is not reached, however, until about the age of five years. The dam bears a single foal following a gestation period of about 11 months. The average life span is 18 years, but individuals as old as 30 are fairly common. Horses, today, are used variously as beasts of burden, for pulling or carrying loads, and riding for pleasure, sport, and transport. There are a wide variety of breeds, varying in size from miniature ponies and horses to the largest draft horses. ***E. c. przevalskii*** (Przewalski's horse; Przevalski's horse). a wild horse, chiefly of Mongolia and Sinkiang, characterized by light yellow coloring and a stiff, upright black mane with no forelock. The only remaining breed of wild horse, it is now endangered and chiefly maintained in zoos. ***E. grevyii*** (Grevy's zebra). a large zebra found in E Africa. It stands 140 to 150 cm (4 1/2 to 5 ft) at the shoulder and weighs about 600 lb (270 kg). It has large, rounded ears and numerous very narrow stripes. ***E. hemionus*** (onager; Asian ass). the wild ass of southwestern Asia that is closely related to *E. kiang*. ***E. h. hemionus*** (chigetai; dziggetai). a mongolian variety of wild ass. ***E. kiang***. a wild ass of Tibet and Mongolia, noted for forming herds of up to a thousand animals. It is swift and has been clocked at speeds of up to 65 km (40 mi) per hour. ***E. zebra*** (mountain zebra; common African Zebra). the most distinctive of equines. Some populations favor rocky hillsides. It usually occurs together with other species of zebras. It has a donkey-like build, long ears, and a characteristic pattern of stripes. It is distinguished from other species of *Equus* by the presence of a dewlap on the throat.

erect. **1:** upright; vertical; perpendicular. **2:** to build; raise; construct.

erectum. erect.

eremacausis. the natural oxidation of organic matter. This process is elevated at higher temperatures. Thus, for example, soils of hot or very warm regions have a very low concentration of humus.

eremium. a desert community.

erg. a unit of energy, the force of 1 dyne acting through a distance of 1 cm.

ergocalciferol. vitamin D_2.

ergodic. pertaining to a process in which any sample, in time or space, is representative of the whole.

ecotone (edge). a zone of transition between two adjacent communities (e.g., between forest and prairie; between woodlot and abandoned field or pasture; between alpine meadow and timberline). **forest ecotone** (forest edge). ecotone that is characteristic of a forest edge, which is often a type of savanna with scattered trees, thickets, grassy or herbaceous plants.

ergosterol. A plant sterol that acts as a that is converted to vitamin D2 ultraviolet radiation.

ergot (secale cornutum; rye ergot; rye smut; clavus). **1a:** the over-wintering mycelial mass (sclerotium) of the ascomycetous fungus *Claviceps spp.* (*q.v.*). It is a small compact, dark purple, spur-like, hardened, horny mass that contains several alkaloids that are poisonous to humans, livestock, and other animals by ingestion. **1b:** the dried sclerotium. **1c:** preparations from the dried sclerotia used in medicine. Usually relatively harmless, ergot derivatives are used to induce or intensify uterine contractions during and immediately following parturition; women have, on occasion, died from its misuse. Ergot preparations are also used to control internal hemorrhage in livestock and humans and to alleviate certain localized vascular disorders (e.g., migraine). While usually relatively harmless, when carelessly applied they can induce hallucinations and even death. **2:** a common name for fungi of the genus *Claviceps*. **3:** A horny growth behind the fetlock of a horse.

ergot poisoning. ergotism.

ergotic. of, pertaining to, or produced by ergot.

ergotism (ergot poisoning; Saint Anthony's fire; mal des ardents; feu de Saint-Antoine; gangrene des Solognots). a disease caused by ingestion of a fungus (*See* ergot). In past centuries, humans and livestock have suffered large losses. Humans usually acquire the disease by ingesting cereal grain (especially wheat or rye) and grain products (e.g., bread) that are infested with the fungus. Livestock may acquire the disease from infested pasturage, hay, grain, or grain screenings. With the exception of newborn pigs, ergotism rarely affects wildlife or livestock other than cattle. Sheep are quite resistant. Ergot preparations used in medical practice today (see ergot) are generally very safe, although clinical disease is sometimes seen from excessive use of ergotamine tartrate in the treatment of migraines. Symptoms of ergotism in humans may include nausea, vomiting, diarrhea, dizziness, weak pulse, thirst, tingling in feet, numbness and coldness of extremities, variable effects on blood pressure, dyspnea, convulsions, and loss of consciousness. Examination may also reveal congestion and inflammatory changes in the gastrointestinal tract and kidneys and gangrene of fingers and toes from persistent peripheral vasoconstriction. Effects of acute ergot poisoning of humans includes thirst, nausea, vomiting, diarrhea, tachycardia, confusion, and coma. **chronic ergotism**. neurotropic ergotism. **convulsive ergotism**. neurotropic ergotism. **gangrenous ergotism** (dry gangrene). one effect of ergot alkaloids is to constrict the smaller capillaries of the circulatory system. This interferes with or stops the circulation of blood in the extremities with ensuing necrosis of affected tissues, especially in humans, of the lower limbs. Signs are similar in cattle. Very small doses of ergot can block circulation in the extremities and dispose the animal to thromboses. The hind feet of cattle are affected first, and the hooves may be sloughed off. As a result, the early stages of this form of ergotism resembles foot-and-mouth disease. The front feet or legs, tips of the ears and tail, and sometimes the tongue may subsequently become necrotic and gangrenous. Secondary infec tion may occur. The comb and beak of the domestic fowl are most affected. **neurotropic ergotism** (convulsive ergotism; chronic ergotism). ergotism that, in humans, induces muscle spasms, hallucinations, and sometimes paralysis; the victim's body feels on fire (hence, the name St. Anthony's fire). When cattle consume relatively large amounts of pasturage or hay infested with ergot, the animals display signs of nervous system injury. These may include convulsions and signs of gastrointestinal distress (signs that do not occur in humans). Early signs are hyperexcitability and sometimes belligerence. Trembling and incoordination follow and running animals exhibit an exaggerated flexure of the forelimbs, and if startled or excited, will often fall. Kicking may alternate with periods of tetanic rigidity and emprosthotonos. Death may occur within three days to a month, although early removal of stock from affected vegetation may result in complete recovery. Neurotropic ergotism is rarely caused, in humans or cattle, by *Claviceps purpurea*. The disease is occasionally seen in animals grazing on *Hilaria mutica* (tobosagrass), which is parasitized by *C. cinerea*. In the United States, neurotropic ergotism of cattle usually results from the ingestion of infected *Paspalum dilatatum* (Dallis grass) or *P. notatum* (Argentine bahia) that are parasitized by *C. paspali*.

ergotized. **1:** affected or poisoned by ergot. **2:** containing ergot.

ergotoxicosis. a mycotoxicosis caused by a species of *Claviceps*.

ergotoxine group. *See* ergotoxine (def. 1).

Erica (heathers). a nearly cosmopolitan genus of heaths (Family Ericacea) containing some 1900 species, some of which, such as blueberry and cranberry, bear edible fruit.

Ericaceae (heath family). a large cosmopolitan family of deciduous and evergreen trees and shrubs that grow on acid soils (or in acid bogs). A variety of mammals have been fatally poisoned by eating leaves or flowers of various species of heaths. Domestic fowl are highly resistant, at least to *Kalmia latifolia*. Humans have been poisoned by honey apparently made from the nectar of several species of heaths. Symptoms of poisoning in livestock include anorexia, repeated swallowing or repeated eructation and swallowing of cud without mastication, profuse salivation, dullness, depression, nausea, and vomiting. The preceding may be accompanied by dyspnea, bloat, frequent defecation, evidence of abdominal pain, and nasal discharge. Autopsy usually reveals a nonspecific gastrointestinal irritation with some hemorrhage and sometimes pulmonary lesions. Animals may eventually exhibit ataxia, prostration, coma, and death. North American genera that include species that are established as poisonous to livestock include *Kalmia*, *Ledum*, *Leucothoe*, *Lyonia*, *Menziesia*, *Pieris*, and *Rhododendron*. A few species have been implicated in serious losses of livestock. Plants in any of these genera can be expected to produce symptoms when they comprise 0.2-1.3% of the diet (based on wet weight of fresh green foliage). Many other heaths may be expected to be toxic. *See also* arbutin, *Arctostaphylos*, *Vaccinium*.

erinose. a disease of plants caused by mites of the genus *Eriophyes*. Plants most frequently affected are grape, walnut, and mountain maple. Development is abnormal with hair-like structures (trichomes) emerging in patches from the leaf surface.

erode. **1:** to abrade or wear away. **2:** to wear away or be worn away by erosion; to wear away or gradually destroy a surface (e.g., soil, a mucosal or epidermal surface) by a physical process such as friction, dissolution, or as a result, for example, of inflammation, injury, contact with a corrosive substance. **3:** to cause to deteriorate; disintegrate. **4:** to corrode. **5:** to destroy tissue by the action of a corrosive substance.

erodible. the quality (e.g., of a soil) of being easily eroded. *Cf.* erosive.

Erodium (heron's-bill; stork's-bill). a genus of annual and perennial herbs (Family Geraniaceae). Some species have been implicated in trefoil dermatitis and rape scald, as have species of *Brassica*, *Medicago*, and *Trifolium*. ***E. cicutarium*** (alfilaria; alfileria; alfillerilla; filaree; pin grass; filaree; red-stem; storksbill). an annual herb (Family Geraniaceae) that is indigenous to southern Europe. It is an important forage plant in the southwestern United States

erosion. **1:** a wearing away (e.g., of a land surface, teeth, intestinal mucosa) or the gradual destruction of a surface. **2:** a state of being worn away or corroded (e.g., by friction, wind or rain, a corrosive chemical). *Cf.* corrosion. **3:** a superficial ulcer in the mucosa of the gastrointestinal tract, one that does not invade the muscularis mucosa. **accelerated erosion**. erosion due to human activity (e.g., that due to plowing, clearing of land, construction of sewers and culverts); the amount or rate of erosion over and above that which occurs naturally. **chemical erosion**. erosion materials including human tissues by certain chemicals such as strong acids and strong bases. **sheet erosion** (sheet-flood erosion). erosion caused by sheets, rather than streams, of running water. **sheet-flood erosion**. sheet erosion. **splash erosion**. that caused by the impact of falling raindrops. **water erosion**. erosion of stone, soil, and other surface features by running or falling water. **wind erosion**. the erosion, transportation, and deposition of topsoil and other surface deposits by the wind. A dust storm is the product of severe wind erosion.

erosion pavement. the layer of pebbles, gravel, or stones that remains on the surface of the ground following the removal of the fine particles by wind or water erosion.

erosive. **1a:** able to erode, abrade, corrode, or wear away as by wind or flowing water. **1b:** characterized by erosion. **1c:** the property of wind or flowing water of sufficient velocity, or of certain chemicals to erode materials. **2:** characterized by erosion. **3:** an agent that erodes tissues (e.g., a corrosive poison) or materials. *Cf.* corrosive, erodible.

error. a mistake; a fault or flaw; inaccuracy; falsity; failure. Error is a widely used but often confusing term. When using this term in the sciences, one must take care to clarify meaning: **1:** the difference between a measurement or experimental result and the "true" value. **2:** the uncertainty of an experimental result. **experimental error**. increased imprecision or inaccuracy in the results of an experiment, presumably due to chance variation or extraneous variables, as opposed to simple errors of observation or measurement. It is also sometimes taken as variation in results that is not accounted for by the hypothesis tested. **random error**. error due to chance such that any deviation in the magnitude and direction of a variable cannot be predicted. **sampling error**. **1:** in statistics, the difference between an observed value of a statistic and the variable it is intended to estimate. **2:** the variation among the observations inany one of an analysis of variance. **3:** any error in selection or collection of samples (sets of specimens) from a population that results in significant differences between the sample and the population that it purports to represent. Such error often results from nonrandom selection of the specimens that comprise the sample. *See also* experimental error. **standard error**. the standard deviation divided by the square root of the number of observations. It is a measure of the variation of a population of means. **type I error**. **1a:** in statistics and experimental research, the rejection of a null hypothesis when it is actually true. **1b:** an error in which an event is predicted which does not, in fact, materialize. **type II error**. **1a:** in statistics and experimental research, the acceptance of a null hypothesis when it is likely to be false and should be rejected. **1b:** failure to predict an event that does in fact occur. Traditionally, types I and II errors have been treated differently. Thus, a 5% probability of type I error in the agricultural sciences (or even a 1% probability) has been considered acceptable, whereas a 20% probability of type II error may be accepted. In some cases, as in risk assessment, however, it is the type II error that one most wishes to avoid.

ERS. Economic Research Service.

eructation. belching; the act of expelling gas brought up from the esophagus or stomach.

erupt. **1:** to emit, discharge, release, gush, well up, issue forth, vent; explode. **2:** to break out, as to break out in a rash.

eruption. **1a:** the act or process of breaking out; emission; explosion; appearing or becoming visible. **1b:** a breaking out of visible skin lesions. **2a:** a rapidly developing dermatosis of the skin or mucous membranes that is usually a local manifestation of systemic disease. The lesions are often marked by prominence, erythema, or both. The shape, size, and condition of the lesions may vary widely. **drug eruption** (medicinal eruption). any rash, eruption or solitary skin lesion caused by the ingestion, injection, inhalation, or insertion of a drug or other xenobiotic substance, but not by skin contact. Such eruptions are most commonly due to allergic sensitization. **fixed drug eruption**. a fixed eruption or hypersensitivity reaction following the administration of a particular drug (e.g., barbiturates, quinine, sulfonamides, tetracycline). Lesions are typically reddish-brown or purple, intensely erythematous, sharply defined macules and are associated with hydropic degeneration of the basal layer. Following involution, the affected areas may erupt again and enlarge if the drug is administered anew. The loci that are most often involved are the extremities including the penis. **fixed eruption**. a circumscribed inflammatory lesion(s) that recur(s) at the same site over a period of time; each attack lasts only a few days but leaves residual pigmentation which accumulates. iodine eruption. an acneform or follicular eruption or granulomatous lesion as part of a systemic reaction to iodine. **medicinal eruption**. drug eruption.

eruptive. pertaining to or characterized by eruption.

Erysimum cheiranthoides (wormseed mustard). a short-lived perennial herb (Family Brassicaceae); plants of this genus are commonly known as blistercress or wallflower. It can cause severe gastroenteritis in grazing livestock, presumably because of the mustard oil content.

erysipelas. a contagious disease of the skin and subcutaneous tissues caused by *Streptococcus pyogenes*. The principle signs are redness and swelling of the affected areas. Among livestock, the disease chiefly affects swine, the joints of which may be affected; breathing is labored.

erysipelotoxin. the toxin of *Streptococcus pyogenes*. It is the causative agent of erysipelas.

erythema (erythema simplex; dermatitis simplex). inflammatory redness or blushing of the skin due to dilation of superficial capillaries. Causes vary (e.g., an allergen, endogenous neurovascular processes, nervous mechanism, sunburn). *See also* acrodynia.

erythema simplex. erythema.

erythematous. pertaining to or marked by erythema; inflamed.

erythredema. acrodynia (def. 2).

erythrocyte (red blood cell; red blood corpuscle; RBC). **1:** a hemoglobin-containing blood cell found in all vertebrate animals. It is compressed, and in cyclostomes and most mammals, it is nearly round in outline; it is oval in other vertebrates. The erythrocytes of most mammals lack a nucleus and thus take the shape of a biconcave disk. They reversibly bind oxygen and carbon dioxide and thus serve in the internal transport and exchange of these gases. **2:** a cell that contains a respiratory pigment, as in some marine worms.

erythrogenic. capable of reddening or causing reddening, as in inflammation.

erythrosine (FD&C Red No. 3; E127; disodium or dipotassium salt of 2,4,5,7-tetraiodofluoroscein). a food color and biological stain,

$$C_{37}H_{36}N_2O_{10}S_3{\cdot}2Na$$

It is used widely, for example, in jams and marmalades, maraschino cherries, pickles and relishes. It is probably not toxic in doses normally encountered. Erythrosine is possibly a weak carcinogen.

Erythrotoxylon. a synonym of *Erythroxylum*.

Erythroxylum (synonym, *Erythroxylon*; coca; hayo; ipado). a large genus of small trees and shrubs (Family Erythroxylaceae) that are indigenous to South America and the West Indies, but cultivated elsewhere. ***E. coca***. (Bolivian coca; Huanaco coca). the leaves contain cocaine and other alkaloids. Mixed with a little alkali, the leaves are chewed by Andean Indians to increase endurance and prevent fatigue and hunger pangs while working or travelling at high elevations. ***E. novagranatense*** (Peruvian coca; Truxillo coca). This species has similar effects and uses as *E. coca*.

esbecythrin. deltamethrin.

escar. a synonym for esker.

eschar. a synonym for esker.

Escherichia coli (formerly *Bacterium coli*; colon bacillus). a species of Gram-negative bacterium (Family Enterobacteriaceae) that are common in the alimentary canal of humans and other vertebrates. It is normally nonpathogenic when present in the gastrointestinal tract, but when free or invasive of other tracts (notably the urinary tract) they are infectious. Some strains are pathogenic and are largely responsible for traveler's diarrhea. When present in milk or water, *E. coli* is an indicator of fecal contamination.

eskar. a synonym for esker.

esker (asar; back furrow; escar; eschar; eskar; osar; serpent kame). a long, low, narrow, sinuous ridge, chiefly of stratified accumulations of gravel and sand formed by a melting glacier or ice sheet. *See also* back furrow.

Esoderm. a trade name for lindane.

essential amino acid. *See* amino acid.

ester. any of a diverse group of naturally occurring and synthetic organic compounds formed by the reaction of an alcohol and an acid (usually a carboxylic or aromatic acid) with the elimination of water.

***S*-ester with *O,o*-dimethyl phosphorothioate**. malathion.

estiv-. aestiv-. a prefix indicating a relationship to summer.

estival. aestival.

estival pond. aestival pond.

estivate. aestivate.

estivation. aestivation.

estradiol-17-β(estra-1,3,5(10)-triene-3,17β-diol; 1,3,5-estratriene-3,17-β-diol). a white or yellow, odorless, crystalline substance or powder, $C_{18}H_{24}O_2$. The β-isomer of estradiol is the most potent naturally occurring ovarian and placental hormone in mammals. It is a steroid synthesized chiefly by the ovary but is also produced by the testis, adrenal gland and placenta, and to some extent by other organs such as liver, depot fat, and skeletal muscle. Estradiol is synthesized metabolically from androstenedione and testosterone. It stimulates and regulates the development of and maintains the secondary female sex characteristics beginning at puberty (e.g., development of pubic hair, growth and development of the vagina, uterus, and fallopian tubes, breast enlargement and shape, roundness of hips, growth and maturation of the long bones). Estradiol-17-β is used in medicine as a steroid hormone and is administered by intramuscular injection or by use of subcutaneous implants. It is a confirmed carcinogen. Other effects, based on research and tests have been reported in the scientific literature. These include teratogenic activity in laboratory animals as well as reproductive and mutagenic effects in humans. *See also* estrogen.

1,3,5-estratriene-3,17-β-diol. estradiol-17-β.

estra-1,3,5(10)-triene-3,17-β-diol. estradiol-17-β.

estrogen. any of a class of female sex hormones (18-carbon steroids) that are elaborated by the ovaries, the adrenal glands, testes, the fetoplacental unit, and certain other peripheral organs and tissues (*See* estradiol-17-β). Estrogens promote the development and maintenance of the reproductive organs and, at the onset of puberty, they promote and regulate the development and maintenance of female secondary sexual characteristics (e.g., pubic hair, breast enlargement and shape, roundness of hips, growth and maturation of the long bones) and sexual behavior. Estrogens also play an essential part in the hormonal control of menstruation, being partly responsible (with progesterone) for the cyclical changes in the lining of the human uterus. Estrogens also have metabolic effects on many organs and tissues not directly related to reproduction.

estrogenic (estrogenous). **1:** of or pertaining to estrogen. **2:** having an action similar to that of an estrogen. Estrogenic compounds include estradiol, chlordecone, coumestrol, diethylstilbestrol, genistein, and zearalinone. **3:** causing estrus.

estrogenous. estrogenic.

estrous. pertaining to estrus.

estrual. pertaining to estrus.

estruation. estrus.

estrum. estrus.

estrus (estrum; estruation; oestrus; heat (in part)). a phase of the estrus cycle (*q.v.*), it is the recurrent, limited period of sexual activity in female mammals other than primates. It is characterized by intense sexual receptivity (heat). *See also* estrus cycle.

estrus cycle (oestrus cycle). the reproductive cycle that occurs in sexually mature, nonprimate, nonpregnant, female mammals either continuously or only during the breeding season. It is accompanied by intense sexual receptivity (*See* estrus). The length of the cycle varies among species from 4 to about 60 days. The cycle is under central control by neuroendocrine secretions periodically released from the hypothalamus which stimulate the secretion of pituitary gonadotrophins. The latter, in turn, promote the production of estrogen and progesterone which are the proximate regulators of the cycle as part of a negative feedback loop. The phases of the cycle are (1) **estrus**: the follicular phase, it is the period of sexual receptivity, during which the Graafian follicles mature, estrogen secretion is at its highest, and the endometrium (lining of the uterus) proliferates. Ovulation and the beginning of corpus luteum formation occur late in this phase in most species. *See* estrogenic. (2) **metestrus**: the postovulatory phase, it is the period when one or more corpora lutea develop and begin to secrete progesterone; estrogen secretion decreases. Mating is no longer sought nor usually allowed by the females of most species. (3) **diestrus**: the luteal phase, in which the corpus luteum is fully functional; it begins to regress in this phase. The growth of new Graafian follicles begins. If fertilization occurs during this phase, the cycle does not progress further until the gestation terminates. This phase is lengthened in some species, if the female does not become gravid following mating, and is termed pseudopregnancy. *See* especially fusarian estrogenism. (4) **proestrus**: the phase during which the corpus luteum fully regresses, progesterone secretion decreases, the next ovarian follicle(s) grows and the secretion of follicle-stimulating hormone and of estrogen begins to increase.

estuarine ecosystem. an ecosystem that lies at the mouth of a river that empties into a salt water body. The character of the system is largely dictated by the interaction between the flowing fresh water and the near-shore ocean waters, chiefly through the action of tides. The water in most parts is brackish. These ecosystems are rich in marine life, euryhaline forms, birds, and wildlife. They are, however, subject to and sensitive to domestic, agricultural, and industrial sewage from point and non-point sources.

estuary. a bay, river mouth, delta, brackish salt marsh, or lagoon that develops at the region of interaction between flowing fresh water from a river and near-shore marine waters.

etaerio. a set of fruits formed from a single flower in which the gynoecium is apocarpous. An etaerio may be comprised of achenes, drupes, or follicles.

2-ethamino-4-isopropylamino-6-methylmercapto-*s*-triazine. ametryn.

ethane dichloride. ethylene dichloride.

ethanedioic acid calcium salt. calcium oxalate.

1,2-ethanediol. ethylene glycol.

1,2-ethanediylbascarbamodithioic acid disodium salt. nabam.

***N,N'*-1,2-ethanediylbis[*N*-(carboxymethyl)glycine]**. edetic acid.

[[1,2-ethanediylbis[carbamodithioato]]-(2—)]manganese. maneb.

ethanoic acid. acetic acid.

ethanol (ethyl alcohol; absolute alcohol; anhydrous alcohol; dehydrated alcohol; ethyl hydrate; ethyl hydroxide; alcohol). a clear, colorless, volatile, flammable liquid, C_2H_5OH, with a melting point of -114°C and a boiling point of 78°C. It is produced by the fermentation and fractional distillation of grain. Other than water, it is probably the most widely used chemical in industry and commerce, e.g., as a beverage ingredient, synthetic chemical, solvent, bacteriostat, germicide, antifreeze, preservative, and gasoline additive.

ethion (ethyl methylene phosphorodithioate). an organophosphate anticholinesterase used as a miticide and insecticide; applied as spray. It is less toxic than parathion, but more toxic than malathion.

ethology. the zoological science that treats animal behavior, especially behavior that occurs under natural conditions.

***o*-(6-ethoxy-2-ethyl-4-pyrimidinyl)-phosphorothioic acid *O,o*-dimethyl ester**. etrimfos.

ethyl carbamate. urethane.

ethyl carbamilate. phenylurethane.

ethyl dimethyl carbinol. *tert*-pentyl alcohol.

***S*-ethyl dipropylthiocarbamate**. EPTC.

ethyl formate (formic ether). a colorless, unstable, flammable liquid, C_2H_5COOH, with a pleasant, distinctive odor. It is used as a yeast and mold inhibitor, a fumigant for raisins and currants, a fungicide for cashew nuts, cereals, tobacco, and dried fruits, and as a food flavoring. Even a brief exposure to the vapor may irritate the skin and eyes and produce narcotic effects. Exposures to high concentrations can be fatal.

ethyl hydrate. ethanol.

ethyl hydroxide. ethanol.

ethyl methyl carbinol. 2-butanol.

ethyl methylene phosphorodithioate. ethion.

ethyl *p*-nitrophenyl benzenethiophosphonate. EPN.

O-ethyl O-*p*-nitrophenyl phenylphosphonothioate (phenylphosphonothioic acid O-ethyl O-(4-nitrophenyl) ester. EPN.

ethyl phenylcarbamate. phenylurethane.

ethyl phenylurethane. phenylurethane.

ethyl pyrophosphate. TEPP.

***S*-ethyl-*N,N*-diisobutylthiocarbamate**. *See* butylate, def. 2.

ethyl urethane. urethane.

ethyl urethan. urethane.

ethylenebis(dithiocarbamate) (EBDC). any of a class of dithiocarbamate fungicides (e.g., maneb, zineb) used on many fruits and vegetables. They are suspected carcinogens.

ethylene bromide. ethylene dibromide.

ethylene chloride. ethylene dichloride.

ethylene dibromide (EDB; 1,2-dibromoethane; ethylene bromide). an extremely toxic, colorless, nonflammable liquid,

$$BrCH_2CH_2Br$$

with a sweetish odor that is miscible with most solvents and thinners. It is an agricultural fumigant and industrial chemical with various uses. It is extremely toxic, is carcinogenic at high doses to laboratory animals, a suspected human carcinogen, and a probable mutagen and teratogen. EDB was banned for most agricultural uses in the United States in 1984.

ethylene dichloride (1,2-dichloroethane; α,β-dichloroethane; *sym*-dichloroethane; dichloroethylene; ethane dichloride; ethylene chloride; 1,2-ethylene dichloride; 1,2-DCE; EDC; Dutch liquid; Dutch oil; glycol dichloride). a toxic, carcinogenic, heavy, clear, colorless, rather viscous, volatile, sweet-tasting, flammable, and moderately explosive liquid,

$$ClCH_2CH_2Cl$$

with a pleasant, chloroform-like odor. It is used as a fumigant of soils and foodstuffs, as an industrial solvent (e.g., of fats. oils, resins, rubber), and in the production of several organic chemicals.

1,2-ethylene dichloride. ethylene dichloride.

ethylene glycol (1,2-ethanediol). a colorless, relatively viscous, nephrotoxic, oxalosis-inducing liquid,

$$HOCH_2CH_20H$$

It is the main component of liquid motor vehicle antifreeze/antiboil formulations and is also used, for example, in hydraulic fluids, condensers, and heat exchangers. It is moderately toxic to humans, although exposure is rather rare because of the low volatility. The lethal oral dose in humans is about 1.4 ml/kg body weight. Domestic cats have a voracious appetite for ethylene glycol and often experience kidney failure when they have access to antifreeze containing this compound.

ethylenebisaminodiacetic acid. edetic acid.

[ethylenebis(dithiocarbamato)]manganese. maneb.

ethylenebis[dithiocarbamic acid] disodium salt. nabam.

ethylenebis[dithiocarbamic acid] manganous salt. maneb.

ethylenediamine (1,2-diaminoethane; EDA). a colorless, alkaline liquid,

$$NH_2CH_2CH_2NH_2$$

It has an ammonia-like odor, is soluble in water and ethanol, slightly soluble in ether, and insoluble in benzene. It is a fungicide and is widely used in the manufacture of chelating agents such as edetic acid. It is also used as a solvent, an intermediate in organic syntheses, an emulsifying agent, an antifreeze inhibitor, and a textile lubricant. EDA is toxic by inhalation and contact.

ethylenediamine tartrate. edetic acid.

ethylenediaminetetraacetic acid. edetic acid.

ethylenediaminetetraacetate. edetic acid.

ethylenediamine-*N,N,N',N'*-tetraacetic acid. edetic acid.

(ethylenedinitrilo)tetraacetic acid. edetic acid.

ethylenedinitrilotetraacetic acid. edetic acid.

1,1'-ethylene-2,2'-di-pyridylium dibromide. diquat.

ethylisocyanate. a liquid, C_2H_5NCO, that is soluble in chlorinated and aromatic hydro carbons. It is used as an intermediate in the production of pesticides. A powerful irritant to skin, eyes, and mucous membranes of the respiratory tract, it is also a chemical warfare agent. When decomposed by heating, it releases sulfur oxides and hydrogen cyanide.

***N*-ethyl-*N'*-(1-methylethyl)-6-(methylthio)-1,3,5-triazine-2,4-diamine**. ametryn.

***N*-[3-(1-ethyl-1-methylpropyl)-5-isoxazoleyl]-2,6-dimethoxybenzamide**. isoxaben.

***S*-[2-(ethylsulfinyl)ethyl]-*O,O*-dimethyl phosphorothioate**. demeton-*O*-methyl sulfoxide.

2-(ethylthio)-ethanethiol *S*-ester + *O,O*-diethyl phosphorothioate. demeton-*S*.

ethylxanthic disulfide. dixanthogen.

etiolation. the production of an abnormal plant by growing it in the dark. Such plants lack chlorophyll and thus become pale yellow or nearly white in color. The stems are typically elongate, and the leaves are small.

etiologic, etiological (aetiological). causal; of or pertaining to etiology.

etiologic agent (aetiologic agent). a causal agent; a chemical or pathogen that causes a specific disease or injury to a living organism.

etiology (aetiology). **1a:** causation. **1b:** the demonstrated causes of a trait, condition, or disease state. **1c:** the origins and course of a disease in terms of pathogens, toxicants, or other contributing causes that might be involved. **1d:** the body of knowledge concerning the causes of disease. **2a:** the study of causality. **2b:** the science of the origins and causes of disease and their mode of operation. *Cf.* pathogenesis.

etrimfos (*o*-(6-ethoxy-2-ethyl-4-pyrimidinyl)-phosphorothioic acid *O,o*-dimethyl ester; *O,o*-dimethyl *o*-(2-ethyl-4-ethoxypyrimidinyl)-6-thionophosphate). a contact and stomach organophosphorus insecticide used against pests of fruit and vegetable crops. It is moderately to very toxic orally to laboratory rats and mice and is also toxic to fish.

Eubacteria. obsolete term for *Bacteria*.

eucaryote. eukaryote.

eukaryote (eucaryote). nucleated unicellular organisms and all multicellular organisms. Eukaryotics are characteristic of most organisms except bacteria. *Cf.* prokaryote.

eukaryotic. that which is characteristic of eukaryotes in which the genetic material is contained within a nucleus that is separated from the cytoplasm by the membranes of the nuclear envelope. Eukaryotes are generally much larger than prokaryotes and are considered more complex and exhibit greater diversity.

eumycetes. the true fung.

eupatorin.a common name of plants of the genus *Eupatorium*.

Eupatorium (eupatorin; snakeroot; thoroughwort; throughwort). a large genus of weedy, many-flowered, usually perennial herbs or shrubs (Family Asteraceae, formerly Compositae). They occur chiefly in tropical and temperate America, with a few species in Eurasia and Africa. Some species are toxic, but the literature on this is confusing, in part because toxicities vary with time and location and because some species are often mistaken for white snakeroot (*E. rugòsum*), which is seriously toxic to both humans and livestock. Toxic concentrations of nitrate have, however, been reported for several species under conditions that favor rapid growth. ***E. ageratoides***. *E. rugosum*. ***E. perfoliatum*** (boneset; thoroughwort). concentrations of nitrate that are toxic to livestock have been reported. ***E. purpureum*** (Joe Pye weed; sweet Joe Pye weed; green-stemmed Joe Pye weed; thoroughwort). Concentrations of nitrate that are toxic to livestock have been reported. Certain other species such as *E. maculatum*, *E. fistulosum*, and *E. verticillatum* are also called Joe Pye weed. ***E. rugosum*** (*E. ageratoides*, *E. urticaefolium*; snakeroot; white snakeroot; white sanicle; fall poison; rich-weed). a showy perennial white-flowered bush that grows to a height of 1 m or more. It commonly occurs in low moist areas from eastern Canada to Saskatchewan, south to eastern Texas, Louisiana, Georgia, and Virginia. The leaves are ovate or heart-shaped, toothed, and are borne oppositely on long stalks. The numerous tiny bright white flowers form clusters at the tips of the branches. It is a widespread and very variable weedy species. White snakeroot is difficult to distinguish from closely related species. All parts of the plant are poisonous. The poisonous principle is tremetol. The syndrome produced in cattle is usually called trembles, and in humans, milk sickness. From the late 1700s to the mid 1900s, this plant was responsible for a condition marked by weakness, tremors, nausea, severe vomition, anorexia, icterus, convulsions, and prostration. The most pronounced lesion is fatty degeneration of the liver. In fatal cases, delirium and coma precede death. The odor of acetone may be noted on the breath of a victim. Kidney damage sometimes results in diminished urine production (*See* oliguria) or anuria. The disease was devastating in some areas, decimating local populations. Milk sickness still occurs occasionally from drinking milk from an exposed family cow. Other animals are also susceptible, and suckling young may be poisoned by tremetol in the mother's milk. Symptoms in cattle and other livestock include weakness, unwillingness to move, stupor, muscular tremors (trembling), and prostration often followed by temporary recovery. In severe cases, the animal ultimately remains prone, becomes comatose, and dies. If made to walk, it abruptly stands still with feet wide apart and trembles. *See* milk sickness, trembles, tremetol. ***E. urticaefolium***. *E. rugosum*.

Eupatorium ageratoides. *Eupatorium rugosum*.

Eupatorium urticaefolium. *Eupatorium rugosum*.

eupeptic. having or pertaining to normal digestion.

Euphorbia (spurges). a large, structurally diverse genus of shrubs and herbs (Family Euphorbiaceae) that are distributed widely in tropical and temperate regions. Most species are poisonous and/or of medicinal value. Some are used for fish and arrow poisons. All parts are toxic and most species have an irritant, milky sap. Some species are violently toxic, others are not known to cause severe poisoning. *E. systyloides*, for example, is used as a remedy for hookworms, but overdosage can cause delirium, convulsions, and death within a few hours. Human intoxications, sometimes severe, have been traced to various species of *Euphorbia* such as *E. lathyris* (caper spurge; mole plant), *E. milli* (crown-of-thorns), *E. peplus* (petty splurge), and *E. pulcherrima* (poinsettia). A number of euphorbia species are ornamentals (e.g. poinsettia, crown-of-thorns, spotted spurge); they may be attractive to children and should thus be considered dangerous. Livestock do not readily eat spurges. The young shoots of some species may be consumed, however, when other forage is unavailable. Some species may be photosensitizing, and concentrations of nitrate that are toxic to livestock have been reported. Symptoms of poisoning may include scours, collapse, and death. In less acute cases, symptoms may be limited to less severe effects such as hair loss, irritation and edema of the mouth and tongue, and symptoms of photosensitization such as dermatitis or edematous enlargement of the head. The following North American species are among those that have been implicated in the poisoning of livestock: *E. corollata* (flowering spurge; tramp's spurge; wild hippo), *E. cyparissias* (cypress spurge, graveyard weed), *E. esula* (leafy spurge, wolf'smilk), *E. helioscopia* (sun spurge, wart weed). *E. maculata* (milk purslane), *E. marginata* (snow-on-the-mountain), and *E. prostrata*. ***E. ipecacuanha*** (ipecac; ipecac spurge; America white ipecac; wild ipecac; Carolina ipecac). a spurge (Family Euphorbiaceae) with emetic properties that is indigenous to the eastern United States. ***E. lactea*** (African milk plant; spotted spurge). a spurge with a fast growing candelabrum form. The sap is toxic and is released when the spine is broken. It is reportedly used by women of certain tribes to rid themselves of evil husbands. This plant is widely cultivated and is often used for hedges in warm climates. ***E. pulcherrima*** (poinsettia). a tropical herb and houseplant commonly sold during the Christmas season.

Euphorbiaceae (the spurge family). a family of dicots that includes poisonous species such as *Hippomane mancinella*, the manchineel tree, *q.v.* The sap of most species contains cucurbitacin, *q.v. See also Euphorbia*.

euploid. descriptive of a cell or cell nucleus that bears an exact multiple of the haploid chromosome number. Euploid cells may be haploid or polyploid with the chromosome number represented by descriptors such as diploid, triploid, tetraploid, etc.

eupnea. normal, unlabored breathing.

Euproctis chrysorrhoea (brown-tail moth). a pest species of moth of the northeastern United States. The caterpillar feeds on tree leaves. *See* dermatitis (brown-tail moth dermatitis).

eurokous. euryoecious.

euroky (euryoky). the ability to tolerate a wide range of environmental conditions.

European alder. *Alnus glutinosa*.

European aspidium. *See* aspidium.

European beachgrass. *Ammophila arenaria*

European beech. *Fagus sylvatica*.

European bison. *Bos bonasus*.

European bittersweet. *Solanum dulcamara*.

European brown tick. *Rhipicephalus sanguineus*.

European cherry laurel. *Prunus laurocerasus*.

European corn borer. *See Pyrausta nubilalis*.

European dairy breed. any breed of dairy cow that had its origin in Great Britain or continental Europe (e.g., Charolais, Simmental, Limousin, Jersey, Holstein, and Guernsey).

European dog tick. *Rhipicephalus sanguineus*.

European domestic cattle. *Bos taurus*.

European false hellebore. *Veratrum album*.

European honeybee. *Apis millifera*.

European moth. *Pyrausta nubilalis*.

European red-berried elder. *Sambucus racemosa*.

European white heath. *Erica arborea*.

European white hellebore. *Veratrum album*.

European wild boar. *Sus scrofa*.

eurybiontic. pertaining to or denoting an organism that tolerates a wide range of a specific environmental factor. *Cf.* stenobiontic.

euryhaline. the ability of an individual aquatic organism, population, or species to tolerate a wide range of dissolved salts in the environment. Salmon are, for example, typically euryhaline. *Cf.* stenohaline.

euryhydric. tolerant of a wide range of moisture levels or humidity. *See also* stenohydric.

euryionic. having or tolerating a wide range of pH. *Cf.* stenionic.

euryoecious (euryoekous; euryokous; eurokous). of or pertaining to a species or population that is tolerant of a wide range of habitats and environmental conditions. *Cf.* amphioecious, stenoecious; one characterized by euroky. *See also* amphioecious, stenoecious.

euryoekous. euryoecious.

euryokous. euryoecious.

euryoky. euroky.

euryphagous. consuming a wide variety of foods; of or pertaining to a heterotrophic organism that uses a wide variety of foods.

eurysubstratic. tolerant of a wide range of types of ecological substrata. *Cf.* stenosubstratic.

eurytropic. said of organisms that respond readily or adapt substantially to changing environmental conditions. *Cf.* stenotropic.

eurytropism. the state or condition whereby an organism responds markedly or adapts to changing environmental conditions. *Cf.* stenotropism.

eustele. *See* stele.

euthanasia (mercy killing). **1:** literally, an "easy death," one that is free from suffering. **2:** the means of providing an easy death. **3:** the practice or deliberate act of painlessly terminating the life of a suffering and incurable human or a wild or domestic animal, usually one who is suffering from intractable pain.

eutrophic. **1:** relating to the process of eutrophication. **2:** relating to a large water body that has undergone substantial eutrophication. A eutrophic water body is rich in nutrients and has a high rate of primary production reflected in well-developed littoral and sublittoral vegetation and dense growth of plankton. There are few fish, and oxygen is depleted during periods of warmth and high insolation. In areas largely free from human influence, eutrophic water bodies are rather old and shallow with well developed vegetation. *Cf.* oligotrophic.

eutrophication. the transition (under otherwise stable conditions) of a body of surface water from an early freshwater stage through intermediate stages of nutrient enrichment to a bog, marsh, or swamp and ultimately a terrestrial system. **anthropogenic eutrophication**. enrichment of large water bodies due to human agency, greatly accelerating the process of eutrophication. The most common proximate causes are due to extravagant use of nitrate fertilizers, the use of detergents that contain phosphorus, the dumping of organic wastes, and runoff containing animal excrement. Eutrophic bodies of water are characterized by large populations of flora and fauna, often leading to oxygen depletion. **natural eutrophication**. the slow, natural, aging process whereby a lake, large river, estuary, or bay evolves into a bog or marsh and eventually disappears. During the later stages of eutrophication, the water body is oxygen poor and choked by abundant plant life as the result of increased amounts of nutritive compounds such as nitrogen and phosphorus. Natural eutrophication occurs slowly as a consequence of the accumulation of silt.

eutrophy. the visitation of a particular flower population or species only by specialized insect pollinators.

evanescent. brief; ephemeral; short-lived; transitory; temporary.

evaporate. **1a:** vaporize; volatilize. **1b:** to undergo or cause to undergo evaporation.

evaporation. **1a:** vaporization; volatilization. **1b:** the change of a substance from a liquid to a gaseous or vapor phase. **2:** the reduction in mass or volume of a liquid by conversion into vapor.

evaporometer. atmometer.

even flow. the planning, scheduling, and management of tree harvests such that the harvest in a given period is equal to that of the preceding period.

evening trumpet flower. *Gelsemium sempervirens*.

evergreen. a term designating those vascular plants that produce and shed leaves throughout the year, such that the branches always bear leaves. An individual leaf may persist for several years. This designation bears no relationship to taxonomic status. *Cf.* deciduous.

everlasting pea. a poisonous pea. *See Lathyrus*.

ewe (yew). a adult female sheep, one that is of a breeding age.

exacerbation. increased severity of a condition, effect, or response, as of a disease, a sign or a symptom.

excavated trail. in forestry, a constructed trail having a width of more than 1.5 m and a mineral soil cutbank height of more than 30 cm.

exceedance. violation of environmental protection standards by exceeding allowable limits.

exchangeable cation. a cation that is held by the adsorption complex of the soil and is readily exchanged with other cations of neutral salt solutions.

excipient. **1a:** any metabolically inert material that is added to a chemical formulation to give a desired consistency or form (e.g., gum arabic, lanolin, soap, starch). **1b:** a substance added to a chemical formulation with the intent of altering or facilitating its effectiveness. *See also* adjuvant.

excise. to remove by cutting out.

excitability. sensitivity to stimulation; irritability.

excitable. **1:** easily aroused or stimulated. **2:** easily agitated, irritated, disturbed, or angered. **3:** able to respond rapidly to a stimulus.

excitant. **1:** stimulant or stimulating. **2a:** that which excites or stimulates. **2b:** a stimulant; any agent that excites the brain or a particular vital function. Such agents are usually named after the organ or function affected (e.g., cerebral excitant, motor excitant). Some are used therapeutically.

excitatory. stimulating; stimulating a process or capable of doing so. *Cf.* inhibitory.

excite. **1:** stimulate; arouse; activate; thrill. **2:** agitate; irritate; anger.

excitoanabolic. stimulating or promoting anabolism.

excitocatabolic. stimulating catabolism

excitoglandular. stimulating glandular secretion.

excitometabolic. inducing or stimulating metabolic changes.

excitomotor. stimulating increased speed of muscular activity.

excitomuscular. inducing or increasing the rate or intensity of muscular activity.

excitor. an agent or stimulus that induces or greatly heightens activity of a living system or system component.

excitosecretory. tending to induce or stimulatle secretion.

excitotoxic effect. *See* effect.

excreta. **1a:** excretion; the excretory products or waste materials eliminated from the body or any cell or tissue of a living organism, often including carbon dioxide, alkaloids, organic acids, nitrogenous compounds, water, and various metabolites of xenobiotics. **1b:** that part of assimilated matter that is eliminated from the body of a living organism as excretion, secretion, or exudation. **1c:** a component of rejecta. **2:** harmful metabolic products of a plant. *Cf.* egesta, excretion, rejecta.

excrete. to separate, concentrate, and eliminate excreta. *See also* excretion.

excretion. **1:** excreta; the excretory products or waste materials eliminated from the body or any cell or tissue of a living organism, often including carbon dioxide, alkaloids, organic acids, nitrogenous compounds, water, and various metabolites of xenobiotics. The term is properly restricted to the elimination of those waste materials that have (in some form) been part of the living protoplasm. **2a:** any behavioral act or physiological process of eliminating waste material from the cells, tissues, or body of a living organism, especially products of metabolism. **2b:** collectively, all processes involved in separating, concentrating, and eliminating excreta. **3:** the elimination by plants of harmful metabolic products.

excretory gland. a gland that separates waste material from the blood.

EXD. dixanthogen.

exo antigen. ecto antigen.

exoantigen. ectoantigen.

exocrine gland. a gland of external secretion (*Cf.* endocrine gland), one in which the secretory products are conveyed by one or more ducts to a free epithelial surface (e.g., a sweat gland). The classification of exocrine glands is based on the structure both of the secretory portion of the gland and the associated system of ducts. **compound exocrine gland**. one with elaborately branched passages and ducts (e.g., the exocrine pancreas). **simple exocrine gland**. a gland with a single secretory passage and a single unbranched duct. **tubular exocrine gland**. a gland within which the secretory portion is a tubule of uniform diameter.

exocytosis. **1:** the presence of migrating inflammatory cells in the epidermis. **2:** (emiocytosis; merocrine secretion; eccrine secretion). the most common type of secretory process, one in which an intracellular vesicle fuses with the plasmalemma (cell membrane) in such a way that the membrane-bound contents of the vesicle are released to the outside in the form of droplets or granules. Neurotransmitters are released from nerve terminals by exocytosis. *Cf.* endocytosis.

exo-1,2-*cis*-dimethyl-3,6-epoxy-hexahydrophthalic anhydride. cantharidin.

***exo*-(±)-methyl-4-(1-methylethyl)-2-[(2-methylphenyl)methoxy]-7-oxabicyclo-[2.2.1]heptane**. cinmethylin.

experimental error. *See* error.

experimental insecticide No. 269. endrin.

exponent. **1:** a person who promotes, represents, speaks for, or advocates (e.g., an idea, program, profession, or organization). **2:** one who expounds, or interprets an idea. **3:** a practitioner of a trade, craft, or profession. **4:** a mathematical function represented by a number, symbol, or expression placed at the upper right of a number (the base number or root). It denotes the power to which the number is to be raised. The power is the number of times the base number is to be multiplied by itself to determine the value represented by the base number with its exponent. Thus, 10^{10} = 10 x 10; $A^{(2y+3)}$ = A x (2y +3).

exponential. **1:** denoting, containing, involving, pertaining to, indicated by, or expressed as a mathematical exponent. **2:** more and more rapid. **3:** explanatory; expository.

exponential growth. sustained growth of an organism, body part, or population (especially a population of microorganisms) that is a simple function of size as measured, for example, by organ or body weight, by the number of cells in population of microorganisms). The larger the structure or population, the faster it grows, constantly doubling within a fixed time period. Such growth is never sustained indefinitely. *See* exponent, exponential growth phase.

exponential growth phase (exponential phase; logarithmic phase). **1a:** a period of maximum sustained (exponential) growth of a population of organisms, an organism, or body part. **1b:** a period during the growth cycle of a population in which the number of individuals in the population increases at an exponential rate (usually taken as doubling within a given time period).

exponential phase. exponential growth phase.

exposure. **1:** revealing, disclosing, displaying. **2:** in proximity to or contact with an agent of disease such that transmissionof the disease or its harmful effects is possible or even likely. **3:** the amount or concentration of a biologically active agent to which an organism is exposed as opposed to the amount that actually enters the body. **4:** exposure to the elements; in the presence of extremes of environment (e.g., high or low temperature; wind) with inadequate protection to a degree that life or health may be impaired.

extension agent. *See* county agent.

extinct. **1a:** extinguished; dead; perished. **1b:** vanished. **2a:** no longer living (e.g., as an organism, a population, a species, a nation); dead; deceased. **2b:** having died out altogether.

extinction. **1:** the act or process of extinguishing or making extinct; the process of becoming extinct. **2:** the death of an entire race, species, or population. In recent decades, a significant number of domestic varieties and species have become extinct because of intensive breeding programs, hybridization, and changing requirements and use patterns with respect to livestock and crops. **3:** the fact of being extinct.

extracellular. outside, or external to, a cell or population of cells.

exudate. **1:** liquid materials that seep from the surface of an organism (e.g., pustular exudates from a sore). **2:** low molecular weight metabolites that seep from plant roots into soil.

exudative calcifying fasciitis. calcinosis.

F

f. femto-

f.p., **fp**. freezing point.

F^+. fluoride ion.

F_1 **generation**. first filial generation (*See* filial generation).

F_2 **toxin**. zearalenone.

2-FAA. *N*-2-fluorenylacetamide.

Fabaceae (pulses; beans). the bean family; a large, heterogeneous, often climbing, nearly cosmopolitan family of herbs. All are leguminous, with nitrogen-fixing microorganisms associated with the roots. They were formerly assigned to Leguminosae (the pea family).

fabism. favism.

face cut. the undercut that is made on the side of the tree opposite the back cut.

facies (plural, facies). **1:** the face, surface, aspect, or shape of a structure. **2:** the countenance of a person or animal, especially an appearance or expression of the face that is indicative of a pathological or morbid condition (e.g., lumpy jaw). **3:** the general appearance, aspect, or composition of a natural grouping, e.g., a population, a type of vegetation, the fauna of a region, habitat, community, ecosystem, geologic formation. **4a:** the appearance or aspect of a particular local modification of an ecological community or a biotope; **4b:** a specialized part, usually local, of a cultural community.

factitious air. nitrous oxide.

factor. **1:** an agent, condition, or an endogenous or environmental variable that contributes to or modifies an action or effect. **2a:** influence, consideration; **2b:** an essential component or ingredient. **2c:** a cause or determinant. **2d:** a vitamin or other essential nutritional element. **3:** a component of a number or expression that is arrived at by multiplication. **4:** a hereditary factor or gene. **abiotic factor**. **1:** any nonliving, nonbiological component of the environment that may directly affect living components (e.g., light and various other forms of energy, temperature, climatic variables, physiographic variables, carbon dioxide, oxygen, trace elements, and water). *See also* biotic factor, biological factor. **absorption factor**. the fraction of a substance to which an organism is exposed that is absorbed. **acetate replacing factor**. thioctic acid. **additive factor**. any of a number of nonallelic genes that affect the same phenotype such that each heightens the expression of the other. **antichromotrichia factor**. *p*-aminobenzoic acid. **anti-gray-hair factor**. *See* *p*-aminobenzoic acid. **antihemorrhagic factor**. a factor that promotes blood clotting. *See* vitamin K. **B-cell replacing factor**. a lymphokine that is essential to the growth and differentiation of B and T cells. **bioaccumulation factor** (BAF). the ratio of the concentration of a substance in the tissue of an organism to the concentration of the substance in the environment, surrounding medium, or the material to which an organism is effectively exposed. *See also* bioconcentration factor, biomagnification factor. **bioconcentration factor** (BCF). the concentration of an environmental chemical in an organism divided by the concentration in the water to which the organism is exposed. *See also* factor, *Cf.*

bioaccumulation factor, biomagnification factor. **biological factor**. any living system, or a part, aspect, product, process, or property of such a system that contributes to or influences a phenomenon or process. A biological factor may have a physical and/or chemical component or components. *Cf.* biotic factor. *See also* physical factor, chemical factor. **biomagnification factor** (BMF). the concentration of an environmental chemical in an organism divided by the concentration in the food of the organism. *See also* factor, *Cf.* bioconcentration factor, bioaccumulation factor. **biotic factor**. **1:** any environmental factor, *q.v.*, that results from the activities of living organisms. *Cf.* biological factor, abiotic factor, modifying factor. **2:** any biological factor that influences the response of a living system. *See also* abiotic factor, biological factor. **blood clotting factor**. any of a number of soluble proteins in vertebrate blood plasma that play essential roles in the complicated process of blood coagulation (e.g., thrombin). *See* dicumarol. **chemical factor**. a factor, environmental or internal to a living system, that is chemical in nature or that has a strong chemical component (e.g., pH, an enzyme, electrolyte content of a medium). **chromotrichia factor**. *p*-aminobenzoic acid. **chronicity factor** (*CF*). the ratio of the acute to the chronic LD_{50} dose. **citrovorum factor**. folinic acid and any of several naturally occurring or synthetic forms of folic acid that support the growth of certain bacteria (e.g., *Leuconostoc citrovorum*). **clearance factor** (C_m). the maximum rate at which any substance is cleared from the blood. *See* clearance. **cobra venom factor** (CVF). a protein constituent of the venom of *Naja naja*, the Indian cobra. It renders properdin factor B responsive to properdin factor D with consequent activation of C3b (complement component 3b, a fragment of the C3 glycoprotein) and other components of complement with lysis of unsensitized erythrocytes. CVF has been injected into laboratory mammals to destroy complement activity for research purposes. *See also* properdin system. **conditioning factor**. a modifying factor that has little or no activity but can significantly alter a response to a stimulus. Thus, sodium can act as a conditioning factor of DOCA activity. **confounding factor** (confounding variable). a variable or factor frequently associated with a process that obscures, confounds, or confuses one's ability to make valid inferences about the variables that may be important to a process or to the output or effects of the process. **density-dependent factor**. any environmental variable that influences the growth of a population (via effects on mortality or fecundity) such that its effectiveness varies with population density (the number of organisms per unit of space occupied). Examples are competition, predation, parasitism, and available food supply. *See also* density dependence. **density-independent factor**. any environmental variable (e.g., extremes of temperature, flood) that affects population growth uniformly with respect to population density. An agent that acts directly and nonselectively upon members of the population is density independent. thus, for example, an agent at low to moderate intensity may act selectively (i.e., removing less fit individuals such as young or diseased individuals), whereas at a higher intensity it may have the same impact on all individuals. **drug-resistance factor**. drug resistance plasmid. **edaphic factor**. any of a large number of soil properties (physical, chemical, biological) that affect the growth and health of plants rooted in the soil and indirectly the growth, health, and structure of the terrestrial community (and usually to a lesser degree, that of nearby aquatic communities) that develops at a given site. edaphic factors may influence the ionic state or the persistence of an anthropogenic chemical at a given location. **elongation factor**. any of several accessory proteins that permit translation to proceed by fa cilitating the correct positioning of aminoacyl-tRNA on the mRNA-ribosome complex. **endogenous factor**. a factor that is naturally internal to a system. **environmental factor**. **1:** any factor or variable in the environment that arises from or constitutes a part of the environment; an environmental variable. **2:** a modifying factor, *q.v.*; any aspect or part of the environment, or any environmental process that contributes to or influences another phenomenon or process. *Cf.* biotic factor. *See also*, modifying factor, teratogen. **exogenous factor** (external factor). a factor external to a system. an environmental factor is an exogenous factor. **external factor**. exogenous factor. **fast death factor**. Any of a number of factors that are responsible for the toxicity of algal blooms. These factors comprise a variable mix of toxic principles produced by diverse populations of organisms associated with a bloom (notably, *Claustridium botulinum* and certain blue-green bacteria). Only certain strains of a given species may produce such a factor. A fast death factor accumulates in the microbial cell and is released via leakage through the cell wall or during cellular decomposition. *See also Microcystis aeruginosa*, algal poisoning, bloom. **growth factor**. any of numerous environmental factors or endogenous principles that variously facilitate growth (e.g., temperature, citrovorum factor). **hereditary factor**. a gene. **internal factor**. a factor, natural or introduced, that is internal to a system. **intrinsic factor**. a variable that is inborn or inherent to an organism, part of an organism, or to a material or object. **limiting factor**. an environmental factor which, by its relative scarcity (e.g., water), excess (salt, water), or unfavorability (e.g., climate or edaphic properties) tends to limit a process (e.g., growth) or the numbers or range of a species, population, or biotic community. **modifying factor**. **1:** any of numerous environmental variables, complex events, and factors in the environment that affect that act to alter the action or influence of other agents on a biological system or process. *Cf.* environmental factor. **2:** any factor that influences a property (e.g., toxicity) or properties of any agent that affects a living system. **3:** a gene that alters the expression of another nonallelic gene. Such modifying genes often have no other known effects. **multiple factor**. two or more pairs of factors (e.g., genes, environmental chemicals) that have similar or complementary cumulative effects on a single response or trait of a living system. **nutritional factor**. a dietary principle (e.g., protein/calorie ratio, a vitamin) that is essential to life and good health. **physical factor**. any factor whether external or internal to a living system, that is a physical (as opposed to chemical or biological) variable (e.g., temperature, moisture, aeration, and pH). **physico-chemical factor**. **1:** a variable that is physical and/or chemical in character as opposed to a biotic factor. **predisposing factor**. an internal or external environmental factor that predisposes a living system to a particular disease, condition or outcome. **properdin factor B** (C3 proactivator; cobra venom cofactor; glycine rich β-glycoprotein II). a blood serum protein (MW 95,000) and component of the properdin system. *See also* cobra venom factor, properdin system. **properdin factor E**. a blood serum protein (MW 160,000) essential to the activation of the third component (C3) of complement by cobra venom factor. *See also* properdin system; cobra venom factor. **pyruvate oxidation factor**. thioctic acid. **risk factor.** single factor (endogenous or exogenous) that is statistically associated with, but not necessarily a cause of, increased risk of injury, damage, morbidity, or mortality (e.g., of a breed, of a herd) by a stated hazard. **transmethylation factor**. choline. **tumor necrosis factor α**. a cytotoxic serum protein produced by numerous types of cells (e.g., macrophages, monocytes, B- and T-lymphocytes) in response to bacterial endotoxin. It kills cells by forming transmembrane channels. It contributes to inflammation; and to wound healing, and the remodeling of tissue by facilitating leukocyte recruitment and the induction of angiogenesis, and by promoting fibroblast proliferation. It can also induce septic shock and anorexia with associated cachexia which may result in death from malnutrition. **uncertainty factor** (UF). **1:** any factor that contributes to the uncertainty of a set of observations or findings. **2:** any factor, arrived at by whatever means to establish a safe dose or concentration of a xenobiotic for a given type or taxon of living organism. **3:** a factor used to determine the reference dose from experimental data. It is intended to account for significant sources of uncertainty such as varying and poorly known sensitivity among members of the target population, uncertainty of measurement, and uncertainty in extrapolating from animal data to humans. **unconscious factor**. a mental process, below the level of awareness, that contributes to a behavior. *See* accident proneness. **vomiting factor**. a general term for the principles (all are trichothecenes) that cause vomiting in ducklings used in bioassays of toxic corn. **Warburg's factor**. a former name for cytochrome oxidase.

factory farm. a large-scale commercial farm or farming operation managed for profit. The owner is likely to be a corporation.

facultative. **1:** capable of, but not confined to, performing a particular action or function. **2:** able to live under a certain set of environmental conditions, but not restricted to those conditions. *Cf.* obligate.

facultative accumulator. *See* accumulator.

fagicladosporic acid. a mycotoxin produced by *Cladosporium epiphyllum*.

fagopyrin. a primary photosensitizing agent found in *Fagopyrum sagittatum* (buckwheat).

fagopyrism. buckwheat poisoning. *See Fagopyrum sagittatum*.

Fagopyrum sagittatum (buckwheat). a fast-growing annual herb (Family Polygonaceae), farmed mostly for its seed which is used to make a characteristically dark flour that is used to make pancakes. This plant, however, is one of a very few species in North America known to cause primary photosensitivity in animals. Buckwheat contains a photosensitizing agent, fagopyrin. *See also Hypericum perforatum*.

Fagus sylvatica (European beech; beechnut). an ornamental deciduous tree with numerous varieties (Family Fagaceae). Humans have occasionally been poisoned by consuming the saponin-containing seeds (beechnuts) and livestock have been poisoned by eating press cakes made from seeds from which the (nontoxic) oil has been expressed. Symptoms are those of weakness and severe gastrointestinal distress.

fahaka. one of numerous common names of fish of the family Tetraodontidae.

Fahrenheit temperature scale. a temperature scale on which the freezing point of water is 32° and the boiling point of water is 212°. *See also* Celsius temperature scale, Kelvin temperature scale.

failure. the loss or inability to function; a state of extreme insufficiency, ineffectiveness or nonperformance. Said of an organ or system. **congestive heart failure**. a condition marked by the inability of the heart to maintain adequate blood flow. It is progressive, and associated with weakness, breathlessness, abdominal discomfort, salt and water retention with edema of the lower parts of the body due to venous stasis, pulmonary congestion. See also cardiomyopathy (congestive cardiomyopathy). **heart failure**. any condition whereby the heart is unable to pump blood normally. Such conditions may be caused directly or indirectly by a poisonous substance, and may be partial or complete, focal or diffuse, temporary or permanent. See also congestive heart failure. **renal failure**. any condition whereby the kidneys are unable to perform adequately. Such conditions may be temporary or permanent; partial or complete; chronic or acute. **respiratory failure**. any condition whereby the lungs are unable to effect the exchange of respiratory gases between blood and ambient air. Proximate causes include lesions of lung tissue and weakness or paralysis of the respiratory muscles. *Cf.* collapse (respiratory collapse).

fainting. *See* syncope.

fair lady. *Atropa belladonna*.

fall. a flock or flight of certain birds (e.g., a fall of woodcocks).

fall crocus. *Cholchicum autumnale*.

fall poison. a common name of *Eupatorium rugosum*.

fallow. **1:** untilled, uncultivated, dormant, idle. **2a:** pertaining to land that is normally used to produce crops but remains unseeded for one or more growing seasons. **bush fallow**. a system of cultivation in which trees and bushes are cleared from virgin land which is then allowed to remain fallow for a period of time before planting.

false acacia. *Robinia pseudoacacia*.

false blackleg. black disease.

false chanterelle. *Omphalotus olearius*.

false hellebore. a common name of herbs of the genus *Adonis*, especially *A. vernalis*.

false indigo. *Baptisia*.

family. 1a: offspring, brood, progeny. **1b:** a group of closely related individuals that interact and live in close association. **2:** a botanic and zoologic taxonomic group comprised of related genera.

family farm. a farm, usually of modest size, that is operated and managed by resident family members. Hired hands are used only under conditions when short-term, intensive labor is required (e.g., during harvests).

famophos. famphur.

famphur (*O,O*-dimethyl-*O*-[*p*-(dimethylsulfamoyl)-phenyl]phosphorothioate (generic); famophos). a slightly water-soluble crystalline powder,

$$(CH_3O)_2P(S)OC_6H_4SO_2N(CH_3)_2$$

It is very soluble in chloroform and carbon tetrachloride. It is a highly toxic organophosphate cholinesterase inhibitor used as an insecticide.

fang. any long, sharp tooth by which an animal seizes, holds, punctures, envenomates, or tears prey. **cheliceral fang**. the sharp, pointed second segment of a spider's chelicera. The venom is introduced into the prey through pores in the tip of the fang. The fangs of most spiders are too short or delicate to penetrate human skin.

farcy. glanders

farm. **1a:** a tract of land or water that is held or leased for the cultivation of crops or rearing of domestic or other animals (e.g., livestock, poultry, mink, fox, trout, shellfish, wildlife) usually with a view to earning a livelihood. **1b:** any parcel or group of parcels of land or water that are managed and farmed as a unit (e.g., a ranch, spread, plantation). **1c:** a tract of land or water devoted to agricultural production (e.g., corn, wheat, cattle) or allied industries such as sylviculture, oyster production, or the production of cultivated or manufactured commodities (e.g., cheese; orchard products such as jams and jellies; honey). Such farms are usually run as commercial ventures. **2:** legal definitions usually stipulate a farm must have a minimum level of economic productivity or profitability. **3:** land on which an industrial activity is conducted (e.g., the drilling or storage of petroleum or the generation of electricity by solar or geothermal power). **4:** an area of land devoted to the storage of a commodity or the emplacement of a group of structures or devices (e.g., a tank farm). **5:** an attributive modifier of another noun (e.g., farm machinery; farm buildings, farm bureau). **6:** in baseball a minor-league team (a farm team, a farm club) that is affiliated with, or owned by, a major league club. **7a:** to cultivate the land. **7b:** to apply the skills and practices of agriculture. **8:** to manage or operate a farm or to contribute substantially through labor, skill, or knowledge to the operations and productivity of one or more farms. **9:** to devote land to agriculture.

farm and home advisor. county agent.

farm size. **1:** total acreage committed, or potentially committed, to production. **2:** the most commonly-used index of farm size is the annual value of gross farm sales. According to the USDA, small farms are those grossing less than $50,000 in sales annually; farms having annual sales of $50,000 or more are classed by the USDA as commercial farms.

Farm Bureau. American Farm Bureau Federation.

farmer. **1:** a person who farms; one who is chiefly engaged in the business of farming. There are numerous types of farmers depending not only on the type of commodities produced (e.g., beef farmer, dairy farmer, hog farmer, corn farmer, wheat farmer, fruit farmer), but also upon the approach taken (organic farmer) and methods employed (catchment farmer). examples are **2:** a person who owns a farm or whose occupation or business is the profitable operation of a farm. *Cf.* agriculturalist. **subsistence farmer**. one who practices or is tied to subsistence agriculture, *q.v.*

farmland. land that is suitable as it stands (including a favorable climate and availability of water either through precipitation or irrigation) for the economically viable production of forage, crops, or rearing of livestock. Included, for example, are cropland, forestland, pastureland, and rangeland.

farmyard. barnyard.

farrow-to-finish. a confinement operation in which pigs are bred and raised to their slaughter weight (ca. 200-250 pounds). *See also* animal feeding operation.

fasciculation. twitching; rapid, often continuous contraction of a skeletal muscle bundle that fails to produce a purposeful movement. *Cf.* convulsion, tremor.

fat. **1:** a compound or mixture of compounds that has a high proportion of saturated fatty acids and glycerol. Fats are important constituents of foods and food stores in many plants and animals. **2:** plump, stocky, obese. **milk fat**. butterfat. **unsaturated fat**. fat that includes fatty acids that have one or more double bonds; not fully hydrogenated.

fat-free lean index. a measure of hog quality that can aid the determination of value. This index was developed by the National Pork Producers Council, an in dustry trade group.

fat-tailed sheep. a class of sheep that has considerable fat by the sides of the tail bones. They are raised for their meat and are widely distributed in southeastern Europe, northern Africa, and Asia.

fatal. **1:** inevitable, destined. **2a:** of, pertaining to, or productive of death. **2b:** causing death; able to cause death. **2c:** inescapably death causing.

fatality. **1:** a disease, condition, accident, or action that results in death. **2:** any instance or individual occurrence of death.

fate. **1b:** destiny; possible or likely future states or ultimate future state of an entity. **1b:** outcome, end result. **chemical fate**. environmental fate. **environmental fate** (chemical fate). a term usually applied to a pollutant (e.g., fertilizer) or to a toxic or hazardous chemical (e.g., a pesticide) meaning the final state (often including pathways and intermediate states) of a chemical at a given location or within a given environment as a result of its movement through the environment and the chemical and physical transformations that it undergoes (e.g., export, mineralization, storage). Fate is largely governed by the equilibrium distribution of the chemical between environmental media (e.g., soil, water, air, organisms), the environmental conditions that obtain (e.g., temperature, pH), and the distribution of the media. Properties of a chemical that help to determine the rate and path of chemical change and the ultimate fate of the chemical include water solubility, vapor pressure, molecular weight, melting point, and the effects of light, heat, air, soil, and water on its behavior.

fatigue. **1a:** tiredness or weariness, it is a common complaint that is usually caused by over-work, lack of sleep. Chronic fatigue is some-times caused by depression or anxiety. See also exhaustion. Alternative causes of fatigue usually produce additional symptoms (e.g., anemia). **1b:** a generalized feeling of tiredness, listlessness, or exhaustion due to prolonged exertion or the effects of psychotropic drugs. There is usually a decrement in performance and a build up of lactic acid in skeletal muscles. **1c:** a condition of discomfort and of muscular weakness such that one works inefficiently and with difficulty. **2:** the state of an organ or tissue whereby the response to a re-peated or long-continued stimulus weakens or ceases. **3:** to induce fatigue as defined above.

fatty acid. any of a group of long-chain, water-insoluble carboxylic acids. They are important components of many lipids (e.g., in glycerides), but most occur naturally only in trace amounts as free fatty acids. The most common naturally occurring fatty acids are oleic (the most widely distributed), palmitic, phenylstearic, and stearic acids. The sodium and potassium salts of fatty acids are water-soluble and are used as soaps (*See* saponification). *See also* aliphatic fatty acid, aromatic fatty acid, essential fatty acid. **aliphatic fatty acid**. any fatty acid (e.g., oleic, palmitic, stearic) that contains no cyclic (aromatic) components. **aromatic fatty acid**. any fatty acid (e.g., phenylstearic) that contains a benzene ring. **essential fatty acid**. any of three unsaturated fatty acids essential to the growth of mammals: linoleic acid, linolenic acid, and arachidonic acid. They are constituents of phospholipids and glycerides, and are thus essential to the formation of cell membranes and fat metabolism. Of these, linoleic acid alone cannot be synthesized and is required in the diet.

fatty tissue. adipose tissue.

fauna. **1:** the entire animal life of a designated locality or region, usually excluding domestic stock that remains under human management. **2:** a faunal work; a list of the animal species of a designated habitat, community, ecosystem, region, or geological period of time.

fava bean. *Vicia faba*.

favism (fabism). an acute form of hemolytic anemia that occurs in Italy following the ingestion of fava beans (e.g., of *Vicia faba*) or inhalation of the pollen. It affects individuals who have an inherited deficiency of glucose-6-phosphate dehydrogenase. Symptoms include fever, vomiting, headache, abdominal pain, diarrhea, severe anemia, and sometimes prostration and coma.

FB/2. diquat.

FD&CA. U.S. Federal Food, Drug and Cosmetic Act.

FD&C Blue No. 1 (Blue 1; 1206 Blue; Aizen Food Blue; C.I. food blue; Cosmetic Blue Lake; D&C Blue No. 4; Food Blue 2; Food Blue Dye No. 1; etc.) a dark purple to bronze powder used as a food coloring in bottled drinks, candy, and baked goods. It can cause chromosomal damage, is possibly carcinogenic, and is banned from French and Finnish foods. Very toxic fumes, NO_x, Na_2O and SO_x, are released on decomposition by heating.

FD&C Blue No. 2 (Blue 2; 1311 Blue; 12070 Blue; C.I. 73015; C.I. acid blue 1; C.I. food blue 1; disodium indigo-5,5-disulfonate; numerous other names) a blue-brown to red-brown powder,

$$C_{16}H_{10}N_2O_8S_2{\cdot}2Na$$

It is soluble in water and concentrated H_2SO_4; slightly soluble in ethanol. It is used as a food coloring in candy, bottled drinks, and pet foods. It is very toxic via intravenous routes; moderately toxic by oral or subcutaneous routes. It can cause brain tumors and is possibly carcinogenic. Its use is forbidden in Norwegian foods. Very toxic fumes, NO_x, SO_x and Na_2O, are released on decomposition by heating.

FD&C Green No. 2 (C.I. 42095; C.I. acid green; C.I. food green 2). a moderately toxic compound, $C_{37}H_{36}N_2O_9S_3{\cdot}2Na$, by ingestion and possibly carcinogenic.

FD&C Green No. 3 (C.I. 42053; C.I. food green 3). a red to brownish-violet powder, $C_{37}H_{36}N_2O_{10}S_3{\cdot}2Na$, that is soluble in water and concentrated sulfuric acid. It is a possible human carcinogen and mutagen.

FD&C Red No. 1 (disodium 3-hydroxy-4-((2,4,5-trimethylphenyl)azo)-2,7-naphthalenedi-sulfonate; disodium 3-hydroxy-4-((2,4,5-tri-methylphenyl)azo)-2,7-naphthalenedisulfonic acid; disodium 3-hydroxy-4-((2,4,5-trimethyl-phenyl)azo)-2,7-naphthalenedisulphonate; C.I. 16155; C.I. food red 6; C.I. food red). a suspected carcinogen, $C_{19}H_{16}N_2O_7S_2{\cdot}2Na$.

FD&C Red No. 3. erythrosine.

FD&C Red No. 40 (C.I. 16035). a red powder, $C_{18}H_{14}N_2O_8S_2Na_2$, that is soluble in water and slightly soluble in absolute alcohol. It is a suspected carcinogen.

FD&C Violet No. 1 (C.I. 42640; C.I. food violet 2). $C_{39}H_{41}N_3O_6S_2{\cdot}Na$, a suspected carcinogen.

FD&C Yellow No. 3 (C.I. 11380; C.I. food yellow 10; C.I. solvent yellow 5). a moderately toxic color by ingestion or skin contact. It is a questionable carcinogen.

feather. an integumental structure that is uniquely characteristic of birds (Class Aves). It is the structural unit of plumage. *See also* integument. Feathers accumulate lead, mercury, and cadmium, for example, and have thus been used for some time as indicators of heavy metal loadings in terrestrial and marine environments. The keratin in bird feathers is not easily degradable, and mercury is probably firmly linked to the disulfide bonds of keratin. Because feathers do not readily biodegrade, retrospective analyses spanning long periods of time are possible.

feather geranium. *Chenopodium botrys*.

febricant. causing fever, febrific. *Cf.* febrifacient.

febricide. **1:** the lowering of body temperature during fever. **2:** an agent that allays fever.

febricity. feverishness; the state of being.

febrifacient. **1:** febrific, febriferous; promoting, inducing, or conveying fever. *Cf.* febricant. **2:** any agent that induces or promotes fever.

febriferous. febrifacient (def. 1).

febrific. febrifacient (def. 1).

febrile (feverish; pyretic; pyrectic). feverish; of or pertaining to fever or feverishness.

febris. fever.

feces. solid or semisolid material, composed of unabsorbed food residues, bacteria, and secretions, that is expelled from the gastrointestinal system. Feces in vertebrate animals are stored in the colon until expelled via the anus. Many xenobiotics or their metabolites leave the animal body via urine or feces (e.g., cadmium). Feces may thus contain poisons or residues of poisons. Monitoring feces for poisons can be misleading because the contents generally reflect only very recent exposures. Furthermore, the composition of feces may be quite variable and may not reflect the actual diet. Fecal material of birds is mixed with excreta from the kidneys. *Cf.* egesta, stool.

fed cattle. beef cattle that are fattened in a feedlot on a high protein ration until ready (usually at a weight of about 1,100 pounds) for sale to a packing plant for slaughter.

federal land. land that is owned by the federal government in the United States and a number of other countries. *Cf.* Crown land.

feed additive. *See* additive.

Feed Grains and Livestock Belt (Corn Belt). a major agricultural region of the United States, once called the Corn Belt because corn, *Zea mays*, was the major crop. It is a great expanse of deep, fertile, well-drained, highly productive soils and long, hot, humid summers. Located in the Midwest and central plains, the heart of region lies in Iowa and Illinois and extends into Indiana, western and southcentral Ohio, southern Minnesota, southeastern South Dakota, eastern Nebraska, northeastern Kansas, and northern Missouri. Large-scale, modern agriculture dominates the region today. While much of the United States corn crop is produced here, considerable diversification has taken place. Soybeans, winter wheat, and alfalfa are also important crops in the region.

feed. **1:** to supply with food or nourishment. **2:** to provide as food. **3:** to satisfy, gratify. **4a:** to be nourished or satisfied. **4b.** to subsist (on). **4c:** to eat or take food; to feed. **5:** to use as pasture. **6:** food, especially that for livestock, fowl, wildlife, and other animals.

feeder calf. *See* feeder cattle.

feeder cattle. steers or cows mature enough to be placed in a feedlot and fattened prior to slaughter. Feeder calves are less than one year old; feeder yearlings are between one and two years old.

feeder yearling. *See* feeder cattle.

feline leukemia-sarcoma virus complex. viruses of cats that cause transmissible leukemia or transmissible fibrosarcoma in kittens.

felt-ringed agaricus. *Agaricus hondensis*.

female fern. *Pteridium*.

femto- (f). a prefix to units of measurement that indicates a quantity of one trillionth 10^{-15} the value indicated by the root. *See* International System of Units.

fenac. chlorfenac.

fenchlorphos. Ronnel.

fenclor. polychlorinated biphenyl.

fenoprop. silvex.

fensulfothion (O,O-diethyl-O-[p-(methylsulfinyl)phenyl]phosphorothioate). a liquid anticholinesterase, $C_{11}H_{17}O_4PS_2$, used as an insecticide and nematocide; also used in some household products to eradicate pests. It is a dangerous substance and is toxic by all routes of exposure, including skin absorption.

fentanyl (phentanyl; *N*-phenyl-*N*-[1-(2-phenylethyl)-4-piperidinyl]propanamide; *N*-(1-phenylethyl-4-piperidyl)propionanilide; *N*-(1-phenylethyl-4-piperidinyl)-*N*-phenylpropionamide). a toxic opioid analgesic, it is a controlled substance in the United States. The citrate (phentanyl) is extremely toxic to laboratory mice.

fenthion (phenthion; phosphorothioic acid *O,O*-dimethyl*O*-[3-methyl-4-(methylthio)-phenyl] ester; *O,O*-dimethyl *O*-(4-methylmercapto-3-methylphenyl) thionophosphate; *O,O*-dimethyl *O*-(4-methylthio-3-methylphenyl) thiophosphate; *O,O*-dimethyl *O*-[4-(methylthio-*m*-tolyl] phosphorothiolate; 4-methylmercapto-3-methylphenyl dimethylthiophosphate). an amber-colored, liquid organophosphate insecticide, acaricide (mainly for use with ornamental plants) and veterinary ectoparasiticide with a slight odor of garlic. Fenthion is a cholinesterase inhibitor. It is very toxic to laboratory rats, and highly toxic to fish, and hymenoptera. *See also* parathion, anticholinesterase.

fenuron (*N,N*-dimethyl-*N*-phenylurea; *N*-phenyl-*N′,N′*-dimethylurea). a white crystalline herbicide,

$$C_6H_5NHCON(CH_3)_2$$

It is used to control deep-rooted perennial weeds and bushes, especially *Convolvulus arvensis* (field bindweed). This compound binds less strongly to soil and is thus effective to greater soil depths than, for example, monuron. It is slightly toxic to laboratory rats. *See also* fenuron TCA.

fenuron TCA. the trichloroacetate of fenuron, it is a white crystalline herbicide used in noncrop areas especially to control woody plants. If hydrolyzed endogenously to aniline, it can cause methemoglobinemia. *Cf.* diuron. *See also* fenuron.

fenvalerate (4-chloro-α-(1-methyl-ethyl)benzeneacetic acid cyano(3-phenoxyphenyl)-methyl ester; α-cyano-3-phenoxy-benzyl α-(4-chlorophenyl)isovalerate; cyano-(3-phenoxyphenyl)methyl 4-chloro-α-(1-methylethyl)benzeneacetate; α-cyano-3-phenoxybenzyl-2-(4-chlorophenyl)-3-methylbutyrate; phenvalerate; pydrin; sumicidin). a broad-spectrum insecticide and veterinary ectoparasiticide. It is neurotoxic, acting on both the central and peripheral nervous systems and an eye and skin irritant. It is very toxic to laboratory rats and highly toxic to fish and hymenopterans. Symptoms of intoxication in humans may include excessive salivation, nausea and vomiting, tremors, incoordination, diarrhea, hypersensitivity to touch and sound, convulsions, and death.

fer-de-lance. *Bothrops atrox*.

feral. **1a**: living normally in the wild. **1b:** referring to plants or animals that were domesticated but have reverted to the wild state. **2:** untamed, undomesticated. **3:** wild or savage. *See also* animal (*Cf.* domestic animal, feral animal, laboratory animal).

ferbam. the iron salt of ethylenebisdithiocarbamate. Its toxicity to animals is low.

fern brake. *Pteridium*.

ferrisulfas. ferrous sulfate.

ferrous gluconate (iron gluconate). a fine yellowish-gray or pale greenish-yellow powder or granules,

$$Fe(C_6H_{11}O_7){\cdot}2H_2O$$

It is used widely as a feed and food additive and in vitamin tablets, and can be dangerous in the home. It contains 33% elemental iron. Up to 10 tablets may produce mild poisoning in a child; up to 20 tablets may prove moderately toxic; and >20 tablets may cause severe poisoning. *See* iron poisoning.

ferrous sulfate (ferrisulfas; copperas; green copperas; green vitriol; iron vitriol; iron sulfate; sal chalybis). a blue-green, water-soluble, crystalline substance,

$$FeSO_4{\cdot}7H_2O$$

It is moderately toxic to laboratory mice when given *per os* (by mouth) and is very toxic if given intravenously. It is toxic to humans, causing gastrointestinal disturbances. Ingestion of large amounts by children can cause hematemesis, tachycardia, liver damage, and peripheral vascular collapse. Ferrous sulfate has a number of applications, including usage as an herbicide and wood preservative, and is used in some pesticides.

fertilization. the fusion of male and female pronuclei, forming a single cell (the zygote) and triggering development.

fertilizer. an organic or inorganic material, natural or synthetic, that is added to the soil to supply one or more primary plant nutrients, such as nitrogen, phosphate (P_2O_5), and potash (K_2O) in a readily available form. Fertilizers may also contain secondary or trace nutrients. Commercially available fertilizers may contain chemicals other than nutrients or may be processed sewage sludge. Application of excess fertilizer or fertilizer attached to eroding soil can be a source of water pollution. **acid-forming fertilizer**. a fertilizer such as ammonium sulfate that adds to the residual acidity of soil. **balanced fertilizer**. a fertilizer that contains each essential nutrient in the proportions necessary to produce a particular crop. **compound fertilizer**. a fertilizer that contains two or more primary plant nutrients (nitrogen, phosphorus, potassium).

fertilizer spike. fertilizer shaped into a spike, which is driven into the ground to slowly supply nutrients to roots of plants.

fertilizer unit. 1% (20 lb) of a short ton.

ferulic acid (3-(4-hydroxy-3-methoxyphenyl)-2-propenoic acid; 4-hydroxy-3-methoxycinnamic acid; 3-methoxy-4-hydroxycinnamic acid; caffeic acid 3-methyl ether). a plant growth inhibitor found in low concentrations in many species of plants. It is usually nontoxic to the producing plant. There are two isomers. The *cis*-form is a yellow oil and the *trans*-form precipitates from water solutions as orthorhombic crystals.

FES. zearalenone.

fescue (fescue grass; festuca). any of various grasses of the genus *Festuca*.

fescue foot (fescue lameness; gangrenous fescue poisoning). a dry gangrene of livestock (mainly cattle; sheep are rarely affected) caused by ingestion of toxic tall fescue, *Festuca arundinacea*. The symptoms and lesions are essentially similar to those of gangrenous ergot poisoning, but fungal sclerotia have not been found in poisonous fescue. Lameness of one or both hind feet usually develops within 10 to 14 days. In severe cases, the front feet and tips of the ears may be affected. Resistance varies among cattle. Some animals recover spontaneously, but the disease usually terminates fatally. An endophytic fungus that infests *Festuca arundinacea* (tall fescue) has been implicated in fescue foot. This fungus was first identified as *Epichloe typhina* and has recently been renamed *Acremonium coenophialum*. It is not clear whether the toxic principle is produced directly by the fungus or whether it is a product of the fungus-plant association. *Cf.* summer fescue toxicosis.

fescue grass. *See* fescue.

fescue lameness. fescue foot.

fescue poisoning. *See* fescue foot.

fescue toxicosis. summer fescue toxicosis.

Festuca (fescue; fescue grass; meadow grass). a genus of some 100 species of Old World grasses (Family Arundinaceae) with wide flat leaves that includes several species of agricultural importance. Some are cultivated for permanent pasturage, hay, or lawns. Fescue may be infested by a endophytic fungus (*acremonium coenophialum*) which contains toxic alkaloids. Livestock ingesting infested grass may develop "fescue foot," *q.v.* Fescue can also accumulate toxic levels of nitrate. ***F. arundinacea*** (*F. elatior arundinacea*; tall fescue; reed fescue). a coarse, deeply rooted perennial bunchgrass (Family Arundinaceae) that is tolerant of a wide range of soil and climatic conditions. It reaches a height of more than a meter. An excellent forage crop, it is sometimes used also in lawns and athletic fields. *F. duriuscula* (hard festuca). a short, perennial bunchgrass that is tolerant of a wide range of climatic and soil conditions. It has a dense, extensive root system that provides for excellent erosion control. ***F. elatior arundinacea***. *F. arundinacea*. ***F. fallax***. *F. rubra commutata*. *F. megalura* (annual fescue; six-weeks fescue). an annual bunchgrass that grows well under relatively dry conditions and in soils of low fertility. Seeding should take place in the autumn or very early spring. ***F. ovina*** (sheep fescue). a widely distributed, perennial bunchgrass having densely clustered stems and short leaf blades, and usually cultivated for lawns and forage. ***F. pratensis*** (meadow fescue). a European fescue, grown for pasture in North America. It is an excellent forage crop. ***Festuca rubra*** (red fescue). a perennial, low-growing, meadow grass of the North Temperate Zone. It is used for erosion control and also to provide cover in orchards and along roadsides. The flower clusters are green, bluish-green, or reddish. ***F. rubra commutata*** (*F. fallax*; chewing fescue; creeping fescue). a hardy, fine-leaved variety grown chiefly in the United States and New Zealand as a lawn grass. The seeds and hay of this plant are sometimes infested by the nematode *Angina agrostis* (*q.v.*) which has caused fatalities in sheep, cattle, and other types of livestock. ***F. viridula*** (green fescue; green leaf fescue; mountain bunchgrass). a perennial grass (family Gramineae); a perennial forage grass found at elevations of 6,500 to 10,000 ft. It is a native North American grass that now occurs from California and Idaho northward to British Columbia and Alberta, Canada.

fetal (foetal). pertaining to, denoting, or having the characteristics of a fetus.

fetal rickets. *See* achondroplasia.

fetal rickets in humans. achondroplasia.

fetation (foetation). the development of the fetus within the uterus.

fetid. malodorous; foul; foul-smelling; putrid; having an offensive or foul or foul odor.

fetor. **1:** foulness, rankness, a very foul odor. **2:** very foul breath; it is sometimes a sign of disease.

fetotoxic. toxic to a fetus; any agent that is toxic to the fetus. Fetotoxic substances are numerous and include ethanol, morphine, salicylates, tetracyclines, thiazides, tobacco smoke, and large doses of vitamin K.

fetotoxicity. **1:** the property, state, or degree of being toxic to the fetus. **2:** injury, retardation, or death of the fetus due to exposure to a toxic agent. Manifestations of fetotoxicity include: lethality, impaired growth, mental retardation, and other dysfunctional states.

fetter. **1a:** to chain, hobble, shackle; confine. **1b:** fetters, chains, shackles; used especially on the feet of a horse to restrain movement.

fetter bush. a common name of plants of the genus *Leucothoe*, and also of *Lyonia lucida* and *Pieris floribunda*.

fetus (foetus). The embryo of a mammal after it has developed the main organs and features of the adult form (e.g., in humans after two to three months of gestation). Prior to this time, it is referred to as the embryo.

feu de Saint-Antoine. St. Anthony's fire.

fever (febris; pyrexia). a condition marked by a body temperature elevated above that which is considered normal for a species or population.

feverish. febrile.

fibrin. the fibrous, insoluble protein that establishes the gel matrix of a blood clot and determines the semisolid character of the clot. fibrin is produced by the action of thrombin on fibrinogen.

fibrinogen. a soluble dimeric protein consisting of three pairs of polypeptide chains that is converted to fibrin by the enzyme thrombin in the presence of calcium ions. It is the inactive precursor of fibrin.

fibrinolysis. The process by whereby sterile blood clots dissolve due to the decomposition of fibrin. It occurs a few hours to several days after clotting; the process is catalyzed by the proteolytic enzyme plasmin.

fibrositis. inflammation of connective tissue of the muscles, joints, ligaments, and tendons.

ficusin. psoralen.

fiddleneck. *Amsinckia intermedia*.

field capacity. the mass or volume of water that remains in a soil following saturation with water and after free drainage becomes negligible.

field mustard. *Brassica rapa*.

filaree. *Erodium cicutarium*.

filial. of or pertaining to children, offspring.

filial generation. as used especially in genetics, pertaining to the offspring of a parent generation or a designated population or subset (e.g., a particular phenotype). **first filial generation** (F_1 genera tion). The first generation produced by the stock of a designated parent generation. Subsequent generations are also designated by number (e.g., second filial generation).

fill bank. an embankment, built with fill, from the outer edge of the travelled portion of a road to the margin of the preexisting or native land surface.

filly. an immature female horse. *See* colt.

find. to give birth to young. Thus a ewe finds a lamb.

fine clayey soil. fine textured soil or soil family, one that contains about 35 to 59% clay. *See* soil family.

fine material. material in thin sections of soil that are composed of particles less than 2 mm in diameter. Such materials are difficult or impossible to resolve by use of a microscope.

fine texture. descriptive of soil that contains about 35 to 59% clay.

fir. a tree of the genus *Abies*.

fire. **1:** combustion, blaze; burning. **2:** combustion of a fuel or other flammable material whereby oxygen combines with material with the release of heat and light (in the form of a flame). **3:** material that is in the process of burning. **4:** conflagration; the destructive burning of a structure of economic or social importance (e.g., of a barn or home, a forest, a town). **5:** a spark or sparks. **6:** to discharge a firearm; the discharge of firearms. **7:** luminescence or brilliance, as or a gem. **8:**f excitement, enthusiasm, or passion. fever or inflammation. **actionable fire**. a fire started or permitted to spread in violation of a law or regulation. **crown fire**. a fire that spreads through the tops of dominant trees or brush. **forest fire**. any wildfire or prescribed fire that occurs in a forest (sometimes taken to include major fires in grasslands, alpine meadows, or tundra). **ground fire**. **1a:** a fire that burns organic materials such as peat or coal within the soil. **1b:** a fire that burns litter and low-growing vegetation. **prescribed burn**. a type of prescribed fire, *q.v.*; the knowledgeable, planned application of fire to a specific unit of land to meet planned resource management objectives. **prescribed fire**. knowledgeable, planned application of fire to natural fuels, including logging debris, grasslands, and/or understory vegetation, such as palmettos, with the intent to confine the fire to a predetermined area. *Cf.* prescribed burn. **prescribed natural fire**. the management practice of allowing naturally ignited fires (e.g., those caused by lightning) to burn under specific management prescriptions without fire suppression. **surface fire**. a fire that burns leaf litter, fallen branches, and other fuels located on the forest floor.

Firemaster FF-1. polybrominated biphenyl mixture.

fish. the common name of poikilothermic ("cold-blooded") aquatic vertebrates of the classes Osteichthyes, Chondrichthyes, and Agnatha. **game fish** (sport fish). an edible fish capable of affording sport and recreation to the angler in its capture and associated activities. **sport fish**. game fish.

fish farming. freshwater commercial aquaculture, fish farming is the most economically important form of aquaculture. This industry is responsible for nearly 20% of world fisheries production. Catfish farming is dominant today in the United States. Products of fish farming also include tilapia (St. Peter's fish), trout, crayfish, oysters, shrimp, salmon, and tropical fish reared for aquaria.

fishing worm. *See* earthworm.

fit. **1:** suitable, appropriate. **2:** competent, capable. **3:** hardy, sound, in good health.

4: to adapt, conform. **5:** to adjust, correct, calibrate. **6:** to equip, outfit. **7:** seizure, convulsion, spasm.

fitness. **1:** soundness, hardiness, health; agreement, suitability, adaptedness. **2:** an ecological concept, fitness is the relative competitive ability of a given genotype as measured by its relative contribution to the genotypic or phenotypic composition of subsequent generations. Fitness is usually measured and expressed in terms of the number of surviving progeny of one genotype compared with that of competing genotypes. **inclusive fitness**. the number of offspring left by an individual compared with the average for the population of which it is a member or with individuals of a different genotype.

fitty. descriptive of a horse that has fits when overheated.

fixation. exposure of plant or animal tissue or whole organisms to a fixative (fixing agent). This is accomplished to preserve and harden the specimen and maintain the shape of cells and other components. Fixation may be employed, for example, with certain museum specimens; specimens preserved for possible examination in the future; and reference specimens. Tissues and small organisms are most often treated with fixatives in preparation for histological examination. The most commonly used fixatives are ethanol, acetic alcohol (a mixture of acetic acid and ethanol), formalin or a formalin-alcohol solution, and osmium tetroxide. Fixation of tissues for microscopic analysis enables a specimen to be cut into thin sections without losing its shape and structural integrity; it also serves to prepare the specimen for staining. The fixative employed depends upon the type of tissue to be examined and the stains, if any, that are to be employed.

fixative. *See* fixation.

flaccid. relaxed, limp, not rigid, not turgid, lacking tone.

flaccidity. the condition of being flaccid.

flagella. plural of flagellum.

flagellate. **1a:** having one or more flagella. **1b:** pertaining to a protozoan with one or more flagella.

flagellum. a hairlike locomotor structure, usually situated at the extreme end of a bacterium or protozoan.

flamingo tailflower. *Anthurium scherzerianum.*

flammable. combustible, inflammable, capable of burning; incendiary, the capacity of a substance or material to ignite and burn readily. *See also* ignition.

flat weed intoxication. flat weed poisoning.

flat weed poisoning (flat weed intoxication). poisoning of an animal by ingestion of dandelion. *See also* stringhalt.

flatpod pea. a poisonous pea. *See Lathyrus.*

flax. *See Linum.*

flaxseed oil. linseed oil.

flaxseed. *See Linum usitatissimum.*

flesh. soft animal tissue, especially meat, the color and sometimes odor (in mammals and birds), sometimes offer clues to disease or condition.

flight. **1:** the act of flying or soaring, e.g., of birds, bats, airplanes. **2:** aeronautics, aviation. **3:** escape; the act of fleeing. **4:** a flock of birds in flight. **5:** a staircase; a set of stairs.

FLIT-MLO. a mosquito larvicide composed

of petroleum-derived aliphatic hydrocarbons.

flock. a collective term applied to a group of certain animals (e.g., birds generally, fowl, waterfowl, sheep).

flood plain. the flat, low-lying area along a watercourse that is subject to flooding and alluviation from time to time. *Cf.* alluvial plain, bottomland. *See also* alluvial community, bottomland.

flora. **1:** plant life, especially of a designated locality or region. **2:** microflora, the set of microbes that populate external and internal surfaces of a normal healthy organism. *Cf.* microflora.

floral. **1:** pertaining to flora. **2:** pertaining to flowers.

Florida arrowroot. a common name of *Zamia integrifolia*.

foothill death camas. *Zigadenus paniculatus*.

flow line. **1:** the position of the surface of a running stream for a designated rate of discharge. **2:** the bed of a body of flowing surface water (e.g., a creek).

Florida velvetbean. *Stizolobium deeringianum*.

flow. **1a:** movement of fluid; fluid stream or current; outpouring. **1b:** to pour or stream; course; gush, circulate. **saturated flow**. the movement of water through saturated soil (*See* soil). **unsaturated flow**. the movement of water through soil that is not fully saturated.

flow line. **1:** the position of the surface of a running stream for a designated rate of discharge. **2:** the bed of a body of flowing surface water (e.g., a creek).

flow rate. **1:** the rate of volume flow of a water body such as a stream or canal. **2:** the rate of let-down of milk by a cow.

flowering glume. lemma.

flowering spurge. *Euphorbia corollata*.

flowers of benzoin. benzoic acid.

fluometuron (*N,N*-dimethyl-*N'*[3-(trifluoromethyl)phenyl]urea; 1,1-dimethyl-3-(α,α,α-trifluoro-*m*-tolyl)urea; *N*-(3-trifluoromethylphenyl)-*N',N'*-dimethylurea). an herbicide with a half-life in soil of about 60 days, it is absorbed mainly through the roots of plants. It is used to control broad-leaved and weedy grasses in cotton. The alkoxyurean herbicides are similar in action, but are more water soluble and less persistent in soil. Fluometuron is very toxic to laboratory rats.

***N*-fluoren-2-yl acetamide**. *N*-2-fluorenylacetamide.

***N*-2-fluorenylacetamide** (2-acetamidofluorene; 2-AAF; 2-acetaminofluorene; 2-acetylaminofluorene; 2-FAA; AAF; acetoaminofluorene; *N*-9*H*-fluoren-2-yl-acetamide; *N*-fluoren-2-yl acetamide). a compound, $C_{15}H_{13}NO$, originally developed as an insecticide, its use has been abandoned. It is moderately toxic by ingestion and intraperitoneal routes. It is a potent carcinogen to some laboratory animals and is a confirmed potent human carcinogen. It releases toxic NO_x fumes when heated to decomposition.

***N*-9*H*-fluoren-2-yl-acetamide. *N*-2-fluorenylacetamide.**

***N*-fluoren-2-yl acetamide**. *N*-2-fluorenylacetamide.

fluoridation. the addition of fluoride to a substance; used especially in reference to fluoridation of drinking water.

fluoride. any salt of hydrofluoric acid, HF, in which fluorine has a valence of -1. Fluorides are major environmental pollutants released into the atmosphere from aluminum reduction, steel manufacturing, and coal-fired power production. Fluoride pollution is associated with extensive damage to livestock, agricultural crops, and timber. Fluoride has a strong tendency to bioaccumulate and can thus be hazardous at very low environmental concentrations. Some plants accumulate tissue concentrations of fluoride that are on the order of 10^6 times that of the ambient air; fluorides have high biological activity that can induce a variety of toxic effects in plants and animals. Some inorganic fluorides are highly toxic; sodium fluoride and sodium monofluorophosphate are, for example, very toxic to laboratory rats. *See* fluoride toxicosis.

fluoride ion (F^+). the highly toxic component of hydrogen fluoride and of soluble fluoride salts (e.g., sodium fluoride). At low concentrations (ca. 1 ppm) in drinking water, it helps to prevent tooth decay, but higher concentrations cause fluorosis, *q.v.* Livestock may be poisoned from the deposition of fluoride from industrial pollution on grazing land. As a consequence, animals become lame and occasionally die. *See also* fluorosis.

fluoride poisoning. *See* fluoride toxicosis, fluorosis.

fluoride toxicosis. systemic, usually chronic (fluorosis), fluoride poisoning. More acute poisoning by fluorides may occur as a result of exposure to hydrogen fluoride or its soluble salts. Symptoms of intoxication may include a peculiar taste, excessive salivation, thirst, nausea, vomiting, abdominal pain, diarrhea, muscular weakness, tremor, central depression, and shock. Death may ensue due to vascular collapse. Typical lesions are those of hemorrhagic gastroenteritis, hypocalcemia, and the inhibition of cellular glycolysis. *See also* fluorosis, fluoride.

fluorine (symbol F)[Z = 9; A_r = 18.998]. a slightly greenish, pale yellow, highly toxic, corrosive, and flammable nonmetallic gaseous or liquid element, F_2 (m.p. -218°C, b.p. -187°C), belonging to the halogen group of elements. Fluorine is the most powerful oxidizing agent and the most electronegative chemical known. It can react directly with nearly all elements including the "inert" gas xenon. Fluorine is extremely dangerous and destructive of skin, mucous membranes, and other tissues. Accidental contact or inhalation should be treated as a medical emergency. Some of its more soluble salts are highly reactive and poisonous. *See also* fluorosis, fluoride toxicosis.

fluoroacetamide (fluoroacetic acid amide; monofluoracetamide). a crystalline solid, freely soluble in water. It is used as a rodenticide and sometimes as an insecticide. It is extremely toxic to laboratory mice and rats.

fluoroacetate. any salt of fluoroacetic acid in which carbon-bound hydrogen atoms are replaced by fluorine atoms. All are rapid-acting and extremely poisonous to humans and other mammals. It may be used as a rodenticide in the United States only by registered pest control operators. Secondary poisoning of cats, dogs, and other animals by consumption of carcasses of fluoroacetate killed rodents is possible. *See also*, aconitase, *Acacia georginae*.

fluoroacetic acid amide. fluoroacetamide.

fluoroacetic acid, sodium salt. sodium fluoroacetate.

fluorosis. chronic fluorine poisoning, often observed in domestic cattle. Common lesions produced by fluoride in grazing livestock include mottling, pitting, hypoplasia, and hypocalcification of the teeth; hyperosteosis, osteoporosis, and osteomalacia. Symptoms may include poor nutrition, stunted growth, lameness, and irregular tooth wear. Fluorosis occurs in many regions of the earth due to the presence of rock phosphate deposits that contain high levels of fluoride, volcanic soils, fluoride pollution (of air and water), and the addition of fluorine-containing minerals (e.g., ground rock phosphate) as supplements to livestock rations. In the United States, the use of defluorinated phosphates is generally an adequate safeguard against fluorosis in livestock. Humans are subject to fluorosis, usually from high levels in drinking water. Symptoms typically include mottling of the teeth (dental fluorosis) and increased bone density (skeletal fluorosis). These effects are not usually serious, but continued intake of excessive levels can soften teeth and produce serious bone abnormalities with crippling. Severe cases are sometimes fatal. *See also* fluoride, fluorine, fluoride toxicosis.

fluostigmine. diisopropyl fluorophosphate.

flush. **1:** to irrigate a canal or cavity with a stream of water. **2a:** to redden. **2b:** a sudden redness of the skin; blush.

flushing. **1:** rinsing out with a stream of fluid. **2:** transient erythema.

fly. common name of any of a variety of winged insects, but given properly only to those insects of the order Diptera (true flies).

fly agaric. *Amanita muscaria*.

fly fungus. *Amanita muscaria*.

fly mushroom. *Amanita muscaria*.

fly-poison. *Amianthium muscitoxicum*.

foal. the young offspring of an equine.

fodder. **1:** food given to livestock, especially straw and hay. **2:** food for livestock that has been "cured" or otherwise processed, but not manufactured.

fodder beet. *Beta vulgaris*.

foetal. fetal.

foetation. fetation.

foetus. fetus.

fog fever. a sometimes fatal condition of cattle marked by acute respiratory distress. It presents about one week following placement on pasturage that has been recently mowed. The exact cause is unknown, but it is thought to be an allergic response of sensitized animals to proteins in grass, pollen, and/or fungal spores; nonimmunologic poisoning has not been ruled out.

foliage beet. *Beta vulgaris*.

folic acid (*N*-[*p*-[[(2-amino-4-hydroxy-6-pteridin-yl)methyl]-amino]benzoyl] glutamic acid). a widely distributed, yellow, water-soluble, crystalline vitamin and essential growth factor in many microorganisms and animals. *See* folinic acid.

follicle-stimulating hormone. FSH.

follitropin. FSH.

fonofos. an extremely toxic phosphonate. a cholinesterase inhibitor used as a soil insecticide and as a pesticide in some household products. It should be considered dangerous when so used. Fonofos is activated by FAD-containing, monooxygenase, and by cytochrome P-450.

food. any substance which, when ingested by a heterotrophic organism, contributes to the maintenance of life. **canned food**. food for humans or domestic animals that is stored in sealed cans. Such food may contain toxins and other toxicants such as pesticides, residual detergent used to clean the food, traces of lead from solder used to seal the can (not allowed in the United States), and additives incorporated into the food at any stage of processing. U. S. law requires only that food manufacturers list on the label those ingredients and additives that are introduced or combined in the final stage of processing. Canned foods imported into the United States are often more hazardous than domestic goods. *See* plant (food plant).

food additive. *See* additive.

food chain. the transfer of energy and nutrients from one organism to the next as they feed on one another. Food chains begin with green plants. Certain nonnutrient materials are also transported through the food chain as well.

food chain transfer. the movement of energy and nutrients with or without metabolic transformation from one level of a food chain usually to the next higher level.

Food Chemicals Codex. a publication that gives test methods and specifications for chemicals used in food.

food color, food coloring. *See* color.

food flavoring (artificial flavor; imitation flavor). any substance added to food to enhance flavor or to imitate natural flavor. More than 1500 such substances derived from petrochemicals are in use. Most are thought to be safe, but some have been linked to hyperactivity and behavioral disturbances in children. See also additive.

food poisoning (bromatoxism). **1:** a vague term that is often used in reference to any illness that results from the ingestion of foods that contain poisonous substances. **2:** poisoning due to ingestion of decaying food that contains toxin-producing microbes. Effects may include vomiting, diarrhea, nausea, and fever. **3:** sitotoxism. See also coturnism. **bacterial food poisoning**. enteritis or gastroenteritis due to the presence of soluble bacterial exotoxins or to multiplication of gastrointestinal bacteria *per se*. This term does not pertain to enteric fevers and dysenteries. A common cause of such poisoning is *Claustridium botulinum*. **true food poisoning**. a term that encompasses poisoning due to ingestion of poisonous organisms (e.g., mushroom poisoning, shellfish poisoning); of plant or animal products (e.g., milk poisoning); or of foods that contain toxin-producing microbes. **staphylococcal food poisoning**. food poisoning due to exotoxins produced by *Staphylococcus spp.*, usually *Staphylococcus pyogenes* var. *aureus*. It is marked by violent vomiting and diarrhea.

food web. the interconnection of food chains in the environment.

Food Blue 2. FD&C Blue No. 1.

Food Blue 2. FD&C Blue No. 1

Food Blue Dye No. 1. FD&C Blue No. 1

Food Blue Dye No. 1. FD&C Blue No. 1.

Food for Peace Program. An informal title denoting the food donation activities carried out overseas under P.L. 83-480 (*See* Agricultural Trade Development and Assistance Act of 1954).

foot-pound-second. a system of English measurement. *See*, for example, acre, acre-foot, acre-inch.

foothill abortion of cattle. epidemic bovine abortion.

foothill death camas. *Zigadenus paniculatus*.

forage poisoning (obsolete). **1:** any poison

ing of livestock associated with the ingestion of fungus-infested forage. *See*, for example, aflatoxicosis. **2:** moldy corn poisoning. **3:** sleepy stage.

forage poisoning. sleepy stage.

forage sorghum. *See Sorghum vulgare*.

forage. **1a:** to search for food or provisions, applied especially to wild and domestic animals. **1b:** to secure by foraging. **2a:** vegetable feed such as hay, grain, and fodder provided to domestic animals. **2b:** the food that wild or domestic animals (especially grazing and browsing animals) secure for themselves.

foramen (plural, foramena). any natural opening or passage in an animal body. Numerous foramina occur in vertebrates, most of which are associated with the skull. **foramen magnum.** the foramen in the base of the skull of vertebrates through which the spinal cord passes.

forb. any broad-leaved, flowering herbaceous plant other than a grass, sedge, or rush. palatable, whose stem is not woody.

force-feed. **1:** to compel to take food, especially by use of a tube introduced into the throat. **2:** to compel an animal or person to take in or absorb nourishment.

forced feeding. forced alimentation.

foreign odor. an odor that is not natural to a farm commodity or agricultural product.

forelle. French meaning trout; also widely used in European countries, especially in restaurants.

forensic. legal; pertaining to, denoting or belonging to law, especially that which concerns courts of justice.

forequarters. **1:** the two front quarters of a domestic mammal such as a horse or cow. **2:** also called fore udder in the cow.

forest. **1:** a plant association predominantly of trees and other woody vegetation that occupies an extensive area of land. **2:** forest land. *See also* woodlot.

forest health. a forest in such condition that it is resilient to stress and damage and such that sustained productivity (under proper management) can be expected. A healthy forest is generally characterized by biodiversity and diversity of habitats that can be expected to sustain biodiversity while meeting present and future resource needs.

forest health agent. any naturally occurring biotic or abiotic influence on a forest, including pests or beneficial organisms and physical factors that exacerbate, ameliorate, counter, or prevent the effects of pests and other harmful agents. Biotic health agents include certain fungi, insects, plants, animals, bacteria, and nematodes. Abiotic agents include snow, frost, snow, fire, wind, sun, drought, and nutrients. Humans are sometimes included in this category.

forest health treatment. the application of any technique that will guard or protect the health of a forest, mitigate damage, or reduce the risk of future damage. Examples are the use of insecticides to control pest species such as the gypsy moth; the spacing of trees to reduce attacks by bark beetles; and maintenance of habitat characteristics favorable to beneficial organisms.

forest land. **1:** any relatively large area of land dominated by trees either native or planted, as one or more contemporary or successive stands, successive crops, or as wilderness. **2:** land, at least 10% of which is occupied by forest trees or was formerly so occupied and has not since been used or developed for non-forest uses.

managed forest land. forest land that is managed, planted, and cropped for purposes of preservation, conservation, or for sustained commercial use and productivity.

forestation. **1:** the planting of forests. **2:** the science of planting and taking care of trees and forests. **3:** the process of establishing and managing forests.

forester. **1a:** one who is trained in forestry. **1b:** one who is engaged in the profession of forestry. **2:** one who oversees the growing and management of timber on an estate. **3:** an officer appointed to protect a forest and preserve the game. **4:** any of various chiefly black moths of the genus *Alypia* and related genera (Family Agaristidae). An important example is the eight-spotted forester (*A. octomaculata*), the larvae of which damage grapevines.

forestry. the application of science, technology, business methods, and conservation practices to the management of forests, forest lands, and forest resources for productivity, economic viability, and human benefit. **sustainable forestry**. the management of a forest and forest resources so that the forest remains healthy and productive indefinitely. This requires stewardship and management based on an understanding of and respect for the forest as an ecosystem. This includes conservation of soil, of biological diversity, of fish and wildlife, surface and groundwater bodies; as well as preservation of water quality. Sustainable forestry also requires rejuvenation and restoration of damaged forest resources.

Forficula auricularia (European earwig). an insect pest (Family Forficulidae) that feeds on blossoms and maturing fruit whereas the nymphs are pests of young shoots, injuring or destroying the leaves.

formic. pertaining to ants or to formic acid.

formic acid. a clear, sour-tasting, colorless, moderately toxic, dangerously caustic, liquid carboxylic acid, H_2CO_2, with a pungent odor. It is the simplest and most acidic of the monobasic fatty acids. Chronic absorption via the skin in humans can cause albuminuria and hematuria. It occurs naturally as a component of ant venom (from which it was originally prepared), in a few other insects, and in certain plants. Formic acid is used in a number of organic syntheses, including the synthesis of acetic acid and certain insecticides and drugs. It is also used to dissolve bones and remove the hair from animal carcasses, and is a component of a number of food flavorings. Formic acid is a strong oxidizing agent, and concentrated solutions are corrosive to tissues; ingestion can prove fatal. Exposure may be via inhalation of the vapor, percutaneous absorption, or ingestion.

formic ether. ethyl formate.

Formicidae. a family of polymorphic colonial hymenopterous insects (ants), 2 to 25 mm long. They are usually black, brown, or reddish in color. Most species are scavengers or predators. Most ants are venomous. The venom is used in defense and to kill insects and other small animals on which some species prey. Envenomation by ants may cause painful reactions, but is rarely fatal to humans. *See also* venom (ant venom), formic acid, myrmicacin.

formonitrile. hydrogen cyanide.

formothion (*O,O*-dimethyl-*S*-(*N*-methyl-*N*-formylcarbamoylmethyl)phosphorodithiolate; phosphorodithioic acid *S*-[2-(formylmethylamino)-2-oxyethyl] *O,O*-dimethyl ester; phosphorodithioic acid *O,O*-dimethyl ester *S*-ester with *N*-formyl-2-mercapto-*N*-methylacet-amide; *S*-(*N*-formyl-*N*-methylcarbamoylmethyl) *O,O*-dimethyl phosphorodithioate). a yellow oil or crystalline substance, it is an anticholinesterase used as a miticide and systemic insecticide chiefly against flies and aphids. It is poisonous to humans and probably also to other vertebrate animals.

formulate. **1:** to state as a formula. **2:** to prepare or compose by a prescribed or established method (a formula).

formulation. **1a:** the process of formulating. **1b:** the product of formulating.

***S*-(*N*-formyl-*N*-methylcarbamoylmethyl) *O,O*-dimethyl phosphorodithioate**. formothion.

fosetyl aluminum (phosphonic acid monoethyl ester aluminum salt; aluminum *tris*(ethyl) phosphite); aluminum *tris*-(*O*-ethylphosphonate), efosite Al, phosethyl Al, LS 74-783, Alliette). a systemic fungicide that is slightly to moderately toxic to laboratory mammals and birds and is carcinogenic to some.

founder effect. *See* effect.

four-o'clock. *Mirabilis jalapa*.

fourth stomach. *See* abomasum.

fowl. in agriculture, any large domestic bird that is raised chiefly for meat.

fowl mannagrass. *Glycera striata*

fowl meadow grass. *Glycera striata*

fowl tick. *Argas persinus*.

foxtail. a common name of *Equisetum*.

fps. foot-pound-second.

fragipan. a hard or very hard restricting subsurface soil horizon when dry, but ruptures rather easily when wet. It is brittle, usually loamy, weakly cemented, and of low permeability. Fragipan is low in organic matter, low to moderate in clay content, but high or very high in silt and/or fine sand.

fragment. **1a:** a piece, portion, particle; remnant. **1b:** to shatter, separate, splinter, divide. **2:** a small lump of soil produced by a disturbance.

fragmentation. in forestry, the conversion of large continuous forest patches into one or more smaller patches surrounded by disturbed areas. This may result from natural agents such as fire, landslides, windthrow, and insect damage. Fragmentation in managed forests also results from the harvesting of timber and associated activities.

fragrance. **1a:** scent, odor, perfume, redolence, bouquet. **1b:** a sweet or agreeable odor. **2:** a perfume or scent; a substance that has a fragrance as in def. 1.

frangula emodin. emodin.

frangulic acid. emodin.

frank. obvious, apparent; often applied to visual signs (e.g., visually obvious blood in the stool) or to lesions.

Franseria discolor (white ragweed). a weedy herb (Family Compositae). Concentrations of nitrate toxic to livestock have been reported.

frass. solid excrement of insect larvae.

Fratol. sodium fluoroacetate.

freehold. real property in the form of land that is held for life.

freely drained. said of a soil through which water percolates freely.

freemartin. the abnormal female member of a pair of twins, where the male is normal, in cattle and some other ungulates. The female is usually an intersex and is sterile. Her condition is thought to have developed due to receipt of male hormones from her twin brother via their joined (anastomosed) placental circulation.

freezing point. **1:** melting point (mp). **2:** the freezing point or melting point of a pure substance is the temperature at which its solid phase is in equilibrium with the liquid phase. Technically, the terms freezing point and melting point can be used interchangeably. If, however, a substance is being cooled, the term freezing point is usually applied to this equilibrium; if a substance is being heated to bring it to this point, the term melting point is usually applied.

French green. cupric acetoarsenite.

Freon [Du Pont]. trademark for a series of compounds (fluorocarbons and chlorofluorocarbons) used in refrigeration and air conditioning equipment, and as fire extinguishing agents, solvents, and cleaning fluids. "Freon" is a clear, water-white liquid or vapor with an etherial odor. In the gas phase is a deadly and insidious hazard and should be handled with care: It is nonirritant and noncorrosive, but is denser than air and can act as a simple asphyxiant. Fatalities from exposure to "Freon" have occurred.

Fresian. a breed of black and white dairy cattle known for its very high yield of milk.

friable. **1a:** any soil (wet or dry) that crumbles easily between the fingers. **1b:** a soil that is easily tilled.

friable. easily crumbled.

friar's cowl. any plant of the genus *Aconitum*.

fricasse. a common name for *Blighia sapida* and its fruit in Haiti.

frost. **1:** a deposit of minute ice crystals on a surface that forms by the condensation of atmospheric vapor when the temperature of the surface is below freezing. **2:** a covering of ice. **3:** the process of freezing, especially of water, or of forming frost. **air frost**. the circumstance where the air temperature is below 0°C, but not at the level of the ground. **ground frost**. the situation in which the air temperature at ground level is below 0°C.

frost hardiness. the ability of a plant to survive frost.

frost pocket. a localized, usually low-lying or exposed area that are more subject to frost than surrounding areas. An area where cold air tends to collect.

frost resistance. frost tolerance.

frost tender. the inability of a plant to survive frost.

frost tolerance (frost resistance). the ability of many plants of mid to high latitudes to withstand freezing temperatures for a substantial period of time. Often called frost resistance, the more appropriate term is tolerance. *See also* frost hardiness.

frosty. covered with frost. freezing weather.

fructivore. an animal that feeds on fruit.

fructivorous. fruit-eating; pertaining to an animal that eats fruit (a fructivore).

frug-. indicating or pertaining to fruit.

frugivore. an animal that subsists chiefly or entirely on fruit.

fruit. **1:** the ripened ovary of a flower together with any accessory parts associated with it. **2:** a vernacular term applied to the fleshy, edible, ripened ovary of a woody plant or vine. **3:** to bear fruit. 4: crop, yield, harvest. **5:** the mature ovary of a plant or animal. **6:** the consequence or outcome of a behavior, labor, or activity. **accessory fruit**. a fruit in which tissue other than that of the ripened ovary forms a conspicuous portion. Apples, figs, and strawberries are examples in which the true fruits (the seeds) lie within, or are embedded within, soft edible tissue. **aggregate fruit**. a fused cluster of several fruits, each of which developed from a single ovary of a single flower, as in a blackberry or magnolia.

fruit wall. pericarp.

fruiting body (conk). the complex reproductive part of a fungus that contains or bears spores. The mature spores are dispersed by the fruiting body. Types of fruiting bodies include apothecia, perithecia, conidiophores, coremia, sporangia, pycnia, aecia, pycnidia, acervuli, and sporodochia.

frutescent. shrublike.

FSH (follicle-stimulating hormone; follitropin; urofolitrophin). a gonadotropic glycoprotein secreted by the anterior pituitary gland that regulates the metabolic activity of the granulosa cells of the mammalian ovary. FSH stimulates maturation of the Graafian follicles of the ovary and the production of estrogens. It is also partially responsible for spermatogenesis. *See* hormone, pituitary gland.

fuel. in forestry, any combustible material found in a forest or associated lands (e.g., standing live and dead trees, shrubs, and other plants; pine needles and other forest litter; peat; coal on or near the surface; structures built by humans). *See* fuel load. **alternative fuel**. fuels other than the traditional fossil fuels (e.g., petroleum and petroleum products, coal products, natural gas). Examples are compressed natural gas, ethanol, methanol, and biodiesel. Extensive use of such fuels could conserve natural fuel sources (e.g., through recycling and use of waste materials), and reduce pollution and dependence on imported petroleum.

fuel load. in forestry, the amount of combustible material, living and dead, that occurs in an area. In the United States this is usually expressed as tons per acre. *See* fuel.

fuelwood. trees that are used for the production of firewood or other wood fuel.

fuller's earth. a naturally occurring, fine-grained, highly absorptive material such as clay (*See* attapulgite) that is comprised mainly of hydrated aluminum silicates.

fulvic acid. a water-soluble, natural organic substance of low molecular weight that is derived from humus. It often occurs in surface waters.

fumes. vapors, or very fine particulates, produced by recondensation in air of a vaporized solid (as when it burns), especially those with a caustic or irritant action.

fumigant. an irritant, usually lethal, smoke, gas, or vapor used (usually in enclosed spaces) to control rodents, insects, or other pests. Substances commonly used as fumigants are acrylonitrile, carbon tetrachloride, ethylene oxide, hydrogen cyanide, and methyl bromide. **soil fumigant**. a pesticidal chemical used to destroy pests in the soil.

fumigate. to apply fumigants.

fumigation. the process of fumigating; the use of a fumigant

functional effect. *See* effect.

fungal. descriptive of or pertaining to fungi.

fungi. plural of fungus.

fungicide. **1a:** a pesticide used to destroy fungi such as molds and mildews. **1b:** a substance that destroys fungi or prevents their growth. Fungicides are secreted by some insects such as ants and also occur in some plants.

fungistat (fungistatic agent). any agent that inhibits the growth of fungi.

fungistatic (mycostatic). **1:** having the capacity to inhibit the growth of fungi. **2:** a fungistat.

fungistatic agent. fungistat.

fungitoxic (mycotoxic). poisonous to, or inhibiting the growth of, fungi.

fungitoxicity. the property of being, or having the capacity to be, fungitoxic.

fungivorous. fungus-feeding.

fungous. of or pertaining to fungi.

fungus (plural, fungi). a heterotrophic, generally nonmotile, non-photosynthesizing, chiefly multicellular organism (Kingdom Fungi) that absorbs nutrients from dead or other living organisms. Formerly considered plants, they are now considered to be an independent line of evolution. Certain mushrooms (e.g., *Amanita*) and many of the lower fungi are dangerously toxic. *See*, for example, ergot.

funicle (funiculus). **1:** the portion of the insect antenna that lies proximal to the antennal club. **2:** the stalk that attaches the ovule to the placenta in the ovary of angiosperms.

funiculus. funicle.

fur. **1:** the hairy covering of a mammal, especially one with soft, thick, close-grown hair near the skin (the underhair or fur proper), with coarser hair growing through it (guard hair). **2:** the pelt; the fur (def. 1), including both the hair and the dried or tanned skin.

fur farm. a farm on which fur-bearing animals such as mink are raised for their pelts.

fur proper. underhair. *See* fur.

7*H*-furo[3,2-g][1]benzopyran-7-one. psoralen.

furo[3,2-g]-coumarin. psoralen.

furocoumarin. any of a group of photoactive antifungal dyestuffs found in plants of the families Apiaceae and Rutaceae that cause primary photosensitization in domestic fowl and livestock. *See* photosensitivity (primary photosensitivity).

fusaric acid (5-butyl-2-pyridinecarboxylic acid; 5-butylpicolinic acid). a mycotoxin produced by fungi of the genus *Fusarium*, especially *Fusarium moniliforme*. It is an antibiotic and wilting agent that causes yellowing of infected plants. Fusaric acid is very toxic to laboratory mice, rabbits, and dogs.

fusariotoxicosis. a condition of poultry due to ingestion of moldy grain. It is characterized by edematous and necrotic oral lesions and is probably due to trichothecene mycotoxins. *Cf.* fusarium toxicosis.

Fusarium. a genus of extremely common ascomycetous fungi (Order Moniales, Family Tuberculariaceae) with sickle-shaped, multicellular conidia. They produce marasmins and fusaric acid. Some are important plant pathogens; some opportunistically (accidentally) infect animals, including humans. *See* fusarium estrogenism.

fusarium estrogenism (*Fusarium* hyperestrogenism; zearalenone hyperestrogenism). a toxicosis caused by zearalenone, *q.v.*, a nonsteroidal estrogen produced by certain toxigenic strains of *Fusarium spp*. Affected animals may be poisoned by fresh or stored feeds or pasturage contaminated with *Fusarium*. The condition is indistinguishable from excessive administration of estrogen. In swine, chiefly weaned and prepubertal gilts are observed with this condition; prominent signs are hyperemia, enlargement of the vulva, and in severe cases, uterine and mammary gland hypertrophy; prolapse of the uterus is seen on occasion. Effects of zearalenone on multiparous sows include diminished fertility, reduced litter size, smaller piglets,, malformations, and probably fetal resorption. Some sows exhibit constant estrus or pseudopregnancy. Lesions in swine include ovarian atrophy and follicular atresia, uterine edema and hypertrophy, and a cystic appearance in degenerative endometrial glands. Dairy heifers may experience weight loss, vaginal discharge, nymphomania, and uterine hypertrophy. Gravid heifers may abort one to three months following conception, usually with multiple returns to service. Ewes receiving 25 mg/day of zearalenone for 10 days exhibited poor reproductive performance due to decreased rates of ovulation. Gravid ewes fed on heavily infested, ensiled corn with a high zearalenone content experienced abortions, premature live births, prostration, and occasionally died. *See Fusarium*, zearallenone.

***Fusarium* hyperestrogenism**. fusarium estrogenism.

fusarium toxicosis. poisoning by trichothecenes secreted by various species of *Fusarium*. This term refers chiefly to skin necroses so produced. A skin test is employed to screen toxic strains of *Fusarium*. *Cf.* fusariotoxicosis.

futures. commodities. not yet produced, to be delivered in the future, *Cf.* actuals.

G

γ. symbol for gamma.

Γ. symbol for gamma.

g. abbreviation of gram.

G. **1:** giga-. **2:** grav; fors; a unit of acceleration equal to the standard acceleration of gravity on Earth, 9.80665 m/sec^2 or approximately 32.174 feet/sec^2.

G6P. glucose-6-phosphate.

G6P dehydrogenase. glucose-6-phosphate dehydrogenase.

G6PD. glucose-6-phosphate dehydrogenase.

gaggle (skein). a flock of geese.

galact- galacto-. combining forms indicating milk.

galactogogue. **1a:** causing the production of milk. **1b:** an agent that induces the production of milk.

galactophorous. milk-carrying.

4-O-β-D-galactopyranosyl-D-glucose. lactose.

galactorrhea. **1:** continued flow of breast milk after weaning. **2:** excessive flow of breast milk during nursing.

4-(β-D-galactosido)-D-glucose. lactose.

galactosis. the production of milk by lacteal glands.

galactotoxin. a toxic substance in milk that is produced by bacteria. *See* galactotoxism.

galactotoxism (galactoxism; galactoxismus). poisoning due to the ingestion of milk that contains toxic substances, especially galactotoxin.

galactoxism. galactotoxism.

galactoxismus. galactotoxism.

Galanthus (snowdrops). a genus of hardy, bulbous, spring-blooming herbs with drooping white flowers (Family Amaryllidaceae). The bulbs are poisonous. *G. nivalis* (snowdrop). a small ornamental plant with nodding flowers introduced to North America from Europe.

gale. a windstorm in which the wind speed is 40 mph (65 km/hr) or more at a height above ground of 32 ft. (9.8 m).

galerina. common name of any fungus of the genus *Galerina*.

Galerina (galerinas). a genus of brown-spored, gilled fungi (Family Cortinariaceae); the edibility of most species is uncertain, but some are poisonous and contain amanitin or other mycotoxins, some of which have yet to be identified. ***G. autumnalis*** (deadly galerina; autumn galerina). a small, brown, innocent-looking but deadly mushroom (Family Cortinariaceae). It is widely distributed throughout North America on well-decayed logs. Symptoms of poisoning, which are similar to those of poisoning by *Amanita virosa*, rarely appear earlier than 10 hours following ingestion. Early symptoms usually include those of acute gastroenteritis with nausea, vomiting, cramps, and diarrhea. Victims may also complain of headaches and pains in the back and joints. Following a remission of variable duration, the victim may develop serious lesions, most notably acute or chronic kidney and liver dysfunction or complete failure, as well as damage to the intestines, genital organs, heart, and nervous system. There may be several remissions during the course of the disease with death occurring several months following exposure. The poisoning is due to the presence of orellanin and possibly other as yet unidentified toxins. ***G. tibicystis*** (sphagnum-bog galerina). this species is difficult to identify and little is known concerning its toxicity. Common to sphagnum bogs, it is one of many moss-inhabiting galerinas. ***G. venenata*** (deadly lawn galerina). a small, deadly mushroom with a moist reddish-brown to buff-colored cap that occurs on lawns and over buried, decomposed wood in Washington and Oregon. It occurs commonly throughout the United States, especially on lawns in the Northwest. Effects are similar to those of *G. autumnalis*.

galitoxin. a toxic resinoid that occurs in the more toxic species of *Asclepias*. It is the principal active agent in asclepias poisoning.

gall. **1:** bile. **2:** an outgrowth or swelling on a plant structure such as a stem, leaf, or root that is produced by the plant in response to the presence of a parasite. **crown gall**. a gall on a plant produced as the result of a bacterial infection.

gallate. any esters of gallic acid, *q.v.*; propyl, octyl, and dodecyl gallates are legally produced antioxidants.

gallbladder. a muscular, sac-like diverticulum of the bile duct in which bile is stored and concentrated. It contracts and empties the bile into the intestines under the influence of the hormone, cholecystokinin. The latter is released into the duodenum in response to the presence of food. The gallbladder is absent in many birds and mammals.

Galleria mellonella (greater wax moth; bee moth; honeycomb moth). an insect pest of honeycombs (Family Pyraliidae, Subfamily Galleriinae), the females of which lay their eggs in masses in crevices within a beehive. When the larvae hatch, they tunnel into the combs, making galleries of silk that remain behind. They also damage the combs and honey cells by puncturing the wax caps, causing honey to leak out and destroying its marketability. These moths usually intrude on abandoned beehives, or active ones in which the bee colony has been weakened by disease or starvation.

gallery. a passageway excavated by an insect under bark or within wood. Such galleries may variously provide food and/or may shelter (as during winter) eggs, larvae, and/or adults, and/or serve as security during egg laying.

gallic acid. the toxic, hydroxyphenol moiety of oak tannin, it is a colorless, water-soluble, crystalline compound. It occurs naturally on the twigs of trees, especially *Quercus* (oak), in response to the presence of gall wasp eggs. Gallic acid and its esters (gallates) are used commercially in the production of antioxidants and dyes and in photography.

galliform. resembling birds of the family Galliformes.

Galliformes. a family of generally large, plump, terrestrial fowl such as the domestic fowl, pheasants, turkeys, quail, partridge, and grouse.

galligenous. of or pertaining to an insect or other organism that produces or induces plant galls.

gallinaceous. **1a:** resembling the domestic fowl, *Gallus domesticus*. **1b:** galliform birds, those assigned to the family Galliformes.

gallotannic acid. tannic acid.

gallsickness. anaplasmosis.

Gallus. a genus of gallinaceous birds including the domestic fowl, jungle fowl, and a number of East Indian species. ***G. domesticus*** (domestic fowl; chicken). descended from various jungle fowl of southeastern Asia and developed in a number of breeds for its flesh, eggs, and feathers. ***G. gallus*** (red jungle fowl). believed to be ancestral to the domestic fowl.

galvano-. a prefix meaning electrical, especially direct current.

gam. a pod of whales.

gamble. gambrel.

gambrel (gamble). a wooden or metal rod inserted into the hocks of swine to support them during butchering. *See also* beef pritch.

game. prey, quarry, especially wild animals hunted by humans either for sport or of necessity.

game birds. a term used variously for all birds of the order Galliformes (gallinaceous, or chickenlike, birds), for certain quarry species within this order, and for a variety of quarry birds of several other orders. In North America the term may include various gallinaceous birds such as quail and turkey, aquatic or quarry birds such as duck and geese, shorebirds such as woodcock, snipe, and plover, and certain nongallinaceous upland birds such as mourning doves.

game fish (sport fish). **1:** an edible fish capable of affording sport to the angler during its capture. **2:** any species of fish categorized as game; licensing and regulations may be imposed upon the angler.

sport fish. game fish.

game law. any law that specifies the regulations governing the hunting or capture of wild game. In most countries today, individuals who obtain a hunting license are allowed to hunt game in accordance with the relevant game laws. Game laws in the United States are designed largely to conserve and to protect wildlife from indiscriminate slaughter by trappers, hunters, and fishermen. Most game laws prohibit lake and river pollution, designate a closed season during which game may not be captured or killed, and limit age, size, and/or sex of legal fish and game.

Gamene. a trade name for lindane.

gametangium. any gamete-producing plant or fungal structure (e.g., an antheridium, oogonium, or archegonium).

gamete. a sex cell (egg or sperm) that contains a single copy of each chromosome.

gametocide. an agent that destroys gametes.

gametocyte. a cell that divides, giving rise to sex cells (gametes: spermatocytes or oocytes).

gametogenesis. **1:** the production of gametes in plants and animals **2:** the specialized series of cell divisions that produce gametes. *Cf.* oogenesis, spermatogenesis.

gametophyte. the gamete-producing, haploid stage or generation in the life cycle of a plant bearing the sex organs that produce gametes. *Cf.* sporophyte.

Gamexan. a trade name for lindane.

gamey. gamy.

Gamiso. a trade name for lindane.

gamma (symbols, γ, Γ). **1:** the third letter of the Greek alphabet. **2:** the symbol for microgram. **3:** a prefix, similar meanings to those of alpha, *q.v.* Gamma is used, for example, specify the location of substituents in a chemical compound or a particular form of an organic chemical such as γ-globulin. **4:** the designated prefix for the most intense form of shortwave radiation, the γ-ray.

gamma globulin (γ-globulin). one of a group of serum proteins in blood plasma with specific electrophoretic mobility. The proteins are produced by the immune system and help to destroy or neutralize infection-causing microbes. The group also includes serum immunoglobulins and other globulins. Gamma globulin, derived and concentrated from the blood of other humans, is used to help provide temporary immunity to certain diseases.

gamma ray (γ-ray). a type of high-energy, ionizing, electromagnetic radiation of extremely short wavelength that can react with atoms and molecules (e.g., of tissue) to eject electrons and produce positively charged ions. *See* ionizing radiation.

gammaglobulinemia (hypogammaglobulinemia). the presence of unusually low levels of gamma globulin in the blood due to a disease of the immune system. The deficiency causes increased susceptibility to many infections by bacteria, viruses, and fungi. *See also* gamma globulin.

Gammalin. a trade name for lindane.

Gammexane. a trade name for lindane.

Gamoxol. a trade name for lindane.

gamy (gamey). **1:** pertaining to or having the flavor or odor of game. **2:** having the flavor or odor of slightly tainted meat. **3:** spirited. **4:** lewd, risque. **5:** unwholesome.

-gamy. **1:** a suffix meaning marriage, union, fertilization, or pollination as denoted by the initial or main element of the word (e.g., allogamy, anemogamy, apogamy, autogamy, exogamy, karyogamy). **2:** a suffix that forms a noun that corresponds to an adjective that ends in -gamous.

gander. an adult male goose.

gang. **1:** a team. **2:** a pack, band. **2:** a herd of elk.

gang bolt. the shaft of a harrow that bears the blades.

gangrene. decay and death of tissue usually due to insufficient vascular supply, followed by bacterial infection and putrefaction; a large mass of tissue may be involved. Untreated gangrene can spread rapidly and is life threatening. It is caused by trauma or disease as in gangrenous ergotism.

gangrene des Solognots. a French name for the ergotism caused by *Clavaceps purpurea*.

gangrenous. pertaining to or affected with gangrene.

gangrenous fescue poisoning. fescue foot.

gap. **1:** breach, hole, hiatus; recess. **2:** a break in the crest of a mountain ridge, sometimes allowing easy passage.

garden angelica. *Angelica archangelica*.

garden huckleberry. *Solanum nigrum*.

garden monkshood. *Aconitum napellus*.

garden net. a net used to cover fruit trees and other productive plants to protect them from birds.

garden pea. *Pisum sativum*.

garden spider. *Argiope*. **banded garden**

der. *A. trivasciata*. **black and yellow garden spider**. *A. aurantia*.

garden truck (garden ware). a colloquial term for vegetables raised for the market.

garden ware. garden truck.

garden wolfbane. *Aconitum napellus*.

garget. **1:** a plant of the genus *Phytolacca*. **2:** ovine gangrenous mastitis.

garlic. *Allium sativum*.

garlic oil. allyl sulfide.

garnish. plants, plant parts, or other edible material used to decorate a plate or serving dish (e.g., parsley, watercress, lemon rinds, croutons).

Gärtner's bacillus. *Salmonella enteritidis*.

gas. **1:** one of the three basic physical forms of matter (gas, liquid, solid) in which the molecules move freely in all directions such that, if unrestrained, it expands indefinitely. A gas can be converted into a liquid and eventually a solid by low temperature (e.g., water vapor) although in most cases, compression at low temperatures is required. **2:** gasoline. **coal gas**. a flammable, explosive, toxic gas comprised chiefly of methane, carbon monoxide, and hydrogen. It is produced by the distillation of coal and used for heating and lighting. **natural gas**. **1:** any gas formed naturally in the earth's crust. Many of these (e.g., hydrogen sulfide) are dangerously toxic. **2:** a naturally occurring gaseous combustible mixture used as fuel. American natural gas is comprised of about 85% methane, 9% ethane, 3% propane, 2% nitrogen, 1% butane, and smaller amounts of higher hydrocarbons. Unprocessed natural gas may contain natural gasoline, sometimes carbon dioxide, hydrogen sulfide, and helium. An occasional well head will yield nearly pure methane. Natural gas, when present in ambient air at high concentrations, is a deadly narcotic and asphyxiant. carbon monoxide results from incomplete combustion of natural gas. **sewer gas**. toxic, usually flammable and explosive gas elaborated by decaying organic matter in sewage.

Gasohol. the registered trade name for a fuel blend consisting of 90% unleaded gasoline and 10% ethanol produced by fermentation. Gasohol emissions contain less carbon monoxide than those from gasoline.

gasoline. a toxic, volatile, flammable liquid petroleum distillate, the most common components being C4-C12 aliphatic hydrocarbons. It typically boils within a range of 30 to 220°C. Applications are mainly as a fuel for internal combustion engines, as a diluent, and as an industrial solvent; it has sometimes been used in museum practice as a defatting and cleaning agent (e.g., to remove fat adhering to the skin of some animal preparations). Toxicity is related primarily to the benzene content. Gasoline is a CNS depressant and an irritant to the skin, conjunctiva, and mucous membranes. Tetraethyllead is sometimes added, which introduces the possibility of lead poisoning from repeated exposure. Combustion of gasoline releases toxic fumes that contain toxic compounds such as carbon monoxide and nitrogen oxides. The vapor is carcinogenic to some laboratory animals. Gasoline is a serious fire hazard and the vapor is highly explosive. *Cf.* ga. *See also* gasoline poisoning.

gasoline poisoning. poisoning by exposure to gasoline, usually by contact, ingestion, or inhalation of fumes from the liquid (not the fumes released by combustion). **acute gasoline poisoning**. early symptoms are flushing, staggering, slurred speech, confusion, and sometimees vomiting. Serious cases can result in un consciousness or coma, cyanosis, and death. Lesions may include pancreatic hemorrhage, splenic congestion, hepatic and renal degeneration, hematopoietic changes, pneumonitis, and pulmonary edema. **chronic gasoline poisoning**. long continued (chronic) exposure to gasoline or its fumes can cause dermatitis with blistering and defatting of the skin. Systemic effects may result from ingestion (as when starting a siphon by mouth), cutaneous absorption, or inhalation. Sniffing of small amounts produces euphoria.

gasping. spasmodic or laborious convulsive breathing with the mouth open.

gasping disease. avian bronchitis (*See* bronchitis).

Gasterophilus intestinalis (horse bot fly; common horse bot fly). a fly (Family Gasterophilidae), the larvae of which parasitize horses. The adult lays its eggs on the hairs of the front legs or shoulders, belly, or hind legs of a horse. The tongue of the horse becomes infested by larvae when it licks itself; the larvae soon burrow into the base of the tongue, then pass to the stomach where they remain for nine to ten months, finally passing out of the horse in the feces. The larvae injure and inflame the lips and tongue and the mucosa that lines the stomach and intestines. The larvae also absorb food, causing malnourishment or even starvation of the host and interfere with glandular secretions. With serious infestations, the host may become gaunt, develop a rough coat, and show signs of gastrointestinal upsets.

gastric. of or pertaining to the stomach.

gastric gland. any of several types of glandular cells situated within the mucosa of the stomach that secrete gastric juice, *q.v.*

gastric juice. a mixture of hydrochloric acid, proteolytic enzymes, and mucus secreted by gastric glands, *q.v.*

gastric mucosal barrier. See barrier.

gastritis. inflammation of the stomach, especially of the mucosa.

gastro-, gastr-. prefixes denoting the stomach.

gastrocolitis. inflammation of the stomach and colon, especially of the mucosa. See gas-trointestinal tract.

gastroduodenitis. inflammation of the stomach and duodenum, especially of the mucosa. See gastrointestinal tract.

gastroentercolitis. inflammation of the stomach, small intestine, and colon. *Cf.* gastroenteritis, gastritis.

gastroenteric. of or pertaining to the stomach and intestines.

gastroenteritis. inflammation of stomach and intestines. There are any number of causes including caustic and irritant chemicals, certain other toxicants, viruses, bacteria, and food poisoning. Any substance that induces vomiting can cause gastroenteritis. Severe gastroenteritis is extremely painful and can be fatal. Depending on the severity of the condition, symptoms may include salivation, nausea, frequent vomiting, diarrhea with watery or bloody stools, abdominal colic, a fall in blood pressure, collaps,e and death.

gastroenterocolitis. inflammation of the stomach, small intestine, and colon. *See* gastrointestinal tract.

gastroenteropathy. any disorder, disease, or pathological state of the gastrointestinal tract.

gastroesophagitis. inflammation of the stomach and esophagus, especially of the mucosa. *See* gastrointestinal tract.

gastrogavage. gavage.

gastrointestinal (GI). **1:** of, related to, or referring to the stomach and intestines of vertebrate animals. **2:** broadly used in reference to the entire alimentary canal. *See also* digestive tract.

gastrointestinal tract (GI tract). **1a:** that portion of the alimentary canal comprised of the stomach and intestines. **1b:** often taken to mean the entire alimentary canal.

gastrolith. stones, or other hard materials used by certain animals to help crush food particles in the gut. *Cf.* grit.

gastropathy. any disorder, disease, or pathological state of the stomach.

gastropod. common name of any mollusk of the class Gastropoda.

Gastropoda. a class of the phylum Mollusca. The head is well developed and bears eyes and tentacles; there is a single shell without a hinge (in some forms the shell is lacking or reduced in size) and a large flattened foot. The mantle cavity, gills, and anus are anterior and the other organs are asymmetrically arranged in most taxa. The visceral hump and shell are spirally coiled and are closed by an operculum.

gastroscope. an optical instrument with a lighted tip used to visually examine the inside of the stomach.

gastroscopy. visual examination of the interior of the stomach by use of a gastroscope.

gastrosis. any disease of the stomach.

gastrospasm. spasm of the stomach.

gastrostolavage. the irrigation or rinsing out of the stomach through a gastric fistula.

gastrotomy. opening the stomach with an incision. It is used clinically, for example, to remove a mass of (e.g., toxic) material from the stomach that cannot be removed by other means such as gavage or induced vomiting.

gaucho. an Argentinian cowboy, especially in the Pampas.

gauge wheel. a wheel mounted on a planter or tillage device to control the working depth.

gavage (gastrogavage). the provision of a liquid or semiliquid food or drugs directly into the stomach by means of a catheter or tube passed through the nares, pharynx, and esophagus or by way of a gastric fistula.

GDP. gross domestic product.

gee. **1:** butter oil. **2:** a voice command to a horse or other beast of burden that directs it to turn to the right. *Cf.* ghee, haw.

gee. butter oil.

geeldikkop. the South African name for "bighead" in sheep.

geese. plural of goose.

geld. **1:** to render an animal sterile, usually by castration. **2:** an animal that is sterile. *Cf.* gelding.

gelding. a gelded (castrated) male horse.

gelsemicine. one of at least three deadly alkaloids in *Gelsemium sempervirens*.

gelsemine. one of at least three deadly crystalline alkaloids in *Gelsemium sempervirens*. It is a mydriatic and CNS stimulant.

Gelsemium (yellow jessamine; trumpet flower). a small genus of twining or trailing woody evergreen shrubby plants or vines (Family Loganiaceae) that occur in woodlands and thickets in Asia and the southern United States. The flowers are yellow, trumpet-shaped, and fragrant. They contain various alkaloids. ***G. sempervirens*** (evening trumpet flower; Carolina jessamine; Carolina yellow jessamine; yellow jessamine; yellow jasmine). a woody, twining, perennial evergreen vine that occurs in dry to wet woods, thickets, and sands of the Coastal Plain and lower Piedmont from Virginia to Florida, Arkansas, and Texas. These plants contain several alkaloids related to strychnine. They are poisonous to all classes of livestock. Children have been poisoned by sucking the nectar from the fragrant flowers.

gelsemoidine. one of at least three deadly alkaloids in *Gelsemium sempervirens*.

gene. a segment of a DNA molecule that encodes the production of a cellular product such as a polypeptide chain or an RNA molecule. Classically a gene is the unit of inheritance defined large by it ultimate expression as a phenotype (e.g., brown eyes).

gene mutation. a point mutation; a change in the nucleotide sequence of a single gene.

gene pool. all alleles of all genes that occur among the individuals that comprise a particular population.

genera. plural of genus.

generic. **1a:** general, prevalent, in general use, indefinite. **1b:** descriptive of an entire group or class. **2:** nonproprietary. **3:** pertaining to or having the status of a biological genus.

generic name. the general or non-trademark name of a product. the generic name of a drug or other substance is usually taken to be the chemical name, a name in common use, or an official name as presented in an official compendium. *Cf.* brand name, trade name, trademark, U.S.N.

-genesis. a suffix indicating manner, kind, or source of origin.

genetic. **1:** pertaining to, caused by, or involving genes. **2:** pertaining to genesis.

genetic engineering. **1a:** the manipulation of an organism's genome by inserting or eliminating specific genes using methods of molecular biology. **1b:** artificial selection as by selective breeding.

genome. the complement of genes that comprise the haploid set of chromosomes.

genotype. the genetic constitution of any individual organism.

genotypic. pertaining to genotype. [RAL-final drf Agric]*

gentle cort. a common name of the deadly mushroom, *Cortinarius gentilis*, which is more commonly called the deadly cort.

genus (plural, genera). a grouping in plant and animal taxonomy that ranks above species and below family. It consists of a number of closely related species that have a number of characteristics in common. They are thus considered to be closely related. "Closely-related" genera are placed into a family (or sometimes a subfamily).

germ cell. a reproductive cell or gamete as opposed to a somatic cell.

gene. The functional unit of heredity, a gene is a locus on a chromosome that codes for a specific protein or several related proteins.

genera. plural of genus.

generic. **1a:** general, common, prevalent, in general use, indefinite. **1b:** collective; descriptive of an entire group or class. **2a:** common to all species of a genus. **2b:** pertaining to or having the status of a biological genus. **3:** universal; standard. **4:** a nonproprietary or nonadvertised product.

generic name. the general or non-trademark name of a product. the generic name of a drug or other substance is usually taken to be the chemical name, a name in common use, or an official name as presented in an official compendium. *Cf.* brand name, trade name, trademark, U.S.N.

genetic. **1:** heritable; pertaining to, caused by, or involving genes. **2:** pertaining to genesis.

genetic drift. the random changes in gene frequencies that are likely to occur in small, isolated populations.

genetic effect. *See* effect.

genetic extinction. **1a:** the selective elimination of a genotype that bears mutant alleles that reduce fitness. *See* death (genetic death). **1b:** the elimination of an undesirable genotype by selective breeding.

genetics. the biological science that treats heredity and variation in relation to the action of DNA.

genin. aglycone.

genital gland. testis or ovary.

genome. **1a:** the complement of genes contained in the cells of a given organism or selected species (i.e., the human genome), including all alleles; often taken as the haploid chromosome set. *Cf.* genotype.

genotoxic. of or pertaining to a substance that is toxic to the genetic material.

genotoxicity. **1a:** the quality of being genotoxic. **1b:** the capacity of a substance to poison genetic material.

genotype. **1a:** the entire gene complement (the genetic constitution) of an organism. **1b:** an exact description of the genetic constitution of an organism with respect to a single trait or to a specified set of traits. **1c:** the entire set of genes transmitted from parent to offspring. *Cf.* phenotype.

genotypic. pertaining to genotype.

genus (plural, genera). a taxonomic group of closely related species of plants or animals. Closely related genera are placed into a family.

geographic information system (GIS). a computer software system that allows users to collect, manage, and analyze large volumes of spatially referenced information and associated attribute attributes.

geologist. a scientist trained in the methods and theory of geology. **agricultural geologist** (agrogeologist). a scientist trained in agricultural geology, *q.v.*

geology. **1:** the physical science that treats the structure, physical processes, and chronology of the earth. This science also contributes very substantially to our knowledge of the origin and evolution of life forms as revealed by fossils and many aspects of the geologic history of the earth. **2:** the geologic aspect, features, formations, and chronology of an area. **agricultural geology** (agrogeology). the division of geology that deals with the character and origin of soils, the behavior of groundwater and surface water reservoirs, the occurrence of mineral fertilizers, and other geological factors that influence agricultural practices. the behavior of underground water.

Geomet. phorate.

germ. **1:** germ cell or gamete of a multicellular organism. **2:** in popular terminology, an organism that causes infection such as pathogenic bacteria, viruses, or fungi.

germ cell. *See* germ.

germ spot. germinal disc.

germicidal. **1:** lethal to pathogenic microorganisms. **2a:** germicide. **2b:** of or pertaining to a germicide.

germicidal agent. germicide.

germicide (germicidal (def. 1); germicidal agent). an agent used to destroy pathogenic microorganisms. Germicides must be registered with the Environmental Protection Agency as pesticides.

germinal disc (germ spot). a reddish spot on the surface of the yolk of an egg. It includes the blastoderm of the fertilized egg and is the site where embryonic development begins.

germinal. of, pertaining to, or influencing gametes (germ cells).

gesarol. DDT.

gestation. synonymous with pregnancy, but most often applied to nonhuman viviparous animals.

gestation period. the period from fertilization to birth. The gestation period for humans is approximately 39 weeks; asses, 360—365 days; cattle, 230—285 days; domestic sheep, 145—150 days; goat, 148—153 days; horses, 330—340 days; swine, 112—115 days.

gestation. synonymous with pregnancy, but usually applied to nonhuman viviparous animals.

get. **1a:** in agriculture, the young of an animal. **1b:** often taken as the offspring of the sire. Get of sire is a show classification in which calves of one sire are rated against calves of others.

Gexane. a trade name for lindane.

GH. somatotrophin.

ghee. an oily semisolid butter prepared in India prepared from the milk of buffalo, cow, goat, or sheep. It is practically 100% butterfat. *See also* buter oil.

GI. gastrointestinal.

GI tract. gastrointestinal tract.

giant arborvitae. *Thuja plicata*.

Giberella zeae (corn root-rot organism). an ascomycete; corn grain on which this fungus has been grown is extremely toxic to rats. *See also* scabby barley poisoning.

giblets. edible parts of the viscera of dressed birds (e.g., gizzard, heart, liver).

gid. staggers (def. 3).

giga- (G). **1:** a prefix that denotes giant size. **2:** a prefix to units of measurement that indicates a quantity of 1 billion (10^9) times the value indicated by the root. See International System of Units.

gilgai (gilgie). **1:** a distinctive relief of microknolls and microbasins that develop on flat clayey soils that exhibit a considerable amount of expansion and contraction in response to wetting and drying. *Cf.* hog-wallow land. **2:** Australian. a small gully or ditch. **3:** Australian. a small pond or pool of water.

gilgie. gilgai.

gill. **1:** any of numerous vertical plate-like structures (lamellae) on the undersurface of the cap of the fruiting body of agaric fungi (mushrooms and toadstools). **2:** a respiratory organ of many aquatic or semiaquatic animals such as fish, crustaceans, and certain amphibians. It is ordinarily a highly vascular outgrowth or projection from the body surface or from a portion of the digestive tract (e.g., of the pharynx in bony fish). Gills are usually thin, highly vascular, membranous, leaf-like, or filamentous organs. They are often major sites of entry or excretion of many toxicants. *See also* branchia, branchial, branchial route, respiratory route. a unit of liquid measure that equals 4 fluid ounces or 0.118 liters. **internal gill**. a gill of fish, derived from endoderm, that projects from the wall of the pharynx. They lie in gill slits (pharyngeal slits) and are protected by a covering, the operculum. **external gill. 1:** a gill, ectodermal in origin and often protected by folds of the body wall as in amphibian larvae. **2:** one of the lamellae on the undersurface of the cap of agaric fungi.

gill-over-the-ground. *Glechoma hederacea*.

gilt. a young female swine that has yet to produce offspring.

ginger. See *Asarum*.

girdling. a procedure in which a hatchet or special tool is used to kill a tree by cutting through the bark and cambium. The process interrupts the flow of nutrients between the leaves and other parts of the tree and the tree dies.

GIS. geographic information system.

githagism. a condition similar to lathyrism, due to poisoning by the seeds of *Agrostemma githago* (= *Lychnis githago*).

gland. an organ, cell, or aggregation of cells that secretes specific substances not related to the normal maintenance metabolism of the organ or tissue itself. In animals, these substances are either secreted onto an epithelial surface (*See* exocrine gland) or directly into the circulating blood (*See* endocrine gland). A gland with the same name and/or function in organisms or differing taxa may not be homologous. Glands, especially those of a highly adaptive nature, that have similar or even "identical" functions may have different evolutionary (or genetic) origins and may depend on different mechanisms.

gland of internal secretion. endocrine gland.

glanders (burr; farcy). a fatal disease of equids that is sometimes transmitted to handlers. Although the disease is now virtually absent in North America and Britain, it still occurs in South America and Asia. The pathogen is *Actinobacillus mallei* (*Malleomyces mallei*; *Bacillus mallei*). Regardless of the initial site of infection, suppurating ulcers develop that spread the pathogen to other parts of the body. Glanders is highly contagious and there is no effective treatment. Affected animals must thus be destroyed to prevent the spread of the disease. **Nasal glanders**. affected animals exhibit swollen submaxillary (salivary) glands and unthriftiness; suppurating, bloody ulcerations of the nasal passages develop. **pulmonary glanders**. a form of the disease marked by dyspnea and unthriftiness. **Skin (cutaneous) glanders**. hard nodules develop under the skin alongside the lymphatic ducts; they enlarge, ulcerate, coalesce, and release a thick fluid that spreads throughout the body via the lymphatics.

glaucophyllum. *Solanum malacoxylon*.

Glechoma hederacea (= *Nepeta hederacea*; ground ivy; gill-over-the-ground; creeping Charlie). a prostrate perennial herb (Family Labiatae) that forms dense, pure stands of ground cover. It occurs throughout most of the United States and Canada. It contains irritant oils and has proven to be lethal to horses if ingested in large amounts, either fresh or in hay.

γ-globulin. gamma globulin.

Gloeocercospora sorghi. a pathogenic fungus that causes copperspot of apples.

glorybind. *See Convolvulus*.

gloss-, gloss-, prefixes denoting or pertaining to the tongue.

***S*-α-(β-D-glucopyranosyloxy)-4-hydroxybenzeneacetonitrile**. dhurrin.

β-D-glucopyranosyloxy-L-*p*-hydroxymandelonitrile. dhurrin.

glucose. the most common simple (six carbon) sugar, it is essential to the energy economy of living organisms. In most vertebrate animals it is stored chiefly in the liver and muscle as glycogen. *See also* sugar.

D-glucose-6-(dihydrogen phosphate. glucose-6-phosphate.

glucose-6-phosphate (D-glucose-6-(dihydrogen phosphate; glucose-6-phosphoric acid; G6P). a normal constituent of resting muscle and an important intermediate in carbohydrate metabolism, $C_6H_{13}O_9P$. Individuals deficient in this enzyme are susceptible to the hemolytic effects of a number of substances (phenylhydrazine, vitamin K). *See also* favism.

glucose-6-phosphate dehydrogenase (G6P dehydrogenase; G6PD). an enzyme that catalyzes the synthesis of phosphogluconate from glucose-6-phosphate. Individuals deficient in this enzyme are susceptible to the hemolytic effects of a number of substances such as phenylhydrazine, and vitamin K; the effects of phenazopyridine are more severe. A heritable deficiency of this enzyme results in a hemolytic anemia induced by the antimalarial pamaquine. *See also* favism.

glucose-6-phosphoric acid. glucose-6-phosphate.

glucoside. **1:** a glycoside in which the sugar moiety is glucose; a compound that yields glucose on hydrolysis. glucosides are *N*-, *O*-, and *S*-derivatives of uridine diphosphate glucose. They occur commonly in plants and insects, but are rare in vertebrate animals. **2:** glycoside (obs.). *Cf.* glycoside.

glume. any of the characteristic chafflike bracts of the inflorescence of grasses, sedges, and similar forms, especially one of a pair of bracts at the base of a spikelet.

glut on the market. a phrase referring to the oversupply of a commodity in commerce, usually with an associated depression of the market.

glutamate. a salt of glutamic acid; often used synonymously with glutamic acid. The term glutamate, however, refers technically to the negative ion. It is a common acceptor of amino acids in mammalian metabolism.

glutamine. a white, crystalline, water-soluble powder that is insoluble in most organic solvents,

$$H_2NC(O)(CH_2)_2CH(NH_2)COOH$$

It is a nonessential amino acid.

β-(γ-L-glutamyl)-aminopropionitrile. the skeleton-deforming principle of at least some herbs of the genus *Lathyrus*, *q.v.* It causes paralysis and skeletal lesions.

glutathione,. a tripeptide consisting of residues of glutamic acid, cysteine, and glycine; it is known to act as a coenzyme in a few enzymatic reactions, but its importance may lie in its role as a nonspecific reducing agent within the cell. It is hypothesized that glutathione serves to maintain the biological activity of certain proteins by keeping selected cysteine side chains in the reduced thiol form, thereby not allowing these residues to oxidize and cross-link with one another to form cystine residues. (Unnecessary cross-links often result in distortions of protein structure.)

gluten. the protein of wheat and other grains that gives the tough elastic character to dough. *See* gluten-induced enteropathy.

gluten-induced enteropathy (celiac sprue; gluten-sensitive enteropathy; nontropical sprue). a condition seen in individuals (mostly Caucasians, occasionally African Americans, but not Asians) who are hypersensitive to cereal grain storage proteins such as gluten or its product gliadin that occur in wheat, barley, and oats. Symptoms include diarrhea, weight loss, bloating, and steatorrhea. Lesions include villous atrophy (a flattening of the mucosal surface) and infiltration of the epithelial layer and lamina propria with inflammatory cells, leading to malabsorption in the small intestine. Gluten is thought to cause this disease via direct toxic action and by recruitment of immunological responses. The disease is usually resolved by administering a diet free of gluten.

gluten-sensitive enteropathy. *See* gluten-induced enteropathy.

glutinous. gummy, viscous, sticky.

Glycera striata (fowl meadow grass; fowl mannagrass). a slender, to loosely tufted, cyanogenic grass (Family Gramineae) that may grow to a height of 1.5 m. It grows in moist ground from Newfoundland to Alberta, Canada, south in the United States to northern Florida, Alabama, and Texas.

glycerol α-mono-chlorohydrin. α-chloroydrin. *See* chlorohydrin.

glycerol α-monochlorohydrin. α-chlorohydrin.

glycine (aminoacetic acid). a nonessential amino acid, it is a common acceptor of amino acids in mammals. See conjugation (amino acid conjugation).

glycine rich β-glycoprotein II. properdin factor B.

Glycine max (cultivated soybean). a normally nontoxic, cultivated plant (Family Fabaceae, formerly Leguminosae). It is, however, potentially goitrogenic. Toxic concentrations of nitrate have been reported. Poisoning of cattle and horses that have consumed trichloroethylene-extracted soybean-oil meal (TCESOM) is also recorded from several countries. Symptoms vary considerably among various species of livestock and laboratory animals.

glycol chlorohydrin. ethylene chlorohydrin. *See* chlorohydrin.

glycol dichloride. ethylene dichloride.

glycoside. **1:** an acetal that yields a sugar (most often a pentose or hexose, but see 3-methyl digitoxose) and a nonsugar (aglycone) on hydrolysis.Glycosides are produced naturally in plants as a result of the chemical combination of diverse hydroxy compounds with various sugars. Glycosides occur more widely and more abundantly in plants than alkaloids. Toxicity varies from nontoxic to extremely toxic. The amount and type of glycosides present in many plants vary greatly due to both endogenous and exogenous modifying factors such as age, growth rate of the plant, climate, and edaphic factors. **2:** glucoside, now obsolete. **cardiac glycoside**. **1:** any of a class of cardiotonic, cardiotoxic glycosides that occur in plants of the genus *Digitalis* (Family Scrophulariaceae), *q.v.* The aglycones are derivatives of cyclopentenophenanthrene; the sugars are unique methyl pentoses. **2:** any plant-derived steroid glycoside with cardiotonic activity. The activity of the heart depends on the presence in the aglycone of an unsaturated lactone ring and a hydroxyl group that have a specific spatial relationship illustrated, for example, by oleandrin. Some 400 such compounds have been identified, mostly in plants of three families: Scrophulariaceae (figworts), Liliaceae (lilies), and Apocynaceae (dogbanes). **coumarin glycoside**. any of a class of glycosides, the aglycone of which is a modification of coumarin. Examples are aesculin and daphnin. **cyanogenetic (cyanophoric, nitrile) glycoside**. a glycoside that yields hydrocyanic acid (HCN), a violent poison, upon hydrolysis. Ruminant animals are more susceptible to cyanide poisoning from plants than are monogastric animals. *See* amygdalin, *Sambucus*. **cyanophoric glycoside** (*See* cyanophoric, def. 2). a glycoside that yields glucose, hydrocyanic acid, and benzaldehyde on hydrolysis. **goitrogenic glycoside**. a goiter-producing thyrotoxic glycoside. **saponic glycoside**. a noncardioactive steroid glycoside; it is a saponin. *See* saponin. **steroid glycoside**. any glycoside in which the aglycone contains a sterol group (e.g., solanine). Their physical properties are similar to those of saponins. While the aglycones are toxic, the toxicity of the intact glycoside is at least partially determined by the sugar moiety on which the solubility of the molecule greatly depends. **tetanic glycoside**. any of a class of tetany producing glycosides. Included are cardiac and saponic glycosides. **triterpenoid glycoside**. a glycoside in which the aglycone is polycyclic and contains 30 carbon atoms. The aglycone is toxic, but the toxicity of the intact glycoside is at least partially determined by the sugar moiety on which the solubility greatly depends.

glycyrrhetinic acid glycoside. glycyrrhizic acid.

glycyrrhiza (licorice; liquorice; licorice root; liquorid). the dried rhizome and roots of varieties of *Glycyrrhiza glabra*. Glycyrrhizin is extracted from it.

Glycyrrhiza glabra (licorice; liquorice; sweet root). a perennial plant (Family Fabaceae, formerly Leguminosae). Licorice should not be ingested by anyone with a cardiovascular disorder. In large doses over extended periods of time it is highly toxic. Licorice contains a saponin-like glycoside, glycyrrhizic acid (*q.v.*), which is much sweeter than sucrose. It has both therapeutic and toxic effects. Licorice root and candy should be consumed with some caution, and should be avoided by hypertensive individuals, those with cardiac or renal disease, and pregnant women who are especially subject to edema.

glycyrrhizic acid (20β-carboxy-11-oxo-30-norolean-12-en-3β-yl-2-*o*-β-D-glucopyranuronosyl-α-D-glucopyranosiduronic acid; glycyrrhizin; glycyrrhizinic acid; glycyrrhetinic acid glycoside). a very sweet saponin-like glycoside extracted from glycorrhiza. It is an adrenocortical stimulant used in the treatment of Addison's disease. It is toxic in large doses, producing headache, lethargy, water and salt retention, potassium secretion, and elevated blood pressure. The depletion of potassium sometimes causes renal failure, cardiac arrest, and death. *See Glycyrrhiza glabra*.

glycyrrhizin. glycyrrhizic acid.

glycyrrhizinic acid. glycyrrhizic acid.

glyphosphate (*N*-(phosphonomethyl)glycine; Roundup™; Kleen-up™). an herbicide of low toxicity to humans, but some domestic dogs and cats develop eye, skin, and upper respiratory effects when exposed during or following an application of this compound to weeds or grass. On occasion, these animals vomit, stagger, and develop muscular weakness of the hind legs when exposed to this chemical on freshly treated foliage.

gm. abbreviation of gram.

Gnaphalium purpureum (purple cudweed). a plant (Family Compositae) in which concentrations of nitrate toxic to livestock have been reported.

gnatho-, gnat-. prefixes indicating or pertaining to the jaw.

GNP. gross national product.

goat. an agile, hardy, usually short-haired, horned ruminant, the domestic forms of which are usually raised for milk and or meat; a few are raised for their wool. Young goats are called kids. **billy goat**. an adult male goat; it has a beard; also called a buck. **nanny goat**. an adult female goat; also called a doe.

goatling. a female goat, 1—2 years of age, which has yet to bear a kid.

goblet cell (beaker cell; caliciform cell; chalice cell). a unicellular (epithelial) gland that secretes mucous. epithelial cell that becomes distended with a large accumulation of mucous. Secretory granules accumulate the apical end, giving it a goblet-like appearance.

goiter. simple goiter is a condition in which an enlarged thyroid produces low levels of thyroxin.

goitrogen. any goiter-inducing substance.

goitrogenic. goiter-inducing; having the ability to cause goiter. Many species of plants contain goitrogenic substances, including, for example, many species and variants of *Brassica*, and food plants in several plant families. Some can be quite dangerous to livestock. *See also* thiocyanate, L-5-vinyl-2-thiooxazolidone, glycoside.

golden false pholiota. *Phaeolepiota*.

golden gram. *Phaseolus aureus*.

golden oats. *Trisetum flavescens*.

goldenrod. any herb of the genus *Solidago*.

goldenweeds. *Oonopsis*.

gonado-, gonad-. prefixes indicating or pertaining to gonads.

gonad. either of the two primary sex organs of higher animals, the ovary or testis. It produces gametes and sex hormones. Its reproductive function can be seriously impaired or destroyed by many toxicants.

gonadal. pertaining to the gonads.

Gonyaulax. a genus of protozoa (Order Dinoflagellata, Class Phytomastigophora) that have yellow to brown chromatophores. They are found in marine, brackish, and fresh waters. ***G. catenella***. a species associated with shellfish poisoning along the Pacific Coast of North America. *G. tamarensis*. the species associated with shellfish poisoning along the Atlantic Coast of North America.

goose grass. common name of *Poa annua*.

goose grass. *Eleusine indica*.

gooseberry. *See Ribes*, currant.

goosefoot. *Chenopodium*. *See also* Chenopodiaceae.

Gossypium barbadense. American Egyptian cotton. *See* cotton, gossypol.

Gossypium hirsutum. upland cotton. *See* cotton, gossypol.

gossypol (1,1′,6,6′,7,7′-hexahydroxy-3,3′-

dimethyl-5,5′-bis(1-methylethyl)[2,2′-binaphthalene]-8,8′-dicarboxyaldehyde; 1,1′,6,6′,7,7′-hexahydroxy-5,5′-di-isopropyl-3,3′-dimethyl[2,2′-binaphthalene]-8,8′-dicarboxaldehyde; 2,2′-bis[1,6,7-trihydroxy-3-methyl-5-isopropyl-8-aldehydonaphthalene]; 2,2′-bis[8-formyl-1,6,7-trihydroxy-5-isopropyl-3-methylnaphthalene]). a toxic, polyphenolic pigment, $C_{30}H_{30}O_8$, found in the seeds of *Gossypium* (cotton). It is a gastrointestinal irritant; large doses cause pulmonary edema, dyspnea, and paralysis in laboratory animals. Some breeds of cattle and other ruminants are resistant to gossypol and may feed on raw seeds, usually with little, if any, effect. Gossypol may also produce temporary sterility in males by lowering the sperm count and has been used in China as an oral contraceptive. It is moderately toxic (e.g., to laboratory rats, domestic fowl) to very toxic (to swine, laboratory rabbits, guinea pigs) to some nonruminant animals. It acts in nonruminant animals by lowering the oxygen-carrying capacity of the blood. *See also* cottonseed.

gossypol poisoning. symptoms of gossypol poisoning in swine, dogs, and Holstein cattle are similar and usually appear abruptly following some weeks of cottonseed meal consumption. They include dyspnea and sometimes frothing (sometimes bloody) at the mouth. Even though the appetite usually remains normal, affected animals (usually livestock or poultry) may become weak and emaciated. Death, sometimes with convulsions, usually intervenes within 2—6 days following the onset of symptoms. Cyanosis develops immediately before death. Nonlethal doses of gossypol in poultry may reduce the hatchability of eggs, depress growth rates, and discolor or alter the consistency of eggs. *See* cottonseed.

gourd. **1:** any of a wide variety of trailing or climbing plants of the family Cucurbitaceae. **2:** the fruit of such a plant. They have hard, durable shells and are used as ornaments or to make simple, colorful utensils such as bowls, cups, ladles, and bottle-like containers.

gousiekte. a sometimes fatal cardiac condition in sheep that have browsed on *Vangueria pygmora*, a poisonous bush. The condition is characterized by myocarditis with dilation of the cardiac muscle and heart failure.

grade. **1:** slope, incline. **2:** degree, grade, level, rank. **3:** condition, status. **4:** quality, rating, value. **5:** any of a number of standards of chemical purity of varying specifications that have been established by various professional associations, organs, or bodies. These include ACS reagent grade (analytical reagent quality or grade); CP (chemically pure, *q.v.*); FCC, a grade that conforms to National Formulary Specifications; NF, a grade of purity conforming to National Formulary specifications; U.S.P, a grade of purity conforming to the specifications of the U.S. Pharmacopeia. A number of other grades exist. *See* purity.

graded effect. *See* effect.

grading. **1a:** sorting and/or classifying. **1b:** flattening, leveling, or smoothing as a roadbed, dike, or embankment. **2:** classifying cotton on the basis of color, foreign matter, and preparation for ginning. **3:** classifying timber, lumber, or logs based on quality or end use.

gradual metamorphosis. *See* metamorphosis.

grain. **1:** seed. **2a:** the caryopsis or dry fruit of a cereal grass such as barley, corn, oats, rice, or wheat. **2b:** the seed-like fruit of certain plants other than grasses such as buckwheat. **3:** any plant that bears seeds or fruit designated as grain.

gram. **1:** the basic unit of mass in the metric system, abbreviated g or gm; it equals 15.432 grains, or 0.035 ounces avoirdupois. **2:** *Phaseolus aureus*.

-gram. a suffix indicating a recording as by an instrument. *Cf*. -graph.

grama grass. *See Bouteloua*.

gramicidin. any of a class of polypeptide antibiotics that contains *D*-amino acids.

gramicidin S. a cyclic peptide produced by *Bacillus brevis* and other *Bacillus* species that acts on Gram-positive bacteria (staphylococci, streptococci).

Gramineae (formerly Poaceae). an agriculturally and ecologically important, nearly metropolitan family of some 9000 species of vascular plants classified among about 600 genera. Grasses are by far the most economically important group of plants. Most are annual or perennial herbs and include both wild and cultivated forms. Grasses are pioneer forms in many ecological communities but, especially in vast areas of low to moderate rainfall, they form the dominant vegetation (e.g., in much of the Great Plains of North America, the steppes of Eurasia, the African veldt the savannas and pampas of South America. The flowers are distinctive and divided into spikelets each of which contains one or more tiny florets. The stems are usually hollow with nodes that are swollen; the roots and rhizomes are usually fibrous. The leaf blade is usually flat and elongate with a culm or sheath at the base that surrounds the stem. The dry fruit is called a grain or caryopsis. *See*, for example, grain, *Agrostis*, *Alopecurus*, *Andropogon*, *Anthoranthum*, *Avena*, *Bambusa*, *Calamgrostis*, *Cynodon*, *Elymus*, *Festuca*, *Hordeum*, *Panicum*, *Paspalum*, *Poa*, *Redfieldia*, *Salvia*, *Sorghum*, *Sporobolus*, *Triticum*, *Trisetum*, *Zea*,.

graminoid. grasslike.

graminology. The science that treats grasses and grass-like plants. *See also* agrostology.

grana. stacked internal membranes (thylakoids) in chloroplasts at which the light reactions of photosynthesis take place.

granivore. a heterotrophic organism that feeds exclusively or predominantly on seeds. **obligate granivore**. an organism that feeds exclusively on seeds.

granular leukocyte. granulocyte.

granulo-. a prefix meaning granular, or indicating a relationship to granules.

granulocyte (granular leukocyte). a type of granular, polymorphonuclear leukocyte that contains distinctively staining cytoplasmic granules. *See also* leukocyte, agranulocyte, basophil, eosinophil, neutrophils.

Granutox. phorate.

-graph. **1:** a suffix indicating a writing, a depiction, or a description. **2:** a recording instrument.

graph-. a prefix indicating a denoting a writing, depiction or description.

grass bacillus. *Bacillus subtilis*.

grass pea. a poisonous pea. *See Lathyrus*.

grass staggers. grass tetany.

grass tetanus. grass tetany.

grass tetany (grass staggers; wetland pasture poisoning; protein poisoning; grass tetanus; lactation tetanus; hypomagnesemia). a syndrome of livestock due to the inability of an animal to absorb adequate amounts of dietary magnesium (*See* hypo magnesemia). This impairment is related to elevated rumen ammonia levels and consequent conversion of dietary magnesium to the relatively insoluble hydroxide. Elevated ammonia suggests poisoning of ruminal microflora since part of the ammonia produced by the digestion of nitrogenous material is normally processed by these organisms. The disease is usually associated with lush pasturage or forage and characterized by hypomagnesemia and frequently by hypocalcemia and other ionic imbalances in serum. Chronic grass tetany may not be accompanied by obvious signs, and is detected clinically by low serum magnesium levels to which affected animals may have adapted. The acute syndrome has a latent period of about 1—26 weeks. Initial signs in cattle and sheep are excitement, anorexia, and progressive incoordination, muscular twitches, bruxing (grinding of teeth), and salivation. More severe signs eventually appear that may include aggression, staggering, and prostration. Tetanic seizures may become general and accompanied by labored breathing and tachycardia, with intermittent remissions, terminating in death. *Cf.* hypomagnesemic tetany.

grav. **1:** *See* G. **2:** abbreviation for gravity or gravitation.

graveyard weed. *Euphorbia cyparissias*.

gravid. pregnant. usually applied to nonhuman viviparous organisms.

gravida. a gravid female; a pregnant woman.

gravity feed. **1:** the provision of fuel, grain, or other materials by force of gravity. **2:** a system or device designed to feed by gravity.

gray. **1a:** a neutral or ashy shade or color. **1b:** dingy, drab. **1c:** cloudy, misty, or foggy. **2:** the color of a mammal's coat due to an intermixture of white and black hairs. **3:** Gy (abbrev.), the International System unit of absorbed dose. It replaces the rad in this system. 1 Gy = 1 joule/kg body weight or 100 rads.

grayanotoxin (andromedotoxin). any of a group of poisonous substances that paralyze the motor nerve terminals in skeletal muscle of humans and laboratory mammals. They also initially stimulate and then paralyze the vagus nerve. Plants that contain grayanotoxins include *Andromeda* (andromeda; bog rosemary), *Kalmia* (laurels), *Ledum* (Labrador tea), *Lyonia* (e.g., stagger bush), *Pieris* (pieris), and *Rhododendron* (rhododendrons, azaleas).

Grayia (hop sage). a genus of plants (Family Chenopodiaceae), the members of which are facultative accumulators of selenium from soil.

graze. to consume standing vegetation, as by many domestic livestock and certain species of wildlife. *Cf.* browse. **2:** to run livestock on pastures, range, or other land suitable for grazing.

grazier. a rancher; one who keeps livestock on range or grazing land.

grazing. **1:** feeding on herbage, algae, lichens, or phytoplankton, by cropping and consuming the above-ground portion of the plants in the case of the herbage, or by consuming entire plants in the case of algae and phytoplankton. **2:** the feed available to grazing livestock on ranges and pastures. **3:** the process of feeding on standing plants other than browse by livestock. *See also* degree of grazing. **accumulated grazing**. Forage retained for later use by livestock. **alternate grazing**. rotational grazing. **deferred rotation grazing**. a management system whereby use for grazing is delayed on various parts of a range during succeeding years. This permits each part of the range to rest for a period during the growing season to foster seed production and an extended period of vegetative growth. **Hohenheim system of grazing**. a system of managing a grazing area by dividing it into a number of paddocks that are grazed in rotation. In this system, a short period of regrowth, facilitated by use of nitrogenous fertilizers, is permitted between grazings. **rotational grazing** (alternate grazing). an intensive system of sustained grazing by moving livestock from one pasture or section of rangeland to another to allow the herbage to reestablish itself at a particular location before grazing resumes. Thus, short periods of intense use is followed by a period of recovery. Forage is thus used more effectively and completely.

greasewood. **1:** *Sarcobatus vermiculatus*, *q.v.* **2:** loosely, any plant of the family Chenopodiaceae. **3:** certain other unrelated small woody shrubs of arid lands such as *Adenostoma fasciculatum* (chamiso, chamisa). a California shrub; and *Ramona polystachya* (white sage). a low shrub of the southwestern United States and northern Mexico. **black greasewood**. *Sarcobatus vermiculatus*.

green fescue. *Festuca viridula*.

greater wax moth. *Galleria mellonella*.

green alga. *See* alga.

green death cap. *Amanita phalloides*.

green dragon. *Arisaema dracontium*.

green gram. *Phaseolus aureus*.

green hellebore. *Helleborus viridis*.

green june beetle. *Cotinus nitida*.

green leaf fescue. *Festuca viridula*.

green thumb. an exceptional aptitude for successful gardening.

green tree retention. the retention of live trees of a specific species and size from harvesting, to achieve a site-specific objective.

green vetch. a poisonous pea. *See Lathyrus*.

green vitriol. ferrous sulfate.

green-spored lepiota. *Chlorophyllum molybdites*.

green-stemmed Joe Pye weed. a common name of *Eupatorium purpureum*.

greenbelt. **1:** a spacious area or strip of largely undeveloped, sparsely settled land, associated with a human community. One that has been set aside to constrain or allow planned development, preserve the character of the countryside and the community, to provide open space, or to provide for recreation or limited agriculture. **2:** a similar area adjacent to a commercial enterprise or establishment usually to prevent intrusion by residential or other commercial properties and to allow for expansion.

greenbottle fly. a brightly colored blowfly.

greenbriar. common name of any plant of the genus *Smilax*.

greenhouse effect. *See* effect.

Grindelia (gumweeds). a genus of herbs of the family Compositae; all are facultative accumulators of selenium from soil.

griseofulvin (7-chloro-2′4,6-trimethoxy-6′-methylspiro[benzofuran-2(3*H*),1′-[2]cyclohexene]-3,4′-dione; 7-chloro-4,6-dimethoxycoumaran-3-one-2-spiro-1′-(2′-methoxy-6′-methylcyclohex-2′-en-4′-one); amudane). a water-insoluble, chlorine-containing, slightly ethanol-soluble, fungistatic antibiotic produced by *Penicillium griseofulvum*, *P. patulum*, and *P. janczewski* (= *P. nigricans*). It is used mainly as an agricultural antifungal agent. It is slightly toxic to humans, usually causing no more than headache and gastrointestinal distress. The toxicity of griseofulvin is generally low, but it is very toxic to laboratory rats, causing liver damage. It is also teratogenic and carcinogenic to laboratory rats.

grist. grain designated for milling (grinding). **2:** the meal or flour produced by such grinding.

grist mill. a building in which grain is ground using large millstones or burrs that are usually turned by a waterwheel served by an adjacent creek or river. Such mills are not now common in the United States.

gristle. cartilage.

grit. **1:** coarse sand, gravel, dirt, dust. **2:** sandstone with a rough surface that includes large, coarse, or angular grains. Grain size may be variable or nearly uniform. **3:** sandstone suitable for grindstones. **4:** gritty material (e.g., oystershell, sand, small stones) ingested by birds to grind feed in their gizzard.

gross domestic product (GDP). The gross national product exclusive of payments on foreign investments.

gross national product (GNP). a quantitative measure of a nation's total economic activity, generally assessed yearly or quarterly. It represents the total monetary value of all goods and services produced in a country during one year.

ground cherry. *Physalis*.

ground hemlock. *Taxus canadensis*.

ground ivy. *Glechoma hederacea*.

ground lichen. *Parmelia molliuscula*.

groundsel. *Senecio*.

groundwater. *See* water.

groundwater level. groundwater table.

groundwater mining. heavy overdrafting, *q.v.*

groundwater surface. groundwater table.

groundwater table (water table; groundwater level; groundwater surface). the upper limit reached by groundwater in a given aquifer. Rock and subsoil voids (with the exception of air pockets) are full of water.

grove. **1:** a small patch, tract, or stand of trees characterized by little herbaceous growth. **2:** a planting of trees, such as an orchard (e.g., an orange, olive, or pecan grove). *Cf.* forest, woodlot.

growth. **1:** increase, especially an increase in size. Growth may take place by accretion (i.e., by the addition, accumulation, or adherence of materials directly from the environment, as in the growth of the shell of some living organisms). **2:** a tumor, cancer, growth, or swelling.

growth hormone. somatotrophin.

growth rate. the rate of positive change in size of a living organism, a component part, or a population. **absolute growth rate**. the actual rate of growth per unit time under a given set of conditions. **relative growth rate**. the rate of growth of one living system under a given set of conditions relative to another living system or to itself at another time or under an alternative set of conditions.

growth substance. auxin.

growthy. in livestock judging, descriptive of an animal that is large in size and well developed for its age.

grugru. **1:** any of several spiny-trunked, tropical feather palms, such as *Acrocomia spp.*, of tropical America. **2:** grugru (grugru grub; grugru worm). the larva of any of several South American billbugs (Family Cucurlionidae) that feeds on the pith of palm trees, sugar cane, etc. It is prized by some as food.

grugru grub. grugru (def. 2).

grulla. a dark, sooty, slate-blue coat color as that of the American quarter horse.

guajillo. *Acacia berlandieri*.

guajillo wobbles. *See Acacia berlandieri*.

guanin. guano.

guano. **1:** a thick terrestrial deposit of the feces of seabirds or bats that is often used as a fertilizer largely because of the high nitrogen and phosphorus content. Extensive deposits of guano are found near the ocean in Peru and form an important industry in which guano is "mined," bagged, and sold as manure, *q.v.* **2:** used without a modifier, this term is usually taken to mean accumulated excrement of birds. **bat guano** (bat manure). Guano produced by bats. Large deposits of guano can be found in certain caves and provide a valuable source of manure.

guaranteed analysis. a list of proximate nutrients, on a feed tag or label, that declares a guaranteed minimum or maximum percentage of each by weight. Fat, ash, crude fiber, and crude protein are usually included.

guardhair. coarse hair that grows through the underhair of many mammals. *See* fur.

Guernsey. a solid, well regarded breed of dairy cattle of that originated on the British island of Guernsey. These cattle are fawn-colored or reddish-fawn and white. They produce a milk of good color and high quality with a high percentage of butterfat.

guideline. a usually unenforced recommendation, direction, or set of instructions put forth by a government agency.

guinea pig. *See Cavia*.

Guineagrass. *Panicum maximum*.

Gulf Coast tick. *Amblyomma maculatum*.

gullet. **1:** esophagus. **2:** the cytopharynx of protozoans. **3:** any cavity through which food passes into an animal body.

gum opium. *Papaver somniferum*.

gumweeds. *Grindelia*.

gust. a sudden breeze, blast, or rush of air.

gut. **1a:** digestive portions of the alimentary tract. **1b:** sometimes used as a synonym for the entire alimentary tract; viscera. **2:** to eviscerate.

Guthion. azinphos methyl.

Gutierrezia (snakeweeds; tansy aster; broomsnakeroot). a genus of glabrous, often glutinous woody shrubs (Family Asteraceae). They are facultative accumulators of selenium from soil. ***G. microcephala*** (*G. sarothrae microcephala*; *Xanthocephalum lucidum*; *Xanthocephalum sarothrae*; broomweed; perennial snakeweed; slinkweed; turpentine weed). a densely branched perennial resinous woody shrub of dry rangelands of the United States from Colorado and Idaho, southern Texas, California, and Mexico. a species that contains the toxic abortifacient, saponin. Losses of livestock, especially sheep and cattle, caused by this plant are sometimes severe in the Trans-Pecos and High Plateau regions of Texas. Adult animals sometimes die from ingesting the plant, but the major impact is due to abortion. Signs of acute poisoning in adult cattle and sheep may include anorexia, sluggishness, diarrhea or constipation, rough coat, mucus in the feces, vaginal discharge, nasal discharge and peeling muzzle (in cattle), mild jaundice (mainly in sheep), and hematuria. Lesions are those of gastroenteritis, nephritis (sometimes severe), and hydropic degeneration of the liver with eventual necrosis. The uterus is edematous and the extraembryonic membranes are hydropic. ***G. sarothrae microcephala***. *G. microcephala*.

guttation. the exudation from plant leaves of liquid water.

guzzler. an area that has been prepared and surfaced to receive and store surface runoff for upland game birds.

gymnosperm. a vascular plant whose seeds are not enclosed in an ovary; a conifer or cycad. *Cf*. angiosperm.

gymnocarpous. angiocarpous.

gypsum. calcium sulfate.

Gymnocladus dioica (Kentucky coffee tree). a large, rough-barked deciduous tree (Family Fabaceae, formerly leguminosae) that is native to moist woods in the eastern and central United States. The raw seeds and fruit pulp are toxic, producing gastrointestinal symptoms such as vomiting and diarrhea with cardiac arrhythmia and coma. The toxic principle is cytisine.

Gymnodinium. a genus of subspherical dinoflagellates (See Dinoflagellata) with numerous variously colored chromatophores. The cell is marked by one annular and one posterior longitudinal groove. Both fresh water and marine species are known. ***G. brevis*** (= *G. breve*, *Ptychodiscus brevis*). a toxic dinoflagellate responsible for red tides, sometimes called gymnodinium blooms, with massive mortality among fish along the west coast of Florida and elsewhere. Millions of fish have been killed by these blooms during the past century and have been responsible for shellfish poisoning of humans. Onshore winds during such blooms may induce transient coughing in humans. Related species are *G. mikimatoi, G. splendens*, and *G. spectabilis*, all of which contain toxins.

gymnosperm. common name for plants of the Gymnospermae. *Cf*. angiosperm. **2:** the naked seed of such a plant (def. 1).

Gymnospermae. a subdivision of the Spermatophyta, gymnosperms are an ancient and diverse group of vascular plants that includes conifers, cycads, ginkos, and yews. The seeds are not enclosed in an ovary, but are borne on a sporophyll or, in the case of pteridosperms, on the frond. *Cf*. Angiospermae.

gynandromorph. an organism, usually an insect, that has both male and female body sectors. Such individuals usually result from the loss of one of the sex chromosomes of a female during cell division. These cells produce a lineage of cells having the XO configuration and thus produce male tissue. Cells that have not lost an X chromosome contain the paired XX configuration and produce female tissues. *Cf* intersex.

gynophyte. a female plant.

gyprock. any rock that a driller finds difficult to drill through.

gypsum (plaster of Paris; land plaster). the hydrated form of calcium sulfate.

$$CaSO_4{\cdot}2H_2O$$

It is a white, odorless, slightly water-soluble compound. Gypsum has numerous uses. It is used, for example, in growing certain legumes, especially peanuts; as a major component of normal superphosphate; and in agricultural irrigation systems, gypsum is used to increase the infiltration rate of water into the soil.

Gyromitra (false morels). a genus of ascomycetous, mostly toxic mushrooms (Family Helvelaceae) with a large, simple, stipitate fruiting body (ascoma) with a folded ascus-bearing cap. Members of this genus are closely related to and easily confused with those of *Helvella*. Poisonous mushrooms of both genera usually produce effects that are similar to those produced by amanitin, but are less severe. Symptoms appear (primarily gastrointestinal) in 6 to 12 or more hours following ingestion. Some species produce hemolysis and central nervous system effects. The active principles vary (See gyromitra toxin). Some humans are resistant. See also helvellic acid, gyromitrin. ***G. (Helvella) esculenta*** (conifer false morel; brain mushroom; beefsteak morel; lorchel; edible false morel; turban top). found throughout North America, but is most common in the north and at high elevations. It may be mistaken for the edible morel, but is deadly poisonous, producing serious effects and death on occasion. It is said to be edible in some locations, but doing so would be unwise. It is tumorigenic in laboratory animals. It contains gyromitrin and methylhydrazine, a volatile pyridoxine antagonist. The toxins can usually be removed (at least partially) by air drying or by boiling, rinsing, and reboiling. Symptoms appear abruptly ca. 6-7 hours following ingestion or inhalation of vapor from cooking; they include headache, malaise, abdominal fullness, and emesis without diarrhea. Recovery without treatment within 2-6 days is common. Fatal hepatic necrosis occurs in some cases. Actual poisonings are rare. See also helvellic acid. ***G. (Helvella) gigas*** (snowbank false morel; giant helvella). found on mountains in Colorado, California and the Pacific, usually near melting snowbanks. It grows to about 1 kg and is gathered and eaten in many locations, but poisonings have been reported. See also gyromitrin. ***G. (Helvella) infula*** (saddle-shaped false morel; hooded helvella). a poisonous species found throughout North America on or by rotten wood or woody debris. It contains helvellic acid. ***G. brunnea*** (*Helvella underwoodii*; gabled false morel). a mushroom with a brownish, convoluted, brain-like cap (about 10 cm wide) on a whitish, branched stalk; the cap may be saddle-shaped or lobed. It occurs throughout much of the eastern United States to Michigan, south and west to Tennessee, Missouri, and Oklahoma. It occurs on humus in deciduous forests. It is poisonous; no deaths have been reported, but the effects of intoxication include severe headaches, vomiting, diarrhea, and blood poisoning. See also gyromitrin. ***G. fastigiata*** (*G. korfi*; thick-stalked morel). closely related to *G. esculenta* and should be considered poisonous. It occurs in northeastern North America, the Rocky Mountains, and locally in the Pacific Northwest on the ground in deciduous or mixed forests. ***G. hydrazine esculenta***. poisonous; contains methyl hydrazine. ***G. korfi***. *G. fastigiata*.

gyromitrin (ethylidine gyromitrin; acetaldehyde- *N*-methyl-*N*-formylhydrazone). an extremely toxic hydrazone of acetaldehyde isolated from various species of *Gyromitra* and *Helvella* (e.g., *G. (Helvella) esculenta*, *G. (Helvella) gigas* and *G. brunnea* (= *H. underwoodii*). It occurs in association with a number of other compounds that are readily hydrolyzed to the proximate toxic metabolite, methylhydrazine, *q.v.*, that appears to be responsible for most or all of the toxic effects of gyromitrin. Symptoms in humans are similar to those produced by ingestion of *Amanita phalloides*.

gyttja. peat, usually formed in stagnant water from highly decomposed plant material, diatomaceous earth, feces, phytoliths, and other fine particles.

H

h. abbreviation of hecto-.

H. the symbol for hydrogen.

H^+. hydrogen ion.

h-. abbreviation for hector-.

habit. **1:** custom, practice, tendency; pattern, routine. **2:** obsession, fixation, addiction. **3:** the form, aspect, or pattern of growth and arrangement of parts characteristic of a plant (or of certain animals).

habitat (ece; oikos; oike). **1a:** the particular place (area) and type of environment in which a particular species or population of organisms (e.g., human, animal, plant, microorganism) naturally occurs, including the nature of the surroundings, both living and nonliving. *Cf.* ecological niche. **1b:** the natural abode of a species of plant or animal. **2a:** the particular abiotic environment in which an organism lives. **2b:** the set of conditions of the physical environment under which a specific type of living system can live and perpetuate itself. The habitat of a biotic community under this definition is termed a biotope. **3a:** habitation, milieux; residence. **3b:** a special environment designed to support life for an extended period of time in an otherwise inhospitable medium or location (e.g., an underwater research vessel; a spaceship. **aquatic habitat**. an underwater habitat where a variety of marine or freshwater flora and fauna occurs throughout much or all of the year (e.g., tide pools, estuaries, bogs, ponds). **critical habitat**. **1:** any habitat is critical to the continued existence of particular plant or animal. **2:** as defined by the Endangered Species Act: any area essential to the conservation of a listed species, though the area need not actually be occupied by the species at the time it is so designated. critical habitat must be designated for all threatened and endangered species under the Act (certain exceptions are specified). Wherever located, the federal government is required to protect such habitat. **critical wildlife habitat**. part or all of a specific location occupied by a particular population or species of wildlife that is recognized as essential to the continued existence of the population. **terrestrial habitat**. a habitat near the soil surface in which soil and air are the predominant media.

habituate. *See* acclimatize.

habituation. **1:** familiarization, acclimatization. **2a:** a gradual diminution or loss of response to a particular stimulus following frequent or continuous stimulation without reinforcement. Habituation can be due to a simple learning process, or can be physiologic in nature. **2b:** the process of becoming accustomed, adapted, or inured to a particular situation or environment (*See*, for example, acclimatization, vernalization).

hacienda. **1:** in the United States, the main dwelling or headquarters on a ranch. **2:** a ranch in the southwestern United States and the spanish-speaking nations of Central and South America.

haematuria. hematuria.

Haemostrongylus vasorum. *Angiostrongylus vasorum*.

hair. 1: a thread, filament, or thin fiber. **2:** a keratinized, filamentous integumental structure that is uniquely characteristic of mammals (Class Mammalia). **3:** pelage; fur; fleece, wool, the hair collectively that covers the body. A number of poisonous substances such as chromium, lead, mercury, selenium, arsenic, and PCB and other halogenated compounds are associated with hair. Such substances are sometimes present in higher concentrations in hair than in any of the internal organs of a poisoned mammal; hair generally contains the highest Hg level of any mammalian tissue. Hair is easily collected and relatively resistant to decay. It can be used to assess the levels and sometimes the history of exposure to halogenated compounds such as PCB, heavy metals such as chromium, lead and mercury, and the metalloid arsenic. The accumulation of heavy metals in the washed hair of mammals exposed to heavy metals can reach values one to two magnitudes greater than in control subjects. Serial segments of human hair have sometimes proven useful in determining the history of exposure, as in the case of the Minimata disaster.

hairy nightshade. *Solanum villosum*.

hairy-spiked grass. *Agropyron dasystachyum*.

half-sib. a half brother or half sister.

halide. **1:** halogen (def. 1b). a compound that contains any of the five highly electronegative elements that comprise group VII of the periodic table (bromine, chlorine, fluorine, iodine, and astatine). *See* halogen; *See also* antiknock agent, dalapon, fluorocarbon, hair, hydrocarbon (chlorinated hydrocarbon, halogenated hydrocarbon), insecticide (organochlorine insecticide), polybrominated biphenyl, polychlorinated biphenyl. **aryl halide**. an aromatic compound with two benzene rings that contain substituted halogens (usually chlorine). The two major classes of aryl halides are those based upon naphthalene and those upon biphenyl. Aryl halides have many uses, which have resulted in substantial human exposure and environmental contamination. Examples are monochlorobenzene, dichlorobenzenes, hexachlorobenzene, and bromobenzene. **hydrogen halide**. a halide such as hydrogen chloride, that has the general formula HX, where X is F (fluorine), Cl (chlorine), Br (bromine), I (iodine), or At (astatine). All are toxic gases.

half-life ($T_{1/2}$). **1:** the average period of time during which the mass, concentration, or activity of a substance decreases to one-half of the initial value in a given system (e.g., a living organism, the atmosphere, a population, community) under specified conditions. **2:** radioactive half-life; the time required for half of the atoms of a radioactive element to undergo decay. See also effective half-life, physical half-life. **atmospheric half-life**. half-life of a specified substance (e.g., a specified air pollutant, vapor, aerosol, microorganism) at a specified location or component of the air environment (e.g., an airshed, a room, a factory). **biological half-life. 1a:** the half-life of a substance within a specified living system as a result of biological processes (e.g., metabolism, excretion, inactivation). **1b:** the half-life of a total dose of a drug, toxicant, radionuclide, or test substance from a living organism. **ecotoxicological half-life**. half-life of a toxicant within a specified ecological system (e.g., a particular population of living organisms, community, or ecosystem). **effective half-life** (effective radioactive half-life). the half-life of a particular radionuclide within a biological system due to radioactive decay and other processes such as excretion. **pharmacologic half-life**. biological half-life as measured by the rate of loss of pharmacologic activity of a substance within a living system or component of a living system. **hamphysical half-life**. the half-life of a radioactive substance through radioactive decay alone. **radioactive half-life**. the half-life of a radioactive substance. See also effective half-life and physical half-life.

Haliotis (abalone). a genus of low, flattened gastropod mollusks, some of which delicacies, others are poisonous.

halite (rock salt). the mineral form (rock salt) of sodium chloride or the rock that consists chiefly of rock salt.

halogen. **1a:** any of the five highly electronegative elements that comprise group VII of the periodic table (bromine, chlorine, fluorine, iodine, and astatine) All are toxic and highly reactive producing a large number of compounds many of which are toxic. **1b:** a halogen-containing compound. These include interhalogen compounds that contain only halogen molecules. *See* halide.

Halogeton (halogeton; barilla). a small genus of herbs and shrubs (Family Chenopodiaceae) found on rangelands in the western United States and other arid regions of the world, they are indigenous to the Mediterranean region and Central Asia. They are poisonous to cattle and sheep because of the high content of soluble oxalates. ***H. glomeratus*** (halogeton, barilla). a small, coarse, annual, extremely toxic herb native to arid, alkaline soils and saline clays of Russia from the vicinity of the Caspian Sea to the steppes of northwestern China and southwestern Siberia. It was introduced from Siberia and has become naturalized in the southwestern United States where it occurs as a weed on disturbed or barren soils. Soluble oxalates may reach extremely hazardous levels in these plants (up to about 35% of dry weight). Livestock fatalities have been confined largely to sheep and are uncommon today. Symptoms are those of oxalate poisoning.

halogeton. common name of any plant of the genus *Halogeton*, especially *H. glomeratus*.

halophillic. descriptive of a plant that thrives in a salty environment. *See* halophyte.

halophobe. a plant that is intolerant of an environment high in salt.

halophobic. descriptive of a halophobe.

halophyte. **1a:** a plant of seashores and estuaries or marshes that have a high salt content. **1b:** a plant that thrives on soils of high salt content.

haloxon. an organophosphate anthelmintic. Clinical signs may occur in hypersensitive sheep 3—5 weeks following treatment, and may include ataxia and a symmetric, spastic partial paresis. The hind limbs in this case are partially flexed, often with the dorsal surface of the hoof on the ground. The condition of the animal may remain in this stage; prognosis is guarded.

hamada. an accumulation of stones on the floor of a desert due to the removal of finer material by wind or water erosion.

Hampshire swine. an American breed of meaty, medium-sized swine that originated in southern England. It is black with a white belt that encircles the front of the body including the front legs and feed. Major improvements of the breed have been made in Kentucky.

hamster. any small murine rodent (Subfamily Cricetinae, Family Muridae) of the genera *Cricetus, Cricelulus, Mesocricetus, or Phodopus*. They are widely used in research and as pets. They are granivorous and phytophagous; all store food and hibernate. They are easily maintained in captivity and can breed throughout the year under laboratory conditions. *Cricetus cricetus* is the most studied species and most common as a pet. It is native to temperate Eurasia, ca. 15 cm long, gray or brown, with round ears and a broad head.

hand-feed. to feed an animal or person by hand, usually at regular intervals. *Cf.* self-feed.

hand feeding (hand rearing). feeding an animal or person by hand, especially an orphaned animal, or one whose parent isincapable of or refuses to feed its young. This is accomplished variously by feeding them a formula or formulation using a bottle, or in some cases a spoon, wetted finger, or similar tools.

hand rearing. hand feeding.

hand. **1:** aid, assistance, help, relief. **2:** a farm worker, laborer; a hired hand. **3:** the standard unit of height of a horse; it is taken from the ground to the withers. One hand is 10.16cm (4 in). **by hand**. work done using only hand tools.

hank. a coiled loop of spun wool that is 560 yards long.

haploid. **1a:** of or pertaining to a cell of a eukaryote that has only a single set of chromosomes. *Cf.* diploid. **1b:** a haploid cell or organism. *See also* monoploid.

haploid cell. a haploid; a eukaryote cell that contains only one set of chromosomes. *See* diploid cell.

haploidy. the state of being haploid.

Haplopappus (synonym *Aplopappus*; rayless goldenrod; jimmy weed; burrow weed). a genus of mostly herbaceous plants (Family Compositaceae) of North and South America. Most species have alternate, rigid leaves and yellow flowers. ***H. venetus*** (coast goldenbush). all parts of the plant are toxic; ingestion by cattle causes trembles. Tremetol, the toxic principle, is excreted in milk. Suckling animals occasionally become poisoned even when the parent is asymptomatic. Humans may also become secondarily poisoned by drinking the milk of affected cattle; the disease in humans is called milk sickness. Toxic concentrations of nitrate have also been reported in this plant. *H. venetus* is closely related to *E. rugosum*.

hapten (haptene; haptin; incomplete antigen; partial antigen). **1:** a molecule that is unable, by itself, to cause the production of antibodies. It may combine (conjugate) covalently, however with an antigenic molecule (carrier molecule). **conjugated hapten** (conjugated antigen). the molecule produced by conjugation (*See* def. 1). Such a molecule may stimulate antibody production. **1: 2:** that portion of a conjugated hapten (def.1) that contains the group on which the antigenic specificity resides (the active site). *See also* antibody.

haptenic. pertaining to a hapten.

hard fescue. *Festuca duriuscula*.

hard liver disease. poisoning by *Amsinckia* of domestic cattle and swine. These animals do not show clear or intense signs f Amsinckia poisoning; consequently, diagnosis is generally made at the abattoir based on the appearance of the liver. See *Amsinckia intermedia*.

hard maple. common name of *Acer nigrum* and *A. saccharum*.

harden. **1a:** to set, become rigid, firm up, become solid, solidify, set, set up. **1b:** toughen, toughen up; temper; inure. **2:** acclimatize.

hardening of the arteries. arteriosclerosis.

Harder's gland. **1:** the deep gland of the semilunar conjunctival fold ("third eyelid") of, for example, swine and deer. **2:** sometimes incorrectly applied to the superficial gland of the semilunar conjunctival fold in *Canis*.

hardy. **1a:** rugged; robust; strong; vigorous. **1b:** denoting a plant or animal that can withstand severe environmental conditions such as low or high environmental temperatures.

hardy annual plant. *See* plant.

harrow. **1:** a farm implement that consists of a heavy frame with a series of teeth, spikes, tines, or upright discs. It is dragged across the surface of a plowed field break up and level off the ground and to destroy weeds, to prepare the seed bed for planting, cover already planted seeds. **2:** to use a harrow. **3:** to torment or cause great distress. **a-harrow**. a harrow with spiked teeth, shaped like a capital A. It is dragged by a hitch from its peak to level and smooth out plowed ground. **brake harrow**. a harrow used to break up large clods of earth left by plowing. **bush harrow**. a kind of light harrow made of bushes, for harrowing grass lands and covering seeds, or to finish the work of a toothed harrow. **disk harrow**. a harrow with two or more gangs of upright disks. It may be used to cut vegetable matter and other debris on the surface of the land or to churn up the top layer of the soil before or after plowing; this, for example, further prepares the ground for planting. Additional uses include to cultivate widely-spaced row crops; to cover broadcast seed; and to prepare the ground for summer fallowing. toothed harrow. a harrow made from pieces of timber or metal into a flat frame set with iron or wooden teeth and drawn over plowed land to level it, or for any of the other purposes indicated above.

Hart's thorn. *Rhamnus cathartica*.

harvested acres. The cropland actually harvested for a particular crop. **planted acres**. the acres planted for a particular crop, whether harvested or not.

harvester. *See* combine.

HAsc. acetylsalicylic acid.

haunch. the hind leg of an animal, especially games animals such as deer.

haw. a voice command to a horse or other beast of burden that directs it to turn to the left. *Cf.* gee.

hay. **1:** grass or other leafy plants that are ready for mowing, especially to be cured for fodder. **2:** dried herbage, such as grasses, alfalfa, or clovers, sometimes including the seeds, that is fed to livestock. **barn-dried hay**. hay dried indoors by blowing air through it. **batch-dried hay**. baled hay dried in groups or batches. In this process, groups of bales are placed on a platform and warm air is blown between the bales. **field-dried hay**. hay allowed to dry in the field after cutting. It is usually less nutritious than barn-dried hay. **grass hay**. hay composed chiefly or entirely of grasses. **green hay**. **1:** uncured hay. **2:** cured hay that retains a green color.

hay bacillus. *Bacillus subtilis*.

hay meadow. a field used to grow hay.

hayo. *Erythroxylum*.

hazard. **1a:** danger, peril, threat. **1b:** exposure to the possibility of loss or injury. **2:** the adverse effect (e.g., cancer, asphyxiation, cardiac arrest) that might result from a potentially harmful component of the environment or from a living system's own behavior. **3:** any environmental component, material, or energy source, (e.g., pollutant or toxic chemical, an explosive substance, ionizing radiation, open flame, heat, noise) that can adversely affect humans, other living systems, the environment, or property. **4:** the hazard of a toxic environmental chemical is a function of exposure and toxicity. This may be defined by the ratio PEC/NOEL, where PEC is the Predicted Environmental Concentration and the NOEL is the No-Observed-Effect Level (NOEL), which is the inverse of the "margin of safety." **imminent health hazard**. a significant threat or danger to health as inferred from evidence sufficient to demonstrate that a product, process, practice, circumstance, or event is of such a danger that immediate correction or cessation is necessary to prevent severe or potentially severe injuries or damage to health. **radiation hazard**. any hazard to public health or to any biotic system due to exposure or potential exposure to ionizing radiation.

hazardous. dangerous, perilous, threatening; exposed to, or concerned with danger.

hazardous material (HAZMAT). any substance capable of endangering life or causing serious injury or damage to humans or other life forms. The term applies broadly to toxic, readily flammable, or explosive materials; to those that emit ionizing radiation; and to any material that is dangerous to handle, use, store, or transport.

hazardous substance. **1:** a substance that can adversely affect living systems by causing illness, injury, malfunction, or a reduced quality of life. **2:** a substance that can cause death or irreparable harm to a living system. **2:** a substance declared hazardous by law, especially a substance that can cause death or irreparable harm. In the United States, the Environmental Protection Agency and the Occupational Safety and Health Administration have, for example, identified numerous chemicals as hazardous.

hazardous tree (danger tree). a tree or part of a tree that is structurally infirm to the point that there is a significant risk of falling and causing damage to persons or property.

hazardous waste. waste that contains hazardous substances of such kinds and amounts that they may cause or significantly contribute to serious illness or death of humans or other living organisms. About 80 million tons of hazardous waste are generated in the United States each year.

HAZMAT. acronym for hazardous material.

HCCH. hexachlorocyclohexane.

HCH. hexachlorohexane; hexachlorocyclohexane.

τ-HCH. **τ-hexachlorohexane** (*See* hexachlorocyclohexane).

HCN. hydrogen cyanide.

head down disease. milkweed poisoning of rabbits. See Asclepiadaceae.

heal. **1:** to return to a normal or healthy state. **2:** to cure, restore, rehabilitate, make whole or healthy.

health. **1:** fitness, vigor, well being; the condition or overall state of being of an organism when it functions normally with no evidence of disease or persistent distress. **2:** in the charter of the World Health Organization (WHO), 'health' is defined as not only an absence of infirmity or illness, but also a state of complete physical, mental, and social well-being.

healthy. in a state of health, fit, well; sound in mind and body; free from disease, pain, or injury; free from any condition of impairment of normal behavior or life functions.

heart. a muscular organ that rhythmically propels blood and maintains its flow through a circulatory system or a vascular network. The single heart of vertebrate animals is composed of cardiac muscle, which contracts rhythmically in response to an intrinsic pacemaker. Nervous control is thus indirect. In most invertebrates that have one or more hearts, cardiac contractions are initiated by the nerves of a heart ganglion.

health effect. *See* effect.

heartwood. the central woody core of a woody branch, stem, or tree trunk, composed of nonliving cells and no longer conducting water and dissolved minerals. It is usually differentiated from the outer layer of wood (sapwood) by its greater density and darker color. As the tree ages, the heartwood increases in diameter, while the sapwood remains about the same thickness. *See also* cambium.

heat. **1a:** warmth. **1b:** excitement. **2a:** estrus (in part). **2b:** the usually intense sexual receptivity of the female mammal that accompanies estrus.

heaves. vomiting (often applied to domestic animals).

heavy-metal tolerance. the ability of certain plants to grow in situations where the concentration of heavy metals is toxic to most plants. Such plants are usually able to detoxate the metal; some, however, do not absorb the metal from the medium in which it grows. Copper tolerance in some species of *Agrostis* is due to the ability of the plant to restrict copper to the roots; it is thus unavailable to the more sensitive shoots.

hector- (h-). a prefix in the SI and metric systems that means one hundred.

hegari. a cultivated grain sorghum; *See Sorghum vulgare*.

Helenium. a genus of annual or perennial shrubs or herbs (Family Compositaceae). A number of species are reported as toxic to livestock: *H. autumnale*, *H. hoopesii*, *H. microcephalum*, *H. nudiflorum*, and *H. tenuifolium*. ***H. hoopesii*** (= *Dugaldia hoopesii*; sneezeweed; orange sneezeweed). A stout, erect, perennial herb, up to 1 m tall, found on mountain slopes and valleys mostly from 2200—3200 m in the Rocky Mountain and Pacific Coast states except Montana and Washington. It contains dugaldin, *q.v. See also* sneezeweed poisoning.

Helianthus annus (common sunflower; wild sunflower). a large, erect, annual herb (Family Asteraceae) that may reach a height 4—5 m with huge flower heads. It is native to and widely distributed throughout the western prairies of North America and is extensively cultivated.

helio-. a prefix denoting or pertaining to the sun.

Heliothis zea (bollworm; corn earworm; tomato fruitworm). an insect (Family Phalaenidae) whose larvae feed on ears of corn, cotton bolls, tomatoes, alfalfa, beans, tobacco, and certain other cultivated plants.

heliotrope. common name of *Heliotropium europaeum*, not the garden heliotrope, *Valeriana officinalis*.

Heliotropium europaeum (heliotrope). a

poisonous annual weed (Family Boraginaceae) that contains hepatotoxic alkaloids of the senecio or pyrrolizidine type (e.g., heliotrine, lassiocarpine, and their *N*-oxides). The oxides are less toxic than the parent alkaloids, but resist digestion, are more soluble and may thus account for the severity of effects. Heliotrope poisoning decreases the metabolic activity of the liver, causing atrophic hepatosis with some fibrosis wherein liver cells increase in size, and are less viable with impaired regenerative ability. The most severe liver damage often occurs on lush, rich pastures. Death in such cases may result from ammonia intoxication due to the high nitrogen content of the forage and the reduced ability of the damaged liver to detoxicate nitrogen. Usually, however, animals may consume large amounts of heliotrope during a whole season with no signs of disease. Hepatic effects will have occurred, however, predisposing the animals to poisoning during the next season. Effects (e.g., loss of condition, changes in metabolism, changes in composition of blood, and death) are cumulative, progressive, and due entirely to liver atrophy. Sheep exhibit a distinctive syndrome with sudden hematogenous jaundice due to altered copper metabolism. It usually presents when sheep pastured for a season on heliotrope are then transferred to pasturage that supports a high intake of copper. Cattle are more sensitive than sheep, but are rarely poisoned. *See also Amsinckia intermedia*.

helitank. a specially designed tank used to transporting and dropping fire suppressants or retardants.

helitanker. a helicopter equipped with a helitank.

helitorch. a specialized drip torch slung and activated from a helicopter. The fuel is a gel.

hellebore. *Helleborus niger*.

helleborism. poisoning by plants of the genera *Veratrum* and *Helleborus*.

Helleborus (hellebore). a genus of European perennial herbs with fibrous roots (Family Ranunculaceae). Several species have been introduced in North America. All parts of the plant are toxic. Active principles are alkaloids such as veratrin. *Cf. Veratrum*. ***H. foetidus*** (stinking hellebore; bear's foot; setterworth). a large branching ornamental plant similar to *H. niger*, but with a bad odor. The powdered leaves have been used as a vermicide for children, but may violently affect or even kill them. ***H. niger*** (Christmas rose; hellebore; black hellebore; true hellebore). a very toxic evergreen plant with white flowers. It is native to Europe and Asia and widely cultivated elsewhere. It contains two highly toxic glycosides, helleborin and helleborein, which are violent irritants and cathartics; other toxic components are helebrin, protoanemonin, and certain saponins. All parts of the plant are neurotoxic and cathartic, especially the leaves. Effects on humans and livestock are similar. Ingestion may cause a burning sensation of the mouth and skin, blistering of the lining of the mouth, paresthesia, nausea, vomiting, gastric distress, hypotension, a weak pulse, and convulsions. It is rarely fatal. ***H. viridis*** (green hellebore; winter aconite). a poisonous evergreen plant similar to *H. niger*, but with greenish flowers tinged at the edges with purple.

helminthagogue. anthelmintic.

helminthic. anthelmintic.

Helminthosporium oryzae. a phytotoxic fungus that infects rice leaves. The toxic principle is ophiobolin A. *See* brown spot of rice.

helmintic. anthelmintic.

hematopoietic gland. a blood-forming or

gan, such as the spleen.

hematuria (haematuria). blood in the urine; often a sign of several nephritis.

heme. a complicated molecule containing iron in the ferrous state, serves as a coenzyme in a variety of biochemical processes. It forms an essential part of the structure of hemoglobin and participates intimately in the uptake and release of oxygen by this protein. Heme is an important part of the cytochromes, enzymes that catalyze the biochemical oxidations and reductions involved in the production of chemical energy in the form of ATP; heme is also associated with the various enzymes that catalyze the cleavage of peroxides.

hemi- a prefix denoting one-half; synonymous with the semi-, demi-.

hemoptysis. emptysis.

hemorrhagic jaundice. human leptospirosis (*See* leptospirosis).

hemorrhagic syndrome of cattle. bovine hemorrhagic syndrome.

hemosozic. antihemolytic.

henbit. *Lamium amplexicaule*.

HEOD. dieldrin.

hepatides. plural of hepatitis.

hepatotoxic (hepatoxic). **1:** toxic to the liver. **2a:** of or pertaining to a hepatotoxicant. **2b:** pertaining to the capacity of an agent to poison the liver, or to the action of such an agent.

hepatotoxicant. liver poison; any chemical that adversely affects the liver.

hepatoxic. hepatotoxic.

heptachlor (1*H*-1,4,5,6,7,8,8-heptachloro-3a,4,7,7a-tetrahydro-4,7-methanoindene). a highly toxic, white to light tan, waxy, water-insoluble, solid, chlorinated, diene insecticide, $C_{10}H_7Cl_7$, making up about 8% of technical grade chlordane. It is nephrotoxic and hepatocarcinogenic to various test animals; it is very toxic to laboratory rats. Effects on humans include irritability, hyperexcitability, tremors, and convulsions with intervening periods of depression, blood dyscrasias, late liver necrosis, and possibly severe gastroenteritis. It causes liver cancer in some laboratory animals. Its use is prohibited in the United States. *See* cyclodiene, heptachlor epoxide.

heptachlor epoxide. a degradation product of heptachlor, $C_{10}H_9CL_7O$, used as an insecticide. It is also the chief metabolite of heptachlor and is carcinogenic to some laboratory animals. *See* heptachlor, cyclodiene.

1*H*-1,4,5,6,7,8,8-heptachloro-3a,4,7,7a-tetrahydro-4,7-methanoindene. heptachlor.

heptenophos (phosphoric acid 7-chlorobicyclo-[3.2.0]hepta-2,6-dien-6-yl dimethyl ester; 7-chlorobicyclo-[3.2.0]hepta-2,6-dien-6-yl dimethylphosphate). a nonpersistent contact and systemic phosphate insecticide used as an ectoparasiticide in veterinary practice.

herb mercury. an herb of the genus *Mercurialis*.

herb. **1:** a nonwoody (nonlignified) annual seed plant. **2:** an aromatic plant or plant part used as a food flavoring, aromatic, or medication.

herb-Christopher. common name of any plant of the genus *Actaea*.

herbaceous. pertaining to or resembling an herb or a nonwoody plant stem.

herbage. the total usable vegetation available to a grazing animal such cattle or deer (e.g., palatable, nutritious, nonwoody herbaceous plants).

herbicide (weedkiller). **1a:** any substance or formulation intended for use as a plant regulator, defoliant, or desiccant. **1b:** a chemical agent (pesticide) used to kill or seriously inhibit the growth of noxious or otherwise undesirable plants. Herbicides are classified as contact or systemic, selective or nonselective, and preemergence or postemergence. Many herbicides are also toxic to animals if swallowed or absorbed through the skin. They are usually less toxic to animals than are insecticides and other pesticides used to control animals. Most of the livestock losses associated with herbicides are the result of improper or careless use. **amide herbicide**. any of a family of herbicides that inhibit seed germination and/or growth of seedlings, probably by blocking protein synthesis in the primary meristems. When applied to foliage, localized or general necrosis is seen at the loci of contact. **anilide herbicide**. any of a family of herbicides (e.g., alachlor, butachlor, propachlor) used as selective pre- and postemergence herbicides to control annual grasses and broadleaved weeds in crops such as maize, beans, cotton, radish, and brassicas. **antiauxin herbicide**. any of a number of growth-regulating compounds that are chemically related to auxins. Included are phenoxy acids such as 4-chlorophenoxy-*iso*-butyric acid (PCIB) and some benzoic acids (e.g., 2,3,5-triiodobenzoic acid (TIBA)). These compounds have effects on cell elongation that are generally opposite to those of auxins. They stimulate root elongation and inhibit that of coleoptile segments. Some, including PCIB, function as auxin antagonists and are thus true antiauxins. **aryl carbamate herbicide**. (aryl carbamic ester). any of a class of herbicides that are absorbed via the roots and have colchicine-like actions. Examples are chlorpropham, diallate, phenylurethan, and propham. They are generally toxic to monocotyledons but not to dicotyledons and are thus used to control grasses in crops such as peas and beets. The most prominent effects are inhibition of oxidative phosphorylation and of RNA and protein synthesis. Photosynthesis is also inhibited and the ATP content of tissues is reduced. **auxin herbicide**. any of a number of synthetic auxins (e.g., 2,4-D) used as herbicides. While these herbicides often cause dedifferentiation and the initiation of cell division in mature cells, they usually inhibit cell division in the primary meristems of intact plants. **carbamate herbicide**. any of a number of carbamates used as herbicides (e.g., propham, chlorpropham, and barban). They act selectively on monocotyledonous seedlings in which they strongly inhibit cell division by interrupting mitotic cycles in root and shoot meristems. **contact herbicide**. one that damages or kills a plant on contact. **hormone-type herbicide**. auxin herbicide. **nitrophenol herbicide**. nitrophenols were the first organic chemicals used in weed control. All are selective herbicides. Some are active on contact, others by absorption through the roots. **nonselective herbicide**. an herbicide that acts on a wide range of plant species. **organoarsenical herbicide**. any organoarsenical used as an herbicide. They cause chlorosis, cessation of growth, and progressive browning followed by dehydration and death. Buds fail to sprout. **phenolic herbicide**. any of a number of phenolic compounds used as herbicides. At high rates of application, they destroy plant membranes. This produces water-logging of the leaves followed by desiccation and death of the plant. **pyrimidine herbicide**. a pyrimidine that acts almost exclusively to inhibit photosynthesis and has little effect on nonphotosynthesizing organisms. **selective herbicide**. an herbicide that acts on one or a few types of plants. **substituted benzoic acid herbicide**. a family of herbicides that includes 2,3,6-TBA, chloramben, and dicamba. Their auxin and herbicidal activity greatly resembles that of the phenoxy auxins. Some of these compounds interfere with geotropic and phototropic curvatures. They also inhibit cell elongation. **substituted urean herbicide**. any of a number of rather selective preemergence herbicides that are absorbed by foliage and roots (e.g., diuron, fenuron, fluometuron, linuron, monuron, neburon, and siduron). Some penetrate deeply into soil and are quite persistent. The half-lives of these compounds in the soil vary considerably. Some penetrate deeply into the soil and are quite persistent. Such herbicides can thus be used to control weeds for extended periods of time. **systemic herbicide**. one that is absorbed and transported internally, killing the plant by acting on tissues that may be remote from the point of contact. **thiocarbamate herbicide**. any of a small family of herbicides with the characteristic group —CS·SS·CS—. They are toxic to germinating seeds. Uptake is via roots and through underground portions of the shoot system. Most are volatile and nonpersistent. Thiocarbamate treatment of grasses causes abnormal emergence of leaves from the coleoptiles. High doses can inhibit leaf emergence completely. Broad-leaved weeds become necrotic, with a consequent suppression of growth. Thio- and dithiocarbamates alter or interfere with a variety of processes such as photosynthesis, respiration, oxidative phosphorylation, protein synthesis, and nucleicacid metabolism. **translocated herbicide**. an herbicide that is distributed throughout the plant following absorption through the leaves or roots. **triazine herbicide**. **1a:** any triazine compound used to broadly inhibit plant growth. The leaves of treated plants become chlorotic, necroses develop, and both photosynthesis and respiration are inhibited with consequent retardation of growth. **1b:** any of a class of herbicides used to control annual grasses and broadleaf weeds. Included are atrazine, prometon, prometryn, propazine, simazine, and triazine; the acute oral LD_{50} in laboratory rats ranges among these herbicides from less than 100 to about 7000 mg/kg body weight. Triazine herbicides have a symmetrical triazine structure. These compounds function as herbicides by inhibiting the Hill reaction in photosynthesis. Overdosage in humans may cause abdominal pain, impaired adrenal function, anemia, dermatitis, diarrhea, eye irritation, mucous membrane irritation, nausea, disturbed thiamine and riboflavin function, or vomiting.

herbicide blue. Agent Blue.

herbicide orange. Agent Orange.

herbicide purple. Agent Purple.

herbicide resistance. *See* resistance (herbicide resistance).

herbicide-resistant. pertaining to plants that continue to propagate when exposed to herbicidal substances. *Cf.* insecticide-resistant, pesticide-resistant; herbicide tolerance.

herbicide tolerance. the inherent or natural ability of an individual organism, population, or species to survive and produce viable offspring following herbicide treatment.

herbicide white. Agent White.

herbivore. a phytophagous animal; one that feeds mainly or exclusively on vegetation.

herbivority. the quality or state of being herbivorous.

herbivorous. phytophagus; feeding on vegetation.

herd. **1:** a group of animals of a particular kind that remain together for a substantial period of time, usually feeding or traveling together. Other terms are often applied to specific types of animals or to special circumstances (e.g., drove, flock, gaggle, raft, remuda). a herd of cattle; a herd of sheep; a herd of zebras. **1a:** the animals, collectively, of a type owned or managed (e.g., on a farm or ranch), as a dairy herd. **2:** sometimes said disparagingly of a large group of people (herd, masses, rabble). **3:** one who cares for a herd (e.g., a cowherd, goatherd, herdsman, shepherd). **4a:** to control or have charge of a large group or herd. **4b:** to tend, lead (cattle, sheep, etc.), or drive (to a destination). **5:** to form a herd or associate as a herd. **mixed herd**. herd; any group or aggregation of wild or domestic mammals of more than one type or species. **trail herd**. a herd of cattle driven along a trail to a destination (especially to market).

herd average. in dairy farming, the average milk and component production per year per cow in a herd or designated herd unit.

herd's grass. *Agrostis alba*.

herding dog. any of a number of breeds of dogs bred originally for herding livestock (e.g., Belgian sheepdog, collie, German shepherd).

heredity. **1:** the transmission or mechanism of transmission of specific genetically based traits from parent to offspring. **2:** the genetic constitution of an individual.

heritage area. any site designated as of historical, architectural, archaeological, paleontological, or scenic significance to the people within a particular governmental jurisdiction (e.g., state, province, nation).

heritage trail. a trail having cultural significance due to established aboriginal use or use by early immigrants.

heron's-bill. *Erodium*.

heteroauxin. former name of 3-indolacetic acid.

Heterobasidiomycetes. *See* Basidiomycetes.

heteroploid. aneuploid.

heteroploidy. aneuploidy.

heterotroph. heterotrophic organism.

heterotrophic (organotrohic; zootrophic). **1a:** pertaining to a mode of nutrition in which nourishment is obtained from exogenous organic material (e.g., plants or plant materials, animals or materials of animal origin, detritus). **1b:** descriptive of organisms that cannot synthesize organic compounds from inorganic substrates (*See* nutrient).

heterotrophic organism. an organism that is incapable of growth and maintenance on purely inorganic media. One that obtains essential nourishment from exogenous organic material. *See* heterotrophic, autotrophic organism.

hexabutyldistannoxane. tributyl tin, tributyltin compound.

τ-hexachlor. *See* τ-hexachlorocyclohexane, lindane.

1,2,3,4,7,7-hexachlorobicyclo[2.2.1]-2-heptene-5,6-bisoxymethylene sulfite. endosulfan.

hexachlorocyclohexane (1,2,3,4,5,6-hexachlorocyclohexane; benzene hexachloride; BHC; HCH; HCCH; TBH). a chlorinated benzene, $C_6H_6Cl_6$, in the form of a white or yellowish, water-insoluble powder or flakes with a musty odor. There are eight stereoisomers having different orientations of the H and Cl atoms; physical properties and toxicity vary among them. Do not confuse with hexachlorobenzene. **τ-hexachlorocyclohexane** (1α,2α,3β,4α,5α,6β-hexachlorocyclohexane; τ-HCH; γ-benzene hexachloride; gamma-benzene hexachloride; gamma-hexachlor; gamma-hexane). an environmentally persistent insecticide and a constituent of a wide variety of agricultural, medical, and veterinary products. It is the most toxic isomer of hexachlorocyclohexane and the chief toxic component of lindane. It is the only isomer that is insecticidal and is also highly toxic to humans and to certain livestock.

1,2,3,4,5,6-hexachlorocyclohexane. hexachlorocyclohexane.

1α2α3β4α5α6β-hexachlorocyclohexane*See* lindane.

hexachlorocyclopentadiene. a very toxic, highly reactive, cyclic alkenyl halide with two double bonds that readily undergoes substitution and addition reactions. It is a light yellow liquid (f.p. 11°C, b.p. 239°C, density 1.7 g/cm^3) with a pungent odor. It was originally used as an agricultural fumigant and as an intermediate in the manufacture of insecticides (e.g., of Mirex and Kepone). Hexachlorocyclopentadiene is an irritant of the skin, eye, and mucous membranes; is strongly lacrimating and is damaging to most vital organs such as the adrenal glands, brain, heart, kidney, and liver.

1,3,4,5,5-hexachloro-1,3-cyclopentadiene dimer. mirex.

hexachlorocyclopentadienedimer. mirex.

1,2,3,4,10,10-hexachloro-6,7-epoxy-1,-4,4a,5,6,7,8,8a-octahydro-*endo, endo*-1,4:5,8-dimethanonaphthalene. endrin.

1,2,3,4,10,10-hexachloro-6,7-epoxy-1,4,-4a,5,6,7,8,8a-octahydro-endo,exo-1,-4:5,8-dimethanonaphthalene. dieldrin.

1,2,3,4,10,10-hexachloro-1,4,4a,5,8,8a-hexahydro-1,4:5,8-dimethanonaphthalene. the active substance in aldrin, *q.v.*

1,2,3,4,10,10-hexachloro-1,4,4a,5,8,8a-hexahydro-1,4:5,8-dimethanonaphthalene. the active substance in aldrin, *q.v.*

6,7,8,9,10,10-hexachloro-1,5,5a,6,9,9a-hexahydro-6,9-methano-2,4,3-benzodioxa-thiepin-3-oxide. endosulfan.

hexachlorohexane (HCH). Several isomers are environmentally persistent with varying pharmacological properties. **τ-hexachlorohexane** (τ-HCH). an environmentally persistent insecticide, it is a constituent of a wide variety of agricultural, medical, and veterinary products. *See* especially lindane.

(1aα,2β,2aα,3β,6β,6aα,7β,7a)-3,4,5,6,9,9-hexachloro-1a,2,2a,3,6,6a,7,7a-octahydro-2,7:3,6-dimethanonaphth[2,3-b]-oxirene. dieldrin.

3,4,5,6,9,9-hexachloro-1a,2,2a,3,6,6a,7,7a-octahydro-2,7:3,6-dimethanonaphth-[2,3-b]-oxirene. endrin.

hexachlorophene (2,2′-methylenebis[3,4,6-trichlorophenol]; 2,2′-dihydroxy-3,3′,5,5′,6, 6′-hexachlorodiphenylmethane; bis-(3,5,6-tri-chloro-2-hydroxyphenyl)methane). a polychlorinated bisphenol, it is a white, crystalline, odorless, water insoluble powder,

$$(C_6HCl_3OH)_2CH_2$$

Hexachlorophene has been used chiefly as a bacteriostat in cosmetics, toothpastes, dermatologicals, and antiinfective germicidal soaps. Hexachlorophene has also been used as an agricultural fungicide and bactericide with applications in veterinary medicine. It is especially effective against Gram-positive bacteria. It is a photocontact allergen and is neurotoxic to laboratory animals and to humans, causing central neurological lesions. It is toxic when applied repeatedly to the skin, and can be lethal, especially in the case of infants. Symptoms of intoxication may include confusion, listlessness, and convulsions. Diffuse status spongiosus of the reticular formation is a common lesion. Hexachlorophene is teratogenic at least in the case of pregnant women who routinely use it.

Hexadon. a trade name for lindane.

hexadrin. endrin.

hexahydro-3a,7a-dimethyl-4,7-epoxyisobenzofuran-1,3-dione. cantharidin.

3,4,5,6,9,10-hexahydro-14,16-dihydroxy-3-methyl-1*H*-2-benzoxacyclotetradecin-1,7(8H)-dione. zearalenone.

2,3,6a,8,9,9a-hexahydro-9a-hydroxy-4-methoxycyclopenta[c]furo[3′,2′:4,5]-furo[2,3-h][1]benzopyran-1,11-dione. aflatoxin M_2

1,2,3,4,5,6-hexahydro-1,5-methano-8H-pyrido-[1,2-a][1,5] ciazocin-9-one. cytisine.

2,3,6aα,8,9,9aα-hexahydro-4-methoxycyclopenta[c]furo-[3′,2′:4,5]furo[2,3-h][1]benzopyran-1,11-dione. aflatoxin B_2.

3,4,7aα,9,10,10aα-hexahydro-5-methoxy-1H,12H-furo[3′,2′:4,5]-furo[2,3h]-pyrano[3,4-c][1]benzopyran-1,12-dione. aflatoxin G_2.

hexahydropyridine. piperidine.

2,3,14,20,22,25-hexahydroxycholest-7-en-6-one. β-ecdysone.

1,1′,6,6′,7,7′-hexahydroxy-5,5′-diisopropyl-3,3′-dimethyl[2,2′-binaphthalene]-8,8′-dicarboxaldehyde. gossypol.

1,3,4,6,8,13-hexahydroxy-10,11-dimethylphenanthro[1,10,9,8-[*opqra*]-perylene-7,14-dione. hypericin.

4,5,7,4′,5′,7′-hexahydroxy-2,2′-dimethylnaphthodianthrone. hypericin.

hexamethylmelamine. altretamine.

***N,N,N′,N′,N′′,N′′*-hexamethyl-1,3,5-triazine-2,4,6-triamine**. altretamine.

τ-hexane. τ-hexachlorocyclohexane.

***n*-hexane**. a colorless, volatile, flammable liquid, with a faint odor. It is a six-carbon alkane,

$$CH_3(CH_2)_4CH_3$$

distilled from petroleum and is used chiefly as a solvent in the extraction of oils from seeds (e.g., of cotton and sunflower seeds), as a paint diluent. Together with other more polar solvents, it is used in the separation of fatty acids. It is moderately neurotoxic. Major symptoms of intoxication in humans with acute, very high rates of exposure are those of CNS depression. Chronic exposure of humans may result in a sensorimotor peripheral axonopathy. The proximate toxicants are metabolites of *n*-hexane (e.g., Hexan-2,5-dione, acetonyl acetone, and methyl *n*-butyl ketone). *See also* hydrocarbon (aliphatic hydrocarbon), axon-opathy.

hexaploid. **1a:** having or pertaining to six homologous haploid sets of chromosomes. **1b:** a hexaploid cell or organism, one that has six sets of homologous chromosomes per cell except of the germ cells.

Hexapoda. Insecta.

2,3,14,20,22,25-hexahydroxycholest-7-en-6-one. β-ecdysone. *See* ecdysone.

Hg. the symbol for the element mercury.

HHDN. aldrin.

high blood pressure. *See* hypertension.

high smut. bunt.

highgrading. removal of only the best trees from a stand, often leaving a residual stand of trees of low quality and value.

highland moccasin. *Agkistrodon contortrix*.

highlead system. a logging system that uses cables rigged to a spar high above the ground so that one end of a log can be lifted during yarding.

Hilaria mutica (tobosagrass). a grass (Family Gramineae) that is sometimes a host to *Claviceps cinerea*. Cattle sometimes contract ergotism from grazing on this grass. *See Claviceps*, ergot, ergotism.

hippo. ipecac (def. 2).

Hirschfeld's bacillus. *Salmonella hirschfeldii*.

histamine (1*H*-imidazole-4-ethanamine; 4-amino-ethylglyoxaline; 2-(4-imidazolyl)-ethylamine; 4-imidazoleethylamine; 5-imidazole-ethylamine; β-aminoethylimidazole; β-aminoethylglyoxa-line). An amine derivative of histidine, it is an organic base, $C_5H_9N_3$, that occurs naturally in living tissues and also as a product of decay. In vertebrate animals, histamine concentrates especially in the lungs, in mast cells and other connective tissue cells, and basophilic blood cells. *See also* ergot.

histo-. a prefix indicating or pertaining to tissue.

histogram. a graph or diagram on which the value of each of a set of variables is depicted by the height of vertical bars.

histohypoxia. an abnormally low concentration of oxygen in a tissue or tissues.

histology (histomorphology; microscopic anatomy). **1:** the science that deals with the identification and study of the microscopic structure of tissues, including the study of intracellular features. It is often of critical importance in the diagnosis of disease. **2:** the organization and cellular structure of tissues.

histomorphology. histology.

histopathological. **1a:** pertaining to the science of histopathology. **1b:** of or pertaining to pathological tissues or their condition.

histopathology. the study of pathologic histology; the microscopic study of abnormal, diseased, or damaged tissue.

histotoxic anoxia. *See* anoxia.

HMM. altretamine.

hog. swine.

hog's potato. *Zigadenus venosus*.

hog-wallow land. a colloquial term denoting flat land that is distinguished by mounds and natural microbasins; relief is often less than 1 m (3 ft).

hog-nosed vipers. snakes of the genus *Bothrops*.

hognose snake. *Heterodon contortrix*.

holiday disease. azoturia.

holism. the conception of living organisms and other complex systems as a functioning whole.

holistic. viewing, relating to, and treating living organisms and other complex dynamic systems as an integrated functioning whole.

holocoenosis. ecosystem.

holocrine gland. a gland whose secretion consists of its own disintegrating cells of the gland itself (e.g., sebaceous gland). *Cf.* merocrine gland.

holocrine secretion. Secretion, as in the case of sebaceous glands, that does not appear to be under nervous control.

Holostei. a subdivision of the Actinopterygii comprised today only of the bowfin and garpike.

home demonstration agent. *See* county agent.

homeostasis. 1: the dynamic steady internal state or equilibrium of the internal environment of a healthy biological system, or a part of such system. **2:** the maintenance of a relatively steady state within a living organism through regulatory mechanisms. In humans, a nearly constant body temperature is maintained by dynamic mechanisms, as are acid-base balance, blood pressure, pH, and numerous other physiological variables. **3:** the tendency to stability of mature ecological systems.

homeotherm. endotherm.

homeothermic. endothermic.

Homobasidiomycetes. *See* Basidiomycetes.

homogenate. a homogenized material.

homogeneity (homogenicity). the state of being homogenous.

homogeneous. comprised of similar ingredients or otherwise having a uniform quality throughout.

homogenicity. homogeneity.

homozoic. pertaining to the same animal or to animals of the same species.

honey. *See Apis mellifera.*

honey mushroom. *Armillariella mellea.*

honeybee, honey bee. *See Apis.*

honeycomb moth. *See Achroia grisella* and *Galleria mellonella.*

hookworm. 1: a parasitic nematode (Superfamily Strongyloidea), especially *Ancylostoma duodenale* or *Necator americanus.* **2:** hookworm disease.

hookworm disease (uncinariasis). **1:** the condition caused by hookworm infestation of the intestinal tract. **2:** ancylostomiasis.

hoose. 1: *Dictyocaulus.* **2:** verminous bronchitis (*See* bronchitis).

hop medic. *Medicago lupulina.*

hop sage. any plant of the genus *Grayia.*

hopclover. *Medicago lupulina.*

Hordeum vulgare (cultivated barley). a cultivated grain (Family Gramineae), often infected by any of a number of fungi. Concentrations of nitrate that are toxic to herbivores have also been reported. Grain infected with *Gibberella saubinetii* (the imperfect stage = *Fusarium graminearum*) is toxic to some animals. Infected seeds are shrunken, may have bluish-black tips, and may confer a darkened color to seed mixtures in which they occur. Even small quantities of infected seed cause intoxication and vomiting in domestic swine. Fatalities among swine are rare, however, because they do not consume heavily infested grain. Horses and mules will not normally eat scabby barley, and ruminants such as cattle can consume heavily infected grain without toxic effects. Laboratory rats are sensitive; guinea pigs are highly resistant.

Toxic concentrations of nitrate have been reported. *See also Claviceps*.

horizon. **1:** edge, vista; sphere, frontier; domain. **2:** soil horizon, *q.v.*

hormone. **1a:** a substance that is produced by one tissue or organ (usually an endocrine gland) directly enters the cardiovascular system, through which it is transported to one or more remote tissues or organs. The hormone stimulates one or more specific physiological effects that are characteristic of the affected target tissue or organ. These effects are often regulatory in nature, and control many vital bodily functions. **2:** phytohormone, *q.v.* Phytohormones include auxins, gibberellins, and cytokinins.

host. **1:** the organism into which a gene from an other organism has been transplanted. **2:** a living organism that is infected by or parasitized by another organism.

hornet stinger. *See* stinger.

horse. **1:** a common name of *Equus caballus* and *E. przewalski*. **2:** a common name of any of a number of hoofed, herbivorous mammals of the genus *Equus*. **3:** a fully mature horse (*E. caballus*); a stallion. **4:** any member of the family Equidae. **5:** to provide with a horse. **6:** a fabricated or natural device or structure (e.g., a hobby horse) that resembles a horse (def. 1) or one on which a person rides or sits as if seated on a horse (def. 1). **7:** trestle; a frame or device, usually made of wood or metal, with four legs, on which materials are mounted or supported (used, for example, in carpentry). **8:** argot for heroin. **9a:** a horse; mounted. **9b:** cavalry; mounted soldiers, as in a troop of horse. **10:** of, pertaining to a horse, or mounted on a horse. **11:** to be in heat; said of a mare. **12:** large or crude (e.g., horse pills). **13:** to joke, to play, horse around. **aged horse**. **1:** commonly defined as a mature horse, one eight years of age or more. **2:** a smooth-mouthed horse; one 12 years of age or older. **American light horse**. pinto horse. **American quarter horse** (quarter horse). a working saddle horse, used especially as a cow pony in the southwestern United States It is large, with a short back, usually weighs more than 1100 pounds, and combines speed and endurance. **blood horse**. a purebred horse. **buttermilk horse**. a roan of a color between red and blue. **cow horse** (cow pony). a horse used to herd cattle and sometimes other livestock, usually on rangelands in the western United States **cow pony**. cow horse. **domestic horse** (modern horse). *E. caballus*. **draft horse** (work horse). any breed of horse that is suited to or used for drawing heavy loads. Draft horses originated in central Europe prior to the Roman invasions. Popular breeds include the extremely large Belgian, Clydesdale, Percheron, and Shire horses, all of which are noted for their endurance, strength, and even temperament. **light horse**. any breed of horse used chiefly for riding or for light work. Such horses arose in the Middle East and North Africa and modern breeds of light horse trace their origins to the Arabian horse, usually through the thoroughbred. Light horses areclassified according to training their training and usage (e.g., racers, trotters, riding horses, and cow horses (cow ponies). Further examples include American saddle horse, Appaloosa, Morgan, palomino, Pinto, quarter horse, and Standardbred, and most ponies. **modern horse**. domestic horse. **pinto horse** (American light horse). a horse with a coat of distinctive, large, irregular color markings, very dark or black and white. This color pattern paint. Pintos were favored by buffalo hunters of the North American Great Plains. Pinto coloring may occur in various breeds, but a pinto breed has been recognized in the United States since 1963. It is a popular general-purpose saddle horse. **quarter horse**. *See* American quarter horse.

saddle horse. a horse bred and/or trained for riding. **wild horse**. **1a:** a vernacular term applied to feral domestic horses, *E. caballus*, or their descendants. **1b:** Przewalski's horse. **work horse**. draft horse.

horse bean. *Vicia faba*.

horse bot fly. *Gasterophilus intestinalis*.

horse briar. a common name of *Smilax rotundifolia*.

horse brush. tetradymia.

horse chestnut. common name of *Aesculus*.

horse malaria. equine infectious anemia (*See* anemia).

horse-biting louse. *Bovicla equi*.

horsebean. *Parkinsonia aculeata*.

horsechestnut. common name of *Aesulus hippocastanum* or the nuts that it produces.

horsefly weed. *Baptisia tinctoria*.

horseguard. *Bembix carolina*.

horseradish. **1a:** *Amoracia lapathifolia*. **1b:** the taproot of *A. lapathifolia*, used as a condiment; it has a sharp flavor.

horsetail. a common name of *Equisetum*.

horsetail fern. a common name of *Equisetum*.

homologous chromosome. either of a pair of chromosomes, in a diploid cell, that have the same linear arrangement of genes. Chromosomes that have the same linear arrangement of genes.

host. an organism that contains or supports another organism.

house fly. any of about 30 species of flies of the genus *Musca*, but more strictly applies to *Musca domestica*, *q.v.*

house rat. *Rattus norvegicus*.

household ammonia. ammonia water.

household bleach. hypochlorite solution.

5-HT. serotonin.

Huanaco coca. *Erythroxylum coca*.

human brucellosis. undulant fever.

humanoid. human-like. *Cf.* anthropoid.

humification. the formation of humus by decomposing organic matter.

humin. the organic matter that remains following extraction of soil with dilute alkali.

humus. **1a:** well-decomposed, relatively stable, dark-brown to black material in the lower organic layer of aerobic soils. It is a complex aggregate of amorphous colloidal organic substances, formed during microbial decomposition and other transformations of plant and animal residues due to the activities of various soil organisms. Humus increases the water-absorbing capacity of soil and contains nutrients that support plant growth. Its chief constituents are derivatives of lignins, proteins and cellulose. **1b:** finished compost. **acid humus**. humus with a pH below 7.0. **raw humus**. a form of humus comprised chiefly of well preserved, though often fragmented plant remains with few fecal pellets.

hundredweight (cwt.). a standard unit of measure of certain agricultural products such as milk, rice, and some livestock. In the United States, a short hundredweight equals 100 lb (45.4 kg); in the United

Kingdom, a long hundredweight weighs 112 lb avoirdupois (50.8 kg).

hunger. **1:** appetite; desire or craving, especially for food. **1:** lack of nutrition; starvation, famine.

hunter's bag. *See* bag limit.

husbandry. **1:** in the broadest sense, farming or agriculture. **2a:** the cultivation and production of marketable crops or livestock. **2b:** the science of raising crops or livestock for food. **3a:** thrift, prudence, frugality, conservation. **3b:** prudent or economically sound management. **4:** management of resources.

husk. **1:** the dry outer covering of a cereal grain. It has little food value and is removed during threshing. **2:** dehusk; to remove the husk of seeds. **3:** a form of bronchitis, *q.v.*, caused by lungworms. *See Dictyocaulus*.

husk tomato. *Physalis*.

hyacinth. *See Hyacinthus orientalis*.

Hyacinthus orientalis (hyacinth). a flowering bulbous plant (Family Liliaceae) that is indigenous to Asia Minor, but is cultivated as an ornamental worldwide. The bulb is toxic, producing severe gastrointestinal distress. It is a source of benzyl alcohol.

hybrid. an offspring of two parents that differ with respect to at least one phenotype. **1b:** the result of a cross between two different cultivars, strains, populations, or species.

hybrid vigor. a frequent phenomenon in which the hybrid produced by crossing two pure bred lines is more vigorous or successful in some way than either of its parents. This is presumably due to increased heterozygosity of the offspring.

hybridization. **1:** the crossing (e.g., by breeding) of genetically different phenotypes to form a hybrid, *q.v.* **DNA-DNA hybridization**. the formation, naturally or artificially, of a duplex DNA molecule by complementary base pairing between two nucleic acid strands derived from different sources. This is the reverse of denaturation (def. 2).

Hydrangea (hydrangea). a genus of deciduous shrubs (Family Saxifragaceae) native to Asia and North and South America. *H. arborescens* (hydrangea; wild hydrangea; seven bark). a North American shrub, growing to more than three 3 meters tall. Several varieties occur in various habitats in the United States including rich woods, banks of streams, and calcareous rocky slopes. It is widely cultivated as a hedge plant. ***H. macrophylla*** (hydrangea, hortensia, hills of snow, snowballs). a common horticultural shrub with large, showy, white, pink, or blue flowers. It is indigenous to Japan, but now occurs in Europe and most of the Western Hemisphere. All parts, especially the flower buds, are deadly toxic. ***H. paniculata*** (paniculate hydrangea). a shrub or small tree with pubescent branchlets. The leaf blade is long and is oblong to elliptic-ovate with elongate panicles.Both ornamental and wild varieties occur in the United States.Wild forms occur chiefly in wooded swamps or along roadsites.All contain hydrocyanic acid and are extremely poisonous to livestock. ***H. quercifolia*** (oak-leaved hydrangea). the leaves are large, coarsely lobed,and have elongate panicles. Both ornamental and wild varieties occur in the United States. They contain hydrocyanic acid and are extremely toxic to livestock. Ornamental forms often escape cultivation.

hydrangea. common name of *Hydrangea spp*.

hydrargyrum. the element mercury.

hydro-, hydr-. prefixes of or pertaining to water.

hydrobiont. an organism that lives chiefly in water.

hydrocarbon. any organic compound composed entirely of carbon and hydrogen atoms (but see derivative hydrocarbon). The carbon atoms may form a straight, or branched, chain or a closed ring. Hydrocarbons are the chief constituents of crude oil, natural gas, and refined petroleum products. Hydrocarbons are prominant among air pollutants; they are released in large amounts into the atmosphere from combustion of fossil fuels used in industrial processes, transportation, and the production of electric power. These hydrocarbons directly injure and often destroy plant and animal populations and even habitats worldwide by means of both routine and large oil spills, especially in marine and coastal environments and by polluting the air. Indirect damage worldwide may include greenhouse effects that results from combustion of fossil hydrocarbon fuels. Many hydrocarbons are toxic and/or carcinogenic to humans, livestock, and wildlife. *See also* herbicide. **acetylene hydrocarbon**. *See* alkyne. **alicyclic hydrocarbon** (alicyclic compound; cyclane; alicyclic). a cyclic aliphatic hydrocarbon consisting of three or more carbon atoms (e.g., cyclohexane) that form a single ring. They combine the properties of straight- or branched-chain aliphatic and cyclic nonaromatic hydrocarbons. *See also* alicyclic. **aliphatic aromatic hydrocarbon**. any aromatic hydrocarbon in which an aliphatic chain replaces one or more hydrogen atoms on an aromatic ring. **aliphatic hydrocarbon**. any straight- or branched-chain hydrocarbon (branches when present are essentially straight). Included are noncyclic alkanes (paraffins), alkenes (olefins; alkadienes), and alkynes, *q.v.* Alkanes are saturated and comparatively unreactive compounds; alkenes are unsaturated compounds that contain one or more double [C=C] bonds; and alkynes are unsaturated, containing one or more triple [C≡C] bonds. The C-1 to C-4 alkanes (e.g., methane and ethane) are simple asphyxiants. The C-5 to C-8 hydrocarbons (e.g., pentane, hexane, heptane, octane) are CNS depressants and *n*-hexane is an especially hazardous neurotoxicant. *See also* terpene. **aromatic chlorinated hydrocarbon**. a class of aromatic hydrocarbons in which one or two hydrogen atoms have been replaced by a chlorine atom. **aromatic hydrocarbon** (arene; aromatic compound; aromatic). any of a class of hydrocarbons (e.g., anthracene, benzene, naphthalene) that includes one or more aromatic (benzene) rings (i.e., one or more planar rings of atoms with delocalized clouds of π (pi) electrons above and below the ring). This distribution of π electrons over several carbon atoms provides a resonance stabilization of the molecule. Aromatic hydrocarbons have an agreeable odor and are further characterized by a low hydrogen-to-carbon atomic ratio; C-C bonds that are quite strong and intermediate in length between those in alkanes and those in alkenes and the tendency to undergo mucleophilic reactions in which hydrogen or other groups on the ring are substituted (addition reactions are uncommon). Benzene is the simplest aromatic hydrocarbon. It forms an equilateral planar six-membered ring. Other aromatic hydrocarbons derive from benzene by substitution of one or more hydrogen atoms by CH_3 or longer carbon chains. Aromatic hydrocarbons are extremely important in the chemical industries, in toxicology, and in all fields of organic and toxicological chemistry. They are the most toxic hydrocarbons found in petroleum, and occur in all crude oils and in many petroleum products. Many are soluble in water, thereby increasing the risk of poisoning to aquatic organisms and to organisms higher in the food chain. Some are environmentally persistent and therefore especially hazardous. **atmos-**

pheric hydrocarbon. a popular (not a chemical) term that encompasses all hydrocarbon compounds that are common air pollutants. They may enter the atmosphere as the result of evaporation, incomplete combustion of fossil fuels, or resuspension by physical means following deposition from the atmosphere. **C-1 to C-4 aliphatic hydrocarbons**. alkanes with one through four carbon atoms (e.g., methane, ethane). These compounds are simple asphyxiants. **C-5 to C-8 aliphatic hydrocarbons**. alkanes with five through eight carbon atoms (e.g., pentane, hexane, heptane, octane) consist of *n*-alkanes. The branched-chain hydrocarbons of this group have many isomers (e.g., there are nine isomers of heptane C_7H_{16}). These compounds are all volatile liquids under ambient conditions. They are used in fuels, as solvents in formulations of a number of commercial products (e.g., glues, inks, varnishes), and in the extraction of fats. They are CNS depressants and *n*-hexane is an especially hazardous neurotoxicant. Exposure to most is chiefly via inhalation. High levels have proven fatal to laboratory animals. Included are a number of environmentally persistent, broad-spectrum insecticides (e.g., DDT, aldrin, dieldrin, heptachlor, chlordane, lindane, endrin, mirex, hexachloride, and toxaphene). **chlorinated hydrocarbon** (organochlorine compound; organochlorine). any halogenated hydrocarbon compound in which one or more hydrogen atoms have been replaced by chlorine atoms. They are broad-spectrum poisons and are toxic to nearly all animals including many vertebrates. Many are carcinogenic, teratogenic, and neurotoxic. About 100 are or have been used as insecticides that degrade only slowly in the environment and accumulate at higher trophic levels. They are lipophilic and accumulate in depot fat (stored fat, as that in adipose tissue). Among the most widely known chlorinated hydrocarbons are aldrin, benzenehexachloride, chlordane, chlordecone, DDT, dieldrin, endrin, heptachlor, hexachloride, PCBs, and toxaphene. Most have been banned or restricted. **cyclic hydrocarbon**. any hydrocarbon with a closed ring. Included are alicyclic hydrocarbons, aromatic hydrocarbons, and cyclic terpenes. **derivative hydrocarbon** (substituted hydrocarbon). any of numerous organic compounds in which one or more hydrogen atoms have been replaced by another element. **fluorinated hydrocarbon**. fluorocarbon. **halogenated hydrocarbon**. a class of derivative hydrocarbons in which one or more hydrogen atoms have been replaced by a halogen atom. Many of these compounds are highly toxic and some are carcinogenic (e.g., vinyl chloride, dichloroethane). Many are used as industrial solvents, pesticides, chemical intermediates, and anesthetics. Examples of the many highly toxic halogenated hydrocarbons are carbon tetrachloride, chloroform, polychlorinated biphenyls, and TCDD. **monocyclic aromatic hydrocarbon**. any aromatic hydrocarbon with a single 6-membered ring. All except benzene have substituent groups. **polychlorinated aromatic hydrocarbon**. any derivative aromatic hydrocarbon that contains more than one or two substituted chlorine atoms. These compounds cause a hyperplastic and hypertrophic lesion of the gastric mucosa in the rhesus monkey and in cattle, but apparently not in the laboratory mouse, rat, and guinea pig. They produce edema and ascites in the domestic fowl, testicular damage in several species, and pancytopenia, especially in monkeys. **polycyclic aromatic hydrocarbon** (PAH; polynuclear aromatic hydrocarbon). any aromatic hydrocarbon having two or more rings. They are formed by the incomplete combustion of other hydrocarbons. Of these, benzo(a)-pyrene is the most studied. PAH compounds occur in engine exhausts, wood stove smoke, cigarette smoke, and charbroiled food; high concentrations of PAH occur in coal tars and petroleum residues. Some metabolites of PAH are car cinogenic (e.g., 7,8-diol-9,10 epoxide of benzo(a)pyrene). **polyhalogenated hydrocarbon**. a substituted hydrocarbon in which more than two hydrogens are replaced by halogen atoms. Such hydrocarbons are highly toxic, immunotoxic, and ubiquitous in the environment, with a high potential for bioaccumulation and biomagnification. *See*, for example, polybrominated biphenyl, polychlorinated biphenyl, polychlorinated dibenzofuran, and polychlorinated dibenzo-*p*-dioxin (PCDD). **polynuclear aromatic hydrocarbon**. polycyclic aromatic - hydrocarbon. **substituted hydrocarbon**. derivative hydrocarbon. **unsaturated hydrocarbon**. any hydrocarbon that has multiple bonds, each involving more than two shared electrons between carbon atoms.

hydrochloric acid (muriatic acid). a colorless, fuming, corrosive, nonflammable, and highly toxic aqueous solution of hydrogen chloride, HCl. concentrated hydrochloric acid (a saturated solution of HCl) contains 36% HCl; it forms a constant-boiling mixture at a concentration of 20% HCl. It has numerous industrial applications. iFor instance, it is used widely as a reagent and in organic syntheses, and in the production of superphosphate fertilizer from phosphate rock. HCl is much less toxic than HFl. Inhalation can cause spasms of the larynx, pulmonary edema, and even death. Hydrogen chloride gas, which constitutes the solute of hydrochloric acid, has a high affinity for water and thus desiccates moist tissues (e.g., eyes and mucous tissue of the upper respiratory tract) that it contacts. If swallowed, HCl causes immediate, intense burning pain, soon followed by vomiting; dark, bloody diarrhea; and a steep drop in blood pressure. Brownish or yellowish stains form in or around the mouth. The throat can swell, suffocating the victim. Peritonitis may occur later if the acid has ruptured the stomach; this is announced by fever and a rigid abdomen. At this stage, a fatal outcome is nearly certain in the absence of immediate medical attention. HCl is much less toxic than hydrofluoric acid. *See also* hydrogen chloride.

hydrocyanic acid. hydrogen cyanide.

hydrocyanic acid poisoning. hydrocyanism.

hydrocyanism (hydrocyanic acid poisoning). poisoning by hydrogen cyanide, *q.v.*

hydrogen cyanide (HCN; hydrocyanic acid; cyanihydric acid; blausäure (German); prussic acid; formonitrile). an extremely toxic, colorless, water-soluble gas or liquid, HCN, that oxidizes readily in air; the vapor is highly flammable. The compressed gas is used by professionals as a fumigant to kill insects and rodents in enclosed spaces (e.g., grain storage bins, warehouses, holds of ships). It is synthesized by a number of processes, but is also produced naturally from cyanohydrins or amino-acid derivatives by more than 800 species of trees in more than 70 families; it occurs at relatively high concentrations in a variety of plants eaten by humans and domestic animals (e.g., pits of wild black cherry, peaches, and apricots, in apple seeds, some varieties of almonds, and lima beans from the tropics). HCN is an extremely toxic cellular asphyxiant that inhibits cytochrome oxidase, with consequent suppression of cellular respiration. Death of mammals from a lethal dose ensues within a few minutes to several hours.

hydrogen sulfate. sulfuric acid.

hydrogeology. the science that treats the geology of groundwater systems, with special emphasis on the chemistry and movement of water. This science is important in estimating the amount of water available for human use, its quality, and rate of replenishment through the hydrologic cycle. Hydrogeologic knowledge is also important to the assessment of the persistence and fate of agricultural chemicals and waste and other anthropogenic substances present at a given location, or in a given aquifer. *Cf.* hydrology.

hydrology. the science that deals with the properties, distribution, and circulation of water near the surface of the earth. This science is important, for example, to the assessment of the transport, persistence, and fate (and hence the potential hazard) of agricultural chemicals and wastes that may find their way into water bodies at a given location or in a given region. *Cf.* hydrogeology.

20-hydroxyecdysone. β-ecdysone. *See* ecdysone.

hydrolysis. **1:** any reaction of water with a salt to yield an acid and a base, one of which is more dissociated than the other. **2:** a decomposition reaction in which a compound reacts with water such that its molecules are cleaved into two new species of lower molecular mass. The hydroxyl group is incorporated into one fragment, the hydrogen atom into another. Hydrolysis is an extremely common type of reaction in living systems. It is important, for example, in the digestion of many foods and in intermediary metabolism. Many hydrolytic reactions are reversible; the reverse reaction is called neutralization, esterification, or condensation. **metabolic hydrolysis**. the metabolic (enzymatic) hydrolysis of organic compounds including many xenobiotics, perhaps most notably organophosphate esters and amides.

hydrolytic. pertaining to hydrolysis.

hydroperitonia. ascites.

hydrophilia. **1:** the state or condition of being hydrophilic. **2:** hydrophilism, the tendency of a tissue to attract and hold water.

hydrophilic. water-loving; having a strong tendency to bind, absorb, or dissolve in water. This term refers, for example, to chemicals that are water soluble or to the regions of a molecule that are polar and therefore attracted to water. Hydrophilic substances include many carbohydrates and complex proteins such as gelatin and collagen. Many toxicants are not hydrophilic in their native form, but become more hydrophilic through metabolism and are thus more readily excreted. Hydrophilic compounds do not diffuse easily through membranes. *See also* hydrophobic, hydrophobia.

hydrophilic amino acid. *See* amino acid.

hydrophilism. hydrophilia (def. 2).

hydrophobia. **1a:** inability to drink. **1b:** a morbid fear of water. **2:** hydrophobia (def. 1) is a characteristic of rabies, which is thus commonly called hydrophobia. *See also* hydrophilia.

hydrophobic. water-fearing; antagonistic to water; unable to dissolve in water. *See also* hydrophilic, hydrophobia. Fats, oils, waxes, many resins, and certain finely divided powders (e.g., carbon black, magnesium carbonate, talc) are hydrophobic.

hydrophyte. a plant that lives in water or in swampy or marsh-like conditions.

hydropic. of or pertaining to dropsy (edema).

hydroponics (chemiculture; soilless gardening; soilless culture; water gardening). a form of aquaculture in which plants are grown in water to which nutrients have been added rather than in soil. This method is used in research, hydroponic home gardens, and commercial cropping operations in many countries. Hydroponically grown plants can be placed closer together than in the field and several crops can be grown in the same tank, thereby conserving space and increasing yields. Hydroponics also reduces weed and pest problems to a minimum. Capital costs are high and attending personnel must be trained. In principle, most plants can be cultivated hydroponically, but size is a limiting factor as physical support becomes a problem. *See also* soilless gardening, soilless culture, chemiculture, and water gardening.

hydrops abdominis. ascites.

hydrops uteri. a condition of livestock characterized by edema of the uterus of pregnant females. The condition may be relieved by draining the uterus.

hydroquinone-β-D-glucopyranoside. arbutin.

hydrous copper sulfate. cupric sulfate.

***p*-hydroxy cinnamic acid**. a phenolic compound secreted by *Adenostoma fasculatum* and other plants of the hard chaparral community (e.g., *Arctostaphylos*). It leaches into the ground apparently inhibiting the growth of nearby grasses.

4′-hydroxyacetanilide. acetaminophen.

***p*-hydroxyacetanilide**. acetaminophen.

5-hydroxy-(2-aminoethyl)indole. serotonin.

6-hydroxy-5-benzofuranacrylic acid δ-lactone. psoralen.

2-hydroxybutane. 2-butanol.

***m*-hydroxycarbanilic acid methyl ester *m*-methylcarbanilate**. phenmediphan.

***p*-hydroxycinnamic acid**. *p*-coumaric acid.

1-(4′-hydroxy-3′-coumarinyl)-1-phenyl-3-butanone. warfarin.

3-hydroxycrotonic acid methyl ester dimethyl phosphate. mevinphos.

8-hydroxydiacetoxyscirpenol (neosolaniol). a trichothecene first isolated from *Fusarium*.

hydroxydimethylarsine oxide. cacodylic acid.

20-hydroxyecdysone. β-ecdysone.

hydroxyethyl trimethyl ammonium hydroxide. choline.

(β-hydroxyethyl)trimethylammonium. choline.

2-hydroxy-3-isobutyl-6-(methylpropyl)pyrazine-l-oxide. aspergillic acid.

H2-hydroxy-3-isobutyl-6(methylpropyl)-pyrazine-1-oxide. aspergillic acid.

p-hydroxymandelo-nitrile-β-D-glucoside. dhurrin.

4-hydroxy-3-methoxycinnamic acid.

3-(4-hydroxy-3-methoxyphenyl)-2-propenoic acid. ferulic acid.

3-hydroxymethyl-chrysazin. aloe-emodin.

4-hydroxy-3-(3-oxo-1-phenylbutyl)-2*H*-1-benzopyran-2-one. warfarin.

6-(10-hydroxy-6-oxo-*trans*-1-undecenyl)-β-resorcylic acid lactone. zearalenone.

***N*-(4-hydroxyphenyl) acetamide**. acetaminophen.

β-[4-hydroxyphenyl]-acrylic acid). *p*-coumaric acid.

4-hydroxyphenyl-β-D-glucopyranoside. arbutin.

(3-(4-hydroxyphenyl)-2-propenoic acid .*p*-

coumaric acid.

α-hydroxytoluene. benzyl alcohol.

2-hydroxy-*N,N,N*-trimethylethanaminium. choline.

[S-(Z,E)]-5-(1-hydroxy-2,6,6-trimethyl-4-oxo-2-cyclohexen-1-yl)-3-methyl-2,4-pentadienoic acid. abscisic acid.

6-(10-hydroxy-6-oxo-*trans*-1-undecenyl)-β-resorcylic acid lactone. zearalenone.

hymenoptera sting. *See* sting.

Hymenoptera (ants, bees, wasps, hornets, yellow jackets, etc.). a large order of endopterygote insects with distinct head, thorax, and abdomen; legs attached to the thorax; usually the thorax and abdomen are separated by a narrow waist. Winged individuals have two pairs of small membranous wings that are coupled by small hooks, have few veins, the anterior pair being the largest. Mouth parts are usually biting, but may be adapted to chewing, lapping, or sucking. All females have a conspicuous ovipositor which may be modified as a sting, saw, or drill. Metamorphosis is complete; the larvae are usually apodous, but some are caterpillar-like with true thoracic legs and abdominal prolegs; pupae are herbivorous or parasitic and in some cases are enclosed in a cocoon. Highly developed social or colonial behavior is common; some species are solitary. The sting or bite may be painful and sometimes lethal, although death of nonsensitive individuals is rare. Common or important genera include *Apis*, *Bombus*, *Nematus*, *Formica*, *Polistes*, *Vespa*, *Vespula*. *See* Aculeata, Hymenoptera sting, hymenopterism, venoms (hymenopteran venoms).

hymenoptera sting. some hymenopterans are not venomous; others can be deadly to humans. The effects of a sting may be largely local and restricted to local symptoms such as slight to intense pain, tenderness, edema, erythema, and itching about the site of the bite. More general reactions are not uncommon, especially in sensitive individuals or to multiple stings. Signs and symptoms often include a generalized urticaria, sometimes with generalized edema, nausea, vomiting, dizziness, malaise, anxiety, wheezing, and thoracic constriction. In severe reactions, further systemic effects such as dyspnea, hoarseness, dysphagia, and confusion may present. Systemic shock, while rare, can be fatal. Signs and symptoms (in addition to any of those above) may include two or more of these: cyanosis, lowered blood pressure, incontinence, unconsciousness, and collapse. Delayed reactions to hymenopteran stings may also occur, and anaphylactoid shock is a risk in subsequent stings. *See* Hymenoptera, venom (hymenopteran venom).

hymenopteran. any insect of the order Hymenoptera.

hymenopterous. of or pertaining to insects of the order Hymenoptera.

Hymenoxys. a genus of small annual or perennial, often weedy herbs (Family Compositaceae). ***H. lemmoni*** (cockerell). a small herb native to California, Utah, northeastern Arizona, and Nevada. It is highly toxic; one half pound per sheep per day is invariably fatal within a few days. The plant is normally grazed by sheep only under poor range conditions. ***H. odorata*** (= *Actinea odorata*; bitterweed; bitter rubberweed). a highly toxic, poisonous, annual weed from a few centimeters to more than 1 m tall. It occurs in semiarid rangelands from southwestern Kansas, western and central Texas, into Arizona, New Mexico, and Mexico. The leaves contain a highly toxic, water-soluble cumulative poison. Sheep may die within a day or two following ingestion of about 1% of their body weight of fresh green leaves. The leaves retain toxicity on drying. Symptoms of acute (experimentally forced, single lethal dose) intoxication include salivation, anorexia, weakness, depression, and vomiting. In natural poisoning of sheep, symptoms are similar but less pronounced. Lesions are variable and sometimes absent. They commonly include, however, pulmonary congestion, inflammation of the abomasum, and hemorrhages of the epicardium. Livestock usually eat bitterweed only under poor range conditions. ***H. richardsonii*** (= *Actinea richardsonii, Hymenoxys floribunda*)(pingue, Colorado rubberweed). a perennial herb with a woody root, up to ½ m tall that is widespread from Canada to Texas and west to Oregon and California. It is somewhat less toxic to livestock than *H. odorata*. Sheep are most often poisoned and have suffered extensive losses from poisoning by pingue throughout portions of the plant's range. Symptoms of acute intoxication include salivation, uneasiness, anorexia, weakness, depression, vomiting, rumen stasis with signs of severe abdominal pain, and prostration.

hyperadiposis, hyperadiposity. an extreme degree of obesity. *See also* hyperliposis.

hypericin (1,3,4,6,8,13-hexahydroxy-10,11-dimethylphenanthro[1,10,9,8-[*opqra*]-perylene-7,14-dione; 4,5,7,4′,5′,7′-hexahydroxy-2,2′-dimethylnaphthodianthrone; hypericum red; Cyclo-Werrol; Cyclosan). a primary photosensitizing agent found in *Hypericum spp.*, it is a reddish-brown fluorescent dianthrone pigment that remains undigested and is chemically unaltered during passage through the liver. Hypericin initiates feeding in the beetle *Chrysolina brunsvicensis*, but can accumulate in mammals that graze on *Hypericum*, causing an intense photosensitization; the symptoms are similar to fagopyrism. *See* photosensitivity.

hypericum red. hypericin.

Hypericum (St. John's wort). a large genus of perennial herbs and shrubs (Family Guttiferae; sometimes given as Hypericaceae). Most species are ornamental (often grown in rock gardens or borders). Some feral, mostly weedy species occur in dry pastures, old fields, woodland borders, and along roadsides. The leaves are characteristically dotted on either surface with translucent pinpoint marks that are black in reflected light or yellowish from transmitted light. The flowers are yellow, numerous, and about 5 cm in diameter. They contain photosensitizing pigments that are unaffected by digestion, are not eliminated by the liver, and reach the capillaries of the skin unchanged. A number of species have poisoned livestock. Common or important species include *H. ascyron*, *H. calycinum*, *H. frondosum*, *H. kalmianum*, *H. patulum*, *H. perforatum* (the most important toxicologically), *H. prolificum*, and *H. punctatum*. ***H. perforatum*** (St. John's wort; Klamath weed; goatweed). a common roadside weed, native to Europe and naturalized elsewhere. It is one of few plant species known to cause primary photosensitivity in animals. Cases are uncommon. It contains the phototoxic pigment, hypericin. Sheep are more resistant than cattle or horses. Effects on sheep include generalized skin irritation, anorexia, and sometimes loss of vision. Because hypericin is resistant to drying and is heat stable, animals can be poisoned by weedy hay. The plant is used in folk medicine for its antibacterial and astringent properties and to soothe the digestive system. Photosensitization, *q.v.*, and other toxic effects also occur in humans. While small doses may not be harmful, it is probably unsafe for human usage. *See*, for example, furocoumarins, *Fagopyrum sagittatum*, *Ammi*, *Cymopterus watsonii*, *Trifolium*, and *Medicago*.

hyperketonemia. Elevated concentrations of ketone bodies in circulating blood plasma. *Cf.* ketonemia, ketosis.

hyperketonuria. elevated concentrations of ketone bodies in urine. *Cf.* ketonuria.

hyperliposis. 1. hyperadiposis. 2. an extreme degree of fatty degeneration.

hyperparasite. a parasite whose host is another parasite.

hyperplasia **1a:** excessive increase in the size of an organ or normal tissue due to an increased rate of division; enlargement of the tissue or organ may occur. Hyperplasia may occur in normal fetal development, wound healing, inflammation, tumor growth, and lymph node activity. **1b:** an abnormally high rate of proliferation of normal cells. An abnormally high rate of proliferation of normal cells that are arranged normally within a given tissue or organ. **2:** hyperplasy; excessive overgrowth of a plant structure. *Cf.* hypoplasia, aplasia.

hyperplastic. of, pertaining to, or characterized by hyperplasia.

hyperplasy. hyperplasia (def. 2).

hyperploid. **1a:** having a chromosome number that is less than an exact multiple of the basic haploid (monoploid) number. *Cf.* aneuploid, hypoploid. **1b:** a hyperploid organism.

hyperpnea. abnormally deep, rapid breathing.

hyperpneic. of or pertaining to hyperpnea.

hyperpyretic. pertaining to hyperpyrexia.

hyperpyrexia. a greatly elevated body temperature in homeothermic animals.

hyperpyrexial. pertaining to hyperpyrexia.

hyperreactive. exhibiting hyperreactivity.

hyperreactivity. a greater than normal responsiveness to stimuli. *Cf.* hypersensisensitivity.

hypersaline. of, pertaining to, or containing unusually large amounts of sodium chloride; highly saline. Most often used with respect to seawater when the salinity is greater than 4.0%.

hypersalinity. the condition of being hypersaline.

hypersensibility. excessive sensibility or sensitivity to a biologically active agent. Do not confuse with hypersensitivity.

hypersensitive. **1:** exhibiting hypersensitivity, *q.v.* **2:** used also to describe organisms within a given population that are killed by a specific toxicant when exposed to a dose equivalent to LD_{05} or below.

hypersensitiveness. **1:** unusual or abnormal sensitivity to the action of a specific agent, as in allergy, anaphylaxis, or hay fever. **2:** obsolete: an increase in an already existing sensitivity to a toxin by repeated exposure. Such cases are, on the contrary, allergic or immunotoxic in character; they are due to a sensitivity to a substance that was formerly innocuous. *Cf.* hypersensitivity, hypersensibility.

hypersensitivity. **1a:** extreme sensitivity to any allergen (e.g., certain drugs, pollens, toxins). Some reactions can be life-threatening, but most are less serious. *Cf.* hypersensibility, hypersensitiveness. 1b: sometimes used as a synonym of allergy (*See also* anaphylaxis). **2a:** an exceptionally strong reaction to an external agent or stimulus of any kind (e.g., noise or touch). **2b:** extreme sensitivity to any agent, chemical or physical. **3:** heightened sensitivity of a plant to infection or infestation by a pathogen such that plant tissue at the site of contact dies, preventing further spread of the disease. plants showing such sensitivity to a par ticular pathogen are highly resistant to infection. Symptoms are limited to minute necrotic lesions. **cell-mediated hypersensitivity**. hypersensitivity (delayed hypersensitivity). **contact hypersensitivity**. a reaction due to a cell-mediated immune response to newly-produced proteins in the skin. This response damages the skin. Examples include reactions to poison oak and poison ivy in humans and reactions of dogs and horses to sensitizing chemicals in plastic collars and plastic food dishes. **cytotoxic hypersensitivity**. cytotoxic reaction (*See* antigen-antibody reaction). **delayed hypersensitivity** (cell-mediated hypersensitivity). an allergic reaction (*See* reaction) characterized by a localized inflammation that appears as much as 24—48 hours following exposure to an antigen (e.g., contact dermatitis). It is produced solely by T-lymphocytes, which secrete lymphokines that direct an assortment of cellular responses. This form of hypersensitivity can be used to assess the integrity of mediated immunity following exposure to suspected immunosuppressiveorimmunomodulatory agents. **immediate hypersensitivity** (type I hypersensitivity). a specific reaction to an antigen that is evident within seconds, minutes or, at most, a few hours following exposure to an antigen. The reaction often results only in a focal inflammation, hemorrhage, and vasculitis at the site of contact with the antigen. There are two classes of immediate hypersensitivity reactions. **class I:** complement-dependent, (i.e., induced by IgG and IgM antibodies). These reactions are equivalent to a type II, cytolytic, or a type III, immune complex-mediated allergic reaction (*See* reaction). Severe cases may present as immune complex diseases including drug-induced systemic lupus erythematosus. **class II:** complement-independent, that is, induced by IgE antibodies, and equivalent to a type I (anaphylactic) allergic reaction (*See* reaction). Severe cases may present as anaphylactic shock or allergic asthma. **induced hypersensitivity**. any reaction that results from exposure to one or more doses of a particular toxicant, as in the case of a severe allergic reaction to an antibiotic. **mediated hypersensitivity**. delayed hypersensitivity. **type I hypersensitivity**. immediate hypersensitivity, class II. **type II hypersensitivity**. immediate hypersensitivity, class I (in part), which may present, for example, as drug-induced hemolytic anemia. **type III hypersensitivity**. immediate hypersensitivity, class I (in part). **type IV hypersensitivity**. delayed hypersensitivity. complex-mediated allergic reaction (*See* reaction). Severe cases may present as immune complex diseases including drug-induced systemic lupus erythematosus. **class II:** complement-independent, that is, induced by IgE antibodies, and equivalent to a type I (anaphylactic) allergic reaction (*See* reaction). Severe cases may present as anaphylactic shock or allergic asthma. **induced hypersensitivity**. any reaction that results from exposure to one or more doses of a particular toxicant, as in the case of a severe allergic reaction to an antibiotic. **mediated hypersensitivity**. delayed hypersensitivity. **type I hypersensitivity**. immediate hypersensitivity, class II. **type II hypersensitivity**. immediate hypersensitivity, class I (in part) which may present, for example, as drug-induced hemolytic anemia. **type III hypersensitivity**. immediate hypersensitivity, Class I (in part). **type IV hypersensitivity**. delayed hypersensitivity.

hypersensitization. **1:** producing or inducing hypersensitivity. **2:** a state of heightened sensitivity to a stimulus of any kind.

hyperserotonemia. an unusually high serum serotonin concentration.

hypersusceptibility. **1a:** a state of abnormally heightened responsiveness, usually in response to foreign agents such as pathogens or chemical substances (e.g., toxicants) which do not affect most individuals within an animal population. **1b:** abnormally heightened response to a biologically active agent. *Cf.* anaphylaxis.

hypertension (high blood pressure). higher than normal blood pressure, especially diastolic pressure.

hypo. sodium thiosulfate.

hypochlorite. **1:** any of a class of salts derived from hypochlorous acid that contain the ion, ClO_3^-. All are highly toxic, caustic, oxidizing, and bleaching agents. They are commonly used, dry or in solution as bleaches, they are dangerous farm and household poisons. They produce active oxygen which is largely responsible for their toxicity. **2:** the hypochlorite ion, ClO_3^-.

hypochlorite poisoning. poisoning by solutions of hypochlorite salts (most often as household or commercial bleaches and related chlorinated products) by ingestion or skin contact. Ingestion produces immediate gastrointestinal distress, with pain and inflammation of the mucous membranes of the mouth, pharynx, esophagus, and stomach; the vomitus has the appearance of coffee grounds. Systemic effects are secondary to local injury and shock. These include confusion, delirium, edema of the glottis, coma, and death due to circulatory collapse. Lesions may include edema of the pharynx, glottis, or larynx and perforation of stomach or esophagus. Inhalation of fumes may produce pulmonary edema.

hypochlorite salt. a metallic salt of hypochlorous acid. *See* hypochlorite.

hypochlorite solution (household bleach). an aqueous solution of a metallic salt of hypochlorous acid (most commonly sodium hypochlorite). Such solutions are strong oxidizing agents and irritants to the skin, eyes, and mucous membranes. They are used to bleach textiles and as antiseptic agents. Ingestion of hypochlorite salts or their solutions usually produces pain and shock locally about the mouth. The human victim may suffer extreme gastrointestinal distress with pain and inflammation of mouth, pharynx, esophagus, and stomach; glottal edema; vomitus the color of coffee grounds; confusion; delirium; coma; circulatory collapse. Lesions in humans commonly include edema of the pharynx, glottis, and/or larynx, with perforation of the stomach or esophagus in severe cases. The fumes can cause pulmonary edema. The systemic effects are secondary to the local irritation and erosion of tissues caused by these substances. Hypochlorous acid and hypochlorites are used chiefly as bleaches and disinfectants, and to purify water. They produce active (nascent) oxygen which, through its oxidizing action, accounts for most of the toxicity of hypochloric acid and hypochlorites.

hypocinesia. hypokinesis.

hypocinesis. hypokinesis.

hypofunction. the functioning of a tissue, organ, or system significantly below that which is normal or usual.

hypogaeic. subterranean, below ground. *See* hypogeal.

hypogeal. in plant science, developing below ground as in the case of fungal fruiting bodies such as truffles; or seed germination in which the cotyledons remain below ground; or fruit development, as in *Arachis* (peanut), where the peduncle becomes positively geotropic and buries its apex in the earth, where the fruit develops.

hypoglossal. beneath the tongue.

hypokinesia. hypokinesis.

hypokinesis (hypokinesia; hypocinesis; hypocinesia; hypomotility) slow or reduced movement.

hypokinetic. pertaining to or characterized by hypokinesis.

hypolethal. of insufficient quantity, power, or intensity to cause death.

hypomagnesemia. 1: reduced serum magnesium levels. **2:** grass tetany.

hypomagnesemic tetany. a condition of domestic cattle and sheep, mostly affecting adult cows and ewes. Affected ewes are usually lactating animals feeding on lush pastures. Cows of any age or condition may be affected, but the disease is most common in beef cattle grazing on cereal crops (especially wheat) or those that are undernourished and exposed to cold, changeable weather. Signs of the disease are irritability, tetany, and convulsions. Severe cases may present abruptly and, in some cases, cattle at pasture may be found dead without prior signs of illness. The condition is similar in sheep and cattle and is strikingly similar to and may be confused with grass tetany, and may be complicated by grass tetany (hypomagnesemia). Furthermore, since it is almost always accompanied by hypocalcemia it is difficult to determine which condition is primary. Hypomagnesemic tetany of calves is clinically identical to grass tetany, *q.v.*

hypomorph. hypomorphic allele (*See* allele).

hypomotility. hypokinesis.

hyponitrous acid anhydride. nitrous oxide.

hypophyseal growth hormone. somatotrophin.

hypophysis. pituitary gland.

hypophysis cerebri. pituitary gland.

hypoplasia (hypoplasty). underdevelopment of a tissue or organ that is less severe than aplasia. *Cf.* hyperplasia, aplasia. **bone marrow hypoplasia**. defective development of bone marrow with serious consequences such as anemia, leukopenia, and lymphocytopenia.

hypoplastic. of, pertaining to, or characterized by hypoplasia.

hypoploid. **1a:** having a chromosome number that is less than an exact multiple of the basic haploid (monoploid) number. **1b:** a hypoploid organism. *Cf.* aneuploid, hyperploid.

hypopnea. abnormally slow, shallow breathing.

hypopneic. of or pertaining to hypopnea.

hypotaxia. a state of poor coordination.

hypotension (low blood pressure). lower than normal blood pressure. In adult humans hypotension is often clinically defined as repeated readings of a systolic pressure below 100 mm Hg and a diastolic pressure below 40 mm Hg. Hypotension may result from shock (*q.v.*), loss of blood, various infectious agents, and any of numerous vasoactive drugs, venoms, or other toxicants. *See*, for example, *Echis*, haloperidol, halothane, *Helleborus niger*, heroin, iron poisoning, kinin peptide, larkspur poisoning, *Ligustrum vulgare*, meprobamate, morphine, nitroglycerine, paraldehyde, parasympathomimetic agent, phenytoin, potassium bromate, potassium permanganate, ricinine, sodium cyanide, tetrodotoxin, thiamylal, toxic shock syndrome.

hypotensive. **1:** characterized by or causing lower than normal pressure, as of blood pressure. **2:** a subject with abnormally low blood pressure. **3:** a hypotensive agent or hypotensor. Toxicants that act on the autonomic nervous system to constrict blood vessels or reduce heart rate, or the force of ventricular contractions, can produce hypotension.

hypothalamic-hypophyseal axis. *See* pituitary gland (anterior lobe).

hypothalamic-pituitary system. *See* pituitary gland (anterior lobe).

hypoxia. a term used variously in the literature of physiology and toxicology. **1a:** less than normal levels of oxygen in inspired air or gases, blood, or tissue, but not so extreme as anoxia. **1b:** a state of reduced oxygen supply to living tissues, even though they are adequately perfused by blood. **2b:** a level of oxygen within living tissue that is below that considered physiologically normal (i.e., less than adequate for normal metabolism). **2b:** a deficiency of oxygen in respiring tissues. **3: anemic hypoxia**. hypoxia due to reduced oxygen-carrying capacity of the blood resulting from anemia, *q.v.* It may result from hemorrhage or poisoning by carbon monoxide, chlorates, nitrites, and certain venoms. *Cf.* hypoxemia. **anoxic hypoxia**. an inadequate supply of oxygen to respiring tissues. It can be due to irritants of the respiratory epithelium, or to poisons that depress respiration. The partial pressure of oxygen in arterial blood is typically low, even though blood oxygen capacity and flow are normal. **arterial hypoxia**. anoxic hypoxia. **diffusion hypoxia**. a sudden transient decrease in alveolar oxygen tension when ambient air is inhaled following nitrous oxide anesthesia. It is due to the diffusion of nitrous oxide from the blood with consequent dilution of alveolar air. **histotoxic hypoxia**. this is not a hypoxia in the strict sense, but is that due to impaired utilization of oxygen by tissue, as in the case of poisoning by cyanide or hydrogen sulfide. *Cf.* anoxia. **stagnant hypoxia**. hypoxia due to decreased blood flow.

hypoxic. of, pertaining to, or characterized by hypoxia.

I

i.p. intraperitoneal.

IA. 3-indolacetic acid.

IAA. 3-indolacetic acid.

iatrogenic. originating from treatment given by a health care professional such as a physician or veterinarian.

iatrogenic disorder. any adverse condition resulting from treatment by a health care professional such as a physician or veterinarian. Improper care, negligence, or less than judicious treatment is usually implied.

ibotenic acid (α-amino-2,3-dihydro-3-oxo-5-isoxazoleacetic acid; α-amino-3-hydroxy-5-isoazoleacetic acid; amino-(3-hydroxy-5-isoxazolyl)acetic acid). a neurotoxic isoxazole which, together with muscimol, is largely responsible for the toxicities of a number of mushrooms including those of the genus *Amanita* (e.g., *A. muscaria* and *A. pantherina*). It is insecticidal and very toxic to small mammals. Ibotenic acid is the precursor of muscimol.

ice. *See* crystalline water.

Iceland poppy. *Papaver nudicaule*.

ichthyophagous. piscivorous.

icteric (icteritious; jaundiced). pertaining to, affected with, or characterized by jaundice.

icteritious. icteric.

icterogenic. causing jaundice (icterus).

icterogenicity. ability to cause jaundice (icterus).

icterohepatitis. inflammation of the liver with marked jaundice (icterus).

icterus. jaundice.

ictrogenic. able to cause icterus (jaundice).

idiosyncratic effect. *See* effect.

illuviation. the deposition of material removed from one soil horizon to another.

imago. The adult, sexually mature stage of an insect.

imagocide. a substance that destroys adult insects (imagos); used especially with reference to mosquitoes.

imbricated snout beetle. *Epicaerus imbricatus*.

1*H*-imidazole-4-ethanamine. histamine.

4-imidazoleethylamine. histamine.

5-imidazole-ethylamine. histamine.

2-(4-imidazolyl)ethylamine. histamine.

imitation flavor. food flavoring.

immediate effect. *See* effect.

immediate response. immediate effect (*See* effect).

immedicable. **1:** pertaining to a disease or condition that is in such a state that it cannot be remedied; incurable in the medical sense. **2:** incurable or irremedial in a social sense.

immigration. movement into a new habitat, community, ecosystem, location, region, or country by an individual or a population on a more or less permanent basis.

immune body. antibody.

immune reaction. immune response (*See* response).

immune response. *See* response.

immunity. **1:** the state or condition of being immune to or resistant to a disease, especially an infectious disease. An animal can acquire immunity to infectious disease by passive or active immunization. **active immunity**. resistance to disease effected by the production of antibodies by an organism, usually in response to antigens peculiar to a specific disease, from a pathogen or disease organism, or by having been vaccinated for the disease. **herd immunity**. the level of resistance of a herd, flock, or other population of animals to a specific disease. **passive immunity**. immunity conferred by transfer of preformed antibodies from a resistant to a susceptible animal. Passive immunity is acquired, for example, from antibodies that pass from the mother to the fetus in utero or from ingestion of the mother's milk. This usually confers an immediate but transient resistance to specific diseases.

immunogen. antigen.

immunogenic. antigenic.

immunogenicity. antigenicity.

-in. *See* -ase.

immunoglobulin. antibody (def.1).

immunology. the science that deals with the immune system.

immunotoxin. antitoxin.

imperial green. cupric acetoarsenite.

impurity. **1:** contamination, adulteration, *q.v.* **2:** a contaminant, an adulterant, *q.v.*

in milk. *See* milk.

in situ. meaning in its original or natural place and position (orientation). This implies, for a substance, that its native physical, chemical, and biological properties are retained intact.

Indian ox. *Bos indicus*.

in vitro. Within an artificial environment, as within a test tube.

in vivo. within a living body.

-in. *See* -ase.

inactivate (deactivate). **1:** to make inactive; deactivate. **2:** to cause a substance, such as an enzyme, to lose a specific type of activity. **3:** to cause an agent of disease (e.g., a microbe) to lose its disease-producing capability. **4:** to transform a substance (reversibly or irreversibly) into an inactive or much less active form.

inactivation (deactivation). **1a:** the state or process of inactivating, deactivating. **1b:** rendering a substance inactive with respect to desired function. **1c:** rendering a substance inert. **2:** elimination of the activity or effects of a biologically active substance by chemical (chemical antagonism) or physical means. *Cf.* chemical antagonism (*See* antagonism).

inactive. not active, dormant, idle, static, inert; stationary.

inadvertent residue. *See* action level.

inanition. debilitation or the process of de

bilitation due to an insufficiency of essential nutrients, as in starvation or malabsorption.

inappetence. lack of an appetite.

inbred. of or pertaining to the result of matings between related organisms.

inbreeding. the mating of very closely related individuals. In animal husbandry this is done to produce and establish purebreeding strains or varieties in which individuals are homozygous for a large proportion of loci or at least for selected loci that produce desired phenotypes. *See also* close breeding.

incisor. a cutting tooth at the front of a mammal's mouth.

incline. **1:** slope, pitch, grade. **2:** bend, lean, tend.

incomplete antigen. hapten.

incomplete metamorphosis. *See* metamorphosis.

incoordination. *See* ataxia (def.1).

increment. **1:** a small positive or negative change in the value of a variable. **2:** a small, often barely perceptible augmentation of a variable. **3:** an increase, or the process of increasing, in number, size, or extent. **4:** in forestry, the increase in diameter, basal area, height, volume, or value of a tree or stand of trees during a given growing season or other designated period.

increment borer. a tool (essentially a hollow auger) used to extract a core of wood from a living tree (*See* increment core).

increment core. the core of wood extracted from a tree by an increment borer. It represents a cross section of the tree and allows one to examine growth rings to determine, for example, tree age and pattern of growth, decay, penetration of preservatives into the wood, and in some cases past pollution episodes.

incubation. **1:** the hatching of eggs by means of natural or artificial heat (as in an incubator). *See* brooding. **2:** the maintenance and stimulation of growth of a bacterial (or other) culture.

Indian aconite. *Aconitum ferox*.

Indian cherry. *Rhamnus caroliniana*.

Indian corn. *Zea mays*.

Indian liquorice. *Abrus precatorius*.

Indian mustard. *Brassica juncea*.

Indian pea. a poisonous pea. *See Lathyrus*.

Indian poke. *Veratrum viride*.

Indian turnip. *See Arisaema*.

indicator. **1:** that which indicates or points. **2a:** a substance that changes color within a specified range of pH or oxidation potential or that renders the completion of a chemical reaction visible. Litmus paper and phenolphthalein solutions are examples of such indicators. **2b:** a sensing instrument which, by detecting a change (e.g., in color), determines the state of a solution (e.g., pH or concentration of a solute). **3:** a tracer. **4a:** bioindicator (*q.v.*); a living organism that is largely restricted to a particular type of environment, or a particular aspect, variable, or condition such that its presence is indicative of a particular environment, aspect, or condition.is **4b:** a living organism that acts as an effects indicator because it is highly sensitive to, and registers an observable response to a particular environ mental variable (e.g., frost, a pollutant) that indicates the active presence of the variable. *See also* indicator species. **effects indicator**. an indicator that registers the effects of an environmental variable such as a pollutant or toxic substance.

indicator species (guide species; index species). a plant or animal species that is a bioindicator. The presence of such a species, for example, in an area, habitat, community, ecosystem, or in a particular soil indicates the presence of a particular substance, process, or environment, or a particular complex of environmental phenomena or conditions. Certain species of lichens, for example, are highly sensitive to sulfur dioxide; their presence indicates low levels of sulfur dioxide pollution at most.

indigenous. native to or originating in a country or region.

indigo. **1a:** any plant of the genus *Indigofera* **2:** A blue dye from *Indigofera spp.*, especially *I. tinctoria* and *I. suffruticosa*, and certain other plants. **creeping indigo**. *I. endecaphylla*.

Indigofera (indigo). A genus of some 700 species of mostly Old World tropical and subtropical herbs and shrubs (Family Fabaceae) used to prepare dyes. Some species are toxic. ***I. endecaphylla*** (creeping indigo). a toxic annual or perennial herb of the Old World tropics. Beta-nitropropionic acid is the toxic principle. The unusually high nitrite content may also contribute to this plant's toxicity. Toxicity varies among the various classes of livestock; swine do not eat this plant. ***I. spicata***. a species thought to cause impairments of calves that are similar to those of crooked calf disease.

indirect effect. *See* effect.

indirect metamorphosis. *See* metamorphosis.

indirect systemic effect. *See* effect.

indisposed. slightly ill; not feeling well; mildly debilitated; ailing.

indium (symbol I) [Z = 49; A_r = 114.82]. a rare, soft, ductile, silvery metallic element of group III of the periodic table found usually in zinc ores. Salts of indium are relatively nontoxic by mouth, but are highly toxic when administered intravenously or subcutaneously. They have proven to be cardiotoxic, hematoxic, hepatotoxic, and nephrotoxic.

indium sulfate. a white or grayish hygroscopic powder that decomposes on heating. It is moderately toxic to rabbits by mouth, but extremely poisonous when administered intravenously.

indole. a toxic alkaloid obtained from coal tar; as a bacterial decomposition product of tryptophan (e.g., in feces or certain bacterial cultures). It is highly toxic and carcinogenic. It is used in highly dilute solutions in perfumes. It also occurs in *Clavipes*, *Gelsemium*, the fruit of *Hippomane mancinella*, and *Peganum* (African rue), and certain other plants.

indole-acetic acid. 3-indolacetic acid.

1*H*-indole-3-acetic acid. 3-indolacetic acid.

indole acetonitrile. a naturally occurring auxin. *See also* 3-indolacetic acid.

3-indolacetic acid (formerly heteroauxin; indole acetic acid; 1*H*-indole-3-acetic acid; IA; IAA; auxin 1AA; indol-3yl-acetic acid). the most widely distributed natural auxin and the chief auxin of higher plants,

$$C_8H_6NCH_2COOH$$

It is also used as a plant growth regulator

and herbicide. Other indole compounds that occur in plants have some auxin-like activity.

indolent. **1:** indisposed to action; inactive; lazy. **2:** sluggish; inactive; not developing. **3:** said of a morbid condition that is painless or nearly so.

indolent ulcer. an ulcer that is slow to heal.

indolizidine-1,2,8-triol. a mydriatic alkaloid that occurs in species of *Astragalus* and *Oxytropis*.

indol-3yl-acetic acid. 3-indolacetic acid.

indoor air. the air within a building, especially one that serves as an abode for humans or domestic animals.

indoor air pollution. chemical, physical, or biological contaminants of indoor air.

indusium. amnion.

inert. **1:** inactive; non-reactive; physiologically inactive. **2:** denoting regions of heterochromatin with inactive genes.

inert ingredient. a component of a pesticide formulation such as a solvent, carrier, or surfactant that is not active against target pests. Such ingredients should not be considered chemically or biologically inert or innocuous.

infectious. *See* contagious.

infectious abortion. brucellosis.

infectious bulbar paralysis. Aujesky's disease.

infectious canine hepatitis. an acute viral disease of canids, especially of domestic dogs and foxes, that is not transmissible to humans. The pathogen is an adenovirus. The incubation period in dogs is six to nine days. Signs include fever, inappetence, congested mucous membranes, and abdominal pain. The case mortality is about 10%, and about 25% of the survivors develop a temporary corneal opacity (hepatitis blue eye). The disease in foxes is marked chiefly as encephalitis. The infection is transmitted mainly by direct contact with a diseased animal. Intravenous fluids, antibiotics, and vitamins are usually given to an infected animal. The disease can be permanently prevented by annual vaccination with a modified live virus.

infectious jaundice. human leptospirosis (*See* leptospirosis).

infestation. invasion of a living organism by parasites.

infiltrate. to penetrate the interstices of a substance, environmental medium (e.g., soil), or tissue. *Cf* perfuse.

infiltration. **1:** permeation, saturation, soaking; penetration. **2:** the penetration into and accumulation of a foreign substance in cells, tissues, or environmental media or the penetration and accumulation of a substance in amounts greater than that which is normally present. **2:** the percolation or seepage of water into soil. Conservation practices such as conservation tillage reduce rates of runoff and increase infiltration rates, thus increasing soil moisture and growth of crops.

inflammation. erythema, heat, pain, and swelling of living tissue that occur as a reaction to chemical or physical irritation (traumatic inflammation) or injury, or to infection.

inflorescence. **1:** an arrangement of flowers in a branching system on the same stem or stalk (peduncle). **2:** the coming into flower of a plant. **cymose inflorescence**.

See cyme. **racemose inflorescence** (indefinite inflorescence). a type of inflorescence in which the meristematic activity occurs at the apex of the main axis and primary laterals and the flowers develop from axillary meristems. An umbel, *q.v.*, is a type of racemose inflorescence.

infrabuccal. below or under the mouth, as an infrabuccal chamber.

infundibulum. *See* pituitary gland (posterior lobe).

ingesta. foodstuffs or any material that is normally consumed by a given species of animal. *Cf.* ingestant, ingestion.

ingestant. any substance that enters the body via the digestive tract. *Cf.* ingesta, ingestion.

ingestion. **1:** the process of taking in food material or liquids through an oral cavity into a digestive tract or food cavity; feeding. **2:** the consumption of food materials. **3:** commonly used for the act of taking any materials into the body of an animal through a mouth.

ingravescent. of gradually increasing severity.

inhibitory. partially or completely restraining, arresting, or inhibiting a process or capable of doing so. *Cf.* excitatory.

inguinal. in or pertaining to the groin.

inguinal canal. the circular passage in the abdominal wall through which a testis descends into the scrotum.

inhalant. a substance (e.g., certain medications, fumigants) that can enter the body by inhalation, *q.v.*

inhalation (inspiration). **1:** the act or process of inspiring, inhaling, or breathing in; drawing air into the alveoli of the lungs by mammals or through the lungs into the air sac system by birds. Inhalation conveys to the lungs and upper respiratory tree not only the natural constituents of air, but other gases, vapors, sprays, or mists, and small particulates (e.g., respirable dusts, ash, smoke, small aerobiota). Some of these, if not cleared from the lungs, may not only cause pulmonary lesions, but may be conveyed to other parts of the body, causing harmful systemic effects. Inhalation is a major route of entry of many microorganisms and toxicants by themselves or borne on gases, vapors, or respirable particles. **2:** a medication or drug in solution that can be introduced by the oral or nasal respiratory route as a nebulized mist or aerosol. Such medications may be designed variously to produce local or systemic effects. **solvent inhalation**. self-intoxication by inhaling the vapors of volatile organic solvents contained, for example, in cleaning fluids, gasoline, glue, lacquer thinners, lighter fluid, and nail polish remover.

inhale (inspire). to take into the pulmonary system (through the mouth and/or nose) by breathing. *See* inhalation.

initial attack. the action taken by the first fire fighting force that arrives on the scene, to halt a fire's spread or potential spread.

initial mature inventory. the portion of the existing total mature forest inventory that is available for harvest. Considerations for inclusion are all management decisions necessary to protect the environment and other forest uses.

injected. **1a:** introduced by injection. **1b:** of or pertaining to a fluid introduced into a body by injection; introduced by injection. **2:** congested. **3:** denoting or referring to blood vessels that are noticeably distended with blood.

injure. to harm, afflict, hurt, impair, or

wound.

injurious. harmful, detrimental; not necessarily destructive or irreversible. injury is generally taken to be less debilitating than damage.

injury. **1a:** harm, hurt, affliction, impairment, wound. **1b:** physical trauma to some part of a living system. This may be limited to physical pain or may include structural impairment. **1c:** physical or functional impairment. **1d:** psychological trauma. **1e:** in toxicology, injury is taken to be less debilitating and less permanent than damage. *Cf.* damage, harm, trauma. **bodily injury**. any impairment of physical condition (e.g., pain, disfigurement, sensory deficits, physical dysfunction of any type). **irreparable injury**. any injury, great or small, that is of constant or frequent occurrence and that cannot reasonably be repaired or receive reasonable compensation under law. **permanent injury**. injury that appears to be beyond the possibility of repair. **serious bodily injury**. severe injury including permanent or protracted loss or impairment of any organ or bodily member; permanent disfigurement.

inkweed. *Drymaria pachyphylla*.

inner gorge. the reach of a stream or stream segment that is bounded by steep side slopes (> 40%) and terminates upslope into more gentle topography.

innocuous (innoxious). harmless, benign, nontoxic; not injurious, damaging, or harmful.

innoxious. innocuous.

inocula. plural of inoculum.

inoculum (plural, inocula). **1:** preconditioned microorganisms placed in compost to initiate or accelerate decomposition. **2:** compost products that contain suitable microorganisms added to new material for composting to assure rapid decomposition.

inoperable lands. in forestry, lands that are assessed as unsuited for timber production now and in the foreseeable future. Factors that contribute to such an assessment include those related to access such as elevation, topography, and steepness of slope, and stability of soils; other factors include size of stands and value of the timber present. Inoperable lands are also those that cannot be harvested without causing serious and irreversible damage to other resources such as soil or water. Lumbering is also incompatible with other designated uses such as in recreation areas, parks, and wilderness areas.

inositol. an aromatic carbohydrate, part of the vitamin-B complex, found in many seeds and cereal grains, vegetables, citrus fruits, and other parts of plants. It also occurs in heart and other muscle, kidney, and liver. Inositol is essential to the required growth of many animals and microorganisms. There are nine isomeric forms. *See also* myo-inositol.

INPC. propham.

insect. any arthropod of the class Insecta, *q.v.*

insect repellent. a compound or formulation that repels insects, usually without killing them. DEET is the most commonly used pesticide in insect repellents today. There may be some danger to the user. It is neurotoxic and can cause brain disorders, slurred speech, difficulty in walking, tremors, and even death. There are documented cases of acute neurotoxicity in children who had been exposed to, or accidentally swallowed, DEET; some terminated fatally. It is moderately toxic to laboratory rats and rabbits, causing chromodacryorrhea, depression, loss of righting reflexes, dyspnea, tremors, coma, terminal convulsions, and death; in at least one study, survivors recovered fully with no evident sequalae.

insect. any arthropod of the Class Insecta.

Insecta (Hexapoda). a very large class of invertebrate animals (Phylum Arthropoda) that contains in excess of one million described species, representing more than 75% of known animals. The sexes are separate and nearly all insects undergo metamorphosis during development. They have three main body parts (head, thorax, and abdomen) and a horny exoskeleton comprised of chitin. The head bears one pair of compound eyes and one or more simple eyes (ocelli). There are three pairs of variously modified mouthparts associated with the diverse modes of ingestion among insects. The digestive system of chewing insects includes a muscular gizzard. There are three pairs of jointed walking legs and a pair of jointed sensory antennae borne on the thorax. Most adult insects (imagos) have two pairs of wings that insert on the second and third thoracic segments; insects are the only flying invertebrates. The abdomen bears terminal appendages, the genitalia. External respiration (breathing) is effected by means of a system of air tubes (tracheae) that open to the outside by means of small, valved, lateral apertures (spiracles). Some primitive insects breathe directly through the body wall and the aquatic larvae of some species breathe by means of external gills). Excretion is by Malpighian tubules. The circulatory system is composed of a tubular heart that pumps blood forward into the head, after which the blood diffuses through the body tissues, returning to the heart.

insectary. a place where living insects are kept, bred, and/or observed.

insecticidal. capable of destroying insects. *Cf.* entomotoxic.

insecticide. **1:** a formulation (pesticide) used especially to kill or control the growth or reproduction of insects. Solvents and emulsifiers are required in most liquid preparations of insecticides and while they are generally of very low toxicity, they must be considered associated causes of insecticidal poisoning. Insecticides should be handled and used with care as they are generally substantially more toxic (some are highly toxic) to birds, other vertebrate animals, and aquatic invertebrates than are fungicides or herbicides. *Cf.* pesticide. **2:** the act of killing insects. **carbamate insecticide**. any salt or ester derivative of carbamic acid used to kill insect larvae, nymphs, and adults on contact (e.g., carbaryl, carbofuran, methomyl, propoxur). carbamate insecticides inhibit cholinesterase at nerve junctions as do organophosphate (OP) insecticides. The inhibition is much less durable, however, and is readily reversed. Carbamate insecticides have gained favor over organophosphates because they break down even more quickly than organophosphates following use and are thus less hazardous to humans and most nontarget organisms. Atropine sulfate injections readily reverse the effects of carbamate inhibition. *See also* carbamate, carbamate poisoning. **chlorinated hydrocarbon insecticide**. organochlorine insecticide. **chlorinated organic insecticide**. organochlorine insecticide. **contact insecticide**. a poison that kills insects or other arthropods by contact with the outer surface of the body. Ingestion is not necessary:such a chemical enters the body either by penetrating the surface or entering the tracheal system. Such insecticides can be useful in controlling phytophagous insects that have sucking mouth parts and are thus not killed by ingestion of insecticides applied to plant surfaces. **household insecticide**. any insecticide produced for use in the home. Many contain pyrethrum and many commercial formulations also contain so-called inert components that are toxic. While pyrethrum is relatively nontoxic to humans, it is toxic to honeybees and many other beneficial insects and also to fish and wildlife. **natural insecticide**. any insecticide traditionally derived from plants. These substances (e.g., rotenone, pyrethrum) have traditionally been regarded as safe for use on animals. Nicotine (as nicotine sulfate) is an exception. Unless carefully used, poisoning may result. Piperonyl butoxide and rotenone are natural insecticides that are quite toxic to mink, especially to pups under eight weeks of age. **organochlorine insecticide** (chlorinated organic insecticide). Any chlorinated hydrocarbon that is a principal ingredient in an insecticide. Many of these insecticides are cumulative poisons that are very toxic to many nontarget organisms, some are carcinogenic, and some are environmentally persistent. They have thus fallen largely into disuse in favor of carbamate and organophosphate insecticides. Symptoms of intoxication in humans may include immediate or delayed vomiting; paresthesia and itching of lips, tongue, and face; headache; sore throat; fatigue; tremors; ataxia; weakness; confusion; convulsions; coma; and death from respiratory failure. There are usually no apparent lesions. **organophosphate insecticide** (organophosphorus insecticide). organophosphate esters (e.g., malathion, parathion, and leptophos) form a very large class of neurotoxic insecticides. Inhibition of the latter enzyme is chiefly responsible for organophosphate poisoning in both invertebrate and vertebrate animals. Some organophosphate insecticides also produce lesions in birds and mammals that are unrelated to their anticholinesterase activity. In these cases, symptoms appear 1—2 weeks following exposure. The hind limbs, and in severe cases the forelimbs also, become paralyzed due to direct damage of nerves. Symptoms of organophosphate poisoning in humans include bronchoconstriction with wheezing and complaints of tightness in the chest; nausea, vomiting, and diarrhea due to stimulation of muscles in the walls of the gastrointestinal tract; twitching; cramps; restlessness, anxiety, and emotional instability; headache; and insomnia. Severe poisoning may further cause respiratory and circulatory depression; convulsions; coma; and death due to respiratory paralysis. Organophosphate compounds are short-lived in the environment and have thus been considered less hazardous to nontarget organisms than many other pesticides (e.g., organochlorines). *See also* carbamate insecticide, organophosphorus compound, organophosphorus-induced delayed neurotoxicity.

organophosphorus insecticide. organophosphate insecticide.

insecticide 3960-X14. Strobane®.

insecticide no. 4049. malathion.

insecticide no. 497. dieldrin.

insecticide-resistant. denoting or pertaining to insects and related pests that continue to multiply when exposed to insecticidal substances. *Cf*. herbicide-resistant, pesticide-resistant.

insectivore. **1:** any eutherian mammal of the order Insectivora (moles, shrews, hedgehogs). **2:** less commonly, any animal that feeds on insects.

insectivorous (entomophagous). **1a:** insect-eating; feeding chiefly or exclusively on insects. **1b:** of or pertaining to an organism that feeds on insects.

inselberg (plural, inselberge). A steep sided hill, composed chiefly of hard rock, that rises abruptly above a plain. Such structures occur mainly in tropical and subtropical areas.

inselberge. plural of inselberg.

inspiration. inhalation.

inspiratory. of, pertaining to, or timed during inhalation, *q.v.*

inspire. inhale.

instar. the stage growth of insects and other arthropods between any two molts.

instinct. **1:** said to be the intrinsic (inherited) cause of behavior in which learning (at least direct, specific learning) does not seem to be involved. It is a much abused term and has been discredited by many, but not all, scientists. **2:** fixed action patterns or stereotypical behavior, evoked by certain stimuli, that is similar in all, or nearly all, individual animals of the same species. Given current theory and state of knowledge, it is better to refer to such behavior patterns as species-specific or species-typical behavior, rather than instinctive behavior. Such behavior is said to be instinctive if it develops in animals that have been reared in isolation from others of the same species. It is assumed by advocates of instinct that learning does not contribute to the development of such behaviors. This is demonstrably untrue in many cases as the "isolated" individual is not isolated from the environment or from itself. Certain so-called instinctive behaviors (or elements of such behaviors) are indeed learned in such situations. The best inference is that all behaviors are the result of inherited, structural, and environmental contributions.

instream use. *See* instream water use.

instream water use. Water use that takes place within the channel of a stream. fish propagation; hydroelectric power generation; navigation; and many recr eational activities are examples. Patterns of instream use locally or regionally bear importantly on water allocation and/or water rights.

integument. the covering of the body. In humans and other vertebrate animals it is comprised of the dermis and epidermis. The integument offers a partial barrier to absorption of xenobiotics, but its effectiveness varies greatly among vertebrates. Included also as part of the **integumentary system** are structures such as hair, feathers, nails, dermal scales, and antlers that arise from the dermis and/or epidermis. Concentrations of environmental chemicals (most notably heavy metals) in hair, feathers, and other integumental structures such as the antlers of deer can often be related not only to temporal and geographic gradients of exposure, but to nutritional deficiencies and disease states as well.

integumentary. pertaining to the integument or a structure arising from the integument.

interdependent association. mutualism.

interferon. A class of small proteins that stimulate viral resistance by cells.

intergrade. a soil that exhibits the properties of two distinctive and genetically different soils.

interior conditions. in forestry, conditions that obtain in a forest, patch, or stand internal to the point where edge effects no longer have a measurable influence on the environment or the biota. Environmental variables generally used to mark this point in temperate regions may include light intensity, relative humidity, temperature, wind, and snow accumulation and snowmelt. Interior conditions in mature forests are usually reached at a distance equivalent to a few times average tree height from the forest edge. *See* edge effect.

intermediate agricultural product. an agricultural product that has a higher per-unit value than a bulk commodity. Such a product is partly processed but not necessarily ready for the consumption. Examples may include soybean meal, wheat flour, vegetable oils, feeds and fodders, animal fats, hides and skins, live animals, and sweeteners such as sugars.

International System of Units (SI; SI Units; Systéme International d'Unités). a system of weights and measures adopted by the International Conference of Weights and Measures in 1960. The system is comprised of (1) seven base units (coherent units), the meter, kilogram, second, ampere, Kelvin, candela, and mole; (2) their decimal multiples and submultiples; (3) other metric units; and (4) derived or supplementary units. The units of one class of derived units (e.g., force, power, frequency) represent the base units to the powers of 10 (e.g., kilometer, millimeter); note that the kilogram, not the gram, was selected as the base unit for mass in this system. Another class of derived units consists of powers of base units (other than 10) and of base units in algebraic relationships (e.g., square meter, velocity, or meters per second). Many derived SI units have individualized names (e.g., joule, hertz, volt). The multiple and submultiple prefixes are tera- (T, 10^{12}), giga- (G, 10^{9}), mega- (M, 10^{6}), kilo- (k, 10^{3}), hecto- (h, 10^{2}), and deca- (da, 10^{1}); the submultiples are deci- (d, 10^{-1}), centi- (c, 10^{-2}), milli- (m, 10^{-3}), micro- (μ, 10^{-6}), nano- (n, 10^{9}), pico- (p, 10^{-12}), and femto-(f, 10^{-15}). The ISU has not entirely replaced older systems of measurements in the natural sciences.

intersegmental. between segments.

intersex. an organism with characteristics that are intermediate between those of males and females or that are indistinct from those of either sex. Some intersexes are sterile, some hermaphroditic. In many species, but not all, intersexes are abnormal individuals. *See* freemartin, gynandromorph.

interspecies effect. *See* interspecific variation.

interspecific. between or among species.

interspecific variation (interspecies effect). variation among species. Variation in structural and functional attributes, even closely related species, can be far reaching and complex. *See also* extrapolation, interspecies extrapolation, grand extrapolation; rodentia; exposure; polymorphism.

interstaminal. an adjective indicating position between stamens.

intertropical hyphemia. *See* ancylostomiasis.

intestinal intoxication. *See* autointoxication (intestinal autointoxication).

intraperitoneal (i.p.). within the peritoneal cavity.

intraepithelial glands. localized accumulations of epithelial secretory cells as those of the urethra.

intrauterine. within the cavity of the uterus.

intrauterine death. death of a fetus while inside the mother's uterus.

intravenous feeding. parenteral alimentation.

introgression. the backcrossing of two plant hybrid populations to introduce new genes into a wild population.

inure. acclimatize.

invasiveness. the ability of an organism to spread beyond its site of origin and become established in new locations.

invertebrate. **1a:** lacking a vertebral column. **2:** an animal that does not possess a vertebral column or backbone.

inverted nipple. blind quarter.

involucre. **1:** a ring of bracts at the base of a compact flower head. **2:** a circlet of leaves that surround the groups of archegonia and antheridia in mosses and liverworts.

io moth. *Automeris io*.

io-moth dermatitis. dermatitis resulting from contact with irritant setae of *Automeris io* (io moth).

ion. an atom, group of atoms, compound, or radical that is electrically charged due to the loss or gain of one or more electrons and thus bears a net positive (cation) or negative (anion) electric charge. Both the chemical and toxicological properties of an ion differ from those of the neutral atom from which it derives. *See* anion, cation.

ioxynil (3,5-diiodo-4-hydroxybenzonitrile; 3,5-diiodo-4-hydroxyphenyl cyanide; 2,6-diiodo-4-cyanophenol). a postemergence contact herbicide with little residual activity. It is used to control broad-leaved seedling weeds in cereal crops and sports turf. Effects appear within 24 hours of application as necrotic foci on leaves. This damage spreads, ultimately killing the plant. Untreated leaves develop chlorosis. Ioxynil is more toxic than bromoxynil.

ioxynil octanoate. a waxy, water-insoluble solid used as an insecticide on cereal grains and sugarcane.

ipado. *Erythroxylum*.

IPC. propham.

ipecac. **1:** *Euphorbia ipecacuanha* (ipecacuanha). **2:** ipecacuanha; hippo; the dried rhizome and roots of *C. ipecacunha*. Formerly used as an emetic, it is now prized chiefly as a source of emetine. It contains the alkaloids cephaeline, emetamine, psychotrine, protemetine, and resin. *See also* ipecac syrup. Ipecac poisoning is termed emetism, *q.v.* **3:** the dried roots or rhizomes of any of several plants having uses similar to those of *C. ipecacunha*. **4:** *Euphorbia ipecacuanha*.

ipecac spurge. *Euphorbia ipecacuanha*.

ipecacuanha. *See Euphorbia ipecacuanha*, ipecac.

ipomea. the dried root of *Ipomoea orizabensis*.

Ipomoea (morning glories). a genus of chiefly herbaceous plants (Family Convolvulaceae); included are both ornamental and weedy species. ***I. batatas*** (sweet potato). a tropical, small-flowered American plant that is now grown worldwide as a food crop. The vines may contain toxic concentrations of nitrate. ***I. orizabensis***. the dried root (ipomea, jalap root; Mexican scammony root) is used as a cathartic.

Iridomyrmes humilis (Argentine ant). an ant (Family Formicidae) of the southeastern United States and California. It is a pest of orchard trees, shade trees, and bee colonies, and is sometimes a household pest.

iris. **1:** common name for certain members of the iris family (Iridaceae), especially those of the genus *Iris*. **2:** the rhizome of *Iris versicolor* (*q.v.*), which was formerly used as a purgative, emetic, and diuretic. **3:** the most anterior part of the vascular tunic of the vertebrate eye. It is a circular pigmented membrane that lies behind the cornea; it is perforated by the pupil.

Iris (iris). a large genus of wild and cultivated plants (Family Iridaceae) named after the Greek goddess of the rainbow. Both wild and cultivated forms contain an irritant substance in the leaves and especially in the thick, fleshy underground rhizome. Ingestion of moderate amounts can produce severe gastrointestinal distress. ***I. versicolor*** (blue flag, poison flag). the common wild iris of North America. It is beardless and rhizomatous bearing blue-violet or perianth white flowers with yellow markings. All parts are toxic, especially the rhizome.

Irish potato. *Solanum tuberosum*.

iron gluconate. ferrous gluconate.

iron sulfate. ferrous sulfate.

iron vitriol. ferrous sulfate.

irreversible toxic effect. *See* effect.

irrigate. to flood, flush out, or thoroughly rinse (e.g., a cavity or wound) with a liquid.

irrigation. **1:** flooding, flushing, or rinsing with water or other liquid, as in flushing out a canal, a cavity, or a wound. *Cf.* lavage. **2:** the process of supplying water artificially to land that supports growing crops. This is usually accomplished by a system of pipes, ditches, or furrows; by flooding; or by the use of catchments or basins.

isehidrosis. anhidrosis.

Isle of Wight disease (acarine disease). a European disease of adult bees, unknown in North America. It is a serious disease caused by *Acarapis woodi*, a mite that invades the tracheae.

isobutane (2-methylpropane; trimethylmethane). an isomer of butane, *q.v.* and an important component of natural gas.

isobutanyl chloride. 3-chloro-2-methylpropene.

***N*-isobutyl-2,6,8-decatrieneamide**. affinin.

***N*-isobutyldeca-*trans*-2-*cis*-6-*trans*-8-trienamide**. affinin.

isocarb. propoxur.

isocyanic acid. methyl isocyanate.

isodrin (endo; endo isomer; Compound 711). an isomer of aldrin, it is an extremely toxic chlorinated diene insecticide (no longer manufactured or used in the United States). Chemical and toxicological behavior is very similar to that of aldrin.

isofluorphate. diisopropyl fluorophosphate.

isoflurophosphate. diisopropyl fluorophosphate.

isoinokosterone. β-ecdysone.

isoniazid (4-pyridinecarboxylic acid hydrazide; isonicotinic acid hydrazide, INH). a tuberculostatic antibacterial used as an antimycotic agent in veterinary practice. Adverse effects of long-term use include liver damage and peripheral neuropathy. Rashes, fever, and CNS effects are also common.

isolate. **1:** to separate a substance or entity from other entities or materials. **2a:** to free from contaminants. **b:** to separate in pure form from a mixture. **3:** that which is separated or isolated (e.g., a chemical isolate, a viable organism separated from its host or culture system). **4:** a plant or animal placed in quarantine.

IsoPPC. propham.

2-isopropenyl-2,3-dihydro-5-acetylbenzofuran. tremetol.

2-isopropoxyphenyl-*N*-methylcarbamate. propoxur.

***o*-isopropoxyphenyl-*N*-methylcarbamate**. propoxur.

isopropyl carbanilate. propham.

isopropyl *N*-(3-chlorophenyl)carbamate. chlorpropham.

***O*-isopropyl *N*-phenyl carbamate**. propham.

***N*-isopropyl-α-chloroacetanilide**. propachlor.

isopropyl-*m*-chlorocarbanilate. chlorpropham.

isopropylfluophosphate. diisopropyl fluorophosphate.

isopropylphosphoramidothioic acid *o*-2,4-(dichlorophenyl) *o*-methyl ester. DMPA.

3-isothiocyanato-1-propene. allyl isothiocyanate.

isothiocyanic acid allyl ether. allyl isothiocyanate.

3-isothiocyanato-1-propene. allyl isothiocyanate.

isoxaben (*N*-[3-(1-ethyl-1-methylpropyl)-5-isoxazoleyl]-2,6-dimethoxybenzamide; benzamizole). a selective preemergent herbicide used for cereal crops. It is slightly toxic *by mouth* to laboratory mice and rats, but the inhaled vapor is extremely toxic to laboratory rats.

isoxazoles. natural insecticides produced by poisonous mushrooms such as *Amanita muscaria*, *A. pantherina*, and *Tricholoma muscarium*. These compounds also have narcotic and psychomimetic effects on humans. Examples are ibotenic acid, muscazone, muscimol, and tricholomic acid. All but ibotenic acid cause visual damage, mental confusion, spatiotemporal dislocation, and memory loss in humans.

Italian broccoli. *B. oleracea* var. *botrytis*.

itchweed. *Veratrum viride*.

Ixodes. a genus of ticks (Family Ixodidae) that are parasitic on mammals. *See* tick, Ixodidae, Ixodoidea.

Ixodidae. a family of arachnids (Order Acarina, Superfamily Ixodoidea) that includes the hard ticks. They are usually parasitic on mammals. *See* tick, Ixodoidea, *Ixodes*, *Dermacentor*.

Ixodoidea (ticks). a superfamily of small biting arachnids (Order Acarina) comprised of species with a conspicuous toothed hypostome. Most are parasitic on mammals. Certain species are venomous (e.g., *Dermacentor andersoni*), producing local erythema, pain, edema, and muscle cramps or serious systemic effects (*See* tick paralysis); others are important vectors of disease (e.g., Rocky Mountain spotted fever, lyme disease); some are chiefly a nuisance to the host. In any case, the wound may become infected. Application of gasoline or heat to the tick usually causes it to loosen its jaws and it can then be removed with forceps (tweezers) and the wound carefully washed. If the jaws remain embedded in the wound, infection or other complications may result. In such cases, a physician should be notified.

J

jack. a male burro.

jack ass. **1:** a burro. *See Equus a. asinus*. **2:** a male burro.

Jack-in-the-pulpit. a common name of several species of *Arisaema* including *A. atrórubens, A. stewardsonii,* and *A. triphyllum*.

Jack o' Lantern. *Omphalotus olearius*.

Jacutin. a trade name for lindane.

jalap root. the dried root of *Ipomoea orizabensis*.

Jamestown weed. *See Datura*.

jaundice (icterus). **1:** a condition characterized by a yellowish skin, sclerae, mucous membranes, deeper tissues, and body fluids; the stool is light and the urine dark. The discoloration is due to the deposition of bilirubin, the concentration of which is increased in blood plasma. Jaundice is seen in a number of disease states and may result from obstruction of the biliary passageways, from hemolysis, or from impaired functioning of hepatic cells. *See also* atrophy (acute yellow atrophy). **2:** to produce jaundice. **toxic jaundice**. that caused by chronic or acute poisoning of the liver by bacterial toxins, or hepatotoxicants such as carbon tetrachloride, phosphorus, or arsphenamine. It is the most common type of jaundice.

jaundiced. icteric.

jake neuritis. Jamaica ginger paralysis (*See* paralysis).

Jamaica ginger polyneuritis. Jamaica ginger paralysis (*See* paralysis).

Java bean. **1:** *Phaseolus limensis*. **2:** the seed of *P. limensis*.

jenny. a female ass.

Jequirity bean. *Abrus precatorius*.

Jerusalem oak. *Chenopodium botrys*.

jessamine. *See* Gelsemium.

jimbrush. *C sordiatus*.

jimmies. a syndrome of livestock, especially sheep grazing on *Notholaena cochisensis* (jimmy fern). Affected animals appear normal until exercised, when they exhibit progressive incoordination, finally stopping in spasm with the back arched. There are associated violent tremors and a pronounced increase in heart and respiratory rates; sometimes prostration. If an affected animal is unmolested, it will usually recover in less than 30 minutes. The attacks recur with repeated exercise. If the poisoning is severe, or further exercise is forced on the animal, death by respiratory failure may ensue. Pregnant animals are most susceptible. Animals usually recover. Domestic fowl s usually recover within a fortnight following removal from range with Jimmy fern.

jimmy fern. *Notholaena*, a fern of the southwestern United States. One race (*cochisensis*) causes a nervous disease in sheep (*See* jimmies) whereas another, *sinuata*, is nontoxic.

jack pine. *Pinus banksiana*.

jimmy weed. common name of any plant of the genus *Haplopappus*.

jimsonweed. *See Datura*.

Job's tears. adlay.

Joe Pye, **Joe Pye weed**. a common name of a number of species of *Eupatorium* (e.g., *E. fistulosum*, *E. maculatum*, *E. verticillatum*), especially *E. purpureum*. It was formerly believed that a Native American physician named Joe Pye used this plant as a medicine in the early colonial days of Massachusetts.

Johnson grass. *Sorghum halepense*.

Jonathan spot. a disease of stored Jonathan apples, in which the fruit have circular lesions.

jointed charlock. *See Raphanus*.

jonquil. *Narcissus jonquilla*.

Juglans nigra (black walnut). a tree (Family Juglandaceae) that poisons plants situated under the canopy. The toxic principle is juglone. *See also* allelopathy, *Salvia leucophylla*, *Artemesia californica*.

K

κ. lowercase kappa.

K. uppercase kappa.

kafir corn. a cultivated grain sorghum. *See Sorghum vulgare*.

kairomone. a biologically active chemical produced and disseminated by living organisms of one species that is advantageous to recipient species. Examples are chemical lures (sometimes), inductants that stimulate adaptation, stimulants (e.g., growth factors), and warning substances (those that signal danger). *See also* allelochemical, allomone, signal.

kale. *Brassica oleracea* var. *acephala*.

Kalmia (laurel; American laurel). a genus of mostly North American evergreen shrubs (Family Ericaceae). Poisonous species occur throughout North America, usually in moist (mesic) areas. They contain grayanotoxins, the toxic principle in most cases being andromedotoxin (grayanotoxin I). All parts of the plant are very toxic to humans, producing lacrimation, rhinorrhea, salivation, nausea, vomiting, dyspnea, hypotension, bradycardia, depression, convulsions, paralysis, and sometimes renal failure. Death may occur in as little as 12 hours. ***K. angustifolia*** (dwarf laurel; sheep-laurel; calfkill; lambkill; sheepkill; wicky). an open woody shrub, growing to a meter or so, usually in abandoned fields in northeastern North America. It is decidedly less dangerous than *K. latifolia*. ***K. latifolia*** (mountain laurel; calico bush; poison laurel; spoonwood; mountain ivy; ivy bush). a dense woody shrub or small tree up to 3½ m tall growing in rocky wooded areas or clearings in the eastern United States. This is probably the best known broadleaf evergreen shrub in North America. The leaves and shoots are highly toxic and were used by American Indians to commit suicide. Mountain laurel is especially dangerous to children; fatalities have occurred. It is, nevertheless, an important browse for deer during winter. ***K. polifolia*** (pale laurel, bog laurel). a small, woody bush, up to 25 cm tall, of wet meadows and bogs in the Rocky Mountains from Alaska southward. At least one variety is poisonous.

Karakul sheep. a breed native to central Asia but raised in a number of countries on several continents. Adults are medium-sized and broad-tailed. They produce a relatively coarse wool that is used to produce heavy fabrics and carpeting. Color varies from black to tan and gray. Late-term fetuses and newborn lambs typically have tightly curled black fleece and provide commercial lambskin (Persian lamb, karakul, broadtail, krimmer, astrakhan). The meat of these lambs is good.

karst. topography characterized by sinkholes, closed depressions, underground caverns, and solution channels.

Karwinskia humboldtiana (coyotillo; tullidora; "buckthorn"). a poisonous, spineless, woody shrub or small tree (Family Rhamnaceae), 1—6½ m tall, that occurs in the deserts of northern Mexico and southwest Texas. The fruit (including seeds and pulp) is very toxic to all forms of livestock and to humans. A paralytic syndrome results from ingestion of the fruit, which contains a mixture of anthracenones ("buckthorn toxins").

Symptoms appear within a few days to several weeks following exposure (1—3 weeks in humans) and include malaise, quadriparesis, and ultimately paralysis of bulbar and respiratory muscles. In humans, the muscles become progressively weaker, beginning with the lower extremities. In livestock, weakness and incoordination first appear in the hind legs, the forelegs often being affected later. Appetite remains normal, but complete prostration can develop in about one month, often followed by death from respiratory failure. Recovery may take up to a year. A second syndrome has been produced experimentally by feeding the less toxic leaves to livestock. This produces nausea, wasting, progressive weakness, and death. *See* buckthorn toxins.

karyogamy. the fusion haploid (N) nuclei in a cell to form a diploid (2N) nucleus.

karyomite. chromosome.

katabatic. of or pertaining to a wind or air current that moves downward or down a slope. *Cf.* anabatic (def. 1).

kcal. kilocalorie.

keg. a small barrel.

Kelvin scale. Kelvin temperature scale.

Kelvin temperature scale (absolute temperature scale; Kelvin scale). the temperature scale in which absolute zero (the temperature at which all molecular motion presumably ceases to exist) = -273° on the Celsius scale. The freezing point of water on this scale is 273°C. and the boiling point of water is 373°C. *Cf.* Celsius temperature scale.

Kentucky coffee tree. *Gymnocladus dioica*.

Kepone. trade name for chlordecone.

Kepone poisoning. chlordecone poisoning.

kerosene (kerosine). a clear, pale yellow, or nearly colorless mixture of liquid petroleum hydrocarbons, mainly of the methane series having 10—16 carbons per molecule. It is used as a reagent and in insecticides, and is occasionally burned for heating or light. Orally it is very toxic to laboratory mice; inhalation of high concentrations may produce drowsiness, coma, and death. Kerosene stoves, unless very well ventilated, may release toxic levels of carbon monoxide.

kerosine. kerosene.

ketogenesis. The production of ketone bodies.

ketone. a class of acidic, partially oxidized hydrocarbons. Ketones are end products of metabolism in animals.

ketone body. any of a class of compounds that include acetone, acetoacetic acid, and beta-hydroxybutyric acid. All are derived from fatty acids and produced chiefly in the liver by the condensation of acetyl CoA, and produced during fatty-acid oxidation. They enter the circulation and undergo further degradation in peripheral tissues. Ketone bodies are normally released into the peripheral circulation and are further oxidized with release of usable energy by peripheral tissues, most notably by skeletal muscles. *See also* acetonemia, ketonemia, ketonuria, ketosis.

ketone propane. acetone.

ketonemia. The presence of detectable concentrations of ketone bodies in circulating blood plasma. *Cf.* hyperketonemia.

ketonuria. the presence of ketone bodies in the urine, as in acetonemia (ketosis) of highly productive dairy cattle.

β-keto-propane, β-ketopropane. acetone.

ketosis. the accumulation of ketone bodies in an organism, often misdiagnosed. In ketosis, large amounts of ketone bodies may be eliminated in the urine. If untreated, ketosis may progress to ketoacidosis, coma, and death. The condition is due to incomplete metabolism of fatty acids, usually as a result of carbohydrate deficiency or poor utilization of carbohydrates, as in starvation, consumption of high-fat diets, pregnancy, during ether anesthesia, and it may also occur in poorly-controlled cases of diabetes mellitus. The cause of ketosis in livestock is the incomplete metabolism of fatty acids due either to a lack of utilization of carbohydrates or a deficiency of carbohydrates in the diet. *See*, for example, lupinosis, trembles. *See also* acetonemia, ketone body, hyperketonuria, ketonuria. **bovine ketosis** (bovine acetonemia; bovine hypoglycemia). ketosis of cattle, which occurs most often in highly productive dairy cattle during peak production of milk or the postparturient period. **primary (bovine) ketosis**. that which occurs when the cow mobilizes body fat faster than the resulting ketones can be eliminated from the body. **secondary (bovine) ketosis**. that which results under conditions in which appetite is depressed (e.g., mastitis, metritis) or from a lack of sufficient nutrients. **ovine ketosis** (pregnancy disease; pregnancy toxemia). that which occurs in ewes during the last weeks of pregnancy. The condition results from the production of excess ketone, which is due mainly to mobilization of fat more rapidly than ketone bodies can be eliminated from the body. Pregnant sheep are especially susceptible because the developing lamb (or more often, twin lambs) are energy demanding. Poor nutrition, change in diet, and other sources of stress such as inclement weather may precipitate the condition. Affected ewes may be withdrawn or indifferent to the flock; they may stand for hours with their heads propped against a wall or fence; they may run into objects, appearing to be blind. Common signs in the later stages of the disease include circling, twitching, muscle spasms, and sometimes convulsive jerking of the head. Periodic convulsions eventually lead to exhaustion and death within 3—4 days. Mortality is high and ewes that have well-developed clinical signs may not respond to treatment.

ketotic. pertaining to ketosis.

key (samara). the two-winged fruit of a maple (*See Acer*).

kg. kilogram.

kiang. a wild ass, *Equus kiang*.

kid. a young goat, of either sex, that is lessthan one year of age.

kidding. birthing (parturition) by a goat.

kidding pen. a pen in which the doe goat is kept when birthing.

kidney beans. *See Phaseolus*.

killer bee. **1:** *Apis millifera dansonii*. **2:** an American hybrid of the African and European honeybees produced by the mingling of domesticated European colonies with an expanding and migrating African colony that escaped from an apiary in Brazil.

kilocalorie (kcal). 1000 calories.

kilogram (kg). 1000 grams. *See also* International System of Units.

kindle. **1:** to light or ignite a fire. **2a:** to arouse, waken, inspire. **2b:** to provoke, incite, stir up. **3:** a group or litter of kittens.

king's gold. arsenic trisulfide.

king's yellow. arsenic trisulfide.

kinin. cytokinin.

kinin peptide. a component of bee venom that can cause smooth muscle contraction and low blood pressure.

kissing bug. assassin bug.

kiwi. **1:** a flightless bird (*Apteryx spp.*) native to New Zealand. **2:** *See Actinidia*.

kiwi fruit. *Actinidia chinensis*.

klapperschlange. any snake of the genus *Crotalus*.

kodo millet. *Paspalum scrobiculatum*.

kohlrabi. *B. oleracea* var. *gongylodes*.

Krebs cycle. tricarboxylic acid cycle.

krubi. common name of plants of the genus *Amorphophallis*.

krysid. ANTU.

Kwell. a trade name for lindane.

kyanol. aniline.

Kleen-up™. *See* glyphosphate.

L

L. symbol for liter (litre).

label. **1a:** a mark, tag, symbol, or emblem; one that identifies, describes, warns, alerts, or informs one about a specific item. **1b:** any printed or written matter (e.g., a note, notice, tag, information, title, warning) attached or appended to a larger writing or to a material or substance, a manufactured item. **1c:** a notice attached to a container that conveys information concerning its contents, intended usage, and the manufacturer; and/or any appropriate or legally required notifications, concerns, or warnings concerning hazards that attend usage, transport, storage, and/or disposal. **2a:** an identifying mark, tag, marker, isotope, etc. **2b:** a substance that has a special affinity for a cell, tissue, organ, or microbe such that it may be introduced and fixed, serving to identify the cell, etc. **3:** an atom or molecule attached to a ligand or to a binding protein that is able to generate a signal that can be monitored in the binding reaction. **3a:** to append or affix a written or printed label. **3b:** to add or introduce and fix a substance into a cell, tissue, organ, or microbe. **warning label**. any label that indicates a danger. The legitimate purpose of a warning label (e.g., on a product) is to inform a user and other parties that might encounter the container of a danger (e.g., presence of a poisonous or otherwise hazardous substance) of which such parties might otherwise remain unaware. Federal laws in the United States require that warning labels be affixed to potentially dangerous products such as drugs, cigarettes, household chemicals, tools, and waste materials. Warning labels may underestimate, or be ambiguous about, the hazard involved. Consequently, items so labeled should be used with considerable caution and stored properly and safely away from access by children. *See also* warning. Examples of warning labels that one may encounter (with sample interpretations in the United States, depending on the nature of the product) include the following:

CAUTION

This label denotes a product that is: (1) harmful if swallowed. It is an irritant. Contact with the eyes and prolonged contact with skin should be avoided. Do not swallow or inhale the vapors. Use in a well-ventilated area. (2) This label usually means that the contents are not lethal unless two tablespoons or more are ingested. (3) Contents are irritant or erosive to the skin, eyes, and mucous membranes.

DANGER POISON
(with skull and crossbones).

This label means that the contents are able to kill an adult if only a tiny pinch or drop is ingested.

Extremely Flammable/
Flammable/Combustible.

This labels warns that the material in the container can ignite if exposed to a flame or spark.

HAZARDOUS WASTE -
HANDLE WITH CARE

This label indicates the presence of a material in the container or conveyance (e.g., rail car, truck) that serves no useful purpose and may be extremely dangerous if it escapes the container. There may be a subsidiary label that indicates the nature of the hazard (e.g., Inhalation Hazard, Marine Pollutant, corrosive substance). The use of such labels is regulated by law and improper disposal of such material is prohibited by law.

Strong Sensitizer.

This label indicates the presence of a substance that may provoke an allergic reaction.

labeled. affixed with a label or tag.

labial gland. *See* salivary gland (minor salivary gland).

Labiatae. the mint family. *See Salvia*.

labriform milkweed. *Asclepias labriformis*.

Laburnum (laburnum). a genus of poisonous shrubs and small trees (Family Fabaceae) that are native to Europe. They have bright yellow flowers, trifoliate leaves, and pendulous racemes. ***L. anagyroides*** (= *Cytisus laburnum*; common laburnum; golden chain; golden chain tree; laburnum). a large, toxic, ornamental shrub or small tree (Family Fabaceae) that grows to 10 m. It is native to the mountains of southern Germany, Switzerland, and France but is cultivated as a garden plant in most parts of the world. The leaves and seed pods are very toxic to humans and to grazing or browsing animals; the active principle is the quinolizidine alkaloid; cytisine, which produces effects similar to nicotine. The seeds are very toxic to livestock and humans; browsing or grazing animals may eat the leaves and pods with fatal results. Symptoms of intoxication include excitement, miosis, nausea, severe vomiting, diarrhea, irregular pulse and breathing, dysphagia, incoordination, convulsions, delirium, and coma; miosis may be noted. Death may result from asphyxiation. Renal damage may occur. The milk of poisoned cattle may be toxic. *See also* cyticism.

laburnum. a plant of the genus *Laburnum*, especially *L. anagyroides*.

lachrymal. lacrimal.

lachrymal gland. lacrimal gland.

lachrymation. lacrimation.

lachrymator. lacrimator.

lachrymatory. lacrimatory.

lacquer tree. *Rhus verniciflua*.

lacrimal (lachrymal). **1:** of, pertaining to, or situated near the lacrimal glands. **2:** characterized by tears or lacrimation.

lacrimal gland (lachrymal gland). one of several acinous glands under the upper eyelids of terrestrial vertebrates that secrete tears, which clean the cornea and keep it moist.

lacrimation (lachrymation). the secretion and discharge of tears (tearing, weeping, crying), especially in excessive amounts.

lacrimator (lachrymator). an agent that stimulates lacrimation on contact with the eyes of a vertebrate animal.

lacrimatory (lachrymatory). of, pertaining to, or able to cause lacrimation.

ß-lactamase. ampicillin.

lactate. **1:** a salt or ester of lactic acid. **2:** a salt of beta-hydroxy propionic acid. **3:** to secrete milk.

lactation. **1:** the production of milk. **2:** lactation period; the period following birth during which the breast continues to secrete milk.

lactation tetanus. grass tetany.

lactational. pertaining to lactation.

lacteal. **1:** milky; pertaining to or similar to milk. 2. a lymphatic vessel that conveys chyle from the intestine.

lactiferous (lactigerous). milk-producing.

lactigerous. lactiferous.

lactimorbus. milk sickness. *See* sickness.

lactose (milk sugar; saccharum lactis; 4-O-β-D-Galactopyranosyl-D-glucose; 4-(β-D-Galactosido)-D-glucose). a sweet-tasting, odorless, white powder or hard, crystalline, water-soluble mass that occurs uniquely in milk:

$$C_{12}H_{22}O_{11}\cdot H_2O$$

It is insoluble in ether and chloroform and is stable in air. It has various uses and can be taken orally as a diluent for substances that would be excessively irritant if given in pure form. Milk from dairy cows contains about 5% lactose.

lactose intolerance. a condition of adult humans in many parts of the world in which lactose is inadequately hydrolyzed in the intestine because of a lack or insufficiency of the enzyme lactase.

Lactuca. a genus of lettuce (Family Compositae). Concentrations of nitrate that are toxic to livestock have been reported in some species such as *L. sativa* (includes salad varieties such as romaine lettuce) and *L. scariola* (prickly lettuce).

lagomorph. an animal assigned to the Lagomorpha.

Lagomorpha. the order of mammals that includes rabbits, hares, and the pika. Skull characteristics and dentition are similar to those of rodents, but are defined by the presence of an additional pair of incisors and by modified hind limbs that permit jumping or hopping.

lamina (plural, laminae). **1a:** in anatomy, a thin, flat layer or plate. **1b:** in horses, a sensitive lamina to which the hoof attaches.

laminae. plural of lamina.

laminitis. **1:** inflammation of any lamina. **2:** inflammation of the lamina of the horse.

Lamium amplexicaule (henbit; dead nettle). a common weed (Family Labiatae) of eastern North America and the Pacific Coast. It can cause staggers (*q.v.*) in sheep, horses, and cattle.

lamium amplexicaule poisoning. poisoning of livestock by *Lamium amplexicaule*. *See* staggers.

Lampropeltis triangulum (milk snake; milksnake). a relatively small, nonvenomous snake (Family Colubridae) that is not dangerous to humans. It occurs in diverse situations throughout much of the eastern United States and portions of the West that lie east of the Rocky Mountains. It is a secretive snake not often seen in the open. In the Northeast it is often mistaken for the copperhead (*Agkistrodon contortrix*), while in the South it is sometimes mistaken for the eastern coral snake (*Micrurus fulvius*). Its common name derives from the mistaken notion that it milks cows at night, often taking prodigious amounts. On occasion it is seen in barns at night, preying on mice (*See* adder). In captivity it has been known to live more than 21 years.

Lamziekte. botulism.

lanceheads. snakes of the genus *Bothrops*.

land capability. the ability of soil resources to support agricultural usage. This quality is usually expressed in terms of land capability classes and subclasses. These categories indicate the suitability of various soils for production of field crops. Soils are thus grouped according to functional properties observed when they are used to grow field crops (liabilities and limitations; risk of damage to the crop; the soils' response to treatment). Capability classes are designated by Roman numerals I through VIII. Class I soils are the best; Class VIII is the poorest.

land capability class. *See* land capability.

land farming (of waste). a process in which toxic or other hazardous wastes are deposited on or in the soil and are degraded naturally by microbes.

land treatment. any action, activity, or project applied to improve the conservation of soil, water, or other resources and to improve productivity of the land.

land-use planning. the process, often embodied in law, whereby decisions are made on future land uses that are thought to best serve the general welfare.

landscape ecology. *See* ecology.

landslide (landslip). **1:** landslip; downslope movement of a large mass of soil or rocks from a mountain or cliff; mud and debris may be included. Such may occur following a torrential rain or frequently repeated precipitation events that eventually saturate the soil, making it more dense and mobile. Landslides may also be fostered or triggered, for example, by road building along slopes, earthquakes, logging, or the undermining action of the ocean along coastal bluffs. **2:** the part of a plow bottom that slides along the wall of the furrow. It steadies the plow, chiefly by helping to offset the pressure exerted by the furrow slice against the moldboard.

landslip. *See* landslide (def.1).

Lantana (lantana). a genus of shrubby plants (Family Verbenaceae), native mainly to Central America northward to the United States regions that border the Gulf of Mexico; also widely used as ornamentals. All varieties can cause hepatogenic photosensitivity in grazing animals. The hepatotoxin is a polycyclic triterpene, rehmannic acid (=lantadene). The original toxic extract of *Lantana* was named lantanin, *q.v.*). The phototoxic agent is phylloerythrin. ***L. camara*** (lantana; red sage; wild sage). A perennial shrub in central and southern Florida; herbaceous farther north. The native plant is usually tall and scraggly, but a number of varieties have been developed as houseplants, greenhouse, and bedding plants. All parts, especially the green berries, are toxic, producing photosensitization of increasing severity with increased exposure to sunlight. In addition to hepatogenic photosensitization, this plant produces symptoms of severe gastroenteritis in livestock. Symptoms of acute poisoning resemble those of belladonna alkaloid (atropine) poisoning. The affected animal is weak and experiences severe gastroenteritis with bloody, watery feces within a day of consumption and dies within 3—4 days. In chronic cases, the animal lives longer and mild to severe symptoms of photosensitization also develop. Complications are common, especially in chronic poisoning. These may include conjunctivitis, severe constipation, jaundice, infestations of bacteria or blow flies, corneal opacity, and blindness. Livestock fatalities are known from Africa, Australia, and the United States. The berries are believed to have fatally poisoned children in the United States.

lantana. *See Lantana*.

lapse rate. **1:** the theoretical rate of decrease in atmospheric temperature with increase of elevation vertically above a given point. **2:** sometimes, the rate of change of any meteorological component with increasing altitude. **adiabatic lapse rate** (adiabetic rate; dry adiabatic lapse rate). the lapse rate of a mass of air moving to a higher altitude through an atmosphere in hydrostatic equilibrium. In such a case, there is theoretically no exchange of heat with the environment. Cooling of the mass is due to expansion as it rises through air of decreasing pressure; the rate of cooling is 5.4^{o}F/1000 ft. of rise.

larch. the common name of any tree of the genus *Larix*.

lard. the rendered fat of swine.

large lungworm. *Dictyocaulus filario*.

larkspur. plants of the genus *Delphinium*.

larkspur poisoning (delphinium poisoning). poisoning by ingestion of plants of the genus *Delphinium*, commonly known as larkspurs. All parts of these plants are toxic, producing the same symptoms as monkshood poisoning. These symptoms may include paresthesia, a burning sensation of the mouth and skin; itching; dermatitis; nausea; vomiting; bradycardia; hypotension; weak pulse; respiratory distress; cyanosis; and convulsions. Symptoms appear almost immediately and death may occur within six hours. The active principles are delphinine and related alkaloids. Sheep are highly resistant and horses rarely eat enough to be gravely poisoned. **larkspur poisoning of cattle**. the indigenous larkspurs of the western United States are an important source of mortality in range cattle. New growth and seeds have the highest concentrations of toxic alkaloids. Both cattle and sheep readily eat larkspur; cattle are often fatally poisoned, but sheep rarely die. Range cattle lose control of motor functions and are eventually unable to stand or regain their feet. Poisoned cattle may choke to death if they vomit while prone. Larkspur-poisoned cattle on hilly ranges often fall with their heads downhill and may be unable to belch; they are thus susceptible to fatal bloat. Death is more often due to respiratory paralysis. In surviving animals, recovery is complete.

larvicidal. destructive to larvae.

larvicide. a chemical that destroys larvae.

lash. **1:** to flail, flog, whip, or scourge; to beat. **2:** to fasten, secure, tether, strap, or tie.

lasher. **1:** one who lashes, beats, or whips. **2:** a length of rope used to lash or secure two or more items together.

Lasso. a tradename for alachlor.

latent. **1:** dormant, inert, quiescent; present, but hidden or inactive; undeveloped. **2:** a period in the course of disease when signs and symptoms have yet to appear; or a period when the pathogen is dormant. **3a:** dormant or seemingly inert, but able to develop under certain circumstances. **3b:** descriptive of characteristics or phenomena that appear or become active under certain circumstances or after some period of time.

latent effect. *See* effect.

latent period (reaction time). a period of inapparent activity between receipt of a stimulus and an appropriate or anticipated response by an irritable tissue or organ.

lateral. **1a:** to the side; farther from the midline. **1b:** pertaining to a side.

laterite. tropical red soils that contain alumina and iron oxides, but little silica due to leaching that occurs under hot, moist conditions.

lathyrism (sweet pea poisoning; lathyrus poisoning). poisoning by plants of the genus *Lathyrus*. Many epidemics have been recorded, although lathyrism in humans is now mainly of historical interest. In many countries the poisonous species are those used as livestock feed. Lathyrism is a paralytic syndrome caused by the ingestion of large amounts of raw or cooked seeds or flour from plants of the genus *Lathyrus*, especially *L. cicera*, *L. clymenum*, *L. latifolius*, *L. odoratus*, and *L. sativus*; *Vicia sativa* may also be involved (*See* vicine). The condition is characterized by partial or total spastic paraplegia usually of the legs, but may also involve the arms in severe cases. Other symptoms may include prodromal pain, cramps, hypothesia or hyperesthesia, and paresthesia. *See also* lupinosis, osteolathyrism, *Lathyrus*.

lathyrus poisoning. lathyrism.

Lathyrus (vetchling; wild pea). a cosmopolitan genus of more than 200 species of annual and perennial herbs (Family Fabaceae). The seeds of a number of species are toxic. Toxic species include *L. cicera* (flatpod pea), *L. clymenum* (Spanish vetch), *L. hirsutus*, *L. odoratus* (sweet pea), *L. latifolius* (sweet pea; perennial pea; everlasting pea; pois vivace (Quebec)), *L. pusillus*, *L. sativus* (grass pea, Indian pea, green vetch), *L. sphaericus*, *L. splendens*, *L. strictus*, *L. sylvestris* (perennial pea; everlasting pea), and *L. tingitanus*. The toxic principles are related to, or derived from, β-cyano-L-alanine. The toxic compound in at least some species of *Lathyrus* that causes paralysis and skeletal lesions is β-(γ-L-glutamyl)-aminopropionitrile. The toxic compound in *L. sylvestris* and *L. latifolius* is L-α,γ-diaminobutyric acid. It causes hyperexcitability, convulsions, and death without skeletal lesions. Both substances have a common precursor, β-cyano-L-alanine. *See also* lathyrism, divicine, vicine.

laughing gas. nitrous oxide.

laurel. *See Kalmia*.

lavage (lavation). **1:** the irrigation or washing out of a cavity or lumen (e.g., that of the stomach or intestines). **2:** to wash out or irrigate (e.g., a cavity or tissue). **blood lavage** (systemic lavage). the washing out of toxic substances from circulating blood by injection of serum into the veins. **gastric lavage**. washing out of the stomach with sterile water, normal saline, 1—5% sodium bicarbonate, or activated charcoal as indicated. This is usually performed to dilute and remove irritants or poisons or to cleanse the stomach prior to or following surgery. In such cases, an early sample of the fluid removed from the stomach is retained for chemical analysis. If appropriate, an antidote is added to the water used to rinse the stomach. Gastric lavage is not used if a corrosive poison has been swallowed because of the risk of the perforation of tissues by the lavage tube. Corrosive acids or alkalis may be diluted by giving large amounts of water or milk to the victim. **intestinal lavage**. dialysis by lavage using a fluid to remove, through the intestinal mucosa, materials that are not being excreted by the kidneys. **systemic lavage.** blood lavage.

lavation. lavage.

law. **1:** a binding rule or body of rules enacted by a governing authority such as a legislature or parliament; that which is ordained or decreed by a ruler or ruling body. **2:** a statement of a universal principle concerning natural phenomena (e.g., chemical, physical, or biological).

LBM. little brown mushroom (*See* mushroom).

leachate. 1a: the liquid product of leaching. **1b:** the solution formed when water (usually rainwater) has filtered through soil or another solid or semisolid matrix in the environment (e.g., toxic waste). The water may be enriched or contaminated as a result of such a process.

leaching (lixiviation). the separation of soluble from insoluble matter by the infiltration of a liquid solvent and the drawing off of the resulting solution. Examples are the percolation of rainwater through soil in the environment or through a hazardous waste site; and the extraction of sugar from sugar beets; the washing of a soluble salts from an insoluble precipitate.

lead tree. *Leucaena glauca*.

leaf (plural, leaves). the chief photosynthetic organ of most green plants, a leaf is a lateral outgrowth of the growing point of a stem. It typically consists of a stalk (the petiole), a thin, flat, expanded portion (the blade or lamina), and the base. The blade is needle-like in most conifers. Stipules that protect the leaf bud appear at the base of the leaves of many plants. Stipules that are green also serve as photosynthetic organs. The leaf blade typically consists of conducting tissues (xylem and phloem) and photosynthetic cells that comprise the mesophyll. The leaf surface is usually covered by a waxy cuticle, which is perforated by pores called stomata. Usually more numerous on the abaxial (lower) side of the leaf, the stomata permit entry and exit of gases and vapor. Leaves are reduced or absent in some green plants.

leaf beetle. any of numerous, often brightly colored beetles of the family Chrysomelidae, that feed on the leaves of plants. The larvae infest the roots, stem, and leaves. *See*, for example, *Oulema melanopus*.

leaf blight. **1:** a characteristic or phase of numerous plant diseases, characterized by necrotic spots or streaks on the leaves and accompanied by seed rot and seedling blight. **2:** any disease that has the characteristics of leaf blight (def. 1). **northern leaf blight** (northern corn leaf blight). a fungal disease of corn caused by *Exsherohilum turcicum*. The leaves of affected plants exhibit elongate tan-gray elliptical spots with subsequent blighting and necrosis.

leaf blotch. **1:** a characteristic or phase of certain diseases (notably fungal) of plants, in which there is a necrotic discoloration of the leaves. **2:** any plant disease having the characterstics of def. 1.

leaf-footed bug (coreid bug). any of numerous plant-sucking or predaceous bugs (Order Hemiptera, Family Coreidae) that normally have leaflike legs. Some are pests of food crops.

leaf lard. lard prepared from the leaf fat of a hog.

leaf lettuce. a leafy type of lettuce (Family Compositae) with loosely clustered, often rippled or curled leaves, sometimes tinged with red.

leaf miner, leafminer). the larvae of any of numerous insects that live in and feed on the parenchyma of leaves.

leaf mold. a compost or layer of soil comprised chiefly of decaying vegetable matter, especially leaves.

leaf roll. a viral disease of potatoes and certain other plants, characterized by chlorosis, an upward rolling of the leaflets, stunting, and necrosis of the phloem.

leaf rust. a fungal disease, especially of cereals and other grasses, characterized by spore-containing, rust-colored pustules on affected leaf blades and sheaths. **brown leaf rust**. a fungal disease of rye caused by the rust *Puccinia dispersa*. The disease is characterized by the presence of long, brown or reddish-brown, granular pustules that may erupt from any part of the plant.

leaf scald. a bacterial disease of sugarcane, most frequently due to *Bacterium albilineans*. The disease is characterized by irregular, whitish streaks on the leaves and defoliation.

leaf spot. **1:** a circumscribed, discolored, diseased area on a leaf that typically includes a central necrotic area. **2:** the name of any disease with such a characteristic as in def. 1. **angular leaf spot**. a disease of plants, caused by any of several species of bacteria, and characterized by angular, watery spots on the leaves and fruit. **angular leaf spot of cucumbers**. a disease in which the leaves initially appear watersoaked and later turn brown and die. The causative agent is the bacterium *Pseudomonas lachrymans*.

leafy spurge. *Euphorbia esula*.

leaky gene. hypomorphic allele (*See* allele).

leash. **1:** a lead, rope, cord, tether, restraint. **2a:** to restrain, control, confine. **2b:** to tether, harness, secure. **3:** group of foxes; a skulk.

leaves. plural of leaf.

lecheguilla. lechuguilla.

lechuguilla (lecheguilla). common name of *Agave lecheguilla* or any of several closely related species in this genus.

lechuguilla fever. lechuguilla poisoning.

lechuguilla poisoning (lechuguilla fever; swellhead). photosensitization of sheep and goats in western Texas, southeastern New Mexico, and northern Mexico due to ingestion of *Agave lecheguilla* (lechuguilla) and subsequent exposure to sunlight. The condition is marked by swelling and crusting of the face and ears. Additional effects include liver damage and consequent icterus, sometimes hemoglobinuria, and often death. Lesions include liver and kidney necroses.

leek. **1:** *Allium porrum*. **2:** common name of several plants related to *A. porrum*.

legend. a code, key, or table such as that which defines or explains, for example, symbols, formations, distance, and patterns that occur on a map.

leggy. **1:** descriptive of a green plant that has unusually long stems or of an animal that has unusually long legs relative to body size, as in very young ungulates.

legume. **1:** any member of the family Fabaceae (= Leguminosae). **2:** a type of fruit produced by a carpel that splits down both sides when ripe (e.g., peas, beans).

leguminous. of or pertaining to legumes or legume-like plants.

Leicester sheep. a breed developed from native stock in Leicestershire, England, in 1755. This breed was first introduced into the United States in colonial times and now raised locally in small numbers. **English Leicester**. a variety with a white face and legs, broad back, and heavy fleece. It is well fleshed and raised chiefly for the mutton. The rams, however, are often cross-bred to Merino ewes; their offspring produce choice wool. **border Leicester**. a variety developed in the border counties of England and Scotland by crossbreeding English Leicesters and Cheviots. The head is bare, the body square, and the carriage alert. Border Leicesters are more common English Leicesters in Canada.

lemma (flowering glume). the lower of a pair of bracts beneath each flower of a grass. *Cf.* glume, palea.

lenacil (3-cyclohexyl-6,7-dihydro-1*H*-cyclopentapyrimidine-2,4-(3*H*,5*H*)-dione; 3-cyclohexyl-5,6-trimethyleneuracil; 3-cyclohexyl-1,5,6,7-tetrahydro-2*H*-cyclopentapyrimidine-2,4(3*H*)-dione). a solid compound, $C_{13}H_{18}N_2O_2$, that is slightly soluble in water and most organic solvents; it is soluble in pyrimidine. A uracil, used as a selective herbicide, it effectively controls many annual and some perennial weeds with little or no damage to a variety of crops. It is absorbed chiefly through the roots. Lenacil is probably only slightly toxic to humans.

lentic. living in standing water such as a swamp, pond, or lake.

leopard spot. appaloosa.

lepto-. a prefix meaning slender.

leptophos (*O*-(4-bromo-2,5-dichlorophenyl)-*O*-methyl phenylthiophosphonate; *O*-(2,5-dichloro-4-bromophenyl)-*O*-methyl phenylthiophosphonate; *O*-methyl-*O*-(4-bromo-2,5-dichlorophenyl)phenyl thiophosphonate; MBCP). an insecticide and pesticide,

$$C_{13}H_{10}BrCl_2O_2PS$$

Leptophos was in use in the United States during the early to mid 1970s. It was banned from further use because of its toxic effects on production and agricultural workers. It is toxic to humans by all routes of exposure. It is neurotoxic, causing severe harm. Symptoms of intoxication are similar to those of multiple sclerosis and may include muscular weakness, staggering gait, tremors, paresthesia, tingling sensations, blurred vision, disorientation and temporary memory loss, auditory hallucinations, and paralysis of the legs. The compound may have teratogenic and mutagenic effects.

Leptospira. a genus of spiral-shaped, motile, aerobic bacteria (family Leptospiraceae). Some species are free living; others are pathogenic to animals. *See* leptospirosis.

leptospirosis. an infection that may affect humans but is chiefly a disease of certain livestock and other domestic animals. It is caused by bacteria of the genus *Leptospira* and is spread by the urine of infected people and animals. **human leptospirosis** (canicola fever; mud fever; swamp fever; hemorrhagic jaundice; infectious jaundice; spirochetal jaundice; swineherds' disease; Weil's disease). The disease tends to infect persons who handle infected animals, their tissues, or the urine of such animals, either directly or in manure, contaminated soil, or water. Signs and symptoms may include headache, fever, chills, and muscle pain; jaundice is not an uncommon feature and subcutaneous bleeding may occur; a rash may also appear. In serious cases, kidney failure and meningitis may occur.

exclosure. an area, such as one that has an easily damaged crop, that physically excludes selected animals (e.g., livestock, deer, or other wildlife).

ethylene chlorohydrin. *See* chlorohydrin.

lesion. **1a:** any circumscribed area of morbid or otherwise pathologically altered tissue. **1b:** focal injury or damage, as a wound. **1c:** a single infected area in a skin disease. **2:** any breakdown or discontinuity in a process or tissue, or loss of function of a part, however small, due to injury or disease.

lesser wax moth. *Achroia grisella*.

lethal. fatal, causing death; deadly.

lethal effect. *See* effect.

lettuce. *See Lactuca*; leaf lettuce.

Leucaena glauca (lead tree). a deep-rooted perennial shrub or tree (Family Fabaceae, formerly Leguminosae). In the United States, it occurs locally in wastelands of the Coastal Plain from Florida to Texas, and in arid lands in the Hawaiian Islands, where it is also cultivated. Ingestion of large quantities of this plant reduces rates of weight gain and results in generally poor condition and loss of hair in livestock. The disease has affected horses, donkeys, mules, rabbits, and swine in Hawaii, but cattle and domestic fowl do very well on *Leucaena* forage. The condition is reversible.

leucocyte. leukocyte.

leukocyte (leucocyte). a white blood corpuscle of which there are two types, granulocytes and agranulocytes, *q.v.* Leukocytes are phagocytic cells that help the vertebrate body to resist infection by ingesting particulate matter (*See also* basophil, eosinophil, neutrophil). While transported by the circulating blood, they can leave the bloodstream and penetrate tissues, apparently in response to signals from injured cells. When leucocytes are destroyed by bacteria, they form pus, which generally cause an abscess if there is no available outlet.

leukoderma. *See* albinism.

leukopathia. *See* albinism.

leukopenia. a condition of reduced numbers of leukocytes in circulating blood; in humans a count of 5000 cells/mm^3 or less. *See* agranulocytosis.

leukopenic. of, pertaining to, characterized by, or able to cause leukopenia.

LH (luteinizing hormone; interstitial cell stimulating hormone; luteotropin; ICSH). a glycoprotein gonadotropin that is secreted by the anterior lobe of the pituitary gland. It stimulates the ovaries to produce progesterone and acts in concert with FSH to stimulate release of estrogens from the Graafian follicles of the ovary. It also induces extrusion of the ripe ovum from the ovarian follicle (ovulation) and subsequent formation of the corpus luteum. *See also* hormone, pituitary gland.

lichen. a distinctive organism with a thallus composed of symbiotically associated fungal and algal cells. The fungus (mycobiont) is usually an ascomycete and is dominant to the algal component (phycobiont). The latter is usually green algae or blue-green bacteria. Many lichens are used as livestock and wildlife feeds, wildlife browse (e.g., *Cladonia rengiferina* - reindeer moss), and emergency food for humans in high latitudes. Several contain mildly toxic chemicals that are removed by boiling in water. Some contain antibiotics. *See Parmelia molliuscula*.

licorice root. *Glycyrrhiza glabra* or glycyrrhiza.

life cycle. **1a:** the set of stages through which an organism passes from origin (e.g., by budding, cell division, or conception) to death. **1b:** life history; the stages through which an organism passes from the beginning of life to maturity, including the reproductive cycle.

life expectancy. the period during which an organism of a given class (e.g., species, age, sex, occupation) can be expected to live on the basis of statistical analysis of existing data, usually in the form of actuarial of life tables.

life history. life cycle (def. 1b).

life root. *Senecio*.

lightning detection system. a network of electronic sensors distributed over a preselected land area and linked electronically to a central computer. The system detects and plots by triangulation, the location of cloud-to-ground lightning strikes in real time.

lignicole. **1a:** an organism that lives on or in wood. **1b:** lignicolous.

lignicolous. living on or in wood.

lignification. formation of wood, a process in which the cell walls of a plant thicken due to deposition of lignin.

lignin. a hard, fibrous polymer present in the walls of cells of xylem and fibers of schlerenchyma. It is comprised of phenylpropane units and serves to stiffen the cell wall. *See* lignification.

lignivorous. descriptive of organisms, especially various insects, that eat wood.

lignocellulose. combined lignin and cellulose, a constituent of wood.

lima bean. **1:** *Phaseolus lunatus*. **2:** the seed of *P. lunatus*.

limber leg. *See Acacia berlandieri*.

limberneck. a common name of botulism in poultry.

lime nitrogen. calcium cyanamide.

lime. **1a:** a citrus fruit. **1b**: the flavor of a lime. **2a:** calcium oxide; calcium hydroxide. **2b:** and salt of calcium. **air-slaked lime**. burned limestone that has been exposed to weathering. **bog lime**. marl.

limonite. *See* bog iron ore.

limicolous. living in mud.

Lincoln sheep. a very large, white-faced, hornless breed with coarse wool, developed in England. It is popular in the United States as the progenitor of other breeds.

Lindafor™. lindane.

lindane ($1\alpha2\alpha3\beta4\alpha5\alpha6\beta$-hexachlorocyclohexane; γ-benzene hexachloride (γ-BHC); gamma-benzene hexachloride; gamma-hexachlor; Aparasin; Aphtiria; Benzahex; Chemhex; ENT 7796; Esoderm; Gammalin; Gamene; Gamexan; Gamiso; Gammexane; Gamoxol; Gexane; Hexadon; Jacutin; Kwell; Lindafor; Lindatox; Lorexane; Pecusanol; Pultox; Quellada; Streunex; Tri-6; Vermaxan; Viton). the legal label name for γ-benzene hexachloride, lindane contains 99% or more of this isomer. It is used as a commercial scabicide, pediculicide, insecticide, and herbicide, and is toxic by all routes of exposure. Lindane is a useful ectoparasiticide for large animals and dogs. It is, however, highly toxic to cats in concentrations necessary for parasite control. Great care must be taken in its use; only cattle in good condition can be expected to tolerate 0.2% lindane applications. Stressed or emaciated cattle have been poisoned by being sprayed or dipped in 0.075% solutions. Horses and hogs appear to tolerate 0.2—0.5% sprays while sheep and swine usually tolerate 0.5% applications. Very young, emaciated, or lactating animals should be treated with extreme caution. The toxic effects of lindane are very similar to those of DDT. Degeneration of kidney tubules, liver damage associated with fatty tissue, and histoplastic anemia have been observed in humans. Exposures to small amounts of lindane can cause headaches, nausea, depression, and respiratory difficulties. Serious acute exposures can cause convulsions, elevated white blood counts, and heart failure. Lindane is fetotoxic and probably carcinogenic. Children are highly susceptible to the toxic effects of lindane. *See also* lindane poisoning.

lindane poisoning. poisoning by lindane, used chiefly in agriculture to control ectoparasites of dogs and larger animals. Cats are, however, poisoned at concentrations below that required for parasite control. Young or lactating animals and those in poor condition have heightened susceptibility to lindane poisoning and should be treated with extreme caution. Poisoning can occur by all routes of exposure. Toxic effects are very similar to those of DDT. **acute lindane poisoning**. symptoms in humans may include dizziness, irritability, restlessness, headache, weakness, nausea, vomiting, dyspnea, tremors, convulsions, and cyanosis. Death from respiratory failure or circulatory collapse may occur. The vapor may irritate the eyes, nose, and throat, and topical application may cause a localized reaction. Lesions may include liver damage associated with fatty tissue, hyaline degeneration of the renal tubules, and histoplastic anemia. Liver damage has occurred in chronically challenged experimental animals. *See also* γ-benzene hexachloride, lindane.

Lindatox. a trade name for lindane.

lingual gland. *See* salivary gland (minor salivary gland).

linseed meal. *See Linum usitatissimum.*

linseed oil (flaxseed oil). an amber to brown drying oil with a characteristic odor and bland taste that is derived from *Linum usitatissimum*, *q.v.* The unboiled oil is usually not toxic if ingested. It is used, for example, in some paints, varnishes, putty, printing inks, oilcloths, soaps, and various pharmaceuticals.

linseed. *See Linum.*

linter. short lint from cottonseed.

Linum (flax; linseed). a genus of narrow-leaved herbs or shrubs (Family Linaceae) that is nearly worldwide in distribution. ***L. usitatissimum*** (common flax; linseed). a casual weed that is grown commercially to produce linen and linseed oil. The entire plant, including leaves and seeds, is highly toxic, producing symptoms, identical to those of cyanide poisoning; the immature seed pods are exceptionally toxic. All classes of livestock have been poisoned by consuming linseed press cake, meal, and flaxseed chaff or screenings. The commercially prepared presscake in the United States rarely contains toxic levels of hydrogen cyanide (HCN). Sometimes small quantities of HCN are following consumption of linseed meal forms goitrogenic thiocyanates in the liver. Linseed meal fed to pregnant ewes has produced serious mortality among their lambs in iodine-deficient areas of New Zealand. *See also* linamarase. ***L. neomexicanum*** (yellow pine flax). an uncommon herbaceous annual or winter annual found in rangelands of New Mexico and Arizona, southward into Mexico. It is poisonous and may have caused livestock losses. *See also* linotoxin.

lipid. any of a group of fats or fat-like substances that are insoluble in water but soluble in organic solvents such as ethanol, chloroform, and ethyl ether. Included are true fats (esters of fatty acids and glycerol), lipoids (e.g., waxes, phospholipids), and sterols.

dl-α-lipoic acid. thioctic acid.

lipotropic factor. choline.

liquid silver. the element mercury.

liquorice. *Glycyrrhiza glabra* or glycyrrhiza.

liquorid. glycyrrhiza.

lithosol. a shallow, humus-rich, stony soil that develops at high altitude on parent materials that resist weathering.

litter. **1:** debris, rubbish, trash. **2:** in ecology and agriculture, organic debris that covers the ground under shrubs or on the floor of a forest. It is usually comprised chiefly of leaves, bark, and twigs. **2:** the offspring that result from a single multiple birth. **leaf litter**. litter, as in a deciduous forest, comprised chiefly of leaves.

litter layer. the layer of litter (*q.v.*) on the forest floor.

liver. **1:** digestive gland; the large digestive organ (or digestive gland) of many invertebrates. **2:** a major organ of vertebrate animals that plays complex roles in digestion, excretion, and detoxication.

livestock. **1a:** animals of any kind that are kept or raised on a farm or agroecosystem for pleasure or use. This definition thus includes wildlife and exotic game animals that meet the above criteria, poultry, and waterfowl as livestock. The term is most often used for domestic mammals raised for meat, meat products, or dairy products (e.g., cattle, swine, horses, goats, sheep). **1b:** domestic mammals raised for meat, meat products, or dairy products, or used as draft animals. (e.g., cattle and other mammals of the genus *Bos*, swine, goats, sheep, horses, asses, and mules). In this use of the term, pets and wildlife are generally not included; llamas are included, but bison are not. **1c:** domestic farm animals other than pets.

livetin. a water-soluble protein of egg yolk.

lixiviation. leaching.

loafing alley. a resting area in a barn.

loafing area. a resting area for cattle such as a loafing alley, loafing barn, or loafing shed.

loafing barn (loafing shed). a barn that provides a resting area for cattle.

loafing shed. loafing barn.

loam. a rich, friable soil comprised of approximately equal amounts of sand and silt plus a smaller amount of clay.

local effect. *See* effect.

local injury. local effect (*See* effect).

loco. **1:** locoweed poisoning; locoweed disease. It is a neuropathy of horses, sheep, and cattle caused by the ingestion of certain species of *Astragalus* and *Oxytropis*. It has caused serious losses of livestock in the western United States. Cattle initially carry their heads somewhat lower than normal; progressive locomotor ataxia soon appears. Common also are head tremors, inability to eat or drink, and abortion. Sheep suffer extreme cerebral depression. Horses are much more susceptible than other livestock; symptoms develop with small amounts of locoweed and the time to onset of symptoms and death is much shorter than in other livestock. Horses become listless, but in the early stages of poisoning are easily excited and dangerous to ride or manage. Goats develop an increasingly severe ascending paralysis. Head tremors and terminal emprosthotonos are usual. Histologically, transitory edematous vacuolization of the brain cells followed by eventual fragmentation of the Golgi apparatus may be seen. **2:** common name of true locoweeds, *Astragalus spp*. and *Oxytropis spp*. **3:** to poison with locoweed.

loco poisoning. *See* loco.

locoism. loco.

"**loco weed.**" marijuana or a cigarette made from marijuana. Do not confuse with locoweed, *q.v.*

locoweed. **1:** *Astragalus*, especially *A. mollissimus*. **2:** any plant of the genera *Astragalus* or *Oxytropis* that causes locoweed poisoning (loco), as opposed to those that may cause selenium poisoning. **3:** loco (def. 2).

locoweed disease. *See* loco.

locoweed poisoning. *See* loco.

loess (löss). a fine-grained, unconsolidated,

wind blown sediment.

log boom. a set of logs bound together to form a raft that can be towed by boat to its destination.

log chain. a steel chain with hooks on each end that is used to secure logs.

log drive. a large set of logs floating down a river.

log mark. a mark of ownership stamped or chopped onto logs.

logarithmic phase. exponential phase.

logging. **1a:** the harvesting of trees by cutting. **1b:** the process or enterprise of harvesting trees and transporting the logs to a sawmill. *See also* harvesting. **cable logging**. a yarding system that uses cables, blocks, and winches. In some regions cable logging is considered conventional, in other places not. **conventional logging**. the felling of trees by any combination of mechanical or hand felling methods and the use of rubber tired or tracked skidding equipment. *Cf.* cable logging.

Lonchocarpus (cube or cubé). **1:** a genus of tropical American shrubs and vines (Family Fabaceae, formerly Leguminosae) used in the preparation of insecticides and fish poisons. The roots contain rotenone, *q.v.* The most important species in this respect are *L. utilis* and *L. urucu*. **2:** a resinous substance (syn: barbasco, timbo) from *Lonchocarpus spp*. which is ground, purified, and used as an insecticide and fish poison.

lone-star tick. *See Amblyomma americanum*.

longleaf pine. *Pinus palustris*.

lopping. chopping branches, tops, and small trees after felling into lengths that it will allow it to lie close to the ground.

lopping and scattering. lopping, *q.v.*, and spreading the slash more or less uniformly over the ground without burning.

Loranthaceae (the mistletoe family). a widely distributed family of toxic shrubby epiphytes with jointed stems that are chiefly parasitic on trees. The flowers are small, the leaves usually broad, and each species produces berries that germinate on the host. These plants are grown and gathered for use as Christmas decorations. Some mistletoes contain toxic amines, notably phenethylamine. Cattle have been seriously poisoned by *Phoradendron serotinum* (American mistletoe), *q.v. See also Arceuthobium pusillum* (dwarf mistleltoe), *Viscum album* (European mistletoe).

Lorexane. a trade name for lindane.

löss. loess.

Lotus tetragonolobus (asparagus pea). a plant grown in southern Europe for the edible pods.

lousicide. pediculicide.

love bean. *Abrus precatorius*.

low blood pressure. hypotension.

low smut. bunt.

low spear grass. common name of *Poa annua*.

low whorled milkweed. *Asclepias pumila*.

Lowry-Brønsted acid. *See* acid.

LS 74-783. fosetyl aluminum.

lubricant. a substance, such as oil, grease, graphite, synovial fluid, or any other liquid, semisolid, or solid that permits free movement of one surface over another without damage by abrasion or by seizing due to unequal expansion caused by heat. *See* lubrication.

lubricate. **1:** to apply oil, grease, or other lubricant to a surface. **2:** to make slippery or reduce friction by application of a lubricant. *See* lubrication.

lubrication. introduction of a lubricant between the contact surfaces of moving parts (e.g., of machinery; of movable joints in an animal) to reduce friction and dissipate heat.

lucern, lucerne. mainly British vernacular for *Medicago sativa*.

lucky bean. *Abrus precatorius*.

lumbago. azoturia.

Lumbricus a genus of earthworms (Family Lumbricidae; Phylum Annelida), a common species of which is *Lumbricus terrestris*.

lumpy jaw (big jaw). **1:** actinobacillosis. **2a:** bovine actinomycosis. **2b:** any actinomycosis with a facies similar to that of bovine actinomycosis.

lungworms. nematodes, chiefly of the family Mastistrongylidae, that invade the air passages of animals. *See*, for example, *Aelurostrongylus*, *Angiostrongylus*, and *Dictyocaulus*.

lupine. a common name of herbs of the genus *Lupinus*.

lupinosis. a mycotoxic disease of grazing animals, marked by liver damage and jaundice and caused chiefly by feeding on sweet lupines (*See Lupinus*). The disease is common in Australia and South Africa, and has been reported in New Zealand and Europe. Sheep, and sometimes cattle and horses, are affected; pigs are susceptible. The causal fungus is *Phomopsis leptostromiformis*, a phytopathogen causing *Phomopsis* stem-blight, especially in white and yellow lupines; blue varieties are very resistant. This fungus produces hepatotoxic secondary metabolites (phomopsins, chiefly phomopsin A) on infected lupine material. Early effects in sheep and cattle are inappetence and listlessness. Complete anorexia and jaundice follow and ketosis is common. Cattle may show lacrimation and salivation. Sheep may become photosensitive. In acute outbreaks, deaths occur in 2—14 days. In acute lupinosis, icterus is prominent. Livers are enlarged, orange-yellow, and fatty. Chronic cases may show small, bronze or tan-colored, fibrous and firm, often distorted livers that are shaped rather like a boxing glove. Copious transudates may be found in the abdominal and thoracic cavities and the pericardial sac. *See Lupinus*, lupinine.

Lupinus (lupine, bluebonnet). a genus of usually low, perennial shrubs (Family Fabaceae, formerly Leguminosae). Some 100 species occur in North America. Toxic plants contain the alkaloid quinolizidine and numerous additional alkaloids (mostly quinolizidine alkaloids), sometimes also nitrogen oxides. The alkaloids are most concentrated in the seeds. Drying does not affect the alkaloids and hay containing *Lupine* may be quite toxic. The lupine alkaloids are highly reactive and minor differences in structure are often reflected in major differences in toxicity. Whereas many lupines are toxic under certain conditions, others may normally provide acceptable forage. Many species and varieties are difficult to place taxonomically. Symptoms produced by lupines vary with species (or variety) and the class of livestock, as does susceptibility. Sensitivity/toxicity also varies seasonally, with varying range conditions. Lupines have proven fatal to a large number of species; sheep are most frequently poisoned, whereas cattle and horses are rarely affected. Symptoms are quite variable. Lu pines are probably responsible for greater losses of sheep in the north central states than any other plant. see also crooked calf disease, piperidine, lupinine, lupinosis, lathyrism. ***L. luteus*** (yellow lupin bean). a European species with fragrant yellow flowers. It is an annual that grows to 0.7 m. *See* sparteine. ***L. niger*** (black lupin bean). *See* sparteine.

luxation. 1. dislocation of a joint.

Lycopersicon esculentum (tomato). a common garden herb and berry (usually thought of as a vegetable) with numerous varieties. The vines and suckers, which contain small amounts of cyanide in the form of the steroid alkaloids solanine and solanidine, have proven toxic to livestock.

lye. **1:** sodium hydroxide. Lye is the vernacular, and often the label term for sodium hydroxide as used in various commercial (e.g., cleaning) formulations. The primary component of drain cleaners, it is an extremely corrosive substance that can destroy any body surface. Even a drop or a dry crystal that comes into contact with damp human skin can cause damage. When ingested, lye quickly erodes the mucosa and other internal tissues and can irreversibly damage part or all of the gastrointestinal tract, with possible lethal consequences. Symptoms are chiefly those of severe gastroenteritis. Death may result from any of a number of complications. **2:** any of a number of strongly corrosive alkalis such as caustic soda, caustic potash, potassium hydroxide sodium hydroxide, as well as sodium and potassium carbonates, oxides, and peroxides. Accidental poisoning of children by these compounds, with burns of the mouth, is not uncommon, but severe, deep esophageal and gastric damage is more common among suicidal adults.

lyme grass. a grass of the genus *Elymus*.

lymph cell. lymphocyte.

lymphadenitis (adenitis). inflammation of a lymphatic gland.

lymphocyte (lymph cell). a type of nonphagocytic leukocyte that usually lacks cytoplasmic granules; the nucleus occupies most of the cell. All lymphocytes produce antibodies. **B lymphocyte** (B-cell; B-cell lymphocyte). a type of lymphocyte, formed in blood marrow, that occurs in blood, lymph, and connective tissue. Certain of these cells produce circulating antibodies when stimulated by specific antigens. **B-cell lymphocyte**. B lymphocyte. **T lymphocyte** (T-cell; T-cell lymphocyte; thymus-dependent lymphocyte; thymus-derived lymphocyte). any of a type of lymphocyte that develops bone marrow stem cells that migrate to and develop in the thymus. The mature cells are transported to other lymphoid tissues (e.g., lymph nodes, spleen), where they interact with B-cells and accessory cells to stimulate antibody production. T-cells also interact directly with antigen-bearing cells, thereby providing cell-mediated immunity. **T-cell lymphocyte**. T lymphocyte.

Lyonia. a genus of North American and east Asian brushy, deciduous and evergreen shrubs (Family Ericaceae). ***L. mariana*** (staggerbush). a deciduous shrub, with nodding white or pinkish flowers, that grows to a height of 2 m. It occurs on peaty or sandy pineland soils and in open woods from southern Rhode Island, New York, New Jersey, and eastern Pennsylvania to western Tennessee and Arkansas, south to Florida and eastern Texas. Poisoning of lambs and calves has been reported. All parts are toxic if ingested, causing salivation, increased tear formation, nasal discharge, vomiting, bradycardia, hypotension, convulsions, and paralysis. The active principle is andromedotoxin.

lytic. pertaining to lysis; having the ability to lyse.

M

μ. lower case mu.

M. upper case mu; mega.

$\mu\gamma$. symbol for micromicrogram (picogram).

μg. a symbol for microgram.

μmg. a symbol for micromilligram (*See* nanogram).

m. abbreviation for milli-.

m.p., **mp**. melting point (*See* freezing point).

mγ. a symbol for micromilligram (*See* nanogram).

mμg. a symbol for micromilligram (*See* nanogram).

Machaeranthera (tansy asters). a genus of plants (Family Compositae) that are facultative accumulators of selenium from soil. ***M. ramosa*** (tansy aster). a species hazardous to livestock because the foliage may contain very high concentrations of selenium.

Macquer's salt. potassium arsenate.

macrobiota. biota that are larger than a few centimeters.

macroelement (macronutrient). an element required in large amounts to ensure normal growth and development of plants. Included are N, P, K, S, Mg, Ca, and Fe. *Cf.* micronutrient.

macronutrient. macroelement.

mad apple. *Datura stramonium*.

mad cow disease. bovine spongiform encephalopathy.

mad itch. Aujesky's disease.

maggot. the worm-like larvae of many insects (e.g., of many dipterans such as the house fly or blow fly). It lacks appendages and a distinct head.

magnetic declination. the angle between true north and magnetic north. Declination varies from location to location and can be set on a compass for a particular location.

maize. *Zea mays*.

mal des ardents. St. Anthony's fire (ergotism).

malaoxon ([*O,O*-dimethyl *S*-1,2-bis(ethoxycarbonyl)ethylphosphorothionate]; MO). an active oxygen analog of malathion, used as a pesticide. Malaoxon, a convulsant, is very toxic to laboratory rats and is carcinogenic, producing both benign and malignant neoplasms. Malaoxon is the active metabolite of malathion and is produced metabolically by the action of cytochrome P-450 on malathion. In higher animals it increases brain phosphoinositide signaling, probably subsequent to inhibition of acetylcholinesterase, and causes convulsions. MO also brings about changes in cerebral lipid composition and may damage several brain regions.

malathion (*O,O*-dimethyl dithiophosphate of diethyl mercaptosuccinate; *O,O*-dimethyl-*S*-1,2-di(ethoxycarbonyl)ethyl phosphorothiolothionate; *S*-(1,2-dicarbethoxyethyl) *O,O*-dimethyldithiophosphate; *S*-1,2-bis(ethoxycarbonyl)ethyl

O,O-dimethylphosphorodithioate; *S*-ester with *O,o*-dimethyl phosphorothioate; [(dimethoxyphosphinothioyl)thio]butanedioic acid diethyl ester; carbofos; insecticide no. 4049; malathon (obsolete); mercaptosuccinic acid diethyl ester; mercaptothion; phosphothion; Cythion). an organophosphorus insecticide and acaride; also used as an ectoparasiticide in veterinary practice. It is neurotoxic and a powerful skin sensitizer, causing allergic dermatitis. malathion may irreversibly inhibit ACH. It is generally less toxic than parathion but produces similar symptoms. It is a significant environmental poison bedause it is relatively persistent in soil. carcinogenic and mutagenic, malathion causes chromosome damage in bone marrow cells and spermatogonia; it induces dominant lethal mutations. *See also* ethion, malaoxon, phosphamidon, insecticide (organophosphate insecticide).

malathon. an obsolete name for malathion.

male hormone. androgen.

maleic hydrazide (MH). a pyridazine herbicide used as a growth retardant for turf grasses, hedges, and trees; to inhibit the sprouting of onions and potatoes; and to prevent the development of suckers in tobacco. It temporarily inhibits growth, often without visible damage to the plant. MH interrupts cell division in both the apical, and subapical meristems of the shoot by specifically inhibiting DNA replication, but not cell extension.

malignant (bovine) anaplasmosis. *See* anaplasmosis.

malignant fever of dogs. canine babesiosis.

Malleomyces mallei. *Actinobacillus mallei*.

mallow. any herb of the genus *Malva*.

malnutrition. **1a:** undernourishment; nutrient deficiency; lack of suitable food substances in the body due, for example, to inadequate diet or to deficient digestion, absorption, or assimilation. **1b:** nutrient excess; excess intake of nutrients or the presence in food of an excess of certain nutrients. **2:** any nutritional disorder, regardless of cause. **3:** the condition of being malnourished. Such a condition may arise from factors mentioned in defs. 1a and 1b. Malnutrition may also arise from poisons present in an otherwise normal diet (e.g., aflatoxins, ergot, pesticide residues, fertilizer, pollutants). *See also* nutrition.

Malta fever. undulant fever.

Malus. **1:** a subgenus of *Pyrus* (*q.v.*) and the appropriate designation for the common (domesticated) apple. **2:** the botanical name for the common apple tree. *See Pyrus malus*.

Malva (mallows). a genus of mostly perennial herbs (Family Malvaceae) that have five-parted, white or rose-colored, silky or papery flowers. ***M. parviflora*** (mallow; cheeseweed). a widely distributed annual weed in North America. This plant is probably responsible for "shivers" in grazing Australian and African livestock. Toxic concentrations of nitrate have been reported.

Malva rotundifolia (button weed). a North American weed commonly found in fields and around outbuildings.

button weed. *Malva rotundifolia*.

mammal. a vertebrate animal of the class Mammalia.

Mammalia (mammals). a class of mostly homeothermic, viviparous, tetrapod vertebrate animals marked by the presence of hair, mammary or milk producing glands, a secondary palate, a four-chambered heart, a muscular diaphragm, and, with rare exception, seven cervical vertebrae. The teeth are in sockets and usually differentiated into incisors, canines, and cusped molars and premolars. All but the marsupials have a chorioallantoic placenta. The class is comprised of three infraclasses — Eutheria, Prototheria, and Marsupialia — only the first two of which contain venomous species.

mammalian. of or pertaining to mammals. *See* Mammalia.

mammalogy (therology). The science that treats all aspects of the biology of mammals.

mammary gland. the milk-producing gland (breast, udder) of all female mammals, that nourishes the young. The presence of one or more pairs of these glands (on the ventral surface of the body) is a defining characteristic of female mammals. Although the state of the gland fluctuates during the estrus cycle, it is usually fully functional following birth of the young. The mature gland consists of several lobes of multiple branching ducts embedded in fatty tissue. The ducts lead from numerous alveoli of milk-secreting cells and in most mammals drain into sinuses that open to the exterior via pores at the nipple. *See also* diethylstilbestrol, factor (milk factor), fusarium estrogenism, Mammalia, progesterone, vinyl chloride.

mammitis. ovine gangrenous mastitis.

manapare. *Bothrops atrox*.

mandelonitrile-β-gentiobioside. amygdalin.

mandioc. *See Manihot*.

mandioca. *See Manihot*.

maneb ([[1,2-ethanediylbis[carbamodithioato]]-(2—)]manganese;[ethylenebis(dithiocarbamato)]manganese;ethylenebis[dithiocarbamic acid] manganous salt; manganous ethylenebis[dithiocarbamate]). a dithiocarbamate fungicide used on foliage; toxicity to animals is low.

manganese (symbol, Mn; manganum)[Z = 25; A_r = 54.94]. a toxic, steel-gray, hard, brittle, lustrous, transition metal that occurs naturally as oxides. It is an essential trace element, but deficiencies have not been demonstrated in humans. *See* manganese poisoning.

manganous ethylenebis[dithiocarbamate]. maneb.

manganum. manganese.

mangel. *Beta vulgaris*.

Mangelwurzel. *Beta vulgaris*.

mangold. *Beta vulgaris*.

Manihot (cassava; manioc; madioc; mandioca). a genus of herbs and shrubs (Family Euphorbaceae) that are indigenous to tropical America but are now widespread throughout the tropics. ***M. esculenta*** (cassava; tapioca). a widely cultivated shrub of tropical America with long, thick, tuberous roots that are the source of a starchy staple (cassava) and of commercially produced tapioca. The raw root is cyanogenic, however, and potentially lethal. ***M. utilissima*** (cassava). the root is toxic, containing a soluble cyanogenetic glycoside, amygdalin, in the sap; amygdalin is present in sufficient quantity to cause cyanide poisoning. Cassava is, nevertheless an important food plant in certain parts of South America. The people using it have learned to remove or hydrolyze the toxins (linamarin and lotaustralin) and to destroy the β-glucosidase that is present in the root prior to cooking.

manioc. *See Manihot*.

manioca. *See Manihot*.

Manitoba maple. *Acer negundo*.

manure. **1:** animal feces usually named for the source (e.g., horse manure, cow manure), usually mixed with other materials such as straw. Manure is used to improve soil fertility. **2**. any type of fertilizer. **bat manure**. bat guano. *Cf*. guano. **horse manure**. horse excrement used as a fertilizer and soil amendment.

map folio. a series of maps in a binder. Such folios are often produced as a set of overlays of mapped information (e.g., on the distribution of soils, fish, water, forests, and wildlife).

maple. a tree of the genus *Acer*.

maple sugar. a food product made by further boiling down maple syrup, *q.v.*

maple syrup. **1:** a sweet syrup with a distinctive flavor produced by partially boiling down the sap of the sugar maple, *Acer saccharum*, or any of several other maple trees during the spring. **2:** a commercial product usually consisting of maple syrup mixed with cane sugar or some other sweetener. *See* maple sugar, *Acer*.

mare. a mature female equine capable of breeding.

mare's-tail. a common name of *Equisetum*.

margosa oil. neem oil.

mariculture (ocean farming; marine aquaculture). a form of aquaculture whereby crops of animals or plants are raised in marine waters either in "pens" or in the open.

marine aquaculture. mariculture.

market gardening. the cultivation of vegetables or flowers on suburban land to supply local markets (e.g., greengrocers, florists, individual consumers).

markry. *Rhus radicans*.

marl (bog lime). soil, essentially clay mixed with carbonate of lime, highly valued as a dressing or fertilizer. It crumbles rapidly and easily. Marl in which the lime is in the form of invertebrate shells is called shell marl. The term is loosely used for a variety of soils, some of which are low in lime content (e.g., the greensand marl of New Jersey). Marling of soil tends to lighten it, correct acidity, and promote nitrification.

marvel of Peru. *Mirabilis jalopa*.

mascagnite. ammonium sulfate.

mass wasting. the movement of soil and surface materials by gravity.

master gland. *See* pituitary gland.

mastitis. a septic or aseptic inflammation of one or more of the mammary glands. **bovine mastitis**. bacterial mastitis of the udders of dairy cattle. It is contracted during milking, whether by machine or by hand; flies are common vectors. The udder becomes inflamed and clotted, and bloody milk is produced. Gangrene may develop, which may lead to death of the animal. Control of mastitis requires segregation of affected animals, use of sanitary facilities, and proper milking procedures. The disease is usually treated with penicillin in combination with other antibiotics and a sulfa drug. *See also* alcohol-alizarin test, bluebag, California mastitis test, *Persea americana*. **ovine gangrenous mastitis** (blue bag; black garget; garget; mammitis). a painful inflammation of the udders of ewes, often caused by *Pasteurella mastitidis*, *Staphylococcus aureus*, and sometimes other pathogens. Affected ewes have a straddling gait; exhibit depression; loss of appetite; solitary habits; and refuse to suckle their lambs. The udders are hard, caked, and marked by dark bluish spots. *See also* blue bag.

mattock. a tool used to loosen the soil in digging. It is similar to a pickax, but one blade is broad instead of pointed. *See also* twibill.

MBCP. leptophos.

MCA. chloroacetic acid.

mcg. microgram.

MCP. MCPA.

MCPA (MCP; (4-chloro-2-methylphenoxy)-acetic acid; 2-methyl-4-chlorophenoxy-acetic acid; (4-chloro-*o*-toloxy)acetic acid). a water-soluble, synthetic phenoxy-acetic acid. It is an auxin used as a potent selective herbicide. It is moderately toxic to laboratory mice.

MCPB. (γ-(4-chloro-2-methylphenoxy)butyric acid). activity and applications are similar to those of MCPA, *q.v.*

meadow. **1:** an area of closed herbaceous vegetation dominated by grasses. **2:** a forest opening dominated by grasses. **3:** any grassy area productive of hay or forage for livestock. **alpine meadow**. meadowland near or above timberline; often used as summer pasturage.

meadow dermatitis. meadow-grass dermatitis.

meadow fescue. *Festuca pratensis*.

meadow grass. *Agrimonia eupatoria*.

meadow grasses. grasses of the genus *Poa*.

meadow pine. a common name of *Equisetum*.

meadow saffron. *Colchicum*.

mechanical site preparation. the preparation of a site for reforestation using mechanized equipment.

medicable. said of a condition or disease, treatable with medication, with a reasonable expectation of a restoration, rehabilitation, or cure.

Medicago (clovers; alfalfa). a genus of herbs and a few shrubs (Family Fabaceae, formerly leguminosae), including some clovers. Some species have been implicated in trefoil dermatitis and rape scald. ***M. hispida*** (bur clover; California bur clover). a forage and cover crop; used for green manure. It is indigenous to Eurasia, and grown in the southwestern United States ***M. lupulina*** (black medic; hop medic; hopclover; blackseed; nonesuch; yellow trefoil). an annual prostrate herb (Family Fabaceae). a valuable forage plant, but considered a weed when present in lawns. ***M. orbicularis*** (button clover; button medic). a forage plant introduced into the United States from Europe. ***M. sativa*** (alfalfa; lucerne (British)). an herbaceous perennial legume and a valuable forage and hay crop and is used also for green manure. It originated in Iran but and wild plants today grow in various locations throughout central Asia, into Siberia. It is the oldest cultivated forage crop in the United States, having been introduced by farmers in 1736. Called the "Queen of Forages," alfalfa is an important component of feed for horses, beef cattle, sheep, dairy cattle, and milking goats. It is palatable and rich in vitamins, minerals, and protein. When cut before the bloom, alfalfa is low in fiber, high in energy, and is thus an important and highly-regarded component of dairy cattle rations. The sprouts are popular as in salads and in sandwiches. Alfalfa contains saponins, is slightly toxic, and is usually mildly photosensitizing. Under conditions of rapid growth, it may contain toxic concentrations of nitrates.

medicament. a medicine or remedy.

medication. **1:** administration of or treatment with a medicine. **3:** a medicament.

medicinal herb; medicinal plant. an herb that has medicinal properties; one that contains substances that have a therapeutic value either in native form or following modification into a partially synthetic product. Many such plants are poisonous.

medicinal plant. *See* medicinal herb.

medicinal squill. *Urginea maritima*.

Mediterranean fever. undulant fever.

Mediterranean squill. *Urginea maritima*.

meiosis (reduction division). the type of division of a primordial germ cell, as it matures into two, then four gametes (e.g., sperm or ova) that contain one-half the number of chromosomes that are normal to the somatic cells of the species.

meiotic. pertaining to meiosis.

Meleagris gallopavo (turkey). the common wild turkey of the Americas (Family Meleagridae), from which domestic strains were developed. Domestic variants of this species are large fowl raised for meat. They are large, terrestrial gallinaceous birds that are confined in the wild to North America. Adult males are called cocks, adult females hens. ***M. g. gallopavo*** (south Mexican turkey). the race domesticated by the Mexican Indians and eventually introduced to New World settlers and into Europe.

Melilotus (sweetclovers). a genus of biennial herbs that occur as roadside weeds throughout much of southern Canada, and the United States. The two principal species are *M. alba* (white sweet clover), and *M. officinalis* (yellow sweet clover). Both species contain coumarin. *See* sweetclover poisoning. Toxic concentrations of nitrate have been reported.

melissophobia. apiphobia.

Melitoxin. dicumarol.

melting point. *See* freezing point.

mendrin. endrin.

***N*-(2-mercaptoethyl)benzenesulfonamide**. bensulide.

***N*-(mercaptomethyl)phthalimide *S*-(*O,o*-dimethyl phosphorodithioate)**. phosmet.

mercaptophos. demeton.

3-mercaptopropanol. 1-propanethiol.

mercaptosuccinic acid diethyl ester. malathion.

mercaptothion. malathion.

merchantable timber. a tree or stand of sufficient size, quality, and/or volume to be suitable for harvesting and marketing.

merchantable volume. the amount of sound wood in a tree or stand that is suitable for harvesting and marketing under a given set of economic conditions.

mercurial. **1:** unpredictable, volatile, or tempermental. **2:** sudden, instantaneous, or immediate. **3a:** any compound that contains mercury. **3b:** any compound or formulation that contains biologically available mercury.

Mercurialis (mercury; dog's mercury; herb mercury). a small genus of slender herbs (Family Euphorbiaceae) native to Eurasia. The dried plants were formerly used as purgatives, diuretics, and antisyphilitics. ***M. annua*** ("boys and girls"). a weak, leafy-stemmed European annual weed (Family Euphorbaceae) introduced into

North America, where it occurs in many waste areas. It has poisoned most classes of livestock. It acts not only as an emetic and purgative, but also as an irritant and narcotic. Affected livestock suffer from gastrointestinal irritation, diarrhea, and hematuria with extreme anemia; death from cardiac depression may intervene. The entire plant contains the toxic principle mercurialine.

mercuric nitrate (mercury nitrate; mercury pernitrate). a hazardous and toxic crystalline compound, $Hg(NO_3)_2$. The crystals are colorless and soluble in water and nitric acid and are insoluble in ethanol, and decompose on heating. It is used to destroy plant lice (*Phylloxera*), in nitrating organic aromatics, and in the manufacture of felt and mercuric fulminate.

mercurous chloride (mercury monochloride; mercury(I) chloride; mercury protochloride; mercury subchloride; mild mercury chloride; calomel; precipité blanc). Probably the best known mercurial, Hg_2Cl_2, it is a very toxic, odorless, tasteless, heavy crystalline powder or white, rhombic crystals, that blackens in air. It is nearly insoluble in water; insoluble in ethanol, ether, and cold dilute acids. Hydrochloric acid, alkaline earth chlorides, and alkalis increase solubility in water, although mercurous chloride decomposes in solutions of alkali bromides, cyanides, or iodides; it also decomposes slowly in alkali chlorides. Calomel is used in small doses as an oral laxative, but if laxation does not result, mercury poisoning is possible (*See also* mercurous chloride poisoning); in this case a saline laxative must be administered as a preventive measure. It is used in powder form as an application to relieve ulcers and skin rashes. Therapeutic uses have also included that of a topical antiseptic, diuretic, and syphilitic. Additional applications are, for example, as a fungicide; in agriculture, to control root maggots on cabbage and onions; in pharmaceuticals; in calomel paper; and in calomel reference electrodes (e.g., the saturated calomel electrode, SCE).

mercurous chloride poisoning ("calomel sickness"). a rare poisoning, the effects of which usually include salivation, epigastric discomfort, and diarrhea. Retention of 30-40 mg./kg body weight of calomel following administration as a laxative may prove fatal. *See also* mercurous chloride, mercuric chloride poisoning.

mercury monochloride. mercurous chloride.

mercury nitrate. mercuric nitrate.

mercury pernitrate. mercuric nitrate.

mercury protochloride. mercurous chloride.

mercury subchloride. mercurous chloride.

mercury. **1:** (symbol, Hg; metallic mercury; elemental mercury; hydrargyrum; liquid silver; quicksilver)[Z = 80; A_r = 200.59]. a toxic transition metal that occurs naturally as the mercuric sulfide (cinnabar); a heavy metal. At ordinary temperatures, the pure metal is a silver, heavy, slightly volatile, non-wetting, mobile liquid. Whereas elemental mercury is usually not very toxic, salts and esters of mercury are. Mercury reduces the rate of photosynthesis by phytoplankton, and produces neurotoxic effects in higher animals. It is a known mutagen, teratogen, and carcinogen. It readily crosses the mammalian placenta and accumulates in milk. Mercury, especially as a methyl mercury (*q.v.*), is a significant environmental poison. Mercury has many applications and is used, for example, in thermometers, barometers, hydrometers, fluorescent lamps, and dental amalgams. It has many scientific applications, and is a component of a number of pharmaceuti cals, fungicides, and agricultural chemicals. **elemental mercury**. **1:** mercury, metallic mercury. **2:** a common name of *Rhus radicans*. **3:** any herb of the genus *Mercurialis*.

mercury(I) chloride. mercurous chloride.

mercy killing. euthanasia.

Merino sheep. one of the best-known breeds of sheep, originating in Spain. It is a hardy, easy to manage, white-faced breed of medium size with fine wool; the rams are horned. It has given rise to several other breeds such as the Rambouilet. There are three strains: A, B, and C (Delaine type). The first two varieties are strongly (A) and moderately (B) wrinkled, while C is much smoother. The quality of wool and meat are better in the Delaine type.

meristem. any region in a green plant that contains actively-dividing or potentially dividing cells. Meristem may occur in a number of locations: apical, axillary, basal, intercalary, lateral, or marginal. **apical meristem**. meristem at the tip of a stem or root. It is typically the dominant meristem of a plant. *See* apical dominance; *See also* abscisic acid. **axillary meristem**. meristem whose cells are capable of active division, but are dormant most of the time. **basal meristem**. meristems at the base of a plant, the cells of which may be dormant or actively dividing. **intercalary meristem**. meristem located some distance from the apical meristem between more or less differentiated tissues. Such meristems facilitate longitudinal growth of a plant organ. **lateral meristem**. meristem situated parallel to the sides of the organ in which it occurs. It brings about peripheral growth. **marginal meristem**. tissue at the edges of a leaf primordium that gives rise to mesophyll and the epidermal tissues of the leaf. **mesophyll**. the ground tissue of a leaf comprised mostly of photosynthesizing cells (the chlorenchyma).

merocrine gland. a gland that produces an acellular secretion. *Cf.* holocrine gland.

merocrine secretion. *See* exocytosis.

merphenyl nitrate. phenylmercuric nitrate, basic.

merryhearts. *Zigadenus nuttallii*.

meso-, mesi-, mes-. combining forms that: **1:** indicate middle or intermediate. **2:** pertain to a mesentery (in anatomy). **3:** secondary or partial (in medicine).

mesobiota. biota comprised of individuals that are large enough to be seen with the naked eye but are smaller than macrobiota, *q.v.*

mesosaprobic. of or pertaining to a polluted aquatic habitat that has a below normal oxygen concentration and a moderately high level of organic decomposition.

metabolic. relating to metabolism.

metabolin. metabolite.

metabolism. the aggregate of all physical and chemical changes that occur within the living substance of an organism, including all energy and material transformations that take place within a cell (cellular metabolism), organ, organ system, or whole organism. Normal metabolism is the underpinning of all life processes. *Cf.* anabolism, catabolism. **ammonotelic metabolism**. that in which ammonia is the end product of nitrogen metabolism, as in many freshwater and marine animals. **basal metabolism**. the energy expended in simple maintenance of the organism at rest. **carbohydrate metabolism**. the catabolic and anabolic processes that involve carbohydrates. Included is the metabolic oxidation of carbohydrates to CO_2 and H_2O with the release of energy. Long, complicated reaction sequences and cycles are components of this process. Carbohydrate biosynthesis also takes place, resulting in the production of storage glycogen. Toxicants can interfere with carbohydrate catabolism and biosynthesis. **ecological metabolism**. respiration by an ecological community or ecosystem. **intermediary metabolism**. the metabolic processes involved in the transformation of food into cellular and tissue components. **lipid metabolism**. complex catabolic and anabolic processes that involve lipids. Disruption of lipid metabolism as by a toxicant can result in pathological accumulations of lipids in the liver (cirrhosis of the liver, fatty liver). **ureotelic metabolism**. that in which urea is the end product of nitrogen metabolism, as in mammals, for example. **uricotelic metabolism**. that in which uric acid is the end product of nitrogen metabolism, as in birds.

metabolite (metabolin). **1:** any product of metabolism, especially a transformed chemical. **2:** any substance involved in metabolism. **active metabolite**. **1a:** a biologically active substance produced by, or modified by, a metabolic process from a precursor. **1b:** any substance that becomes more active than the parent compound as the result of metabolic processes. **reactive metabolite**. a metabolite that is more reactive than the parent compound. Such metabolites mediate many types of toxic effects (e.g., carcinogenesis, cell necrosis, mutagenesis, teratogenesis, hypersensitivity reactions). **stable metabolite**. a stable, normally nonreactive metabolite. **toxic metabolite**. a reactive metabolite that is more toxic than the parent compound. **proximate toxic metabolite**. a stable, toxic metabolite produced in a tissue remote from the target organ. **ultimate toxic metabolite**. a toxic metabolite produced within the target organ by metabolic activation, from a proximate or intermediate metabolite.

metabolizable. capable of being chemically transformed by metabolism.

metacentric. of or pertaining to a chromosome or chromatid in which the centromere is in the median position such that the arms of the chromosome are of equal length. *Cf.* acentric, acrocentric, telocentric.

metacetaldehyde. metaldehyde.

metal. **1a:** any element that is electropositive and reactive (e.g., alkali metals, alkaline earth metals), tending to lose electrons on reaction. Most are further marked by luster, malleability, ductility, and conductivity of electricity and heat. Their oxides and hydroxides are basic. **1b:** the transition elements which are less reactive, amphoteric, and tend to form complexes. The distinction between metals and nonmetals is not sharp. **2:** in vernacular terms, any lustrous, malleable, ductile, solid element or alloy that readily conducts electricity and heat and which forms positive ions in aqueous solution. Not all of these properties need be present in all forms of the substance to be popularly designated as a metal (e.g., cast iron is not malleable or ductile, mercury is a liquid at normal ambient temperatures, some metals are poor conductors). **heavy metal**. a variously defined term and often used in the literature without clear definition. Definition 1b is most useful in toxicology. **1a:** a metal with a density greater than 5 g/cm^{-3}. This definition is arbitrary and includes 53 naturally occurring elements and 16 synthetic elements with widely divergent chemical properties. **1b:** any element with a density greater than 5 g/cm^{-3} that (1) that binds preferentially to ligands that contain nitrogen or sulfur as opposed to those that preferentially bind to ligands that contain oxygen, and (2) those that are borderline with respect to ligand selection. This definition includes those metals that are poisonous as mercury, chromium, cadmi um, arsenic, and lead (See also def. 1c, below). They are extremely important environmental poisons and are the most extensively investigated chemically and toxicologically. They are environmentally persistent, are toxic even at low concentrations, and tend to bioaccumulate. See also heavy-metal tolerance. **1c:** in plant nutrition, any metal of moderate to high atomic number (e.g., Cu, Zn, Ni, Pb) present in certain soils (e.g., due to a mine spoil or outcrop) that inhibits plant growth except in the case of a few tolerant species and ecotypes. **2:** in electron microscopy, a metal of high atomic number used to introduce electron density into a biological specimen by staining, shadowing, or negative staining. **3:** often loosely applied to any metal or metalloid of relative high atomic mass, including some metals that do not fit the chemical definition of a heavy metal (e.g., copper and zinc) and some (e.g., arsenic, a metalloid) which are not true metals.

metallic. of, pertaining to, or resembling metal.

metalloid. any chemical element having properties that are intermediate between those of metals and non-metals. The metalloid elements are boron, silicon, germanium, arsenic, antimony, tellurium, and astatine. See also metal (heavy metal).

metaldehyde (metacetaldehyde). a white, crystalline substance, $(CH_3CHO)_n$, it is a toxic, flammable polymer of acetaldehyde. Metaldehyde is a commonly used molluscacide (snail bait) and herbicide. It is usually blended with bran and is palatable to dogs and farm animals. Metaldehyde is hydrolyzed in stomach acid to acetaldehyde polymers that readily enter the brain. It is a major cause of poisonings in small animals. *See also* metaldehyde poisoning. Metaldehyde is a strong irritant.

metaldehyde poisoning. poisoning due to ingestion of metaldehyde. The effects are similar in all mammalian species. Neurologic signs dominate and initially include hyperesthesia, muscle weakness, muscle tremors, and incoordination, followed by continuous tonic convulsions and emprosthotonos. Additional signs include hypersalivation (frothing at the mouth), polypnea, dyspnea, tachycardia, and pyrexia (with profuse sweating in horses), cyanosis, and loss of consciousness. Convulsions frequently occur. In severe cases, early death from respiratory failure may intervene; survivors may suffer liver failure. Lesions include congestion and edema of liver, kidney, and lungs, intestinal hemorrhage. Opening of the stomach or rumen on autopsy may reveal a formaldehyde-like odor.

metallic mercury. elemental mercury.

metallophyte. a plant that tolerates substrates (e.g., soils, bottom sediments) that have very high levels of heavy metals. **facultative metallophyte**. pseudometallophyte. **obligate metallophyte**. a metallophyte that is restricted to substrates that contain relatively high concentrations of heavy metals. **pseudometallophyte** (facultative metallophyte). a plant that can tolerate high concentrations of heavy metals but one not confined to soils or other media that contain heavy metals.

metaminodiazepoxide. chlordiazepoxide.

metastrongyle. metastrongylid.

metastrongylid (metastrongyle). **1:** common name of any worm of the family Metastrongylidae. **2:** of or pertaining to the family Metastrongylidae.

Metastrongylidae. a large family of parasitic nematodes such as *Angiostrongylus*, *Dictyocaulus*, and *Crenosoma*.

Metazoa. a subkingdom comprised of multicellular animals that are generally com posed of many specialized cells organized into tissues, organs, and organ systems. They possess a coordinating nervous system. Included are all animals except the Protozoa and Parazoa (sponges).

meteorology. the branch of science that deals with the behavior of a planet's atmosphere, especially that of the earth. Applied meteorology is importantly devoted to the analysis and prediction of weather, locally and regionally. **agricultural meteorology** (agrometeorology). the science that treats the relationships between meteorological variables and agricultural practices and concerns.

metestrus. *See* estrus.

β-methallyl chloride. 3-chloro-2-methylpropene.

2-methylallyl chloride. 3-chloro-2-methylpropene.

methallylchloride. 3-chloro-2-methylpropene.

β-methallylchloride. 3-chloro-2-methylpropene.

methamidophos (*O,S*-dimethyl phosphoramidothioate). a highly dangerous pesticide, miscible with water, that is used to protect cole, cotton, lettuce, and potato crops. It is used also in some household products to eradicate pests.

methane (marsh gas). a colorless, odorless, nontoxic, flammable gas, CH_4, that forms explosive mixtures with air. Methane is the principal component of most natural gas and a major component of coal gas. It is formed by the anaerobic decomposition of organic matter, as in marshes.

methanearsonic acid (methylarsonic acid; methylarsinic acid; monomethylarsinic acid). a biomethylated arsenic herbicide, CH_5AsO_3, prepared from sodium arsenite and methyl iodide. It is a skin and eye irritant and, as the monosodium salt, is moderately toxic by ingestion to laboratory rats. Methanearsonic acid and cacodylic acid are the two organoarsenic compounds most likely to be encountered in the environment. *See also* biomethylated arsenic, monosodium methanearsenate, disodium monomethanearsenate.

methane-carboxylic acid. acetic acid.

methanesulfonic acid tetramethylene ester. busulfan.

methanol (methyl alcohol; wood alcohol; wood naphtha). the simplest of the primary aliphatic alcohols, it is a clear, colorless, mobile, strongly polar, flammable liquid, CH_3OH, with a characteristic odor. It is miscible with ether, ethanol, and water and is used widely in paint and varnish removers, as an industrial solvent, as a synthetic intermediate, and in certain types of antifreeze fluids; mixed with ethanol and soap, it is used as a solid canned fuel. It is also used as a denaturant of ethanol. Methanol is highly toxic to humans and damaging to the kidney. It is readily absorbed by the skin and gastrointestinal mucosa. The vapor is also toxic if inhaled, increasing breathing rate. It can cause blindness and death.

methemoglobin (MetHb; ferrihemoglobin). a form of hemoglobin that does not bind oxygen reversibly. The iron of MetHb has been oxidized to the ferric state, with essentially ionic bonds. Methemoglobin is normally present in circulating blood of vertebrates, but its concentration is increased under certain circumstances such as poisoning by nitrates and certain other chemicals. Significant increases result in cyanosis. *See* methemoglobinemia, nitrite poisoning.

methemoglobinemia. the presence of unusually large amounts of methemoglobin in the blood with resulting cyanosis due to lack of oxygen in the peripheral tissues. The condition may be caused by nitrates and certain other toxic chemicals; by injury; or due to defective NADH methemoglobin reductase (a recessive autosomal trait); or may result from an abnormality in hemoglobin M, *q.v.* *See* methemoglobin, nitrite poisoning.

methiocarb (3,5-dimethyl-4-(methylthio)-phenol methylcarbamate; methylcarbamic acid 4-(methylthio)-3,5-xylyl ester; 4-(methylthio)-3,5-xylyl methylcarbamate; 4-methylthio-3,5-dimethylphenyl *N*-methylcarbamate; mercaptodimethur; metmercapturon). a molluscicide and an insecticide used on vegetable and fruit crops.

4-methoxybenzenemethanol. anise alcohol.

methomyl. a carbamate insecticide, it is a cholinesterase inhibitor; injection of atropine sulfate readily reverses inhibition. Symptoms of poisoning are similar to those of organophosphate poisoning. The oral LD_{50} in rats is 17 mg/kg body weight. Cattle have reportedly been poisoned by ingesting forage accidentally sprayed with this compound.

methoxy-DDT. methoxychlor.

1-methoxycarbonyl-1-propen-2-yl dimethyl phosphate. mevinphos.

methoxychlor (1,1′-(2,2,2-trichloroethylidene)bis[4-methoxybenzene]; 2,2-bis(*p*-methoxyphenol)-1,1,1-trichloroethane; methoxy-DDT; DMDT). a white, crystalline, chlorinated hydrocarbon,

$$Cl_3CCH(C_6H_4OCH_3)_2$$

Methoxychlor is incompatible with alkaline materials. It is soluble in water, insoluble in ethanol. An insecticide used to destroy mosquito larvae and house flies, especially in dairy barns; it is considered relatively safe to use. Methoxychlor is nephrotoxic and hepatotoxic. Dairy calves tolerate 265 mg/kg body weight; 500 mg/kg is mildly toxic; and 1 g/kg produces rather severe poisoning. Sheep are little affected by methoxychlor. One dog was given 990 mg/kg daily for 30 days with no signs of poisoning.

methoxydiuron. linuron.

3-methoxy-4-hydroxycinnamic acid. ferulic acid.

(3-(3-methoxy-2-methyl-3-oxo-1-propenyl)-2,2-dimethylcyclopropane-carboxylic acid 2-methyl-4-oxo-3-(2-propenyl)-2-cyclopenten-1-yl ester. allethrin II.

methoxyphenol. *See* butylated hydroxyanisole.

2-*p*-methoxyphenylmethyl-3-acetoxy-4-hydroxypyrrolidine. anisomycin.

[2R-(2α,3α,4β)]-2-[(4-methoxyphenyl) methyl]-3,4-pyrrolidinediol 3-acetate. anisomycin.

methyl bromide (bromomethane). a colorless, transparent, easily liquefied gas, CH_3Br, with a chloroform-like odor and burning taste. It is used chiefly in drug manufacture, as a soil fumigant, and as a fumigant in grain storage warehouses. This compound is also used as a pesticide against fruit flies in shipments of citrus fruit. It is a strong irritant to the skin and is poisonous by all routes of exposure. Methyl bromide is carcinogenic to laboratory animals. Exposed workers may experience irritation of the eyes, skin, mucous membranes, and/or pulmonary system. Presenting symptoms are seen one-half to six hours after exposure and may include visual problems, headache, nausea, vomiting, and tremor. Coma may ensue and the victim may die from respiratory or circulatory failure. Survivors may experience permanent kidney or brain damage. Low vapor concentrations above threshold that are undetectable by taste or odor may cause chronic poisoning that affects the CNS Symptoms include disturbances of speech and vision, confusion, lethargy, and muscle pains.

methyl-5-butyl-2-benzimidazole carbamate. parabendazole.

methyl1-(butylcarbamoyl)-2-benzimidazolecarbamate. benomyl.

methyl carbamic acid 1-naphthyl ester. carbaryl.

methyl carbamic acid 2,3-dihydro-2,2-dimethyl-7-benzofuranyl ester. carbofuran.

methyl demeton. demeton-*S*-methyl.

methyl 3-(dimethoxyphosphinyloxy)crotonate. mevinphos.

methyl 1-(dimethylcarbamoyl)-*N* (methylcarbamoyloxy)thioformimidate. oxamyl.

2-methyl4,6-dinitrophenol. dinitrocresol.

methyl ester. methyl isocyanate.

methyl esters. the accepted chemical name of biodiesel, *q.v.*

methyl ethyl carbinol. 2-butanol.

methyl isocyanate (MIC; isocyanic acid, methyl ester; isocyanatomethane; methyl mustard oil). a colorless, slightly water-soluble, highly volatile liquid, CH_3NCO, used in the synthesis of pesticides such as Sevin Carbaryl; it is the active ingredient in certain soil fungicides and nematocides. MIC is highly flammable and a significant fire hazard. Both the liquid and the vapor are very toxic and are strong irritants and sensitizers of the eyes, skin, and mucous membranes. Systemic effects in humans can result from all routes of exposure. Initial symptoms are usually tearing and damage to the cornea, as evidenced by the opacity. If inhaled, MIC causes immediate constriction of the respiratory passages and lungs and the victim gasps for breath. It can inflame the pulmonary tissue (bronchitis) with accumulation of fluid, resulting in labored, painful breathing and possibly death. At high concentrations, the vapor can also cause blindness, pulmonary fibrosis, emphysema, and gynecologic effects. Severe acute exposure can result in sudden death. Chronic exposure may cause irreversible damage to the eyes, liver, and kidneys; chronic bronchitis; and may exacerbate the symptoms of asthma. Atmospheric moisture causes MIC to decompose rapidly in the environment.

methyl parathion (phosphorothioic acid *O,O*-dimethyl *O*-(4-nitrophenyl) ester; *O,O*-dimethyl-*O-p*-nitrophenylphosphorothioate). a white, crystalline, slightly water-soluble solid when pure,

$$(CH_3O)_2P(SO)OC_6H_4NO_2$$

It is an extremely toxic organophosphate inhibitor of cholinesterase that is commonly used as an insecticide. Methyl parathion is also used in some household products to eliminate pests. It is an explosion hazard when heated and is toxic by all routes of exposure.

methyl phthalate. dimethyl phthalate.

methyl Systox. demeton-*S*-methyl.

methyl viologen (2+). paraquat.

methylacetal. acetone.

β-methallyl chloride. 3-chloro-2-methylpropene.

2-methylallyl chloride. 3-chloro-2-methylpropene.

β-methallylchloride. 3-chloro-2-methylpropene.

(±)-2-*exo*-(2-methylbenzyl-oxy)-1-methyl-4-isopropyl-7-oxabicyclo[2.2.1]-heptane. cinmethylin.

***O*-methyl-*O*-(4-bromo-2,5-dichlorophenyl)-phenyl thiophosphonate**. leptophos.

3-methyl-2-butanoic acid 2-*sec*-butyl-4,6-dinitrophenyl ester. binapacryl.

methylcarbamicacid4-(dimethylamino)-*m*-tolyl ester. aminocarb.

methylcarbamic acid 2,3-(iso-propylidene-dioxy)phenyl ester. bendiocarb.

methylcarbamic acid 2,3-(isopropylidene-dioxy)phenyl ester. bendiocarb.

2-methyl-4-chlorophenoxyacetic acid. MCPA.

3-methylcrotonic acid 2-*sec*-butyl-4,6-dinitrophenyl ester. binapacryl.

3,3′-methylenebis[4-hydroxy-2H-1-benzo-pyran-2-one]. dicumarol.

2,2′-methylenebis[3,4,6-trichlorophenol]. hexachlorophene.

2-(1-methylethoxy)phenol methylcarbamate. propoxur.

***o*-(1-methylethyl)phosphoramidothioic acid *o*-(2,4-dichlorophenyl) *o*-methyl ester**. DMPA.

methyl-2-isopropoxyphenylcarbamate. propoxur.

4-methylmercapto-3-methylphenyl dimethylthiophosphate. fenthion.

methylmercaptophos. demeton-*S*-methyl.

methylmercury. **1:** either of two compounds that contain the methyl group: dimethylmercury, $(CH_3)_2Hg$, and monomethylmercury, CH_3Hg^+. Both compounds are notorious as poisonous environmental pollutants. The methylmercuries are water and lipid soluble, and are thus bioavailable and readily absorbed by aquatic organisms. Once formed in sediment, they may ultimately become widely distributed in both aquatic and terrestrial food webs. Fish tissues may contain more than 1000 times the concentration of mercury as the surrounding water in surface waters contaminated by these organomercurials. **2:** in the literature, the term methylmercury is often used indiscriminantly, and without clarification, to indicate one or the other of these two compounds as well as common salts of monomethylmercury.

methylmercury neurotoxicity. Minamata disease.

1-methyl-4-(1-methylethyl)2,3-dioxabicyclo[2.2.2]oct-5-ene. ascaridole.

2-methyl-2-(methylthio)propanal*O*-[(methylamino)carbonyl]oxime. aldicarb.

2-methyl-2-(methylthio)propionaldehyde *O*-(methylcarbamoyl)oxime. aldicarb.

(3-methylphenyl)carbamic acid 3-[(methoxycarbonyl)amino]phenyl ester. phenmedipham.

3-methyl-1-phenyl-5-pyrazolyl dimethylcarbamate. pyrolan®.

2-methylpropane. isobutane.

2-methyl-2-propanol. *tert*-butanol.

***N*-(2-methylpropyl)-2,6,8-decatriene-amide)**. affinin.

2-(1-methylpropyl)-4,6-dinitrophenol. dinoseb.

1,10-methyl-2-(3-pyridyl)-pyrrolidine sulfate. nicotine sulfate.

3-(1-methyl-2-pyrrolidinyl)pyridine. nicotine.

(*S*)-3-(1-methyl-2-pyrrolidinyl)pyridinesulfate (2:1). nicotine sulfate.

1,3-(1-methyl-2-pyrrolidyl)pyridine sulfate. nicotine sulfate.

methylthionine chloride. methylene blue.

methylthionium chloride. methylene blue.

methyl-3-(*m*-tolylcarbamoyloxyphenylcarbamate. phenmedipham.

methylene blue (methylthionine chloride; 3,7-bis(dimethylamino)phenothiazin-5-ium chloride, C.I. basic blue 9, methylthionium chloride; tetramethylthionine chloride; 3,7-bis(dimethylamino)phenazathionium chloride; Swiss Blue; C.I. 52015; Solvent Blue 8; Urolene Blue). a dark, lustrous green, water- and ethanol-soluble compound that is toxic by ingestion.

$$C_{16}H_{18}N_3SCl{\cdot}3H_2O$$

It is used as a reducing agent, a bacterial stain, and a dye. Methylene blue is used therapeutically in human medicine as an antimethemoglobinemic, and an antidote to cyanide poisoning. In veterinary practice it serves as an antiseptic, disinfectant, and an antidote to cyanide poisoning. It is sometimes effective in ameliorating the course of nitrite poisoning by reducing the methemoglobin to the normal ferrous form.

metobromuron (3-(*p*-bromophenyl)-1-methoxy-1-methylurea). an herbicide used mainly to control weeds in carrots, groundnuts, peas, soybeans, and potatoes. It is moderately toxic to laboratory rats.

metritis. an inflammation of the uterus usually acquired at the time of parturition or by excessive matings during estrus. *Escherichia coli* is the most common causative pathogen, though *Streptococcus spp*, *Staphylococcus spp*, and other pathogens are sometimes isolated.

mevinphos (3-[(dimethoxyphosphinyl)oxy]-2-butenoic acid methyl ester; 3-hydroxycrotonic acid methyl ester dimethyl phosphate; 1-methoxycarbonyl-1-propen-2-yl dimethyl phosphate; methyl 3-(dimethoxyphosphinyloxy)crotonate; O,O-dimethyl-1-carbomethoxy-1-propen-2-yl phosphate; α-2-carbomethoxy-1-methylvinyl dimethyl phosphate; 2-carbomethoxy-1-methylvinyl dimethyl phosphate; 2-carbomethoxy-1-propen-2-yl dimethyl phosphate; CMDP). a yellow, liquid phosphate ester,

$$\alpha\text{-2-}(CH_3O)_2P(O)OC(CH_3){:}CHCOOH_3$$

Mevinphos is an extremely poisonous, extremely dangerous systemic poison used chiefly to control pests such as mites, aphids, and houseflies. It is a cholinesterase inhibitor and is toxic by all routes of exposure. The *cis*-isomer is about 100 times more toxic than the *trans*-isomer. The commercial product is a mixture of *cis*- and *trans*-isomers. When heated to decomposition it releases toxic phosphate fumes. *See also* Phosdrin®.

mevalonic acid (3,5-dihydroxy-3-methylpentanoic acid; 3,5-dihydroxy-3-methylvaleric acid; β,δ-dihydroxy-β-methylvaleric acid; hiochic). an organic acid,

$$CH_2OHCH_2C(OH)(CH_3)CH_2COOH$$

It is an intermediate in the biosynthesis of squalene, cholesterol, and coenzyme Q in plants and animals.

mexacarbate (4-(dimethylamino)-3,5-dimethylphenol methylcarbamate (ester);

methylcarbamic acid 4-(dimethylamino)-3,5-xylylester;4-dimethylamino-3,5-xylyl methylcarbamate). a colorless, crystalline, water-insoluble substance, $C_{12}H_{18}N_2O_2$, that is soluble in ethanol, acetone, and benzene. A carbamate pesticide used mainly to control snails and slugs it is extremely toxic to laboratory rats.

Mexican horse. a breed of horse that originated from Andalusians and Arabians. It was introduced into Mexico in the sixteenth century by the Spaniards.

Mexican Jack (Spanish jack). a small, familiar burro that is seldom used as breeding stock in the United States

Mexican moccasin. *Agkistrodon bilineatus*.

Mexican scammony root. the dried root of *Ipomoea orizabensis*.

Mexican tea. *Chenopodium ambrosioides*.

Mexican whorled milkweed. *Asclepias mexicana*.

MF. milk fat.

mg. milligram.

Mg. the symbol for mercury.

mg/m^3. milligrams per cubic meter. It is a common standard used to express the number of allowable milligrams of a substance in a cubic meter of air.

mgm. for milligram.

MH. maleic hydrazide.

Michaelmas daisy. *See Aster*.

MIC. methyl isocyanate.

micro- (μ). **1:** a prefix to units of measurement that indicates a quantity of 1 millionth (10^{-6}) the value indicated by the root. *See* International System of Units. **2:** often used in the vernacular to mean very small.

microbe. microorganism.

microbial pesticide. any microbe used as a pesticide. *See Bacillus thuringiensis*.

microbial toxin. any toxin produced by a microorganism. Pathogenic microorganisms generally normally harm of kill a host organism by the production of toxins rather than by cell invasion. The virulence of such a microorganism is thus due to the ability of the microbial toxins to damage cells that are important to the vitality or viability of the host. See endotoxin, exotoxin.

microbicidal. destructive to microbes.

microbicide. microbicidal; an agent that destroys microbes.

microbioassay. an assay to determine the potency of a nutrient, a toxicant, or active substance or factor by observing its effect upon the growth of selected microorganisms. Such tests are considered fundamental by some professionals; others believe they are generally very limited in application and interpretation.

microbiota. the microscopic (unicellular and multicellular) organisms of a habitat, ecosystem, or specified region. *Cf.* macrobiota, microorganism.

microclimate. the climate, for example, of a localized area or a specific portion or segment of an area, locality, habitat, or ecosystem. Specific examples of a microclimate are: the climate near the ground; that on the south slope of a hill or mountain; that in the canopy of a forest; that on the windward side of a lake; and that near the surface of a lake or pond. A microclimate may contrast sharply with the climate taken as a whole for an ecosys tem, locality, or region. Thus, the climate adjacent to the north side of a home, that of a portion of a farm that is influenced by a pond or small lake, that on the leeward side of a windbreak, that of a forest canopy, and that of the herbaceous layer on the forest floor would be considered microclimates).

Microcystis. a genus of toxic cyanobacteria. *See also* Annie, Fannie, and Mike. ***M. aeruginosa***. a poisonous species of cyanobacteria. At least one toxic principle, a "fast death factor" that is algal in origin, has been isolated from nearly pure laboratory cultures. Highly toxic to humans and livestock, this factor is a cyclic polypeptide that contains aspartic acid, glutamic acid, D-serine, valine, ornithine, alanine, and leucine. *See* factor (fast death factor).

microelement. a trace element that is required in very small amounts to ensure normal plant growth and development. Included are Mn, B, Cl, Zn, Cu, and Mo. *Cf.* micronutrient, macroelement.

microgram (mcg; symbols, γ and μg). one-millionth of a gram (10^{-6} g); one-thousandth of a milligram (10^{-3} mg.).

micrometer (symbol, μm). *See* micron.

micromilligram. nanogram.

micron (symbol, μ; micrometer [symbol, μm]). 10,000 Å units or one-millionth (10^{-6}) meter.

micronutrient. a naturally occurring chemical that is essential to normal growth and development of an organism but is required only in small amounts. Micronutrients include trace elements (microelements) and vitamins. *Cf.* macronutrient, microelement.

microorganism (microbe). **1:** any minute one-celled organism, not distinguished as plant or animal in nature. **2:** a minute living organism (unicellular, or in some cases multicellular), especially a bacterium or protozoan, most of which are invisible to the naked eye. Many microorganisms are of agricultural and/or toxicological interest. *Cf.* bacterium, microbiota.

micropropagation. a form of cloning or propagation in which a single cell of the meristernatic tissue is used to produce a new plant.

microscopic anatomy. histology.

Micruroides euryxanthus (formerly *Micrurus euryxanthus*; Arizona coral snake; Sonoran coral snake). a small, rare snake (Family Elapidae) that inhabits semidesert areas of the southwestern United States and northwestern Mexico. It is the only species in the genus. Body length rarely exceed 50 cm. It is a glossy snake with a small head that is indistinct from the neck; the snout is rounded; the canthus is indistinct. The eyes are small with round pupils. The body is slender, elongate, does not taper; the tail is short. As opposed to *Micrurus*, *q.v.*, the head is solid black to the angle of the jaw, terminating in a straight edge across the parietal scales. The posterior tips of these scales mark termination of the head which is bordered by a yellow or whitish neck band succeeded by a red ring. *Micruroides* is distinguished from the nonvenomous sand snakes (*Chilomeniscus*) and shovel-nosed snakes (*Chionactis*) by the elongate body, black snout, and unmodified rostral plate. The red bands on the body are bordered by yellow or white in contrast to king snakes (*Lampropeltis*), which have red bands bordered by black. The body is ringed by alternating bands of red and black rings, each separated by light rings; the tail is banded by alternating black and light rings. It is secretive and not usually regarded as dangerous but the venom is highly toxic and this snake should be approached with caution.

Micrurus (American coral snakes). a genus comprised of some 40 species of usually small, slender, brightly colored, venomous, New World snakes (Family Elapidae). They occur from North Carolina to Texas in the United States and from Coahuila and Sonora in Mexico to Argentina and Bolivia in South America. They are distinguished by a small head with small eyes, no distinct canthus, a rounded snout, and no evident neck; a long, slender untapered body with a short tail; a rounded snout, and small eyes with a round pupil. There are nine scales on the crown; the nasal abuts on the single preocular; there is no distinct canthus. The only teeth rooted in bone are the rather large maxillary tubular fangs. Nearly all species are brightly colored with complete alternate rings of black, yellow (sometimes white), and red (usually). The venom is highly toxic and all are dangerous. The bite is sometimes painless, but a bite with envenomation is typically accompanied by localized pain (which may be severe), edema, and sometimes paresthesia of the affected area. This may be followed by slow onset of paralysis and impaired breathing, that very often terminates in death if untreated. Additional effects may include excessive salivation, weakness, drowsiness, slurred speech, difficulty in swallowing, ptosis, cardiovascular effects, and sometimes hypotension, a weak pulse and other signs of shock. Death is rare in individuals receiving antivenin therapy and mechanical respiration.

mild mercury chloride. mercurous chloride.

mildew. **1:** any fungus that grows on plant material. Downy or powder mildews are ascomycetous fungi that cause plant disease. **2:** any fungal disease of plants in which the fungus is seen on the surface of the plant; any disease produced by mildew (def. 1).

mildewcide. a substance that destroys mildew.

milk. a heterogeneous liquid, produced by the mammary glands of female mammals, that nourishes the young. the milk dairy animals (e.g., cattle, goats, sheep), their milk serves also as an important commodity used by humans. **bloody milk**. that from an animal with ruptured blood vessels and injured inner gland tissues. Such milk should be disposed of. **bulk milk**. milk stored on the farm in tanks rather than stored in cans. *See* bulk milk tank. **can milk**. milk kept in 10-gallon cans on the farm to cool and for transport to a dairy product manufacturing plant. **cow's milk**. the milk of domestic cattle contains approximately 87% water, 3.8% emulsified fat and fatty acids, 3% casein, 5% lactose, and small amounts of nutritional components to include calcium, copper, iron, magnesium, and a number of vitamins. The composition of milk varies widely among mammals, especially with respect to the fat and protein content. **in milk**. said of a lactating female. **manufacturing grade milk**. milk produced under regulations that are less stringent than those for producing grade A milk. *See* manufactured milk products; *Cf.* manufacturing milk. **manufacturing milk**. milk used in the manufacture of dairy products. **raw milk**. unprocessed milk as it comes from a dairy cow, goat, or other mammal. **reconstituted milk**. the product (e.g., reconstituted skim milk) that results when a dried milk is rehydrated with water. **skim milk**. processed milk with a fat content of less than 0.5% (usually less than 0.1%). Skim milk has about half as many calories as whole milk and is nearly cholesterol free, but the levels of protein, lactose, minerals, and water-soluble vitamins are not reduced. **surplus milk**. the quantity of grade A liquid milk in excess of that needed for Class I purposes.

milk fat (MF; butterfat). the lipid component of milk. It is a complex mixture of triglycerides that contains numerous fatty acids. Milk fat is sometimes simply referred to as fat. Electronic photometry and infrared absorption are generally used in rapid tests for milkfat.

milk fever (parturient apoplexy; parturient fever; parturient paresis; parturient paralysis). an endemic paralytic disease of cattle near parturition. It is a metabolic disease accompanied by hypocalcemia.

milk flow. the amount of milk produced by a dairy cow or herd in a designated period of time.

milk purslane. *Euphorbia maculata*.

milk serum. the nonfat compartment of milk.

milk shed. a geographic region where dairy milk is produced for a given milk marketing area.

milk snake, milksnake. *Lampropeltis triangulum*.

milk sugar. lactose.

milk toxins. aflatoxins M (*See* aflatoxin).

milk vetch. *See Astragalus*.

milking. **1:** the withdrawal of milk by hand or by use of a milking machine, from an animal's mammary gland. **2:** the quantity or quality of milk produced by an animal's udder at a given time. **3:** pertaining to milking (def 1).

milking machine. a piece of equipment used to improve the efficiency of milking cows. Components include a vacuum pump, teat cups, and a milk line or milker pail. Milk flow is induced by alternate application of the vacuum and atmospheric pressure.

milking parlor. a clean, enclosed, protected area in which dairy animals are normally milked.

milksnakes. *See* adder.

milkweed. any plant of the Family Asclepidaceae, especially those of the genus *Asclepias*.

milkweed family. Asclepiadaceae.

milkweed poisoning of rabbits. this type of poisoning is caused by feeding hay that contains woolly pod milkweed. *Asclepias eriocarpa*, reported only from the southwestern United States. It is sometimes called head down disease, inasmuch as affected rabbits develop paralysis of the neck muscles and loss of coordination. If an animal has not consumed too much of the weed and the paralysis has not progressed too far, an attempt may be made to treat it. The head of the rabbit is held so that it can drink water and consume food. Leafy greens and carrots should be fed. The poisonous principle is a resinoid; consumption of the green plant equalling approximately 0.25% of an animal's weight is lethal.

milli- (m) a prefix to units of measurement that indicates a quantity of 0.001 (10^{-3}) times the value of the root. *See* International System of Units.

milligamma. nanogram.

milligram (mg; mgm) a unit of mass in the cgs (centimeter-gram-second) system, taken as 10^{-3} gram, which equals 0.015432 grains or 0.000035 ounces.

millimicrogram. nanogram.

millimole (mM). one-thousandth (10^{-3}) mole (def.3).

milo. a cultivated grain sorghum. *See Sorghum vulgare*.

Minamata disaster. an occurrence in which

thousands of persons living near Minamata Bay in Japan during the period 1953—1960 were poisoned by eating fish contaminated by mercury wastes dumped into the bay from several chemical plants More than 100 persons died. The proximate poisons were alkyl mercuries, notably monomethylmercury and dimethylmercury. The family of symptoms experienced by these people became known as Minamata disease, *q.v.*

Minamata disease (methylmercury neurotoxicity). a severe toxic neuropathy first recognized as a result of the Minamata disaster, *q.v.* The syndrome is caused by alkyl mercury poisoning (chiefly by methylmercury, *q.v.*), which may result in permanent paralysis, other neurologic and mental disabilities, or death. Symptoms may include constriction of the visual field (tunnel vision), cerebellar ataxia, paresthesias, progressive paralysis, deafness, and dysarthria. Lesions include loss of neurons, especially from the cerebral cortex and cerebellar cortex. CNS development in the young may be seriously affected. At risk are those who eat seafood contaminated with organomercury compounds. *See also* Minamata disaster, methylmercury, dimethylmercury, monomethylmercury.

miner's disease. ancylostomiasis.

mineral. an inorganic substance other than water (e.g, an ore, a mineral acid). **accessory mineral**. any mineral that occurs in small amounts in rock such that its presence or absence does not affect the properties of the aggregate rock.

mineral green. cupric acetoarsenite.

mineralization. the change of an organic compound into inorganic compounds with the release of elements and inorganic compounds by oxidative mechanisms, weathering, or the action of microbes. **biological mineralization**. the decomposition of organic compounds with the liberation of inorganic minerals by the action of microbes.

mintweed. *Salvia reflexa*.

miosis. constriction of the pupil of the eye.

miotic. **1:** of, pertaining to, characterized by, or producing, miosis (def.1). **2:** an agent that causes the pupil to contract.

Mirabilis. a genus of common perennial garden herbs (Family Nyctaginaceae) that are native to the American tropics. About ten species are cultivated in the warmer parts of the United States including *M. jalapa* (four-o'clock; marvel of Peru), an herb with fragrant yellow, red, or white flowers. All parts of these plants are poisonous.

mirex (1,1a,2,2,3,3a,4,5,5,5a,5b,6-dodecachlorooctahydro-1,3,4-metheno-1*H*-cyclobuta(c,d)pentalene; dodecachloropentacyclodecane;dodecachloropentacyclo(3,2,2,-$0^{2,6}$,$0^{3,9}$,$0^{5,10}$)decane; hexachlorocyclopentadienedimer; 1,3,4,5,5-hexachloro-1,3-cyclopentadiene dimer; perchloropentacyclo(5.2.1.$0^{2,6}$.$0^{3,9}$.$0^{5,8}$)decane). an intensely white, odorless, crystalline substance, $C_{10}Cl_{12}$, that is insoluble in water, soluble in benzene and dioxane. It is an insecticide, that is especially dangerous because it is highly persistent in the environment, bioaccumulates, and is poisonous to nontarget animals including humans. Mirex is highly toxic to humans by ingestion and moderately toxic by inhalation and skin absorption.

miscarriage. **1a:** failure, malfunction, misapplication. **1b:** mismanagement, misapplication; a blunder. **2:** abortion (*q.v.*), especially spontaneous abortion.

miscarry. **1:** to fail, blunder, malfunction. **2:** abort (def. 2).

mistletoe family. *See* Loranthaceae.

mite. a minute arachnid (Order Acarina). Some are parasitic and/or vectors of disease.

miticide. a pesticide used to destroys mites and other small arachnids such as the fruit-infesting European red mite and the common red spider. *See also* acaricide.

mitigate. to moderate; allay, palliate, render milder.

mitigated. diminished in severity.

Mitis green. cupric acetoarsenite.

mitosis. a continuous process of cell replication to form two daughter cells with identical sets of chromosomes. This process is classically divided into four sequential stages: prophase, metaphase, anaphase, and telophase. The period between divisions is called interphase or interkinesis. During interphase, DNA is synthesized and prior to the onset of prophase, the chromosomes replicate, forming pairs of identical sister chromosomes called chromatids.

mixed gland. **1:** a gland that houses both serous and mucous secretory cells. **2:** a complex gland that is both exocrine and endocrine in character (e.g., the pancreas).

mixture. any heterogeneous combination of two or more substances without chemical union (e.g., air, alloys, blood, concrete, petroleum, sea water, many venoms). A mixture cannot, therefore, be represented by a chemical formula. The components may or may not be uniformly dispersed and in many cases can be separated into the component materials by mechanical means. **complex mixture**. a mixture comprised of numerous substances (e.g., crude petroleum). Operationally, a complex mixture is one in which the component parts cannot be separated out by convenient or cost-efficient means. Such mixtures, if toxic, (petroleum, many petroleum distillates, liquid synfuels) are extremely difficult to assess. In particular, during the course of testing or experimentation, the makeup of such a mixture may change (e.g., due to differential volatilization, rates of decay, and reactivity). Furthermore, such mixtures may produce complex and confusing combinations of signs and symptoms. The set of symptoms may vary complexly with concentration and age of the mixture due to differential stabilities, solubilities, toxicities, and mode of action and other properties of the component materials. Many animal venoms are complex mixtures.

mM. millimole; millimolar.

Mn. the symbol for manganese.

Mo. symbol for molybdenum.

MO. malaoxon.

molar gland. *See* salivary gland (minor salivary gland).

moldering. *See* decay.

moldy corn poisoning (equine nervous syndrome; choking staggers; cornstalk disease; stomach staggers). a nervous disease especially of horses due to ingestion of moldy forages and feeds, especially corn. Symptoms vary but usually include excitement and paralysis. Autopsy reveals degenerative changes (most commonly necrosis and liquefaction) in the white matter of the brain and upper spinal cord. There are three clinical types: lethargic, nervous, and paralytic moldy corn poisoning. It is likely that early outbreaks designated as moldy corn poisoning were caused chiefly by aflatoxins. *Cf.* aflatoxicosus, forage poisoning, moldy corn toxicosis. *See also* encephali tis (virus equine encephalitis). Poisoned cattle experience a severe, usually fatal gastroenteritis.

mole death. strychnine.

mole-nots. strychnine.

mole plant. *Euphorbia Lathyris*.

molecular biology. biochemistry.

molinate. a thiocarbamate herbicide.

mollusc. mollusk.

molluscacidal. destructive to mollusks.

molluscacide (molluscicide). an agent that is destructive to mollusks.

molluscicide. molluscacide.

mollusk (mollusc). any member of the Phylum Mollusca. Most venomous mollusks are either gastropods (univalve mollusks) or cephalopods (octopuses and squids). Bivalve mollusks (clams, oysters, and mussels), however, may accumulate and even concentrate in their tissues, poisons (e.g., chromium) from the medium, from contaminated sediments, or their food. *See also* shellfish poisoning.

from the medium (e.g., air, soil, water) in their tissues

Molten's disease. Pictou disease.

molting hormone. ecdysone.

molybdenum (symbol, Mo)[Z = 42; A_r = 95.94]. a transition element that occurs naturally as the sulfide in molybdenite. Mo occurs as a dark-gray to black powder with a metallic sheen or as a silver-white mass. It is an antagonist of copper in animal nutrition. Certain species of plants, in soils with high concentrations of molybdenum, can accumulate this element to concentrations that are not toxic to the plants, but are poisonous to grazing animals. Legumes generally accumulate much higher levels of molybdenum than most other plants under a given set of conditions. Soils with normal copper content and low concentrations of molybdenum can similarly poison grazing animals. On the other hand, soils with normal copper content and high levels of molybdenum (or with low copper and moderate molybdenum levels) may produce symptoms of copper deficiency in grazing animals. Molybdenum toxicity to livestock is strongly influenced by the amount of inorganic sulfate in the forage consumed. Even modest amounts of sulfates in the soil can effectively control molybdenum intoxication of livestock.

monalide. a post-emergent anilide herbicide that decomposes rapidly in the soil. It is used mainly to destroy weeds of umbelliferous crops such as carrots, celery and parsley.

Monday disease. azoturia.

Monday dyspnea. byssinosis.

Monday fever. byssinosis.

Monday morning disease. azoturia.

Moniliaceae. a family of fungi that includes the genus *Aspergillus*.

monkey's bread. the fruit of *Adansonia digitata*, *q.v.*

monkshood. **1a:** *Aconitum*. **1b:** aconite.

mono-. a prefix denoting one.

monochasial cyme. *See* cyme.

monochasium. *See* cyme.

monochloroacetic acid. chloroacetic acid.

α-monochlorohydrin. α-chlorohydrin. *See* chlorohydrin.

monocrotophos. a highly toxic organophosphate insecticide; also used in some household products to eradicate pests.

monoculture. an agricultural production unit (e.g., a field, a farm) that cultivates and produces an entire crop that consists of genetically similar organisms (e.g., a single species or cultivar).

monocyte. a large, mononuclear, phagocytic leukocyte that has more cytoplasm than lymphocyte.

monofluoracetamide. fluoroacetamide.

monoploid. **1a:** haploid; having one set of chromosomes. **1b:** a cell or organism that has a single set of chromosomes.

monosodium methanearsonate (MSMA). the monosodium salt of methylarsonic acid, CH_4AsNaO_3. MSMA, a herbicide, is somewhat less toxic to laboratory rats than the disodium salt, disodium monomethanearsonate (also a herbicide).

monosomy. *See* aneuploidy.

Monterey pine. *Pinus radiata*.

monuron (*N'*-(4-chlorophenyl)-*N*,*N*-dimethylurea; 1,1-dimethyl-3-(*p*-chlorophenyl)-urea; 3-(*p*-chlorophenyl)-1,1-dimethylurea; CMU). an environmentally persistent substituted urea used as a somewhat selective preemergence herbicide and soil sterilant. It is used mainly to control broadleaved weeds and grasses in non cropland areas and in crops such as cotton, sugar cane, pineapple, and asparagus.

morbid. **1:** diseased; of, or pertaining to, disease. **2a:** unhealthy or unwholesome; gruesome.

morbidity. **1:** the state of being diseased or morbid. **2:** the rate of illness in a population; morbidity rate.

morbidity rate. the number of cases of a specific illness or disease in a defined population, herd, flock, or community during a specified period of time (usually a month or a year) per unit (usually 1000, 10,000, or 100,000) of population.

morbific. pathogenic; pertaining to or causing disease or morbidity.

morbigenous. pathogenic.

morbus. disease.

Morerastrongylus costaricensis. *Angiostrongylus costaricensis*.

Morfamquat. a highly selective bipyridylium quaternary salt that has been used as a contact postemergence herbicide to control a variety of annual weeds in cereals; grasses are highly resistant. *Cf.* diquat, paraquat.

moribund. **1:** dying; in a dying condition or state; in a condition of impending death, often characterized by a deepening stupor or coma. **2:** dormant. **3:** a dying person or animal.

morphactin. any of a class of derivatives of fluorene-9-carboxylic acids that powerfully regulate plant growth and development. They inhibit shoot elongation and cause morphological abnormalities in intact plants. This is accomplished by blocking endogenous auxin transport. Morphactins also promote the cell divisions that give rise to lateral primordia in the root pericycle. The organization of such primordia is, however, greatly disturbed and further growth is inhibited.

morning glory. any plant of the genus *Ipomoea*.

morphologic effect. *See* effect.

mortality. **1:** the state or quality of being mortal (subject to death). **2:** a heavy loss of life; death in large numbers. **3:** the total number of deaths in a given population or community at a given time. **4:** death rate; **a:** the proportion of deaths to population. **b:** the ratio of deaths that occur to those expected. **consequential mortality**. death that is secondary to the action of a given agent (e.g., an insecticide). **delayed mortality**. death that occurs some time following trauma or removal of the source of lethality. **perinatal mortality**. mortality that occurs at about the time of birth, or during the period when life *ex utero* is considered possible; sometimes taken to include stillbirths and deaths during the first days of life.

Morus (mulberry tree). a genus of hardy, long-lived Asiatic trees with milky sap (Family Moraceae). They are grown in the United States for their berry-like fruit. The unripe fruit and shoots are toxic. The principal species are *M. alba* (white mulberry), which is native to China, and *M. nigra* (black mulberry), which is native to Asia.

mosquito. a small, long-legged insect (Order Diptera; Family Culicidae). Mosquitos are true flies, the females of which usually have piercing and sucking mouth parts and must presumably feed at least once upon mammalian (or in some cases avian) blood before their eggs can mature. Males typically feed on fruit or plant juices. The eggs are laid singly or together in the form of a raft, usually in stagnant water (e.g., ponds, pools, open containers). The pupae are aquatic and have gills but frequently come to the water's surface, where they breathe air through tubes on the thorax. Mosquitoes may thus be effectively controlled by spreading oily substances on the water's surface, thus suffocating the pupae. *See Aedes*, *Anopheles*.

mouse killer. mouse poison (*See* poison).

mount. to climb on or get onto, for example, a horse or motorcycle.

mountain balsam. *Abies lasiocarpa*.

mountain bunchgrass. *Festuca viridula*.

mountain mahogany. *See Cercocarpus*.

mousebane. any plant of the genus *Aconitum*.

MSMA. acronym of monosodium methanearsonate.

mucilage. a slimy material comprised of complex derivatives of carbohydrates that is produced by many plants and animals.

mucin. A glycoprotein that is the chief constituent of mucus.

muck. greatly decomposed, wet, organic soil.

muckiness. **1a:** having the characteristics of muck. **1b:** comprised of muck.

mucopurulent. comprised of mucus and pus.

mucosa. a moist mucous membrane that lines a hollow organ or body cavity such as the stomach and intestines. It consists of three layers: an inner glandular layer; a middle layer of loose connective tissue (lamina propria) that contains lymph, blood vessels, and the bases of glands; and an outer muscular layer (mucosa muscularis).

mucosal disease. bovine viral diarrhea.

mud fever. human leptospirosis (*See* leptospirosis).

mug-wort. *Artemisia*.

mulberry tree. any tree of the genus *Morus*.

mule. the F_1 hybrid of a male domestic ass and a female horse.

mule killer. common name of any of a number of arthropods in the southern United States (e.g., whip scorpions, mantids, walking sticks) alleged to kill livestock by stinging, biting, or when swallowed.

muley. a polled cow.

multiparous. pertaining to or descriptive of a female mammal that has borne more than one offspring.

multiple allelomorph. multiple allele (*See* allele).

multiple cropping. the consecutive production of two or more crops of similar or dissimilar commodities on the same land within the same year.

multiple use. a system of resource use where the resources in a given land unit serve more than one purpose (e.g., recreation, forestry, grazing of livestock, wildlife conservation, watershed management, fishing, water supply). Such a system can be managed in any of three ways: 1. different uses are assigned to different, usually adjacent subareas, which together form a composite multiple use land unit; 2. alternation in time of different uses on the same areas; 3. more than one use of an area at one time. In the latter case, conflicts among users may occur; these may be resolved if one use (or one resource) is considered to be primary such that other (secondary) uses (resources) are supported only if they are compatible with the primary use.

mung bean. *Phaseolus aureus*.

murder. a flock of crows.

muriate of ammonia. ammonium chloride.

muriatic acid. name for commercial grade hydrochloric acid.

Musa sapientum. a large tropical plant, several varieties of which are cultivated on plantations in Central and South America, the West Indies, the Canary Islands, and the Philippines. All varieties produce well only in areas of high temperature, moisture, and deep soil. ***Musa paradisiaca* var. *sapientum*** (banana). a perennial, herbaceous, tree-like plant indigenous to southeastern Asia but also cultivated in other tropical areas. *See* banana.

muscicide. an agent that destroys flies.

muscoid. see unabridged.

mushroom. **1:** any of a large variety of fungi (mostly of the subdivision Basidiomycetes) that usually have a fleshy, usually aerial, umbrella-like, fruiting body (actually a sporophore); it arises from an underground mycelium. The mushroom is typically differentiated into a stipe (the stalk) and pileus (the cap). The spores develop in the folds (gills) or pores on the undersurface of the pileus. In some species, the edge of the cap is united with the stalk by a membrane, the partial veil. This ruptures in the mature mushroom, exposing the underside of the cap and leaving a ring (the annulus) about the top of the stalk. Many mushrooms are edible, others are unpalatable or toxic, and some are deadly poisonous. In some cases, certain populations of a species may be edible while other populations are poisonous. **2:** any macroscopic fungal fruiting body. *Cf.* toadstool. **little brown mushroom** (LBM). certain of the brown-spored gilled mushrooms are edible, others such as *Cortinarius gentilis* and *Cortinarius orellanus* are deadly poisonous. Mushroom hunters are thus well advised not to gather or eat LBMs.

musquash root. *Cicutum maculata*.

mustard. *See Brassica.*

mustard family. Brassicaceae.

mustard greens. *See Brassica juncea.*

mustard oil. allyl isothiocyanate.

muster (ostentation). a group of peacocks, especially if displaying.

mutafacient. mutagenic.

mutagen. an agent that induces mutagenesis; one that brings about an increase in the natural rate of mutation within a population.

mutagenesis. **1a:** the origin, development, or production of a mutation in heritable material. **1b:** the induction of heritable changes in genetic material within living cells. While mutagenesis is a normal process that plays an important role in adaptive evolution of plants and animals, most mutations are harmful.

mutagenic (mutafacient). able to induce mutations or to increase the rate of mutations in a population. *See* mutagen.

mutant. **1a:** any organism, gene, or character that has undergone a mutational change. **1b:** of, pertaining to, or designating a gene that has undergone a mutation. **1c:** an organism or cell carrying one or more mutant genes. **1d:** the character or phenotype produced as the result of gene mutation. **2:** any gene that is rare in a population, or produces a phenotype that differs from the wild type.

mutate. **1:** to undergo mutation. **2:** to effect a change in the genetic constitution of a cell by altering its DNA.

mutation. **1:** any sudden or discontinuous, potentially heritable, change in the genetic constitution of a cell or organism. Most mutations are recessive and deleterious to the affected organism. If a mutation is advantageous, it may eventually become dominant through natural selection.

mute. **1a:** silent, speechless. **1b:** silent, unable to talk or call out. **2a:** to silence, quiet, hush. **3a:** to deaden, muffle, or suppress. **4:** a pack of hounds.

mutualism (interdependent association). symbiosis in which both organisms benefit. The relationship may be one of complete interdependence.

mw. milkweed. *See Asclepias.*

mycetismus (mycetism; mushroom poisoning; mushroom mycotoxicosis). poisoning due to ingestion of a fungus, especially a poisonous mushroom. In North America alone, more than 70 species of mushrooms are known to be poisonous to humans. Of these, *Amanita phalloides* causes most of the confirmed fatalities. Diagnosis is often difficult, due in part to: (1) the time between exposure and presentation of symptoms; (2) the diversity of signs and symptoms that may appear; (3) failure to secure reliable identification of species eaten; or (4) inability to identify the toxin present in blood or excreta of the victim.

mycoplasma. any of numerous parasitic microorganisms (Class Mollicutes]. All lack a cell wall, can survive without oxygen, and are the smallest self-reproducing prokaryotes. These microbes are common causes of pneumonia and urinary tract infections.

Mycoplasma. a genus of parasitic bacteria. Most species have no cell wall. *See also* mycoplasma.

mycorrhiza. a symbiotic association between a fungus and the roots of a plant.

mycosis (plural, mycoses). any disease caused by a fungus.

mycostat. fungistat.

mycostatic. fungistatic.

mycotic. relating to or caused by a fungus; relating to a mycosis.

mycotoxic. fungitoxic.

mycotoxicosis. **1:** poisoning by fungal or bacterial toxins, usually in contaminated food or animal feeds. **2:** poisoning due to ingestion of fungi (e.g., *Claviceps* spp., poisonous mushrooms). **mushroom mycotoxicosis**. mycetismus.

mycotoxin. **1:** any toxic chemical that is produced naturally by fungi. Numerous mycotoxins have been identified. They are active at very low concentrations, sometimes making detection difficult, and mechanisms of action are remarkably varied. *See*, for example, ochratoxin, patulin, penicillic acid, *Penicillium*, toxin (e.g., T-2 toxin). **2:** sometimes restricted to any poisonous substance produced by fungi that infest food or animal feed. **anthraquinoid mycotoxin**. toxic derivatives of anthraquinone (*q.v.*) that occur in fungi. *See*, for example, luteoskyrin, rubroskyrin. **bisfuranoid mycotoxin**. any of a class of mycotoxins that contain the dihydrofurobenzofuran system. All metabolites of these mycotoxins characteristically include either the unsaturated 7,8-dihydrofurano(2,3-b)furan or the more reduced 2,3,7,8-tetrahydrofuro(2,3-b)furan. They are hepatotoxic and tumorigenic to various animals. Examples are sterigmatocystin and aflatoxins.

mydaleine. a poisonous ptomaine in putrefying viscera. Symptoms of intoxication include salivation, miosis, intermittent fever, and arrest of the heart in diastole.

mydatoxin (mydatoxine). a deadly ptomaine, $C_6H_{13}NO_2$, from putrefying viscera and flesh or from human intestines maintained at low temperatures for a long time.

mydatoxine. mydatoxin.

mydriasis (pupillary dilation). excessive, prolonged, or abnormal dilation of the pupil of the vertebrate eye. Causes vary and may include fear, hysteria, coma, anesthesia, botulism, drugs and certain other chemicals, and irritation of the cervical sympathetic nerve.

mydriatic. **1:** causing mydriasis. **2:** a drug that causes mydriasis.

myel-, **myelo-**. a prefix referring to or indicating **a:** bone marrow, **b:** spinal cord.

myochaemoglobinuria. azoturia.

myoinositol. a stereoisomer of inositol (*q.v.*) it is the precursor of phosphatidylinositol. The latter is an important constituent of animal membranes, muscle, and brain tissue.

Myrica cerifera (bayberry). a fragrant, perennial, evergreen shrub (Family Myricaceae) suspected of being carcinogenic.

Myristica fragrans (nutmeg). a tropical tree (Family Myristicaceae) with entire leaves and flowers that produce a fleshy fruit, nutmeg, *q.v. See also* nutmeg, oil of nutmeg.

Myristicaceae (nutmeg family). a family of trees with unisexual flowers and monadelphous stamens; the seeds are arillate.

myristicin. a toxic, crystalline phenolic ether, $C_{11}H_{12}O_3$, with a strong odor. It is the chief toxic substance in the ripe seeds of *Myristica fragrans* (nutmeg). Ingestion of large amounts of the spice (5-15 g) causes symptoms similar to those of atropine poisoning. It reduces the levels of cytochrome P-450 and thus inhibits monooxygenations catalyzed by this enzyme. Myristicin occurs also in a number of essential oils.

Myxophyta. Cyanobacteria.

N

N. symbol for nitrogen.

NAA. 1-naphthaleneacetic acid.

nabam (1,2-ethanediylbascarbamodithioic acid disodium salt; ethylenebis[dithiocarbamic acid] disodium salt; disodium ethylenebis[dithiocarbamate]). the sodium salt of dimethyldithiocarbamate,

$$NaSSCNHCH_2CH_2NHCSSNA$$

Nabam is a very toxic agricultural fungicide. Its toxicity to animals is low.

Naematoloma fasciculare (sulfur tuft). a bitter-tasting poisonous mushroom that occurs throughout much of North America on stumps and logs or on soil over buried wood. Whereas some individuals in North America have suffered gastric distress for a few hours with complete recovery, fatalities have been reported in Asia and Europe.

naked ladies. *Cholchicum autumnale*.

naked lady. *Amaryllis belladonna*.

naled. a colorless, liquid, organophosphate insecticide, $C_4H_7Br_2C_{12}O_4P$, that is slightly soluble in aliphatic hydrocarbons and very soluble in aromatic hydrocarbons. A cholinesterase inhibitor, it is dangerous to both production workers and applicators. It produces multiple symptoms, ranging from minor eye irritation, cramps, and weakness to paralysis, convulsions, and hypotension.

nano- (n). a prefix to units of measurement that indicates a quantity of 1 millionth (10^{-9}) the value indicated by the root. *See* International System of Units.

nanogram (millimicrogram; micromilligram; milligamma; symbols, mγ, μmg, mμg, ng). a unit of mass equivalent to 10^{-9} gram or 10^{-3} microgram.

nanometer (nm). 10^{-9} meter or 10 Ångstrom units.

nap-at-noon. a common name of plants of the genus *Ornithogalum*.

α-naphthaleneacetic acid. 1-naphthaleneacetic acid.

1-naphthaleneacetic acid (NAA; α-naphthaleneacetic acid; naphthylacetic acid). a synthetic auxin-type plant growth regulator and selective herbicide related to auxin 1AA. It is moderately toxic to laboratory rats. It is, for example, lethal to *Brassica kaber* (= *B. arvensis*) (charlock) and *Beta vulgaris* (sugarbeet), but is essentially nontoxic to cereal crops. NAA is moderately toxic to laboratory rats.

1-naphthalenol methylcarbamate. carbaryl.

1-naphthyl *N*-methylcarbamate. carbaryl.

naphthylacetic acid. 1-naphthaleneacetic acid.

2-(naphthyloxy)propionamide. napropamide.

α-naphthylthiourea. ANTU

1-naphthylthiourea. ANTU.

1(naphthyl)-2-thiourea. ANTU

napropamide (*N,N*-diethyl-2-(1-naphthalenyloxy)propanamide; *N,N*-diethyl-2-(1-naphthyloxy)propionamide**1**; 2-(naphthyl oxy)propionamide). an herbicide of low toxicity to mammals.

naptalam. a colorless, crystalline amide, $C_{10}H_7NHCSNH_2$, that is soluble in alkaline solutions and sparingly soluble in ethanol, acetone, and benzene. It is used as an analytical reagent. The sodium salt is used as a soil-applied preemergence herbicide that is absorbed through the roots. It may persist in the soil for as much as 2 months. Naptalam closely resembles a morphactin in its physiological action in that it promotes the elongation of isolated coleoptile segments but inhibits the elongation of intact coleoptiles. It inhibits root elongation. Morphactins, naptalam, and certain substituted benzoic acids interfere with geotropic and phototropic curvatures. *Cf.* alachlor, diphenamid, propachlor.

Narcissus (narcissus). a genus of bulbous plants (Family Amaryllidaceae). Many of these are favored as house and garden plants. Certain hardy forms are commonly referred to as daffodils. The use of common names leads to confusion (see daffodil, jonquil). These names should be used professionally with care, if at all. Nomenclature is further complicated by the numerous wild and cultivated hybrids that are known. The bulb is toxic. Common or garden species include *N. jonquilla* (jonquil), native to southern Europe and Algeria and the only species that may properly be called the jonquil; *N. poeticus* (poet's narcissus), a species native to southern Europe, especially Greece; *N. tazetta*, a species with unusually large bulbs; and *N. triandrus* (angel's tears), originally native to Spain and Portugal.

narcissus. a plant of the genus *Narcissus*.

Narthecium ossifragum (bog asphodel). a European bog herb (Family Liliaceae) with linear leaves and greenish-yellow flowers. It causes hepatogenic photosensitization in livestock. There is a similar species in the United States, *N. americanum* (also known as the bog asphodel). *See also* phylloerythrin.

natality (birth rate). **1a:** the ratio of the number of births in a target population during a stated period of time to the total number of individuals in the population. **1b:** operationally, the number of offspring produced per female or per individual in a specified population per unit time. **1c:** treated as an emergent property at the population level or organization, natality is the rate of production of new individuals by birth, germination or fission. **2.** the production of new individuals (by whatever means) in a specified population. *See also* mortality. This usage is not recommended.

nebulization. the reduction of a solid or liquid to a mist or spray made up of very fine particles.

neburon (*N*-butyl-*N'*-(3,4-dichlorophenyl)-*N*-methylurea; 3-(3,4-dichlorophenyl)-1-methyl-1-*n*-butylurea). an herbicide used, for example, to control annual weeds in certain evergreens and to control *Stellaria media* (common chickweed) and *Cerastium vulgatum* (mouse-ear chickweed) in crops such as *Medicago sativa* (alfalfa), tomatoes, and strawberries. Neburon is only slightly toxic to laboratory rats.

neck rein (bearing rein). **1a:** a rein used to signal a horse to turn by pushing against its neck toward the direction of turn. **1b:** to use a rein in the manner of a neck rein to direct a horse to turn.

necrose. **1:** to cause necrosis. **2:** to undergo necrosis; to become necrotic.

necroses. plural of necrosis.

necrosis (plural, necroses). **1a:** tissue death; death of a circumscribed (focal) portion of tissue. Necrosis in plants often results from fungal infection, and the shape of necrotic areas often indicates the specific cause.

necrotic. denoting, pertaining to, or concerned with, or characterized by necrosis.

necrotizing. causing or able to cause necrosis.

neem oil (nim oil; margosa oil). an aromatic oil from the fruit and seeds of *Melia azadirachta*. It is an anthelmintic, a pesticide in certain formulations, an insect repellent, antifeedant, and an ethanol denaturant.

nekton. organisms that swim actively in water.

Nemathelminthes. Aschelminthes.

nendrin. endrin.

neonatal. of or pertaining to a newborn mammal.

neonate. a newborn mammal.

neonicotine. anabasine.

Nepeta hederacea. *Glechoma hederacea*.

nephritic. **1:** of or pertaining to nephritis. **2:** pertaining to the kidneys. **3:** a drug or agent used in the treatment of renal disease.

nephritides. **1:** plural of nephritis. **2:** a collective term, used in reference to all types of nephritis.

nephritis (plural, nephritides). acute or chronic inflammation of the kidney that may be a focal, or a diffuse proliferative, or destructive process that may involve the glomeruli, tubules, or interstitial tissue. Causes vary but include bacteriotoxins, streptococcal infections, septicemia, drugs, heavy metals. **nephrotoxic nephritis**. that produced in a vertebrate animal by the injection of antikidney antibody from a different species that had earlier received a preparation of glomerular basement membrane from an animal of the same species as the one to be inoculated. **saturnine nephritis**. that seen in chronic lead poisoning. **tartrate nephritis**. an acute form of nephritis induced by subcutaneous injection of racemic tartaric acid.

nephritogenic. causing, or able to cause, nephritis.

nephron. the functional unit of the kidney of reptiles, birds, and mammals. While nephrons vary in structural detail, all contain a structure (the Malpighian corpuscle) that contains a knot of capillaries (a glomerulus) that receives a filtrate of circulating blood and an associated tubule that reabsorbs parts of the filtrate into the circulation. The remainder is excreted.

nephropathia. nephropathy.

nephropathic. of, pertaining to, characterized by, or productive of nephropathy.

nephropathy. any disease or morbid condition of the kidneys. *Cf.* nephrosis.

nephrosis. **1a:** any disease or lesion of the kidney that is accompanied by degenerative lesions of the renal tubules with attendant, noninflammatory edema, albuminuria, and decreased serum albumin. See also nephrotic syndrome. **1b:** a clinical syndrome characterized by edema, excess albumin in urine, and cholesterol in blood.

nephrotoxic agent. a poison that is harmful or damaging to the kidney of vertebrates. There are numerous such agents including many antibiotics, biologicals, botanicals, herbicides, oxalosis-inducing agents (e.g., oxalic acid), pesticides, solvents, all nonsteroidal antiinflammatory drugs, and most, if not all, heavy metals.

Nerium (oleander). a genus of evergreen shrubs (Family Apocynaceae) native to southern Europe and Japan. ***N. indicum*** (oleander). a common species found in Hawaii. It contains several toxic glycosides, including the steroid glycoside, odoroside. Symptoms and toxicity are similar to those of *N. oleander*. ***N. oleander*** (oleander; rosebay). an ornamental evergreen shrub or bush native to Europe that is widely cultivated as a house and garden plant. All parts of the plant are toxic; the leaves are extremely poisonous. They contain toxic cardiac glycosides, the most important of which are oleandroside and nerioside. Symptoms in humans and other animals include discoloration of the mouth, nausea and vomiting, weakness and sometimes dizziness, a rapid pulse, sweating, cold extremities, mydriasis, anorexia, abdominal pain, and bloody feces. In fatal cases, the heartbeat is weak and irregular; dyspnea and coma are common. See also oleandrin, oleandrin poisoning.

nervous syndrome of cattle and sheep (rape blindness). a disease, reported from Canada and the United Kingdom, that results from ingestion of rape, *Brassica napus*. Affected animals initially stand alone or wander, often bumping into solid objects, signaling blindness. The disease often arrests at this point or the animal becomes aggressive. Specific lesions are absent and the eye exhibits no abnormality except the absence of a pupillary reflex. Blindness may be permanent, but with good feed and care, vision usually returns gradually.

nervous system. a highly organized, complex system that is present in all multicellular animals. All such systems have a highly organized receptor-conductor-effector system comprised of cells (nerve cells or neurons) that generate and transmit information, in the form of electrical impulses, by means of electrochemical mechanisms. Additional types of cells that support or provide nutrients to the neural tissue may also be present. The nervous system receives information in the form of both external and internal stimuli, interprets it, conveys it, and coordinates a rapid and suitable (adaptive) muscular and/or glandular response. In higher animals, the nervous system coordinates (together with the neuroendocrine system) the adjustments and reactions to stimuli such that behavior is appropriate to both internal and environmental conditions. The nervous systems of most, if not all, metazoans also stores information and have some capacity to modify the processing of information based on experience (learning).

nettle. (stinging nettle). **1:** any plant of the family Urticaceae, especially those of the genus *Urtica*. **2:** any of numerous unrelated plants with stinging hairs (e.g., *Hesperocnide*, *Laportea*, *Urtica*).

nettle syndrome. a disease of hunting dogs caused by the stinging hairs of *Urtica chamaedryoides*. It is characterized by excessive salivation, emesis, pawing or wiping at the muzzle, a slow irregular heartbeat, and muscular weakness. *See also* syndrome.

neuritis. **1:** inflammation of a nerve or nerves that is often associated with a degenerative process. Neuritis is attended by pain and tenderness over the nerves; anesthesia and paresthesia; paralysis; wasting; and elimination of reflexes. Causes vary, but include compression or trauma; direct infection of the affected nerves; various disease states; metabolic disorders; toxins; heavy metal poisoning; ethanol; carbon tetrachloride. **2a:** sometimes used to denote noninflammatory lesions or disease of the peripheral nervous system. **2b:** often used interchangeably with neuropathy.

neuropathy. **1:** any disease, inflammation, disturbance, damage, or pathological change in the peripheral nervous system, regardless of etiology (e.g., arsenical neuropathy, diabetic neuropathy, ischemic neuropathy, traumatic neuropathy). *Cf.* polyneuropathy, encephalopathy, myelopathy, neuritis. **2:** noninflammatory lesions in the peripheral nervous system, as opposed to neuritic inflammatory lesions. **alcoholic neuropathy** (polyneuritis potatorum; alcoholic neuritis). neuropathy due to thiamine deficiency in chronic alcoholism. **buckthorn polyneuropathy**. a neuropathy that results from ingesting the fruit of shrubs of the genus *Rhamnus* (buckthorns). It is a symmetrical ascending neuropathy beginning in the lower limbs and terminating in the brain stem with consequent respiratory paralysis and death. Survivors recover slowly, but usually completely. There is no specific therapy. **delayed neuropathy**. organophosphorus-induced delayed neurotoxicity.

neurotropic (polioclastic). having an affinity for nervous tissue, or exerting its principal effect upon the nervous system.

neutral ammonium fluoride. ammonium fluoride.

neutrophil. **1:** staining readily with neutral dyes. **2:** a granular leukocyte that stains readily with neutral dyes. These cells ingest bacteria and small particles.

nib. *See* beak.

Nicotiana (tobacco, nicotiana). a genus of plants (Family Solanaceae), all or nearly all of which contain nicotine and a number of carcinogenic hydrocarbons. Accidental or intentional use or misuse of tobacco produces frequent intoxication or other effects on humans, livestock, and other animals. Species known to have poisoned livestock in the United States include *N. attenuata* (wild tobacco, coyote tobacco), *N. glauca* (tree tobacco), *N. tribonophylla* (wild tobacco, desert tobacco), and *N. tabacum*. All of these produce similar, mainly neurologic effects. Signs may include muscular weakness, tremors, staggering, and collapse. Violent heart action may be observed, but the pulse is weak and rapid and the extremities become cold. Labored breathing is common. Time to death varies from a few minutes to several days. In at least one case, pregnant swine that ingested tobacco stalks developed arthrogryposis.

nicotiana. *Nicotiana*.

nicotine (3-(1-methyl-2-pyrrolidinyl)pyridine; β-pyridyl-α-*n*-methylpyrrolidine). an extremely poisonous alkaloid. The pure compound is a thick, colorless, nearly odorless, flammable oil,

$$C_5H_4NC_4H_7NCH_3$$

It has a sharp, burning taste, is miscible with water, and turns brown in air. Nicotine occurs in all parts of tobacco plants (*Nicotiana*) but is most concentrated in leaves. It is one of the most toxic and addicting of all drugs. Heavy users (cigarette smokers), however, acquire a pronounced tolerance to nicotine. It is also used as a contact insecticidal fumigant in closed spaces. Nicotine is apparently the addictive component of cigarettes.

nicotine sulfate (1,10-methyl-2-(3-pyridyl)-pyrrolidine sulfate; (S)-3-(1-methyl-2-pyrrolidinyl)pyridine sulfate (**2:**1); 1,3-(1-methyl-2-pyrrolidyl)pyridine sulfate). a poisonous substance,

$$C_{20}H_{26}N_4{\cdot}O_4S$$

It is used as an insecticide. Animals poisoned by nicotine or by nicotine sulfate show tremors, incoordination, nausea, and disturbed respiration, and finally dark bloody hemorrhages in the heart and in the lungs, and congestion of the brain. Mildly affected animals recover rapidly and spontaneously. When heated to decomposition, nicotine sulfate emits toxic gases.

nicouline. rotenone.

nide (bouquet). a nest of pheasants.

nightshade. common name of any of a number of plants of the genus *Solanum* (Family Solanaceae) and of some other plants of the Family Solanaceae. *See also* *Atropa belladonna, Phytolacca americana*.

nightshade family. Solanaceae.

nim oil. neem oil.

nine-banded armadillo. *Dasypus novemcinctus*.

niter. potassium nitrate.

nitraphen. nitrofen.

nitrate. a salt or ester of nitric acid. Nitrates as such are only moderately toxic, but if they are not promptly absorbed during digestion, they can be reduced to nitrites by microflora in the intestines. The latter are extremely toxic to humans. They may also combine with amines to produce carcinogenic nitrosamines. Nitrates are among the most dangerous food additives. Livestock and other animals have been poisoned by ingesting plants that contain high levels of nitrates. These include certain crops such as corn, oat hay, sorghum, and weedy pasture plants from a number of families (e.g., Amaranthaceae, Chenopodiaceae, Brassicaceae, and Solanaceae). Other possible sources of nitrate poisoning of animals are pond water or well water with high nitrate content, and even the ingestion of fertilizer, machine oil, and silage. Livestock and wildlife that graze on the above-ground portions of plants that contain toxic levels of nitrates, including such crops as corn, oat hay, and sorghum, and a variety of weedy pasture plants may be seriously poisoned; poisoning occurs mainly under conditions and in seasons when plant growth has been especially rapid. *See also* nitrite, nitrite poisoning.

nitrate reduction. the conversion of nitrate to reduced forms of nitrogen. **assimilatory nitrate reduction**. reduction of nitrate, usually to ammonium, for the synthesis of amino acids and proteins.

(nitrato-*O*)phenylmercury. phenylmercuric nitrate, basic.

nitre. potassium nitrate.

nitrification. the oxidation of ammonia to nitrite by aerobic microorganisms. **autotrophic nitrification**. the oxidation of ammonium to nitrate by the sequential action of two chemoautotrophic organisms, one of which produces nitrite from ammonium and the other oxidizes the nitrite to nitrate.

nitrite. a salt or ester of nitrous acid. These compounds depress motor centers of the spinal cord; have an antispasmodic action; and, as vasodilators, they lower blood pressure. Symptoms of intoxication in humans may include headache, vertigo, palpitations, visual disturbances, flushed skin, nausea, vomiting, diarrhea, methemoglobinemia and, especially in infants, may be followed by coma, and death. They are used in the treatment of angina pectoris. Nitrite is added to 60—65% of all pork produced in the United States, as well as some other meats, poultry, fish, and cheeses. It is especially prevalent in processed meats such as bacon, sausage, luncheon meats, and hot dogs, where it is used to preserve the pink color and inhibit the growth of bacteria that cause botulinum poisoning. Nevertheless, most known cases of nitrite poisoning of humans in the United States are those occur in infants who have been given well water contami nated with nitrites. Nitrites are also of environmental significance because they react with some amines, yielding carcinogenic nitrosamines. *See also* nitrite poisoning, nitrate.

nitrite poisoning. nitrites are released during the digestion of food materials that contain nitrates, *q.v.*, They are substantially more toxic than nitrates and are the most common proximate cause of intoxication from the ingestion of nitrates. Nitrites impair the oxygen-carrying capacity of the blood by oxidizing ferrous hemoglobin (Hb) to the ferric methemoglobin. Symptoms appear rapidly. They include cyanosis, trembling, weakness, and severe dyspnea. The blood becomes characteristically dark, chocolate-brown in nitrite poisoning. In some cases, the dominant symptoms are diuresis, pronounced vasodilation with concomitant reduction in blood pressure, increased cardiac activity, and coma. In these cases, vasoconstrictive drugs can be effective. In other cases, anoxia dominates. The anoxia is responsive to methylene blue. In active animals, acute intoxication results in death by asphyxiation when about 50% or more of the hemoglobin has been converted to methemoglobin. *See also* nitrate.

2-nitro-1,1-bis(*p*-chlorophenyl)butane. Bulan.

1,1′-(2-nitrobutylidene)bis[4-chlorobenzene]. Bulan.

nitrofen (2,4-dichloro-1-(4-nitrophenoxy)-benzene; 2,4-dichlorophenyl*p*-nitrophenyl ether; nitraphen; nitrophen; nitrophene). a moderately toxic, selective, pre- and postemergent herbicide. It should be placed on the soil surface and not incorporated into soil. Nitrofen is a probable carcinogen.

nitrofene. nitrofen.

nitrogen (symbol, N)[Z = 7; A_r = 14.007]. a very electronegative, odorless, diatomic, relatively inert, gaseous element that condenses to a liquid at -195.79°C (77.36° K). It comprises about 78% by volume of the atmosphere. Nitrogen is an important component of living matter.

nitrogen cycle. the biogeochemical cycle of nitrogen by means of nitrogen fixation, nitrification, decomposition by putrefaction (ammonification), and denitrification.

nitrogen dioxide. a heavy, yellow-brown or reddish-brown, toxic, oxidant gas that normally exists in equilibrium with other nitrogen oxides. A component of smog, it is produced by a number of industrial processes and is also produced during the fermentation of fodder to form silage (*See* silo-filler's disease). *See also* nitrogen oxide red.

nitrogen dioxide poisoning. nitrogen dioxide intoxication of humans may result from exposure to silage (See silo-filler's disease) or to a number of industrial processes. Symptoms vary as a function of concentration and duration of exposure. Acute exposures may have a fatal outcome within a few hours to a few days as a result of pulmonary edema. Presenting symptoms often include muscular weakness, dyspnea, and coughing which may persist for 2-3 weeks. Following this period, the victim may experience fever (with or without chills), severe dyspnea, heavy ineffective coughing, and cyanosis. Treatment may be ineffective and subjects often die within 3½-6 weeks following exposure. Victims occasionally recover spontaneously. Lesions in such cases include discrete pulmonary nodules and pronounced neutrophilic leukocytosis. Symptoms in milder cases are similar to those of bronchopneumonia.

nitrogen fixation (dinitrogen fixation). the conversion of molecular nitrogen, N_2, to ammonia and subsequent forms that are useful in biological processes. **associative nitrogen fixation**. an interaction between a green plant and a free-living diazotroph that accelerates the rate of nitrogen fixation.

nitrogen oxide (NO_X). **1:** any of the gaseous oxides of nitrogen. These include nitric oxide (NO); nitrogen dioxide (the monomer, NO_2, and the dimer, N_2O_4); nitrogen oxide red; nitrogen pentoxide; nitrogen trioxide; nitrous oxide. All of these gases are pulmonary irritants and can cause acute pulmonary reactions when inhaled. The most hazardous of these oxides are NO and NO_2. **2:** nitrous fumes. **3:** nitrous oxide.

nitrogen oxide poisoning. poisoning by inhalation of NO_2 or its dimer, N_2O_4. It is severely irritating to the pulmonary epithelium and associated tissues. Unless the vapor concentration is very high, there may be no symptoms near the time of exposure with the possible exception of a transient cough, nausea, and fatigue; symptoms usually present within 5-72 hours. Acute exposure may cause pulmonary edema and fatal bronchiolitis fibrosa obliterans. Inhalation for even very brief periods of time of air containing 200 700 ppm of $N0_2$ can be fatal. $N0_2$ disrupts certain enzyme systems such as lactic dehydrogenase and may also cause lipid peroxidation. These two gases are usually the chief component of most fumes that contain nitrogen oxides and cause most of the lung damage in such cases. See also nitrogen oxide. [RAL-final drf Agric]*

nitrogen oxide red. a nitrogen dioxide, N_2O_4, it is often produced together with nitrogen dioxide during the fermentation of fodder to form silage (*See* silo-filler's disease).

nitrogenase. an enzyme that catalyzes N_2 fixation.

nitrolime. calcium cyanamide.

nitrophen. nitrofen.

***N*-(4-nitrophenyl)-*N'*-(3-pyridinylmethyl)urea**. pyriminyl.

nitrosamine. any of a group of *N*-nitroso derivatives of secondary amines, $R_2{=}N{-}N{=}O$ (where R is an alkyl or aryl group), formed by the combination of nitrates with amines or amides present in the organism. They occur in many foods, whiskey, cosmetics, herbicides, rubber factories, tanneries, iron foundries, and a number of other industrial settings. Some are strongly carcinogenic to experimental animals. See nitrate. Nitrosamines are the most studied of the *N*-nitroso compounds.

nitroso-. a prefix indicating the presence of the nitroso radical.

***N*-nitroso compound** (NOC). any of a class of compounds that contain the nitroso radical. They are carcinogens that can cause tumors in many different organs in experimental animals of various species.

nitroso radical. a reduction product, -N:O, of nitrite (NO_2).

***N*-nitrosodimethylamine** (dimethylnitrosamine; DMN; DMNA). an oily, flammable, yellow, water- and ethanol-soluble liquid, $(CH_3)_2NNO$. It is an extremely toxic, carcinogenic nitrosamine and is toxic by all routes of exposure, including skin contact. Reaction time in humans is usually a few hours. Early symptoms include nausea, vomiting, headache, abdominal cramps, diarrhea, muscular weakness, and fever. The major lesion is liver damage which may cause death. Tests on laboratory animals indicate linkage to liver, lung, and kidney cancer. When added as a preservative to stale herring meal, sodium nitrite is converted into a compound which is very toxic to mink, causing hepatic degeneration, ascites, and extensive internal hemorrhage. DMN is a probable human carcinogen. It is used as an industrial solvent, a rubber accelerator, an intermediate in the production of rocket propellants, a gasoline additive, antioxidant, and pesticide.

nitrosohemoglobin. a form of hemoglobin in which the nitroso radical, -N:O, binds to the heme moiety of hemoglobin. It is toxicologically similar to carbon monoxide hemoglobin. *See* nitrosohemoglobin poisoning.

nitrosohemoglobin poisoning. a condition similar to that of carbon monoxide poisoning. It has caused extensive mortality among cattle especially in the central United States. Symptoms are also similar to nitrite poisoning, but the blood has a normal coloration and contains no methemoglobin.

nitrous oxide (dinitrogen monoxide; laughing gas; hyponitrous acid anhydride; factitious air). a colorless, sweet-tasting, ethanol-soluble, slightly water-soluble gas, N_2O. It is used as an inhalation anesthetic in dentistry and surgery. See also nitrogen oxide. *See also* acetylene-block assay, anesthetic gas, denitrification.

nodding thistle. *Carduus*.

NO_X. nitrogen oxide.

NOC. *N*-nitroso compound.

Nolina texana (sacahuista; sacahuiste; beargrass). a perennial herb (Family Liliaceae) of the foothills and rangelands of western Texas, Arizona, and Mexico. Grazing animals normally feed on the buds, blooms, and mature fruit, which contain no chlorophyll. If the animals also graze on other plants or plant parts containing chlorophyll, photosensitization will usually develop. Symptoms initially include apathy and anorexia, followed within a day or two by generalized and progressively severe jaundice. Affected animals remain, when possible, near water and in the shade. The urine may be red, due to hemoglobinuria, or yellow. Discharges from nostrils and orbit are common. Declining weight, strength and increasing debility leading to death within a week following onset of symptoms is usual. *See also* phylloerythrin.

nonesuch. *Medicago lupulina*.

nontropical sprue. gluten-induced enteropathy.

1,4,5,6,7,7-hexachloro-5-nor-bornene-2,3-dimethanol cyclic sulfite. endosulfan.

norbormide. a toxic, white, water-insoluble solid that is soluble in dilute acid used as a rodenticide.

normo- a prefix meaning normal, usual; conforming to the rule.

northern corn leaf blight. northern leaf blight (*See* leaf blight).

Northern white cedar. *thuja occidentalis*.

Norway pine. *Pinus resinosa*.

Norway rat. *Rattus norvegicus*.

nosogenic. pathogenic.

nosopoietic. pathogenic.

notalian. antiboreal.

Notholaena sinuata (Jimmy fern; cloak fern). a neurotoxic perennial, evergreen, upright fern that occurs on dry, rocky hills especially those with limestone soils in western Texas, Arizona, New Mexico, and Mexico. It is responsible for jimmies" of range livestock, especially sheep.

nourishing. providing or able to provide nourishment.

nourishment. **1a:** nutrition. **1b:** food, foodstuffs, edibles; any substance that nourishes and supports life. **2:** the act or process of nourishing or of being nourished.

nucleotide. an organic monomer that combines with other nucleotides in the formation of nucleic acids. The most important nucleotides are those derived from the bases adenine, guanine, cytosine, thymine, and uracil.

nutmeg. **1:** ***Myristica fragrans.*** **2:** the ground, dried kernels of the seeds of *M. fragrans*, used as a spice. Large amounts (5—15 g) are poisonous, causing CNS excitation, euphoria, hallucinations, delirium, flushing of the skin, tachycardia, and lack of salivation.

nutmeg family. Myristicaceae.

nutrient. **1a:** any substance that provides an organism with the elements that are essential to normal metabolism; those substances that promote growth and development and support life processes. **1b:** often defined operationally as any substance assimilated by a living organism that promotes growth (nitrogen and phosphorus in wastewater can be considered nutrients under this definition, although they are sometimes environmentally damaging). **2:** imparting, serving as, or providing nourishment.

nutrient deficiency. *See* malnutrition.

nutrient excess. *See* malnutrition.

nutriment. a nutritious substance; that which nourishes.

nutrition (nourishment). **1:** all processes involved in the acquisition and utilization of foods that support growth, development, and support metabolism, vital activities, and reproduction. **2:** the science that treats processes involved in nutrition as defined above. *See also* malnutrtion.

nutritional. pertaining to nutrition.

nutritional deficiency. any suboptimal amount, concentration, or proportion of a particular nutrient in a particular diet can be regarded as a deficiency. A diet may be nutritinally deficient if certain nutrients are not present in correct proportions. Operationally a diet is nutritionally deficient if the organism maintained on the diet fails to thrive, grow, and achieve reproductive competence.

nutritious (nutritive). nourishing, wholesome, sustaining; affording nutriment.

nutritive. nutritious.

nux vomica (quaker buttons; bachelor's buttons; poison nut; dog buttons; vomit nut). the dried ripe seed of *Strychnos nux-vomica*. It contains several toxic alkaloids, most importantly strychnine and brucine. It has been used in medicine as a bitter tonic and CNS stimulant and is used in veterinary medicine as a bitter tonic and to treat chronic indigestion, inappetence, and atony of the rumen.

Nyssa sylvatica (black tupelo; black gum; sour gum; pepperidge; tupelo). a deciduous tree (Family Nyssaceae) that is indigenous to North America. It is horticulturally important largely because of its beautiful red autumn foliage. The timber is used for a variety of purposes (e.g., as pulpwood, to make veneer and to make cabinets and furniture).

O

O. the symbol for oxygen.

oak. **1:** any tree of the genus *Quercus*. **2:** wood or lumber from an oak tree.

oak poisoning (quercus poisoning). a toxic syndrome to which many species of mammals are susceptible. It is usually acquired by ingestion of buds and immature leaves of most, perhaps all, species of oak in North America and Europe. Animals are usually poisoned from browsing on scrub oak or upon the foliage of felled trees. Gallic acid, present in the tannin oaks, is the active principle. Signs are chiefly those due to gastrointestinal distress and a distinctive renal dysfunction. Clinical signs include anorexia, nasal discharge, rumen stasis in ruminants, polydipsia, polyuria, constipation, and abdominal tenderness and pain. Poisoning has been prevented experimentally by adding calcium hydroxide to tannic acid. If calcium hydroxide is added to the ration to make up 15% of the bulk, the syndrome may be prevented in livestock, and may relieve symptoms if given early in the course of the disease. The syndrome is similar to acorn poisoning, *q.v.*

oak-leaved hydrangea. *Hydrangea quercifolia*.

oats. *Aven fatua*; *A. sativa*.

obese. overweight; excessively fat.

obesity. the condition of being obese. *Cf.* steatosis.

obligate. **1:** to compel, constrain, require. **2:** confined to performing a particular action or function. **3:** used in reference to a living organism that can survive only under a restricted set of environmental conditions. *See also* metallophyte, accumulator, selenophile. *Cf.* facultative.

obligate accumulator. *See* accumulator.

ocean farming. mariculture.

ochratoxin. any of a number of hepatotoxic mycotoxins (e.g., from *Aspergillus ochraceus*, *A. sulphureus*, *A. melleus*, and *Penicillium viridicatum*). They sometimes contaminate cottonseed, corn, other stored grains, and peanuts. **ochratoxin A**. the main ochratoxin; a chlorine-containing mycotoxin, $C_{20}H_{18}ClNO_6$, produced by a number of fungi including *Aspergillus ochraceus* and *Penicillium viridicatum* that contain an α,β-unsaturated lactone moiety. It is hepatotoxic and nephrotoxic and known to be extremely toxic to laboratory mice, rats, monkeys, dogs, domestic fowl, ducklings, and trout.

octachlorocamphene. toxaphene.

1,2,4,5,6,7,8,8-octachloro-3a,4,7,7a-tetrahydro-4,7-methanoindan. chlordane.

1,2,4,5,6,7,8,8-octachloro-2,3,3a,4,7,7a-hexahydro-4,7-methano-1H-indene. chlordane.

1,2,4,5,6,7,8,8-octachloro-4,7-methane-3a,4,7,7a-tetrahydroindane. chlordane.

***n*-octyl alcohol, primary**. 1-*n*-octanol.

oculopathy. ophthalmopathy.

oestrus. estrus.

oestrous cycle. estrus cycle.

Ohio buckeye. common name of *Aesulus glabra* or the nuts that it produces.

-oid. suffix meaning resembling or similar.

Oignon sauvage. *Arisaema atrórubens*.

oike. habitat.

oikos. habitat.

oil garlic. allyl sulfide.

oil of myristica. oil of nutmeg.

oil of nutmeg, volatile (oil of myristica). a steam-distilled oil from the kernels of ripe seeds of *Myristica fragrans* (nutmeg). Ingestion of large amounts are narcotic, causing delirium and death.

oil of vitriol. sulfuric acid.

olefin. alkene.

oleovitamin D_2. vitamin D_2.

olig-, oligo-. a prefix meaning few, very little, or deficient.

omnivore. an animal that eats a broad range of food materials of both plant and animal origin.

omnivorous. pertaining to an animal that is able to subsist on or has the ability to consume a wide variety of foods, including both plants and animals or their products. Whereas *Homo sapiens* is omnivorous, many individuals (strict vegetarians for personal, religious, or social reasons) and populations (e.g., those in the far north that subsist on a few types of animal prey) are not omnivorous.

Omphalotus illudens. See *Omphalotus olearius*.

Omphalotus olearius (Jack o' Lantern, orange Jack o' lantern; false chanterelle). an orange to yellowish-orange, mild to moderately poisonous mushroom (Family Tricholomataceae) with sharp-edged gills that usually occurs in clusters at the base of stumps or buried roots of deciduous trees in eastern North America and California. Some experts consider this part of a species complex, in which case the eastern North American form, should be designated *O. illudens* and its western counterpart, *O. olivascens*. All parts are poisonous. If ingested, it typically causes nausea, vomiting and/or diarrhea, gastric distress, sensory disturbances, and muscle relaxation that persist for a few hours to 2 days. See mycetismus gastrointestinalis.

on the hoof. denoting a living meat animal.

onager. *Equus hemionus*.

onion. *Allium cepa*.

Onoclea sensibilis (sensitive fern). a neurotoxic perennial fern found in open woods, meadows, and old fields throughout the eastern United States, west to Texas, and north into Canada. This plant produces neurologic effects in horses that graze on it.

oogonia. plural of oogonium.

oogonium (plural, oogonia). the unicellular female reproductive organ of certain algae and fungi. *See* oosphere.

Oonopsis. (goldenweeds). a genus of plants (Family Compositae), all of which are obligate accumulators of selenium from soil.

oosphere. a large, nutrient rich, nonmotile female gamete or ovum produced in an oogonium, *q.v.*

open. a vernacular term denoting a nonpregnant farm mammal.

ophthalmopathy (oculopathy). any disease

of the eyes.

ophi-, ophio-. a prefix that **1:** refers to or denotes snakes. **2a:** indicates a thing that suggests a snake or is snake-like. **2b:** denotes the quality of being or resembling a snake; used in relation to a specified structure or quality.

ophic. ophidic.

ophidian. **1:** snakelike. **2:** a snake.

ophidiasis. ophidism.

ophidic. of, pertaining to, caused by, or obtained from snakes.

ophidism (ophidiasis; ophiotoxemia; ophitotoxemia). **1:** any condition that results from envenomation by a snake. **2:** snake venom poisoning.

ophiotoxemia. ophidism.

ophitotoxemia. ophidism.

opium. **1:** gum opium; crude opium. A highly toxic, addictive substance, it is a mixture that contains morphine, codeine, thebaine, papaverine, and narcotine. The crude resin is a gummy substance obtained from the unripe seed capsules of the opium poppy, *Papaver somniferum*; it also occurs in the fruit and sap of this plant. **2:** *Papaver somniferum*.

opium poppy. *Papaver somniferum*.

optimal. **1a:** of or pertaining to the optimum; most favorable; ideal. **1b:** pertaining to the intensity (temperature, concentration) or amount (heat, dosage) of an environmental factor that most favors growth and reproduction of a living system. Some essential substances have an optimum range of concentration or dosage above which toxic effects occur and below which deficiencies may occur. **1c:** the most efficient; the most cost-effective (e.g., use of a mechanism that optimizes the efficiency or effectiveness of energy acquisition; use of a mechanism that minimizes energy expenditure of a life process) **2:** of or pertaining to any property or condition of a living system (e.g., a behavioral trait such as optimal foraging; optimal reproductive strategy) or its environment that favors production of the largest possible number of viable offspring.

optimum. **1a:** that condition of an environment (e.g., surroundings, habitat, locale, region) that is conducive to an organism's or population's best or ideal (*See* optimal) functioning, including the realization of optimal reproductive function and the closest approach to biotic potential. **1b:** that degree (intensity) or amount of any environmental factor or set of factors that supports or allows full development and reproduction by a living system.

orange argiope. *Argiope aurantia*.

orange Jack o' lantern. *Omphalotus olearius*.

orange sneezeweed. *Helenium hoopsii*.

orchard. land used for growing fruit trees. Most orchards today are large and operated by commercial growers. Orchards are no longer a frequent feature of farms.

order. **1a:** a group of related families of plants or animals. **1b:** a taxonomic category subordinate to a class that contains one or more families that share certain characteristics and are deemed to be closely related. Names of orders typically end in *-ales* in plants and *-a* in animals.

Oregon fir. *Pseudotsuga menziesii*.

Oregon pine. *Pseudotsuga menziesii*.

Oregon water hemlock. *Cicuta vagans*.

organ (organum). a structure within the

body of most multicellular organisms that is comprised of one or more tissues that interact to perform a single, usually specialized, coherent, function or set of functions. **critical organ**. the organ that first reaches the critical concentration of a toxicant under specified conditions of exposure for a given species or population. *See also* critical concentration, effect (critical effect).

organic. **1:** integral to; central to. **2:** having an organized structure. **3a:** living; alive; biotic; biotic in origin. **3b:** arising from an organism. **4:** pertaining to substances derived from living organisms. **4:** of or pertaining to an organ or organs. **5:** of or pertaining to chemical substances that contain carbon. **6:** of, pertaining to, or cultivated by the use of fertilizers from animal or vegetable sources as opposed to the use of synthetic chemicals.

organic decay. *See* decay.

organism. any individual living being, including humans.

organochlorine. **1:** of or pertaining to chlorinated hydrocarbons. **2:** an organochlorine compound.

organochlorine compound (organochlorine). chlorinated hydrocarbon. *See* hydrocarbon.

organochlorine insecticide (chlorinated organic insecticide). any chlorinated hydrocarbon that is prominent in the formulation of an insecticide. Symptoms of intoxication in humans may include immediate or delayed vomiting; paresthesia and itching of lips, tongue, and face, headache, sore throat, fatigue, tremors, ataxia, weakness, confusion, convulsions, and coma. Death from respiratory failure may result. There are usually no pathological findings.

organomercurial pesticide. any of a number of compounds of three types: alkylated mercury compounds, aromatic mercury compounds, and alkoxy-alkylmercury compounds.

organophosphorus. organophosphorus compound.

organophosphorus compound (organophosphorus). an organic compound that contains bound phosphorus. Many of these compounds are powerful acetylcholinesterase inhibitors used as insecticides. Those developed for use as a nerve gas, *q.v.*, are among the most toxic synthetic compounds ever produced; they are lethal to humans in minute amounts. There are many types of organophosphorus compounds, most notably phospholipids, which are widely distributed in nature, esters of phosphinic and phosphonic acids which are economic poisons, pyrophosphates which are extremely toxic cholinesterase inhibitors used as insecticides, and phosphoric esters of glycerol, glycol, etc., which serve as components of fertilizers. *See also* organophosphorus ester, pesticide, organophosphorus poisoning.

organophosphorus-induced delayed neurotoxicity (OPIDN; organophosphorus ester-induced delayed neurotoxicity; organophosphorus-induced delayed neuropathy). a neurologic condition in humans, certain other mammals, especially cattle, (but not laboratory rats and mice), domestic fowl, and turkeys (but not Japanese Quail) due to exposure to certain organophosphate esters. The condition is unknown in poikilothermic vertebrates and invertebrates. The disease is progressive, with signs of ataxia appearing after 8-14 days; the victim becomes weak and unsteady, with eventual flaccid paralysis of the extremities. The characteristic lesions are those of a Wallerian-like axonal degeneration which appears at about the time of onset of symptoms. Certain sensory and motor pathways of the spinal cord are affected as are large fibers of the peripheral nerves (often only the sciatic nerve). *See also* neurotoxic esterase, tri-*o*-cresyl phosphate, test (organophosphorus-induced delayed neurotoxicity test).

organophosphorus poisoning. acute poisoning by an organophosphorus compound. Precise symptoms of organophosphorus poisoning vary considerably among vertebrate and invertebrate taxa. *See* insecticide (organophosphate insecticide), organophosphorus compound.

organoselenium compound (organoselenium). an organic compound that contains one or more selenium atoms. Such are produced synthetically (e.g., 1,4-diselenane) or by the action of various organisms (e.g., dimethylselenide, dimethyldiselenide and dimethylselenone). All are poisonous and are of considerable environmental importance. Some organisms convert inorganic selenium to dimethylselenide. This type of biomethylation is efficiently carried out by several genera of fungi and is accompanied by the very strong "ultragarlic" odor of the product. The biotransformation of organic and inorganic selenium to dimethylselenide and dimethyldiselenide occurs in animals (e.g., laboratory rats) and the volatile compounds are exhaled. Dimethylselenone is produced by bacteria. *See also* selenium.

organotellurium compound. an organic compound of tellurium. Even though tellurium is a rather rare element, its compounds are of some environmental significance. They can be produced synthetically and by microorganisms. Dimethyltelluride, for example, can be produced by fungi from inorganic tellurium compounds. Tellurium compounds are toxic, but less so than their selenium analogs.

organotin compound, **organotin**. any organic compound that contains tin. Many have the general formula, RnSnX4-n, where n = number, R = a hydrocarbon group, and X = the organic or inorganic moiety linked to a tin atom by a carbon atom. In general, the toxicity of a series of organotins is maximal where n = 3. In most cases, toxicity is conferred largely by the R group. Most organotins are readily absorbed through the skin. The principal mode of action, especially of R3SnX molecules involves bonding to proteins, presumably by the sulfur on the cysteine and histidine residues. Mitochondrial function is disrupted by any of several biochemical processes. Annual worldwide production of organotins may exceed 40,000 metric tons. Applications are numerous, including use as fungicides, acaricides, disinfectants, heat stabilizers, catalytic agents, and in anti-fouling paints.

organotrophic. heterotrophic.

organotropism. the state or condition of having a special affinity for particular tissues or organs. Said of chemicals and pathogenic agents.

organum. organ.

oriental poppy. *Papaver orientale*.

origin. *See* center of origin.

Ornithodorus (pajaroello). a genus of soft ticks that parasitize a variety of mammals; many are disease vectors. They are indigenous to Mexico and the southwestern United States. The bite of some species causes a local vesiculation, pustulation, rupture, ulceration, and eschar, with varying degrees of local swelling and pain.

Ornithogalum (star-of-Bethlehem; nap-at-noon). a genus of bulbous plants (Family Liliaceae) that is common to roadsides, meadows, and grasslands. It is an Old

World genus with two North American species, *O. umbellatum* and *O. nutans*. *O. umbellatum* is one of the most dangerous lilies in the United States. The bulbs are toxic and have caused the deaths of children. Livestock have been killed by bulbs of wild-growing plants brought to the surface by plowing, frost heaving, or rooting by swine. ***O. umbellatum*** (star of Bethlehem; nap-at-noon; snowdrop). an Old World onionlike perennial herb and garden plant (Family Lilaceae) introduced to North America from the Mediterranean region and naturalized, occurring in grasslands, thickets, and along roadsides from Newfoundland to Ontario and from New England to North Carolina, Kansas and Nebraska. The bulbs are toxic and have caused losses of sheep and cattle. Children have been poisoned by eating the bulbs of related species.

orpiment. arsenic trisulfide.

Orthene. acephate.

Ortho 12420. acephate.

orthoarsenic acid. arsenic acid.

Oryza. a genus of annual grasses and cereal crops (Family Gramineae). ***O. rufipogon*** (red rice). an annual grass, the grain of which is red. It is a serious pest in fields planted in white rice, *O. sativa*, as it cannot be readily separated from the white rice. As a result it can lower the yield and value of threshed rice. It is very difficult to control or eliminate. ***O. sativa*** (rice; white rice). an annual marsh grass (Family Gramineae), that has been cultivated for millennia for its starchy seeds (rice grains). Several varieties of two types are cultivated, lowland and upland rices; the former requires irrigation.

oryzalin (4-(dipropylamino)-3,5-dinitrobenzenesulfonamide; 3,5-dinitro-N^4,N^4-dipropylsulfanilamide). a preemergence herbicide that is particularly effective in orchards and vineyards. It is carcinogenic to some laboratory animals.

osmose. to pass through a semipermeable membrane by osmosis.

osmosis. the diffusion molecules of a pure solvent through a semipermeable membrane (a membrane that selectively prevents the passage of solute molecules but is permeable to the solvent) that separates two solutions of differing solute concentrations. The transfer is from a region of relatively low solute concentration to one of relatively high solute concentration. In living systems, the solvent is water and the migration tends to equalize solute concentrations on either side of a cell membrane. The net direction of osmosis is determined by the relative osmotic pressures of solutions within the cell and those without (See endosmosis, exosmosis).

osmotic. of or pertaining to osmosis.

Osteichthyes. the class of vertebrate animals that contains the bony fishes characterized by the presence of an endoskeleton chiefly of bone. They are the dominant fishes, and considered the most highly evolved, invading all types of waters. The single external gill opening is covered by an operculum, and in modern forms the spiracle is greatly reduced or absent. Most have an air bladder (swim bladder) and a covering (operculum) over the gills.

osteomalacia (osteomalacosis). a disease marked by increasing softness of bones due to loss of calcium salts. The condition is accompanied by rheumatic pains in the limbs, vertebral column, and chest, anemia, and progressive muscular weakness. *Cf.* osteopetrosis. *See also* fluorosis, renal failure.

osteomalacosis. osteomalacia.

osteopetrosis. a disease characterized by

excessive calcification of bone and spontaneous fractures. It is considered inherited in humans, but in the case of certain herbivores, the disease can result from prolonged calcium intake or poisoning by vitamin D_3 analogs that are present in certain plants such as *Trisetum flavescens* and *Cestrum diurnum*. *See also* enzootic calcinosis.

@@@

osteosclerosis congenita. achondroplasia.

OTBE. tributyl tin; tributyl tin compound.

Oulema melanopus (cereal leaf beetle). an Old World leaf beetle, introduced into North America in 1962. It is a serious pest of oats, cereal grasses, and certain other small grains.

outcross. **1a:** a mating between individuals of the same species, variety or breed who are not otherwise closely related. **1b:** an individual produced by such a mating.

outdoor air. *See* ambient air.

outface. *See* slope (dip slope).

ova. plural of ovum.

ovalbumin. albumen.

ovariectomy. surgical removal of an ovary.

ovary (plural, ovaries; ovarium). **1:** any female organ in plants or animals that produces an ovum. **2:** in zoology, one of the female reproductive glands. It contains the ripening ova prior to fertilization. In vertebrates, there is a pair of ovaries that produce steroid hormones as well as ova. **3:** in botany, the enlarged, ovule- or seed-bearing lower part of the pistil in angiosperms that ripens into a fruit.

ovary wall. pericarp.

overdominance. the condition in which a heterozygote is expressed in a more extreme phenotype than either of the corresponding homozygotes.

overdrafting. the removal of more groundwater from an aquifer (e.g., a spring or well) than is naturally replenished. This can cause the water table to fall and may result in increased costs of pumping, subsidence of land which reduces the future capacity of the aquifer, intrusion of saltwater, reduced stream flows in interconnected aquifers and surface-water systems, and ultimately exhaustion of groundwater reserves. Overdrafting occurs in the Great Plains and other parts of the western United States Heavy overdrafting, especially of large aquifers, is sometimes termed groudwater mining.

ovine. of, from, or pertaining to sheep (genus *Ovis*).

ovine anaplasmosis. *See* anaplasmosis.

Ovis. *See* sheep.

ovum (plural, ova; egg). the mature female gamete or germ cell.

ox (plural, oxen). in the strict sense, an ox is a mature castrated male bovine used as a draft animal.

oxamyl. a white crystalline compound used as a pesticide with tobacco, fruits, other crops, and ornamental plants.

7-oxabicyclo-(2,2,1)-heptane-2,3-dicarboxylic acid. endothall.

oxalemia. the presence of an excess of oxalates in circulating blood.

Oxalidaceae. the wood-sorrel family. *See Oxalis*.

Oxalis (sorrel). a genus of small, rather delicate herbs (Family Oxalidaceae).

Some species sometimes contain lethal concentrations of soluble oxalates. Species of the genus *Rumex* are also commonly referred to as "sorrel," thus causing occasional confusion in reports of livestock losses in the United States. ***O. cernua***. *O. pes-caprae*. ***O. corniculata*** (creeping lady's sorrel). a hardy species that can contain lethal amounts of soluble oxalates. ***O. pes-caprae*** (*O. cernua*; Bermuda buttercup; Bermuda oxalis; soursob; sorrel). a species that occurs as an introduced weed in California and Florida. It sometimes contains lethal concentrations of soluble oxalates. It has caused sheep fatalities in Australia. Symptoms and lesions are those of oxalate poisoning, *q.v.*

oxalosis. a generalized deposition of calcium oxalate in renal and extrarenal tissues, including the formation of urinary calculi. Oxalosis-inducing agents include oxalic acid, methoxyflurane, ethylene glycol, antirust agents, all of which are nephrotoxic. Some garden plants and many plants of arid and semi-arid regions contain oxalates. *See* calcium oxalate. *See also* *Arisaema*, *Calaldium*, *Portulaca*, *Rhamnus*, *Rheum*, *Rumex*, *Salsola*, *Sarcobatus*, *Symphytum*,

oxamyl ((2-(dimethylamino)-*N*-[[(methylamino)-carbonyl]oxy]-2-oxoethanimidothioic acid methyl ester; *N',N'*-dimethyl-*N*-[(methylcarbamoyl)oxy]-1-thiooxamimidic acid methyl ester; *N,N*-dimethyl-α-methylcarbamoyloxy-imino-α-(methylthio)acetamide; methyl 1-(dimethylcarbamoyl)-*N*-(methylcarbamoyloxy)thioformimidate; thioxamyl). a white crystalline compound used as a pesticide with tobacco, fruits, other crops, and ornamental plants.

oxen. plural of ox.

Oxford sheep. a rather large-bodied, hornless breed developed by crossing Hampshire and Cotswold sheep. It was selected for size and productivity.

oxidation. **1:** formerly, the chemical combination of oxygen with another substance. **2a:** any reaction in which electrons are transferred from one substance (oxidation) to another. The substance that gains electrons is termed the oxidizing agent, whereas the substance that loses electrons is termed the reducing agent. *See also* reaction (detoxification reaction, oxidation reaction, redox reaction), oxidase. **2b:** partial loss of electrons, as when electrons are displaced within a molecule, is also considered oxidation. **3:** dehydrogenation, the removal of one or more hydrogen atoms from a molecule.

oxidative phosphorylation. a critical metabolic process in aerobes whereby ATP is formed from ADP and orthophosphate. The process driven by the flow of electrons along the electron transport chain, with O^2 as the final acceptor. *See*, for example, respiration (aerobic cellular) respiration, AMP, ADP, ATP, arsenic, bromethalin, herbicide (thicarbamate herbicide).

oxybiotic. aerobic.

oxygen (symbol, O)[$Z = 8$; $A_r = 15.9994$]. a colorless, odorless, tasteless, neutral, nonmetallic, diatomic gas that supports combustion, and the most abundant and widely distributed chemical element on Earth. It combines with most other elements to form oxides and is essential to animal and plant life. Pharmaceutical-grade oxygen contains not less than 99.0%, by volume, of O_2. It is used in carbon monoxide poisoning, to relieve hypoxia, in cryosurgery, and at hyperbaric pressures to conduct cardiac and other surgery. Therapeutic overdosage can cause convulsions. **active oxygen**. highly reactive species of oxygen such as superoxide, hydroxyperoxyl radical, hydrogen peroxide, and the hydroxyl radical. They destroy microsomal cyto chromes by initiating autoxidation. Active oxygen species are formed in the animal body by successive additions of an electron (e-) and a hydrogen ion (H+) to a molecule of molecular oxygen, yielding $H0_2$-.

oxyntic gland. acid gland (def. 1)

oxyosis. acidosis.

oxyphilous. acidophilic.

oxyphyte. a plant that grows under acidic conditions.

oxytocic. of or pertaining to oxytocin or to a substance (e.g., sparteine) with endocrinologic properties similar to those of oxytocin.

oxytocin. a hormone secreted by the posterior lobe of the pituitary gland. It controls uterine contractions and milk letdown.

Oxytropis (locoweeds; point locoweeds; pointvetches). a genus of perennial herbs and small shrubby plants (Family Fabaceae, formerly Leguminosae) that occur chiefly in mountainous areas of Europe, Asia, and North America. Examples are *O. lambertii* (white loco, white point loco, stemless loco), *O. saximontana*, and *O. sericea* (white loco, white point loco). There is a intergradation of characteristics, which is probably due to hybridization, among the locoweed species of this genus. No doubt many of the specific names have been used falsely in the toxicological literature. Toxic species contain indolizidine-1,2,8-triol. *See also Astragalus*.

P

p. abbreviation of pico-.

PABA. *p*-aminobenzoic acid.

Pachysandra procumbens (box). a hedge plant (Family Buxaceae) that sometimes escapes cultivation. It contains alkaloids and certain other biologically active compounds. Numerous animals have been killed by eating clippings from box hedges. The leaves are very toxic; a small amount can be lethal. Symptoms are those of severe gastrointestinal irritation, sometimes producing bloody stools.

Pacific poison oak. *Rhus diversiloba*.

pajaroello. any soft tick of the genus *Ornithodorus*.

palea. the upper of a pair of bracts beneath each flower of a grass. It is usually thinner and more delicate than the lower pair (lemma). *Cf.* glume, lemma.

palm family. Palmaceae (Arecaceae; Palmae).

palm vipers. snakes of the genus *Bothrops*.

Palmaceae (Arecaceae; Palmae). some 2000 species of mainly tropical, woody trees and some vines and shrubs (Family Palmaceae). They usually have a tall columnar trunk that lacks cambium and a crown of large feathery leaves. *See Areca catechu*.

Palmae. Palmaceae.

palmotoxin B_0. a mycotoxin from *Aspergillus flavus*. It is extremely toxic (*i.p.*) to rats.

palsy. **1a:** a condition of uncontrollable tremor of the body or of one or more body parts. **1b:** paralysis. **2:** a weak or enfeebled condition. **3:** to become palsied or paralyzed. *See* paralysis.

pancytopenia. a condition in which the concentration of erythrocytes, blood platelets, and all types of white cells, in circulating blood are substantially reduced.

panic grasses. *Panicum*.

panicle. a racemose inflorescence (*See* inflorescence) in which the flowers form on stalks (peduncles) that arise alternately or spirally from the main axis.

paniculate hydrangea. *Hydrangea paniculata*.

Panicum. (panic grasses) a widely distributed genus of grasses (Family Gramineae) that occupy a diversity of habitats. They can produce hepatogenic photosensitivity in grazing animals. *See also* phylloerythrin. ***P. capillare*** (witchgrass). in addition to photosensitization, concentrations of nitrate toxic to livestock have been reported. ***P. maximum*** (Guineagrass). a coarse, perennial, drought-resistant, bunchgrass. It is indigenous to Africa, but was introduced into the New World via the West Indies. It is cultivated locally in Florida and Texas, but has become an important forage grass in parts of tropical South America.

Papaver (poppies). a genus of about 50 plant species (Family Papaveraceae) that are nearly worldwide in distribution, although few species occur naturally in North America. Most contain a variety of toxic alkaloids, sometimes in high concentra tions. ***P. nudicaule*** (Iceland poppy). a species, sometimes cultivated, that has been implicated in the poisoning of livestock. ***P. orientale*** (oriental poppy). a common garden poppy that contains some alkaloids but is generally considered nontoxic. ***P. rhoeas*** (corn poppy; red poppy). occasionally cultivated, it has been implicated in the poisoning of livestock. ***P. somniferum*** (opium poppy; gum opium; poppy seed; poppy opium). an herb introduced to North America from Europe, it has escaped from cultivation, often occurring near dwellings, along roadsides, and in abandoned fields from Newfoundland to North Dakota, Puget Sound, and southward. It contains a large number of isoquinoline alkaloids including codeine, morphine, and papaverine. The unripe seed capsule is very toxic, and seeds used as animal feed following removal of the oil have proven lethal. *See also* nontoxic, opium.

Papaveraceae (poppy family). a family of herbs, or sometimes shrubs, many of which are common garden plants and nearly all of which produce toxic alkaloids or other poisonous principles. *See*, for example, *Sanguinaria canadensis*, *Chelidonium majus*, *Papaver spp.*, *Argemone mexicana*, *Dicentra*, and *Corydalis*.

papaw. *Asimini triloba*.

paperflower. any herb of the genus *Psilostrophe*.

Papilio palyxenes asterius (black swallowtail;). a large, beautiful, deep indigo-black butterfly (Family Papilionideae) that is indigenous to the United States east of the Rocky Mountains. feed on carrots, celery, caraway, dill, parsley, parsnips, and similar plants of economic importance. They are often called celery worms, parsley worms, and similar vernacular names.

para-. **1:** prefix meaning positioned beside, situated near, surrounding, accessory to, beyond, apart from, etc. **2:** in chemistry, *para-* or its abbreviation, *p-*, indicates the substitution of 2 atoms, each bonded to a carbon atom or situated on opposite sides of a benzene ring.

parabendazole ((5-butyl-1*H*-benzimidazol-2-yl)carbamic acid methyl ester; methyl-5-butyl-2-benzimidazole carbamate; 5-butyl-2-(carbomethoxyamino)benzimidizole). a veterinary anthelmintic with nematocidal action,

$$C_{13}H_{17}N_3O_2$$

It is moderately toxic to laboratory rats and mice.

paracasein. a synonym for casein (def. 2).

paraesthesia. paresthesia.

paraffin. **1:** alkane. **2:** a white translucent mixture of hydrocarbons that is obtained from petroleum; it is solid at room temperature. Mild exposure causes chronic dermatitis and boils. Intense exposure is associated with cancer of the scrotum.

paralyses. plural of paralysis.

paralysis (plural, paralyses). **1:** palsy (in part); a temporary or permanent impairment, inhibition, or loss of voluntary motor function. **2:** loss, impairment, or reduced ability to perform any bodily function (e.g., secretion, sensation). **ascending paralysis**. a paralysis that progresses from the lower to the upper (or posterior to anterior) portions of the body in humans and other vertebrate animals or from the periphery toward the center of the body. **anesthesia paralysis**. that following anesthesia. **arsenical paralysis**. that due to arsenical poisoning. **Chastek paralysis**. paralysis of foxes or mink due to thiamine deficiency caused by feeding on raw fish that contain an enzyme destructive to thiamine. Effects include loss of appetite, emaciation, and eventual paralysis and death. **ginger paralysis**. Jamaica ginger paralysis. **global paralysis**. paralysis that involves the entire body of the animal. It is usually rapidly fatal. **immunological paralysis**. lack of specific antibody production following exposure to large doses of the antigen. Elimination of antigen relieves the condition. **jake paralysis**. Jamaica ginger paralysis. **Jamaica ginger paralysis** (jake paralysis; jake neuritis; ginger paralysis; Jamaica ginger polyneuritis). paralysis of the extremities, especially of the legs, due to imbibing Jamaica ginger as a beverage. **lead paralysis**. that caused by lead poisoning. It is due to a peripheral neuritis and marked by wristdrop. **mixed paralysis**. combined motor and sensory paralysis. **motor paralysis**. loss or impairment of muscular contraction. **narcosis paralysis**. that which occurs during anesthesia. **sensory paralysis**. that due to loss or impairment of sensory function. **tick-bite paralysis**. a progressive, ascending, flaccid motor paralysis observed in humans (mostly children) and domestic animals. It is due to the bites of certain ticks, notably those of the genera *Dermacentor* and *Ixodes*. The condition is presumably due to a venomous saliva. The victim usually recovers following removal of the tick. **toxic paralysis**. that due to exposure to a toxic substance. **traumatic paralysis**. that due to trauma.

paralytic. **1:** pertaining to paralysis, the capacity to cause paralysis, or to suffering from paralysis. **2:** a chemical that paralyzes.

paralytic haemoglobinaemia. azoturia.

paralyze. to bring about a state of paralysis; to render an organism or part of an organism unable to move.

paralyzed tongue. a condition of domestic cattle that results from long-term consumption of *Descurainia pinnata*, *q.v.* Partial or complete blindness is the first indication of poisoning. Subsequently, the affected animal loses the use of its tongue and cannot swallow. The animal becomes progressively thinner and weaker but the symptoms gradually disappear with improved feed and the daily administration of 2-3 gallons of water.

paralyzer. **1:** that which causes paralysis. **2:** any substance that inhibits a chemical reaction.

paraoxon (*0,O*-diethyl-*O*-(p-nitrophenyl) phosphate; phosphoric acid diethyl 4-nitrophenyl ester; diethyl *p*-nitrophenyl phosphate; phosphacol; E-600). an extremely toxic activation product of parathion. It is formed by the desulfuration by microsomal monooxygenases of parathion. Paraoxon is a potent anticholinesterase and other serine esterases and, as such, finds wide usage as a pharmacologic tool.

paraquat (1,1′-dimethyl-4,4′-bipyridinium; *N,N′*-dimethyl-γ,γ′-dipyridylium; methyl viologen (2+)). a yellow, solid, water-soluble, bipyridylium quaternary ammonium salt, used as a rapid-acting, nonspecific contact herbicide. It is better known than other members of this group (e.g., diquat, chlormequat, and difenzoquat). It disrupts plant membranes, causing wilting and desiccation. It is highly toxic and has caused large numbers of human deaths. Intoxication of humans and other animals may occur by all routes of absorption (suicidal hypodermic injections are known). Systemic effects in such cases are independent of the route of exposure and the formulation encountered. Paraquat is, nevertheless generally considered safe to use if suitable precautions are taken. Contamination of food and drinking water are serious concerns that must be monitored and guarded against. Poisoning by paraquat is complex and devastating to humans and other animals. Chronic effects due to long-continued exposure to low concentrations are not well known; paraquat is carcinogenic to some laboratory animals. Acute exposure of test animals to aerosols causes pulmonary fibrosis, and the lungs are affected even when exposure is through nonpulmonary routes. Acute exposure may cause focal erosion of tissue and extensive systemic effects including pulmonary fibrosis, variations in the levels of catecholamine, glucose, and insulin, and severe damage to a number of organs. Presenting symptoms include burning sensations in the mouth and throat, vomiting, dyspnea, and sometimes diarrhea, followed within a few days by dyspnea, cyanosis, and signs of impairment of renal, heart, and liver function. Progressive interstitial pneumonia may develop. Long-term nonlethal exposure can cause skin irritation, nosebleeds, and severe eye injury. Death is usually due to pulmonary insufficiency. A constant lesion in fatal cases is pulmonary fibrosis, often accompanied by pulmonary edema and hemorrhaging.

parasite. an organism that lives on or within, and at the expense of, another organism (host).

parasiticidal. lethal to parasites.

parasiticide. 1: lethal to parasites. **2:** an agent that kills parasites.

parasympathetic. **1:** of, pertaining to, functioning like, or identifying a division of the autonomic nervous system of vertebrate animals. **2:** cholinergic.

parasympathetic nervous system (cholinergic nervous system). a division of the autonomic nervous system of vertebrates. This system generally promotes activities associated with normal bodily functions. It thus controls contraction of the smooth muscles of the blood vessels and gastrointestinal tract as well as glandular secretions (e.g., from the salivary glands, bladder, genital organs, and rectum). It acts, for example, to constrict pupils, lower blood pressure, slow heart rate, constrict bronchi, contract the urinary bladder, and stimulate digestion. *Cf.* sympathetic nervous system.

parathion (phosporothioic acid *O,O*-diethyl *O*-(4-nitrophenyl; PA) ester; *O,O*-diethyl *O-p*-nitrophenylphosphorothioate;diethyl-*p*-nitrophenyl monothiophosphate; DNPT). a highly toxic, highly dangerous, phosphorothioate organophosphate insecticide and acaracide,

$$(C_2H_5O)_2P(S)OC_6H_4NO_2.$$

Parathion is a dark brown to yellow liquid, that is soluble in alcohols, aromatic hydrocarbons, esters, ketones, ethers, and oils and nearly insoluble in water, kerosene, and petroleum ether. Parathion is highly toxic to a wide range of insects but should not be used in the home or any shelter that houses animals because it is toxic also to mammals. Parathion is a highly dangerous, supertoxic compound (as little as 120 mg is known to have killed an adult human; and 2 mg has killed a child). Most accidental poisonings of humans are due to cutaneous absorption; some are due to inhalation. Symptoms of parathion poisoning in humans first appear several hours following exposure. Symptoms may include tearing, nausea, vomiting, muscular twitching, diarrhea, abdominal cramps, salivation, headache, vertigo, runny nose, miosis, respiratory distress, generalized and profound muscular weakness, confusion, jerky movements, convulsions, and coma. Death may result from respiratory failure due to CNS paralysis.

parathyroid gland. any of two pairs of small glands that control calcium and phosphorus metabolism in vertebrate animals and regulate concentrations in circulating blood and bones. They are located at the base of the neck within or near the thyroid gland. See also gland.

parenchyma. relatively unspecialized tissue. **ground parenchyma**. parenchyma frequently forms a ground tissue, given this name, in which other tissues are embedded.

paresis. **1:** partial or complete muscle paralysis; sometimes reported as muscular weakness. **2:** dementia paralytica; a disease of the brain that is syphilitic in origin.

paresthesia (paraesthesia). **1a:** an abnormal, morbid, or perverted sensation, or the condition of having such sensations (e.g., numbness, prickliness, burning, tickling, or tingling), with or without additional symptoms. **1b:** numbness and tingling in an extremity which often occur with the return of the blood flow to a nerve following temporary blockage of or mild injury to a blood vessel. **2:** beyond (normal) sensation.

Paris green. cupric acetoarsenite.

Parkinsonia aculeata (horsebean). often used as a hedge plant, it is a small evergreen tree (Family Fabaceae, formerly Leguminosae) with clusters of fragrant yellow flowers and feathery pendulous branches. Toxic concentrations of nitrate have been reported in the foliage.

Parmelia molliuscula (ground lichen). a lichen that contains usnic acid. It is at most moderately toxic to cattle and sheep but nevertheless should be considered potentially lethal. Most poisonings occur during winter when forage is scarce. Most cases are mild; symptoms include ataxia especially of the hind legs. Total paralysis beginning in the hind legs may occur in severe cases and the animal is then unable to rise. Affected animals usually remain alert and appetite is normal.

parrot green. cupric acetoarsenite.

parsleyworm. a common name of the larva of *Papilio palyxenes asterius*.

parthenogenesis (unisexual reproduction). the production of offspring from an egg that has not been fertilized.

partial antigen. hapten.

partition. screen, separation; to devide or separte; the tendency of a substance to exhibit an affinity for one material over another. See partition coefficient.

partition coefficient (PC). a measure of the relative solubility of a chemical in either of two non-miscible liquids when placed in a container holding both liquids. The PC of a substance is the ratio of its concentration in one of the liquids to its concentration in the other. **octanol/water partition coefficient**. the ratio of the solubility of a test substance in a 1-*n*-octanol (the lipid phase) to that of an aqueous phase. It is a measure of the ease with which a substance is transferred from water to lipids. This value is very helpful in estimating the ability of a xenobiotic to cross cell membranes. In practice, it helps toxicologists to infer the bioavailability or uptake (especially via the gill tissue of fish) of a given toxicant since those with high octanol/water PCs are lipophilic and are usually much more readily absorbed than those with low coefficients.

parturient apoplexy. milk fever.

parturient fever. milk fever.

parturient paralysis. milk fever.

parturient paresis. milk fever.

parturition. **1:** passage and delivery of a newborn animal through the terminal portion of the female reproductive tract or birth canal. **2:** childbirth.

paspalism. paspalum staggers.

Paspalum. a large, widely distributed genus of mostly perennial grasses (Family Gramineae). One or more flattened racemes with hard, papery bracts surround the flowers. The seed heads are millet-like. Some species, including the following, are sometimes infested with *Claviceps paspali*, *q.v.* ***P. dilatatum*** (dallis grass). a South American grass that now grows wild in the United States and elsewhere; ***P. notatum*** (Argentine bahia; bahia grass). a rapidly growing, poorly seeding, forage grass that is also used to control erosion. ***P. scrobiculatum*** (kodo millet), a species indigenous to India. *See* paspalum staggers, ergotism (neurotropic ergotism).

paspalum staggers (paspalism; dallis grass poisoning). poisoning of cattle grazing on the millet-like seed heads of grasses of the genus *Paspalum*, *q.v.*, when infested by the fungus *Claviceps paspali*. Horses and sheep are sometimes poisoned. The condition is usually marked by tremors, incoordination, a spastic pelvic limb gait, and falling. It is due to mycotoxins produced by the fungus. A large, single dose can produce signs that persist for several days. Animals display continuous trembling of the large muscle groups; the gait is jerky and limb movements are uncoordinated. Animals fall in awkward positions if they attempt to run. Prolonged exposure can result in complete paralysis. Recovery follows removal of the animals to feed not contaminated with sclerotia of *C. paspali*. Topping of the pasture to remove affected seed heads has been effective in controlling this condition.

pathobiology. a science that is closely allied to the science of pathology considered broadly inclusive of subdisciplines such as plant pathology (plant pathobiology), and animal pathology (animal pathobiology) with emphasis on biological processes and mechanisms as opposed to medical or legal aspects.

pathogen. **1:** any disease-producing agent; any virus, organism, or substance (e.g., a toxin) that can cause disease. **2:** in common usage, this term is often restricted to a disease-causing living organism, usually a microorganism.

pathologic, pathological. **1a:** morbid or diseased. **1b:** indicative of or caused by a morbid condition. **2:** pertaining to pathology.

pathogenesis (pathogeny; pathogenesy). **1:** the course of development of a disease or morbid state especially with respect to cellular and tissue-level processes. **2:** the pathologic, physiologic, or biochemical mechanisms resulting in the development of a disease or morbid process.

pathogenesy. pathogenesis.

pathogenetic. pathogenic.

pathogenic (pathogenetic; morbific; morbigenous; nosogenic; nosopoietic). of, pertaining to, or causing pathogenesis; giving rise to disease, pathologic conditions, or morbid symptoms.

pathogenicity. the state or quality of causing or being able to cause disease.

pathogeny. pathogenesis.

pathologist. a specialist in the practice or study of pathology.

pathology. **1:** the largely descriptive study of the causes and nature of disease and dysfunctional states, especially the study of morphological and functional changes in tissues as a consequence of disease, whether caused by pathogenic organisms or by exogenous or endogenous chemical or physical agents. Tools used range from those of microscopy to gross and subgross anatomy, and from those of biochemistry to physiology. While largely descriptive in nature, both quantitative and experimental studies are also important, as is the study of toxic substances. **2:** the structural and functional characteristics of disease, especially at the tissue and organ level.

pawpaw. *Asimina triloba*.

Paxillus involutus (poison paxillus). a reputedly immunotoxic mushroom (Family Paxillaceae) which gradually sensitizes a subject following several exposures, ultimately causing kidney failure. They occur on or near wood in mixed decidu ous evergreen forests throughout much of North America.

Pb. the symbol for lead (plumbum).

PBB. polybrominated biphenyl.

PCB. polychlorinated biphenyl.

PCNB. pentachloronitrobenzene.

PCP. pentachlorophenol.

PCP poisoning. pentachlorophenol poisoning.

PCPA. chlorophenoxyacetic acid.

PDB. *p*-dichlorobenzene.

pea flower locust. *Robinia pseudoacacia*.

peach tree. *Prunus persica*.

pear tree. any tree of the genus *Pyrus*.

peat moss (sphagnum moss; bog moss). **1:** a moss from which peat may form, especially one of the genus *Sphagnum*. **2:** a dried moss (def. 1) used mainly as a mulch or seedbed to help acidify the soil.

peba. *Dasypus novemcinctus*.

Pecusanol. lindane.

ped. **1:** a naturally-formed aggregate of soil particles. *See also* soil aggregation.

pediculicide (lousicide). **1:** lethal to lice. **2:** an agent that kills body lice; usually applied to the skin.

penchlorol. pentachlorophenol.

penetrant. any agent (e.g., dimethyl sulfoxide (DMSO)) that increases the speed and ease with which a liquid penetrates or permeates a membrane, tissue, or other material. Penetrants may be used to facilitate absorption of toxicants or other pharmacologically active agents by cells or tissues. See, for example, . **2:** See nematocyst.

penetrate. to pierce, enter; invade; infiltrate, permeate; to pass deeply into (e.g., cells, tissues, a body cavity).

penetrating. **1a:** invasive, piercing, cutting, or entering deeply, as in a penetrating wound. **1b:** able to pass deeply, e.g., into tissue. **2:** sharp, e.g., a penetrating odor.

penetration. **1a:** piercing, entering, ingress. **1b:** invasion. **1c:** absorption, q.v., infiltration, permeation **2:** intelligence, insight, mental acumen. **3:** depth of focus of a lens.

penis. the male copulatory organ, especially of mammals, certain birds, and some invertebrate animals.

Penstemon (beard tongue). a large genus of North American herbs (Family Scrophulariaceae) that are facultative accumulators of selenium from soil.

penta. pentachlorophenol.

penta-. a prefix indicating five, five of, grouped in fives, etc.

pentachlorin. DDT.

pentachloronitrobenzene (PCNB). a color

less, musty-smelling, crystalline compound,

$C_6Cl_5NO_2$

PCNB is used as a pesticide, soil fungicide, and has been used as a seed dressing. It is an irritant and is known to cause cancer in certain strains of laboratory mice.

pentachlorophenol (penta; penchlorol; PCP). a toxic, fungicidal, bactericidal, light-brown, solid, chlorinated hydrocarbon, C_6Cl_5OH, prepared by catalytic chlorination of phenol. It is used as a wood preservative, protecting it from bacteria, fungi, and slime. It is also used for termite control and as a general herbicide and defoliant. Cattle have become ill from licking the wood or breathing the fumes from fences made from PCP-treated wood. Pentachlorophenol is also fetotoxic and teratogenic. *See* pentachlorophenol poisoning. Livestock fed in troughs made of PCP-treated lumber may salivate excessively and display irritation of the oral mucosa.

pentachlorophenol poisoning (PCP poisoning). poisoning marked by signs that may include nervousness, weakness, rapid pulse, rapid respiratory rate, muscle tremors, fever, convulsions, and death. Chronic poisoning may result in CNS depression, lightheadedness, sleepiness, tremors, nausea, loss of appetite and weight loss, fatty liver, nephrosis, and cancer. There is no known antidote. Livestock fed in troughs made of PCP-treated lumber may salivate excessively and display irritation of the oral mucosa.

3-pentadecylcatechol (3-pentadecyl-1,2-benzenediol; 3-pentadecylpyrocatechol; tetrahydrourushiol; hydrourushiol; dihydrohengol; 3-PDC). a toxic white powder, C_6Cl_5OH, that is soluble in acetone, benzene, ethanal, and ethyl ether. It is used as a chemical intermediate, a diagnostic aid, and as an algicide, bactericide, and herbicide. It is a constituent of urushiol.

(2β,3β,5β,22R)-2,3,14,20,22-pentahydroxycholest-7-en-6-one. ponasterone A.

2,3,14,22,25-pentahydroxycholest-7-en-6-one. α-ecdysone.

2,4-pentanedione. acetylacetone.

pentanochlor. a postemergent anilide herbicide that rapidly degrades in the soil. It is used mainly to weed tomatoes and umbelliferous crops such as carrots, celery, and parsley.

pepper spot. a disease of clover, characterized by numerous black specks on the leaves, caused by a fungus, Pseudoplea trifolii.

pepper. **1:** a plant of the genus *Piper*. **2a:** the fruit of plants of the genus *Piper*. **2b:** any of a number of products obtained from plants of the genus *Piper*. **3:** the fruit of certain other plants other than those of the genus *Piper* that bear spicy fruits with a sharp pepper-like flavor. **4:** the fruit of a variety of unrelated plants that do not have a spicy, pepper-like flavor (e.g., green pepper). **5a:** to sprinkle as one sprinkles pepper; to apply pepper. **5b:** to make spicy. **black pepper**. **1:** *Piper nigrum*. **2:** a widely used spicy condiment from *Piper nigrum*. It is usually not toxic, unless inhaled in mass. **Java pepper**. *Piper cubeba*. **tailed pepper**. *Piper cubeba*. **true pepper**. any plant of the genus *Piper*.

pepperidge. a common name of *Nyssa sylvatica*.

pepsin. a protein-digesting enzyme secreted by the gastric glands of vertebrates.

perchloropentacyclo(5.2.1.0^{2,6}.0^{3,9}.0^{5,8})-decane. mirex.

percolate. **1:** to strain or filter a liquid through porous material. **2:** to trickle slowly through a porous substance, as rainwater percolates through soil. **3:** a liquid that percolates.

percolation. **1a:** the slow movement of a liquid (usually water or a solvent) through a porous material such as soil). **1b:** the extraction of soluble materials from the material through which a solvent percolates. **2a:** the movement of water (e.g., rainwater) downward and radially through subsurface soil layers. The excess of water over that amount which is sufficient to saturate the soil will usually continue downward into the groundwater. Percolation may play an important role in the transport of environmental chemicals (e.g., fertilizer and pesticide residues) to new locations. In some cases (as in the case of improperly situated land fills or hazardous waste dumps), percolation can extend pollution to nearby streams and in extreme cases may cause irreversible pollution of groundwater aquifers. **2b:** the flow of groundwater due to gravity through pore spaces in rock or soil.

perennial pea. a poisonous pea. *See Lathyrus*.

perfusion. **1:** the pouring of a liquid over or through a material. **2:** artificial passage of a fluid through blood vessels of an organ or whole animal. *Cf.* infiltration, suffusion. **organ perfusion**. the maintenance, by perfusion, of an organ isolated from any vascular supply either *in situ* or isolated from the body. The organ is perfused with a suitable fluid driven mechanically through its own vascular system. Such preparations offer excellent opportunities to test the behavior and effects of a pharmaceutical or other xenobiotic and its interactions with the test organ.

perianth. a collective term for the sepals and petals of a flower.

pericarp (ovary wall; fruit wall; seed vessel). the wall of the ovary of green plants.

perinatal. of or pertaining to the time near birth. See perinatal period.

perinatal period. **1a:** near the time of birth. **1b:** the time when viability *ex utero* is considered possible. **1c:** the last third of the gestation period. In *Homo sapiens* the perinatal period is usually taken to begin at the 20th week of gestation. For some purposes the perinatal period in humans may include the first postnatal week.

periodic. recurring at regular intervals in time or space.

peripheral. **1:** superficial, marginal; pertaining to or situated at or near the margin or periphery. **2a:** outlying; located away from the center of a structure or away from a central structure. **2b:** nearer to the periphery than some designated point or structure.

peripheral nervous system (PNS). **1:** that portion of the nervous system of an animal that lies outside the central nervous system (that part which includes a "brain"). **2:** in vertebrate animals, the PNS includes the nerves and ganglia that lie outside of the brain and spinal cord. *See also* central nervous system.

periphloic. amphiphloic.

periphyton. a complex matrix of aquatic algae and heterotrophic microbes attached to submerged surfaces. It is a major source of food for numerous aquatic invertebrates and herbivorous fishes. Periphyton can bioconcentrate environmental chemicals such as mercury and cadmium by several orders of magnitude over concentrations that occur in the surrounding water.

peristalsis. a series of involuntary, usually rhythmic contractions that move the con tents along in a tubular organ with muscular walls (e.g., a digestive tract). In portions of a digestive tract it may also produce a churning action that mixes the contents and aids digestion.

peritoneal cavity. the potential space that lies between the lining (peritoneum) of the body wall and that of the viscera. This space contains a small amount of fluid.

peritrichous. descriptive of microorganisms that bear cilia or flagella over their entire surface.

perixylic. amphixylic.

permafrost. permanently frozen subsoil.

permeability. **1:** the quality of being permeable. **2a:** the relative ability of a membrane, cell, soil, or other potential barrier to permit passage of fluids or dissolved materials. **2b:** the rate or relative rate of passage of liquids or dissolved materials through such barriers.

permeable. pervious, penetrable; permitting the passage of fluids or dissolved materials (said of a membrane or other potential barrier). **selectively permeable**. a property of a living cell membrane whereby permeability is regulated.

permeate. **1:** to penetrate, infiltrate, saturate, or pass through, as through a filter or semipermeable membrane. **2:** any component of a solution or suspension that can permeate (def. 1). *Cf.* permeant.

permeation. The process of spreading through, infiltrating, saturating, or penetrating, e.g., a substance, cell, tissue or organ.

pernicious. **1:** noxious, destructive, deadly, fatal, tending to be fatal **2:** of or pertaining to a severe, often fatal disease.

pernicious vomiting. continuing uncontrollable vomiting.

Peronospora tabicini. *See* blue mold of tobacco.

1,4-peroxido-*p*-menthene-2. ascaridole.

Persea americana (avocado). a tree (Family Lauraceae), the leaves, fruit, bark, and seeds of which are toxic. Livestock, various mammals and birds, and probably fish have been poisoned under natural conditions. In cattle, horses, and goats milk flow is reduced and a noninfectious, sometimes fatal mastitis develops. The mastitis can be controlled but milk flow never returns to normal. The ripe fruit is toxic to canaries. Varieties of avocado range in toxicity from weakly to highly toxic.

Persian insect powder. pyrethrum flowers.

Persian tick. *Argas persinus*.

persic oil. *See* apricot-kernel oil.

persistence. **1a:** stability, endurance, tenacity; durability; resistance to degradation. **1b:** durability or stability of a characteristic structure or behavior in the face of opposition or adverse environmental conditions. **2a:** chemical stability of a substance, usually under a stated set of conditions. **2b:** of or pertaining to the length of time a chemical compound will remain unaltered under specified conditions or in designated situations or environments. **absolute environmental persistence**. a term that applies to those chemicals (e.g., heavy metals) that never undergo dissolution in the environment. Burdens of such chemicals are reduced at a given location only via transport mechanisms. **aquatic persistence**. chemical stability of a substance in a water body. **environmental persistence**. stability, or the tendency to stability, of a substance following introduction into the environment. The most hazardous toxicants in the environment are generally those that persist; they are not necessarily the most toxic.

persistent. **1a:** having the quality of persistence, *q.v.* **1b:** lasting, enduring, durable, long-lived, resistant to degradative forces or to decomposition, remaining.

pest. **1a:** any organism considered to be disruptive to human activities, especially one that interferes with human well-being, comfort, the value or well-being of pets or domestic animals and plants, or other human resources. **1b:** most commonly applied to any organism that has a negative economic impact, such as one that damages crops or reduces their yield or marketability, one that irritates or injures pets or livestock, one that damages or reduces the value or pleasure one takes in house or garden, or one that spreads disease. An organism that, furthermore, has no appreciable redeeming features in the context in which it is considered a pest. **emergent pest**. a special case of faunal imbalance whereby alternative species become pests by virtue of some environmental change such as an imbalance in, or even extinction of, the population of an another pest (e.g., DDT control of the codling moth has resulted in a burgeoning of mite populations to the point where they became serious pests).

pesticidal synergist. a compound (e.g., piperonyl butoxide) added to an insecticide or other pesticide to increase its effectiveness. *See also* synergist.

pesticide. any compound or formulation such as an herbicide, insecticide, miticide, or rodenticide, that is used to destroy plant and animal pests or to inhibit their activity. Virtually all are toxic to humans, some seriously so. **contact pesticide**. a chemical that kills pests through contact with the body surface, i.e., ingestion or inhalation is not required. **household pesticide**. any pesticide used in the home (e.g., fly spray, ant and roach baits, fly spray, insect repellents). In the United States alone, about 2.5 million humans are thought to be affected annually by such substances; 70% of those affected are children under the age of five. Symptoms of pesticide exposure in the home vary but often include nausea, coughing, dyspnea, eye irritation, dizziness, weakness, blurred vision, muscle spasms, and convulsions. Most are lipid-soluble cumulative poisons. **persistent pesticide**. a pesticide such as DDT that decomposes slowly, if at all, or remains active following one growing season. While such pesticides may give longer protection against the targeted pests than non-persistent pesticides, they may contaminate the harvest or may bioaccumulate, poisoning wildlife, livestock, pets, and even humans (examples are DDT and related organochlorine pesticides, MCPA, and PCB).

pesticide tolerance. the amount of pesticide residue permitted under law to remain in or on a harvested food crop. In the United States, the Environmental Protection Agency attempts to set these at levels well below the point where the chemicals might be harmful to consumers.

pesticide-resistant. pertaining to pests that are not easily killed or damaged by pesticides but continue to multiply when exposed to pesticides.

petiole. the stalk of a leaf.

pestilent. epidemic, widespread, and destructive; devastating; epidemic.

Petit Prêcheur. *Arisaema atrórubens*.

petroleum (crude oil). a naturally occurring complex mixture of tens of thousands of liquid hydrocarbons, usually with small amounts of sulfur, varying amounts of heavy metals, and traces of nitrogen and oxygen compounds. Some crude oils contain as much as 10% sulfur. Full chemical characterization of any crude oil has proven impossible. Petroleum is toxic and a skin irritant. Oil pollution of the marine environment (and of certain freshwater environments) produces heavy mortality of fish and wildlife worldwide. In addition, large spills have caused environmental disasters of major proportions. *See* petroleum fuel.

petroleum distillate. any of a wide variety of materials of varying toxicity distilled from petroleum (e.g., kerosene, gasoline, paint thinner). Petroleum distillates cause numerous deaths from accidental poisoning, especially of very young children. In the United States, nearly one-fourth of all deaths from accidental poisoning result from exposure to these substances. *See* petroleum fuel.

petroleum fuel. any of a wide range of petroleum products used as liquid fuels for heating equipment or engines. They may be broadly classified as distillate fuels (*See* petroleum distillate) and residual oils. Residual oil is the heavy oil that remains following removal of distillates in a refinery. Included are some diesel fuels and fuel oils. All are polluting, some to a high degree, and most will emit carbon monoxide fumes if burned under less than optimal or prescribed conditions.

pg. picogram.

pH. symbol for the negative logarithm (base 10) of the hydrogen ion concentration of an aqueous solution. It is a measure of the acidity or alkalinity of the solution. A solution with a pH of 7.00 is neutral; one with a pH of 7.0-14.0 is alkaline; one with a pH lower than 7.00 is acidic.

Phacochoerus (wart hogs). a genus of wild swine (Family Suidae). Included is the African wart hog. It is characterized by greatly enlarged curved tusks and three large wart-like prominences on each side of the face.

Phaeophyta (brown algae). a plant phylum containing the largest of the algae. They are multicellular, almost exclusively marine, and are filamentous or composed of leaf-like structures (thalli). They sometimes provide major forage for livestock.

phage. bacteriophage.

phage particle. bacteriophage.

phagocyte. any cell (e.g., certain leukocytes) that is able to ingest and destroy particulate matter and other cells such as bacteria, protozoa, colloids, and debris.

phagocytic. of or pertaining to phagocytes or phagocytosis.

phagocytosis. ingestion and digestion of particles by a cell. *See* phagocyte.

Phalaris arundinacea (reed canarygrass). a wild North American grass (Family Gramineae) that may become heavily ergotized. *See Claviceps*.

pharmaco-. an affix denoting relationship to a drug or medication.

pharmacology. the study of biologically active chemicals, especially of medicinal drugs, their origins, nature, chemistry, action, and applications. **biochemical pharmacology**. the subdiscipline that deals with the biochemical processes and mechanisms responsible for drug actions. **clinical pharmacology**. that division concerned with the use of therapeutic agents in the prevention, treatment, and control of human disorders. **veterinary pharmacology**. the study and application of the principles and practices of pharmacology to the concerns of veterinary medicine. *See also* phytopharmacology.

Phaseolus (kidney beans; bush beans; lima beans; etc.). a genus of warm-climate, cyanogenic, twining, annual or perennial herbs (Family Fabaceae, formerly Leguminosae). A number are grown for their edible pods and seeds. ***P. aureus*** (mung bean; black gram, golden gram, green gram, gram, mungo bean). an annual herb (Family Fabaceae) with green or yellow seeds, that is indigenous to Asia. Both the seeds and sprouts are edible. It is used for forage or green manure. ***P. coccineus*** (scarlet runner bean). an ornamental vine with pods that are considered edible, but has a record of toxicity in Europe. ***P. limensis*** (= *P. lunatus* in part) (lima bean; Java bean; Burma bean; Sieva bean; etc.). there are numerous varieties of this bean, all of which contain the cyanogenic glycoside, linamarin (= phaseolunatin). The lima beans commonly cultivated in the United States and other countries in the temperate zone are essentially nontoxic. They are small, light green to nearly white in color, easily recognized, and essentially non-toxic. An exception is the black lima bean of Puerto Rico, which has proven toxic to humans. Tropical varieties are often quite toxic. They are usually small, plump beans that are solid-colored or sometimes spotted. Many instances of cyanide poisoning are known from New Guinea due to eating raw lima beans (cooking is preventive). Lima beans imported into Canada or the United States for use as food or feed must by law be analyzed for cyanide. Specimens from the United States contain less than 0.01% hydrocyanic acid (HCN), whereas tropical varieties contain more than 0.01% HCN, frequently much more. Some tropical varieties contain as much as 0.3% HCN Plants containing 0.02% HCN are potentially hazardous to livestock. ***P. lunatus***. *Phaseolus limensis*.

pheasant's eye. a common name of herbs of the genus *Adonis*.

phenmedipham ((3-methylphenyl)carbamic acid 3[(methoxycarbonyl)amino]phenyl ester; 1*m*-hydroxycarbanilic acid methyl ester m-methylcarbanilate; methyl-3-(*m*-tolylcarbamoyloxy)phenylcarbamate). a selective carbamate herbicide. It is slightly toxic to laboratory rats.

Phenoclor. polychlorinated biphenyl.

phenol carbinol. benzyl alcohol.

10*H*-phenothiazine. phenothiazine (def.1).

phenothiazine. **1:** 10*H*-phenothiazine; thiodiphenylamine; dibenzothiazine. A photocontact allergen and antiemetic, $C_{12}H_9NS$, used as an insecticide and anthelmintic. It also lowers body temperature. **2:** any of a class of antipsychotic drugs such as chlorpromazine. They are derivatives of the phenothiazine nucleus.

phenotype. **1:** any individual discernable or measurable trait (structural or functional) of an organism (or type of organism) as opposed to genotype. A phenotype is the result of interactions between genotype and environment. **2:** the sum of all such traits of an individual organism or population.

phenotypic. referring to or denoting phenotype.

phenoxyacetic acid. **1:** a light tan, flammable powder,

$$C_6H_5OCH_2COOH$$

It is used, for example, as an intermediate for dyes, pesticides, and pharmaceuticals, and as a precursor in antibiotic fermentations. **2:** any of a class of substituted compounds of phenoxyacetic acid (def. 1), some of which show auxin-like activity. Various chlorine-substituted species disrupt plant metabolism especially at the meristems and have been widely used as selective herbicides. These compounds are especially effective as weedkillers because they are immune from the plant's endogenous IAA oxidizing system, which usually inactivates excess auxins. 2,4-D, 2.4.5-T, and MCPA are phenoxyacetic acids.

phentanyl. fentanyl.

phenthion. fenthion.

phenvalerate. fenvalerate.

***N*-phenyl isopropyl carbamate**. propham.

3-α-phenyl-β-acetylethyl-4-hydroxycoumarin. warfarin.

phenylamine. aniline.

phenylcarbamic acid 1-methylethyl ester. propham.

phenylcarbamic acid ethyl ester. phenylurethane.

phenylcarbamide. 1-phenyl-2-thiourea.

phenylcarbinol. benzyl alcohol.

***N*-phenyl-*N'*,*N'*-dimethylurea.** fenuron.

***N*-(1-phenylethyl-4-piperidinyl)-*N*-phenylpropionamide**. fentanyl.

***N*-(1-phenylethyl-4-piperidyl)propionanilide**. fentanyl.

phenylformic acid. benzoic acid.

phenylmercuric acetate ((acetato)phenylm ercury; acetoxyphenylmercury; phenylmercury acetate; PMA; PMAC; PMAS). a white to cream-colored, crystalline compound,

$$C_6H_5HgOCOCH_3$$

It is slightly soluble in water and soluble in benzene, ethanol, and glacial acetic acid. It is a strong irritant and is used as a bacteriostatic preservative, antiseptic, fungicide, and herbicide (especially for crabgrass), a mildewcide for paints, and a slimicide in paper mills.

phenylmercuric nitrate, basic ((nitrato-*O*)-phenylmercury; merphenyl nitrate). a fine, white, crystalline substance or light gray powder,

$$C_6H_5HgNO_3{\cdot}C_6H_5HgOH$$

It is insoluble in ether, slightly soluble in water and ethanol, and moderately soluble in glycerol. It is used as a germicide, a fungicide, and is sometimes used as a preservative in drug or toxicant formulations. It is toxic to humans.

phenylmercury acetate. phenylmercuric acetate.

phenyl-methanol. benzyl alcohol.

1-phenyl-3-methyl-5-pyrazolyl dimethylcarbamate. pyrolan®.

***N*-phenyl-*N*-[1-(2-phenylethyl)-4-piperidinyl]propanamide**. fentanyl.

phenylthiourea. 1-phenyl-2-thiourea.

1-phenylthiourea. 1-phenyl-2-thiourea.

1-phenyl-2-thiourea (1-phenylthiourea; phenylcarbamide; phenylthiourea; PTU). a compound with bitter, needle-like crystals, $C_7H_8N_2S$, that is soluble in water, ethanol, and aqueous ether. It is a selective toxicant and rodenticide. While much more toxic to rodents than to humans, it is extremely hazardous to humans and is poisonous by ingestion and by intraperitoneal routes. Toxicity to other animals is largely unknown.

phenylurethan. phenylurethane.

phenylurethane. (phenylurethan; phenylcarbamic acid ethyl ester; ethyl phenylcarbamate; ethyl phenylurethane; ethyl carbamilate). a white crystalline aryl carbamate,

$$C_6H_5NHCOOC_2H_5$$

It is soluble in common organic solvents and hot water. It has an aromatic odor and clove-like taste. Phenylurethane is an herbicide that interferes with photosynthesis. It inhibits the growth of barley at concentrations that do not affect the spindle. It delays the germination of oats and wheat with ensuing abnormal growth.

pheromone. a chemical, usually a sex attractant, produced by some animals that attracts others usually of the opposite sex of the same species

Philanthus triangulum (bee wolf). a solitary European wasp that captures honey bees with which it stocks its nest. The bees serve as food of the wasp larvae. This wasp is a significant pest to beekeepers in certain locations.

phloem. the primary tissue that transports sugars, amino acids, and certain other materials through a vascular plant. It is composed of hollow, elongated vessels that extend throughout the plant. These are the sieve tubes or angiosperms or the sieve cells of ferns and gymnosperms. Phloem also contains supporting fibers (bast). *See* xylem.

Pholiota. a mushroom genus of brown-spored agarics of North America and Europe (Family Strophariaceae). They have an annulus and grow on open ground or decaying wood. ***P. autumnalis***. former scientific name of *Galerina autumnalis*. ***P. squarrosa*** (scaley pholiota). a mushroom that grows in clusters at the base of trees, most commonly aspen and birch, and on logs and stumps. It causes gastric distress in some individuals.

phomin. cytochalasin B.

Phomopsis paspalli. a fungus that produces the toxic macrolide mycotoxins kodo-cytochalasin-1 and -2. It commonly infests kodo millet, *Paspalum scrobiculatum*

Phomopsis vexans. *See* eggplant blight.

Phoradendron serotinum (formerly *P. flavescens*; American mistletoe; false mistletoe). a yellowish-green shrubby epiphyte (Family Loranthaceae) that contains chlorophyll. It grows on various deciduous trees in eastern North America, from Florida to eastern Texas, north to southern Ontario, southern Indiana, southern Illinois, southern Missouri, and southeastern Kansas. The leaves are broad, thick, and greenish; the fruit is a globose, pulpy drupe. All parts but especially the small white berries are toxic by ingestion, producing gastrointestinal symptoms and bradycardia. Leaves have been used for tea and a coffee substitute, apparently with no ill effects.

phorate (phosphorodithioic acid *O,o*-diethyl *S*-[(ethylthio)methyl] ester; *O,o*-diethyl *S*-(ethylthio)methylphosphorodithioate; *O,o*-diethyl *S*-ethylmercaptomethyl dithiophosphate; thimet; Agrimet; Geomet; Granutox; Rampart; Thimenox). a clear, slightly water-soluble liquid used as a broad-spectrum, systemic organophosphate insecticide. It is a neurotoxicant that acts via acetylcholinesterase inhibition. It is highly dangerous and toxic by all routes of exposure but is used in some household products to eradicate pests.

Phormia regina (black blowfly). a fly (Family Calliphoridae) that lays its eggs chiefly on animal carcasses. It also lays eggs in soiled wool about the rump of sheep on which the maggots feed. The larvae may also attack the horns, causing the skin about the horn to fester; the wool may loosen and molder.

phorophyte. a plant upon which an epiphyte grows.

phosalone (phosphorodithioic acid *S*-[(6-chloro-2-oxo-3(2H)-benzoxazolyl)methyl] *O,o*-diethyl ester; phosphorodithioic acid, *O,o*-diethyl ester, *S*-ester with 6-chloro-3-(mercaptomethyl)-2-benzoxazolinone; 3-(*O,o*-diethyldithiophosphorylmethyl)-6-chlorobenzoxazolinone; 6-chloro-3-(*O,o*-diethyldithiophosphorylmethyl)benzoxazolone;*S*-(6-chloro-2-oxobenzoxazolin-3-yl)-methyl diethyl phosphorothiolothionate). a toxic, organic compound,

$$C_{12}H_{15}ClNO_4PS_2$$

The colorless crystals are soluble in alcohols and aromatic solvents. It is used mainly as an insecticide and molluscicide. It is a component of some household products used to eradicate pests; such products should be considered dangerous and handled with care.

Phosdrin®. a mixture of organophosphates that contains more than 60% mevinphos and less than 40% chemically related insecticides. *See also* mevinphos.

phosethyl Al. fosetyl aluminum.

phosmet (phosphorodithioic acid S-[(1,3-dihydro-1,3-dioxo-2*H*-isoindol-2-yl)methyl] *O,o*-dimethyl ester; phosphorodithioic acid *O,o*-dimethyl ester *S*-ester with *N*-(mercaptomethyl)phthalimide; *O,o*-dimethyl *S*-phthalimidomethyl phosphorothionate; *N*-(mercaptomethyl)phthalimide *S*-(*O,o*-dimethyl phosphorodithioate); phthalophos). a colorless, crystalline, partially water-soluble, organophosphate, $C_{11}H_{12}NO_4PS_2$. It is a dimethyl ester of phosphorodithioic acid. It is an anticholinesterase used as an insecticide and acaricide.

phosphacol. paraoxon.

phosphamidon (2-chloro-2-diethylcarbamoyl-1-methylvinyldimethylphosphate). a highly toxic, colorless, liquid, organophosphate insecticide, $(CH_3O)_2P(O)OC(CH_3):C(Cl)C(O)N(C_2H_5)_2$, that is soluble in water and organic solvents. It is more toxic than malathion but less toxic than parathion. Effects are due to cholinesterase inhibition. It is a component of some household products designed to eradicate pests. Phosphamidon is toxic to humans by all routes of exposure.

phosphate. any of the three series of salts of phosphoric acid. **orthophosphates** include all of the naturally occurring phosphates, the anion of which is HPO^-; all are acidic. **metaphosphates** are those with the anion HPO^{2-}; all are neutral. **pyrophosphates** are those with the anion PO^{3-}; all are alkaline. Phosphates vary enormously in toxicity, some being virtually harmless and others deadly in trace amounts. Toxic phosphates kill primarily by the reversible inhibition of acetyl cholinesterase, although they affect many other enzyme systems also. In general, an exposed animal dies or recovers completely from the action of phosphates.

phosphine. **1:** hydrogen phosphide; phosphoretted hydrogen; PHI. A colorless, very toxic, hydride of phosphorus, H_3P, with a characteristic foul garlicky (or dead-fish) odor and is soluble in ethanol, ether, and cuprous chloride solution. It is produced in small amounts during putrefaction of phosphorus containing organic matter and is synthesized from white phosphorus and aqueous alkali hydroxide. Among many uses, PHI is as an active agent in some rodenticides. Phosphine is a very strong pulmonary tract irritant and central nervous system depressant. Inhalation affects the eyes and causes dilation of the pupils. A high concentration in the air is rapidly fatal, producing fainting, lowered blood pressure, pulmonary damage, convulsions, paralysis, and coma. Somewhat less serious exposures in humans may cause fatigue, vertigo, vomiting, a sensation of coolness, bronchitis, edema, dyspnea, and associated pain in the region of the diaphragm. It is a potential hazard in industrial processes and in the laboratory because of its inadvertent production in chemical syntheses involving other phosphorus compounds. **2:** any of a class of organophosphorus compounds derived from phosphine by substitution of organic groups for H atoms. **3:** an orange-yellow basic dye; Its assigned name in the Colour Index is orange 15 but a common commercial name is phosphine.

phosphonic acid monoethyl ester aluminum salt. fosetyl aluminum.

***N*-(phosphonomethyl)glycine**. glyphosate.

phosphoric acid. a colorless, highly toxic, corrosive liquid. It is used, for example, in the manufacture of fertilizers, detergents, animal feed, drugs, and soft drinks; it is used also in sugar refining, metal cleaning, engraving, and lithography. It can severely damage eyes and skin on contact and is moderately toxic if ingested. Under certain conditions it can cause respiratory difficulties, especially as an aerosol of mist.

phosphoric acid 2-chloro-1-(2,4-dichlorophenyl diethyl ester. chlorfenvinphos.

phosphoric acid 7-chlorobicyclo-[3.2.0]hepta-2,6-dien-6-yl dimethyl ester. heptenophos.

phosphoric acid 2,2-dichloroethenyl dimethyl ester. dichlorvox.

phosphoric acid 2,2-dichlorovinyl dimethyl ester. dichlorvos.

phosphoric acid diethyl 4-nitrophenyl ester. paraoxon.

phosphoric acid dimethyl ester, ester with *cis*-3-hydroxy-*N,N*-dimethylcrotonamide. dicrotophos.

phosphoricacid3-(dimethylamino)-methyl-3-oxo-1-propenyl dimethyl ester. dicrotophos.

phosphorodithioic acid *S*-[2-chloro-1-(1,3-dihydro-1,3-dioxo-2*H*-isoindol-2-yl)ethyl] *O,O*-diethyl ester. dialifor.

***O,o*-phosphorodithioic acid *S*-[(6-chloro-2-oxo-3(2H)-benzoxazolyl)methyl] *O,o*-diethyl ester**. phosalone.

phosphorodithioic acid *S*-[[(4-chlorophenyl)thio]methyl] *O,o*-diethyl ester. carbophenothion.

phosphorodithioic acid, *O,o*-diethyl ester, *S*-ester with 6-chloro-3-(mercaptomethyl)-2-benzoxazolinone. phosalolne.

phosphorodithioic acid *O,O*-diethyl *S*-[2-(ethylthio)ethyl] ester. disulfoton.

phosphorodithioic acid *O,o*-diethyl *S*-[(ethylthio)methyl] ester. phorate.

phosphorodithioic acid *O,O*-diethyl ester *S*-ester with *N*-(2-chloro-1-mercaptoethyl)phthalimide. dialifor.

phosphorodithioic acid *S*-[(1,3-dihydro-1,3-dioxo-2*H*-isoindol-2-yl)methyl] *O,o*-dimethyl ester. phosmet.

phosphorodithioic acid *O,o*-dimethyl ester *S*-ester with *N*-(mercaptomethyl)phthalimide. phosmet.

phosphorodithioic acid *O,O*-dimethyl ester, ester with 2-mercapto-*N*-methylacetamide. dimethoate.

phosphorodithioic acid *O,O*-dimethyl ester *S*-ester with *N*-formyl-2-mercapto-*N*-methylacet-amide. formothion.

phosphorodithioic acid *O,O*-dimethyl *S*-[2-(methylamino)-2-oxoethyl] ester. dimethoate.

phosphorodithioic acid *S*-[2-(formyl-methylamino)-2-oxyethyl] *O,O*-dimethyl ester. formothion.

phosphorodithioic acid *O,O*-diethyl ester *S*-ester with *N*-(2-chloro-1-mercapto-ethyl)phthalimide. dialifor.

phosphorodithioic acid *S*-[2-chloro-1-(1,3-dihydro-1,3-dioxo-2*H*-isoindol-2-yl)-ethyl] *O,O*-diethyl ester. dialifor.

phosphorodithioic acid *S,S'*-1,4-di-oxane-2,3-diyl *O,O,O',O'*-tetraethyl ester. dioxathion.

phosphorofluoridic acid bis(1-methylethyl) ester. diisopropyl fluorophosphate.

phosphorofluoridic acid diisopropyl ester. diisopropyl fluorophosphate.

phosphorothioic acid *O,O*-dimethyl *O*-[3-methyl-4-(methylthio)phenyl] ester. fenthion.

phosphorothioic acid *O,O*-dimethyl *O*-(4-nitrophenyl) ester. methyl parathion.

phosphorothioic acid *O*-[2-(ethylthio)ethyl] *O,O*-dimethylester mixed with *S*-[2-(ethylthio)ethyl]. demeton-*S*-methyl.

phosphorothioic acid *O,O*-diethyl *O*-(3,5,6-trichloro-2-pyridinyl) ester. chlorpyrifos.

phosphorothioic acid *O,O*-diethyl *O*-[2-(ethylthio)ethyl] ester mixture with *O,O*-diethyl *S*-[2-(ethyl-thio)ethyl]-phosphorothioate. demeton.

phosphorothionate ester. a class of esters in which the oxygen of the phosphate group has been replaced by sulfur. These compounds, which include the insecticides chlorothion, diazenon, parathion, and tributylphosphorothionate, thus contain an O_3P:S group. These compounds are often more toxic to insects relative to mammals than their oxygen-containing analogs.

phosphorothionate insecticide. *See* phosphorothionate ester.

phosphorus (symbol, P)[Z = 15; A_r = 30.975]. a reactive, solid, highly combustible, nonmetallic allotropic element that is widely distributed in nature, always in a combined state. There are no stable isotopes, but several radioactive isotopes. It occurs as a phosphate in all living cells. Elemental phosphorus and its fumes have a characteristic fishy odor and are extremely poisonous. It occurs in three main forms: black, red, and white. Elemental phosphorus causes severe burns or intense inflammation and fatty degeneration of the liver and other viscera; repeated inhalation of the fumes often causes a necrosis of the lower jaw (phosphonecrosis). The largest single use of phosphorus compounds, many of which are caustic or toxic, is in fertilizers. Cupric sulfate is used therapeutically as an antidote. See also phosphoptomaine, radiophosphorus. **black phosphorus**. a black, lustrous crystalline substance that resembles graphite or coal. It is stable in air, does not ignite spontaneously, is electrically conductive, and is insoluble in most solvents. **ordinary phosphorus**. white phosphorus. **red phosphorus** (amorphous phosphorus). a dark violet-red amorphous powder that is infusible, insoluble in most solvents including carbon disulfide, and much less reactive than white phosphorus. Pure red phosphorus is much less toxic than white phosphorus. It may explode from friction or on contact with an oxidizing agent; large masses may ignite spontaneously. It reacts with oxygen and water vapor to emit phosphine fumes. Red phosphorus is used in pyro technics, in the manufacture of fertilizers, pesticides, and various organic syntheses, and in the manufacture of incendiary shells, cartridges, and smoke bombs. **white phosphorus** (ordinary phosphorus; yellow phosphorus). an extremely poisonous white or transparent crystalline, wax-like solid that darkens and yellows on exposure to light, and ignites spontaneously in moist air at about 30°C. It must thus be stored under water away from heat or dissolved in carbon disulfide. White phosphorus is toxic by ingestion or inhalation and contact with the skin can cause severe burns. Ingestion of as little as ca. 50-100 mg can cause severe gastrointestinal irritation with bloody diarrhea, skin eruptions, liver damage, oliguria, cardiovascular collapse, convulsions, coma, and death. The effects of chronic systemic intoxication may include phosphonecrosis, weight loss, spontaneous fractures, and anemia. It is used in the manufacture of rodenticides, smoke screens, and analytical chemistry. **yellow phosphorus**. white phosphorus.

phosphorylation. the addition of a phosphate group to a molecule. Metabolic phosphorylation of proteins at specific sites is a common means of modifying their activity.

phosphothion. malathion.

phosporothioic acid *O,O*-diethyl *O*-(4-nitrophenyl; PA) ester. parathion.

photo-. a prefix indicating a response, sensitivity, or relationship to light.

photoactivation. the activation of a chemical, a chemical reaction, or certain processes of living organisms by light (e.g., the activation of chlorophyll by sunlight during photosynthesis).

photoactive. chemically reactive to light in the visible or ultraviolet portions of the spectrum.

photoallergic. of, pertaining to, or characterized by an intensified, delayed contact-type sensitivity to light, especially sunlight.

photoallergic dermatitis. *See* photoallergy.

photoallergy. **1a:** an allergic sensitivity to light. **1b:** photoallergic dermatitis; a type of delayed contact dermatitis produced by the interaction of light and certain chemicals. *See also* dermatitis, photosensitivity, photocontact allergen. Light produces the sensitivity reaction.

photoautotroph. photoautotrophic organism.

photoautotrophic organism (photoautotroph; photolithotrophic organism; photolithotroph). an autotrophic organism whose energy is derived from sunlight. Included are all green, and some purple plants, algae, and a few types of bacteria. *See also* autotrophic organism, chemoautotrophic organism.

photochemical. **1a:** pertaining to the chemical action or activity of light. **1b:** chemically reactive in the presence of light or other radiation. **1c:** influenced or initiated by light (said of a chemical or chemical reaction). *See* photochemical oxidant.

photochemical oxidant. any air pollutant produced by the action of sunlight on oxides of nitrogen and hydrocarbons. *See* photochemical air pollutant.

photodermatitis. a rapidly developing dermatitis due to exposure to ultraviolet light. It results from the action of ultraviolit light. light (usually that in sunlight) following exposure of a subject to a phototoxic substance or photoallergen. *Cf.* photodermatosis, photosensitivity. **photoallergic photodermatitis** (photoallergic dermatitis). a rapidly developing allergic dermatitis of sun-exposed areas of the skin known mainly from livestock and humans. **phototoxic photodermatitis**

(phototoxic dermatitis). a rapidly developing dermatitis with hyperpigmentation of sun-exposed areas of the skin following exposure to agents containing photosensitizing substances (e.g., coal tar, some perfumes, drugs, or plants that contain psoralens), followed by exposure to sunlight. *See* photodermatitis.

photodermatosis. a morbid condition of the skin resulting from exposure to light.

photosensitive. **1:** highly sensitive to or reactive to light. *See* photosensitivity, photosensitization. **2:** highly responsive to photoperiod or changes in photoperiod.

photosensitivity. **1:** sensitivity of an organism or a chemical (e.g., a pigment) to light. **2:** the property or condition of a biological clock such that it is reset daily by the photoperiod. **3a:** any abnormal response to sunlight (usually a skin reaction) due to the presence of a sensitizing agent in the body. **3b:** hypersensitivity of an animal to light due to the presence of any of a number of photodynamic pigments in the peripheral circulation. In such cases, exposure to sunlight produces erythema and pruritis of lightly pigmented, unprotected skin followed by edematous suffusions and usually necrotic skin lesions. Thus, cattle are most affected on white or unpigmented areas of the skin and white sheep are most affected about the head. *See also* photoallergy, phototoxicity. **hepatogenic (hepatogenous) photosensitivity**. a condition in which the photodynamic pigment is a normal breakdown product of digestion that is eliminated by the healthy liver. Thus, the condition is the immediate result of liver dysfunction which is usually due to hepatotoxic principles present in the photosensitizing plant. The condition may be accompanied by pronounced general jaundice. *See also Agave lecheguilla, Brassica napus, Lantana spp., Nolina texana, Panicum spp., Tetradymia spp., Tribulus terrestris*, phylloerythrin. **ictrogenic photosensitivity**. a term sometimes applied to hepatogenic photosensitivity. The generalized jaundice and intensely-pigmented livers result from an inability of the affected animal to excrete pigment molecules. **primary photosensitivity.** the case in which the photodynamic pigment, unaltered by digestion, enters the general metabolism. Only two North American plants, *Fagopyrum sagittatum* and *Hypericum perforatum*, are known to cause primary photosensitivity.

photosensitization. a condition in which the skin reacts abnormally to light, especially to ultraviolet rays or sunlight, due to the presence of a phototoxic pigment in the skin. Light energy is absorbed by the pigment inaugurating a signal to receptor molecules that rapidly initiates chemical reactions that are injurious to the skin. Included are the production of reactive oxygen intermediates or changes in cell membrane permeability.

photosynthesis. a process in which the energy of sunlight is converted to chemical energy. It occurs in green (chlorophyll-bearing) plants and phototrophic bacteria that contain a similar pigment, bacteriochlorophyll. In a sequence of reactions, carbohydrates are manufactured from oxygen, carbon dioxide, and water in the presence of sunlight. The net reaction can be summarized as:

$$6CO_2 + 6H_2O + \text{light energy} \rightarrow C_6H_{12}O_6 + 6O_2$$

net photosynthesis. total or gross photosynthesis less respiration. It is usually determined by measuring the net uptake of carbon dioxide by the leaf.

phreatic. **1a:** pertaining to a well. **1b:** pertaining to groundwater that is reachable or probably reachable by drilling.

phreaticole. an organism that inhabits phreatic groundwater.

phreatophyte. a deep-rooted plant that absorbs water from the permanent water table or from the soil layer immediately above it. *See also* phreatic.

phthalic acid dimethyl ester. dimethyl phthalate.

phthalophos. phosmet.

phycology. algology.

Phyllanthus abnormis (spurge). a small annual or perennial herb (Family Euphorbaceae) of western and southwestern Texas. It is nephrotoxic and hepatotoxic to most types of livestock. Toxicity varies geographically.

phylloerythrin. a normal product of the digestion of chlorophyll by vertebrate animals, it is a photosensitizing porphyrin pigment. It is an anaerobic breakdown product of chlorophyll in the forestomachs of ruminants. Normally it is rapidly excreted via the bile. In certain cases of liver dysfunction or bile duct occlusion, it may reach the peripheral circulation where it absorbs sunlight and initiates a phototoxic reaction. It is nearly always the photosensitizing pigment in hepatogenous photosensitization. This has been verified in many cases as in copper poisoning or poisoning by *Lippia rehmanni*, *Lantana camara*, *Myoporum laetum* (ngaio), *Narthecium ossifragum* (bog asphodel), *Panicum* spp., *Tribulus terrestris* (puncture vine), or *Myoporum laetum* (ngaio).

Phylloxera (plant louse). a genus of plant lice (Family Phylloxeridae) that are destructive to many commercially important plants. See also mercuric nitrate.

phylum (plural, phyla). **1:** a group of related classes of animals, representing the highest (and broadest) category within the animal kingdom; roughly equivalent to the "division" in botany. **2:** used also by some plant taxonomists in preference to "division."

phyone. somatotrophin.

Physalis (ground cherry, husk tomato). a genus of annual or perennial herbs (Family Solanaceae). The seeds and unripe fruit are toxic. The toxic principle is an extremely toxic protein, physalitoxin.

physiogenic disease. abiotic disease.

physiologic, physiological. **1:** pertaining to physiology or the phenomena of physiology. **2:** of or pertaining to normal (not pathologic) vital processes. **3:** of or pertaining to a process or biologic entity that is identified by its functional activity or effects rather than its anatomic structure (e.g., a physiologic sphincter). **4:** of or pertaining to a dosage (or the effects) of a xenobiotic, at a concentration, amount, or potency that falls within the normal range of the naturally occurring endogenous substance (e.g., a hormone or neurotransmitter) that it mimics or resembles.

physiological saline solution. *See* saline (physiological saliane).

physiological salt solution. *See* saline (physiological saline).

physiology. the biological science that treats the chemical and physical processes in the functioning of cells, tissues, organs, organ systems, and whole organisms. While this discipline does not treat structure per se, it is strongly rooted in histology and anatomy. **cell physiology**. the study of the physiology of cells. **comparative physiology**. the study and comparison of the physiology of different taxa. **general physiology** the study of the general principles and scientific bases of physiology. **pathologic physiology**. the study of pathological states and processes. **plant physiology**. the branch of physiology that treats the physicochemical pro cesses of plants. **special physiology**. the physiology of the special senses.

Phytar. cacodylic acid.

phytoalexin. any chemical constituent of a plant that repels, inhibits the growth of, or destroys insects or pathogens, especially fungi. They are usually formed in response to chemical stimuli from the pathogen. Examples are pisatin and ipomeamarone.

phytocidal. destructive or lethal to plants.

phytocide. a chemical used to destroy and control plants. *Cf.* herbicide.

phytodermatitis. dermatitis caused by exposure to plants. The condition can be due to any of a number of mechanisms and modes of exposure (e.g., mechanical or chemical irritation, toxicity, allergy, or photosensitization). *See also* phytophotodermatitis, *Urtica*.

phytoecdysone. any plant substance with ecdysone-like activity. Some are extremely potent. The principal ecdysone of plant extracts is β-ecdysone. *See also* ponasterone.

phytogenous. of plant origin; produced by plants.

phytohormone (plant hormone). any substance normally produced by a plant at a location removed from the site of action within the plant and is active in minute amounts. Phytohormones control plant functions such as growth. There are three types: auxins, cytokinins, and gibberellins.

phytokinin. cytokinin.

Phytolacca (pokeweed; pokeberry; scoke; garget; pigeonberry). a genus of perennial poisonous herbs, shrubs and trees (Family Phytolaccaceae) of tropical and warm areas. ***P. americana*** (synonym, *P. decandra*; pokeweed; scoke; poke; garget; pigeonberry). a tall, perennial herb with smooth stems and leaves and an unpleasant odor. It is a common plant that grows on low ground often in barnyards, clearings and along roadsides from Florida to Texas, northward to New England, southern Quebec, New York, and southern Ontario. The ovaries are green, and the berries, when ripe in autumn, are dark purple. All parts are extremely toxic. Animals are rarely poisoned. Swine, however, have been poisoned by rooting out and eating the roots. Even now people are sometimes killed when they eat the extremely toxic, dark purple berry or the root which may be mistaken for wild parsnip or horseradish. Gloves should be worn when handling this plant.

Phytolaccaceae (pokeweed family). a family of plants with alternate entire leaves and perfect flowers and similar in general characteristics to plants of the family Chenopodiaceae. *See Phytolacca*.

phytopathogenic. pathogenic to plants; of or pertaining to a substance or an organism that is pathogenic to or productive of disease in plants.

phytopathology (plant pathology). the study of dysfunctional and pathological states and processes in green plants. *See* pathology.

phytopharmacology. the study of the effects of biologically active substances on plants, especially on growth. *See also* pharmacology.

phytophlyctodermatitis. meadow-grass dermatitis.

phytophotodermatitis. phototoxic dermatitis due to the ingestion of parts of plants such as certain species of *Brassica*, *Erodium*, *Euphorbia*, *Medicago*, *Poa*, and *Trifolium* with subsequent exposure to light.

See also phytodermatitis.

phytophysiology. the science of plant physiology.

phytoplankton. plankton comprised of plants or that component of plankton comprised of plants.

phytotoxicologist. a scientist concerned with poisonous plants.

phytotoxicology. a science that deals with poisonous plants.

Picea mariana (black spruce; double spruce; eastern spruce; bog spruce). a slow growing, often stunted but hardy evergreen conifer (Family Pinaceae) that is indigenous to North America. It grows in moist, acidic soils as in peat bogs. The wood is used for a range of products including pulpwood, planing-mill products, and furniture.

picketline-bound. *See* agoraphobia.

picloram (4-amino-3,5,6-trichloro-2-pyridinecarboxylic acid; 4-amino-3,5,6-trichloropicolinic acid). a crystalline solid or colorless powder,

$$C_6H_3Cl_3N_2O_2$$

It is a pyridine herbicide and defoliant with the odor of chlorine. It is toxic also to higher animals and humans by ingestion and inhalation and is a suspected occupational carcinogen. Picloram and its salts are absorbed by leaves and roots and translocated to new growth. It is used to control broadleaved weeds (except Cruciferae) in grasses. Even at low application rates it alters leaf morphology and inhibits the growth of leaves. With normal usage, terminal growth ceases. Following high application rates, it may persist in the soil for nearly a year.

pico- (p). a prefix to units of measurement that indicates a quantity of 10^{-12} or one-trillionth of the value indicated by the root. *See* International System of Units.

picogram ($\mu\gamma$; pg; microgamma; micromicrogram). a unit of mass equal to 10^{-12} or one-trillionth of a gram.

picry. *Rhus radicans*.

Pictou disease (Molten's disease; Winton disease). a disease of domestic horses and cattle in Nova Scotia marked by cirrhosis of the liver. It results from ingestion of *Senecio jacobeus*.

pie plant. *Rheum rhabarbarum*.

Pieris (pieris). **1:** a genus of butterflies (Family Pieridae). The best known and most widespread species of which is *P. rapae*, the cabbage butterfly; the caterpillar feeds on cabbage and other Cruciferae. *See* sinigrin, allyl isothiocyanate. **2:** a genus of evergreen shrubs or small trees of eastern North America and Asia (Family Ericaceae).

pieris. common name of butterflies plants of the genus *Pieris* (and plants of the genus *Pieris*).

pig. **1:** a young swine. **2:** *See Sus*.

pigeon tick. *Argas reflexus*.

pigeonberry. a plant of the genus *Phytolacca*.

pigment. **1:** any colorant, natural or synthetic; a dye, stain, tint, coloring, or paint. *See also* color. **2:** any of numerous substances or formulations in the form of a solid, crystal or powder that imparts a specified positive color to another substance or material. By this definition, talc, whiting, barytes, and clays are not pigments but carbon black can be considered a pigment. Most are insoluble in water or organic solvents. Whereas a pigment is usually dispersed in a vehicle or substrate, it is physically and chemically unaffected by the vehicle. Pigments are classified in any of numerous ways not treated here. But see color. In some classifications, white is considered a color. **3:** a tincture; any medicinal preparation that is painted onto the skin (tincture of iodine). *Cf.* color, colorant, dye. **animal pigment**. any pigment that occurs naturally in animals. Examples are rhodopsin and melanin. **biological pigment**. any pigment that occurs in a living organism. Pigments play many diverse roles in life processes (photoreception; diverse photochemical reactions; photosensitization; the provision of cryptic and warning coloration; smoke screens; internal respiration; enzyme activity; detoxication). Examples of biological pigments are hemoglobin, chlorophylls, cytochromes, ferrichromes, gossypol. All of these examples contain iron. **chromate pigment**. chromates used as pigments. All are toxic. *See also* chromate. **iron pigment**. pigments that contain iron. Many play critical biological roles; many are toxic. Examples are hemoglobin, cytochromes, ferrichromes, iron oxides. **lead pigment**. any pigment that contains lead compounds. All are potentially toxic if ingested. Examples are lead carbonate, lead molybdate, lead antimonite, and lead nitrate. **photosensitizing pigment**. any pigment, usually a plant pigment, that causes photosensitization in animals (usually herbivores) when ingested and the animal is subsequently exposed to sunlight. *See*, for example, hypericin, phylloerythrin. *See also* photodermatitis, photosensitivity, photosensitization. **phytotoxic pigment**. a toxic plant pigment. Photosensitizing pigments are phytotoxic. **plant pigment** (vegetable pigment). any of a large variety of organic pigments (after def. 2, above) that are produced by plants. Examples are carotenoids such as carotene and xanthophyll, flavanoids such as catechins, and flavones. Some can be prepared synthetically. **respiratory pigment**. a pigment that reversibly binds oxygen and carbon dioxide. *See* hemoglobin. **synthetic pigment**. any pigment that is synthesized and not found in nature.

pigweed. *Chenopodium*. *See also* Chenopodiaceae.

piloerection. the erection or apparent stiffening of body hair due to stimulation and contraction of the erector pili muscles. It is usually accompanied by dilation of the pupils. It is often due to excitement, especially fear.

Pimenta acris. the bay-rum tree. *See* bay leaf.

pine. any tree of the family Pinaceae.**true pine**. any tree of the genus *Pinus* (Family Pinaceae), *q.v.*

pink nose. acorn fruit.

pinkeye. equine viral arteritis.

pinna. **1:** the outer, funnel-like structure of the ear, as in humans, that gathers sound waves. **2:** a leaflet of a fern frond or of any compound pinnate leaf.

2-pinnate. bipinnate.

pinnate. situated along a central axis as in a compound leaf with leaflets on either side of a common petiole.

Pinus (pines; true pines). a genus of about 80 species of coniferous trees (Family Pinaceae), it is the largest and most widely-distributed genus of the family. The genus is best represented at lower altitudes in the North Temperate Zone but some species even occur on slopes of high mountains in the tropics. The leaves occur as single needles or clusters of two to five which (together with length of the needles) offers a means of, or aids in, identifying many species. Many of the pines are economically important, providing, for example, lumber of varying type and quality, resins, balsams, pitch, turpentine, and rosin. The seeds of several American and Mediterranean species produce edible seeds. ***P. banksiana*** (jack pine). a North American species thrives on poor and sandy soils as on the dry slopes of the Rocky Mountains. It does not provide prime lumber as the trunk is often gnarled but it is much used as a source of pulpwood. It is also used as a "pioneer" species to colonize areas prior to the introduction of more commercially valuable species. ***P. caribaea*** (slash pine). a species that is widely cultivated on sandy soils in tropical areas. ***P. palustris*** (longleaf pine; Southern yellow pine). a tree with hard, highly resinous wood that is used for heavy construction and a source of pulpwood and naval stores such as pitch, resins, and turpentine. ***P. pinaster*** (cluster pine). a tree that is widely-distributed in Spain and southern France. The cluster pine is the chief source of turpentine in Europe. ***P. ponderosa*** (ponderosa pine; western yellow pine; blackjack pine). a large to very large dominant coniferous tree that grows to about 60 m under favorable conditions. It is widely distributed at moderate elevations in western North America from southern California, New Mexico, and trans-Pecos Texas to southern British Columbia to North Dakota. The needles and buds are embryotoxic and abortifacient to cattle. The wood is hard and second only to Douglas fir in importance as a commercial timber tree in North America. ***P. radiata*** (Monterey pine). native to California, this species has been used extensively in reforestation in Chile and New Zealand. ***P. resinosa*** (Norway pine; red pine). a species that has suffered from thoughtless, intense harvesting and has a current range similar to that of *P. strobus*. The wood is moderately heavy and used chiefly in rough carpentry and general construction. This tree is often used in reforestation. ***P. sabiniana*** (digger pine; grayleaf pine). a true pine of California that produces an edible nut. The resin or balsam of this tree is used to produce abietene. ***P. strobus*** (white pine). a tree with a straight-grained soft wood with little resin that is easy to work and is chiefly in interior trim and cabinetwork. It was once abundant from Newfoundland to Manitoba and in much of the Eastern United States. It has been greatly depleted over much of its former range by extensive overcutting and by attacks of white-pine blister rust. ***P. sylvestris*** (Scotch pine). a widely distributed species, ranging from Scotland to Siberia is one of the most valuable timber trees in Europe. It is a popular Christmas tree in the United States. ***P. virginiana*** (Virginia pine). a tree of the Appalachians and the Piedmont of the northeastern United States It is a popular Christmas tree.

Piper (true peppers). a large genus of tropical plants (Family Piperaceae), the true peppers. ***P. cubeba*** (tailed pepper; Java pepper). a tropical shrub of southern Asia. *See* cubeb, cubebism. ***P. nigrum*** (black pepper). a woody vine, indigenous to the East Indies, it is the source of white and black pepper, which are widely used condiments. The fruit contains safrole. *See* piperism.

piperidine (hexahydropyridine). a highly toxic, soapy feeling, liquid alkaloid with the characteristic odor of pepper; it is a completely saturated six-membered ring that contains 1 nitrogen atom,

$$CH_2CH_2CH_2CH_2CH_2NH$$

used chiefly as an insecticide. It occurs in some species of *Lupinus* and in small amounts in *Piper nigrum*. *See also* piperine, piperism, pyridine.

3-(2-piperidinyl)pyridine. anabasine.

piperine (1-[5-(1,3-benzodioxol-5-yl)-1-oxo-2,4-pentadienyl]piperidine; 1-piperoylpiperidine). a crystalline, slightly soluble alkaloid from *Piper nigrum* (black pepper) and related species such as *P. clusii*, *P. geniculatum*, *P. longum*, and *P. retrofractum*. It yields piperidine and piperic acid on hydrolysis. Piperine provides the pungent taste of brandy; it is used as an insecticide and is more toxic to *Musca domestica* (the house fly) than is pyrethrum. *See* piperism.

piperism. poisoning by black pepper. *See* piperine.

1-piperoylpiperidine.piperine.

Piroplasma. Former generic name of *Babesia*.

piscivorous (ichthyophagous). feeding on fish; said of an animal that feeds chiefly or exclusively on fish.

Pisum sativum (garden pea; English pea). a popular, widely-cultivated annual herb (Family Fabaceae) that yields an edible seed. It is indigenous to Eurasia.

pitch. slope, grade, decline, or descent.

pithomycotoxicosis. facial eczema.

pituitary growth hormone. somatotrophin.

pityriasis rubra. exfoliative dermatitis.

placenta. a specialized organ of eutherian mammals, formed from the chorion and uterine tissue following implantation of the fertilized egg in the uterus. It is the organ of metabolic exchange between the mother and the embryo or fetus. The placenta is also the organ of transport of toxicants from maternal blood to that of the fetus. *See also* barrier (placental barrier).

placental barrier. *See* barrier.

placental membrane. placental barrier (*See* barrier).

plains whorled milkweed. *Asclepias pumila*.

plankton. microscopic plants and animals that live in water bodies, fresh water, brackish, or marine water. *See also* nekton, phytoplankton, zooplankton.

plant. any member of the kingdom Plantae. The responses of plant populations to a xenobiotic chemical vary as a function of phenological stage and effects of environmental variables (past and current) such as nutrient availability, moisture, and interactions with other plants. In most cases, plant tissue concentrations of toxic chemicals do not bear a quantitative relationship with exposure concentrations. Effects can be indirect in the sense of impact on soils or plant species found in association with the affected plant. **accumulator plant**. **1a:** historically, any plant species, the individuals of which accumulate, from the environment, relatively high concentrations of a specific chemical element (e.g., a heavy metal) in their tissues. **1b:** any plant that is an accumulator (def. 1a). **annual plant**. a plant that completes its life cycle in one growing season, a term usually applied only to herbaceous plants. **autotrophic plant**. any plant that synthesizes all organic compounds from inorganic substrates, usually by photosynthesis. *Cf.* autotrophic, holotrophic, holozoic. *See also* autotrophic organism, photoautotrophic organism. **bee plant**. *Cleome serrulata*. **biennial plant**. a plant that matures and dies after two seasons; usually applied only to herbaceous plants. *See also* annual plant, perennial plant. **food plant**. any plant, some part of which is edible by wild or domestic animals or by humans (with or without preparation). Many food plants have poisonous parts, poisonous life stages, or require special preparation to assure safety. Avocado leaves, apple or pear seeds, rhubarb leaves, apricot, cherry, peach, and plum pits, tomato leaves, and the outer green husks of walnuts can all have toxic effects if enough is in gested. Even a small amount of green potato sprouts can be very toxic. *See also* garden plant, houseplant. **garden plant**. any plant, other than an agricultural crop, that is typically cultivated in a garden or out-of-doors as an ornamental, for use in landscaping, or occasionally for food. Some or all parts of numerous garden plants are pharmacologically active or poisonous when ingested or in some cases touched. Many are deadly. **green plant**. any plant that contains chlorophyll; nearly all are autotrophic. **hardy annual plant**. an annual outdoor plant that grows readily from seed even during cold winters and whose seeds can survive cold winters or significant periods of drought. **hedge plant**. **1:** a tree, shrub, and sometimes even a rapidly growing annual plant that can form a continuous close growth or hedge. A number of these plants are poisonous. *See*, for example, *Taxus* (yew), *Pachysandra procumbens* (box), *Hydrangea* (hydrangeas), and *Ligustrum* (privet). **2:** *Ligustrum vulgare*; *Dieffenbachia seguine*. **house plant**. *See* houseplant. **perennial plant**. any plant that lives more than 2 years. *See also* annual plant, biennial plant; usually applied only to herbaceous plants. Some tend to live indefinitely, others just a few years. **photoautotrophic plant**. **pie plant**. *Rheum rhabarbarum* (rhubarb). **weather plant**. *Abrus precatorius*.

plant hormone. phytohormone; auxin.

plant louse. *Phylloxera*.

plant pathology. phytopathology.

plant-derived pesticide. botanical pesticide *See* botanic.

plant-derived insecticide. botanical insecticide. *See* botanic.

Plantae. the plant kingdom; usually defined to include all green plants (i.e., all organisms that contain chlorophyll). In many classifications, the fungi and bacteria are also included. Organisms included in this kingdom are usually easily distinguished from members of the animal kingdom (Animalia). Most are autotrophic; they are typically attached to a substrate and unable to move freely about (exceptions are fungi and certain flagellate algae). In addition to a semipermeable cell membrane, most plant cells are surrounded by a cell wall of cellulose; starch is a common storage polysaccharide.

plasmid. a circular piece of DNA in which the two ends of the original DNA strand are covalently bound. It is not essential to growth.

pleurisy root. *Asclepias tuberosa*.

plumeless thistle. *Carduus*.

PMA. phenylmercuric acetate.

PMAC. phenylmercuric acetate.

PMAS phenylmercuric acetate.

pneusis. breathing.

PNS. peripheral nervous system.

Poa (meadow grasses; speargrasses; bluegrasses). a genus of grasses (Family Gramineae) native to North America that may become heavily ergotized. *See Claviceps*. ***P. ampla*** (big bluegrass). a grass (Family Gramineae) found in Yukon, southward to California and New Mexico. It is used to reseed rangeland. ***P. annua*** (annual bluegrass; annual meadow grass; early meadow grass; low spear grass; annual poa; winter grass; goose grass). an annual forage grass (Family Gramineae) that is indigenous to Eurasia. It is considered a pest plant when it occurs in good turf. ***P. bulbosa*** (bulbous bluegrass; winter bluegrass). a perennial grass, native to Europe, sometimes found in pas tures, that has bulb-like growth.

Poaceae. *See* Gramineae.

podzol (podsol). a generally poor type of nutrient-poor, acid soil in which the top layer has a bleached washed-out appearance due to the loss of iron, aluminum oxides, and/or humus by leaching into a lower layer which is brightly colored due to the presence of iron.

poet's narcissus. *Narcissus poeticus*.

poikilotherm. ectotherm.

poikilothermic. ectothermic.

poinsettia. *Euphorbia pulcherrima*.

point locoweeds. any plant of the genus *Oxytropis*. 2

pointvetch. any plant of the genus *Oxytropis*.

pois vivace (Quebec)). a poisonous pea. *See Lathyrus*.

poison. **1:** a substance that injures, impairs, or kills a living system, a system component, or disrupts the normal maintenance and perpetuation of the system by virtue of its chemical or physical interaction with the metabolic processes of the system. **2:** any substance which, in small amounts, is injurious to health or endangers the life of a living organism when in contact with, absorbed by, or developed within the body of the organism. All chemicals are poisonous at some level of exposure. **3:** any substance that endangers the fitness of a species by virtue of its chemical or physical interactions with their metabolic processes. **contact poison**. any poison that acts on contact with the skin or outer surface of a living organism and does not require another exposure route to have its full effect. *See also* insecticide (contact insecticide). **cumulative poison**. a toxicant that is not normally eliminated from the body or is so slowly eliminated that continued or repeated exposure to nontoxic amounts or concentrations may result in a build-up to toxic levels. In some cases, there may be cumulative subclinical effects that eventually result in damage, even though the poison is normally eliminated. **ecosystem poison**. any toxicant such as certain herbicides and pesticides (e.g., chlorinated hydrocarbons) that can damage or impair ecosystems or key ecosystem components. Such substances (organochlorine pesticides such as DDT) often affect a broad spectrum of organisms. **environmental poison**. any toxicant, usually of anthropogenic origin, that becomes dispersed in the environment in quantities at least large enough and dispersed enough to poison or burden large or significant components of the environment such as an air shed, watershed, or one or more water bodies (e.g., an aquifer, a lake, or a stream). **mitotic poison**. a poison that interrupts or disrupts cell division in plants or animals. **mouse poison** (mouse killer). any of a large variety of poisons used to poison *Mus musculus* (the house mouse). Mouse and rat poisons are among the most harmful pesticides available for use in the home. They may contain arsenic, strychnine, or phosphorus, all of which are rapid-acting and lethal when ingested. **nerve poison**. any toxicant that acts directly on nervous tissue. **pesticidal poison**. a poison whose toxic properties are commercially exploited to kill pests. **plant poison**. any poison of plant origin. **rat poison**. any of a large variety of poisons and special concoctions (e.g., warfarin) that are used to poison the domestic or Norway rat, *Rattus norvegicus*. They may contain arsenic, strychnine, or phosphorus, all of which are rapid-acting and lethal when ingested. Rat and mouse poisons are among the most dangerous pesticides available for home use. **stomach poison**. a poison (usually a pesticide) that is active only via the oral route (e.g., stomach insecticide). **systemic poison**. **1a:** a poison that produces generalized or whole-body effects. **1b:** a poison that produces toxic effects in major parts of an organism that are remote from the site of contact between the organism and the toxicant. *Cf.* toxicosis. **vascular poison**. a poison that acts on vessels of the blood vascular system.

poison ash. *Rhus vernix*.

poison bean. a plant of the genus *Sesbania*.

poison creeper. *Rhus radicans*.

poison dogwood. *Rhus vernix*.

poison elder. *Rhus vernix*.

poison hemlock. *Conium maculatum*.

poison ivy. **1:** the common name of *Rhus radicans*. **2:** correctly applied only to *R. radicans* but often used interchangeably for *R. diversiloba* (Pacific poison oak), *R. radicans* (poison ivy), and *R. toxicodendron* (poison oak). *See Rhus*. **3:** *Rhus* dermatitis (poison ivy dermatitis).

poison oak. **1:** the common name of *Rhus toxicodendron*. **2:** often used interchangeably for *R. diversiloba* (Pacific poison oak), *R. radicans* (poison ivy), and *R. toxicodendron* (poison oak). *See Rhus*. **3:** *Rhus* dermatitis (poison oak dermatitis).

poison paxillus .*Paxillus involutus*.

poison suckleya. *Suckleya suckleyana*.

poison sumac. *Rhus vernix*.

poison vine. *Rhus radicans*.

poisonvetches. *See Astragalus*.

pokeberry. a plant of the genus *Phytolacca*.

pokeweed. a plant of the genus *Phytolacca*

pokeweed family. Phytolaccaceae.

polioclastic. neurotropic.

pollen. microspores produced in vast numbers in the anthers of flowering plants. They are essential to sexual reproduction. They form a fine dust in the atmosphere and are considered a natural or background pollutant. They are the cause of pollinoses.

pollen sac. any of the cavities in an anther in which pollen is produced.

pollenosis. pollinosis.

pollinosis (plural, pollinoses; pollenosis). congestion of the mucous membranes of the nasal passages due to contact with pollen; hay fever. *See also* hypersensitiveness.

pollutant. any substance or form of energy, of natural or anthropogenic origin, that is present in the environment in greater than natural concentrations or amounts due to human activity and which has a net detrimental or undesirable effect on humans, the environment, or upon something of value to humans in the environment. Most are a matter of concern because of their toxicity. **air pollutant**. any pollutant in any form or chemical phase that occurs in air. **water pollutant**. any pollutant that occurs in fresh or marine water bodies (lakes, streams, reservoirs, oceans) regardless of the route or mode of entry.

polybrominated biphenyl (PBB; polybromobiphenyl). any of a class of brominated hydrocarbon compounds used as fire retardants and components of heat resistant plastics. They are chemically related to polychlorinated biphenyls. Massive livestock poisonings occurred in Michigan in 1973 due to the accidental addition of about two tons of PBB flame retardant to livestock feed during its formulation. Some 500 cattle farms and 150 dairy farms were placed under quarantine. As a precautionary measure, 30,000 cattle and swine were killed. Signs exhibited by cattle exposed to the contaminated feedstock usually included salivation, lacrimation, diarrhea, inappetence, dehydration, weight loss, alopecia, abnormal growth of hooves, decreased milk production, and teratologies; gravid cows aborted. Autopsy revealed lesions of the kidney, liver, gallbladder, and thymus. Failing and moribund animals suffered extensive subcutaneous edema and hemorrhage.

polybromobiphenyl. polybrominated biphenyl.

polybrominated biphenyl mixture (Firemaster FF-1). this mixture has been shown to be carcinogenic in some laboratory animals.

polychlorinated biphenyl (chlorobiphenyl; chlorinated diphenyl; chlorinated diphenylene; polychlorobiphenyl; PCB; Aroclor; Clophen; Fenclor; Phenoclor; Pyralene; Santotherm). any of a series of about 200 nonflammable, highly persistent organic liquids derived from biphenyl (a compound with two linked benzene rings). Acute toxicity is generally low but PCBs are of concern because of their persistence in the environment and in body fat. They are very stable, have become widespread in the environment, and considerable concern has been expressed over their effects on human health, wildlife, and the environment. Effects on animals due to ingestion of various PCBs include hepatic neoplasia, hepatic porphyria, alterations in mixed function oxidase systems, liver hypertrophy, increased bile flow, reproductive dysfunction (in birds), and reproductive failure (in mink, probably in fish). PCBs cause eggshell thinning in birds and have thereby contributed to the extinction or near extinction of certain birds.

polychlorobiphenyl. polychlorinated biphenyl.

polychlorocamphene. toxaphene.

Polygonaceae (buckwheat family). a family of widely distributed shrubs, trees, and erect or climbing herbs. A number of species are cultivated as ornamentals. *See Fagopyrum sagittatum*, *Polygonum*.

Polygonum (smartweeds). a large, widely distributed (rare in the tropics) genus of annual or perennial herbs and small shrubs (Family Polygonaceae). At least some species are photosensitizing. Toxic concentrations of nitrate have been reported for some.

polyp glucosaminidase. chitinase.

pome. a fleshy fruit such as an apple or pear with a fleshy outer portion and a papery core with seeds.

ponasterone. any of four polyhydroxylated steroids isolated from *Podocarpus nakaii* (Family Podocarpaceae). They exert potent ecdysone activity on houseflies, silkworms, and other insects. Ingested plant material containing as little as 1 μg/kg ponasterone can be lethal to caterpillars of the cecropia moth (*Samia cecropia*). *See also* ponasterone A, phytoecdysone.

ponasterone A ((2β,3β,5β,22R)-2,3,14,20,-22-pentahydroxycholest-7-en-6-one; 25-deoxyecdysterone). the first of four ponasterones (A, B, C, D) to be isolated. It is highly toxic to caterpillars of the cecropia moth (*Samia cecropia*). *See* ponasterone.

ponderosa pine. *Pinus ponderosa*.

poor Annie. *Veratrum viride*.

poppy. any plant of the genus *Papaver*.

poppy family. Papaveraceae.

poppy opium. *Papaver somniferum*.

poppy seed. *Papaver somniferum*.

porcine amelia. *See* amelia.

pork belly. a side of fresh pork, used to produce bacon.

positive electrode. anode.

positive taxis or tropism. a tendency to move or grow toward the source of the stimulus.

post-. a prefix indicating location behind or occurring after.

potassium acid arsenate. potassium arsenate.

potassium arsenate (potassium acid arsenate; potassium dihydrogen arsenate; Macquer's salt). a toxic, colorless, water-soluble, ethanol-insoluble crystalline compound, KH_2AsO_4. It is used as an insecticide, an analytic reagent, in textile printing, and in the preservation of hides. It is a strong irritant and is toxic by ingestion or inhalation.

potassium chloride (potassium muriate). a colorless, salty tasting, water-soluble crystalline compound, KCl. It is used for example in pharmaceuticals, as a fertilizer, and a food additive. Large oral doses can cause gastrointestinal and cardiovascular problems. Near-fatal poisonings are known.

potassium cyanate. a very toxic, white, water-soluble, crystalline powder, KOCN, used as an herbicide.

potassium dihydrogen arsenate. potassium arsenate.

potassium iodide (KI). white, water- and ethanol-soluble crystals, granules, or powder, KI, with a bitter, salty taste. It is used, for example, in feed additives for livestock and as a dietary supplement in table salt at a concentration of up to 0.01%. KI is teratogenic and a cause of congenital goiter.

potassium muriate. potassium chloride.

potassium nitrate (niter; nitre; saltpeter). a colorless or white, slightly hygroscopic, crystalline powder, KNO_3, with a pungent salty taste. It is a strong oxidizing agent and there is considerable risk of fire or explosion if it is shocked, heated, or comes into contact with organic matter. Potassium nitrate is used, for example, in fertilizers.

potato. *Solanum tuberosum*.

potato family. Solanaceae.

potato poisoning. poisoning due to the ingestion of potatoes (*Solanum tuberosum*) or potato parts (the skin or green sprouts) that contain high levels of solanine, *q.v.* Potatoes normally contain about 7 mg of solanine per 100 g of potato; 20-25 mg/100g is toxic. Symptoms of intoxication may include headache, abdominal pain, nausea, vomiting, diarrhea, fever, and neurologic disturbances such as restlessness, drowsiness, apathy, confusion, stupor, visual disturbances, and hallucinations.

poultry husbandry. the management of poultry production, especially for meat.

prayer bean. *Abrus precatorius*.

precatory bean. *Abrus precatorius*.

precipité blanc. mercurous chloride.

pregnant. **1:** a synonym of gravid but applied most often to humans and sometimes domestic mammals. **2:** of or pertaining to a woman, or to the condition of a woman, who carries within her a developing embryo or fetus.

preparation K. dixanthogen.

prickly cucumber. *Echinocystis lobata*.

primary producer. a green plant or any microorganism that can convert light or chemical energy into organic matter.

prime mover. *See* agonist (def. 1b).

pro-. a prefix meaning prior to, previous to, the precursor of, in front of, for, or forward.

procambial strand. procambium.

procambium (procambial strand). meristematic tissue from which vascular bundles develop.

procaryote. prokaryote.

procedure. a particular methodical way, usually consisting of several steps, whereby a desired result may be accomplished.

process. **1:** a mechanism, means, method, procedure, or mode of action used to achieve a certain result. **2:** the progress of a disease. **3:** an anatomical projection or outgrowth as from an organ or tissue such as bone.

prokaryote (procaryote; prokaryotic cell). a simple unicellular organism (protist) in which there is no true nucleus; the genetic material is thus not separated from the cytoplasm by a membrane. They have very few subcellular structures and are further characterized by a high rate of reproduction (cell division). Included are bacteria and cyanobacteria. They are used in a variety of mutagenicity tests. *See also* eukaryote.

producer. *See* autotrophic organism, primary producer.

2-propenal. acrolein.

proestrus. *See* estrus.

progesterone (pregn-4-ene-3,20-dione; Δ^4-pregnen-3,20-dione; corpus luteum hormone; luteohormone; progestational hormone; progestin). a steroid hormone produced naturally by the corpus luteum of the ovary, the adrenal gland, and the maternal placenta. It is also produced synthetically. In mammals generally, it functions in the maintenance of gestation, suppresses ovulation, and stimulates development of the mammary glands and that of the maternal placenta following implantation. In veterinary practice, progesterone has been used to control habitual abortion and to suppress or synchronize estrus.

progestin. progesterone.

prognosis. a forecast, especially of the probable future course of a disease including the prospect for recovery or other outcome.

prokaryotic. of, pertaining to, or resembling prokaryotes; lacking the organelles found in more complex cells (eukaryotes).

prokaryotic cell. prokaryote.

prometon. an atrazine used only as a nonselective herbicide for the control of most annual and perennial broadleaved and grass weeds on non-crop areas. *See* triazine herbicide.

prometryne. an atrazine herbicide, similar in biological properties to atrazine and simazine but absorbed through foliage as well as the roots. *See* triazine herbicide.

prongbuck. *Antilocapra americana*.

pronghorn. *Antilocapra americana*.

pronghorn antelope. *Antilocapra americana*.

propachlor (2-chloro-*N*-(1-methylethyl)-*N*-phenylacetamide; 2-chloro-*N*-isopropylacetanilide; *N*-isopropyl-α-chloroacetanilide). a tan powder, $C_{11}H_{14}NOCl$, that is soluble in ethanol and benzene. It is a preemergence selective herbicide that is absorbed through the roots and inhibits root elongation by inhibiting cell division by influencing RNA and protein synthesis. It is moderately toxic to laboratory rats. Other soil-applied amides with similar action include alachlor, diphenamid, and naptalam.

propanethiol. 1-propanethiol.

1-propanethiol (*n*-propyl mercaptan; 3-mercaptopropanol; propyl mercaptan; propanethiol). a highly flammable oil, C_3H_8S, with a disagreeable odor and a boiling range of 67-73°C, used as an herbicide. It is moderately toxic, orally, to humans, mildly toxic when inhaled, and is a severe eye irritant. Toxic fumes, SO_x, are released on heating to decomposition.

propanil (*N*-(3,4-dichlorophenyl)propanamide; 3′,4′-dichloropropionanilide; *N*-(3,4-dichlorophenyl)propionamide). a postemergent contact anilide herbicide; applied to the foliage, it rapidly degrades in the soil. It is used mainly to weed umbelliferous crops such as celery, carrots, and parsley and also in rice and potatoes. Damage is in the form of localized or general necrosis, depending on the dose applied. It also inhibits growth but this is probably a secondary effect. Propanil is moderately toxic to laboratory rats.

propanone. acetone.

2-propanone. acetone.

propenenitrile. acrylonitrile.

2-propenenitrile. acrylonitrile.

2-propenyl isothiocyanate. allyl isothiocyanate.

propham (phenylcarbamic acid 1-methylethyl ester; carbanilic acid isopropyl ester; *N*-phenyl isopropyl carbamate; isopropyl carbanilate; *O*-isopropyl *N*-phenyl carbamate; INPC; IPC; IsoPPC). an aromatic carbamate herbicide applied to soil by spraying. It is generally effective against monocotyledons only, and does not control crabgrass (*Digitaria spp.*). It is moderately toxic to laboratory rats. It causes contraction of the plant chromosomes, is an inhibitor of gibberellin-induced α-amylase synthesis in barley endosperm, and thus probably inhibits gene derepression (i.e., the onset of metabolic changes) but not processes that are already in progress. It causes mitotic aberrations (e.g., anaphase bridges, blocked metaphases, nuclear fragments, giant vesiculate nuclei, increased numbers of chromosomes) in certain root and shoot cells of *Avena* and *Allium*. *See also* chlorpropham.

propoxur (2-(1-methylethoxy)phenol methylcarbamate; *o*-isopropoxyphenyl-*N*-methylcarbamate; 2-isopropoxyphenyl-*N*-methylcarbamate; methyl-2-isopropoxyphenylcarbamate; aprocarb; isocarb). a toxic, white, crystalline powder,

$$(CH_3)_2CHOC_6H_4OOCNHCH_3$$

It is soluble in water and most polar solvents. It is a dangerous carbamate insecticide, molluscicide, and is a component of some household products to eradicate pests. It is toxic by ingestion and inhalation. It inhibits cholinesterase but injection of atropine sulfate readily reverses the inhibition. Symptoms of empoisonment are similar to those of organophosphate poisoning. Oral toxicity may vary considerably. The oral LD_{50} in laboratory rats has been reported as 95 mg/kg body wt but >800 mg/kg in domestic goats. *See also* insecticide (carbamate insecticide).

propyl mercaptan. 1-propanethiol.

***n*-propyl mercaptan**. 1-propanethiol.

protein. any of a large variety of complex nitrogenous macromolecules, essential for life, that occur in all living cells and in biological fluids such as blood plasma. They are composed of one or several long polypeptides composed of α-amino acids, R—CH(NH_2—COOH), that are connected by peptide linkages (—CO·NH—) which are formed by elimination of H_2O between the NH_2 group and COOH group of successive amino acid residues. R is hydrogen or a ring or chain of carbon atoms.

proton number (symbol, Z; atomic number). the number of protons in the nucleus of an atom. The electron structure, which determines the nature of chemical bonding, depends upon the degree of electrostatic attraction of the nucleus. The latter is a function of the proton number which is thus the prime determinant of an element's chemical behavior.

Prunus. a genus of mostly deciduous, small trees and shrubs (Family Rosaceae) of the North Temperate Zone and the Andean Mountains of South America. Most species have edible fruits and many are cultivated. ***P. amygdalus*** (synonym, *Amygdalus communus*; common almond). an edible almond but see *P. amygdalus amara*. ***P. amygdalus amara*** (synonym, *Amygdalus communis amara*, bitter almond). a variety of common almond that has a very bitter seed; it contains amygdalin and benzaldehyde. The latter compound confers a characteristic odor to the nut. *See* amygdalin. ***P. armeniaca*** (apricot). a tree grown for its edible fruit. The pits contain amygdalin. ***P. caroliniana*** (cherry laurel). a lovely, shiny-leaved, poisonous wild cherry which may grow to a height of more than 40 feet. The leaves, seeds, and bark contain amygdalin. ***P. laurocerasus*** (European cherry laurel). a species that is very similar to the American species. The leaves, seeds, and bark contain amygdalin. ***P. persica*** (peach tree). a widely cultivated fruit tree native to China. The roots contain large amounts of amygdalin. ***P. serotina*** (black cherry; wild black cherry; wild cherry; black choke). an important timber tree, the bark, leaves, and especially the seeds of which are toxic. Symptoms of intoxication from chewing the seeds include stupor, vocal cord paralysis, twitching, convulsions, and coma. Amygdalin is the source of the hydrocyanic acid that produces these effects. This species is also known to be a cause of arthrogryposis in terrestrial vertebrates.

prussic acid. hydrogen cyanide.

pseudoanaphylactic. anaphylactoid.

Pseudomonas tabaci. a bacterium (Family Pseudomonadaceae) that contains wildfire toxin and is the causative organism of wildfire disease in tobacco plants. *See* toxin (wildfire toxin).

pseudorabies. Aujesky's disease.

Pseudotsuga menziesii (douglas fir; douglas pine; Douglas spruce; Oregon fir; Oregon pine). a coniferous tree of western North America, often reaching a height of more than 200 ft. (60 m). The bark is reddish-brown bark, the needles flattened and narrow, cones are light-brown cones. The timber produced is strong and durable. It is the state tree of Oregon.

Psilostrophe (paperflowers). a genus of small, toxic, bushy, wooly herbs with a perennating taproot (Family Compositaceae) that occur in open dry rangelands of the western United States. They have caused substantial losses of sheep. Afflicted sheep begin to show some incoordination when running. They become sluggish and anoretic and often cough violently, with or without vomiting. The animals become progressively depressed and emaciated and eventually die. Lesions are generally limited to some albuminous degeneration of renal tubules.

psilostrophe. common name of a plant of the genus *Psilostrophe*.

psoralen (7*H*-furo[3,2-g][1]benzopyran-7-one; 6-hydroxy-5-benzofuranacrylic acid δ-lactone; furo[3,2-g]-coumarin; ficusin). one of a class of furocoumarin phytoalexins, $C_{11}H_6O_3$, that occur in many species of plants of the families Rutaceae (e.g., bergamot, cloves, limes), Apiaceae (e.g., celery, parsnips), Fabaceae (e.g., *Psoralen coryfolia*), and Moraceae (e.g., figs). Psoralins are defensive against attacks by insects and fungi. They are photosensitizing and phototoxic not only to insects but to laboratory animals and humans also. They have been used in photochemotherapy to treat psoriasis, vitiligo, and mycosis fungoides.

Pteridium (bracken; bracken fern; female fern; fern brake). a genus of moderately toxic perennial ferns (Family Polypodiaceae) that occur chiefly in open woods, upland pastures, or abandoned fields on dry, sandy soils. ***P. aquilinum*** (eastern bracken; brake fern; eagle fern). a species largely of the northeastern United States and Canada but also widely distributed over the earth. Bracken is sometimes used as fodder or bedding for livestock, even though it is poisonous. Poisoned cattle exhibit a high temperature, dyspnea, excessive salivation and hemorrhage. Found chiefly in northeastern United States and Canada. Also called brake fern, eagle fern, bracken. *See also* thiaminase.

Pteridium. a perennial fern that grows widely on acid soils. It occurs mainly on heaths and upland areas.

PTU. 1-phenyl-2-thiourea.

ptyal- bial-, sialo-, siali-. prefixes denoting saliva.

Puccinia dispersa. a widespread rust (Family Pucciniaceae) that causes brown leaf rust (*See* leaf rust) in rye.

puff adder. *Heterodon contortrix*. *See also* adder.

pulmonary mite. *Cytodites nudus*.

pulses. *See* Fabaceae.

Pultox. lindane.

pump. **1:** a device that transfers fluids from one location to another by creating a pressure differential between the two locations (e.g., by suction). **2:** to force a fluid along a particular pathway by creating a pressure differential along the pathway.

puncture vine. *Tribulus terrestris*.

pupillary dilation. mydriasis.

purgation. catharsis.

purgative. cathartic.

purine. an organic base present in the nucleotides and nucleic acids of plant and animal tissue. Adenine and guanine are purines that are nearly universally present in living systems.

purple cockle. *Agrostemma githago*.

purple cudweed. *Gnaphalium purpureum*.

***Purshia tridentata* (bitterbrush)**. a valuable forage plant (family Rosaceae) of the high mountains of the western United States

putrefaction. *See* decay.

pyrazon. an herbicide, it is a plant growth inhibitor with a tendency to produce abnormal plants. The most prominent biochemical effect is a significant reduction in the rate of photosynthesis.

Pyrausta nubilalis (European moth; corn borer moth). a moth, native to Europe, and now common in many areas of eastern North America. The larvae (corn borers) feed on and destroy corn and certain other plants such as beans and potatoes. *See* corn borer.

pydrin. fenvalerate.

***N*-3-pyridylmethyl-*N'*-*p*-nitrophenylurea**. pyriminyl.

Pyralene. polychlorinated biphenyl.

pyrazon. a pyridazine herbicide, it is a plant growth inhibitor with a tendency to produce abnormal plants. The most prominent biochemical effect is a significant reduction in the rate of photosynthesis.

pyrectic. febrile.

pyrethrin (pyrethrum (in part); pyrethrum insecticide; pyrethrum flowers; pyrethrum extract; Dalmatian insect powder; Persian insect powder). **1:** a botanical comprised of several esters extracted from the flowers of three species of chrysanthemum (Family Asteraceae) using hot carbon dioxide under high pressure. Pyre-thrin is rapidly degraded photochemically in the environment. The most intense reactions in humans and other terrestrial vertebrates result from inhalation but poisoning may occur through ingestion or absorption through the skin. In serious acute poisoning, the first sign is dermatitis, which may appear within half an hour of exposure; the skin becomes red and burns and itches at the site of contact or more generally; the cheeks soon swell. The initial effects may be followed by progressive hyperexcitability, loss of coordination, tremors, nausea, vomiting, diarrhea, convulsions, and muscular paralysis. Death from respiratory paralysis may intervene. Less intense exposure may result in headache, gastrointestinal distress, and paresthesia of the lips and tongue. When removed from exposure, the victim's symptoms may require two days to two weeks to subside. Commercial sprays used as a plant insecticide usually contain about 0.1% pyrethrin and may include other active agents. Though seldom used now, pyrethrin has been used in concentrated form as a dust to kill fleas on animals. **2:** any of six insecticidal allethrin analogs extracted from pyrethrum flowers, *q.v.*: cinerins I and II, jasmolins I and II, and pyrethrins I and II, *q.v.* Pyrethrins I and II comprise 71% of the extract. All of these compounds are highly toxic (to the target organisms), fast-acting paralytic, broad-spectrum contact insecticides. They are only moderately toxic to laboratory rats, *by mouth*. They act primarily by slowing the opening and closing of sodium ion channels in nerve membranes with a consequent increase in Na^+ permeability and depolarization. Symptoms of acute intoxication in humans are those of gastroenteritis and CNS disturbance and may include diarrhea, nausea, vomition, tinnitis, headache, numbness of tongue and lips, syncope, hyperexcitability, incoordination, convulsions, prostration, and death due to respiratory paralysis. These substances are also hepatotoxic and nephrotoxic and can cause severe allergic dermatitis systemic allergic reactions. *Cf.* pyrethrum.

Pyrethrum. a former genus (Family Asteriaceae), the species of which are now included in the genus *Chrysanthemum*. *See* pyrethrum flowers. *Cf. Anacyclus pyrethrum.*

pyrethrum (in part). pyrethrin.

pyrethrum extract. pyrethrin.

pyrethrum flowers (Dalmatian insect powder; Persian insect powder). **1:** the dried flowers of *Chrysanthemum (Pyrethrum) cinariifolium* and *C. coccineum* (= *C. roseum*), both indigenous to Dalmatia, Montenegro, and western Asia, and from those of *Anacyclus pyrethrum* which is native to Morocco. The powdered flowers or the extract (pyrethrum) are used in insecticidal preparations, as a scabicide, and in veterinary practice as an ectoparaciticide. **2:** pyrethrin. *See also* pyrethrum.

pyrethrum insecticide. pyrethrin.

pyretic. febrile.

pyrexia. fever.

β-pyridyl-α-*n*-methylpyrrolidine. nicotine.

2-(3-pyridyl)piperidine. anabasine.

pyrimidine. the parent compound of a number of heterocyclic nitrogenous compounds, $C_4H_4N_2$, including the bases cytosine, thymine, and thymine.

pyriminyl (N-(4-nitrophenyl)-*N'*-(3-pyridinylmethyl)urea; *N*-3-pyridylmethyl-*N'*-*p*-nitrophenylurea; pyrinuron). an extremely hazardous substance,

$$C_{13}H_{12}N_4O_3$$

pyriminyl is used as a rodenticide. It is extremely toxic to humans, laboratory rats, and feral Norway rats by ingestion. Systemic effects include distorted perceptions, hallucinations, muscular weakness, nausea and vomiting. When heated to decomposition, toxic NO_X fumes are released.

pyrinuron. pyriminil.

pyroacetic ether. acetone.

pyrolan® (dimethylcarbamic acid 3-methyl-1-phenyl-1*H*-pyrazol-5-yl ester; 3-methyl-1-phenyl-5-pyrazolyl dimethylcarbamate; 1-phenyl-3-methyl-5-pyrazolyl dimethylcarbamate; 3-methyl-1-phenyl-5-pyrazolyl dimethylcarbamate). a crystalline substance,

$$C_{13}H_{15}N_3O_2$$

It is a reversible cholinesterase inhibitor that was formerly used as an insecticide. pyrolan® is very to extremely toxic by ingestion to humans and other higher animals. Though more toxic, the effects of pyrolan® on humans are presumably similar to those of carbaryl.

Pyrus (pear trees). a genus of mostly deciduous trees (Family Rosaceae) native to cool-temperate regions; included are a number of valuable fruit trees and ornamentals. The leaves of some species contain arbutin. ***P. malus*** (apple). sometimes given as the scientific name for the common apple. The presumed ancestral species of the modern common apple tree are *Malus communis*, *M. pumila*, and *M. domestica*. *Malus* is the botanical name for all apples. Scientifically, in this author's view, *Malus* should be treated as a subgenus of *Pyrus*. Furthermore, because of the long history of domestication and hybridization, a species designation for the common (domesticated) apple is not warranted.

Q

q.v. (Latin for *quo vide*). which see.

quack grass. *Agropyron spp*.

quadrat. **1:** a small square or rectangular plot of land, usually 1 meter on a side that is marked off and used, for example, to as one of a number of sampling plots to study the distribution of small plants and animals.

quadruped. a four-footed animal.

qualitative. **1a:** pertaining to kind or type. **1b:** descriptive in non-numerical terms. In general, qualitative differences, determinations, or assessments are those based on two or more discrete values or states of a system that cannot be represented numerically. Thus, large dose/small dose, large amount/small amount, highly concentrated/moderately concentrated are qualitative comparisons; large overdose, nontoxic are qualitative judgements or assessments. *Cf.* quantitative.

qualitative analysis. chemical analysis (*q.v.*) to determine those elements and compounds that are present in a sample of material of unknown composition.

quality grain. good quality grain, usually indicated by a high specific weight.

quantal effect. *See* effect.

quantitative. **1a:** pertaining to amount or degree. **1b:** descriptive in numerical terms. A quantitative observation, measurement, estimate, or inference is one based on precise numbers. To be quantitative an observation must be measurable, based on measurements, or based on ratios or other numerical values. This term is generally used in reference to a property that can be characterized or defined in terms of measurements, counts, ratios, or other numerical values. The amount or concentration of a substance given in precise or exact numerical terms is quantitative. *Cf.* qualitative.

quantitative analysis. chemical analysis (*q.v.*) to determine the amount by weight of each element or compound present in a sample of material of quantitatively unknown composition.

quarantine. **1:** isolation of a person, plant or animal from others of the same population, specied, herd, etc. for a period of time to ensure that such are free from transmissible disease. **2:** with respect to agriculture, quarantine is used **a:** as a restraint on importing specified plants or animals from areas of endemic disease of agricultural pests. **b:** to isolation of animals or plants suspected of carrying pests or diseases to prevent their spread to others. In the United States, both states and the federal government may impose quarantines or operate quarantine facilities. **3:** to effect a quarantine as defined above.

quart (abbreviation, qt). **1a:** a unit of liquid measure equal to 1/4 gallon or 0.9468 liters. **1b:** a dry measure used, for example, to measure dry goods such as grain. It holds somewhat more than the fluid measure. **imperial quart**. a unit of liquid measure that contains about 1.1359 liters.

quarter evil. *See* blackroot (def. 1).

quarter horse. *See* horse (American quarter horse).

quebracho. *Aspidosperma*.

queen. a female domestic cat of breeding age.

"**queen mother of poisons**." *See* aconite, *Aconitum*.

Quellada. a trade name for lindane.

Quercus (oak). a genus of perennial, woody, deciduous trees and shrubs (Family Fagaceae). *See* oak poisoning, acorn poisoning.

quercus poisoning. oak poisoning. *See also* acorn poisoning.

quicklime. calcium oxide.

quicksilver. the element mercury.

***N'*-(2-quinoxalinyl)sulfanilamide**. sulfaquinoxaline.

qt. quart.

R

®. the symbol for a registered trademark.

r. former abbreviation for Roentgen.

R. abbreviation for gas constant; electrical resistance; radical (usually an alkyl or aryl group, thus ROH is an alcohol); respiration; respiratory exchange ratio; Roentgen; remainder of a chemical formula; the unit of resistance in the cardiovascular system.

rabbit brush. tetradymia.

rabbitbrush. common name for plants of the genus Chrysothamnus.

rabies. an often fatal notifiable viral disease of mammals, especially carnivores. It is transmitted to humans by infected animals. Animals commonly infected, other than humans, include bats, cats, dogs, foxes, raccoons, skunks, and cattle. The disease is usually transmitted by biting but is also known to be transmitted to humans by bats without a direct bite. In this case, transmission is assumed to be via inhalation of virus-laden aerosols.

race (subspecies; variety). a subspecies or a geographic subdivision of a species. A geographically defined group of breeding populations that differs from other similar groups in the frequency of one or more heritable traits.

raceme. a rather elongated and slender inflorescence in which the pedicels attach to a simple central axis.

racemose gland. any gland that would have the appearance of a bunch of grapes if viewed three dimensionally as in a reconstruction.

racemose inflorescence. *See* inflorescence.

radical. **1:** a plant part that arises from a root close to the ground such as a basal leaf or the stem of a flower. **2:** an ionic group with one or more positive or negative charges (e.g., OH^-, NH_4^+, SO_4^{--}). Such groups do not exist in the free state but as a unit in a compound. **free radical**. a usually short-lived, highly reactive fragment formed by the splitting of a molecular bond. It has one or more unpaired electrons. Such a species is symbolized in chemical formulas by a dot indicating that each has an unpaired electron ($(C_2H_5)^{\bullet}$, $Cl^{\bullet}$). Even though short-lived, free radicals initiate many kinds of chain reactions. Such a reaction occurs when free radicals attack a substrate and are regenerated during the progress of the reaction. Free radicals form, for example, in tissues exposed to ionizing radiation and are thus a proximate cause of radiation damage that may occur. *See also* active oxygen, superoxide.

radicle. an embryonic root.

radio-labelled. pertaining to a substance to which a radioactive tracer has been added.

radioactive. of, pertaining to, or having the quality of radioactivity. **radioactive substance**. any element, compound, or mixture that exhibits radioactivity. Many of these substances are extremely toxic or otherwise hazardous. Those who generally incur the greatest risks from exposure to radioactive substances are uranium miners, radiologists, and workers in the nuclear industry. Other potential sources of serious or even catastrophic exposure are nuclear accidents (e.g., core meltdown at a nuclear power plant), radioactive wastes, and nuclear warfare. **radioactive waste**. any waste that contains radioactive materials. Sources of such waste include (1) high level waste produced during the reprocessing of radioactive fuel following usage in a nuclear reactor, (2) transuranic waste resulting from the reprocessing and fabrication of plutonium to make nuclear weapons, and (3) low level wastes contained in the solutions, rags, clothes, and tools used to clean and decontaminate nuclear reactors. Nuclear wastes remain hazardous for extremely long periods of time. For example, plutonium 239, the fuel of breeder reactors, remains extremely hazardous for 24,000 years. Secure, long-term storage of radioactive wastes is a problem that has yet to be fully resolved.

radiotoxemia. an obsolete synonym of radiation sickness. *See* sickness.

radish. *See Raphanus*.

Radopholus similis. a nematode (Family Tylenchidae) that attacks a variety of economically important plants, including banana, citrus, sweet potato, and sugarcane.

rafter. to plow the ground so that the furrow is turned over onto the adjacent unplowed land.

ragwort. *Senecio*.

rain drop. appaloosa.

rain gun. a mechanism used to apply irrigation water under pressure to grasslands or sugar beets. It consists of a series of widely-spaced fixed or mobile nozzles.

rain lily. *Zephyranthes atamasco*.

rainfall. rain; the amount of precipitation from the atmosphere, usually during a specified period of time.

rainwater. water that has fallen as rain and has been collected and used for irrigation.

rale. any abnormal respiratory sound detected by auscultation.

Ralstonia. *See* bacterial wilt of corn.

ram. an adult male sheep or goat; one that has not been castrated.

Rambouillet sheep. a breed valued for high-quality wool, they are of medium to large size and are the largest and strongest of the fine-wool sheep; the rams are horned. The breed was developed in France from the Spanish Merino sheep. It has become very popular in the United States and is the foundation of most flocks of the western rangelands; the ewes are frequently bred to rams of coarser-wooled breeds. While prized for their wool, Rambouillets also provide quality mutton.

Ramona polystachya (white sage). *See* greasewood.

Rampart. phorate.

ranch. **1:** a large farm, usually on rangeland, that specializes in the production of cattle, sheep, or horses. **cattle range**. a large grassland farm on which cattle are raised.

ranch hand. *See* rancher.

rancher. the owner of or a worker (ranch hand) on a ranch.

rancid. having a disagreeable taste or odor due to partial decomposition. Used especially with respect to butter and certain other fats.

random error. *See* error.

range. **1a:** roughly equivalent to rangeland;

any land that supports vegetation suitable for grazing by wildlife or domestic livestock including grasslands, woodlands, shrublands, and forest lands. **1b:** sometimes restricted to open, extensive grasslands or scrub; open space.

range enhancement. any treatment, development, or structure implemented to achieve or maintain the desired plant community of a particular range, grassland, woodlands, shrubland, or prairie area that supports grazing animals.

rangeland. any of a broad category of land usually marked by the presence of native plant communities that are often associated with grazing animals. Rangelands are managed by ecological, not agronomic methods. *Cf.* range.

Ranunculus (buttercup; crowfoot). a very large genus of toxic herbs, the type genus of the family Ranunculaceae. All species contain a moderately toxic vesicant oil, protoanemonin, in varying amounts. All parts of the plant are toxic. Ingestion may produce a severe gastroenteritis in which salivation, diarrhea, and signs of abdominal pain are typical. Oral tissues may be inflamed. All forms of livestock are susceptible. Buttercups are distasteful to most livestock and poisoning is rare. Protoanemonin is unstable and buttercup hay is therefore nontoxic. *See also* anemonin, dermatitis (primary contact dermatitis), Ranunculaceae.

rape. **1:** *Brassica napus*. **2:** a usually violent crime associated with heterosexual or homosexual sex. In general usage and in law, this term is used in reference to humans. Some vertebrate animals, however, sometimes show similar behavior.

rape blindness. nervous syndrome of cattle and sheep, *q.v.*.

rape poisoning. poisoning by ingestion of *Brassica napus*, q.v. Symptoms in grazing ruminants include ruminal atony, adipsia, aphagia, and death unless suitable food and care are given.

rape scald. a phototoxic syndrome of livestock. Some species of *Brassica*, *Erodium*, *Medicago*, and *Trifolium* have been implicated.

rapeseed meal. a feed concentrate used extensively during World War II. It is toxic to livestock but its effects can be partly eliminated by hot water extraction of the meal or supplementation with iodine. *See Brassica napus*.

rapeseed oil. the oil expressed from the seeds of *Brassica campestris*. It is used in the manufacture of margarine, soaps, and lubricants.

rappel crew. an initial response crew that rappels from a helicopter to access fires.

rat. a common name that includes a wide variety of medium-sized rodents, such as members of the following genera: *Rattus, Sigmodon,* and *Dipodomys. See also* Cricetidae, Muridae. **albino rat**. **1:** a phenotype that sometimes occurs in various populations of *Rattus*. The fur is white and the eyes are pink due to a lack of pigment. **2:** any strain of laboratory rat inbred for albinism. **laboratory rat**. any rat bred and reared in captivity, originally derived from a small population of *Rattus norvegicus*, the Norway rat, for use in laboratory studies. There are many strains and lines of such rats, all of which are now genetically far removed from the wild type. While they can be considered congeneric with *Rattus*, they should not be characterized as *R. norvegicus*, nor referred to as the Norway rat. Laboratory rats no longer have any counterparts in nature. Laboratory rats are often used as test species in toxicology but often respond differently to a specific chemical than other mammals including humans and other rodents. Thus, for example, the adult laboratory rat is highly resistant to the action of cardiac glycosides and to the fetotoxic effects of thalidomide. **Wistar rat**. an inbred strain of laboratory rat that is homozygous at most loci. This strain is the result of strict brother-sister inbreeding over many generations to develop animals of closely similar genetic composition that are suitable for research.

rat killer. rat poison. *See* poison (rat poison).

rate-of-cut. the proportion of watershed area allowed to be cut each year.

ratsbane. arsenic.

rattlebox. a common name of herbs of the genus *Crotalaria*.

rattlebrush. a plant of the genus *Sesbania*.

rattlesnake. any snake of the genus *Crotalus*.

rattle-weed. *Baptisia tinctoria*.

rattleweed. *Crotalaria sagittalis*.

Raphanus (radishes). a genus of herbs that includes both wild and cultivated radishes. ***R. raphanistrum*** (wild radish; jointed charlock). an annual plant (Family Brassicaceae) that is dangerous to livestock in North America and Europe. It has been implicated in poisonings of cattle and sheep, including lambs. The animals may develop severe gastroenteritis, apparently because of the mustard oil content. *See also* raphania. ***R. rusticanus***. an outdated scientific name for *Armoracia rusticana* (horseradish). ***Raphanus sativus*** (radish; wild radish). an herb (Family Brassicaceae) from which the cultivated radish was developed. The seeds of this plant are the cause of raphania. Toxic concentrations of nitrate have also been reported.

Rattus. a genus of omnivorous rats (Family Muridae). ***R. norvegicus*** (Norway rat; house rat; barn rat; wharf rat; domestic rat; brown rat; sewer rat). the most common and widely distributed rat in nature. It is a pest and disease carrier in many human communities; strains originating more than a century ago from poorly defined populations of *R. norvegicus* stock are the most important subjects used in many fields of biological, agricultural, and veterinary research (e.g., in genetics, immunology, neurobehavioral sciences, pathology, oncology, pharmacology, physiology, toxicology, transplantation). Such are also extremely important in toxicological research and testing, especially to predict the effects of chemicals on humans. *See also* rat (laboratory rat). ***R. rattus*** (black rat). it is smaller and darker in color than the Norwegian, sewer, or brown rat *(Rattus norvegicus)* and has longer ears and tail. It is much less common and appears to be driven out wherever it is in contact with *R. norvegicus*. The black rat is the species that ordinarily transmits plague to humans by means of the flea, *Xenopsylla cheopis*.

γ-ray. gamma ray.

ray fungi. bacteria of the genus *Actinomyces*.

ray-finned fishes. *See* Actinopterygii.

rayless goldenrod. common name of any plant of the genus *Haplopappus*.

RBC. erythrocyte.

realgar; red orpiment. arsenic disulfide.

recharge. refill, resupply.

Reclamation, U. S. Bureau of. *See* Bureau of Reclamation.

recrudescence. the growth or increase of a process or condition such as the recrudescence of the gonads in certain animals (e.g., most birds) as a new breeding season approaches.

red alga. *See* alga.

red arsenic. arsenic disulfide.

red arsenic glass. arsenic disulfide.

red arsenic sulfide. arsenic disulfide.

red baneberry. *Actaea rubra*.

red blood cell. erythrocyte.

red blood corpuscle. erythrocyte.

red cedar. *Thuja plicata*.

red clover. *Trifolium pratense*.

red cohosh. *Actaea rubra*.

red fescue. *Festuca rubra*.

red fox. *Vulpes fulva*.

red jungle fowl. *Gallus gallus*.

red maple. *Acer rubrum*.

red pine. *Pinus resinosa*.

red poppy. *Papaver rhoeas*.

red sage. *Lantana camara*.

red squill powder. a fairly specific neurotoxic rodenticide prepared from *Urginea maritima*. It contains cardiac glycosides and is readily eaten by house mice and Norway rats. It is especially effective against rats and mice because they lack the ability to vomit.

red-stem filaree. *Erodium cicutarium*.

red water. any of several diseases of certain domestic animals indicated by reddish urine.

red-mouth bolete. *Boletus subvelutipes*.

red-tailed hawk. *See* 3-chloro-*p*-toluidine.

Redfieldia flexuosa (blowout grass). a grass (Family Gramineae) of the central and southern Great Plains of North America that grows well on deep, loose sands. It is used to reseed rangeland, establish pastures, and control blow-outs.

redox reaction. oxidation-reduction reaction.

redroot. *Amaranthus*.

Redskin. allyl isothiocyanate.

redtop. *Agrostis alba*.

reduction division. meiosis.

reduviid. assassin bug.

Reduviidae (assassin bug family). a family of insects, most species of which prey on other insects, from which they suck body fluids; a few species suck blood from vertebrates. A few are vectors of disease to humans. The forefemora are powerful structures used for grasping and holding prey.

redwater disease. a urinary syndrome of cattle and sheep.

redwater fever. bovine babesiosis.

reed fescue. *Festuca arundinacea*.

reentry interval. the period of time following application of a pesticide at a given location during which workers are not permitted to re-enter the area.

Reglone. diquat.

regulation. **1:** the control or maintenance of structural, and functional integrity in the face of often changing environmental conditions. **2:** a rule or requirement promulgated by the government department or agency that is responsible for implementing a particular law. The purpose of a regulation and the requirements stated therein is to bring those subject to the law into compliance. All regulations are published in the Code of Federal Regulations.

rein. the non-metal part of a bridle. It is a long, narrow strap, usually of leather, that attaches to either side of the bit, *q.v.*. It is used to control a horse.

renal. of or pertaining to the kidneys.

renal dropsy. renal failure.

renal failure (renal dropsy). failure of the kidney to maintain water and electrolyte balance, with consequent progressive systemic acidosis. The causes of acute renal failure include trauma, vascular insults, uncontrolled hypertension, or any condition that impairs blood flow to the kidneys.glomerulonephritis, acute obstruction of the urinary tract). Certain substances such as heavy metals and certain bacterial toxins can cause renal failure. Without treatment, acute renal failure progresses to chronic or end-stage renal failure. This stage is marked by acidosis and the toxicity of specific electrolytes, which leads to volume overload, hypertension, congestive heart failure, osteomalacia, secondary hyperparathyroidism, and death.

renal insufficiency. reduced ability of the kidney of vertebrates to maintain water and electrolyte balance. *See* renal failure.

rennet casein. a synonym for casein (def. 2).

rennet curd. a synonym for casein (def. 2).

rennet stomach. the abomasum.

rent. payment to an owner or owner's agent for use of land or a domicile. **cash rent**. rental payment in cash to the landlord rather than a share of the crop produced.

The landlord thus shares no risk with the tenant and has no beneficial interest in the commodity produced. **crop share rent**. rental payment to the landlord by a tenant farmer with a share of the crop rather than cash. This offers the tenant a mechanism for sharing risks with the landlord. And the landlord has a beneficial interest in the value of the commodity produced.

repellent. **1:** able to drive off or repel; repulsive. **2:** an agent that repels noxious organisms (e.g., insect pests, ticks, mites). **3:** an agent that reduces swelling.

reproductive potential. the theoretical maximum ability of a living organism to produce offspring. A value never achieved, in large measure, because of environmental resistance, *q.v.* *Cf.* biotic potential.

reptile. any vertebrate animal of the class Reptilia.

Reptilia. a class of amniote, poikilothermic, air breathing, predominantly terrestrial, usually pentadactyl tetrapod vertebrates. They are characterized by a bony skeleton, a cranium that bears a single occipital condyle, a three-chambered heart (four-chambered in the case of crocodilians), a dry horny skin that bears epidermal scales, plates, or scutes, and internal fertilization. The lungs are functional throughout life. Most are oviparous (egg-laying) but some are ovoviviparous in the sense that the eggs hatch internally and the young are born live. The egg shell is leathery. Included in this class are lizards (Order Squamata, Suborder Sauria) and snakes (Order Squamata, Suborder Serpentes).

reregistration. the reevaluation and relicensing of existing pesticides in the United States that were originally registered prior to current scientific and regulatory standards.

residual effect. *See* effect.

resilience. **1:** elasticity; adaptability; ability to rebound. the ability of an ecosystem to maintain its structural and functional integrity following disturbance.

resin. **1:** any of a chemically diverse group of complex nitrogenous compounds that occur in some plants and share certain physical properties. They are solid or semisolid at room temperature, brittle, easily melted, and flammable. They are insoluble in water. Some toxic resinoids are among the most virulent of poisons to humans and other animals. They are direct irritants of the nervous system or of muscles. **2:** rosin. **3:** any of a variety of organic monomers (e.g., acrylic resin, autopolymer resin) that are insoluble in water. *Cf.* resinoid.

resinoid. any substance that contains or resembles a resin. Some resinoids are among the most virulent of poisons. *See* resin.

resinous. pertaining to or derived from a resin.

resistance. **1:** insusceptibility; invulnerability; immunity. **2:** obstinancy. **3: antibiotic resistance**. resistance to antibiotic challenge by a microorganism due to the production of a protein that disables an antibiotic or prevents its passage into a cell. **herbicide resistance**. the inherited ability of a cultivated plant to survive and reproduce following exposure to a concentration or dose of an herbicide that is normally lethal to the wild type. Such resistance may occur naturally, by the selection of cultivars produced by tissue culture, or may be induced by means of genetic engineering. **pesticide resistance**. The inherited capacity of pests exposed to a pesticide to develop resistance to that pesticide.

resmethrin. a semisynthetic mixture of pyrethrin-like pyrethroid isomers. Resmethrin (under brand names such as Bioresmethrin, Chryson, Cismethrin, and Synthrin) is used largely in homes, gardens, and greenhouses to control a wide variety of insects. It is moderately toxic to mammals including humans. Acute dermal LD_{50} in laboratory rats is about 2.0 g/kg; acute oral toxicity varies among rat strains from about 1 to 4 g/kg body weight. The mode of action is via an effect on the sodium channels, leading to hyperexcitability. The various brands, however, vary considerably in toxicity. Bioresmethrin is essentially nontoxic when administered orally to laboratory rats, whereas Cismethrin (due to the *cis*-component) is very toxic, 90 to 170 mg/kg body weight.

resource base. *See* agricultural resource base.

resource industry. any industry that uses the primary resources from agricultural production, fisheries, forestry, or mining.

resource recovery. the process of recovering and using discarded materials to obtain useful matter or energy.

respirable. **1:** able to be respired or suitable for respiration. **2:** inhalable; able to be inhaled or breathed in (e.g., respirable dust). *See* dust.

respirable particles. respirable dust. *See* dust.

respiration. **1a:** an ATP-generating process in which either inorganic or organic compounds serve as electron donors (becoming oxidized), and inorganic compounds (mostly oxygen but also sulfate, and carbonate) serve as the ultimate electron acceptors (becoming reduced). **1b:** the interchange of oxygen and carbon dioxide between an aerobic organism and the environment or between individual living cells and the surrounding medium. **2:**

ventilation or breathing as by gills or lungs (*See* external respiration). **aerobic cellular respiration**. cellular respiration in which free oxygen is the final acceptor of hydrogen atoms from a series of reactions that produce ATP. The general equation for this process is:

$$C_6H_{12}O_6 + 6O_2 = 6CO_2 + 6H_2O + 36\ ATP + heat$$

This is accomplished by a series of reactions that transfer energy from the organic substrate to ATP. Glycolysis in which glucose is converted to pyruvate within the cytoplasm initiates the process. In aerobic respiration, the pyruvate so formed is converted to acetyl coenzyme A which enters the tricarboxylic acid cycle (*q.v.)* in the mitochondria; it is converted to carbon dioxide and hydrogen atoms with the release of electrons to the electron transport chain (respiratory chain). In summary, these reactions reduce a number of pigmented enzymes (cytochromes, flavoproteins) and yield molecular oxygen which combines to form water. By coupling with the oxidative phosphorylation of AMP and ADP, the energy released at each step of this process is stored in ATP. **anaerobic respiration**. cellular respiration in which glucose is converted, in the absence of molecular oxygen, to lactic acid with the release of energy to other metabolic processes. During this process, the organic substrate is not completely oxidized. **audible respiration**. abnormal or abnormally discernable respiratory sounds of any lung-breathing animal, particularly mammals or birds (e.g., wheezing, rales). **cellular respiration** (internal respiration). the oxidation of end products of glycolysis with the storage of considerable energy in ATP. The oxidant in the respiration of eukaryocytes (*See* aerobic respiration) is molecular oxygen. The oxidant in some bacteria (*See* anaerobic respiration) is sulfate rather than oxygen. **Cheyne-Stokes respiration**. the breathing pattern characteristically observed in coma due to affection of the CNS centers of respiration. It is marked by a rhythmic decline and gradual increase in depth of breathing, and sometimes in rate, to a maximum, followed by a decrease with regularly recurring periods of apnea. The cycles normally range from 30 seconds to 2 minutes in duration, with 5 to 30 seconds of apnea. **external respiration**. the exchange of respiratory gases (oxygen and carbon dioxide) between the external environment and the cells of an organism. **internal respiration**. cellular respiration. **irregular respiration**. that in which there is no definite cycle or rate of breathing. **labored respiration**. labored breathing; forced or difficult, usually irregular breathing.

respiratory. **1a:** of or pertaining to respiration of any type. **1b:** of or pertaining to the process of breathing or of gas exchange across cellular membranes (as in gills or lungs) **2:** Pertaining to lungs, gills, or other organs of breathing.

respiratory arrest. cessation of spontaneous breathing.

respiratory center. the structure or portion of the brain stem that regulates the depth and rate of breathing.

respiratory chain. *See* electron transport system.

respiratory distress. *See* dyspnea.

respiratory syndrome of cattle. pulmonary emphysema of cattle. *See* emphysema.

respiratory tree. *See* upper respiratory system.

respire. **1:** to breathe. **2a:** exchange of respiratory gases across membranes at the cell, tissue, or organism level; **2b:** the exchange by an organism of oxygen and carbon dioxide with the environment as part of the process of respiration.

response. **1a:** any activity (behavioral, metabolic, etc.) of an organism that occurs in response to a stimulus. **1b:** classically, any observed or measured action or movement due to the application of a stimulus. **2:** an action or actions taken to minimize the impacts of a release of contaminating materials to the environment. Also termed response action. *Cf.* reaction, reflex. **allergic response**. allergic reaction. *See* reaction. **immune response** (immune reaction). the response of the immune system following exposure to a specific antigen. Such a response may be manifested in the production of antibodies, mediated immunity, or immunological tolerance. **reflex response**. reflex.

restricted use. when a pesticide is registered in the United States, some or all of its uses may be restricted if the pesticide requires special handling because of its toxicity. Restricted-use pesticides may be applied only by trained, certified applicators or those under their direct supervision.

resuspension. the suspension of materials that have previously been taken out of a liquid or gaseous medium. The term is often applied to pollutants that have been taken up by bottom sediments. In the case of aluminum, for example, acidification of the water column by acid rain may solubilize toxic aluminum salts present in bottom sediment causing them to reenter the water column. Similarly, some pesticides applied to crops or soil can be resuspended in the atmosphere by air currents, creating a hazard for nontarget animals.

retaining wall. a wall that resists lateral pressure other than wind pressure, especially one that helps to prevent earth or rock slides. *See also* abutment.

retch. **1a:** an involuntary attempt to vomit. **1b:** to attempt, involuntarily, to vomit. *Cf.* retching.

retching (dry heaves; dry vomiting; vomiturition). an involuntary attempt to vomit, sometimes repeatedly.

reversible effect. *See* effect.

reversible toxic effect. *See* effect.

Reye's syndrome. a striking, often fatal disease of children mainly in southeast Asia, India, and Africa. The etiology is by no means well established. All of the following have been implicated: consumption of aflatoxin-contaminated food, increased use of aspirin, and viral infections such as chicken pox and influenza in association with aspirin use. Symptoms include fever, vomiting, diarrhea, convulsions, coma, and death. Lesions include fatty degeneration of heart, liver and kidneys, cerebral edema, neural degeneration, and lymphocytolysis. *See* aflatoxin,

RHA. rolling herd average.

rhabarberone. aloe-emodin.

Rhamnaceae. the buckthorn family. *See Rhamnus*.

Rhamnus (buckthorn). a genus of Asiatic perennial evergreen and deciduous shrubs and small trees (Family Rhamnaceae). Most occur in the temperate regions of the Northern Hemisphere. The leaves and fruit are toxic. Species that occur in North America include *R. caroliniana* (Carolina buckthorn; Indian cherry), *R. cathartica* (common buckthorn), *R. frangula* (buckthorn; alder buckthorn), and *R. purshiana* (cascara buckthorn, bearberry, bearwood, coffeeberry). ***R. cathartica*** (common buckthorn; Hart's thorn, waythorn; Rhineberry). indigenous to Eurasia, this plant was imported into the United States as a hedge plant. It has since escaped and occurs in hedge rows and pastures in the eastern United States All parts of the plant are toxic and contain a purgative. It has a bitter taste, however, and are seldom eaten by livestock. ***R. frangula*** (alder buckthorn). an ornamental shrub cultivated for the beauty of its shiny green leaves that turn bright yellow in autumn. ***R. purshiana*** (cascara buckthorn; bearberry; bearwood; coffeeberry). a small tree of the Pacific Coast of North America. The dried bark (cascara sagrada) contains emodin and is used as a mild laxative. ***R. rhabarbarum*** (*R. rhaponticum*; rhubarb; pie plant). a widely grown garden vegetable with a leaf stalk of rhubarb that has a sour but pleasant taste, due chiefly to the presence of nearly nontoxic malic acid. The leaf also has a sour taste due to the presence of citric and oxalic acids. The content of oxalic acid and soluble oxalates in the leaf is usually sufficient to produce severe effects and even death in humans.

rheum emodin. emodin.

Rhineberry. *Rhamnus cathartica*.

rhinotracheitis. inflammation of the upper respiratory tract, especially including the trachea and nasal passages. **infectious bovine rhinotracheitis** (red nose). a viral disease of cattle in which the respiratory tract is congested and its tissues inflamed.

red-berried elder. *Sambucus canadensis* and *S. pubens*.

red nose. infectious bovine rhinotracheitis.

red squill. *Urginea maritima*.

red rice. *Oryza rufipogon*

Rhipicephalus appendiculatus. *See* east coast fever.

Rhipicephalus sanguineus (brown dog; European brown tick; European dog tick). a hard-backed tick that infests ruminants and dogs. It is a vector of piroplasmosis; also called European dog tick, European brown tick.

rhizome. an underground, often horizontal, stem of plants such as *Acorus calamus*, *Iris*, and *Sanguinaria*.

rhodanate. thiocyanate.

rhododendron. *See Rhododendron*.

rhubarb. *Rheum rhabarbarum*.

Rhus. a genus of toxic deciduous and evergreen shrubs, vines, and small trees (Family Anacardiaceae) that are widely distributed in the temperate zone and subtropics. Most have compound leaves which turn a brilliant scarlet color in the autumn. The flowers are usually small and occur in large panicles. Some species are virulently poisonous and may prove fatal to humans. All North American species produce a severe allergic dermatitis (*Rhus* dermatitis) in sensitive individuals that is marked by itching and burning of the skin and edema. On rare occasions, serious cases can be life-threatening. Poisonous members of the genus with smooth fruits and leaves are sometimes placed in the genus *Toxicodendron*. ***R. diversiloba*** (syn. *Toxicodendron diversilobum*; Pacific poison oak; Western poison oak). a shrub or vine with three or five toothed or lobed leaflets and growing to 2.5 m. It occurs in western North America from southern British Columbia to northern Baja California. The toxic principle is urushiol. ***R. quercifolia***. *Rhus toxicodendron*. ***R. radicans*** (syn. *T. radicans*; formerly *R. toxicodendron*; poison ivy; poison vine; cowitch; "poison oak"; poison creeper; picry; mercury; markry). a climbing vine, small shrub, or, rarely, a small tree with three-lobed leaflets; most common in shaded woodlands, woodland glades or borders, or along roadsides. It is toxic to humans and closely related primates throughout the year. Birds eat the fruit in winter and are unaffected. The active principle is urushiol. *R. radicans* and *R. diversiloba* cause more cases of delayed contact sensitivity than all other known sensitizers combined. Their effects are quite similar: dermal contact produces a mild to severe self-limiting dermatitis which develops within 12-14 hours following reexposure of a sensitized individual. The condition is characterized by a cutaneous erythema with edema, itching, and burning of the skin. Blistering (with discharge of serum) and pruritus with exudation may develop. Ingestion can seriously injure the mucous linings of the mouth and gastrointestinal tract. Except in hypersensitive individuals, healing is complete in about 10 days. Sensitivity slowly diminishes in the absence of further exposures. Contrary to common belief, immunity is not conferred by chewing the leaves or fruit of the plant. Perhaps 30% of North Americans have an antigen-specific tolerance to these plants. ***R. succedanea*** (wax tree). a poisonous tree of Japan that grows to about 10 m. It has compound leaves with 9-15 lustrous leaflets. The whitish fruit contains a wax-like substance used in candlemaking. ***R. toxicodendron*** (syn. *Toxicodendron quercifolium*; *R. quercifolia*; poison oak; Eastern poison oak). **1:** a shrub of the scrub oak-pine savannahs of the Atlantic and Gulf coastal plains of North America. It grows to a height of 2 m, with leaves that are three-lobed or rounded and hairy on both surfaces. It is easily confused with *R. radicans*, and its effects on humans are identical. The active principle is urushiol. **2:** earlier literature treated *R. radicans* as a variant of *R. toxicodendron*. ***R. verniciflua*** (varnish tree; lacquer tree). a poisonous, ornamental tree that is indigenous to China and Japan. It grows to a height of about 20 m. The leaves are compound, with 11 to 15 large leaflets and long, hanging clusters of whitish flowers. It yields the lacquer used in Japan to protect and to put a beautiful high gloss on woodenware. ***R. vernix*** (syn. *Toxicodendron vernix*; poison sumac; poison dogwood; poison elder; poison ash; swamp sumac; thunderwood). a shrub or small tree up to 5 m high. It occurs in damp, marshy situations from Quebec to central Florida. All parts of the plant are toxic, producing an allergic contact dermatitis (Rhus dermatitis). The active principle is the resin urushiol.

Ribes (gooseberries; currants). a genus of ornamental plants (Family Saxifragaceae) that produce popular fruits (gooseberries, currents).

rice. *Oryza*.

rich-weed. a common name of *Eupatorium rugosum*.

ricin (ricine). **1:** either of two extremely poisonous glycoprotein toxins (RCL III and RCL IV) found in the castor bean plant, *Ricinus communis*. The two toxins are present in highest concentration in the seeds. They are violent irritants and hemorrhagins. They act by inhibiting ribosomal protein synthesis. Each of these two nearly identical lectins is composed of two polypeptide chains connected by a disulfide bridge. When ingested, one chain binds to cells of the intestinal wall, allowing entry of the other chain into the cytoplasm of a cell. A single molecule of ricin is sufficient to kill the cell. They can induce labor in pregnant women, and very small amounts are lethal to humans. **2:** ricin mixture. **3:** the individual lectins. This usage is confusing and, with current knowledge, unacceptable. *Cf.* ricinine. *See also* ricin agglutinin, ricin mixture.

ricin mixture. a mixture of several different proteins, of which two agglutinins (RCL I and RCL II) and two toxins (RCL III and RCL IV) have been isolated. Toxicity resides largely in the two toxins (*See* ricin). The minimum lethal oral dose in some laboratory animals may be as low as 0.01 mg/kg body weight. When injected, the minimum lethal dose is about 0.1

ng/kg. Even so, castor bean press cake, a livestock feed, can be detoxified simply by heating. Individual variation in sensitivity is high but on the average, horses are most vulnerable and poultry are most resistant to ricin poisoning. Very low doses have antitumor activity. *See also* ricin, ricin agglutinin.

ricine. ricin.

Riddell's groundsel. a common name of *Senecio riddellii* and *Senecio plattensis*.

rigidity. **1a:** stiffness or inflexibility; tenseness. **1b:** rigor; abnormal or morbid stiffness or inflexibility.

rill. a small channel in soil produced by splash erosion. Rills are normally filled in during the course of cultivation.

rinderpest. an acute, usually fatal infectious disease of cattle, sheep, etc., caused by a paramyxovirus of the genus Morbillivirus and characterized by high fever, diarrhea, and lesions of the skin and mucous membranes. Also called cattle plague.

ringless honey mushroom. *Armillariella tabescens*.

riparian buffer. buffer strip.

riparian doctrine. the precept or law that holders of riparian property have the right to use the resources found on, or that pass through, such property.

riparian right. a right under law, such as fishing or use of water (e.g., for irrigation), held by a person who owns riparian property.

risk. **1:** the probability of occurrence of a specified undesirable or adverse effect (e.g., loss, harm, injury, death) as a function of hazard and exposure. As the probability of experiencing the event approaches zero, so does risk. Risk is based on the probability that the undesired event (e.g., a specific transmissible disease) will occur and the severity of the outcome (e.g., mortality rate; loss of production) if it does occur. **1b:** the probability that harm will result to a specified population (e.g., a herd, a breed, a crop, a cultivar) under specified conditions. Risk is expressed in various terms (e.g., per person, per animal, per unit of time (incidence), per event). **1c:** the probability that a substance, action, process, or event will cause harm (usually injury, damage, or death). **2:** the probability of occurrence of an adverse effect combined with the probability that the effect, if it does occur, will be at an unacceptable (or specified) level of severity. *See also* factor (risk factor).

risk assessment (risk analysis). the process of estimating the magnitudes and probabilities of occurrence of adverse effects of specified processes, activities, accidents, or catastrophes, either natural or as a consequence of human activities (e.g., on farming operations or yield). While risk assessments can be helpful in identifying risks, even the most sophisticated assessments suffer from irreducible uncertainties. As a consequence, (1) a large array of diverse, ever-changing methodologies have been used to estimate the probability of undesirable events, and (2) risk assessments offer only one element in making decisions (e.g., regulatory decisions) regarding risk in any given case.

risk avoidance. the reduction of risk by altering, changing the timing, location of, or curtailing a process or activity such as the time of planting or the planting of a tolerant or resistant plant.

risk characterization. the summary output of a risk assessment. It may be qualitative or quantitative.

risk communication. any purposeful ex

change, formal or informal, of information concerning health, environmental, or other risks among interested parties.

risk control. the control of risk by society, whether through social, legal, scientific, or engineering methods and activities.

risk management. the process of judgment and evaluation that uses the results of risk assessment to evaluate alternative regulatory and/or nonregulatory options to reduce risk to an acceptable level and selecting among them. In addition to scientific/technical input, this process necessarily takes into account legal, economic, political, and social factors.

robin. a powerful phytotoxin found in the roots, bark, leaves, and seeds of *Robinia pseudoacacia*. The antitoxin of robin is termed antirobin.

Robinia pseudoacacia (black locust; bastard acacia; black acacia; false acacia; pea flower locust). a large, common, rough-barked, native tree (sometimes a large shrub; Family Fabaceae) of eastern North America, especially from Pennsylvania through Georgia and the Ozark Mountains. It sometimes grows to 25 m. The branches bear spines. The bark, foliage, seeds, and sprouts are very toxic due to the presence of a powerful phytotoxin (robin). Ingestion of relatively small amounts of the toxic plant material can produce symptoms that include vomiting, diarrhea, dullness, depression, a weak pulse, and coldness of the extremities. Toxicity, at least of the leaves, may be seasonal. Horses are much more sensitive than cattle.

rock dove. *Columba livia*.

rock maple. common name of *Acer nigrum* and *A. saccharum*.

rock pigeon. *Columba livia*.

rock salt. *See* halite.

Rocky Mountain fir. *Abies lasiocarpa*.

Rocky Mountain wood tick. *Dermacentor andersoni*.

rodent. any mammal of the order rodentia.

Rodentia (rodent family). the largest order of placental mammals, containing more than 6400 known species and subspecies. Included are mice, voles, rats, cavies, guinea pigs, hamsters, nutria, pacas, porcupines, squirrels, and beavers. All have one pair of continuously growing, rootless, chisel-like upper incisors that are used in gnawing. There are no canines; the molars and premolars are flat-crowned and used in grinding. The lower jaw is mobile and able to move not only up and down but also foreward, backward, and laterally. Rodents are extremely diverse behaviorally, physiologically, and nutritionally. They range in size from a few grams to about 60 kg, are nearly worldwide in distribution but are absent from the Australian region. Rodents occupy a diversity of habitat types. Most are omnivorous or herbivorous; a few, however, are carnivorous. Many species are important agricultural pests.

rodenticide. **1:** any substance or formulation that kills rodents (e.g., warfarin, 6-mercaptopurine, malachite green, rubratoxin, scilliroside). **2:** any chemical compound or formulation produced for the purpose of killing, repelling, or preventing rodents from damaging food, crops, or valuable material. Most are applied to the control of the house mouse, *Mus musculus*, or the Norway rat, *Rattus norvegicus*. Rodenticides are of considerable toxicological and agricultural concern because their use often results in direct or secondary poisoning of nontarget animals (e.g., pets, foxes, swine, certain omnivorous birds, birds of prey). Rodenticides are also responsible for accidental and suicidal poi sonings of humans. **anticoagulant rodenticide**. a rodenticide such as Warfarin that causes massive internal bleeding by virtue of its action as an anticoagulant. These rodenticides are potentially hazardous to all mammals and birds.

rodentine. rodent-like; of or pertaining to rodents.

roe. **1:** eggs of fishes. The roe of many bony fishes (Class Osteichthyes) are edible but that of most marine species is at times poisonous, sometimes extremely so. **2:** the swollen ovaries or extruded eggs of decapod crustaceans. **3:** Old World deer of the genus *Capreolus* (roe, roe deer).

roe deer. *See* roe.

Roentgen (röntgen; symbol, R formerly, r). a unit of measurement of an exposure dose of gamma- or X-ray radiation defined by the ability to ionize air. It corresponds to that amount of ionizing radiation that, in dry air at STP, is sufficient to liberate electrons and positrons carrying a total charge of 2.58 $x10^{-4}$ coulombs/kg of air, or to produce two ionizations per cubic μm of water or living tissue.

Roentgen intoxication. radiation sickness. *See* sickness.

röntgen. Roentgen.

rolling herd average (RHA). the average milk production per herd per year based on 365 days. The test record of the same period during the preceding year is subtracted from the new record, and a new rolling 365-day average is calculated.

Roman vitriol. the pentahydrate salt of cupric sulfate.

Ronnel (phosphorothioic acid; O,O-dimethyl O-(2,4,5-trichlorophenyl)ester; dimethyltrichlorophenyl thiophosphate; fenchlorphos). a systemic organophosphate, Ronnel is a cholinesterase inhibitor,

$$C_8H_8Cl_3O_3PS$$

It is used in households and on farms as an insecticide and parasiticide. It is an eye irritant to humans and can cause liver and kidney damage; it is also a possible human carcinogen and teratogen. Therapeutic doses given orally to pregnant blue foxes, *Alopex lagopus*, reduce the average litter size fell per vixen from 9.5 in controls to 1.2 in test animals. Effects on the fetus included incomplete ossification of cranial bones, cleft palate, and hydrocephalus.

root. **1:** the underground portion of a plant. **2:** the proximal end of a nerve. **3:** a portion of an organ imbedded in tissue.

Rosaceae. (the rose family). a family of herbs, shrubs, and trees of various growth forms that are widely distributed throughout the Northern Temperate Zone. Many are cultivated as ornamental garden or house plants, for their fruit, or (in a few cases) for their medicinal value.

rosary bead. *Abrus precatorius*.

rosary pea. *Abrus precatorius*.

rose family. Rosaceae.

rosin. the solid, amber, translucent resin from various pine trees (*Pinus* spp.). It is used chiefly to prepare plasters and ointments.

rotation. **1:** planned rotation among years of successive cuttings of timber in a forested area. **2:** planned rotation of land committed to a particular use: grazing, cultivation of particular crops. *See* grazing (deferred rotation grazing; Hohenheim system of grazing; rotational grazing). **crop rotation**. an agricultural practice of sys tematically varying the crops planted on a particular piece of land from time to time with a view to saving or increasing the mineral or organic content of the soil, improving soil structure, reducing erosion, increasing crop yields, or reduceing or eliminating insects, weeds, and plant diseases. Thus, for example, a cultivated crop such as corn may be alternated with a legume such as a clover. The latter adds nitrogen to the soil. *See* agriculture (sustainable agriculture).

rotenone ([2R-(2α,6aα,12aα)]-1,2,12,12a-tetrahydro-8,9-dimethoxy-2-(1-methylethenyl)-[1]benzopyrano-[3,4-b]furo[2,3-h][1]benzopyran-6(6aH)-one; Derrin; nicouline; tubatoxin). a dangerously toxic, white crystalline compound, $C_{23}H_{22}O_6$, that is soluble in ethyl ether and acetone but insoluble in water. It decomposes on exposure to light. It has been used as an insecticide, a scabicide, a fish poison (currently the most important application), in fly sprays, flea collars, and in mothproofing agents. Fatalities from overexposure have occurred, although it is safe for most mammals as normally used. It is poisonous to humans and other higher vertebrates by ingestion and possibly other routes of exposure; swine are especially susceptible; it is very toxic to fish.

rotten neck of rice. brown spot of rice.

roundleaf. a common name of *Smilax rotundifolia*.

Roundup™. *See* glyphosphate.

roundworm. any animal of the phylum Nematoda.

Roxarsone. a trademark name for 4-hydroxy-3-nitrophenylarsonic acid. It is marketed primarily as a growth promoter added to animal feed and as an antibacterial for domestic fowl. It is a pentavalent organic arsenical with a relatively high acute toxicity to rats and dogs.

rubber weed. *Actaea richardsoni*.

rubification. **1:** reddening. **2:** the development of red color in soil; soil reddening.

rubratoxin. either of two mycotoxins (rubratoxins A and B) produced by *Penicillium rubrum* and *P. purpurogenum* which often infest cereal grains. Both appear to be extremely toxic and are known from outbreaks of toxicoses in the United States. Rubritoxin B is extremely toxic to laboratory rodents, rabbits, cats, and dogs. It is hepatotoxic, embryotoxic, and teratogenic. Effects typically include anorexia, dehydration, somnolence, diarrhea, and jaundice; internal hemorrhage is common.

Rubus macraci (akala). a shrub (Family Rosaceae) cultivated in the Hawaian islands for its large, red, edible fruit.

ruby arsenic. arsenic disulfide.

Rudbeckia hirta (black-eyed Susan). a common wildflower, the yellow petals of which radiate from a large, dark central disk.

rumenitis. inflammation of the rumen.

Rumex (dock; "sorrel"). a nearly cosmopolitan genus of hardy, mostly biennial and perennial, often weedy herbs and shrubs (Family Polygonaceae). Most species are native to, and widely distributed within, north temperate regions. Many are common to pastures, meadows, roadsides, and other situations typified by dry, acidic, or poor soil. Some species are frequently referred to in the literature as sorrel. This causes some confusion in reports of livestock losses in the United States. Many species are toxic due to the presence of sometimes lethal amounts of soluble oxalate in the plant tissues. Ex amples of species that have caused losses among livestock in various countries: *R. crispus* (yellow dock), *R. acetosella* (sheep sorrel, common sorrel), and *R. acetosa* (garden sorrel). Those species with fleshy leaves are often sour but edible and are used especially in salads. Toxic concentrations of nitrate have been reported for various species under conditions of rapid growth. *See also* emodin.

ruminant. an animal that regurgitates and rechews food (the cud) from the first stomach (the rumen); examples are antelope, cow, deer, sheep.

run-off. runoff.

runner. *See* stolon.

running spiders. spiders of the genus *Chiricanthium*.

runoff (run-off; environmental runoff). that part of precipitation, snow melt, or irrigation water that is not retained by the soil but runs freely off the land into streams or other surface waters. It can carry dissolved or suspended pollutants from the air and land (e.g., fertilizers, pesticides, other polluting agrochemicals, and industrial wastes) into the receiving waters.

rural. **1a:** agrarian; agricultural; of; related to; or pertaining to farming; agricultural areas. **1b:** sylvan; rustic; provincial.

rush. a common name of plants of the genus *Equisetum*.

Russian thistle. *Salsola pestifer*

rutabaga. *Brassica napus* var. *napobrassica*.

rye. *Secale cereale*.

ryegrass staggers. *See* staggers.

rye ergot. ergot.

rye smut. ergot.

S

σ. *See* sigma. For terms beginning with this prefix, see the root word.

Σ. *See* sigma. For terms beginning with this prefix, see the root word.

S. the symbol for sulfur.

sac. a bag or pouch-like structure.

sacahuista. *Nolina texana*.

saccharin sorghum. *See Sorghum vulgare*.

Saccharomyces malei (apple cider yeast). a yeast (Family Saccharomycetaceae) that occurs naturally in fruits. It is used in the production of sparkling hard cider.

saccharum lactis. lactose.

saccular gland. acinar gland.

sacculus. a little sac.

safety. **1:** security; protection. (note: safety is always relative or uncertain). **2a:** the reciprocal of risk, or the probability that harm will not occur under a set of specified conditions. **2b:** the approximate opposite of hazard; the improbability of exposure to possible loss or injury. *See also* margin of safety. **3:** the expectation that a biologically active agent will cause relatively little or no harm under a given set of circumstances. *See also* margin of safety. **4:** a mechanism prevent, helps to prevent accidents as the safety on a firearm that prevents accidental discharge.

sage. *See Artemisia*, *Salvia californica*. **annual sage**. *Salvia reflexa*. **hop sage**. *Grayia*. **sand sage**. *Artemisia filifolia*.

sage sickness. a rapidly developing disease of horses unaccustomed to foraging on sage. The animals become nervous and tend to fall when forced to move due to a partial paralysis of the forelimbs. Within a week or two the animals recover and are able to consume large quantities of sage without recurrence of symptoms. The species responsible for the disease is *Artemisia filifolia* (and sometimes *A. spinescens*).

sagebrush. *Artemisia*.

sago palm. common name of plants of the genus *Cycas*.

Saint Anthony's fire. ergotism.

sal ammoniac. ammonium chloride.

sal chalybis. ferrous sulfate.

Salientia. Anura.

saline (salty). **1:** salty; salt-like; of the nature of a salt; of, pertaining to, or containing salts or a salt, especially sodium chloride. **2:** saline solution; a salt solution. **3:** pertaining to water or soil that is rich in soluble salts. **physiological saline** (physiological saline solution; physiological salt solution). an isotonic aqueous solution of salts at a concentration that is isotonic with the cells or body fluids of a particular type of organism. Such solutions are used chiefly to bathe and temporarily maintain living cells *in vitro*, or as the carrier of drugs or other substances that are administered by injection. Human physiological saline solutions contain 0.9% sodium chloride.

saline solution. *See* saline.

salinity. **1a:** saltiness; a measure of the total concentration of dissolved sodium chloride in water. **1b:** a measure of the total weight of dissolved salts per unit volume of solvent. Thus, the salinity of sea water is taken as the total amount by weight (in parts per thousand, ppt) of dissolved solids following the conversion of all bromide and iodide to chloride; of all carbonate to oxide; and the complete oxidation of all organic matter. The salinity of ocean water averages 35 ppt. *See also* chlorinity.

salinization. **1:** the process whereby soluble salts accumulate in soil or other environmental matrices. This often results in a reduction in arable land due to the toxicity of the salinized soil to plants. **2:** excessive salinity, especially of soil.

saliva. the secretion of salivary glands. Such glands are present and open into or near the mouth in many vertebrate and invertebrate animals. The usual function of saliva is the moistening and lubrication of food; in some cases it may also serve an excretory role (*See* excretion). The saliva of humans and most mammals is a clear, viscous fluid (secreted also by mucous glands of the mouth) that contains water, mucin, organic salts, and ptyalin. In some tetrapods and insects, saliva contains amylases (e.g., ptyalin in mammals) which may initiate the digestion of starch. The saliva of insects contains various digestive enzymes depending upon the nature of the diet. In blood-sucking animals such as leeches and mosquitos, saliva contains anticoagulants. The saliva of many animals is poisonous, often serving as a venom.

salivary (sialine; sialic). of or pertaining to saliva or to the secretion of saliva.

salivary gland. any of several glands of multiple evolutionary origin in the digestive system of many vertebrates and certain invertebrates (e.g., biting insects, cephalopods) that secretes saliva. **internal salivary gland**. the sublingual and submandibular glands treated as a single gland. **major salivary gland**. the three largest pairs of salivary glands (parotid, submandibular, sublingual) of the oral cavity treated as a unit. **minor salivary gland**. any of the smaller, largely mucous-secreting, exocrine glands (buccal, labial, lingual, molar, palatine) of the oral cavity of vertebrates. **sublingual salivary gland** (sublingual gland). one of three pairs of vertebrate salivary glands, located near the base of the tongue. **submandibular salivary gland** (submandibular gland). one of three pairs of vertebrate salivary glands, located below the mandible. **submaxillary salivary gland** (submaxillary gland). either of a pair of small glands near the jaw that secrete saliva into the mouth.

salivation. the secretion of saliva, including ptyalism.

salivatory. pertaining to the secretion of saliva.

Salmonella. a genus of pathogenic, toxin-secreting, aerobic and facultatively aerobic Gram-negative bacteria (Family Enterobacteriaceae). ***S. cholerae-suis***. a species that infects swine; it sometimes also causes an acute gastroenteritis and enteric fever in humans. ***S. enteritidis***. (Gärtner's bacillus). a cause of gastroenteritis in humans. ***S. hirschfeldii*** (Hirschfeld's bacillus). a cause of enteric fever in humans. ***S. schotmülleri*** (Schotmüller's bacillus). a cause of enteric fever in humans. ***S. typhi*** (=*S. typhosa*; typhoid bacillus). the cause of typhoid fever; exposure of humans is usually via contaminated water and food. ***S. typhimurium***. this species secretes a ptomaine, which causes the most common form of food poisoning in humans. Victims are afflicted with flu-like symptoms and sometimes die from the effects of the toxin. *See also* *Salmonella*/microsome test. ***S. typhosa***. *S. typhi*.

salmonellosis. any disease caused by bacteria of the genus *Salmonella.*

salmonid. a fish of the family Salmonidae.

Salmonidae. a family of bony fishes that includes trout and salmon. The mouth is relatively large with well-developed dentition; the scales are small; and there is a small, dorsal, adipose fin anterior to the base of the tail. Most species inhabit cool water (freshwater and/or marine water) in the northern hemisphere.

Salsola pestifer (Russian thistle). a bushy annual herb (Family Chenopodiaceae) that is most commonly seen in the prairie regions of the United States and in waste areas. It may contain hazardous concentrations of soluble oxalates and probably salsoline (which has been isolated from *S. richteri*). Concentrations of nitrate that are toxic to livestock have also been reported.

salsoline. a moderately toxic crystalline substance isolated from *Salsola richteri*, $C_{11}H_{15}NO_2$.

salt. **1:** the compound formed by replacement of the hydrogen of an acid by a metal or an equivalent. In general, the reaction of an acid with a base yields a salt plus water. Most salts ionize in aqueous solution. **2:** sodium chloride.

salty. saline.

salt block (salt lick). a solid block of salt, about 10 in. on a side, sometimes medicated, and placed in a location available to livestock and sometimes game animals.

salt lick. salt block.

salt poisoning (brine poisoning). a lethal disease of animals, notably pigs, fed on garbage, due largely to excessive intake of sodium chloride (table salt). If sufficient quantities of fresh drinking water are available, poisoning does not usually result.

salt water intrusion. the invasion of fresh surface- or groundwater by salt water. If the salt water comes from the ocean it may be called sea water intrusion. Such invasion may, if the rate of recharge by fresh water is low, create a semipermanent or permanent toxic environment, even a virtual desert at and about the site, that precludes use of the site for crop production.

saltbushes. plants of the genus *Atriplex*.

saltpeter. potassium nitrate.

Salvia (sage). a large genus of shrubs or bushy plants (Family Labiatae), often aromatic, that occur in warm and temperate regions. Many of the species that occur in the United States occupy relatively dry situations. Certain Old World species occur occasionally in North America but rarely persist. ***S. laciniata***. *S reflexa*. ***S. leucophylla*** (sage). this shrub (Family Labiatae) and certain others such as *Artemisia californica*, that comprise the soft chaparral communities of southern California, produce volatile terpenes that are apparently adsorbed by the soil and inhibit the growth of grasses (Family Gramineae), sometimes producing bare areas of one or a few meters wide about these communities. ***S. mellifera*** (black sage). a perennial herb of the Pacific Coast of the United States It is sometimes used as forage; the flowers are favored by bees. ***S. reflexa*** (syn. *S.laciniata*; annual sage; mintweed). an annual, often bushy weed (Family Labiatae) named for the turned-back (reflexed) leaves. It is relatively common in dry, open areas from Wisconsin to Montana and south to Mexico; it is uncommon in the eastern United States. It has caused severe losses among livestock in Australia and cattle in the United States. Australian material is high in nitrate, and the symptoms are those of nitrate poisoning. Symptoms differ in the United States, however, where the poisoned cattle quickly develop muscular weakness and symptoms of gastrointestinal distress.

Salzburg vitriol. the pentahydrate salt of cupric sulfate.

samara. key. *See Acer*.

Sambucus (elders). a genus of large, deciduous shrubs or small trees that have compound leaves (Family Caprifoliaceae). Documented poisonings of humans by members of this genus are rare. Nevertheless, most species should be considered dangerous, especially to children. ***S. canadensis***. (common elder; elderberry; black elder; red-berried elder; scarlet elder). a shrub 1-4 m tall that occurs in the northern United States and Canada. The stems have a white pith and are only slightly woody. The twigs bear 5-11 elliptic to lanceolate, sharply serrated leaflets. All parts of the plant are poisonous, including the raw berries. The leaves, shoots, bark, and roots should be considered very toxic. Nevertheless, limited intake of raw berries is not likely to cause harmful effects. Cooked elderberries are often used to make pies and jams and as such are safely eaten. The active principle is sambunigrin, a cyanogenic glycoside. Symptoms of intoxication, which usually appear several hours following ingestion of the plant material, include dizziness, headache, nausea, vomiting, tachycardia, and possible death. ***S. pubens*** (American red elder; red-berried elder; stinking elder). a large shrub up to 8m tall, with warty bark and brown pith, pubescent branches, and 5—7 ovate-lanceolate leaflets which are downy or glabrous beneath. The ripe, cooked berries are edible. The roots, stems, leaves and possibly the flowers are toxic and cathartic. The leaves and roots contain cyanogenic glucosides, and high con-concentrations of nitrate that may affect livestock have been reported. ***S. racemosa*** (European red-berried elder). a hardy species that grows to a height of about 4 m. Toxicity and effects are similar to those of *S. canadensis*.

sample. **1a:** any subset of a population. **1b:** a representative subset of a population. **1c:** a subset of a population collected with a view to estimating some characteristic of the population by observation or measurement of the sample. The specimens that form the sample are selected and collected in such a way that the sample can be regarded as representative of the population with respect to the characteristic(s) to be investigated. *Cf*. specimen. **grab sample**. a sample, usually of a single specimen, presumably taken randomly (often not). In air pollution science, a grab sample of air is considered a type of short-period sample. **random sample**. a sample in which every member of the population from which it is drawn has an equal and independent chance of being included. **representative sample**. **1a:** a representative part of a larger unit used to study the properties of the whole. **1b:** a true random sample of sufficient size such that the distribution of values about a suitable measure of central tendency (e.g., average or mean value) is identical (or sufficiently similar) so that sample observations or measurements can be used to make confident predictions about the population as a whole. *See also* population. **short-period sample** (spot sample). a sample of air collected over a short period of time and usually taken to a central laboratory for analysis. The WHO Expert Committee on Atmospheric Pollutants (1963) recommended that for purposes of international comparison of routine measurements, a short-period air sample, should be defined as one taken over a period of 30 minutes. *See also* grab sample.

spot sample. short-period sample.

sampling error. *See* error.

sand dune. dune.

sand viper. *Heterodon contortrix*. *See* adder.

sand-crack. a crack in the hoof of a horse, usually occurring on the inside of the forefoot (quarter-crack) or the forepart of the hindfoot (toe-crack). The horse may be lamed if the crack is deep enough to expose sensitive tissue.

quarter-crack. *See* sand-crack.

sandcorn. *Zigadenus paniculatus*.

sangui-, sanguin-, sanguino-. prefixes that indicate blood, bloody, or bloodiness.

sanitarian. a qualified professional who is skilled in sanitation and public health.

sanitary. clean (as a clean environment), germ-free, healthful, hygienic, sterile, conducive to health.

sanitation. the process of making sanitary, preventing disease, or promoting health, especially with respect to freeing the environment of disease-causing organisms and other components that are unfavorable to health.

sanitize. to make something clean, often in the sense of destroying microbes.

santonica. the unexpanded flower heads of various species of *Artemisia*. It is the source of α-santonin, *q.v.*

α-santonin. an odorless and initially tasteless white powder, $C_{15}H_{18}O_3$, that turns bitter and that turns yellow when exposed to light. It is only slightly soluble in water but is soluble in ethanol, chloroform, and in most fatty and volatile oils. Alpha-santonin has a tricyclic structure and is the inner anhydride or lactone of santoninic acid. It is extracted from santonica. It has been used as a nonirritating, powerful anthelmintic, especially against *Ascarus lumbricoides*, and in the treatment of urinary incontinence but because of its toxicity it is generally no longer used. It causes disturbances of vision, especially affecting color vision in humans and sometimes causes disorders of taste, hearing, and smell. Large doses stimulate the CNS causing nausea, vomiting, headache, epigastric pain, muscle twitching, confusion, diarrhea, tonic-clonic convulsions, intermittent periods of depression, and even coma. Death from cardiopulmonary collapse may intervene.

Santotherm. polychlorinated biphenyl.

sap. **1:** any naturally occurring fluids of a living organism or tissue. **2:** the sugary fluid transported through the phloem of a vascular plant. **3:** the cytoplasm (ground substance) of a cell.

Sapindaceae (the soapberry family). a family of tropical trees and shrubs that bear leathery fruits that contain a saponin that is used for cleaning by natives. The most notorious of these plants is *Blighia sapida*, *q.v.*

Saponaria (soapworts). a genus of hardy annual and perennial European herbs (Family Caryophyllaceae) with white, red, or pink flowers. They contain poisonous saponins. ***S. officinalis*** (bouncing bet; cow cockle). a stout perennial plant of Asia but widely naturalized and common in many places as a roadside plant. The flowers are large and rose colored. It contains saponins and is toxic to livestock, although actual instances of poisoning are rare. ***S. vaccaria***. a coarse annual plant with pale red flowers. It is toxic to livestock, although poisonings are rare.

saponin (saponic glycoside). any of a class of bitter-tasting, steroidal glycosides of plant origin. The aglycone, a sapogenin, is a steroid alcohol. Saponins form nonal kaline colloids in water that characteristically produce a froth or foam when shaken. Many are toxic to animals and have been used as fish poisons. Some (e.g., diosgen) are used in the commercial production of steroid hormones. They may function in the plant as a deterrent to herbivores. Toxic saponins are gastrointestinal irritants and, when present in sufficient concentration in plant tissue, may produce severe effects. They also have hemolytic effects. Saponins are not readily absorbed through the intact gastrointestinal tract and the most serious effects occur in association with other irritant substances that are able to injure the wall of the digestive tract. *See Phytolacca americana*, asterosaponin.

saprist. organic soil in which most of the plant matter has decomposed and the original tissue cannot be identified.

sapro-, sapr-. combining forms denoting rotten, putrid, or decayed.

saprobe. **1:** a saprophytic organism. **2:** an organism that flourishes in water rich in organic matter.

saprobic (saprobiont). pertaining to water that is rich in decaying organic matter.

saprobic system. an ecological classification of polluted water bodies that are undergoing self-purification. The system is based on relative levels of pollution, oxygen concentration, and types of indicator microorganisms present.

saprobiont. saprobic.

saprophyte. a heterotrophic organism such as a bacterium or fungus that externally breaks down dead organic matter before absorbing the products.

saprophytic. of or pertaining to saprophytes, *q.v.*

saprophytism. the mode of nutrition of saprophytes, *q.v.*

saprozoic. of or pertaining to organisms that feed on solutions of organic material rather than on solid organic matter. Saprozoic organisms include the flagellate protozoan *Polytoma uvella* and many gastrointestinal parasites.

sapwood (alburnum). *See* wood.

sarco-. a combining form denoting muscle or a resemblance to flesh.

Sarcobatus (greasewood). a genus of oxalate-containing plants (Family Chenopodaceae) of the southwestern United States. *See* greasewood. ***S. vermiculatus*** (black greasewood). a low, spiny, woody, deciduous shrub that contains lethal amounts of soluble oxalates. Mass mortality of sheep has occurred in a number of locations but cattle are rarely poisoned. Symptoms are those of oxalate poisoning. It is considered useful forage by most ranchers assuming that moderate amounts of other forage are available.

SARE. Sustainable Agriculture Research and Extension Program.

Sarothamnus scoparius. *Sparteum junceum*.

sarsa. sarsaparilla.

sarsaparilla (sarsa). the dried root of *Smilax aristolochiaefolia* and a number of other species of the genus. It is used as a flavoring agent in beverages and in the treatment of psoriasis.

Sauria. a suborder of the order Squamata (Class Reptilia), comprised of lizards. They are mostly terrestrial, pentadactyl quadrupeds with long, slender bodies and a long tail. A few lizards, e.g., ***Anguis*** (slow worm), are limbless. A movable quadrate bone provides a flexible joint between the upper jaw and the skull, increasing the gape of the jaw; the teeth are fused to the jaws. Most have moveable eyelids, are oviparous, and insectivorous; some are herbivorous.

sausage poisoning. allantiasis.

savanna. savannah.

savannah (savanna). the tropical and subtropical grassland biome of Africa and South America. It is characterized by extensive grasslands with scattered trees, and usually by annually recurring wet and dry seasons. The term is applied mainly to extensive plains in South America and Africa.

scabby. pertaining to grain (e.g., oats, wheat, corn, barley) that is infected with fungi.

scabby barley. barley infested with the fungi, *Gibberella* or *Hordeum vulgare*.

scabby barley poisoning. a disease of livestock caused by ingestion of barley infested with the fungi, *Gibberella* or *Hordeum vulgare*, *q.v.* Ingestion of small amounts of scabby barley induces vomiting in swine but cattle are unaffected even following ingestion of large quantities.

scarlet elder. *Sambucus canadensis*.

scat (dropping; fecal dropping). the fecal dropping of an animal, especially that of a terrestrial mammal. Scats are fairly easy to collect for some animals, whereas the animals themselves may be difficult to observe or capture. They often offer the opportunity to identify some of the animals present in an area and to gain information regarding the recent diet of an individual or population of animals and the hazards encountered thereby (e.g., deficiencies, toxicants).

scabby barley. barley infested with the fungi, *Gibberella* or *Hordeum vulgare*.

scallion. a common name for any edible plant of the genus *Allium* that has a reduced bulb (e.g., the leek and shallot).

scarlet maple. *Acer rubrum*.

scarlet runner bean. *Phaseolus coccineus*.

scato-. a combining form denoting feces. *See also* copro-, sterco-.

scavenger. **1:** an organism that feeds predominantly on carrion or offal. **2:** a substance added to another substance, mixture, or other system to remove pollutants or other impurities.

scent gland. a cutaneous gland that produces odoriferous secretions (pheromones or recognition odors; they may be located on different parts of the body, e.g., under the chin (rabbit); between the digits); on the medial surface of the metatarsus (deer); in the preorbital fold (antelope); in the occipital region of the head (camel); on the flank (hamster); in the region about the anus (perianal region) and on the dorsum of the tail base (carnivores). *See also* stink gland.

Scheele's green. cupric arsenite.

schisto- a combining form indicating a split or cleft. *See also* schizo-.

schizo-, schiz-. combining forms denoting split, cleft, or division. *See also* schisto-.

school (shoal; draught). a party or group of fish.

Schotmüller's bacillus. *Salmonella schotmülleri*.

Schweinfurt green. cupric acetoarsenite.

scilliroside. an extremely toxic cardiac glycoside isolated from the bulbs of

Urginea maritima. It is used as a rodenticide.

scillism. poisoning by *Urginea maritima* (squill).

scillitic. pertaining to *Urginea maritima* (squill).

sclera (white of the eye). the white fibrous outer layer of the eyeball in vertebrate animals.

sclero-, scler-. combining forms denoting hardness (induration), sclerosis, relationship to the sclera.

sclerotia. plural of sclerotium.

Sclerotinia cepivorum. *See* white rot.

sclerotium (plural, sclerotia). a dormant stage of various true fungi in the form of a usually dark, compact, hardened mass of mycelium. When mature, it detaches from the host and begins to grow, sending out hyphae or producing fruiting bodies, when conditions become favorable. *Claviceps purpurea*, an ascomycete fungus parasitic on rye that produces sclerotia known as ergots in place of the ovary of the host grass flower. Ergots are overwintering structures that germinate the following spring to produce stalked stroma in which the perithecia are formed. *See* ergotism.

Sclerotium. a genus of fungi that may infest corn or sweet potatoes. Ingestion of infested feed produces severe respiratory distress and anorexia. The course of the disease is rapid, terminating fatally within a week.

scoke. a plant of the genus *Phytolacca*.

scoparius. the dried tops of *Sparteum junceum* (=*Cytisus scoparius*) used as a cathartic and diuretic. although not authorized, seeds and dry flowers are included in such preparations.

-scopy. a suffix denoting an action or activity involving use of a viewing instrument.

Scotch broom. *Sparteum junceum*.

Scotch pine. *Pinus sylvestris*.

scouring rush. *Equisetum hyemale*.

scouring. *See* scours.

scours (scouring). diarrhea or dysentery in livestock, notably in the newborn. **peat scours. 1a:** that due to molybdenum poisoning of grazing cattle. **1b:** molybdenum poisoning of cattle.

scrapie. a disease of sheep and goats, considered to be a variant of bovine spongiform encephalopathy, *q.v.* Meat scraps from diseased sheep used in cattle feed may have introduced the disease to cattle.

screef. to expose a soil surface for planting.

scrota. *See* scrotum.

scrotal. pertaining to the scrotum. *See* scrotal sac.

scrotal sac. either of two sacs that contain the descended testes of male mammals. Together they comprise a scrotum.

scrotectomy. excision of part or all of the scrotum.

scrotitis. inflammation of the scrotum.

scrotum (plural, scrota or scrotums). an external musculocutaneous pouch that contains the descended testis and part of the spermatic cord of male mammals.

scrubber. a pollution control device that uses a spray of water, a reactant, or a dry process to trap pollutants in emissions before they leave the source.

scurs. horn tissue that regrows following the dehorning of cattle. It appears as small horns attached to the skin, not to the skull.

scutch grass. *Cynodon dactylon.*

Se. symbol for selenium.

sea lyme grass. *Elymus arenarius.*

sea onion. *Urginea maritima.*

sea salt. sodium chloride recovered by evaporation from marine waters.

sebaceous (sebaceus). oily; fatty; having or pertaining to an oily secretion, especially sebum or a sebum-like secretion.

sebaceous gland. any of a number of epidermal exocrine glands in the skin of mammals that secretes sebum, *q.v.* One or more of these glands, each of which contains large fat granules, open near the surface of the skin by each hair follicle. Sebaceous glands are also present in areas that lack hair (e.g., the lips) but are absent from the palms and soles (or pads) of the feet.

sebaceus. sebaceous.

sebo-, seb-, sebi-. prefixes indicating sebum or sebaceous.

seborrhea. excessive secretion of sebum by the sebaceous glands.

oil gland. **1:** a sebaceous gland. **2:** any gland that produces an oily secretion such as a uropygial gland.

sebum. a lipoidal secretion of the sebaceous glands of the skin of mammals. It is a mixture that contains, for example, varying amounts of triglycerides, fatty acids, waxes, steroids, and cellular debris. this fluid lubricates the hair and skin, acting to waterproof the skin and to restrict desiccation. Sebum from the ears is wax-like and called cerumen; that from the foreskin of the penis is called smegma. *See* skin, sebaceous.

***sec*-butanol**. 2-butanol.

secale cornutum. ergot.

Secale cereale (cultivated rye; rye). a hardy annual grass (Family Grammineae) and agricultural crop grown for grain and straw. It has been heavily ergotized at times (*See Claviceps*, ergotism). Toxic concentrations of nitrate that are toxic to livestock have been reported at times, especially during periods of unusually rapid growth.

9,10-secoergosta-5,7,10(19),22-tetraen-3-ol. vitamin D_2.

secondary parasitism. autoparasitism.

sedge. **1:** any of a number grasslike herbs (Family Cyperaceae) that commonly occur in well-drained soils or marshy areas. They are of little nutritional value for livestock. **2:** a flock of cranes.

sediment. particulate matter that has been transported by wind, water or ice and later deposited in an area or has been precipitated from water. Sediments have been repeatedly shown to accumulate toxic chemicals. Under favorable conditions sediments provide useful records of pollution chronology and can demonstrate the sources of specific pollutants. **bottom sediment**. that which has settled onto the bottom of a water body such as a lake, pond, or stream. Sediments that accumulate beneath bodies of standing or slow-moving water play a very important role in freshwater ecology. They contain nutrients and often a rich biota as well as toxic chemicals. Organic compounds in sediments may be taken up by humans and a wide variety of organisms, sometimes with disastrous consequences (e.g., methyl mercury). **suspended sediment**. parti culate matter that remains in air or water either because of its small size or due to resuspension by wind or water currents. Resuspension of sediments is often a continuing source of environmental poisons. *Cf.* seston.

sedimentation. the process of forming sediment.

seed. **1a:** typically the ripened, fertilized ovule of a plant. Some plants, however, produce seeds asexually (agamospermy). **1b:** the embryonic plant within a protective coat.

seed beetle (seed weevil). any of several species of beetles (Family Bruchidae) that infest the seeds of legumes.

seed coat. the outer covering of a seed.

seed corn. ears or kernels of corn set aside for use as seed.

seed dressing. an antifungal or insecticidal substance applied to protect seeds from infestations. *See*ds so treated are toxic to animals and cannot be used in feeds. *See*ds so treated have, at times, devastated wildlife (especially bird) populations. *See also* organomercurial fungicide.

seed leaf. cotyledon.

seed plant. **1:** a seed-bearing plant; spermatophyte. **2:** a plant cultivated for its seed.

seed vessel. pericarp.

seed weevil. seed beetle.

seedling. **1:** any young plant, especially one grown in a nursery for transplanting. **2:** a tree grown from a seed extending from the time of germination to the sapling stage. **3a:** a tree having a diameter at breast height of 1 cm or less. **3b:** a tree less than 1 m tall. **bareroot seedling**. a seedling whose roots are exposed at the time of planting. The seedlings are grown in a nursery and removed from the soil in which they were grown prior to planting in the field. **container seedling**. a seedling grown in a small container under controlled environmental conditions. **plug seedling**. a seedling grown in a small container under carefully controlled (nursery) conditions. The soil remains bound up in its roots when it is transplanted.

seedlot. **1a:** a quantity of cones or seeds of the same species, source, year of collection, and quality. **1b:** a quantity of seeds from a single collection at a particular time and location. Such a seedlot can represent those from a single tree collection or from a pool collected at one time from a number of trees.

seepage. the emergence of water from the ground (or seepage into the ground) over an extended surface or line in contrast with emergence from a single location such as a spring.

seep. **1:** of a liquid: to leak, drip, trickle, ooze; soak. **2:** any area where water seeps out of the ground. *Cf.* seepage, seepage zone.

seepage zone. an area on the slope or at the base of a hill where water often or continuously seeps to the surface.

segmented worm. *See* Annelida.

seige. a flock of cranes (also sedge).

seizure. convulsion.

selective. **1:** discerning; discriminating; exhibiting a high degree of specificity with respect to activity, action, or choice. **2:** descriptive of herbicides that kill some plants and not others. Broadleaved plants are usually unaffected by such herbicides. Such herbicides are thus valuable for applications on cereal crops and lawns. Some selective herbicides (e.g., Simazine) kill germinating plants only. **3:** descriptive of the degree to which a drug produces the desired effect in relation to adverse effects that may occur; pertaining to drug selectivity.

selenide. a compound of selenium, with another element or radical.

seleniferous. containing selenium, as in seleniferous soils.

selenium (symbol, Se)[Z = 34; Ar = 78.96]. an allotropic nonmetallic (metalloid) element that is chemically similar to sulfur. Selenium is an essential trace element in the diet. It is, nevertheless, highly poisonous to domestic animals, wildlife, and humans at concentrations in the diet somewhat above that required (ca. 50-70 μg/day). Chronic selenium poisoning (selenosis) has occurred in humans who consumed about 40 μg/kg body weight on a daily basis. The median lethal dose ranges from about 1.5-6 mg/kg body weight, depending upon the species and compound tested. Selenium causes reproductive effects, liver and kidney damage, and is tumorigenic and teratogenic in experimental animals. Many soils contain selenate and selenite compounds. In some cases, soil concentrations are such that selenophilic plants (e.g., *Astragalus pectinatus*) can accumulate selenium to concentrations that are toxic to livestock. Alkali disease or blind staggers is an acute or subacute condition of livestock due to the consumption of grain or accumulator plants that contain high concentrations of selenium (usually above 5mg/kg of total mass).

selenium poisoning (selenosis). chronic poisoning of horses, cattle, and swine characterized by emaciation, hair loss, anemia, and bone and liver damage. It is caused by ingesting grains and forage raised on soils high in selenium. **acute or subacute selenium poisoning** (in animals). blind staggers.

selenophile. any plant that accumulates selenium from the surrounding medium (usually soil). **facultative selenophile.** any selenophile that grows on soils that do not contain selenium as well as soils that do. **obligate selenophile.** any plant that grows only on seleniferous soils and contains selenium at some stage of growth, e.g., *Astragalus*.

selenosis. selenium poisoning.

self digestion. autolysis.

self-feed. to supply food to animals in a way that allows them to eat as much and as often as they want. *Cf.* hand-feed, *ad libitum*.

self-fertilization (autogamy). pollination by pollen from the same flower.

semi-. a prefix denoting one-half; synonymous with demi-, hemi-. These prefixes are sometimes compounded, usually in this order, hemidemisemi-. This sample compound prefix denotes one-eighth.

seminole bead. *Abrus precatorius*.

semiochemical. **1:** any substance produced by a living organism that conveys information to other organisms, especially that which stimulates or helps to regulate behaviors (e.g., social behavior, aggregation, sexual stimulation, trail following). **2:** of or pertaining to semiochemicals or to semiochemical interactions.

semiochemical interaction. any interaction that involves the transmission of information between organisms by means of chemicals elaborated by some or all of those that interact. *See* allelochemical, semiochemical.

semolina. a coarse wheaten flour produced

after the fine flour has been milled. It is used to make pasta and puddings. *See Triticum durum*.

Senecio (groundsel; ragwort; squaw weed; life root; senecio). a semicosmopolitan genus of more than 1200 species of herbs with many-flowered, chiefly yellow heads (Family Compositaceae), of which more than two dozen are known to be poisonous. They contain alkaloids that cause hepatogenic photosensitization. Most cases of senecio poisoning involve cattle and horses. Sheep and goats are more resistant but have been poisoned experimentally. Examples of poisonous North American species are *S. glabellus* (bitterweed), *S. integerrimus*, *S. longilobus* (thread-leaf groundsel), *S. plattensis*, *S. riddellii* (Riddell's groundsel), *S. spartioides* (broom groundsel), and *S. vulgaris* (common groundsel). *See also* senecio poisoning.

senecio. *Senecio*.

senecio poisoning. a sometimes fatal poisoning of herbivores (usually cattle and horses) and sometimes of humans due to ingestion of herbs of the genus *Senecio*. In livestock, onset of the disease is usually abrupt, typically following a latent period of several months. The animal is initially sluggish, uneasy, and anorectic, becoming rapidly weak and emaciated with signs of abdominal pain and reduced sensibility. Death may intervene in as little as a week. The mucous membranes take on a yellow or muddy coloration and the skin (and milk of cattle) develops a distinctive sweetish and unpleasant odor. In the later stages, the animal may become progressively weak, dying quietly. Some animals become restless and walk aimlessly into or through structures such as fences. Death is therefore sometimes accidental. In cattle, the coat roughens and the muzzle becomes dry and scaly; tenesmus with constipation or diarrhea is common. **acute senecio poisoning**. Such is rare in humans and livestock. Post mortem examination of affected animals may reveal specific liver lesions, the exact nature of which depends chiefly on the duration of the condition. Thus, necrosis and severe hemorrhage into the lobules are characteristic of acute intoxications. *See Senecio*. **senecio poisoning, chronic**. human intoxications in Africa usually stem from ingestion of bread made from contaminated flour. In the West Indies, poisoning may result from the native custom of brewing a medicinal tea from various indigenous plants. Signs and symptoms in humans are similar to those of venous occlusive disease and include ascites, headache, hepatomegaly, nausea, vomiting, abdominal pain, apathy, and emaciation. The vomitus and stool may be bloody. Lesions include those of hepatic fibrosis or cirrhosis. In cattle, the gall bladder may also be greatly enlarged with an accumulation of bile; nephropathies may be observed.

senecoicacid 2-*sec*-butyl-4,6-dinitrophenyl ester. binapacryl.

sensitinogen. any antigen that sensitizes an animal or produces hypersusceptibility (e.g., an anaphylactogen, an allergen).

sensitive fern. *Onoclea sensibilis*.

sensitize. **1:** to render sensitive. **2:** to render an individual sensitive to a chemical. **3:** to immunize. *See also* sensitized antigen.

sensitized. rendered sensitive.

sensitizer. **1:** antibody. **2:** a substance that causes a substantial proportion of exposed animals to develop an allergic response following additional exposure to the same or to closely related substances.

separation. **1:** a division, split, breach, or rupture. **2:** isolation; segregation; keeping apart; **3:** isolated, being kept apart. **4:** dividing into parts.

separation layer. abscission layer.

sepsis. **1a:** the presence in circulating blood and other tissues of disease-producing microorganisms (bacteria or viruses) or their toxins. **1b:** the condition resulting from def. 1a, which may include abscesses throughout the body or septicemia.

sept-, septo-. combining forms meaning septum.

septic. **1:** infected; any condition that produces pus and is caused by sepsis. **2:** heavily polluted; said of a habitat or freshwater zone that is rich in decomposing organic matter, high in carbon dioxide, and very low in dissolved oxygen.

septicemic. of or pertaining septicemia or its effects.

septicemic infection. an infection due to invasion of the bloodstream by pathogenic organisms.

septicine. a ptomaine produced from hexylamine and amylamine during the putrefaction of red meat (flesh).

septico-, septic-, sept-. prefixes denoting sepsis, septic; putrefaction.

septicopyemia. pyemia and septicemia occurring at the same time.

septicopyemic. pertaining to septicopyemia.

septis. poisoning caused by the products of putrefaction.

septum. a membranous or fleshy wall that separates two cavities in a plant or animal.

serotonin (5-hydroxytryptamine; 5-HT; 5-hydroxy-(2-aminoethyl)indole; 3-(2-aminoethyl)-1*H*-indol-5-ol; 3-(β-aminoethyl)-5-hydroxyindole; 5-hydroxy-3-(β-aminoethyl)indole; enteramine; thrombocytin; thrombotonin). a catecholamine with neurotransmitter properties. In vertebrate animals, it is concentrated in the hypothalamus within the granular synaptic vesicles of nerve terminals. Serotonin nerve terminals also occur in nuclei of the parasympathetic and sympathetic nervous systems and in the pineal gland. It is synthesized in the brain from tryptophan. Serotonin is associated with the control of wakefulness, mood, and pain sensation (e.g., in mammals); it is a chiefly inhibitory CNS neurotransmitter and a potent constrictor of smooth muscle, especially of the larger blood vessels, thus elevating blood

5-hydroxytryptamine. serotonin.

serpent kame. a synonym for esker.

serum disease. serum sickness. *See* sickness.

serum reaction. serum sickness. *See* sickness.

service (serving; covering). a term used in breeding domestic animals and large wildlife species. It indicates the mating of male to female. **natural service**. natural mating of livestock as opposed to artificial insemination.

service sire. The sire to which a female is currently being bred.

serving. *See* service.

sesame cake. the presscake that remains after the oil has been expressed from sesame seeds. It is a protein-rich feed for livestock. *See also* sesame oil, *Sesamum*.

sesame oil. a valuable oil that is expressed from sesame seeds by crushing.

Sesamum (sesame). an erect herb that attains height of about 2 m (6 ft). It is indigenous to Africa but is cultivated widely in tropical and subtropical regions. It is an important crop in China, India, Nigeria, the Sudan, and warmer areas bordering the Meditteranean Sea. *See also* sesame oil, sesame cake.

sesame. a plant of the genus *Sesamum*.

sesbane. a plant of the genus *Sesbania*.

Sesbania (coffeeweed; coffeebean; rattlebox; bagpod; rattlebrush; sesbane; poison bean). a genus of warm climate, annual or perennial herbs, shrubs, and small trees (Family Fabaceae). The three closely related species named below occur in the Gulf Coastal Plain of the southern United States and are toxic to livestock. Symptoms vary somewhat among types of stock but are similar for all three species of *Sesbania*. The acute condition is characterized by gastroenteritis that may produce necrotic lesions, diarrhea (sometimes constipation in cattle), and may be rapidly fatal. ***S. drummondii*** (= *Daubentonia drummondi*, *D. longifolia* (in part); coffeebean; rattlebrush; rattlebox, etc.). ***S. platycarpa***. *Sesbania vesicaria*. ***S. punicea*** (= *Daubentonia punicea*, *D. longifolia*, in part)) (purple sesbane, purple rattlebox, etc.). ***S. vesicaria*** (= *S. platycarpa*, *Glottidium vesicarium*) (bagpod, bladderpod, coffeebean, etc.).

setterworth. *Helleborus foetidus*.

sewage sludge. *See* biosolid.

sewer rat. *Rattus norvegicus*.

sex. **1:** gender. **2a:** coitus, copulation, intercourse. **2b:** procreation.

sex-linked disease. X-linked disease.

shallot. *Allium ascalonium*.

shear. **1a:** to clip, crop, or cut. **1b:** to shave wool, to fleece; to trim. **2:** the distortion of a body by two opposing, parallel forces whereby adjacent parts slide over one another in a plane within the body that is parallel to the planes of force. **3:** in culturing Christmas trees, to prune the branches to encourage dense foliage and to shape the tree.

shears. scissors, usually large and strong, used to shear.

sheep. a ruminant of the genus *Ovis* (Family Bovidae). Sheep are raised for their wool, meat, and/or milk. Most domestic sheep are raised where extensive range or grazing conditions are available. Sheep, when not carefully managed, can destroy large tracts of grassland. **2:** leather made from the skin of sheep.

sheep-dip. a wash or lotion applied to the skin or fleece of sheep, usually by immersing animals in vats, in order to destroy ectoparasites.

sheep fescue. *Festuca ovina*.

shell. a hard outer covering or layer, as that of a nut.

shellfish. common name of shelled mollusks and crustaceans, especially those used as human food; sometimes restricted to mollusks that have shells.

shigellosis. bacillary dysentery (*See* dysentery).

shivers. a disease of Australian and African livestock (cattle, sheep and horses) probably caused by consumption of large amounts of *Malva parviflora*. It is a nervous syndrome characterized by severe muscular tremors which are intensified by exercise. The appetite remains normal and mild cases recover quickly if the animals are kept quiet. *M. nicaeensis* (= *M. borealis*) has been implicated in a similar syndrome in livestock in the United States. *Malva spp.*, when ingested by laying fowl, produce a discoloration of yolk in stored eggs. Malvalic and sterculic acids are responsible for this effect.

shoal. a school of fish.

shorthorn (beef shorthorn). a breed of compact cattle with short horns. There are three varieties, the dairy shorthorn, beef shorthorn, and Lincoln shorthorn. Shorthorns may be red, red and white, white, or roan in color.

showy milkweed. *Asclepias speciosa*.

Shropshire sheep. a medium-sized, fast-growing, mutton breed of sheep with a dark face, developed in Shropshire and Staffordshire, England, and recognized as early as the middle of the 19th century. It is an excellent breeder and produces both a good wool and a good grade of mutton. It is the most popular sheep of English origin in the United States and is widely distributed throughout the world because of its ability to adjust to a broad range of conditions.

SI. International System of Units.

sialic. salivary.

sialine. salivary.

sialo-, siali-, bial-, ptyal-. prefixes denoting saliva.

sick. **1:** unwell; ill. **2:** the state of feeling unwell; nauseous; nauseated. **3:** disturbed or mentally ill.

sickness. **1a:** a state of being ill; a disease. **1b:** any condition marked by pronounced deviation from a healthy state; illness; an ailment, disease, disorder, or malady. **1b:** a feeling of being ill; nausea. **African horse sickness** (AHS). an insect-borne, viral disease of equids. The disease is marked by fever, swelling of the heart, and internal hemorrhage; the case fatality rate is high. **calomel sickness**. mercurous chloride. **car sickness**. a form of motion sickness. **compressed-air sickness**. bends. **decompression sickness**. bends. **green tobacco sickness**. a brief, recurrent, occupational illness of tobacco harvesters. Symptoms include headache, dizziness, vomiting, and prostration. **Jamaican vomiting sickness**. ackee poisoning. **lambing sickness**. a disease of domestic ewes that is nearly identical to milk fever of cattle. **milk sickness** (lactimorbus). trematol poisoning of humans, a disease caused by ingesting meat or contaminated milk from cattle poisoned by *Eupatorium rugosum*, *q.v.* A nearly identical condition is produced by *Haplopappus*. Milk sickness is a devastating disease that was common from the late 1700s to the mid 1900s. Even now, this disease occurs occasionally when milk from an exposed family cow is drunk. Symptoms include severe vomiting, dyspnea, delirium, convulsions, coma, and death. In non-lethal cases, recovery is slow. *See* trembles, trematol poisoning. *See also* trematol, trembles. **motion sickness**. nausea induced by motion under certain conditions of motion, especially where the eyes, middle ear, and neck proprioceptors are receiving different or inconsistent signals, hence sometimes called air sickness, car sickness or sea sickness, depending on mode of transportation. **radiation sickness** (actinotoxemia; Roentgen intoxication; radiotoxemia (obsolete)). **1a:** the original definition prior to the development and deployment of the atomic bomb, it is the illness produced by whole-body x-irradiation of humans and laboratory animals at a level exceeding about 100 rem. The severity and character of the disease are dose dependent. In mild cases, the usual symptoms are nausea, vomiting, anorexia, debility, and leukopenia. More serious cases are marked by a reduction or total loss of platelets with bleeding, anemia due to reduced capacity to produce new erythrocytes, and an increased risk of intercurrent infection due to severe loss or disappearance of leukocytes. **1b:** illness produced by exposure to any type or source of ionizing radiation. It is essentially indistinguishable from def. 1a, above. In its acute form, it is usually referred to as acute radiation syndrome. **sage sickness**. a rapidly developing disease of horses unaccustomed to foraging on sage. The animals become nervous and tend to fall when forced to move due to a partial paralysis of the forelimbs. Within a week or two, the animals recover and are able to consume large quantities of sage without recurrence of symptoms. The species responsible for the disease is *Artemisia filifolia* and sometimes *A. spinescens*. **sea sickness**. a form of motion sickness. **serum sickness** (serum disease; serum reaction). an immune complex disease that presents several days following administration of a foreign serum protein, especially horse serum. It is marked by both focal and systemic effects that may include urticaria, edema, fever, general lymphadenopathy, joint pains, and sometimes albuminuria. *See* anaphylaxis, anaphylactic shock, antivenin. **spewing sickness**. sneezeweed poisoning. **Western duck sickness**. *See* alkali disease.

side effect. unintended effect (*See* effect).

Sieva bean. **1:** *Phaseolus limensis*. **2:** the seed of *P. limensis*.

sieve tube. *See* phloem.

Sigmodon (cotton rat). a genus of medium-sized herbivorous rats with a rough coat. They are sometimes pests on farms or ranches. Cotton rats occur in grasslands and brushy areas in the southern United States, Central America, and northern South America.

sigmoido-, sigmoid-. combining forms denoting sigmoid, usually the sigmoid colon.

sign. an abnormality of structure or function that can be identified upon direct or clinical examination of a diseased or ailing animal or human. It is an objective measure of an animal's condition as opposed to a symptom described by humans.

silage. feed for livestock (fodder) prepared by storing and fermenting green forage plants (often chopped stalks and leaves of corn) in a silo. *See* silo-filler's disease.

silkweed. common name of a plant of the genus *Asclepias*.

silo. a large structure, usually located near a barn, used to store grain or silage. **bunker silo** (horizontal silo). a wide trench, 6-8 ft deep and usually 50-100 ft long, in the side of a hill from which surface water has been diverted. 50-100 movable self-feeders can be constructed in either or both ends of such a structure to make silage available to livestock. **horizontal silo**. bunker silo.

silo-filler's disease. a rare poisoning of agricultural workers due to the inhalation of nitrogen oxides (usually nitrogen dioxide and nitrogen tetroxide) while collecting silage in poorly ventilated areas such as a silo or a room at the base of a silo. The gases tend to collect in the bottom of the silo during fermentation, sometimes exposing workers to high concentrations of these gases. Concentrations of nearly 60,000 ppm of nitrogen dioxide have been recorded from silos housing fermenting silage. A breath or two under such conditions can produce acute intoxication. Symptoms may not appear until several hours following exposure and typically include respiratory distress, pulmonary edema, sometimes loss of consciousness, and rarely death. The release of nitrogen oxides by silage often reduces the possibility that the consuming livestock will suffer nitrate poisoning. *See also* nitrate.

silver vine. *Actinidia polygama*.

silverleaf nightshade. *Solanum eleagnifolium*.

silvex (2-(2,4,5-trichlorophenoxy)propionic acid; fenoprop; 2,4,5-TC; 2,4,5-T/-silvex). the α-methyl derivative of 2,4,5-T, it is a selective pre- and postemergence herbicide and is superior to 2,4,5-T on certain woody species such as *Quercus* (oak). Silvex is generally somewhat less toxic than 2,4-D. It is moderately toxic to laboratory rats and is irritating to skin, eyes, and mucous membranes. In mammals, most is excreted unchanged in the urine. The use of this herbicide on orchards, rice fields, sugarcane, rangeland, and other noncrop sites was terminated in the United States by the EPA in March, 1985. Silvex is carcinogenic to some laboratory animals.

silvicide. an herbicide used to destroy woody shrubs and trees.

Silybum marianum (variegated thistle). a thistle-like herb (Family Asteraceae) with spiny, white-spotted leaves. Concentrations of nitrate that are toxic to herbivores have been reported.

sinigrin (1-thio-β-D-glucopyranose 1-[*N*-(sulfooxy)-3-buteninidate]monopotassium salt; sinigroside; myronate potassium; potassium myronate; allyl glucosinolate). a crystalline β-glucopyranoside and glycoside of allyl isothiocyanate, $C_{10}H_{16}KNO_9S_2$, isolated from various species of Cruciferae (mustard family, crucifers), most notably from the seeds of *Brassica nigra* and the root of *Alliaria officinalis*. It decomposes in a reaction catalyzed by myrosinase when a plant is damaged, releasing the mustard oil, allyl isothiocyanate, and potassium hydrogen sulfate. The isothiocyanate is a potent vesicant and an irritant to many species of animals, thus tending to inhibit grazing. This substance, however, attracts butterflies of the genus *Pieris*, *q.v.*

sinigroside. sinigrin.

sinistro-. combining form indicating left or toward the left.

sink. **1:** a basin or depression in a flat surface. **2a:** a basin, pool, or pit used to deposit waste or sewage. **2b:** any repository used for long-term storage of matter or energy. **2c:** any reservoir large enough to receive or absorb energy or matter without undergoing significant change. Thus, body fat is a sink for lipophilic compounds (e.g., organochlorines). **3a:** any component of a living system that removes and retains, for a time, wastes or other materials that are inimical to other system components or to the system as a whole. Thus, metallotheionein may act as a sink in the removal of toxic metals from the systemic circulation. **3b:** any living system or system component (metabolite, cell, organ, tissue) that is a net importer, storage depot, and end-user of a metabolite or other resource. The root of a plant is a sink that receives sugars that were synthesized in the leaves and converts them to a storage form (polysaccharides). **4:** in air pollution science, any area or environmental component that, by whatever process, removes one or more pollutants from the air. Moist ground and green foliage are important sulfur dioxide sinks. *See also* storage, metallotheionein.

Sinox. dinitrocresol.

siphonostele. *See* stele.

Siphunculata. Anoplura.

sire. a male parent, e.g., of a horse and certain other mammals.

six-weeks fescue. *Festuca megalura*.

skein. a flock of (flying) geese. *See also* gaggle.

skin (cutis). **1:** the outer covering or integument of an animal, especially of a vertebrate animal. The vertebrate skin is comprised of two layers: (1) the epidermis, comprised of epithelium, and (2) the dermis, comprised of connective tissue with associated structures. The skin of fishes and some reptiles is covered by dermal or epidermal scales or dermal plates. The outermost layer of the epidermis (stratum corneum) of terrestrial vertebrates is cornified and variously modified as spines, nails, claws, beaks, and other structures. The skin is an organ of breathing in many animals, especially those that are aquatic or semiaquatic (e.g., Amphibia). In so-called warm-blooded animals (homeotherms), the skin is important in temperature control; feathers in birds serve this function, as do hair and sweat glands in mammals. The skin of mammals has follicles that produce hair, sebaceous (oil) glands, and usually sweat glands. *See* dermis, epidermis. **true skin**. *See* dermis.

skulk. a leash of foxes.

skunk cabbage. *Veratrum californicum. See also Symphytum foetidus.*

slaked lime. calcium hydroxide.

slash. the residual vegetation remaining on the ground due to tree-felling and other forest practices or activities. *See* lopping and spreading.

slash burning. broadcast burning.

slash pine. *Pinus caribaea.*

sleepy stage (forage poisoning; stomach stage). a disease of horses of unknown etiology, usually associated with feeding on moldy hay and grain.

sleuth (sloth). a group of bears.

slickenside. the shiny surface formed when two peds rub against each other because of expansion due to wetting.

slickspot. a small area of surface soil that becomes slick when wet due to alkalinity or a high concentration of exchangeable sodium.

slide. **1:** to slip, skid, or drift. **2:** a mass movement due to a slope failure along one or more slip surfaces. The mass usually breaks up and travels downslope. If enough water is present, the flow of debris may become a torrent, causing considerable damage to structures below. **3:** microscope slide. **microscope slide**. a rectangular glass plate on which an object is placed for examination under a microscope.

slime flux. a slimy, tacky substance that oozes from cracks in the bark of a tree or from injuries (e.g., due to pruning, insect damage). **alcoholic slime flux**. a white, foamy substance that seeps from cracks in the bark or from sapwood near the base of a tree. This material, which ultimately kills the tree, contains a high concentration of sugars, starches, and proteins. **brown slime flux**. a slimy, brown substance with an offensive odor that oozes from cuts in shade trees, usually those of declining vigor (most commonly birch, elm, and maple). The clear sap contains nutrients that attract bacteria, fungi, and insects that take up and metabolize the nutrients, releasing a complex of wastes that become rancid and turn brown on exposure to air. *See also* bacterial wetwood.

slinkweed. *Gutierrezia microcephala.*

slip. said of a domestic animal, to abort or give birth prematurely. Said of a domestic animal.

sloe. the fruit (e.g., plums, cherries) of trees of the genus *Prunus*. Whereas the fruit of most species is edible, the pit usually contains amygdalin.

slope. **1a:** an incline, grade, slant, or pitch; the deviation of a surface from the horizontal, measured as a percentage, a numerical ratio, or degrees. **1b:** any inclined surface. **1c:** a natural incline of the ground, as on the side of a hill. **1d:** in agriculture and forestry, a slope is the inclined surface of any part of the earth. Slope is determined by the angle formed by the surface in question and the horizontal plane. *Cf.* angle. **2:** the side of a hill, mountain, or similar natural structure. **3a:** in mathematics, the tangent of the angle between a given straight line and the x-axis of a system of Cartesian coordinates. **3b:** the derivative of the function whose graph is evaluated as a curve at a specified point. **4:** to move on an incline or in an oblique direction with respect to a horizontal or vertical plane. **5:** to deviate from the horizontal or vertical; to incline. **6:** to build or form a slope (e.g., to slope an embankment). **adverse slope**. an uphill incline used to haul or skid logs or other loads. **cut slope**. the face (slope) of an excavated bank at the side of a roadway (or drainage ditch, etc.) required to lower the natural ground line to that desired for the roadway. **back slope**. dip slope. **dip slope** (back slope; outface). a slope of the land's surface that conforms roughly to and is determined by the underlying rocks. **fill slope**. the upper slope of an excavated bank at the side of a roadway built on a hillside.

slope failure. the failure of a slope to remain intact and in place. *See* slide, slump.

slope processes. all processes, operations, and events that change the configuration of a slope, chiefly those by which rock, surface materials, and soil move downslope under the dominating influence of gravity.

slope stability. resistance of a slope to erosion and slides.

sloth. **1:** sleuth. **2:** a type of herbivorous arboreal animal of the order Edentata, *q.v.*

sludge. any solid, semisolid, or mud-like residue generated from the wastewater purification process of a municipal, commercial, or industrial wastewater treatment plant or from an air pollution control facility (exclusive of the treated effluent from a wastewater treatment plant). It results from screening, chemical treatment, coagulation or sedimentation, filtering, pressing, bacterial digestion, and oxidation. The disposal of sludge is regulated under the Clean Water Act. **activated sludge.** sludge that results when primary effluent is mixed with bacteria-laden sludge, then agitated and aerated to promote microbial growth and activity, thereby accelerating the breakdown of organic matter in sewage undergoing secondary waste treatment. **primary sludge**. sludge produced by a sedimentation process. **secondary sludge**. sludge resulting from microbial digestion.

slump. a mass movement in which slope failure occurs on a usually curved slip surface and the mass moves downslope as an intact block. Slumps often occur in place, while slides usually break up and travel downslope.

slurry. a thin semifluid mixture of a fine insoluble material.

small intestine. the division of the alimentary canal that lies between the stomach and the cecum. It absorbs digested food and nutrients.

small Jack-in-the-pulpit. *Arisaema triphyllum.*

smartweed. herbs and small shrubs of the genus *Polygonum.*

smegma. *See* sebum.

smilax. **1:** common name for any plant of

the genus *Smilax*. **2:** common name for *Asparagus asparagoides*.

Smilax (greenbriar). a genus of widely distributed, mostly woody, climbing plants (Family Lilaceae; placed in a separate family, Similaceae by some botanists), which are sources of sarsaparilla. The starchy root of several species was used as food by American Indians. ***S. rotundifolia*** (greenbriar; greenbrier; bullbriar; catbrier; devil's hopvine; horse briar; roundleaf). a prickly, green-stemmed, wiry species that is indigenous to North America. It is a weed that is often difficult to eradicate, once established.

Smith-Lever Act. *See* Cooperative Extension Service.

smog. **1:** naturally and artificially produced mixtures of smoke, fog, hydrocarbons, and other air pollutants dispersed in the atmosphere. **2:** air pollution associated with oxidants, especially the nitrogen oxides, NO_X, and ozone; photochemical smog. **3:** originally defined as a mixture of smoke and fog dispersed in the atmosphere. **oxidant smog**. photochemical smog. **photochemical smog** (smog; oxidant smog). a highly irritating, usually dense, lacrimatory haze produced by photochemical reactions that occur under the influence of strong sunlight in air polluted by various sources but most importantly by automobile exhaust gases under temperature inversion conditions.

smoke. **1:** in chemistry, an aerosol of solid or liquid particles generated by combustion, thermal decomposition, or evaporation. **2a:** a suspension in the atmosphere of small particles produced by combustion. **2b:** visible emissions (other than vapor) from a smoke stack or chimney. **2c:** particles in the atmosphere as a result of the incomplete combustion of fossil fuels or other organic material. **3a:** a cigarette. **3b:** a cigarette made from marijuana. **secondary smoke**. tobacco smoke that is inhaled by a nonsmoker who is in a location where smoking occurs. **side-stream smoke**. smoke produced by a cigarette and released at the distal tip. Studies indicate that such smoke poses a risk to nonsmokers (and presumably smokers). Only about 4% of the total smoke produced by a cigarette is inhaled by the smoker. The remainder is side-stream waste in which the concentration of pollutants is about twice that inhaled by the smoker. **tobacco smoke**. in general smoke produced by tobacco when a person lights a tobacco product such as a cigarette, cigar, or pipe tobacco. **wood smoke**. smoke produced by the combustion of wood as in a fireplace or wood stove. Wood smoke contains fine, respirable particles that can damage the lungs, cause lung cancer, or, when dense, can cause asphyxiation.

smoke management. in conducting a program of prescribed burns in forestry practice, predetermined burning prescriptions and firing techniques are commonly employed to minimize the adverse effects of the smoke produced in areas where smoke presents a hazard to humans or other resources. *See* smoke-sensitive area.

smoke-sensitive area. an area in which smoke density or accumulation present a hazard to public health or safety, private or public resources, or to esthetic elements in the environment (e.g., obstructing scenic views).

smut. **1:** dirt; grime; soot. **2:** a fungus disease of cereal grains caused by a species of *Ustilago*. **3:** any of a group of basidiomycete fungi (Order Ustilaginales). They are normally facultative parasites, often causing substantial damage to agricultural crops. Infested grains may poison livestock. *See*, for example, smutty oat disease.

smut ball. bunt.

smutty oat disease. smutty oat poisoning.

smutty oat poisoning (smutty oat disease). a disease of cattle and horses known from a few cases in the 19th century. Mortality was high among horses feeding on oat grain infested with smut. The disease in cattle was associated with hay that had been cut before the (already-infested) grain was ripe.

snag. a standing dead tree.

snake (serpent; ophidian). common name of any of about 3000 species of limbless reptiles (Order Squamata, suborder Serpentes). The blood of most is mildly toxic. **nonvenomous snake**. any snake that lacks fangs and a venom gland. **venomous snake**. any snake that has fangs and venom glands, which are the only anatomic attributes that distinguish venomous from nonvenomous snakes. There are more than 400 species of venomous snakes worldwide; about 360 are terrestrial; less than 200 are considered dangerous to humans. Most fatalities from snakebite occur in Asia. Collectively, venomous snakes are called thanatophidia or toxicophidia. *See*, for example, antitoxin, antivenin, *Agkistrodon*, *Bothrops*, *Coluber*, *Crotalus*, *Heterodon*.

snake pipes. a common name of *Equisetum*.

snake venom poisoning. poisoning due to the venom received from a snakebite. The amount and potency of the venom delivered varies widely, and not all bites of venomous snakes introduce venom. Just as dangerous, in many cases, is the possibility of tetanus (as in the bite (or clawing) of almost any animal) or other infectious disease. A person bitten by any snake (venomous or nonvenomous) should seek medical treatment.

snakeberry. *Actaea rubra*.

snakebite. **1:** the bite or penetration of teeth, especially fangs, with or without envenomation, when a snake strikes an animal. Most snakebites inflicted on humans worldwide are by nonvenomous species (such bites may become seriously infected). Most snakes strike or bite an extremity. The bite of many types of snakes are distinctive or nearly so and should be noted before treatment, especially with antivenin. Envenomation does not always result from the bite of a poisonous snake. Symptoms may rapidly appear following envenomation. **2:** the condition resulting from the bite or strike of a snake, especially in the region of the wound. Symptoms vary according to the family, genus, or species of snake involved, with the depth and location of the wound, the amount of venom delivered, and other factors including the condition of the victim. Nevertheless, the bite of a venomous snake usually soon produces unambiguous symptoms. *See* snake venom poisoning.

snakeroot. *See Eupatorium*.

snakeweeds. *See Gutierrezia*.

sneezeweed. *Helenium hoopsii*.

sneezeweed poisoning (spewing sickness). poisoning of sheep, and perhaps other grazing livestock, by certain plants of the genus *Helenium*, *q.v.* Symptoms include dullness and depression, sometimes nausea and vomiting which may be accompanied by sialism with frothing, and belching, a rapid and irregular pulse, and rapid breathing. Postmortem examination may reveal a variety of lesions that include gastrointestinal irritation, necrosis, and sometimes perforation of the intestinal wall where it comes in contact with ingested plant material; hepatic and renal damage, and usually necrotic lesions of the lungs.

snow on the mountain. *Euphorbia marginata*.

snowdrop. any herb of the genus *Galanthus*.

soap plant. *Zigadenus venosus*.

soapberry family. Sapindaceae.

soda niter. sodium nitrate.

sodium antimonyl tartrate. antimony sodium tartrate.

sodium arsenite (sodium metaarsenite). a whitish or light grayish powder that is soluble in water, slightly soluble in ethanol, and absorbs atmospheric carbon dioxide. It is used, for example, as an antiseptic, insecticide, a contact herbicide (it causes rapid wilting), and by taxidermists in arsenical soaps. It is a confirmed human carcinogen. It appears to be deadly poisonous by all routes of exposure, although most human poisoning is by ingestion. This salt has been shown to inhibit respiration and growth, and to upset normal mitosis in affected plants.

sodium azide. an extremely toxic, unstable, explosive, white crystalline compound, N_3NA, that is soluble in water and liquid ammonia. It is used to make lead azide explosives, as a propellant used to inflate automotive safety bags, as an herbicide, agricultural pesticide, and to control fruit rot. Sodium azide is a potent vasodilator and has been used therapeutically as an antihypertensive. Symptoms of intoxication can include hypotension, tachycardia, palpitations, tachypnea, severe headache, hypothermia, and unconsciousness, usually followed by rapid recovery. Overdosage sometimes, however, causes convulsions and death by respiratory arrest or heart failure. It is a mutagen in plants and bacteria and a suspected carcinogen.

sodiumcacodylate ([(dimethylarsino)oxy]sodium As-oxide; sodium dimethylarsonate). the sodium salt of cacodylic acid, it is a white, deliquescent, powdery or amorphous, water and ethanol soluble, crystalline substance,

$$(CH_3)_2AsOONa{\cdot}3H_2O$$

It is used as an herbicide and, in veterinary practice to improve the quality of blood by increasing erythrocyte count and hemoglobin concentration. Sodium cacodylate is toxic to terrestrial vertebrates by ingestion and inhalation. This salt is rapidly acidified in the vertebrate stomach with the liberation of inorganic arsenic, probably as arsenic acid, which is then reduced to arsenous oxide (As_2O_3).

sodium chlorate. a colorless, odorless, flammable, crystalline substance, $NaClO_3$, with a saline taste; it is a strong oxidant and bleach. It is a widely used herbicide and defoliant and has numerous other applications. Plants treated with sodium chlorate and contaminated clothing worn by the applicator are highly combustible and constitute fire hazards. Poisoning of livestock by sodium chlorate has occurred both from the ingestion of treated plants and from accidental consumption of feed to which it was mistakenly added as sodium chloride. Cattle are sometimes attracted to foliage treated with sodium chlorate. Considerable quantities must be consumed before signs of toxicity appear. Ingestion results in the conversion of normal hemoglobin to methemoglobin.

sodium chloride (salt; common salt; table salt; sea salt; halite (rock salt)). colorless, transparent, somewhat hygroscopic, salty tasting, crystals or white crystalline powder, NaCl, that is soluble in water and glycerol and slightly soluble in ethanol. An essential nutrient, it is generally considered moderately toxic. It is used as a condiment, in curing foods, and as an analytical reagent. Overindulgence may cause stomach irritation and vomiting; the most common systemic effect from ingestion of table salt is hypertension, and existing hypertension can be exacerbated with renal injury. Acute poisoning by in gestion may also cause vomiting and dehydration. Intraperitoneal administration can terminate pregnancy. Teratogenic and reproductive effects have been induced in laboratory animals and there is some indication of mutagenicity.

sodium cyanate (cyanic acid sodium salt). a very toxic, white, crystalline powder or colorless, monoclinic crystals, NaOCN, soluble in water, ethanol, and ethyl ether. It is used as a nonselective herbicide, a chemical intermediate, in the study of sickle cell anemia, the preparation of certain medications, and in heat treatment of steel.

sodium dimethylarsonate. sodium cacodylate.

sodium fluoride. a white, crystalline powder, NaF, with a salty taste that is soluble in water and very slightly soluble in ethanol. When used as an insecticide, it is often dyed blue. It is toxic to humans and other terrestrial vertebrates by ingestion and inhalation. Overexposure may cause conjunctivitis, nausea, retching or vomiting, eventual weakening of the heart, disturbances of renal function, and poor coagulability of blood.

sodium fluoroacetate (sodium monofluoroacetate; fluoroacetic acid, sodium salt; compound 1080; 1080; Fratol). a fine, white, odorless, tasteless, hygroscopic, nonvolatile, highly water-soluble, extremely toxic powder, $CH_2FCOONa$. It is apparently highly toxic to all terrestrial vertebrates, causing convulsions and ventricular fibrillation. Formerly used as a rodenticide in rural situations and as a repellant against rodents and predatory mammals such as the coyote (*Canis latrans*) and Arctic fox (*Alopex algopus*). It has often accidentally killed wildlife. Symptoms of poisoning may include vomiting, irregular breathing, auditory hal lucinations, anxiety, paresthesia of the face, facial tics, irregular pulse, convulsions, and coma. Death may result from cardiorespiratory collapse as a result of pulmonary edema or ventricular fibrillation.

sodium fluosilicate. sodium hexafluorosilicate.

sodium hexafluorosilicate (sodium fluosilicate; sodium fluorosilicate; sodium silicofluoride). a very toxic, white, granular powder, Na_2SiF_6, used as an insecticide, rodenticide, mothproof for woolen fabrics, and as a pediculicide in veterinary practice. It yields 2NaF and SiF_4 in aqueous solution. Humans have died from exposure to this compound. It is sometimes used as an insecticide.

sodium hyposulfite. sodium thiosulfate.

sodium metaarsenite. sodium arsenite.

sodium monofluoroacetate. sodium fluoroacetate.

sodium nitrate (soda niter). a colorless, transparent, odorless, crystalline, granular powder, or white or nearly white, opaque fused masses or sticks, $NaNO_2$. It is an oxidizing agent, a fire hazard near organic material, and may explode when shocked or at high temperature (537°C). The natural, impure form is Chile saltpeter, and the impure industrial form is called caliche. Sodium nitrate is used, for example, in soldering flux and pyrotechnics and as a color fixative, fertilizer, and reagent. It is used to enhance the taste of cured and various processed meats (e.g., bacon, bologna, frankfurters) and fish and as a preservative to prevent the growth of *Claustridium botulinum* in meats.

sodium pyroborate. sodium borate. *See* borate.

sodium selenate. a toxic, very water-soluble, decahydrate, white crystalline substance,

$$Na_2SeO_4 \cdot 10H_2O$$

It is used as a horticultural insecticide on nonedible plants. Anhydrous sodium selenate is extremely poisonous.

sodium silicofluoride. sodium hexafluorosilicate.

sodium subsulfite. sodium thiosulfate.

sodium TCA. sodium trichloracetate.

sodium tetraborate. sodium borate. *See* borate.

sodium thiosulfate (sodium subsulfite; hypo; sodium hyposulfite). a powder or white, translucent crystals,

$$Na_2S_2O_3 \cdot 5H_2O$$

Also available as the anhydride, it is soluble in water and oil of turpentine, nearly insoluble in ethanol, and is deliquiscent in humid air. It has a cool taste and a bitter aftertaste. Among many uses are the removal of chlorine from solution and the bleaching of wood pulp (as antichlor). Toxicity is low and it is poorly absorbed from the gastrointestinal tract.

sodium trichloroacetate (sodium TCA). the sodium salt of trichloroacetic acid, it is a water-soluble, very corrosive compound,

$$CCl_3COONa$$

used as an herbicide and pesticide. It is absorbed by the roots of plants and is used chiefly to control grasses (Family Gramineae) such as *Agropyron repens* (couch grass) and *Avena fatua* (wild oats) that infest crops such as *Beta vulgaris* (sugarbeet), peas, and *Brassica oleracea* (kale). It is used also as a protein precipitant and a histological decalcifier and fixative.

soft detergent. any cleaning agent that readily breaks down in nature.

soil. a mixture, usually stratified, of mineral matter from weathered rock together with organic components due to the decomposition or partial decomposition of vegetation. Both classes may form a soil *in situ* or, more commonly, may be transported by wind or water to a particular site. Soil often exhibits extreme variability in composition, chemical and physical properties, and gross structure even within a small area. A simple description or classification of soil to be sampled, e.g., based on soil profile analysis, is insufficient to ascertain its toxicologically relevant properties. Variations in properties such as pH within soil types can decisively alter their behavior with respect to persistence, mobility, extractability, and bioavailability of the environmental chemicals contained therein. In nearly all studies that involve chemical analyses of soils, the variability among results is high. This generally reflects true variability or inadequate sampling rather than chemical analytical errors or limitations. Furthermore, whereas biota usually make up well-defined statistical units, soils rarely do. Beyond this, concentrations of toxic chemicals in soils may bear little immediate relationship to their ecotoxicological significance or bioavailability. Thus, knowledge of a chemical's presence or concentration in soil alone is unlikely to provide useful information regarding any aspect of environmental toxicology, risk assessment, or management. **acid soil**. Soil with a pH value lower than 6.6. **alderwood soil**. a soil formed from glacial till and characterized by a highly compacted, sometimes impermeable subsoil horizon. **alkaline soil**. soil that has a pH value of less than 7.3. **alluvial soil**. any soil that has developed on a fairly recent alluvium. **anthropic soil**. a soil, no longer considered a natural *in situ* soil, due to a long history of cultivation. **argillaceous soil**. a soil that contains a large amount of clay particles. **basaltic soil**. *See* basalt. **brashy soil**. a soil comprised of rough, angular rocks, the interstices of which are filled with fine soil particles. **buried soil**. soil, originally on the surface, that has been buried at a considerable depth by geologic deposits such as volcanic ash, by flooding, or glacial deposits laid down during interglacial periods. **alluvial soil**. a usually fertile soil derived from alluvial deposits. *See* alluvium. **early soil**. a well-drained, usually sandy or sandy-loamy soil that warms early in the year. **seleniferous soil**. a soil with a high selenium content; such soils are encountered in various parts of the world. Farming on such soils has often resulted in the production of highly toxic crops and animal feedstocks. Occasionally, when animal feeds contain only moderate levels of selenium, the addition of protein supplements (e.g., linseed meal), sulfur compounds, or inorganic arsenicals to the animals' diets may counter some or all of the toxic effects of selenium. The efficacy of food additives is, however, extremely undependable. **sodic soil**. soil that is high in exchangeable sodium.

soil acidity. the hydrogen ion concentration in the soil solution expressed as pH. Soil with a reaction below pH 6.5 is generally considered acid. *See* acidity, soil reaction.

soil aggregation. aggregation of groups of soil particles by cohesion into masses such as clods or peds. The degree of aggregation is inversely related to the degree of solubility of components.

soil amendment. *See* amendment.

soil horizon. a layer in a soil profile, *q.v.* **A horizon** (A-horizon). the top layer of a soil profile, *q.v.* It usually contains humus. **B horizon** (B-horizon). the horizon that lies immediately below the A-horizon. It contains deposits of organic matter leached from the A horizon. **C horizon** (C-horizon). that which lies beneath the B-horizon and above bedrock. It consists of weathered rock.

soil profile. a vertical section of soil from the soil surface to the parent rock. *See* soil horizon.

soil reaction. a measure of the acidity or alkalinity of soil as a function of pH. Thus, soil with pH above 9 is considered very strongly alkaline; 8.5-9.9 is strongly alkaline; 7.9-8.4 is moderately alkaline; 7.4-7.8 is slightly alkaline; 6.6-7.3 is neutral; 6.1-6.5 is slightly acid; 5.6-6.0 is moderately acid; 5.1-5.5 is strongly acid; 4.5-5.0 is very strongly acid; a pH below 4.5 is extremely acid.

soil series. a group of soils, the horizons of which, except for the texture of the surface portion, have similar characteristics and arrangement in the soil profile. They are given proper names after certain place names within the regions where they occur (e.g., Norfolk, Miami, and Houston).

soil sterilant. a pesticidal agent used to temporarily or permanently destroy all biota in soil. Soil sterilants must be registered with the EPA as pesticides.

soil-to-plant transfer factor. the ratio of radionuclide activity per unit mass (usually on the basis of wet weight) of plant tissue to activity per unit mass of soil.

soilless culture. hydroponics.

soilless gardening. hydroponics.

solan. an amide herbicide applied to foliage. It produces a localized or general necrosis, depending upon the dosage. It also inhibits growth.

Solanaceae (nightshade or potato family). a

family of herbs and shrubs that includes about 85 genera and nearly 2000 species. A number of genera, such as *Aconitum* (monkshood), *Atropa* (nightshades), *Hyoscyamus* (henbanes), *Mandragora* (mandrakes), *Aconitum* (mandrake), *Datura* (jimsonweeds), *Nicotiana* (tobaccos), and *Solanum* (nightshades, potato) are dangerously toxic.

solanaceous. of or pertaining to plants of the family Solanaceae, the alkaloids they contain, or the pharmacologic agents derived from then.

solanaceous alkaloid. any alkaloid that occur in plants of the family Solanaceae; most are toxic. Examples are aconitine, scopolamine, hyoscyamine.

solanine (solatunine). an extremely toxic alkaloid and steroid glycoside,

$$C_{45}H_{73}NO_{15}$$

It occurs in a number of species of *Solanum* and *Lycopersicon*. Solanine, which is not readily absorbed, yields a neurotoxic steroid alkamine (solanidine), the biologically active aglycone, and a sugar (solanose) on hydrolysis. Solanidine is considered the most toxic moiety. Symptoms and lesions of poisoning by solanine are similar among various mammals but vary specifically with the relative amounts of the irritant glycoalkaloids and the released alkamines that reach the animal's system. Livestock have been killed by grazing on vines of solanine-containing plants and humans have died from ingesting potatoes (i.e., the tubers) with "sun burned" spots or with sprouts. Spoiled potatoes may contain more solanine than normal. Solanine has been used in agriculture as an insecticide and was formerly used to treat bronchitis, epilepsy, and asthma. *See* solanum poisoning.

solanine poisoning. poisoning by ingestion of plant materials that contain solanine. The chief proximate poison is solanidine, a product of the hydrolysis of solanine. *See also* solanine, solanum poisoning. Symptoms vary with source (as other toxicants may be involved). *See* especially potato poisoning, solanum poisoning.

Solanum (nightshades). a largely tropical genus of herbs, shrubs, trees, and vines (Family Solanaceae); some species are highly toxic. Many are ornamental, some are vegetables, such as potatoes and eggplants; others have medicinal value. Both fresh and dry plant material are toxic, thus hay containing poisonous *Solanum* spp. is hazardous to livestock. Symptoms and lesions produced by the various species are similar, but vary with the relative amounts of the particular irritant glycoalkaloids and alkamines that reach the animal's system. Concentrations of nitrate that are toxic to livestock have also been reported. *See also* solanum poisoning. ***S. dulcamara*** (European bittersweet; climbing nightshade). a shrubby vine native to Europe; introduced and widely naturalized in North America. Leaves and berries are toxic. Ingestion produces a prodromal scratchy sensation in the oropharynx followed by abdominal pain, sometimes headache, anorexia, nausea, vomiting, diarrhea, shock, and respiratory depression. Cattle, horses, and sheep have been poisoned by ingestion of this plant. The berries, red when ripe, are toxic and have caused fatalities in children. ***S. eleagnifolium*** (silverleaf nightshade; white horse nettle). a plant of dry, open woods, prairies, and disturbed areas, mostly in the southwestern United States. Cattle have been severely poisoned by ingestion of as little as 1 g/kg body weight of this plant, and it has, at times, caused substantial losses among cattle. Sheep are more resistant; goats are even more resistant. ***S. esuriale***. ingestion of this plant causes calcinosis in herbivores. *See* enzootic calcinosis. ***S. intrusum*** (garden huckleberry; wonderberry). a cultivated species that is very closely related to

S. nigrum. The fruit appears to be non-toxic. ***S. malacoxylon*** (glaucophyllum). a plant indigenous to Argentina and Brazil, it causes calcinosis in herbivores. ***S. melongena***. a species that bears lavender flowers and purple fruit. ***S. m. esculentum*** (eggplant). a food plant, the vines contain solanaceous alkaloids and may be toxic to livestock. *See* eggplant blight. ***S. nigrum*** (black nightshade; deadly nightshade; common nightshade; garden huckleberry). an annual weak-stemmed plant that often lies on the ground but sometimes grows erect to 2/3 m. The leaves are highly toxic as are the berries, especially when green. Symptoms of intoxication following ingestion are chiefly those of gastrointestinal poisoning and include abdominal pain, nausea, vomiting, diarrhea, and circulatory and respiratory depression. Numerous poisonings have been reported, though some, perhaps many, of these poisonings in the United States were due to *S. americanum*. ***South torvum***. contains substances that can induce calcinosis in herbivores. *See* enzootic calcinosis. ***S. triflorum*** (three-flowered nightshade; cutleaf nightshade). an annual herb, native to the prairies of North America, now a weed common to disturbed soils and waste areas of British Columbia and most of the United States. It contains solanaceous alkaloids and is known to be toxic to livestock. ***S. tuberosum*** (potato; Irish potato). a tropical and subtropical perennial herb now cultivated widely in temperate regions as a vegetable. The vines, sprouts, peelings, and sunburned or spoiled potatoes have poisoned livestock on many occasions; human poisoning has also been validated. These plants contain solanine. The highest concentrations occur in the green parts, especially sprouts and the sun-greened skin of the tuber. Symptoms and lesions are generally those of acute solanum poisoning, severe GI symptoms, headache, cold, clammy skin, and circulatory and respiratory depression. ***S. villosum*** (hairy nightshade). a toxic annual weed that is common in the eastern United States and locally in the west.

solanum poisoning. poisoning by any of various species of plants of the genus *Solanum*, *q.v.* It is essentially a form of solanine poisoning. Poisoning is almost always acute but not always fatal; chronic poisoning is known, however. **solanum poisoning, acute**. symptoms and lesions are similar for the various species of plants and include the effects of the gastroirritant glycoalkaloids and the neurotoxic effects of one or more alamines such as solanidine. Symptoms of gastrointestinal irritation range from slight to severe. They include nausea, vomition, anorexia, abdominal pain, constipation, or diarrhea, sometimes with bloody feces. Neurotoxic effects may include apathy, drowsiness, and even stupefaction, tremor, dyspnea, salivation, and progressive debilitation (weakness, paralysis, prostration, and coma). Death (from paralysis) or recovery occurs within a few hours. **solanum poisoning, chronic**. the effects of chronic poisoning of cattle (established for *Solanum carolinense*) include anorexia, emaciation, constipation, and ascites.

solatunine. solanine.

solenostele. *See* stele.

Solidago (goldenrod). a genus of erect herbs (Family Asteraceae), usually with golden (occasionally white) composite flowers. There are many species, among which hybridization is common. Some species are considered toxic to livestock, due at times to the presence of toxic concentrations of nitrates. ***S. virgaurea***. an aromatic species that occurs in Europe and North America. It has a diuretic action.

solubility. **1a:** the capacity or tendency of one substance (the solute) to mix or blend uniformly with another (the solvent). **1b:**

the extent of such mixing under specified conditions. The solubility of a given solute varies among solvents and is affected by factors such as temperature and pH.

solum. that portion of a soil situated above the relatively undisturbed or unaltered matter.

Solvent Blue 8. methylene blue.

somatotrophin, **somatotropin** (adenohypophyseal growth hormone; anterior pituitary growth hormone; growth hormone; GH; hypophyseal growth hormone; phyone; pituitary growth hormone; somatotropic hormone; STH). a species-specific anabolic protein hormone produced by the acidophile cells of the anterior lobe of the pituitary gland. It promotes body growth, stimulates protein synthesis, promotes lipid mobilization and milk production, and inhibits glucose utilization.

somatotropic hormone. somatotrophin.

Sonoran coral snake. a common name of *Micruroides euryxanthus*

soot. any brown to black dust or powder with a high carbon content produced by the incomplete combustion of any carbonaceous material. It is a mixture containing variable amounts of organic tars (sometimes more than 50% by weight), colloidal carbon, and refractory inorganic materials depending upon the composition of the combusted material and the conditions of combustion. Soot has physical and chemical characteristics that distinguish it from carbon black. Long exposure to soot via direct contact or by inhalation can result in pulmonary tumors and cancer of the skin or scrotum.

sophorine. cytisine.

sorbefacient. **1:** promoting absorption. **2:** an agent that promotes absorption.

sorghum. a common name of grasses of the genus *Sorghum*.

sorghum forage poisoning. poisoning of livestock by certain species and varieties of *Sorghum*, *q.v.*

Sorghum (sorghum; Johnson grass; Columbus grass; sudan grass). a genus of grasses (Family Gramineae) that have been cultivated in Africa and Asia since antiquity. Hundreds of varieties are known but these are not well characterized genetically. They are thus usually divided into two categories: (1) the cultivated (grain) sorghums (*See S. vulgare*) and sudan grass which are annuals, and (2) Johnson grass, a weedy perennial. Sudan grasses produce hydrocyanic acid in varying concentrations, depending upon the strain and growing conditions. Conditions that do not support normal growth usually increase cyanide formation. Soils high in nitrogen may yield a plant with highly toxic levels of cyanide. Livestock pastured in fields with stunted or immature sudan grass, especially if it is dark green in color, may be at risk. ***S. almum***. a perennial sorghum originating in South America. A forage crop, it has sometimes caused serious mortality in livestock. This species hybridizes with *S. halepense*, producing very toxic seeds. ***S. halepense*** (Johnson grass; Egyptian millet). native to Eurasia, it is a locally common noxious weed in old fields and waste areas in many parts of the United States. It has sometimes produced serious mortality of cattle, especially in a number of southern states. Elsewhere it is sometimes used as a forage and hay crop. ***S. vulgare*** (cultivated (grain) sorghums). a species that includes a large variety of cultivated sorghums, many of which are unnamed. Included are the saccharin (forage) sorghums (sweet or cane sorghums, sorgo), grain sorghums (including hegari, kafir corn, and milo), and broom corn. Many varieties are unnamed. Sorghum toxicity of the seeds and foliage is due to the cyanogenetic glycoside, dhurrin. Signs of poisoning in livestock are those of cyanide poisoning. Cattle suffer the most fatalities among livestock. The saccharin sorghums may also accumulate toxic levels of nitrates, and sudan grass has been implicated in photosensitization of sheep. ***S. v. sudanense*** (sudan grass). a variety that is sometimes toxic; a photosensitizer.

sorghum forage poisoning. poisoning of livestock by certain species and varieties of *Sorghum*, *q.v.*

sorgo. a type of forage sorghum. *See Sorghum vulgare.*

sorrel. a common name most appropriately applied to *Oxalis pes-caprae* but often used in reference to dock and certain other species of *Rumex*. The latter usage has caused some confusion in reports of livestock losses in the United States.

sour gum. a common name of *Nyssa sylvatica.*

soursob. *Oxalis pes-caprae.*

south Mexican turkey. *Meleagris g. gallopavo.*

South African bont tick. *Amblyomma hebraeum.*

Southdown sheep. a popular breed of sheep used for mutton. It is small, but thickset, and is highly valued for its lamb and mutton. The breed was developed on the South Downs of Sussex and is one of England's oldest breeds. It is now worldwide in distribution, finding much use in improving other breeds. The wool is a fine grade of medium length.

southern yellow pine. *Pinus palustris.*

sow. **1a:** an adult female swine, one that has produced at least one litter. *Cf.* gilt. **1b:** the adult female of certain other mammals, e.g., bears.

soybean. *See Glycine max.*

sp. abbreviation for species, singular.

Spanish broom. *Sparteum junceum.*

Spanish jack. Mexican jack.

Spanish tea. *Chenopodium ambrosioides.*

Spanish vetch. a poisonous pea. *See Lathyrus.*

spanon. chlordimeform.

sparteine([7*S*-(7α,7aα,14aα,14aβ)]-dodecahydro-7,14-methano-2*H*,6*H*-dipyrido[1,2-a:1′,2′-e][1,5]diazocine; *l*-sparteine; lupinidine). a toxic, viscous liquid, tetracyclic oxytocic alkaloid extracted especially from the tops of *Sparteum junceum* (=*Cytisus scoparius*) but also from *Anagyris foetida*, *Lupinus luteus*, and *L. niger*. It is cardiotoxic with a digitalis-like action. **sparteine sulfate**. a crystalline compound used therapeutically as an oxytocic and formerly as a cardiotonic drug. *See also* broom poisoning, cytisine.

***l*-sparteine**. sparteine.

Sparteum junceum (*Cytisus scoparius*; *Sarothamnus scoparius*; common broom; weavers' broom; Spanish broom; Scotch broom). an attractive, rather hardy, leguminous European shrub. It has been naturalized in parts of North America where it is often planted as cover along roadsides or other dry gravelly embankments. The raw seeds and leaves of are toxic and contain two toxic principles, sparteine and cytisine. Symptoms of intoxication include a weak pulse, hypotension, intestinal paralysis, and muscular weakness. The active principle is sparteine.

speargrasses. grasses of the genus *Poa*.

species. **1:** a group of entities (e.g., of organisms or coins) that are acknowledged as distinct from other such entities. **2:** a taxon that forms the unit of biological classification. It falls below genus in the hierarchy of classification. **3:** biologically, a species of sexually reproducing organisms is a population of individuals that do not normally breed with other individuals of other such populations.

species aggregate. *See* aggregate.

species-dependent. of or pertaining to a characteristic, behavior, or response that is specific to a given species, although it may not be invariably present or observed in the species. *See also* species-typical, specific (species-specific).

species-specific. of or pertaining to a characteristic behavior, or response of an organism that is observed in all members of a given species under the same conditions.*See also* species typical.

species-specific behavior. **1a:** any behavior pattern that is performed by all members of a given species under the same conditions and which is not modified by learning. **1b:** behavior that is specific to and characteristic of a given species. Few, if any behaviors are truly species-specific. Species-typical is usually the preferred term.

species-typical. of or pertaining to a characteristic, behavior, or response that is typical of a given species; one that is neither restricted solely to the species nor invariably observed in the species. *Cf.* specific (species-specific). *See also* species-dependent, specific (species-specific).

specific cholinesterase. acetylcholinesterase.

speckled loco. *Astragalus lentiginosus*, a true locoweed.

sperma-, spermato-, sperm-. combining forms denoting semen or spermatozoa.

Spermatophyta (seed plants). A division of the Kingdom Plantae that includes all seed-bearing, vascular plants. It is comprised of two classes, the Gymnospermae and the Angiospermae.

spermatozoid. antherozoid.

sphagnum moss. peat moss.

sphagnum-bog galerina. *Galerina tibicystis.*

sphere-. a prefix indicating spherical, a sphere.

sphincter. a ring-shaped smooth (involuntary) muscle that, upon contraction, constricts or closes a passage, an opening, or the lumen of a tubular structure (e.g., the anal sphincter or the sphincter muscles in the distal end of a cow's teat).

sphygmo-, sphygm-. prefixes denoting pulse.

spider. with the exception of two small groups, all spiders are venomous. Fortunately, the fangs of most species are too short or fragile to penetrate human skin. Nevertheless, at least 60 species in the United States alone have been implicated in bites to humans. Species that are dangerous include the widow spiders, *Latrodectus mactans*, and related species; the brown or violin spider, *Loxosceles reclusa* (sometimes called the brown recluse), and related species; the jumping spiders, *Phidippus* species; the tarantulas, *Aphonopelma* and *Pamphobeteus*; the trap-door spiders, *Bothriocyrtum* and *Ummidia* species; the so-called banana spiders, *Phoneutria* and *Cupiennius sallei, Lycosa* (Wolf spider), and *Heteropoda*; the crab spider, *Misumenoides aleatorius*; the running spiders, *Liocranoides* and *Chiracanthium*; the orbweavers, *Neoscona vertebrata, Araneus* species, and *Argiope aurantia* (or ange argiope); the running or gnaphosid spiders, *Drassodes*; the green lynx spider, *Peucetia viridans*; and the comb-footed or false black widow, *Steatoda grossa*. *Pamphobeteus, Cupiennius,* and *Phoneutria* are not native to the United States but may be brought into the country on produce or other materials.

spider lily. a common name usually applied to bulbous plants of the genus *Hymenocallis* (*q.v.*) and locally applied to plants of the genus *Lycoris*. *See Hymenocallis*.

spilanthol. affinin.

spin-, spino-. combining forms denoting (1) the spine, and (2) spinous.

spinach. *Spinacia oleracea*.

Spinacia oleracea (spinach). a common short-season garden herb. It contains moderate amounts of oxalic acid. Has produced death from calcium deficiency in experimental animals maintained on diets that just meet minimum calcium requirements.

spir-, spiro-. prefixes that denote (1) a coil; coil-shaped. (2) breathing.

spirit level. a tool used by carpenters and others to determine whether a surface is sloping or horizontal. It is comprised of a slightly bent transparent tube, mounted in a frame, and partially filled with ethanol, ether, or a similar fluid; space for a small bubble remains. If the bubble remains in the middle, the surface is horizontal. A spirit level may contain two tubes at right angles to one another, allowing carpenters, for example, to register deviations from the vertical as well as the horizontal.

spirochetal jaundice. human leptospirosis *See* leptospirosis.

spirochetosis. any infection caused by spirochetes.

splanchno-, splanchn-, splanchni-. combining forms denoting the viscera. *See also* viscero-.

spondylo-, spondyl-. combining forms denoting the vertebrae.

spongio-. combining form denoting sponge, spongelike, spongy.

spontaneous behavior. **1:** behavior that is (1) involuntary, automatic, or unconscious or (2) immediate, unrehearsed, impetuous, or unplanned.

spontaneous movement. autonomic movement.

spoon-feed. **1:** to feed with a spoon; to cause to be fed with a spoon. **2:** fed with a spoon. **3:** pampered; given excessive care. **4:** given no opportunity to act or think independently.

spor-, sporo-, spori-. combining forms denoting seed, spore.

sporangium (plural, sporangia). a structure in which spores develop.

spore. **1:** a minute structure other than a seed that is capable of developing into a new individual. **2:** a reproductive body in nonseed plants.

sporicidal. able to destroy spores.

sporicide. an agent that kills spores.

sporidesmin. a mixture of chlorine-containing mycotoxins (A-H and J) secreted by *Pithomyces chartarum* (= *Sporidesmium bakeri*) (Class Deuteromycetes), a saprophyte of decaying grasses. It causes facial eczema and liver damage in sheep and cattle in New Zealand and Australia. It is the causative agent of facial eczema in domestic sheep. *See* eczema (facial eczema). **sporidesmin G**. a possible cause of facial eczema in humans.

Sporobolus airoides (alkali sacaton; alkali dropseed; tussock grass). a grass (Family Gramineae) indigenous to North America that is used to reseed grazing lands.

sporophyll. a modified leaf that bears sporangia.

spotted spurge. *Euphorbia lactea*.

spreading adder. *Heterodon contortrix*.

spring vetch. *Vicia sativa*.

springhalt. stringhalt.

sprouting broccoli. *B. oleracea* var. *botrytis*.

spurge family. Euphorbiaceae.

spurges. certain plants of the family Euphorbaceae, especially those of the genus *Euphorbia*, *q.v.*; sometimes applied also to other unrelated plants such as those of the family Thymelaeaceae. *See Daphne*, *Phyllanthus abnormis*.

sputum. a secretion of the lungs coughed up in large amounts in some lung diseases.

squash. *Curcubita maxima*.

squaw weed. *Senecio*.

squirrel corn. a common name of *Dicentra spp*.

squirrel food. *Zigadenus venosus*.

St. Anthony's fire (mal des ardents; feu de Saint-Antoine). **1:** now obsolete, it was the original name (feu de Saint-Antoine) for ergotism, based on the first recognized case in Dauphiné, France in 1039 AD, where St. Anthony's remains were housed. The cause of the condition was unknown until 1676. **2:** neurotropic (convulsive) ergotism. **3:** *Claviceps purpurea*.

St. John's wort. *Hypericum*.

stachbotryotoxicosis. a hemorrhagic mycotoxicosis of horses and cattle following ingestion of oats, hay, or fodder that has been overgrown by the fungus *Stachbotrys atra*. Humans may be poisoned if exposed to infested forage either by inhalation or percutaneous absorption of the toxins. Typical effects in humans are skin rash, pharyngitis, and mild leukopenia. *See also* trichothecene.

Stachybotrys atra. the cause of stachbotryotoxicosis. The active principle is the mycotoxin satratoxin.

stafftree family. Celastraceae.

staggerbush. *Lyonia mariana*.

staggergrass. *Amianthium muscitoxicum*.

staggers. **1:** a common term for any of a number of mostly unrelated neurotoxic diseases of livestock, e.g., atamasco poisoning, bracken staggers, grass staggers (*See* grass tetany), blind staggers (subacute selenium poisoning), *Lamium amplexicaule* poisoning, annual ryegrass staggers, paspalum staggers, perennial ryegrass staggers, choking staggers (*See* moldy corn poisoning), and stomach staggers (*See* moldy corn poisoning). **2:** a type of decompression sickness characterized by mental confusion, vertigo, and muscular weakness. **3:** gid, a disease of sheep, marked by swaying and uncertain gait, caused by the presence of the larva of *Taenia multiceps*. **annual ryegrass staggers**. an often fatal disease of livestock of any age or sex due to grazing on *Lolium rigidum* (annual ryegrass). It is prevalent at times in western and southern Australia and South Africa from November through March. **ryegrass staggers**. either of two related neurotoxic diseases.

staggerweed. **1:** *Dicentra eximia*. **2:** *See Delphinium*.

stagnant anoxia. *See* anoxia.

stalk. a stem-like structure that forms the (usually supportive) base of a flower or leaf (used in this dictionary in place of pedicel, peduncle, and petiole).

stall. a barn, shed, stable, or boxed off area within a barn used to house livestock, usually a single animal, especially a horse or cow.

stall-feed. **1:** to feed and maintain an animal in a stall. **2:** to fatten an animal for slaughter by retaining and feeding in a stall.

stallion. a mature male horse capable of breeding.

stamen. that part of the flower that produces pollen; the pollen-bearing organ of a flower composed of the anther (pollen sac) and filament (stalk).

stand (stand of trees). a community dominated by trees that are distinguishable as a group from growth on adjoining areas. A stand is relatively uniform in species composition, age, structure, and condition, thus forming a silvicultural or potential management unit. **age stand**. the average age of the trees that form a stand. **all-aged stand**. an uneven-aged stand that contains all age classes. **even-aged stand**. a stand of trees comprised of one or two age classes. Such stands are often the result of fire, clearcutting, or shelterwood. **forest stand**. a stand that occurs within a forested area. **free-growing stand**. a stand of healthy, commercially valuable trees, the growth of which is not impeded by competition with other vegetation. **mixed stand**. a stand comprised of two or more tree species. **natural forest stand**. an unmanaged stand that has developed under natural conditions. **pure stand**. a stand comprised of a single species of tree. **uneven-aged stand**. a stand of trees containing three or more age classes. In a balanced uneven-aged stand, each age class is represented by approximately equal areas, providing a balanced distribution of diameter classes.

standard deviation. the square root of the sum of the deviations from the mean squared, divided by one less than the number of observations. A measure of the variation within a population of individuals. *See also* coefficient of variation.

standing crop (standing crop biomass; standing stock). the total mass of organisms that make up a population, a specified component of a population, or that within a given area or community. Standing crop may be measured in terms of volume, mass, or energy content. *See also* biomass.

standing crop biomass. *See* standing crop.

standing stock. standing crop.

Staphylococcin. any bacteriocin produced by *Staphylococcus spp*. *See also* bacteriocin.

Staphylococcus. a genus of Gram-positive, nonmotile, facultatively anaerobic, chemoorganotrophic bacteria. Most are halogen tolerant. Some are pathogenic. *See also See also* bacteriocin, mastitis, metritis.

star-of-Bethlehem. a common name of plants of the genus *Ornithogalum*.

starvation. **1:** deprivation or withholding of food to an organism for a relatively long period of time. **2:** the suffering or condition resulting from lack of food or the inability to take in, digest, absorb, or metabolize nutrient substances. Such a condition is marked by loss of biomass and many changes in behavior and metabolism. Unless remedied, death is the ultimate outcome.

starwort. *See Aster*.

-stat. a suffix indicating an agent intended to keep something from changing or moving.

stearo-, stear-. combining forms denoting fat. *See also* steato-.

steato-. combining form denoting fat. *See also* stearo-.

steatogenesis. **1a:** biosynthesis of lipids. **1b:** commonly used to denote lipid accumulation in the testes of nonmammalian vertebrates following spermatogenesis.

steatosis. **1:** obesity; corpulence. **2:** fatty degeneration, as of the heart or liver.

steer. **1:** a male bovine castrated prior to the appearance of secondary male sex characteristics. Steers finish faster and are less aggressive than bulls. In many countries they are used as draught animals. **2:** sometimes also called a bullock.

stelar. pertaining to a stele.

stele (vascular bundle; vascular cylinder). the central core of the stems and roots of a vascular plant. It is comprised of vascular tissue (xylem, phloem), ground tissue (e.g., pith, medullary rays), and an outermost layer, the pericycle. There are several types of stele. **atactostele**. a stele typical of monocotyledons. The bundles of vascular tissue are distributed unevenly throughout the ground tissue. **eustele**. a stele found in dicotyledons and gymnosperms. It is comprised of a ring of discrete primary vascular bundles situated within a single ring of endodermis and the pericycle. **perforated stele**. a stele with gaps that are not associated with leaves. **polycyclic stele**. an arrangement of two steles, one within the other. **polystele**. a stelar arrangement in which one or more parallel, sometimes anastomosing, vascular systems occur. **protostele**. a stele having no leaf gaps. **siphonostele**. medullated protosteles and solenosteles treated collectively as a single type of stele. **solenostele**. a siphonostele that contains a core of pith internal to the xylem and also a cylinder of phloem internal to the xylem. *See* stele.

Stellaria media (chickweed). a perennial weed (Family Caryophyllaceae) that grows to about 0.6 m. It is endemic to Europe but has become established in much of North America. Toxic concentrations of nitrate that are toxic to herbivores have been reported.

stem turnip. *B. oleracea* var. *gongylodes*.

stemless loco. *Oxytropus lambertii*.

stenecious. stenoecious.

steno-. combining form denoting narrowness or constriction.

stenobiontic. pertaining to or denoting an organism that is intolerant of a variable or unstable habitat. *Cf.* eurybiontic.

stenoecic. stenoecious.

stenoecious (stenoecic; stenecious; stenoecious; stenokous). of or pertaining to a living organism that is restricted to or tolerant of a narrow range of habitats and environmental conditions; one characterized by stenoky. *See also* amphioecious, euryoecious.

stenoekous. stenoecious.

stenohaline. of or pertaining to an aquatic organism that tolerates only very small variations in the salinity of the water; true of certain jelly fish and numerous fishes. *Cf.* euryhaline.

stenoionic. tolerating or characterized by a narrow range of pH. *See also* euryionic.

stenoky. the quality or condition of being stenoecious, whereby a species or population of organisms tolerates a relatively narrow range of ecological conditions. *See also* euroky.

stenophagous. pertaining to an organism that subsists on only one or a few kinds of food.

stenosis. a narrowing or constriction of a tube, duct, or vessel within an organism.

stenotropic. said of organisms that exhibit a limited response or fail to adapt to changing environmental conditions. *Cf.* eurytropic.

stenotropism. the state or condition whereby an organism responds little or fails to adapt to changing environmental conditions. *Cf.* eurytropism.

stereo- 1: a combining form that denotes a solid, or a solid condition or state. **2:** a prefix that denotes spatial qualities, three-dimensionality.

sterilant. **1:** an agent that renders an animal sterile or incapable of reproduction. **2:** an agent that destroys microbial activity (e.g., toxicants, steam, ultraviolet light, high velocity electron bombardment). **chemical sterilant**. a chemical that eliminates the viability of a microbe. *Cf.* chemosterilant.

sterile. **1a:** of or pertaining to sterility. **1b:** unable to produce offspring; barren; apogenic. **2:** aseptic; not producing microorganisms; free from living microorganisms. **3:** axenic; pertaining to cultures of a particular microorganism that are uncontaminated by other microorganisms.

sterility. **1:** infertility; apogeny; in general, the inability to conceive or produce viable offspring (female sterility), fertilize, or induce conception (male sterility). **2:** a condition of being aseptic or free from living microorganisms and their spores.

sterilization. **1:** the use of chemical or other means (e.g., steam, UV light, high-velocity electron bombardment) to destroy all microorganisms in water or on surfaces. *Cf.* disinfection. **2:** the inability to reproduce. **3a:** any process (e.g., surgery, exposure to a xenobiotic) whereby the capacity of an individual organism to reproduce is eliminated. **3b:** in pest control, the use of radiation or chemicals to severely impair or kill body cells that are essential to reproduction. The destruction of all living organisms in water or on the surface of various materials.

sterno-, stern-. combining forms denoting the sternum.

steroid. **1:** any of a large family of polycyclic lipids that contain the tetracyclic cyclopenta[a]phenanthrene nucleus. Included are many hormones (e.g., adrenocortical hormones, gonadal steroids), precursors of certain lipid-soluble vitamins, certain active principles in toad venoms (e.g., bufanolide), and a variety of drugs (e.g., synthetic estrogens). **2:** pertaining to steroids (def. 1). **3:** sometimes used for any of a number of substances that are structurally related to steroids (def. 1). Included are bile acids, sterols, cardiac glycosides, saponins, and a number of drugs.

stetho-, steth-. combining forms denoting the chest.

Stewart's disease. bacterial wilt of corn.

STH. somatotropic hormone (*See* somatotrophin).

stheno-. combining form denoting strength, force, power.

stimulate (plural, stimuli). **1a:** to provoke a response. **1b:** to excite an organ or structure to functional activity or to bring about an increase in activity.

stimuli. plural of stimulus.

sting. **1a:** a sharp, usually brief pain or burning sensation. Most result from a puncturing of the skin by the mouthparts or stinger of any of numerous species of arthropods, including insects, myriapods, and arachnids. Other invertebrates such as jellyfish, sea urchins, some sponges, some mollusks, and several species of venomous fish, such as stingray, toadfish, rabbitfish, and catfish. *Cf.* bite. **1b:** an injury due to the combined effects of the mechanical injury incurred and the venom injected into the wound. Except for the associated trauma, which may be slight to severe, stings are not usually serious. Nevertheless, poisoning, infection, or rarely anaphylactic shock must always be considered possible and preventive measures taken (*See*, for example, hymenoptera sting). **2:** a stinger, *q.v.* **3:** to introduce venom by stinging. **hymenoptera sting**. some hymenopterans are not venomous, while others can be deadly. In mammals and birds, the sting usually produces some tenderness, apparent pain and/or itching, redness, and edema at the site of a sting and sometimes intense pain. Stings may also result in more generalized symptoms that include itching, urticaria, weakness and malaise, and sometimes anxiety, nausea, vomiting, dizziness, edema, wheezing, difficult breathing, and abdominal pain. Severe cases may also exhibit some of the following: hoarseness, thickened speech, dysphagia, confusion, and despondency. Hypersensitive individuals may develop deep, systemic (anaphylactic) shock with any or all of the following: hypotension, cyanosis, incontinence, collapse, and unconsciousness. Delayed anaphylactoid reactions may occur and are a risk in subsequent stings.

stinger. a general type of organ of envenomation. It is a hard, piercing organ of certain animals such as bees, wasps, and stingrays. It may consist of a chitinous spicule, bony spine, other hard, sharp, piercing structures and a venom gland, sac, or scattered cells that secrete venom. *Cf.* sting. **bee stinger**. the stinger of the worker bee is a sharp, barbed, modified ovipositor situated in the posterior wall of the abdomen. It is connected to the venom sac. The bee only stings once because when the bee drives the stinger into a victim, the barbs anchor it and the stinger is torn from the bee (who later dies) as it flies off. The stinger of the queen bee lacks barbs. Drones have no stinger. **hornet stinger.** the stinger is similar to that of a bee but is retained following envenomation. **wasp stinger**. the stinger is similar to that of a bee but is retained following envenomation.

stinking elder. *Sambucus pubens*.

stinging nettle. *See* nettle.

stink gland. a scent gland, *q.v.*, that secretes a foul-smelling substance, as in the case of certain insects (e.g., *Cimex lectularius*, cockroach). *Cf.* odoriferous gland.

stinking hellebore. *Helleborus foetidus*.

stinking smut. bunt.

stinkweed. *Datura stramonium*.

Stirifos. tetrachlorvinphos.

Stizolobium deeringianum (deering velvetbean; Florida velvetbean). a tropical vine (Family Fabaceae) indigenous to Asia. It is grown as a cover crop, for green manure, for use as a stock feed, or for its purple flower clusters.

stocking. a measure of the area occupied by trees relative to a desired or optimal area. The measure is usually based on the basal area of well-spaced trees per acre or hectare.

stolon. 1: a vegetatively propagating plant structure. It may consist of a runner that extends along the surface of the ground, taking root at the nodes where a cluster of shoots arise and send out new runners. If a stolon is short, taking root only at the tip, it is termed an offset. **2:** a stemlike or stalklike structure in some invertebrates; new individuals bud off from these structures.

stoma (plural, stomata). **1:** a minute opening or pore. **2a:** any of a large number of small apertures in the epidermis of portions of the aerial parts of green plants (e.g., leaves, stems, flowers), in some underground rhizomes, and in the leaves of some aquatic plants. The exchange of gases and loss of water by transpiration take place through these openings. Each stoma is bordered by a pair of guard cells, which control its opening and closing. **2b:** this term is sometimes taken to include the aperture and the guard cells combined.

stomach. **1:** the most anterior or superior organ of the vertebrate digestive tract (the gastrointestinal tract), it is usually situated between the esophagus and intestine. There is structural and functional variation among various taxa. Thus, the stomach may be simple or compartmentalized. In birds, the posterior end (the gizzard) is muscular and grinds food; in ruminant herbivores, the stomach is comprised of several chambers which facilitate the digestion of cellulose by commensal bacteria. The stomach typically functions in the temporary storage of food, digestion (mechanical and chemical), and absorption with translocation to other parts of the body. The food is broken up by the churning action of the muscular walls (muscularis mucosa) of the stomach. Glandular cells in the mucosa secrete digestive (gastric) juices that contain hydrochloric acid, which lowers the pH of food, and pepsin, which initiates protein digestion (a hydrolytic process); most protein digestion takes place in the stomach. Some substances that are ionic at a pH near 7 and above are neutral in the stomach and are thus readily absorbed. Stomach contents, other than HCl, may include food particles, gastric mucin, gastric lipase, and pepsin. A pyloric sphincter at the posterior or inferior end of the stomach regulates the passage of food into the duodenum. **2a:** any structure in an invertebrate that is analogous to the vertebrate stomach; such organs are common. **2b:** the entire digestive cavity of an invertebrate.

stomach stage. sleepy stage.

stomach staggers. moldy corn poisoning.

stomachal. stomachic; pertaining to the stomach.

stomachic. pertaining to the stomach.

stomal. pertaining to a stoma.

stomas. a plural of stoma, *q.v.*

stomata. a plural of stoma, q.v.

stomatal. pertaining to a stoma.

stomatitis. inflammation of oral mucosa.

stom-, **stomat-**, **stomato-**. prefixes denoting an opening or mouth.

stomato-, stom-, stomat-. combining forms denoting mouth.

stomatopathy (stomatosis). any disease or lesion of the oral cavity.

stomatosis. stomatopathy.

stork's-bill. *Erodium*.

storksbill. *Erodium cicutarium*.

strabismus. crossed eyes.

stramonium. *Datura stramonium*.

Streptococcus. a genus of Gram-positive, non spore-producing, facultatively anaerobic bacteria. Some are pathogenic. *See also* erysipelas, metritis, yogurt.

Streptomyces avermitilis. *See* avermectin.

stressor. an agent capable of producing stress. **systemic stressor**. a nonspecific agent that both induces stress and stimulates defense mechanisms in a living organism. At certain concentrations, or amounts, or intensities, virtually all environmental factors become stressors. **topical stressor**. an agent that induces a localized stress response (e.g., inflammation).

Streunex. a trade name for lindane.

Streunex. lindane.

stringhalt (springhalt). a myoclonic affliction of one or both hindlimbs, chiefly of horses. It presents as spasmodic overflexion of the joints. Lesions of a peripheral neuropathy have been observed, involving the sciatic, peroneal, and tibial nerves. While of unknown etiology, severe forms of the disease have been attributed by some to lathyrism in the United States and to flat weed (dandelion) intoxication in Australia. Flexion may be mild, with only a spasmodic lifting and grounding of the foot. In severe cases, the foot may be drawn sharply up, touching the belly, and then struck violently on the ground. Serious cases may involve atrophy of the lateral thigh muscles or even the quarters. The condition may be progressive, and euthanasia may become necessary.

springhalt. stringhalt.

strip. **1a:** to pare; to peel; to remove a covering. **1b:** to uncover; undress. **2:** a belt or ribbon of land.

strip farming. a method of farming in which strips of land across contours are planted in different crops.

strip grazing. an approach to grazing in which animals, allowed access to a small part of a field, and separated from the rest of the field by a temporary (usually electric) fence.

strip-cropping. a means of cultivation in which plowing follows contours of the land in order to reduce erosion.

stripper's asthma. byssinosis.

Strobane® (terpene polychlorinates; Dichloricide aerosol; Dichloride Mothproofer; insecticide 3960-X14). an amber liquid mixture of polychlorinated terpenes used as an insecticide. It is carcinogenic to some laboratory animals.

strontium (symbol, Sr)[Z = 38;Ar = 87.62]. a soft, low-melting, silvery-white metallic element of the alkaline earth series. Its chemical and biological properties are similar to those of calcium, and it is thus readily absorbed and stored in bone. The anions of various strontium salts (e.g., bromide, iodide, lactate) are used therapeutically. **strontium-89** (^{89}Sr). a radioactive isotope of strontium, a beta emitter, T½ = 51 days. It is used as a tracer in studies of strontium absorption by living tissues and systems (e.g., bone). **strontium-90** (^{90}Sr). a radioactive isotope of strontium, a beta emitter, T½ = 29 years. It is very dangerous and because of the amounts produced, its long half-life, and selective incorporation (with slow turnover) into bone tissue, ^{90}Sr poses a serious hazard to vertebrate animals. Relatively low doses cause bone cancer in many vertebrates, although sensitivity varies. swine, for example are much more resistant than dogs.

strychnia. strychnine.

strychnidin-10-one. strychnine.

strychnine (strychnidin-10-one; strychnia; dog button, mole-nots, mole death). **1:** a colorless, extremely bitter, nearly water-insoluble, crystalline powder,

$$C_{21}H_{22}N_2O_2$$

It occurs in plants of the genus *Strychnos*. It is a fast-acting, convulsant alkaloid that is extremely toxic to mammals, acting chiefly by blocking the inhibition of nerve conduction by glycine in the spinal cord. Strychnine is used mainly as a rodenticide; because of this, it is a significant cause of poisoning of small animals, including pets. In medicine and allied fields, it is used almost entirely as a research tool. **2:** to poison by strychnine.

strychnine poisoning. strychnic poisoning; strychninism.

strychninism (chronic strychnic poisoning; chronic strychnine poisoning). a toxic condition due to the misuse of strychnine. Early signs are apprehension, tremors, and twitching of the face and arms. The condition may advance to include severe convulsions and respiratory arrest. Symptoms are chiefly those of central nervous system stimulation and may also include increased acuity of hearing, vision, touch, taste, and smell, followed by tonic convulsions and vomiting. *See also* strychnine, strychnic poisoning.

substupor. **1:** a state of partial (lethargy, daze, stupefaction) or nearly complete unconsciousness such that the victim can be aroused only briefly by repeated, intense stimulation. *Cf.* coma. **2:** a disorder characterized by diminished response to stimuli.

stuporose. stuporous.

stuporous (stuporose). pertaining to or characterized by stupor.

sturgeon family. Acipenseridae.

styloid. elongate; having an elongated shape.

Suaeda fruticosa (burroweed). a common weed (Family Chenopodiaceae) that has reduced available forage for grazing animals in large areas of the western United States

burroweed. *Suaeda fruticosa*.

sub-. a prefix meaning beneath; less than normal; inferior. *Cf.* hypo-.

subarid. moderately arid.

sublethal effect. *See* effect.

sublingual gland. sublingual salivary gland.

submandiular gland. submandibular salivary gland.

submaxillary gland. submaxillary salivary gland.

subspecies. race.

substituted hydrocarbon. derivative hydrocarbon (*See* hydrocarbon).

substrate. **1:** a specific substance or class of substances on which an enzyme of ferment acts catalytically. **2a:** any solid surface on which a layer or film of another substance is deposited. **2b:** substratum; a material that underlies and supports another (e.g., the soil on which a plant grows; a layer of rock beneath the soil surface). **suicide substrate**. a substrate that complexes with an enzyme to yield a reactive metabolite that inhibits further activity of the enzyme.

subtenolin. an antibiotic produced by *Bacillus subtilis*.

subterranean clover. *Trifolium subterraneum*.

subtilin. an antibiotic produced by *Bacillus subtilis*.

succession. progression; sequence; series.

allogenic succession. the replacement of one biotic community by another as a result of changes in the external environment. Reverse (allogenic) succession is a common result of chronic pollution. *Cf.* autogenic succession. **autogenic succession**. the replacement of one biotic community by another as a result of biological activities within the community. Toxic biological secretions often play a role in this form of succession. *Cf.* allogenic succession. **ecological succession**. the series of stages whereby a biotic community gradually changes in a particular area under a given set of environmental circumstances. This may be perceived as the appearance of a sequence of different communities over a period of time, usually with transitional stages, until a community (the climax community) results that can remain stable for a long period of time. Autogenic succession and reversed allogenic succession are of toxicological interest. **primary succession**. succession that begins with pioneer species (e.g., algae, lichens) that become established in an area devoid of life (e.g., a recent lava flow; an excavation to bare rock; a sand dune). **reversed allogenic succession**. allogenic succession in which the train of communities is reversed over that which naturally occurs; regression toward an earlier stage of succession. Reversal of succession is a fairly common effect of pollution at the community level. **reversed succession**. reversed allogenic succession. **secondary succession**. the establishment of a new community on a site previously occupied by an established community (e.g., a clearing in a forest, an abandoned field). *See also* allelopathy, sere.

succinic acid 2,2-dimethylhydrazide. daminozide.

Suckleya suckleyana (poison suckleya). a rare but locally abundant cyanogenic herb (Family Polygonaceae) of the Great Plains from Montana to Colorado and New Mexico. Symptoms and lesions are those of cyanide poisoning. At times, suckleya is nontoxic. It has caused heavy losses among cattle in Colorado and New Mexico.

sudan grass. a common name of grasses of the genus *Sorghum*.

sudor-. combining form denoting sweat, perspiration.

sudoriferous gland. sweat gland.

Suffolk sheep. developed in England, it is a relatively large hornless breed with a black face, bare head, and bare black legs. They grow rapidly, the mutton is of high quality, they are excellent breeders but the fleece is rather light. The rams are often used for crossbreeding to produce lambs for slaughter.

sugar beet. *Beta vulgaris*.

sugarbeet. *Beta vulgaris*.

Suidae (swine). a family of omnivorous artiodactyls that includes domestic and wild swine of the genera. *Babirussa*, *Phacochoerus*, and *Sus*.

sulf-, sulfo- 1. as a prefix to a chemical compound, it indicates the presence of a sulfur atom.

sulfa drug. sulfonamide.

sulfabenzpyrazine. sulfaquinoxaline.

sulfallate (diethylcarbamodithioic acid 2-chloro-2-propenyl ester; diethyldithiocarbamic acid 2-chloroallyl ester; 2-chloroallyl diethyldithiocarbamate; CDEC; CP 4742; Vegadex). an amber liquid herbicide, $C_8H_{14}ClNS$, that is soluble in water and most organic solvents. It is an irritant to the skin, eyes, and mucous membranes. It is carcinogenic in some laboratory animals and a suspected human carcinogen.

2-sulfanilamidoquinoxaline. sulfaquinoxaline.

sulfaquinoxaline(4-amino-*N*-2-quinoxalinylbenzenesulfonamide; *N'*-(2-quinoxalinyl)-sulfanilamide; 2-sulfanilamidoquinoxaline; *N'*-(2quinoxalyl)sulfanilamide; sulfabenzpyrazine). a sulfonamide, $C_{14}H_{12}N_4O_2S$, used chiefly as an antimicrobial and coccidiostat in veterinary practice. It upsets normal blood-clotting mechanisms of mink and causes extensive internal hemorrhage, which results in serious losses.

sulfate. a salt or ester of sulfuric acid.

sulfate of ammonia. ammonium sulfate.

sulfathiazole (4-amino-*N*-2-thiazolylbenzenesulfonamide; N^1-2-thiazolylsulfanilamide). a sulfonamide, $C_9H_9N_3O_2S_2$, the use of which is now confined to use in veterinary practice as an antimicrobial because of serious unintended effects on humans.

sulfocyanate. thiocyanate.

sulfonamide (sulfa drug; sulphonamide). any of a class of antimetabolites with the general formula, $R{\cdot}SO_2{\cdot}NH_2$. Included are sulfanilamide and its derivatives (e.g., sulfapyridine, sulfathiazole, sulfadiazine). They are active against bacteria; some are herbicidal (e.g., oryzalin). Bacteria that do not require folic acid or that are able to use preformed folic acid are not affected by these drugs. All are toxic. Some tend to cause injury to the urinary tract by precipitating in the kidney.

sulfur dioxide (sulfurous anhydride; sulfurous oxide). a colorless, nonflammable, irritant gas or liquid, SO_2, with a strong, acrid, suffocating odor that is soluble in ethanol, water, and ether. It is a powerful reducing agent used to prevent oxidative decay of food and medicinal products; as a bleach; as an intermediate in the production of sulfuric acid, sodium sulfite, glass, disinfectants, fumigants, wine, and in brewing. It has numerous additional uses. It forms sulfurous acid, H_2SO_3, in aqueous solution. Sulfur dioxide is a by-product of coal and fuel oil combustion and is a global air pollutant of considerable significance and is chiefly responsible for acid rain. Atmospheric sulfur dioxide corrodes metals and contributes to the decay of some building materials. When inhaled at low atmospheric concentrations most sulfur dioxide is removed in the upper respiratory tract. It is nevertheless a powerful irritant of the eyes, upper respiratory tract and bronchi. It can cause rapid, acute irritation of the eyes with tearing and inflammation; its action on the respiratory tract causes coughing, dyspnea, and spasm of the larynx; it can also cause pulmonary edema and respiratory paralysis. High and lethal concentrations are asphyxiating.

sulfuric acid (hydrogen sulfate; oil of vitriol; battery acid; electrolyte acid). the most widely used industrial chemical, it is a dense, colorless, oily, highly corrosive, highly reactive, oxidizing, dehydrating liquid, H_2SO_4. It is an important laboratory reagent and is used in the synthesis/-manufacture of many chemical compounds including the manufacture of fertilizers, nitrate explosives, artificial fibers, dyes, drugs, detergents, glue, paint, paper, furs, and food products. Additional uses include those of a dehydrating agent, an electrolyte in wet-cell batteries, the pickling of steel. Sulfuric acid dissolves most metals. It erodes, oxidizes, dehydrates, and/or sulfonates most organic materials. The dilute acid is an irritant to the skin, eyes, and mucous membranes. In concentrated form (commercial grade acid ranges from 93-98% H_2SO_4), however. At such concentrations it is exceedingly corrosive to tissues causing erosion and necrosis. It readily penetrates skin and mucous membranes to erode underlying tissues with charring and necrosis. Ingestion causes severe gastroenteritis; perforation of the esophagus or stomach may produce rapidly fatal results. Concentrated sulfuric acid is also phytotoxic, readily and rapidly penetrating leaf tissue and other soft tissues. Although cell walls are apparently unaffected, cytoplasm is rapidly destroyed; and chlorophyll and chloroplasts are apparently destroyed. Sulfuric acid combines with the magnesium atom of chlorophyll *in vitro,* destroying this pigment. The primary action, otherwise, on plant tissue may be via dehydration, taking up bound water. *See also* sulfoxism.

sulfuric acid, copper(2+) salt (1:1). cupric sulfate.

sulph-, **sulpho-**. the British equivalent of sulf-, and sulfo-.

sulphonamide. sulfonamide.

sumac family. Anacardaceae.

sumicidin. fenvalerate.

summer fescue toxicosis (summer toxicosis; tall fescue toxicosis; fescue toxicosis; summer syndrome). a warm season condition of cattle, sheep, and horses marked by a reduction in weight gain or in milk production. Animals are afflicted during the summer when eating *Festuca arundinacea* (tall fescue) forage or seed. The severity of the condition varies from field to field and from year to year. It is caused by toxic alkaloids in forage or seed associated with an endophytic fungus, *Acreminium coenophialum*, that infests the fescue. Reduced performance may appear within 1-2 weeks after fescue feeding is started. Further signs of intoxication include elevated body temperature, increased breathing rate, rough hair coat, lower serum prolactin levels, and excessive salivation; the animals seek wet spots or shade. Lowered reproductive performance and agalactia have been reported for both horses and cattle. A thickened placenta and birth of weak foals have been reported in horses.

summer syndrome. summer fescue toxicosis.

summer toxicosis. summer fescue toxicosis.

sun spurge. *Euphorbia helioscopia*.

sunflower. *See Helianthus annus*.

sunflower family. Asteraceae.

sunscald. a form of photosensitization that affects swine in the southeastern United States during autumn and winter when pastured on *Avena sativa*. It is especially common on wet pasturage. Symptoms include intense itching with necrosis and sloughing of skin. The ears may be lost in acute cases.

super-. a prefix indicating an excess; above, superior. Super- often has the same usage as supra-.

superparasitism. autoparasitism.

superphosphate. a fertilizer comprised mainly of calcium phosphates and calcium sulfate. It is manufactured by treating calcium phosphate ("phosphate rock"), $Ca_3(PO_4)_2$, with sulfuric acid. Fluorides, when present in rock, release hydrogen fluoride and other gaseous fluorine compounds, which can be hazardous in the absence of efficient absorbing equipment.

supra-. a prefix denoting a position above the part indicated by the word to which it is joined. *Cf.* super-.

suprarenal gland. adrenal gland.

Sus (swine). a genus of omnivorous mam

mals (Family Suidae) including four wild species that occur in Europe and Asia as well as a number of domestic breeds. ***S. scrofa*** (European wild boar; wild boar). a species characterized by the upturned tusks of the male.

suspension. **1:** a fluid that includes fine, insoluble, particulate matter that is evenly dispersed throughout. **2:** a temporary stoppage, delay, or interruption as in suspension of the use of a substance (e.g., a pesticide) when deemed necessary by the responsible governmental authority in order to prevent an imminent hazard that may result from continued use.

Sustainable Agriculture Research and Education Program (SARE). a competitive grants program authorized under the Food Security Act of 1985. It supports farmer—scientist teams that conduct on-farm experiments in the use sustainable agricultural methods such as the use of less chemically intensive methods of pest control and the maintenance of soil fertility. Cooperative Extension personnel also work with farmers to encourage adoption of sustainable practices.

swallow worts. *See Asclepias.*

swamp. a community or ecosystem dominated by woody plants. The soil of such a system is saturated with, and is commonly or intermittently covered by standing water but there is no accumulation of peat at the surface.

swamp-cypress. a common name of *Taxodium distichum.*

swamp fever. **1:** equine infectious anemia (*See* anemia). **2:** human leptospirosis (*See* leptospirosis).

swamp maple. *Acer rubrum.*

swamp milkweed. *Asclepias incarnata.*

swamp sumac. *Rhus vernix.*

swarm. **1a:** a crowd; horde. **1b:** a large number of entities, especially if in motion. **2:** a hive or colony. **3:** to teem, abound, crawl, or stream. **4:** to plague, infest, or overrun. **absconding swarm**. a flight bees that have abandoned a hive due to damage to the hive or to a disturbance.

sweat gland (sudoriferous gland). a type of tubular exocrine gland in the skin of most mammals that secretes sweat. Sweat is transported to the skin via a duct that originates in the dermal secretory portion of the gland. In many mammals, these glands are restricted to the pads of the feet. In humans and higher primates, they are widely distributed and often play an important role in evaporative cooling.

swedish turnip. *Brassica napus* var. *napobrassica.*

sweet cane. calamus; *Acorus calamus.*

sweet clover poisoning. a pernicious hemorrhagic disease of animals that consume toxic quantities of spoiled (moldy) sweet clover hay or silage (*See Melilotus*). All herbivores are susceptible but cattle are the most common victims, while sheep, swine, and horses are sometimes poisoned. All clinical signs are due to hemorrhage resulting from impaired blood coagulation. The appearance of clinical disease following consumption of toxic sweet clover varies greatly as a function of dicumarol content of the feed, the amount consumed, and the age of the animals affected. In some cases, clinical signs may appear only after several months of feeding on contaminated feed; onset of symptoms may be abrupt. As a consequence, death from massive hemorrhage may occur following trauma of any kind with little or no warning. Initial signs may include subcutaneous swellings due to subdermal hemorrhage, stiffness or lameness from bleeding into muscles and joints, epistaxis, the appearance of hematomas, and gastrointestinal bleeding. The blood has a reduced capacity to coagulate and bleeding may become massive, leading rapidly to death. Lesions revealed on autopsy typically include large subcutaneous and internal extravasations of blood. Neonates may be affected but rarely die unless the dam exhibits clinical signs of the disease.

sweet flag. *See Acorus calamus*.

sweet grass. *See Acorus calamus*.

sweet Joe Pye weed. *Eupatorium*.

sweet Joe Pyeweed. *Eupatorium Purpureum*.

sweet pea. a poisonous pea. *See Lathyrus*.

sweet pea poisoning. lathyrism.

sweet potato. *Ipomoea batatas*.

sweet root. *Glycyrrhiza glabra* or glycyrrhiza.

sweet sorghums. forage sorghums (*See Sorghum vulgare*).

sweetclovers. biennial herbs of the genus *Melilotus*.

swellhead. lechuguilla poisoning.

Swift's disease. acrodynia (def. 2).

Swiss Blue. methylene blue.

swine. pig; hog; common name of any mammal of the family Suidae, especially those of the genus *Sus*, *q.v.*

swine fever. *See* African swine fever.

swineherd's disease. human leptospirosis (*See* leptospirosis).

Swiss chard. *Beta vulgaris*.

swooning. syncope.

sym-. *See* syn-.

symbion. symbiont.

symbiont (symbiote; symbion). an organism that participates in a symbiosis.

symbiosis. **1a:** the close association of two or more dissimilar organisms or populations. **1b:** all relationships that occur between members of two different species or intraspecific relationships, especially where more than one phenotype is represented within the species (e.g., different color phases). **1c:** mutualism; sometimes used for close associations where both receive an advantage from the association.

symbiote. symbiont.

symbiotic. of or pertaining to symbiosis.

sympath-, sympatheto-, sympathico-, sympatho-. combining forms relating to the sympathetic part of the autonomic nervous system.

sympathetic. **1:** pertaining to, caused by, or exhibiting sympathy, compassion, or understanding. **2a:** of or pertaining to the sympathetic nervous system of vertebrates or any of its components. **2b:** of or pertaining to segmental nerves that supply the spiracles of insects. **2c:** of or pertaining to coloration of an organism that imitates that of the surroundings.

sympathetic nervous system. that part of the autonomic nervous system concerned mainly with processes involving homeostasis, the utilization of energy, and stress or emergencies. Adrenalin, the chief neurotransmitter of this system, is liberated in heart and visceral muscle, glands, and internal vessels. Acetylcholine also plays a role as a neurotransmitter at ganglionic synapses and at sympathetic terminals in skin, skeletal muscle, and the walls of blood vessels. The sympathetic system controls heat loss, which is accomplished in mammals, for example, by regulating blood vessel tone, the secretion of sweat, and erection of hair, etc. It also plays a critical role in preparing an animal to respond to an emergency by, for example, accelerating the heart rate, causing blood vessels other than those to the heart and skeletal muscle to constrict, mobilizing glucose (or glycogen in the case of birds) from the liver, and stimulating the secretion of adrenalin. The latter prepares the animal for 'flight or fight,' as appropriate. The actions of the sympathetic system tend to oppose those of the parasympathetic system.

sympathicolytic. antiadrenergic.

sympatholytic. antiadrenergic.

sympathoparalytic. antiadrenergic.

Symphytum (comfrey). a genus of perennial European herbs (Family Boraginaceae). Certain species were formerly thought to have medicinal value. ***S. foetidus*** (skunk cabbage). a toxic perennial herb (Family Araceae) with heavy roots and broad leaves that is native to low, swampy areas throughout the eastern United States. The leaves and rhizomes are toxic. The raw leaves can produce an intense burning sensation, irritation, and edema when chewed and can cause severe gastroenteritis if swallowed. The active principle is calcium oxalate. *See also Veratrum californicum.*

symptom. **1a:** any bodily condition or feeling experienced (and noticed) by a person (e.g., pain, nausea, dizziness, anxiety, depression) that is indicative of a disease, abnormality, or morbid phenomenon. *Cf.* sign.

syn-, **sym-**. prefixes indicating together, with, joined; sym- appears before b, p, ph, or m.

syncope (fainting; swooning; a fainting spell; a swoon). a sudden, temporary loss of consciousness due to a generalized cerebral ischemia.

syndrome. **1a:** a concurrence of signs or symptoms. **1b:** a set of associated signs and/or symptoms that are associated with a particular anatomic, physiologic, behavioral peculiarity or with a particular disease, type of disease, or morbid state. Taken together they provide a portrait that may help to diagnose or treat the condition. *See also* sign, symptom.

synergistic effect. *See* effect.

synovia (synovial fluid). a clear lubricating fluid of joints, tendon sheathes, and bursae. It is secreted by the synovial membrane of a joint.

synovial. denoting or pertaining to synovia.

synovial fluid. synovia.

synovial membrane. a membrane that lines the capsule of a joint.

synovitis. inflammation of a synovial membrane.

synthetic ammonia. anhydrous ammonia.

syrupy. **1a:** pertaining to syrup. **1b:** having the consistency of syrup.

Systéme International d'Unités. International System of Units.

systemic. **1:** of or pertaining to a system. **2a:** of, pertaining to, or involving the whole organism. **2b:** involving parts of the body of an organism that are remote from the site of exposure to a chemical such as a certain insecticides. **3:** translocated; descriptive of a chemical that is absorbed by and transported throughout the tissues or a living organism. Examples are systemic fungicides and pesticides applied to a plant in order to render plant tissues toxic to fungi, insects, or other pests. **4:** descriptive of that part of the circulatory system that serves the general circulation excluding the gas-exchanging surfaces of lungs or gills.

systemic effect. *See* effect.

Systox. demeton.

T

$T_{1/2}$. *See* half-life.

™. trademark.

2,4,5-T (2,4,5-trichlorophenoxyacetic acid). a synthetic, light tan or gray solid chlorophenoxy auxin with three substituted chlorine atoms,

$$C_6H_2Cl_3OCH_2CO_2H.$$

Originally developed as a chemical warfare agent in World War II, it has been widely used as a selective postemergence herbicide and defoliant and also as a bactericide and fungicide. A component of Agent Orange, 2,4,5-T is a highly toxic irritant of the skin and gastrointestinal tract, is hepatotoxic, a probable abortifacient, and is thought to be carcinogenic, teratogenic, and fetotoxic. Livestock (sheep) in areas of usage have demonstrated nephritis, hepatitis, and enteritis at autopsy. In the late 1970s, where forests were sprayed in the United States, the fetuses of many local pregnant women miscarried, a number of mammals sickened and died, and many species of birds disappeared. Because of trace contamination with dioxin, its use on all food crops except rice was banned in the United States in 1970 by the Department of - Agriculture. Because of its probable potency as an abortifacient, its use was further restricted in 1979, by the U.S. Environmental Protection Agency to use mainly as an herbicide on highway borders, rangelands, rice, and sugarcane fields. In 1985, its use on all noncrop sites was banned by the EPA.

2,4,6-T (2,4,6-trichlorophenol). a solid, nonflammable chlorinated phenol that forms yellow flakes and is soluble in ethanol, acetone, and ethyl ether. It is used as a fungicide, herbicide, and defoliant.

2,3,6-TBA (2,3,6-trichlorobenzoic acid). a substituted benzoic acid, TBA is a potent synthetic auxin. TBA and its aldehyde are all-purpose, non-selective, preemergence and early postemergence herbicides. It is used to control weeds in *Zea mays* (corn, maize) at doses of 0.5-1.0 kg/ha. It is absorbed by plant leaves and roots and can persist in soil for long periods with high residual toxicity.

T cell. T lymphocyte (*See* lymphocyte).

2,4,5-T/silvex. silvex.

T-cell lymphocyte. T lymphocyte (*See* lymphocyte).

Tabac du diable. *Veratrum viride*.

table salt. sodium chloride used in the home.

taeniacide. teniacide.

tailflower. *See Anthurium*.

tall fescue toxicosis. summer fescue toxicosis.

summer kill. a partial or complete kill of a fish population in a pond or lake during the warm season. Such kills may result from excessively warm water, depletion of dissolved oxygen, release of toxic substances from a decaying algal bloom, or by a combination of such factors.

tall fescue. *Festuca arundinaceae*.

Tanacetum (tansy). a genus of annual and

perennial herbs (Family Asteraceae). All parts of the plant contain a toxic oil (cedar leaf oil, *q.v.*), the toxic component of which is thujone (formerly termed tanacetin from this source). Small amounts of flowers and young leaves are used as a seasoning but can cause violent reactions and death if consumed in large quantities. The thujone content of these plants varies from none to approximately 95%. ***T. vulgare*** (common tansy). a highly toxic perennial herb, growing to a height of 1 m. It is indigenous to Europe but has become naturalized elsewhere, occurring widely in the United States, especially in the eastern states and the Pacific Northwest. The leaves, flowers, and stem contain a toxic oil, tanacetin.

tanbark. *See* butt cut.

tannate. any salt or ester of tannic acid.

tannic. of or pertaining to tan (tan-bark) or tannin.

tannic acid (gallotannic acid). **1:** a toxic glucoside, $C_{76}H_{52}O_{46}$, that yields gallic acid and glucose on hydrolysis. It occurs in many plants, especially in the bark of oak trees and related species (Family Fagaceae). **2:** tannin. *See also* oak poisoning.

tannin. **1:** any of a group of astringent, acidic materials derived from gallic acid. They occur in the leaves, vascular tissues, bark, unripe fruits, galls, and seed coats of many higher plants. They are found in vacuoles, cytoplasm, and sometimes cell walls or in specially enlarged tannin cells or systems of cells that appear as granular masses or yellow, red, or brown bodies. They are used to tan animal hides to produce leather, in photography to prepare the inks and dyes, and as clarifying agents for beer and wine. **2:** tannic acid. **3:** any of a number of substances that contain tannin. *See also* gallic acid, oak poisoning, acorn poisoning.

tansy. *Tanacetum*.

tansy aster. *See Gutierrezia*.

tansy mustard. *Descurainia pinnata*.

Taphrina epiphylla (alder witches' broom). a fungus (Family Taphrinaceae) of European origin that infests alders, causing greatly excessive branching.

tapioca plant. cassava.

tapioca. *See Manihot*.

target gland. a gland that is stimulated by the internal secretion of another gland or by some other specific external stimulus.

tarweed. *Amsinckia intermedia*.

taxa. plural of taxon.

Taxaceae (the yew family). a family of resinous evergreen trees and shrubs, the yews, distributed throughout the warmer regions of the earth. Some species are cultivated for their ornamental value. A number of forms are known to be toxic; *See* especially *Taxus*.

taxes. plural of taxis.

Taxidea taxus (badger). a white-faced, black-and-gray-furred, burrowing, nocturnal carnivorous mammal (Family Mustelidae). It is stout-bodied with short legs and is fairly common in many rural areas in North America.

badger. common name of *Taxidia taxus*.

taxine, taxine alkaloids. a mixture of cardiotoxic alkaloids isolated from the berries and leaves of the yew, *Taxus baccata* (Family Taxaceae) and certain other members of the genus. Taxines are pharmacologically poorly understood cardiac depressants. In taxine poisoning, the heart is slowed, eventually stopping in diastole. Mule deer seem impervious to these alkaloids at the concentrations present in at least some cultivars of *T. baccata* (personal observation by the author). *See also Taxus*.

taxis (plural, taxes). a directed response or orientation by a motile organism toward the source of a weak or "positive" stimulus (e.g., a food source) or away from a "negative" or strong stimulus (e.g., strong light, a strong irritant). *See* teleorganic, approach withdrawal, adience, appetitive.

Taxodiaceae. redwood family. *See Taxodium distichum*.

Taxodium distichum (baldcypress; swamp cypress). a deciduous tree of swamps (Family Taxodiaceae). It is grown as an ornamental and, because the wood is very resistant to moist conditions, it is used in outdoor construction.

taxon (plural, taxa). any defined taxonomic group of organisms of any rank (e.g., species, family, order, phylum).

taxonomic. of or pertaining to taxonomy.

taxonomy. the science of naming and classifying organisms. It is a complex and fundamental field of biological science that provides a sound and necessary basis for all biology-based sciences, basic and applied. It is the study of the theory, procedures, and rules of classification of organisms, usually according to phenotype and inferred evolutionary (genetic) relationships. Taxonomic knowledge is important for nearly any aspect of agriculture in which breeds, cultivars, breeding, or hybridization programs are involved. **classical taxonomy**. this discipline identifies homologous morphological structures and serological and biochemical data to describe and classify plants and animals. **experimental taxonomy** (biosystematics). a subdiscipline of taxonomy that relies largely on experimental breeding to establish or clarify relationships among closely related species or genera and to define taxa.

Taxus (yew). a genus of poisonous evergreen trees and shrubs (Family Taxaceae) that are native throughout the Northern Hemisphere. While the outer fleshy part of the fruit (the aril) is generally edible, the seeds are poisonous. Human intoxication by most species is very rare. *T. baccata* has been the most extensively studied. ***T. baccata*** (English yew). a poisonous woody plant that grows to a height of more than 12 m. There are a large number of cultivars; toxicity might vary among them. All parts are toxic, producing severe gastrointestinal distress, diarrhea, mydriasis, muscular weakness, coma, convulsions, and cardiac and respiratory depression. *T. baccata* is readily eaten by livestock and has been responsible for numerous fatalities among cattle. ***T. canadensis*** (American yew, Canadian yew, ground hemlock). a hardy, low, spreading bush that sometimes grows to a height of 1.4 m. It occurs in rich woodlands and thickets from Newfoundland, Nova Scotia, and New England to portions of Virginia, Kentucky, Indiana, Illinois, and Iowa and is often used as a hedge plant. The pulp is edible but seeds and wilted foliage contain taxine and are poisonous and notoriously lethal to livestock.

2,3,6-TBA (2,3,6-trichlorobenzoic acid). a potent synthetic auxin, TBA and its aldehyde are all-purpose, non-selective, preemergence and early postemergence herbicides. It is used to control weeds in *Zea mays* at doses of 0.5-1.0 kg/ha. It is absorbed by plant leaves and roots and can persist in soil for long periods with high residual toxicity.

TBH. hexachlorocyclohexane.

TBT. **1:** tributyl tin **2:** tributyl tin compound.

TBTO. **1:** tributyl tin **2:** tributyl tin compound.

2,4,5-TC. silvex.

TCA. trichloroacetic acid.

TCA cycle. tricarboxylic acid cycle.

TCDD (2,3,7,8-tetrachlorodibenzo[b,e]-[1,4]dioxin;2,3,7,8-tetrachlorodibenzo-*p*-dioxin; 2,3,7,8-tetrachlorodioxin; "dioxin"). a compound formed as a by-product in the manufacture of certain chlorinated benzene compounds, especially trichlorophenol and its derivatives, some of which are widely used as herbicides; it may also occur as a contaminant in such products. It is one of the most toxic substances known, and is also teratogenic and a potent carcinogen to laboratory animals. Effects of exposure may include edema, hemorrhage, severe weight loss, thymic atrophy, kidney anomalies, lowered gestation index, decreased fetal weight, increased ratio of liver to body weight, and increased fatty deposition in the liver. TCDD has often been used erroneously as a synonym for dioxin, but it is only one of the class of compounds termed dioxins; it is, however, the most toxic.

***m*-TCPN**. chlorothalonil.

TDE. 1,1-dichloro-2,2-bis(*p*-chlorophenyl)-ethane.

team. **1a:** a pair; a set. **1b:** a crew or gang; a squad. **2a:** a pair of horses, or other draft animals; especially a set of individuals matched in some way or that work together, as in pulling a cart or coach. a yoke of oxen; a span of mules; a brace or team of ducks.

teart. **1:** soil or plants that contain high concentrations of molybdenum. **2:** molybdenum poisoning of livestock that graze on vegetation grown in soil that contains high concentrations of molybdenum.

teartness. a synonym for molybdenum poisoning in the United Kingdom. *See* teart.

Teleostei. a subdivision of the Actinopterygii that includes all modern bony fish except holosteans and chondrosteans.

telocentric. of or pertaining to a chromosome or chromatid in which the centromere is located at the extreme end of the chromosome. *Cf.* acentric, acrocentric, metacentric.

temephos (Abate). a colorless, crystalline, pesticidal cholinesterase inhibitor, $[(CH_3O_2)PSOC_6H_4]S$ It is toxic by ingestion and inhalation.

Temik™. aldicarb.

tempeh poisoning. bongkrek poisoning.

temper. **1:** nature; disposition; temperament. **2:** anger; rage; passion. **3a:** to make temperate or moderate. **3b:** to allay or soothe. **4:** to adjust; adapt; modify. **5:** to inure, toughen; acclimatize. **6:** to add or remove moisture from grain to optimize grinding. **7:** to soften leather. **8:** to harden, toughen, or make resilient as in the tempering of steel. **9:** the hardness and strength of a rolled metal. **10:** the nominal carbon content of steel. **11:** the degree of hardness or resiliency conferred on steel by tempering. **12:** to anneal or toughen glass. **13:** to soften and increase the resiliency of hardened steel or cast iron by reheating to a temperature considerably lower than used in hardening or manufacture of the original metal or well below that from which quenching for hardening was done. **14:** to reheat and harden steel or cast iron.

tendo-. combining form denoting a tendon.

teniacide (taeniacide; tenicide). **1:** lethal to tapeworms. **2:** an agent that is lethal to tapeworms.

tenicide. teniacide.

TEPP (ethyl pyrophosphate; tetraethyl pyrophosphate). a clear, white to amber, hygroscopic, liquid organophosphate, $(C_2H_5)_4P_2O_7$, that quickly hydrolyzes on contact with water. It is miscible with water and all organic solvents except aliphatic hydrocarbons. It has been used as an insecticide and rodenticide. It is a potent irreversible anticholinesterase. TEPP is no longer in use as an insecticide because it is supertoxic to humans and extremely toxic also to other mammals. It appears to ba cumulative poison by all routes of exposure, and is rapidly absorbed through the skin. One herd of cattle of varied ages was accidently sprayed with 0.33% emulsion of TEPP; all reportedly died within 40 minutes.

tera-. **1:** a prefix used in the SI and metric systems to signify one trillion. **2:** a combining form that denotes a teras.

teras (plural, terata). a malformed individual or morphologic anomaly (a monster) that arises during development and growth. Such monsters may be exemplified by the presence of redundant, defective, misplaced, or grossly malformed parts or the absence of critical parts. In the case of mammals, a teras is a grossly malformed fetus.

terata. plural of teras.

teratogen. any chemical or environmental factor that can cause serious malformations during growth and development of plants or animals. In vertebrate animals, teratogens are those agents that act from fertilization to hatching, birth, or during metamorphosis. Teratogens are mechanistically diverse and include, for example, malnutrition, various drugs and environmental chemicals (e.g., thalidomide), ionizing radiation, and virus infections (e.g., rubella).

teratogenesis (teratogeny). **1a:** the production of monstrous growths or fetuses. **1b:** abnormal development or growth that produces monsters (teras), monstrous growths, or fetuses, whether viable or nonviable. *Cf.* dysmorphogenesis. **2:** the set of processes involved in the production of a monster or teras.

teratogenetic. pertaining to teratogenesis.

teratogenic. **1:** of, pertaining to, or having the action of a teratogen. **2:** able to or tending to produce developmental malformations and monstrosities.

teratogeny. teratogenesis.

teratology. **1:** the study of malformations, monstrosities, or critical deviations from that which is morphologically and functionally normal in developing and growing organisms. **2:** the study of the origin and nature of terata and the processes involved in their initiation and development. Much of this field is weighted toward abnormalities caused by radiation, viruses, and xenobiotics, including drugs. Teratology generally combines knowledge from developmental biology, pathology, and specific knowledge of the physiology and biochemistry of the particular organisms studied. Branches of this science include phytoteratology and zooteratology.

terbacil. a selective pyrimidine herbicide absorbed mainly through the roots. It is used to control many annual and some perennial weeds without damage to a number of crops.

terminal residue. that amount of a pesticide residue that remains in an agricultural product at the point of human ingestion. *See also* maximum residue limit.

terminated. **1:** complete; concluded; finished; completed; ended (as of life). **2:** aborted; discontinued; stopped. **3:** killed. **4:** adjudicated.

termiticide. a substance that destroys termites.

terpene polychlorinates. Strobane®.

terpene. **1:** any of a large class of naturally occurring saturated hydrocarbons produced by plants that are formed completely by 5-carbon isopentyl (isoprene) C_5 units with the general formula, $(C_5H_8)_n$. The smaller terpene molecules occur in plants (e.g., geraniums, mints, etc.), giving them their characteristic odors. Examples of terpenes with larger molecules are carotenoids, lanosterol, phytol, squalene, vitamin A, and natural rubber. Some terpenes are allopathic agents. *See also* aplysin, aplysinol, roridin A, fusidic acid. strobane, Artemesia californica. Some are allopathogenic agents (*See Artemisia*, *Salvia*). **2:** a moderately toxic, unsaturated hydrocarbon, $C_{10}H_{16}$, found in plant oleoresins and essential oils.

terpenism. poisoning by a terpene; the condition is marked by nausea, vomiting, convulsions, unconsciousness, pulmonary edema, and tachycardia.

terpenoid. any compound with an isoprenoid structure similar to that of terpenes; many are alcohols. Terpene derivatives such as camphor, menthol, terpineo, borneol, and geraniol are terpenoids.

***tert*-butyl hydroxide**. *tert*-butanol.

tertiary butanol. *tert*-butanol.

test (*See also* toxicity test). **1a:** to examine or verify. **1b:** an assay, trial, examination, or experiment. *See also* alternative test method, assay, auditory startle, BALB/3T3, bioassay, color (FD&C color, for example), concentration (lethal concentration, etc.), conditioning, cosmetic, dose (lethal dose, etc.), double blind, experiment, extrapolation, microbioassay, no-effect level, paralysis time, partition coefficient (octanol/water partition coefficient), prothrombin time, sodium benzoate, toxicity test. **1c:** a criterion, gage, standard, or norm. **2:** a chemical reaction or reagent that has clinical significance (*See* chemical test). **3a:** a usually hard, often rigid, protective covering (e.g., of certain protozoans, echinoderms of the class Echinoidea). **3b:** the exoskeleton of *Limulus*. **3c:** testa; the protective outer covering of a seed. **allergenicity test**. any direct test of the allergenicity of a substance (e.g., dermal sensitization tests). *See also* radioallergosorbent test. **Bettendorff's test**. a test for arsenic in which hydrochloric acid is mixed with a test fluid and stannous chloride is then added. If the resultant fluid contains arsenic, a piece of tin foil placed in the fluid will form a brown precipitate. **carcinogenicity test**. any test of the potential of a chemical to induce cancer. Such are usually long-term tests, most frequently tests of laboratory rats and mice that are conducted throughout most of the life of the test animal (2 years or more for rats; 1.5-2.0 yrs for mice). Depending on the nature of the test chemical and the probable route of exposure of humans, the test chemical may be delivered via food, gavage, drinking water, dermal application, or inhalation. **chemical test**. **1a:** the process of examining a substance, tissue, or other material to determine its chemical composition or structure (e.g., Bettendorff's test, Fleitmann's test, Reuss' test, Wormley's test, Yvon's test). **1b:** the reagent or substance (e.g., barium hydroxide, diphenylamine, picramic acid) used in such a test. **delayed neuropathy test**. organophosphorus-induced delayed neurotoxicity test. **field test**. any test conducted out of doors or in a situation not under the control of the investigator. The natural environment (and even the human indoor environment) is usually far more complex and more variable than that which is obtained in a laboratory situation. Even so, test results from field tests may be more realistic in some cases.

Fleitmann's test. a test for arsenic in which hydrogen is generated in a test tube containing the test liquid. The material is heated and filter paper treated with silver nitrate solution is placed over the top of the test tube. Arsenic is present if the moist paper is blackened. **Gutzeit's test**. a test for arsenic in a liquid. A small amount of sulfuric acid and a small piece of zinc are added to the suspected liquid, which is then boiled. Filter paper treated with silver nitrate solution is placed in the vapor stream above the test tube. If the paper turns yellow, arsenic is present. **laboratory test**. any test conducted in a laboratory, under at least partially controlled environmental conditions. *See*, for example, electrolyte measurement. **organophosphorus-induced delayed neurotoxicity test** (delayed neuropathy test). the potential of an organophosphate to produce this condition is usually tested by observing the appearance of paralysis of the leg muscles of treated mature domestic fowl or examination, at autopsy, for motor nerve degeneration. The clinical response of these animals is similar to that of humans. **photoallergy test**. *See* phototoxicity test. **phototoxicity test**. any test that evaluates the dermal effects of light, especially ultraviolet light, on the test organism subsequent to treatment with the test chemical. Such tests may be employed to assess phototoxicity and photoallergenicity of a test chemical. In general, the chemical is applied to a portion of the skin of control subjects and experimental subjects; the area is then irradiated with ultraviolet light in the experimental (but not in the control) groups. **picrate test**. a sensitive test for cyanide that is suitable for use in the field. Filter paper dipped in a 5% aqueous solution of sodium carbonate and 0.5% picric acid is allowed to dry slightly and then suspended in a test tube over a biological sample (e.g., macerated plant material, minced liver, rumen contents) to which a few drops of chloroform or dilute acetic acid have been added. **Reinsch's test**. a test for arsenic in which a copper strip is placed in the test liquid, which is then acidified and boiled with hydrochloric acid. The presence of arsenic is indicated by the appearance of a gray deposit on the copper. **Rimini's test**. a test for the presence of formaldehyde in liquids (e.g., milk, urine) using a dilute solution of phenylhydrazine hydrochloride, sodium hydroxide, and sodium nitroprusside. **significance, test of**. a statistical test that provides a criterion to help determine whether a difference between the expected (hypothetical) and observed values is small enough to be reasonably due to chance. **toxicity test** (toxicity assay). any of extremely numerous and varied tests used to establish the toxicity of a substance to any of a wide variety of living systems (e.g., cells, organisms, herds, communities) tested under various exposure regimes and for various purposes, often to help establish regulations or to control or treat the problem. **Wormley's test**. a test for alkaloids in which the test solution is treated with picric acid or a dilute iodine/potassium-iodide solution. The presence of alkaloids is indicated by a color reaction. **Yvon's test**. **1:** a test for alkaloids in which the test solution is added to an aqueous mixture of bismuth subnitrate, potassium iodide, and hydrochloric acid. The presence of alkaloids is indicated by a red color reaction. **2:** a test for acetanilid in which a urine specimen is treated with chloroform and heated with yellow nitrate of mercury. The presence of acetanilid in the urine is indicated by a green color reaction.

testis (plural, testes). the male gonad or organ of generation; the primary sex organ that produces sperm. In vertebrates, there is a pair of testes that produces both spermatozoa and steroid hormones under the control of the pituitary gonadotrophins.

tetano-, tetan-. combining forms denoting tetanus, tetany.

TETD. disulfiram.

tetra-. a prefix meaning four.

2,3,5,6-tetrachloro-1,4-benzenedicarboxylic acid dimethyl ester. chlorthal-dimethyl.

2,4,5,6-tetrachloro-1,3-benzenedicarbonitrile. chlorothalonil.

2,4,5,6-tetrachloro-3-cyanobenzonitrile. chlorothalonil.

2,3,7,8-tetrachlorodibenzo[b,e][1,4]dioxin. TCDD.

2,3,7,8-tetrachlorodibenzo-*p*-dioxin. TCDD.

2,4,5,6-tetrachloro-1,3-dicyanobenzene. chlorothalonil.

2,3,7,8-tetrachlorodioxin. TCDD.

tetrachlorodiphenylethane. 1,1-dichloro-2,2-bis(*p*-chlorophenyl)ethane.

tetrachloroisophthalonitrile.chlorothalonil.

***m*-tetrachlorophthalodinitrile**. chlorothalonil.

2,3,5,6-tetrachloroterephthalic acid dimethyl ester. chlorthal-dimethyl.

tetrachlorvinphos (2-chloro-1(2,4,5-trichlorophenyl)vinyl dimethyl phosphate; 2-chloro-1-(2,4,5-trichlorophenyl)ethenyl phosphoric acid dimethyl ester; Stirifos). a moderately toxic, organophosphate insecticide used to control pests on food crops and livestock. It is one of the least toxic to vertebrates of the cholinesterase inhibitors (due probably to its low solubility in water). No fatal cases of human exposure are known. This compound is sold as wetable powders and dusts, emulsifiable concentrates, and in a special formulation for cattle which is mixed with the feed. At high concentrations, technical grade tetrachlorvinfos causes hyperplastic thyroid lesions and adrenal cortical adenomas in laboratory rats and is carcinogenic to laboratory rats and mice.

tetracycline [*Lederle*]. a yellow, odorless, crystalline powder,

$$C_{22}H_{24}N_2O_8{\cdot}0\text{-}6H_2O$$

Tetracycline is nearly insoluble in water, chloroform, and ether, slightly soluble in ethanol, and very soluble in dilute hydrochloric acid and solutions of alkali hydroxide. Tetracycline is a broad-spectrum antibiotic obtained from certain bacteria of the genus *Streptomyces*; it can also be synthesized. It acts by interfering with protein synthesis in prokaryotes. Tetracycline is often prescribed to treat urinary infections, streptococcal infections, pneumonia, brucellosis, and rickettsial diseases such as typhus. It may also be used to treat bacterial infections in individuals who are sensitive to penicillin. Bacteria may become resistant to this antibiotic but this does not usually occur with a short course of treatment. It is very toxic to laboratory rats. Prolonged use in humans may result in dental discoloration, fatty liver, and renal damage. Tetracycline is teratogenic and may cause severe damage to teeth and cataracts in the fetus if administered to a prospective mother.

Tetradymia (horse brush; rabbit brush). a common hepatotoxic, photosensitizing bushy plant of the western United States. It causes more livestock losses from photosensitization than any other plant in the United States. Bighead appears in sheep within 16-24 hours following ingestion of a toxic dose. If a sufficient amount of *Tetradymia* forage is consumed, death can result from liver damage alone. *See also* phylloerythrin.

tetraethyl pyrophosphate. TEPP.

tetraethylthioperoxydicarbonic diamide. disulfiram.

tetraethylthiuram disulfide. disulfiram.

tetrahydro-. a prefix denoting the attachment of four hydrogen atoms as in tetrahydrofolate (H_4folate).

[2R-(2α,6aα,12aα)]-1,2,12,12a-tetrahydro-8,9-dimethoxy-2-(1-methylethenyl)-[1]benzopyrano-[3,4-b]furo[2,3-h][1]-benzopyran-6(6aH)-one. rotenone.

2,3,6a,9a-tetrahydro-9a-hydroxy-4-methoxycyclopenta[c]furo[2,3-h][1]-benzopyran-1,11-dione. aflatoxin M_1

3,4,7aα,10aα-tetrahydro-5-methoxy-1H,-12H-furo[3′,2′:4,5]-furo[2,3-h]pyrano[3,4-c][1]-benzopyran-1,12-dione. aflatoxin G_1.

2,3,-6aα,9aα-tetrahydro-4-methoxycyclopenta-[c]-furo-[3′,2′:4,5]furo[2,3-h]-[1]benzopyran-1,11-dione. aflatoxin B_1.

1,3,4,5-tetrahydroxycyclohexanecarboxylic acid. chlorogenic acid.

6′,7′10,11-tetramethoxymetan. emetine.

tetramethyl thiurane disulfide. thiram.

tetramethylene bis(methanesulfonate). busulfan.

tetramethylthionine chloride. methylene blue.

tetramethylthiuram disulfide. thiram.

tetraploid. **1a:** having four homologous sets of chromosomes. **1b:** a tetraploid cell or organism, one that has four sets of homologous chromosomes per cell, except of the germ cells.

teturamin. disulfiram.

texan-. *See* tetano-.

Texas armadillo. *Dasypus novemcinctus*.

Texas fever. *See* babesiosis.

thalamo-, thalam-. combining forms pertaining to the thalamus.

thallitoxicosis. thallotoxicosis.

thallium (symbol, Tl)[Z = 81; A_r=204.37]. a highly toxic, soft, malleable, bluish- or grayish-white, metallic element of group III of the periodic table. Thallium and certain of its salts occur in lead and cadmium ores as well as in flue dust and during zinc smelting. It readily forms toxic compounds on contact with water. Thallium is toxic by all routes of exposure; avoid contact. It was formerly used as an insecticide and rodenticide. Symptoms of acute poisoning by thallium and many of its salts commonly include nausea, vomiting, severe abdominal pain, tremors, delirium, convulsions, dyspnea, paralysis, coma, collapse, and death. Chronic poisoning may cause alopecia. Lesions include hemorrhagic gastroenteritis and encephalopathy.

thallium sulfate (thallous sulfate). an extremely toxic, colorless, water-soluble, crystalline substance, Tl_2SO_4. Thallium sulfate is used as a rodenticide, in ant baits, and as a reagent in analytical chemistry. It is similar toxicologically to thallium, *q.v.*

thallium poisoning. thallotoxicosis.

thallotoxicosis (thallitoxicosis; thallium poisoning). poisoning by thallium or thallium-containing substances. It usually occurs following accidental ingestion of a rodenticide.

thallous sulfate. thallium sulfate.

thanatopsy. autopsy.

Theileria parva. *See* east coast fever.

therapeutic effect. *See* effect.

therio-. combining form denoting animals.

therm-, thermo-. combining forms denoting heat.

therology. mammalogy.

thiabendazole. (2-(4′-thiazoyl)benzimidazole; 4-[2-benzimidazolyl]thiazole). a white to tan crystalline substance, $C_{10}H_7N_3S$, used as a systemic fungicide on citrus fruits. Rats and dogs tolerate a relatively high dose.

thiaminase. an enzyme that catalyzes the hydrolysis of thiamine (vitamin B_1) into pyrimidine and a thiazole derivatives. It occurs in raw fish, certain bacteria, and an occasional plant (e.g., *Dryopteris felix-mas* (male fern), *Equisetum* (horsetails), and *Pteridium acquilinum* (eastern bracken)). It is the only known plant enzyme that is toxic to animals. *See*, for example, bracken staggers, *Pteridium*, equisetosis, *Equisetum*.

thiamine (vitamin B_1; 3-(4-amino-2-methylpyrimidyl-5-methyl)-4-methyl-5,β-hydroxy-ethylthiazolium chloride). also called the antineuritic vitamine, thiamine is a water-soluble vitamin,

$$C_{12}H_{17}ClN_4OS$$

It is essential for growth and the prevention of beriberi. It is a precursor of the coenzyme thiamine pyrophosphate, which plays a role in carbohydrate metabolism in the decarboxylation of α-keto acids. Many organisms are unable to synthesize thiamine and it therefore has to be supplied from an external source. Thiamine deficiency in the technologically advanced nations is uncommon (but see thiamine shock). Thiamine occurs, for example, in wheat germ, whole grains, enriched breads and cereals, and pork. Thiaminase is responsible for thiamine deficiency in certain poisonings, especially of livestock (*See*, for example, bracken staggers, equisetosis). Some poisons can cause disturbances of thiamine metabolism (*See*, for example, herbicide (translocated herbicide)).

***N*¹-2-thiazolylsulfanilamide**. sulfathiazole.

2-(4′-thiazoyl)benzimidazole. thiabendazole.

Thimenox. phorate.

thimet. phorate.

3,3′-thio-bis-1-propene. allyl sulfide.

thio-TEPA. thiotepa.

thioallyl ether. allyl sulfide.

3,3′-thio-bis-1-propene. allyl sulfide.

thiocarbamide. thiourea.

thiocarbonate. any compound that contains the thiocarbonate group, --CS.SS.CS--. *See also* herbicide (thiocarbamate herbicide).

thioctic acid (6,8-thioctic acid; dl-α-lipoic acid; 1,2-dithiolane-3-pentanoic acid; 1,2-dithiolane-3-valeric acid; 6,8-dithiooctanoic acid; 5-(1,2-dithiolan-3-yl)valeric acid; 5-(1,2-dithiolan-3-yl)pentanoic acid; acetate replacing factor; pyruvate oxidation factor; POF). a crystalline substance, $C_8H_{14}O_2S_2$, that is nearly insoluble in water but soluble in fat solvents. It is a growth factor in many bacteria and protozoa. It is used in the treatment of liver disease and *Amanita* poisoning. It occurs naturally in yeast and liver and is prepared synthetically.

6,8-thioctic acid. thioctic acid.

thiocyanate (rhodanate; sulfocyanate). any

compound that contains the radical —SCN; organic thiocyanates are derivatives of thiocyanic acid, HSCN, in which the hydrogens have been replaced by hydrocarbon moieties (e.g., CH_3). They were the first synthetic organic insecticides, killing on contact. The volatile, lower molecular mass thiocyanates (methyl, ethyl, isopropyl) are rapid-acting potent poisons used as fumigants in the control of insects. Thiocyanates are thyrotoxic to livestock, causing thyroid hyperplasia and symptoms of hypothyroidism. Effects on lambs are similar to those of L-5-vinyl-2-thiooxazolidone (*q.*v.). Thiocyanate glycosides have been isolated from plants that have caused symptoms of hypothyroidism in livestock. *See* isothiocyanate.

thiodemeton. disulfoton.

thioimidodicarbonic diamide. 2,4-dithiobiuret.

thioperoxydicarbonic acid diethyl ester. dixanthogen.

thiophosphorus insecticide. any sulfur-containing organophosphosphate insecticides (e.g., methylparathion, parathion). Metabolic activation requires desulfuration of these compounds to their oxon analogs (e.g., methylparaoxon and paraoxon).

thiotepa (triethylenethiophosphoramide; tris-(1-aziridinyl)phosphine sulfide; thio-TEPA). an extremely toxic cytotoxic alkylating agent related to triazone, *q.v.* It is used as an insect sterilant and antineoplastic agent. It causes delayed reproductive effects in human females and is a human carcinogen. Inhalation or contact with thiotepa particles is hazardous.

thiourea (thiocarbamide). a sulfur analog of urea with white, lustrous, bitter-tasting crystals, $(NH_2)_2CS$, that has been used as a rodenticide and is moderately to highly toxic to humans. It is a skin irritant and affects bone marrow, causing anemia. Thiourea is a known carcinogen and may not be used in food products in the United States and is known to cause liver and thyroid cancer in experimental animals.

thioxamyl. oxamyl.

thiram (tetramethyl thiurane disulfide; tetramethylthiuram disulfide; bis-(dimethylthiocarbamyl) disulfide; thiuram; TMTD). a white or yellow, crystalline powder,

$$[(CH_3)_2NCH]_2S_2$$

Thiram has a characteristic odor and is water-insoluble but soluble in a number of organic solvents. It is a dithiocarbamate fungicide that is used also as a disinfectant of seeds, nuts, and fruits, to control bacteria in edible oils and fats; and as an ingredient in antiseptic sprays and soaps and suntan lotions. It is toxic by ingestion and inhalation and an irritant of skin, eyes, and mucous membranes, causing sneezing, coughing, and skin and eye irritations; the reaction is especially severe if a subject drinks alcohol while exposed to thiram. These properties led to the development of disulfiram (Antabuse).

thiuram. thiram.

Thlaspi arvense (fanweed). a toxic, widely distributed temperate zone weed (Family Brassicaceae) that causes severe gastroenteritis in livestock when ingested, presumably because of the mustard oil content.

thoraco-, thorax-, thoracico-. combining forms denoting the chest (thorax).

thorn. a woody, spine-like structure with a sharp point that is continuous with the vascular system of the plant.

thornapple. *See Datura*.

thorny-headed worm. any worm of the Acanthocephala.

thoroughwort (formerly throughwort). a common name of *Eupatorium* species.

thread lungworm. *Dictyocaulus filario*.

thread-leaf groundsel. a common name of *Senecio integerrimus* and *S. longilobus*.

three-flowered nightshade. *Solanum triflorum*.

thrombo-, throb-. combining forms pertaining to or denoting blood clot.

thrombocytin. serotonin.

thrombotonin. serotonin.

throughwort. *Eupatorium*.

throwback. *See* atavism.

thuja (thuya). **1:** any tree of the genus *Thuja*. **2:** the wood of the sandarac tree.

Thuja. a genus trees (Family Cupressaceae, the cypress family) with very small cones that have scalelike leaves borne by flattened, fan-like branchlets. There are numerous cultivated varieties including many dwarf forms. Only two of the several species of *Thuja* are indiginous to North America, the rest are Asian. The wood of the North American species (see below) is soft but highly resistant to decay and thus are important in supplying fence posts. ***T. occidentalis*** (arborvitae; white cedar; northern white cedar). a tree of eastern North America; it is a popular hedge tree that has many cultivated forms. ***T. plicata*** (giant arborvitae; red cedar; Western red cedar). a tree of western North America that is much larger and substantially more important as lumber than *T. occidentalis* and is used chiefly to make shingles and shakes.

thunderwood. *Rhus vernix*.

Thurber loco. *Astragalus thurberi*, a true locoweed.

thuya. thuja.

thyme-, thy-, thymi-. **1:** combining forms denoting the thymus. **2:** combining forms denoting a relationship to the mind, soul, or emotions.

-thymia. a suffix denoting relation to the mind, soul, or emotions. *See also* thyme- (def.2).

thymus (thymus gland). an unpaired, bilobed, lymphpoid organ of vertebrates that develops from the third pharyngeal pouch. It lies ventrally in the region of the lower neck and anterior to, or in bipeds, superior to the heart. The thymus is essential to the development of immunity and controls cell-mediated immunity, homograft rejection, and allergic aspects of the immune response in mammals. It is essential to the development of T-cells. At sexual maturity, the thymus atrophies in most vertebrates other than birds. It is active only if the pool of T-cells is depleted.

thymus gland. thymus.

thymus-dependent lymphocyte. T lymphocyte (*See* lymphocyte).

thymus-derived lymphocyte. T lymphocyte (*See* lymphocyte).

thyreo-. an obsolete spelling for the affix thyro-.

Thyridopteryx ephemeraeformis (bagworm). **1a:** a moth (Family Psychidae) that occurs widely throughout the eastern United States and areas near the Gulf of Mexico. Within this large area, they infest nearly all species of trees. These pests are controlled largely by the use of insecticides or by handpicking the cocoons (*See* Psychidae) before the eggs hatch at the end of May. **1b:** the larvae of this moth. They spin a protective silken bag about themselves.

bagworm. *Thyridopteryx ephemeraeformis*.

thyro-, thyr-. combining forms pertaining to or denoting the thyroid gland.

thyroid storm. *See* crisis (thyroid crisis).

TIBA (2,3,5-triiodobenzoic acid). an anti-auxin herbicide that inhibits cell division in primary meristems. This compound and certain other substituted benzoic acids have similar but weaker antitropistic properties to those of morphactins.

tick fever. bovine babesiosis.

tick. common name of any representative of the arachnid superfamily, Ixodoidea, *q.v.* Bites produce itching sensations and local skin irritation. Some species are disease vectors (e.g., of typhus, Rocky Mountain spotted fever). The bite of poisonous species produces pain, erythema, swelling, muscle cramps and, in some cases, tick paralysis, *q.v.* The application of gasoline or heat to the tick should cause it to loosen its jaws. It can then be removed with forceps (tweezers). Do not leave jaws embedded in the wound, which should be carefully washed. If there is any question of infection or complications, a physician should be notified.

till. 1: boulder clay; clay soil that contains rocks of varying size as occurs in glacial deposits. 2: to turn and break up the soil for planting crops as by plowing, harrowing, turning with a shovel. 3: soil prepared by tilling.

tillage. the process of tilling or husbanding the soil. **brand tillage**. a farming strategy in which old land is allowed to lied fallow and regain its natural cover while new land is cleared, tilled, and planted. This practice is common in the tropics where soils are easily exhausted and continuing use of the land is otherwise impossible.

tilth. 1: a fine-textured surface soil that results from cultivation; the condition of such a soil. 2: to work the soil to form a fine tilth.

timberland. 1: forested land that can annually produce a minimum amount of wood of commercial value. This term does not apply to lands where timber cutting is prohibited. 2: land where commercial cutting of timber is active or proposed. *See* timber (def. 2).

time recording device or monitor. a mechanical or electronic device that monitors and displays the interval between milkings and together with the starting and ending time of the preceding milking.

tincal. sodium borate (*See* borate.

TMTD. thiram.

TNF-α. tumor necrosis factor α.

toadstool. **1:** any of numerous fungi with an umbrella-like cap (pileus). This term is essentially synonymous with mushroom in both the narrow and broad senses but is more often used for inedible species. **2:** todesstuhl; death's stool; a vernacular term applied to any unappealing, unpalatable, or poisonous mushroom. *Cf.* mushroom, mycetismus.

tobacco. 1: the dried, usually processed, leaves of *Nicotiana* spp., as that in cigars, cigarettes, snuff, pipe tobacco, or chewing tobacco. It is carcinogenic to humans in all forms. Tobacco is teratogenic and a cause of arthrogryposis in livestock. *See also* cigarette, cigarette smoke. 2: *Nico-*

tiana spp., *q.v.*

tobosagrass. *Hilaria mutica*.

toco-. a combining form denoting childbirth.

toe-crack. *See* sand-crack.

tolerance (toleration). 1: the ability to endure pain or hardship. 2: the ability to endure exposure to a large amount (or a high concentration) of a substance (e.g., food, drug, toxicant) without adverse effects. This term may apply to an individual organism, a type of organism, a population or species, or even to an ecological community. 3: the capacity of an organism to show decreasing sensitivity to a particular substance with subsequent exposure. *See also* intolerance. drug tolerance. showing a progressive decrease in sensitivity to a drug with subsequent doses. herbicide tolerance. the inherent or natural ability of a species to survive and reproduce following herbicide treatment. immunologic tolerance. a state of immunologic responsiveness to an antigen such that an amount that would ordinarily induce an immune reaction does not do so. law of tolerance. that the distribution of an organism is limited by its tolerance to the fluctuations of a single factor. This is actually a working hypothesis that may hold in some instances and not others. Nevertheless, microdistribution of some species is limited by toxic plant secretions. For example, we may be facing reductions and/or redistributions of certain species of certain populations of organisms due to toxic chemicals in the environment. pesticide tolerance. the permissible residue levels for pesticides in raw agricultural produce and processed foods. In the United States, when a pesticide is registered for use on a food or a crop, a tolerance (or exemption from the tolerance requirement) must be established. These are established by the EPA and are enforced by the Food and Drug Administration and the Department of Agriculture. self tolerance. *See* horror (horror autotoxicus).

tolerance level. that concentration of a chemical residue in food or feed above which adverse health effects are possible and corrective action should be taken.

tolerance limit. 1a: the greatest amount, concentration, or intensity of a biologically active agent or stimulus to which a living system can be exposed for a specified period time and (1) not give evidence of distress, injury, or damage, or (2) still survive. *See also* limit, limit value, exposure limit, maximum allowable concentration, ceiling level, ceiling value. 1b: in toxicology this term usually refers to the concentration or dose of a substance that a living organism can endure without ill effects. median tolerance limit (TL_{50}). the concentration of a substance in a suitable diluent that 50% of a test population can survive for a specified period of exposure.

toleration. *See* tolerance.

tolguacha. *See Datura*.

α-toluenol. benzyl alcohol.

tom turkey. the male turkey.

tomato fruitworm. *Heliothis zea*.

tomato. *Lycopersicon esculenta*.

-tome. a suffix that indicates or denotes 1: a cutting instrument, the first element in the compound usually indicating the part that the instrument is designed to cut. 2: segment, part, or section.

tomigoff. *Bothrops atrox*.

ton. **1b:** any of a number of units of weight. The most commonly used are given here. The British spelling of ton is tonne. **freight ton**. a common unit of volume of sea freight equal to 40 cubic feet. **long ton**. a common unit of weight in the United Kingdom. It equals 2240 pounds (1016.0469088 kg). **short ton**. a common unit of weight in the United States. It equals 2000 pounds (907.18474 kg).

tonka bean camphor. coumarin.

tonne. *See* ton.

top-, topo-. combining forms that denote location, place; topical.

torciopelo. *Bothrops atrox*.

torsion. 1a: the act of twisting. 1b: the condition of being twisted or turned. 2: a type of mechanical stress that causes the deformation of an object when one end is twisted in one direction and the other end is held motionless or twisted in the opposite direction.

toxaphene (chlorinated camphene; octachlorocamphene; polychlorocamphene). the generic name for a complex insecticidal mixture of chlorinated camphenes, chiefly octachlorocamphene, which has the approximate empirical formula, $C_{10}H_{10}Cl_8$. Toxaphene is an extremely toxic, amber, waxy, solid mixture of more than 170 compounds with a mild odor of chlorine and camphor. The individual compounds vary widely in toxicity. The insecticidal activity, however, resides in just a few of these compounds. Toxaphene was used not only to control insects of cotton and other crops but also to control insects on livestock and to remove unwanted fish from freshwater lakes. Symptoms of intoxication in exposed mammals include nausea, confusion, agitation, tremors, epileptiform convulsions, and unconsciousness. The oral LD_{50} in laboratory rats is 69 mg/kg. Toxaphene is carcinogenic to some laboratory animals and is a suspected human carcinogen. Toxaphene was banned from use in 1982 in the United States by the Environmental Protection Agency, except for on livestock and in emergencies. All registrations for toxaphene in the United States have been cancelled.

toxemia (toxaemia; toxicemia; toxicohemia; toxinemia; blood poisoning). 1: any of a variety of intoxications, pathological conditions, or disease syndromes caused by the presence of poisonous substances, especially bacterial toxins, in circulating blood. 1a: septicemia; a generalized systemic intoxication due to dissemination of bacterial toxins via the circulatory system from a focal infection. 1b: clinical manifestations of 1a; a functional or organic disturbance or pathological condition due to dissemination of toxicants derived from protein catabolism. *See*, for example, ophidism. 2: injury to a plant by insect or other poisons. alimentary toxemia. a form of autointoxication due to the absorption of poisonous substances from the alimentary canal that were produced therein. toxemia of pregnancy (eclampsia and preeclampsia). a poorly defined term used in reference to metabolic disorders of pregnancy in humans that are marked by edema and serious disturbances of renal function (with albuminuria), central nervous system function, and blood pressure. **eclampsia**. a metabolic disorder of late pregnancy that may occur from the 20th week of pregnancy until a week after delivery. It is characterized by an intensification of the symptoms described for preeclampsia, above, plus the development of muscle spasms, seizures, and coma. The etiology is unknown. Both of these toxemias usually diminish and disappear within a week following childbirth but are sometimes fatal. **preeclampsia**. a metabolic disorder of late pregnancy that may occur from the 20th week until a week after delivery. It is marked by edema and serious disturbances of renal function (with albuminuria), central nervous system function, and blood pressure. It is sometimes fatal. If convulsions and coma also occur, the condition is termed eclampsia. The condition may be mild (mild preeclampsia) with a significant rise in blood-pressure, a puffiness of the face, hands and feet, proteinuria, and excessive weight gain during the last trimester. In severe cases (severe preeclampsia), the above symptoms are more pronounced and the victim may have headache, blurred vision, abdominal pain, and irritability.

toxemic jaundice. copper poisoning.

toxic effect. *See* effect.

toxic pulmonary emphysema. *See* emphysema.

toxic pulmonary emphysema of cattle. *See* emphysema.

Toxic Substances Control Act (TSCA). a U.S. federal law passed in October 1976. This act authorizes the EPA to develop and implement means to minimize dangers from toxic substances. Pesticides, drugs, tobacco products, nuclear materials, cosmetics, foods, and food additives are excluded from compliance with this act or the regulations promulgated therefrom.

toxic systemic effect. *See* effect.

toxic waste. 1: any waste material that is toxic. 2: any waste material that contains poisonous substance(s) as designated under law such that their use, transportation, and disposal is regulated. Several laws apply in the United States. These are The Resource Conservation and Recovery Act of 1976, as amended; The Toxic Substances Control Act of 1976, as amended; the Comprehensive Environmental Response Compensation and Liability Act of 1980 (CERCLA). Many governmental jurisdictions have similar laws that regulate the handling and disposal of such wastes.

toxicity. **1:** the state or quality of being poisonous; the capacity to poison living systems; having a toxic or poisonous quality. This is sometimes erroneously taken to be an inherent property of a substance. The toxicity of a substance is always due, to a greater or lesser extent, to certain interactive properties of the substance, the specific target (e.g., species, strain, sex, age, organ, prior exposure to the toxicant), and the environment. **2:** potency, virulence, deadliness, lethality, or the degree of virulence of a poison or of a toxic microbe. The degree to which a toxicant is injurious, lethal, or poisonous to plant or animal life. **3:** toxicity is often operationally defined in terms of the amount (or concentration) of a particular toxicant required to produce a stated detrimental effect in a given type or species of living organism under specified conditions of exposure for a given end point (e.g., mortality, morbidity, eggshell thinning, loss of fecundity); the toxic dose is often given in terms of unit body weight of the organism of concern (e.g., mg of toxicant per kg body weight). *See also* factor (modifying factor), nontoxic, toxic, toxicity rating. **absolute effluent toxicity**. the toxicity of a waste stream or effluent without consideration of dilution. **acute toxicity**. **1:** the capacity of a biologically active agent such as a pesticide or herbicide to cause serious harm, injury, or death to a living system following a single dose or a single brief exposure. **2:** harm to a living system that is manifested within a relatively short time following exposure to a toxicant. **3:** the amount of pesticide that will seriously affect or destroy a test animal in a single dose. *See also* toxic. **aquatic toxicity**. the capacity of a substance in a body of water (usually taken to be a freshwater system — a stream, lake, reservoir, or groundwater body) to poison those organisms that normally live in such a body. *Cf.* marine aquatic toxicity. **biogenic toxicity** (allelopathy). the chemical inhibition of one organism by another (allelopathy); sometimes taken to also include autotoxicity. **chronic toxicity**. **1:** the capacity of a biologically active agent such as a pesticide or herbicide to produce toxic effects following long-continued or numerous brief exposures. Trustworthy quantitative information on chronic toxicity is rarely available. **2:** any poisonous effect that results from long-continued or numerous brief exposures to a substance. **3:** occasionally used synonymously with delayed toxicity. This usage is not recommended. **comparative toxicity**. the capacity of a toxicant to poison a particular type or species of organism relative to its toxicity to other types of taxa of the same rank (e.g., a genus, a species, a family, dairy bulls). If comparisons are made among a set of species, all should be taxonomically related (e.g., all are members of the same genus, a set of closely related genera, or the same family). This term is sometimes confused with relative toxicity, *q.v. See also* toxicology (comparative toxicology). **delayed toxicity** (latent toxicity). the appearance of clinical toxic effects following a relatively extended latent period (e.g., tumor development and accompanying symptoms; poisonings by certain mushrooms). This term is not synonymous with chronic toxicity. **dermal toxicity**. the ability of a toxicant to poison people or other vertebrates by contact with the skin. **indirect toxicity**. the capacity of a substance or organism to exert toxic effects (1) only following metabolic transformation to another chemical species (the proximate poison) or (2) by virtue of contamination by or accumulation of a poison or poisons. *Cf.* secondary toxicity. **latent toxicity**. delayed toxicity. **marine aquatic toxicity**. the toxicity of a substance in a saltwater body to those organisms that normally live in such a body. *Cf.* aquatic toxicity. **overdosage toxicity**. the toxic effect that predictably occurs when the dosage of a drug that is administered to a subject exceeds the therapeutic range. **relative toxicity**. **1:** the toxicity of one toxicant or class of toxicants relative to another for a specified target under specified conditions. This term is often confused with comparative toxicity, *q.v.* Factors that affect relative toxicity in addition to the inherent properties of the toxicants under consideration at a given time are numerous and may include dosage, route of exposure, physical nature or phase of the toxicant, temperature of the toxicant, ambient temperature, humidity, health and condition of the subject (individual, population, system), and interactions with other chemicals (poisonous and nonpoisonous). **2a:** the toxicity of a chemical determined for a given effect relative to that for another effect of the same chemical. This is a special case of relative toxicity. **2b:** the toxicity of two or more chemicals with respect to one specified adverse effect. This is a special case of relative toxicity. **3a:** the toxicity of an effluent discharge (e.g., from a manufacturing plant) relative to that following mixture with the receiving waters. **3b:** the toxicity of an effluent relative to a dilution using water that has a composition similar to that of the receiving waters. **secondary toxicity**. the state, quality, or capacity (which may be temporary) of an ordinarily nonpoisonous organism, material, or substance to poison living systems. The situation may arise by accumulation or contamination with toxicants from the environment. Cadmium, copper, lead, and fluorine are examples of chemicals that may accumulate to dangerous levels in certain plants. *Cf.* indirect toxicity. **selective toxicity**. **1a:** the state whereby the action of a given toxicant is restricted to one or a few structures, processes, species, or higher taxon of organism. Only rarely do poisonous compounds used to control pests and weeds possess selective toxicity. **1b:** as applied to pesticides and herbicides, selective toxicity is the nature of a poison such that it produces injury to one kind of living system (the uneconomic form) without injuring another (the economic form), even though the two systems are in intimate contact. Very few substances exhibit selective toxicity.

toxico-, tox-, toxi-, toxo-. combining forms denoting poison, a toxin, or poisoning.

toxicon (obsolete). toxin.

toxin (toxinum; biotoxin; toxicon (obsolete)). defined variously by different specialists, sometimes without regard to existing usage. Definitions 1a and 1b are preferred. **1a:** true toxin; biotoxin; a usually proteinaceous poison, produced metabolically by a living organism, that can induce the production of antitoxins when injected into the tissues of an animal, especially a vertebrate animal. True toxins are produced by certain bacteria, a few higher plants, and by certain venomous animals (e.g., stinging hymenoptera, venomous snakes); they are variously named according to source (e.g., bacteriotoxin, microbial toxin, mycotoxin, phytotoxin, or zootoxin). **1b:** in the original sense, any poison of high molecular weight, usually a polypeptide or protein, synthesized by a microorganism, especially a bacterium. *See* microbial toxin. **2:** in pathology and medical jurisprudence, a toxin is usually considered to include any diffusible alkaloidal substance (e.g., ptomaines, abrin, brucin); snake venoms and wasp venoms; microbial toxins. **3:** often synonymous with any poison including mixtures (e.g., snake venoms). This usage is not recommended. ***Amanita* toxin**. amanitatoxin. **animal toxin**. zootoxin. **anthrax toxin** (bacillus anthracis toxin). a culture filtrate of *bacillus anthracis* that contains at least three different substances: an edema factor, a lethal factor, and a protective antigen. **bacterial toxin**. bacteriotoxin. **botulinum toxin** (botulin; botulinus toxin; botulismotoxin). the generic name of a set of at least eight extremely toxic, antigenically distinct protein neurotoxins produced anaerobically by the bacterium, *Claustridium botulinum*. Humans are usually exposed by consuming inadequately sterilized canned or bottled food. The botulinum toxins resist decomposition by gastric and intestinal juices. *See* botulism. **botulinus toxin**. botulinum toxin. **botulismotoxin**. botulinum toxin. **cholera toxin.** choleragen. **diphtheria toxin**. the causative agent of diphtheria. It is an antigenic protein toxin secreted by *Corynebacterium diphtheriae* that interferes with protein synthesis in eukaryotic cells and also destroys tissue. **dysentery toxin**. the exotoxin of various species of *Shigella*. **ectotoxin**. an obsolete synonym for endotoxin. **exotoxin** (ectotoxin (obsolete); extracellular toxin; soluble toxin). any pathogenic proteinaceous toxin produced by bacteria, usually Gram-positive bacteria, that are released into the surrounding medium. Many are highly active enzymes, are more heat labile and generally more toxic than endotoxins, and cause more severe effects. Exotoxins are detoxified (but retain their antigenicity) by treatment with agents such as formaldehyde (formol toxoid) that do not affect endotoxins. Exotoxins (e.g., botulinum toxin) are the most toxic substances known and include cardiotoxins, hemolysins, cardiotoxins, diphtheria toxin, neurotoxins, plague toxin and various cell-disrupting enzymes such as lecithinase and collagenase). Ectotoxin is an obsolete synonym. *See* botulinum toxin, endotoxin, toxemia. **extracellular (bacterial) toxin**. exotoxin. **F_2 toxin**. zearalenone. **intracellular (bacterial) toxin.** endotoxin. **microbial toxin**. an extremely toxic poison of high molecular mass synthesized metabolically by a microorganism. It is, by and large, the virulence of the toxins produced that determines the virulence of a particular microbe. Microorganisms also produce a variety of toxic substances (e.g., acetaldehyde, formaldehyde) that are usually not termed toxins. *See also* bacteriotoxin, endotoxin, exotoxin, toxinologist. **milk toxin**. any of the aflatoxins G. **mushroom toxin**. any toxin produced by any of the numerous species of poisonous mushrooms. The known mushroom toxins have been variously named and classed such that casual examination of the relevant literature can be confusing and misleading. normal toxin. a toxin solution that holds exactly 100 lethal doses per ml of diluent. phytotoxin. any toxin produced by a plant. Most are alkaloids and produced by fungi; toxins are less common in higher plants (examples are juglone, ricin, abrin). plant toxin. phytotoxin. soluble toxin. exotoxin. tetanus toxin (tetanotoxin). **1:** the neurotropic, heat-labile exotoxin of *Clostridium tetani*; it is the cause of tetanus. It is a crystalline protein (molecular weight 67,000) that consists of two parts: a neurotoxin (tetanospasmin) and a hemolysin (tetanolysin). It is one of the most toxic substances known. The major action is apparently the blocking of inhibitory synaptic impulses. true toxin. toxin (*See* def. 1a, above). A term that encompasses exotoxins produced by certain bacteria and toxins produced by a few fungi, a few higher plants, and by venomous animals such as stinging Hymenoptera and venomous snakes. wildfire toxin (*N*-[2-amino-4-(3-hydroxy-2-oxo-3-azetidinyl)-1-oxobutyl]-L-threonine). a colorless, hygroscopic, fungal phytotoxin,

$C_{11}H_{19}N_3O_6$

It is soluble in water and methanol and moderately soluble in ethanol. Wildfire toxin is produced by *Pseudomonas tabaci*, the bacterium that causes wildfire disease of tobacco. It is highly toxic to diverse organisms including algae, bacteria, animals, and higher plants.

***N*-[2-amino-4-(3-hydroxy-2-oxo-3-azetidinyl)-1-oxobutyl]-L-threonine**. wildfire toxin (*See* toxin).

toxolysin. antitoxin.

TR-1. verruculogen.

trace element. **1:** an element that occurs in living organisms in trace or very low concentrations. Most are essential to normal growth, development, and functioning of an organism but are present in and used by the organism only in minute quantities. A total lack of certain of these eventually causes death but concentrations that are higher than normal (e.g., of chromium, cobalt, copper, fluorine, iodine, selenium) can be toxic. **2:** an element present in a material only in minute or trace quantities (e.g., trace elements in coal); these sometimes have toxicological significance, as in the burning of fossil fuels. essential trace element. a trace element that forms part of the structure of metabolically important molecules (e.g., enzymes), or is an important cofactor or catalyst of metabolic processes. The total (or near total) lack of which eventually causes death; excessive amounts are toxic. The essential trace elements of humans and most vertebrates include Zn, Mg, Cu, Ni, Mn, Co, Cr, I, and Se; those of plants include Cu, Zn, Mg, Mn, and B.

tracheo-, trache-. combining forms denoting the trachea.

tractor. a heavy-duty farm vehicle that is usually powered by diesel fuel. The rear wheels are large, sometimes huge. It has numerous important uses on a farm (hauling heavy loads; plowing, harrowing). It has largely replaced the horse as the main provider of muscle on a modern farm.

tragacanth. a demulcent used in veterinary practice as a vehicle for insoluble medications or to mask the taste of medications.

tramp's spurge. *Euphorbia corollata*.

transgene. a foreign gene present in both somatic and germ cells of a plant or animal. It is expressed in one or more tissues and is inherited by progeny by normal Mendelian inheritance. *See* transgenic animal, transgenic plant.

transgenic. pertaining to an organism in which a foreign DNA gene (transgene) is incorporated into its genome early in development.

transgenic animal. a genetically engineered animal or the offspring of a genetically engineered animal. Such an animal usually contains genetic material from one (or more) unrelated organisms (e.g., a virus, plant, or another animal). *See also* transgenic, transgenic plant, transgenosis.

transgenic plant. a genetically engineered plant or the offspring of a genetically engineered plant. Such a plant usually contains genetic material from one or more unrelated organisms, such as a virus or other plant. *See also* transgenic, transgenic animal, transgenosis.

transgenosis. the transfer of genes from one organism to another unrelated organism, where they are incorporated into the genome and expressed phenotypically.

transmethylation factor. choline.

trefoil dermatitis. trifoliosis.

trefoil. any herb of the genus *Trifolium*.

trembles (tremetol poisoning of cattle). a disease primarily of cattle caused by consuming tremetol-containing plants such as *Eupatorium rugosum*, *E. urticaefolium*, and *Haplopappus venetus*. Humans have been secondarily poisoned by drinking milk from, and occasionally eating the meat of, an affected animal. The condition in humans is called milk sickness. Symptoms in cattle may include weakness and anorexia, conspicuous trembling, stiffness of the limbs, frequent falling especially upon forced exertion, and gaseous abdominal distension. In acute cases, the animal may lose consciousness and die within a few days. Chronic cases are typified by periods of remission and exacerbation. Symptoms in humans are similar to those in cattle but also include vomiting, complaints of vague pains, listlessness, abdominal distress, and constipation. The proximate cause of the weakened condition in humans and cattle is probably hypoglycemia; death results from ketosis and acidosis. Conspicuous lesions are fatty degeneration of the liver, kidneys, and sometimes muscle. Death in the chronic form of trembles usually results from starvation or intercurrent infection.

tremetol (2-isopropenyl-2,3-dihydro-5-acetylbenzofuran; tremetone). a lipophilic alcohol, it is a toxin produced by certain plants such as *Eupatorium rugosum*, *E. urticaefolium*, and *Haplopappus heterophyllus*. It causes muscular weakness and trembles, *q.v.*, in cattle and in humans who drink contaminated milk or who occasionally eat the meat of an affected animal. It causes milk sickness, *q.v.*

tremetol poisoning of cattle. trembles.

tremetone. tremetol.

tri-. a prefix indicating three.

Tri-6. a trade name for lindane.

triallate (bis(1-methylethyl)carbamothioic acid S-(2,3,3-trichloro-2-propenyl) ester; diisopropylthiocarbamic acid S-(2,3,3-trichloroallyl) ester; 2,3,-trichloro-2-propen-1-thiol diisopropylcarbamate; S-(2,3,3-trichloro-2-propenyl) bis(1-methylethyl)carbamothioate). a thiocarbamate herbicide, $C_{10}H_{16}Cl_3NOS$, closely related to diallate. It is used chiefly to control wild oats and blackgrass in cereal crops. It is a mitotic poison in young shoots but its major effect is suppression of cell elongation. In wild oat seedlings exposed to triallate, cell divisions are fewer and mitotic abnormalities can be observed.

triarylated organotin compound (triaryltins). an organotin compound with three aromatic groups. These compounds are biocidal, immunotoxic agents, used mainly as fungicides and insecticides.

triaryltins. triarylated organotin compounds.

triazine. 1: any of three isomers, $C_3H_3N_3$, that contain a 3-membered carbon ring, three double bonds, and three nitrogen atoms. Triazine is used as an herbicide but is very or even extremely toxic to sheep, cattle, and horses. These animals have been fatally poisoned when grazing on triazine-treated pastures as much as a week following spraying. Signs of intoxication may include anorexia, weight loss, tenseness and general depression and muscular weakness. 2: any of various structurally related derivatives of these compounds. *See* atrazine, herbicide (triazine herbicide).

triazine compound. triazine derivative.

triazine derivative (triazine compound). 1: any of a number of compounds that are structural derivatives of triazine, a number of which are used as herbicides. *See* atrazine, ametryn, herbicide (triazine herbicide). 2: triazine (def. 2).

1*H*-1,2,4-triazol-3-amine. 3-amino-1,2,4-triazole.

1,2,4-triazol-3-ylamine. 3-amino-1,2,4-triazole.

Tribulus terrestris (puncture vine; caltrop). a prostrate annual weed (Family Zygophyllaceae), introduced to the southern United States from southern Europe. It occurs commonly on dry soils along roadsides and deserts mainly in the southern and southeastern states. It causes hepatogenic photosensitization in grazing animals. It is a causative agent of bighead in sheep. The hepatotoxin inhibits the conjugation of pigments, which normally occurs prior to excretion via the bile; it also inhibits excretion. Concentrations of nitrate that are toxic to livestock have also been reported. *See also* phylloerythrin, bighead.

tributyl tin, tributyl tin compound (bis(tributyltin) oxide; TBT; TBTO; OTBE; Biomet TBTO; Butinox; hexabutyldistannoxane). an alkyltin compound,

$$[(C_4H_9)_3Sn]_2O$$

It has bactericidal, fungicidal, molluscicidal, and insecticidal properties. TBTs are of particular environmental importance because of their growing use as industrial biocides, e.g., in the control of slime, fungi, and bacteria. A major use is that of TBT paints used to coat maritime ships to prevent the growth of fouling organisms. It is very or extremely toxic to laboratory rats by ingestion, and is considered very toxic to humans. Mechanisms of toxic action are unknown and the direct primary effects are poorly known. It appears to be neurotoxic and immunotoxic and is a suspected teratogen. Exposed humans may experience headaches, blurred vision, stomach aches, dizziness, fatigue, and chronic dermatitis.

tricalcium arsenate. calcium arsenate.

1,2,3-tricarboxylic acid. citric acid.

tricarboxylic acid cycle (citric acid cycle; Krebs cycle; TCA cycle). an important series of metabolic reactions in aerobic respiration in plants and animals. In this cycle, acetyl coenzyme A formed from pyruvate produced during glycolysis is oxidized completely to carbon dioxide and water through a sequence of interconversions of various carboxylic acids (oxaloacetate, citrate, ketoglutarate, succinate, fumarate, and malate). The indirect result of this process is the synthesis (reconstitution) of ATP, although some ATP is produced by direct coupling with cycle reactions. The two molecules of pyruvate that enter a cycle from glycolysis produce 30 ATP molecules. *See* citric acid, respiration.

1,4,5-trideoxy-1,4-imino-5-(4-methoxyphenyl)-D-xylo-pentitol 3-acetate. anisomycin.

-trichia. a combining form denoting condition or type of hair.

2,3,6-trichlorobenzoic acid. 2,3,6-TBA.

trichloroacetic acid (TCA). a colorless, extremely toxic, hygroscopic solid with colorless, deliquescent crystals,

$$C_2HCl_3O_2$$

It is a strong acid and an oxidative metabolite of trichloroethylene (TCE). It is used as an herbicide (sodium trichloroacetate) and an intermediate in the manufacture of pesticides. As in the case of other haloaliphatic acids, TCA inhibits plant growth, and causes leaf chlorosis and necrosis. It affects numerous additional plant processes. TCA is used in the laboratory to precipitate protein. It is corrosive to skin, eyes, and mucous membranes and is carcinogenic to some laboratory rodents.

trichloroacetonitrile (trichloromethylnitrile). an insecticide, C_2Cl_3N, and a strong irritant of the eyes and skin. It is very toxic to laboratory rats.

1,2,4-trichlorobenzene (*unsym*-trichlorobenzene). a colorless, flammable, stable, water-insoluble liquid, $C_6H_3Cl_3$, with a pleasant odor similar to that of *o*-dichlorobenzene; it is miscible with most oils and organic solvents. It is used chiefly as a solvent in the manufacture of chemicals, in dyes and intermediates, herbicides, and insecticides. It is toxic by contact, ingestion, and inhalation. The principle lesions found in experimental exposures of laboratory animals are those of the liver, kidney, and lung. Humans are exposed predominantly by runoff from precipitation and drinking water. It is hepatotoxic and a strong irritant of the skin, eye, and upper respiratory tract. Symptoms of intoxication may include drowsiness, lack of coordination, and unconsciousness.

***unsym*-trichlorobenzene**. 1,2,4-trichlorobenzene

2,3,6-trichlorobenzeneacetic acid. chlorfenac.

2,3,6-trichlorobenzoic acid. 2,3-6-TBA.

1,1,1-trichloro-2,2-bis(*p*-chlorophenyl)ethane. DDT.

1,1′-**(2,2,2-trichloroethylidene)bis[4-chlorobenzene]**. DDT.

1,1′-**(2,2,2-trichloroethylidene)bis[4-methoxybenzene]**. methoxychlor.

tri**chloromethylnitrile**. trichloroacetonitrile.

2,4,6-trichlorophenol. 2,4,6-T.

2,4,5-trichlorophenoxyacetic acid. 2,4,5-T

2-**(2,4,5-trichlorophenoxy)propionic acid**. silvex.

2,3,6-trichlorophenylacetic acid. chlorfenac.

2,3,-trichloro-2-propen-1-thiol diisopropylcarbamate. triallate.

***S*-(2,3,3-trichloro-2-propenyl) bis(1-methylethyl)carbamothioate**. triallate.

trichochromogenic factor. *p*-aminobenzoic acid.

Trichodectes climax. *Bovicola caprae*.

Trichodectes parumpilosus. *Bovicola equi*.

Trichodectes scalars. *Bovicola bovis*.

Trichodectes sphaerocephalus. *Bovicola ovis*.

trichothecene; trichothecene mycotoxin. any of a class of more than 40 sesquiterpenoid mycotoxins produced by a variety of fungi that grow chiefly on grains (e.g., *Cephalosporium, Fusarium, Myrothecium, Stachybotrys, Trichoderma, Trichothecium*, and *Verticimonosporium*). Many of these compounds have bacteriocidal, fungicidal, and insecticidal activity and are highly toxic to various plants and animals. They invade numerous plants and many agricultural products, the ingestion of which can cause serious poisonings of humans and other animals. An outbreak of trichothecene poisoning in Siberia in 1944 provided considerable information on the effects and toxicities of these substances on humans. Because of a shortage of food related to World War II, many people ate moldy grains such as barley, millet, and wheat. The skin of those who did became inflamed and they suffered gastrointestinal disorders, including vomiting and diarrhea, and multiple hemorrhage. About 10% of those affected died. All trichothecenes are dermatotoxic; some (e.g., T-2 toxin) are neurotoxic; the crude toxins may be immunosuppressive. Examples are roridin, verrucarins, deoxynivalenol (Rd-toxin), deoxynivalenol-monoacetate (Rc-toxin), diacetylnivalenol, diacetoxyscirpenol, monoacetylnivalenol (fusarenon-X), nivalenol, trichothecin, and T-2 toxin.

2,4,5-trichlorophenoxyacetic acid. 2,4,5-T.

1,4,5-trideoxy-1,4-imino-5-(4-methoxyphenyl)-D-xylo-pentitol 3-acetate. anisomycin.

triene. alkene.

trietazine. an atrazine herbicide, introduced in 1972 as a component of herbicide formulations for weed control in potatoes and peas.

triethylenethiophosphoramide. thiotepa.

α,α,α-trifluoro-2,6-dinitro-*N,N*-dipropyl-*p*-toluidine. trifluralin.

1,1,**1-trifluoro-2,6-dinitro-*N,N*-dipropyl-p-toluidine**. trifluralin.

***N*-(3-trifluoromethylphenyl)-*N′,N′*-dimethylurea**. fluometron.

trifluralin (1,1,1-trifluoro-2,6-dinitro-*N,N*-dipropyl-p-toluidine; 2,6-dinitro-*N,N*-dipropyl-4-(trifluoromethyl)benzenamine; α,α,α-trifluoro-2,6-dinitro-*N,N*-dipropyl-p-toluidine; 2,6-dinitro-*N,N*-dipropyl-α,α,α-trifluoro-p-toluidine; 2,6-dinitro-*N,N*-dipropyl-4-trifluoromethylaniline; *N,N*-dipropyl-2,6-dinitro-4-trifluoromethylaniline). a yellowish-orange solid,

$$F_3C(NO_2)_2C_6H_2N(C_3H_7)_2$$

It is insoluble in water and soluble in ethanol, acetone, and xylene. An herbicide used to control annual grass and weeds in a number of crops (chiefly cotton) by incorporation into the top 2-5 cm of soil. Trifluralin inhibits cell division in root meristems, arresting cells in prophase within a few hours of treatment. This is usually followed by other mitotic aberrations, the enlargement of some meristematic cells which have become multinucleate, and a progressive decrease in the amount of meristematic tissue. It may completely inhibit lateral root development, especially in cotton. It is very toxic to laboratory rats and is carcinogenic to some laboratory animals.

trifoliosis (dew poisoning; alsike poisoning; trefoil dermatitis; clover disease). alsike-induced photosensitization that occurs in livestock due to ingestion of any of several types of clover and alfalfa (*See Brassica*, *Erodium*, *Medicago*, and *Trifolium*). In sensitized horses, the skin reddens from sunlight and a dry necrosis of the skin or edema with a serous discharge from the affected areas ensues. The most affected parts of the body are usually the muzzle and feet which are most exposed to moisture when on pasture (hence "dew poisoning"). Symptoms may also include excitement or depression and sometimes colic, stomatitis, and diarrhea which may be severe, in which case the affected animal becomes emaciated.

Trifolium (clover; trefoil). a genus of annual, biennial, or perennial herbs (Family Fabaceae, formerly Leguminosae). Various species have been implicated in trifoliosis and rape scald, as have species of *Brassica*, *Erodium*, and *Medicago*. *T. alexandrinum* (berseem; berseem clover; Egyptian clover). a variety of kohlrabi, it is a common fodder crop in Egypt and India. It is also grown for forage in the southwestern United States *T. hybridum* (alsike clover). originating in the Alsike parish in Sweden, it is a tall, slender, short-lived perennial with white flowers. Alsike poisoning has affected cattle, horses, sheep, and swine. Animals are photosensitized and usually succumb on alsike pasturage during sunny weather. Nevertheless, this clover may be acceptable for hay and as pasture under properly controlled irrigation or on dry land with an effective annual precipitation that equals or exceeds 45 cm (18 in.). *T. pratense* (red clover). a low-growing herb, often used as a forage crop. Hay containing substantial amounts of second cuttings or late-season cuttings of red clover may seriously affect cattle, horses, and sheep. The syndrome is initially characterized by slobbering but other symptoms soon appear and become increasingly severe unless contaminated hay is removed from the diet. Symptoms include bloating, a stiff gait, reduced milk flow, diarrhea, emaciation, and abortion. Pasturage in fields with red clover may produce photosensitization within a few days, with consequent edema and necrosis of tissues surrounding the eye and sometimes impaired vision or even blindness. *T. subterraneum* (subterranean clover). a species that sometimes accumulates lethal concentrations of copper from the soil. The growing foliage contains an estrogen that has caused extensive damage to flocks of sheep in Australia. Symptoms of poisoning in sheep include infertility, abnormal lactation, dystocia, and prolapse of the uterus.

Triglochin (arrowgrass). A semicosmopolitan genus of grass-like perennial plants (Family Juncaginaceae) that grow in clumps. They have long, linear leaves which, unlike grasses, are thick. These plants contain widely variable amounts of HCN and have been responsible for intoxication of livestock. Symptoms and lesions are those of cyanide poisoning.

1,3,8-trihydroxy-6-methyl-anthraquinone. emodin.

1,3,8-trihydroxy-6-methyl-9-10-anthracenedione. emodin.

4,5,7-trihydroxy-2-methyl-anthraquinone. emodin.

2,3,5-triiodobenzoic acid. TIBA

1,2,4-trimethoxy-5-(1-propenyl)benzene. asarone.

trimethyl carbonol. *tert*-butanol.

trimethyl methanol. *tert*-butanol.

trimethylammonium ion. choline.

trimethylmethane. isobutane.

trip. **1:** a journey or expedition. **2:** to activate or trigger, as a switch. **3:** to stumble or to fall over an object. **4:** to gamble or lope. **5:** a herd of goats.

triploid. **1a:** having three homologous sets of chromosomes. **1b:** a triploid cell or organism, one that has three sets of homologous chromosomes per cell except the germ cells.

2,4,6-tris-(dimethylamino)-*s*-triazine. hexamethylmelamine. altretamine.

tris(1-aziridinyl)phosphine sulfide. thiotepa.

Trisetum flavescens (golden oats; yellow oats; yellow oat grass). a tall grass (Family Gramineae), growing to nearly 1 m in height, with flat leaves and a terminal, rudimentary flower. It is indigenous to middle Europe but has been naturalized elsewhere (e.g., in North America). It contains substances that can induce calcinosis accompanied by excessive formation of dense bone with eventual obliteration of marrow spaces in herbivores. *See* enzootic calcinosis.

Triticum. a genus that includes a variety of cereal grasses that are broadly adaptable to a range of climates. Included are the most important and widely-cultivated cereals in many temperate areas (e.g., Europe, North America, and Australia). All species have 2-5 flattened, flowered spikelets within a terminal cylinder spikelet. *T. durum* (durum). a hard wheat grown in southern Europe and in the United States It is used to make semolina. *T. aestivum* (*T. vulgare*). an annual cereal grass and the most widely-cultivated of all wheats. It is known only as cultivated varieties. The grain, when milled, yields a fine white flour. *T. vulgare*. *T. aesticum*.

troop. 1: a band, corps, company, or legion. 2: a group of soldiers, especially those mounted on horses (a troop of cavalry). 3: a group of kangaroos, baboons or certain other animals.

-trophic. a suffix indicating a relationship to food or nutrition.

tropho-, troph-. combining forms indicating a relationship to food or nutrition.

-trophy. a suffix meaning food or nutrition.

-tropic. a suffix indicating a turning toward or an affinity for.

tropical bont tick. *Amblyomma variegatum*.

tropical hyphemia. *See* ancylostomiasis.

true aster. *See Aster*.

true cholinesterase. acetylcholinesterase.

true hardy aster. *See Aster*.

Temik™. aldicarb.

true hellebore. *Helleborus niger*.

true pine. *See* pine; *Pinus*.

trumpet flower. *See Gelsemium*.

trunk (bole). 1: bole; the main stem of a tree. 2: the body of a quadruped and certain marine mammals, exclusive of appendages. 3: an elongate proboscis as that of an elephant. 4: the main stem of an nerve or vessel.

Truxillo coca. *Erythroxylum novagranatense*.

trypsin. a proteolytic enzyme that hydrolyzes peptide bonds on the carboxyl side of the amino acids arginine and lysine.

TTD. disulfiram.

tuba root. derris root.

tubatoxin. rotenone.

tuber water hemlock. *Cicuta vagans*.

tubular gland. a gland composed of one or more tubules that terminate blindly.

tubuloacinar (acinotubular gland). any gland, the secretory units of which are elongated acini.

tubuloacinar gland (tubuloalveolar gland). a tubular gland terminating in a blind sac.

tuff. a rock composed of the fine volcanic particles that have usually been fused together by heat.

tullidora. a common name of *Karwinskia humboldtiana*.

tung nut. *Aleurites*.

tung oil (Chinawood oil). a yellow, flammable drying oil that is soluble in other oils, chloroform, ether, and carbon disulfide. It is extracted from the nuts of *Aleurites*, chiefly from *A. fordii*. The nuts and leaves also contain saponins. Ingestion of a single nut is sufficient to produce vomiting and severe abdominal pain within half an hour, followed somewhat later by diarrhea. Severe poisoning may end in exhaustion and collapse. Livestock that ingest prunings or fallen branches of *Aleurtes* present similar symptoms. Little, if any, tung oil is now produced in the United States.

tung oil tree. *Aleurites*.

tung tree. *Aleurites*.

tunnel disease. ancylostomiasis.

tupelo. a common name of *Nyssa sylvatica*.

turkey. *See Meleagris gallopavo*.

turkey corn. *See Dicentra*.

turkey X disease. 1a: aflatoxicosis. 1b: a mycotoxicosis of domestic turkeys caused by aflatoxins B from *Aspergillus flavus*, *q.v. See* aflatoxicosis.

turnip. *B. rapa* var. *rapifera*.

turnout. a widening in a narrow roadway where a vehicle may pull off to permit other vehicles to pass or sometimes to turn around.

turpentine weed. *Gutierrezia microcephala*.

tussock grass. *Sporobolus airoides*.

twibill. a mattock in which one blade is similar to that of an adz and the other like that of an ax.

twice-pinnate. bipinnate.

tympanic body. tympanic gland.

tympanic gland (tympanic body). any of the mucous glands of the mucosa of the tympanic cavity.

type I error. *See* error.

type II error. *See* error.

typhoid bacillus. *Salmonella typhi*.

Tyson's gland. unicellular gland.

U

udder. the suspended bag-like mammary gland of many quadrupeds. It terminates in teats from which the young suckle. fore udder. the forequarters of a cow.

UDMH. 1,1-dimethylhydrazine.

ulexine. cytisine.

ultra-. a prefix denoting excess, exaggeration; beyond.

umbel. A type of racemose inflorescence (*See* inflorescence) in which the flowers occur in flat-topped clusters on their pedicels that arise from the same point on the main axis like the ribs of an umbrella.

umbo. a rounded or blunt projection arising from a surface, as on the scale of a pine cone.

uncinariasis. hookworm disease.

undercut. the first cut made in felling a tree. *See* back cut.

underfur. underhair. *See* fur.

underhair (underfur, fur proper). the hair of a mammal that lies close to the skin as opposed to guardhair. *See* fur.

undernourishment. *See* malnutrition.

underwater agriculture. aquaculture.

undulant fever (Mediterranean fever; Malta fever; brucellosis; human brucellosis). brucellosis in humans, a disease caused by bacteria of the genus *Brucella* (*See* brucellosis), that is acquired by direct contact with diseased animals or by ingestion of infested meat, milk, or cheese. Brucellae are also present in aborted fetuses and uterine secretions of infected animals. Veterinarians, farmers, and slaughterhouse workers run the highest risk of infection. There is no effective vaccine for human brucellosis. The disease is treated, with antibiotics. If untreated the disease usually persists for months but is rarely fatal to humans.

uneven-aged. descriptive of a forest or stand of trees that contains intermingled trees that differ markedly in age. By convention this is generally taken as a minimum range of 10 to 20 years. *See* all-aged; *See also* stand (uneven-aged stand).

ungulate. 1: any hoofed mammal. 2: hoofed; having hoofs.

uni-. a prefix denoting single, unpaired.

Unified National Strategy for Animal Feeding Operations. a joint program of the USDA and U.S. Environmental Protection Agency. This program is intended to minimize water pollution from large animal confinement facilities and that due to land application of manure. This is accomplished by the application of site specific comprehensive nutrient management plans. *See also* animal feeding operation.

unintended effect. *See* effect.

unsaturated fat. fat that includes fatty acids that have one or more double bonds; not fully hydrogenated.

unsoundness. unhealthiness; shakiness. In livestock (especially horses), unsoundness is any departure from normal that renders the animal unfit for service. Unsoundness may be temporary or permanent and may have environmental (e.g., disease) or hereditary components.

unstable. **1:** erratic; liable to change; changeable; not constant. **2:** reactive; explosive. **3:** readily decomposes or disintegrates.

unsteady gait. an erratic manner of walking.

unsymmetrical. asymmetric.

unthriftiness. poor weight gain of an animal or human relative to that expected from the quantity and quality of the diet.

uptake. *See* absorption.

urano-, uranisco-. combining forms relating to the hard palate.

ure-, urea-. prefixes of or pertaining to urea or urine.

urea. a white crystalline solid or powder,

$$CO(NH_2)_2$$

it is the primary nitrogenous waste of elasmobranchs, amphibians, and mammals where it occurs in blood, lymph, and urine. Urea and carbon dioxide are the end products of protein metabolism. It is produced in the liver chiefly from ammonia released during the deamination of amino acids; it is also formed from ammonia compounds in the body and can be formed directly from arginine. Urea normally comprises 80-90% of total urinary nitrogen in humans. Urea was the first organic compound to be synthesized in the laboratory. Aside from its metabolic importance, urea is produced industrially in high volume. Industrial applications include, for example, the manufacture of plastics and resins, herbicides, fertilizers, animal feeds, pharmaceuticals, and adhesives.

ureter-. a combining form denoting the ureter.

urethan. urethane.

urethane (carbamic acid ethyl ester; ethyl carbamate; ethyl urethane; ethyl urethan; urethan). a moderately toxic, synthetic white powder or crystalline ester amide,

$$CO(NH_2)OC_2H_5.$$

It is odorless, tastes like saltpeter, and forms solutions that are neutral to litmus. It is the repeating unit of polyurethane resins. It is used as an antineoplastic agent, a solubilizer and cosolvent for fumigants and pesticides, and as a veterinary anesthetic. Its use is incompatible with alkalies, acids, antipyrine, camphor, chloral hydrate, menthol, salol, or thymol. Urethane is moderately toxic to laboratory mice. It is a hepatotoxicant, causing liver necrosis, an immunosuppressive agent, and a direct-acting carcinogen.

urginea. the bulbs of *Urginea indica* (Indian squill) and *U. maritima*. It is the source of squill.

Urginea. a genus of Old World bulbous herbs (Family Lilaceae), chiefly of the Mediterranean region; discovered in Algeria. *U. maritima* (= *Scilla maritima*) (Mediterranean squill; red squill; medicinal squill; sea onion). a bulbous herb (Family Lilaceae) of southern Europe and northern Africa. The bulbs are cardiotoxic and widely used in commercial rodenticides (as red squill powder) because they are readily consumed (and assimilated) by rats, whereas they are distasteful to humans and species other than rats. The bulbs contain several cardiotoxic glycosides, including proscillaridin, scillaren A, scillaren B, and scilliroside.

uric-. *See* uris-.

2,6,8-trioxypurine. uric acid.

uric oxide. uric acid.

uric acid (2,6,8-trioxypurine; lithic acid; uric oxide). an odorless, tasteless, white, crystalline compound, CHNO, that is water soluble but insoluble in ethanol and ether; soluble in hot, concentrated sulfuric acid; and soluble in glycerol, aqueous solutions of alkali hydroxides, sodium acetate, and sodium phosphate. When heated, it emits toxic fumes of hydrogen cyanide. Uric acid is the primary nitrogenous waste of birds, reptiles, and certain invertebrates, especially insects. It is the end product of purine metabolism in primates including humans and in some dogs and certain other mammals. If not excreted, uric acid remains in the body unchanged and may form renal calculi ("kidney stones") and gouty concretions.

uric oxide. uric acid.

uricotelic. 1a: of or pertaining to the excretion of nitrogen in the form of uric acid. 1b: pertaining to animals that excrete most excess nitrogen in the form of uric acid. Birds are uricotelic. *Cf.* ureotelic.

urin-. a combining form relating to urine. *See also* use-, uris-.

urinary syndrome of cattle and sheep (redwater disease). this disease is a hemolytic anemia resulting from consumption of rape, *Brassica napus*. The symptoms and lesions are almost indistinguishable from copper poisoning. Affected animals stand apart and do not feed. The urine is dark colored and foams on striking the ground. Visible mucous membranes and the carcass are pale or icteric. Blood hemoglobin levels are very low; the liver is unusual in appearance and necrotic; the kidney is congested and dark appearing. In recently postpartum dairy cattle, this syndrome is indistinguishable from puerperal hemoglobinuria (*See* hemoglobinuria).

urinary. pertaining to urine.

urine-. *See* uris-.

uris-, uric-, uric- . combining forms relating to uric acid.

uris-, urine-. combining forms relating to urine.

urofolitrophin. FSH.

Urolene Blue. methylene blue.

uropygial gland (preen gland). a compound alveolar gland of birds located at the root of the tail, dorsal to the pygostyle. It secretes an oily mixture of fatty acids and wax from a papilla that opens onto the dorsal surface at the base of the tail. Birds apply this to the feathers with their beaks when preening. The secretion aids grooming and waterproofs the plumage. A few birds lack a uropygial gland.

ursin. arbutin.

Urtica (nettle). a widely distributed genus of annual and perennial herbs (Family Urticariaceae) introduced to North America from Eurasia. They bear stinging hairs that penetrate the skin and inject irritant chemicals that include formic acid and histamine. *U. chamaedryoides* (nettle). a slender nettle with stinging bristles, it occurs in rich woods, bottomlands, and waste areas from Florida to Texas and Mexico and northward to West Virginia, Kentucky, Missouri, and Oklahoma. It is the cause of nettle syndrome, *q.v.*

us-. a prefix **1:** meaning not. **2:** indicating removal, reversal, release, or deprivation. **3:** expressing an intensive action.

USID. Agency for International Development.

USDA. U. S. Department of Agriculture.

utero-, uter-. combining forms pertaining to or indicating a relationship to the uterus. *See also* hystero- (def. 1), metra-, metro-.

V

vaccine. a preparation of dead or weakened pathogens, or of derived antigenic determinants, that is used to facilitate the production of antibodies or immunity against the pathogen.

virulent. **1a:** toxic, deadly, able to cause disease. **1b:** pertaining to an organism that is able to cause disabling or deadly disease.

virulence. the degree of ability of an organism to cause disabling disease or death.

virus. an infectious particle, having some but not all of the properties of a living organism. It is comprised of a protein capsule surrounding a core of nucleic acid. A virus requires a host organism to replicate.

vaccine. a suspension of antigen (e.g., dead or weakened pathogenic bacteria or viruses; proteinaceous toxins from pathogenic bacteria) that will, when introduced into animal tissues, provoke the formation of antigen-specific antibodies. *Cf.* antivenin. *See also* vaccination.

Vaccinium (blueberry; cowberry; cranberry). a genus of widely distributed deciduous, evergreen shrubs (Family Ericaceae). The leaves of a number of species contain arbutin [e.g., *V. angustifolium* (low bush blueberry), *V. corymbosum* (high bush blueberry), *V. macrocarpon* (cranberry) *V. vitis-idaea* (cowberry)].

vagina. a sheath-like passage of the female reproductive tract in mammals that extends from the vulva to the cervix (neck of the uterus). Vaginal absorption is an uncommon, but sometimes important route of entry of poisons. *See* potassium permanganate.

vaginitis. inflammation of the vagina.

vascular cylinder. stele.

veal. meat from calves fed on milk.

vector. **1:** an organism, often an insect or rodent, that transmits disease. *See also* transvector. **2:** an object that transports genes into a host cell (a plasmid, virus, or other bacteria). A gene is placed in the vector which then "infects" the host.

Vegadex. sulfallate.

vegetal. a common name for *Blighia sapida* and its fruit in Cuba and Puerto Rico.

vegetation. **1:** the total plant life or plant cover of an area taken as a whole (e.g., the vegetation of the Russian steppes). **2:** the assemblage or inventory of plant species in an area. **3:** plant life in general.

vegetative control. any nonpoint source pollution control practice that involves the planting of cover crops to minimize the loss of pollutants from a site by reducing erosion.

vegetative propagation (vegetative reproduction). **1:**. nonsexual reproduction of plants by the growth and detachment of part of a plant (the parent plant) that becomes established as an independent plant body. Differentiated organs that may undergo vegetative propagation include bulbs, corms, rhizomes, stolons, and tubers. **2:** nonsexual reproduction of multicellular animals by budding as in the case in *Hydra*.

vegetative reproduction. vegetative propagation.

velvet bean. *Pithecolobium lobatum.*

velvetbean. *See Stizolobium deeringianum.*

venom. **1:** any of numerous, usually complex, chemically and pharmacologically diverse toxic secretions of animals that are transmitted to other animals by means of a bite, sting, puncture, or in rare instances by "spitting." Venoms were his torically classified on the basis of the type of tissue presumed to be affected (e.g., hemotoxin, neurotoxin). Most venoms, however, appear to affect nearly any cell or tissue which they contact and the specific effects observed are due chiefly to the degree of accumulation at a particular locus in the victim's body. **2:** unacceptable usage: a poison of animal origin. **ant venom**. a venom produced by an ant. Constituents vary but usually include enzymes and one or more organic acids including formic acid. Formic acid predominates in the venoms of red ants, *Formica spp.*, whereas the venom of *Myrmicaria natalensis* contains approximately 35% acetic acid, 31% isovaleric acid, and 22% propionic acid. Envenomation may be painful. **arthropod venom**. any venom secreted by an arthropod. Most spiders and many insects are venomous. Both the venom and the apparatus of envenomation vary considerably among arthropods. **bee venom**. the venom of bees (Order Hymenoptera). It is more complex and contains a greater variety of proteinaceous materials than do wasp and hornet venoms. These venoms produce neurotoxic, hemolytic, and strong histamine-like effects. Honey bee venom, for example, contains histamine, small peptides, mellitin, apamin, phospholipase, proteins (five of which are antigenic to humans), free amino acids, a histopeptide, simple sugars, and several phospholipids. *See also* apamin, mellitin, mast cell degranulating peptide. **black widow venom**. *See* latrodectus venom. **cobra venom**. the venom of true cobras, those of the genus *Naja*. It is a complex mixture of toxins and enzymes, the most important of which are cobra venom factor, cobra hemotoxin, and cobrotoxin. The latter is a powerful neurotoxin and is usually the dominant toxin of cobra venom. In some populations or individual snakes, the venom is high in necrotizing factor but low in neurotoxin. **crotalid venom**. the venom of a crotalid snake. They are generally rich in proteolytic enzymes. The venoms of New World crotalids alter peripheral resistances, often damage blood vessels, and may alter blood cells and blood coagulation mechanisms, as well as cardiac and pulmonary dynamics, and nervous system functions. The venom is extremely toxic but intensity of effects varies as a function of the amount and kind of venom injected and other factors. Symptoms in humans may include local pain, inflammation, hemorrhage and necrosis, dizziness, motor and sensory depression, hypotension (may be severe), and collapse. Death may occur, usually 18-32 hours following the bite. **heloderma venom**. the venom (sometimes called lizard venom) of the two lizards of the genus *Heloderma*, *H. suspectum* and *H. horridum*. It contains serotonin, amine oxidase, phospholipase A, hyaluronidase, and proteolytic activity. Injection of a large dose into a laboratory animal lowers systemic arterial pressure with a concomitant decrease in circulating blood volume, respiratory distress, and tachycardia. A lethal dose reduces ventricular contractility. **hemotoxic snake venom** (e.g., of *Crotalus*, *Agkistrodon*). any snake venom that contains cytolysins that enzymatically destroy cell membranes and tissues, with consequent hemorrhage from various organs. Such venoms may cause considerable pain, swelling, discoloration, hemorrhage, and necrosis that spreads rapidly from the area of the bite. Hematuria usually occurs and hemorrhaging from the mouth, nose, eyes, and gastrointestinal tract is common. Nausea, vom ition, circulatory collapse, and death may ensue. Hemorrhage is due to one of three types of hypofibrinogenemia: (1) without fibrinolysis or thrombocytopenia; (2) with fibrinolysis but no thrombocytopenia; (3) with thrombocytopenia and no fibrinolysis. **honey bee venom**. *See* bee venom. **hornet venom**. wasp and hornet venoms are distinguished from bee venoms by their lower content of peptides. They do contain kinin peptide, which may cause smooth muscle contraction and lowered blood pressure. Two biogenic amines in wasp or hornet venoms (serotonin and acetylcholine) lower blood pressure and cause pain. Acetylcholine may cause malfunction of heart and skeletal muscles. **hymenoptera venom**. hymenopteran venom. **hymenopteran venom** (hymenoptera venom). any venom produced by a hymenopteran. They are usually concentrated mixtures of water-soluble, nitrogen-containing chemicals. The major types of chemical species present are biologically synthesized (biogenic) amines, peptides and small, nonenzymatic proteins, and enzymes. The most common of the biogenic amines is histamine, found in the venoms of bees, wasps, and hornets (*See also* 5-hydroxytryptamine). Wasp and hornet venoms contain serotonin, and hornet venom contains the biogenic amine acetylcholine. Among the peptides and low-molecular-mass proteins in hymenopteran venoms are apamin, mellitin, and mast cell degranulating peptide in bee venom; wasp kinin, and hornet kinin. Enzymes contained in bee, wasp, and hornet venom include phospholipase A and hyaluronidase. Phospholipase B occurs in wasp and hornet venom. **insect venom**. any venom produced by an insect. The composition of insect venoms varies greatly among taxa and within the class. **krait venom**. the venom and its effects are similar to that described for cobra venom but while the bite is usually less painful than that of a cobra, the clinical manifestations may be more severe. Systemic effects include shock, often with conspicuous respiratory depression and coma; abdominal pain may be intense, and chest pain is sometimes reported. **latrodectus venom**. an extremely toxic, neurotoxic venom of *Latrodectus mactans,* the composition of which remains unclear. It is comprised of about 5 or 6 proteins (some are enzymes, including hyaluronidase) and a neurotoxic principle that appears to be a polypeptide with a MW of perhaps 130,000. It acts on nerve endings or myoneuronal junctions, destroying peripheral nerve endings or producing an ascending paralysis. The known pharmacological properties of the crude venom are inconsistent with the effects of this venom on humans. Early effects of envenomation in humans usually include muscle fasciculations, muscular weakness, sweating, tender, painful, and often enlarged lymph nodes, lymphadenitis, and hypertension. Severe cases often exhibit lower back and abdominal pain with muscular (especially abdominal) rigidity, and sometimes severe paroxysmal muscle cramps and arthralgia. Additional symptoms that may occur include nausea, vomition, conjunctivitis, edema of the eyelids, hyperemia, skin rash, and pruritis. *See also* black widow venom poisoning. **lizard venom**. *See Heloderma* venom. **loxosceles venom**. a largely cytotoxic, hemotoxic, necrotizing, and coagulant venom produced by spiders of the genus *Loxosceles*, most notably that of *L. reclusa* (brown recluse). The bite produces an unusual and characteristic necrotic lesion in humans; systemic effects may include disseminated intravascular coagulation. The venom is comprised of 10-12 proteins with enzyme activity that is greater than that of latrodectus venom. Two toxins (mw about 24,000) have been isolated from the protein fraction of *L. reclusa* venom. This venom contains a number of enzymes, and is necrotizing and lethal to mice and rabbits. No principle has been isolated that produces the lesion characteristic of *Loxosceles* bites. **neurotoxic snake venom** (e.g., of *Mi-*

crurus, *Naja*). any snake venom that contains one or more neurotoxins that dominate the effects produced by envenomation. Type A neurotoxins act primarily on the higher cardiorespiratory centers of the CNS; type B neurotoxins generally act on the myoneural junctions. The human victim of a neurotoxic snake venom may experience muscular weakness, paralysis of the mouth and throat, and ultimately of the respiratory muscles. Additional effects seen in serious cases of envenomation may include drowsiness, optic ptosis, muscular weakness, difficulty in swallowing, trismus, occasionally convulsions (e.g., from cobra venom), bradycardia, a weak pulse, dyspnea, with eventual cardiac and respiratory collapse and death. **rattlesnake venom**. a venom produced by the venom glands of snakes of the family Crotalidae, especially those of the genera *Crotalus* and *Sistrurus*. The venom is complex and contains many poisonous constituents including hemorrhagins. The lung appears to be the major target organ, where increased vascular permeability causes pulmonary congestion and hemorrhage. The venom may cause local bleeding, severe pain, superficial edema of rapid onset, and hemorrhagic bleb formation. **scorpion venom**. any venom produced by a scorpion. In the amount delivered by a sting, although usually painful, the venom of relatively few species produces effects that require medical attention. The most toxic venoms are produced by scorpions of the family Buthidae. The buthid venoms contain extremely toxic neurotoxins which are small basic proteins with a mw of about 7000. They also contain 5-hydroxytryptamine, which probably does not significantly contribute to toxicity. The buthid venoms act to enhance the release of acetylcholine by motor neurons and by postganglionic autonomic neurons with increased axonal Na^+ permeability. Symptoms in humans may include focal pain and inflammation, nervousness, debility, cardiac arrhythmia, and respiratory distress. All cases of severe poisoning must be treated with antiscorpion serum as soon as possible. Small children are especially at risk from scorpion stings. **snake venom**. a venom produced by a snake. Such venoms are complex mixtures that do not lend themselves to complete analysis. Included are polypeptides, enzymes (e.g., phospholipases, peptidases, proteases, collagenases, phosphoesterases, acetylcholinesterase) and other toxic proteins, biogenic amines, carbohydrates, and metal ions. Among the most important components are the smaller (nonenzymatic) polypeptides (those containing 60-70 amino acid residues), many of which are neurotoxic; some are more toxic than the crude venom. Proteolytic enzymes are also prominent constituents, often causing widespread tissue damage and necrosis. In terms of effects, snake venoms are comprised of variable amounts of blood coagulants, anticoagulants, agglutinins, cytolysins, proteolysins, antibactericidin, and neurotoxins. Most components appear to occupy specific chemical and physiologic receptor sites. Envenomation by a snake may be complicated by the release of autopharmacologic substances (e.g., histamine, serotonin) that can make diagnosis and treatment difficult. Individual snake venoms produce multiple effects, yet they are often classified on the basis of what is considered by some as the primary toxic effect (e.g., cardiotoxic, hemotoxic, myotoxic, neurotoxic). The use of such classification has sometimes caused serious errors in the treatment of snakebite. *See also* enzymes of snake venom, snake venom poisoning. Snake venoms sometimes cause severe anaphylaxis and nearly instantaneous death in humans or animals (especially livestock) that have experienced snakebite in the past. **spider venom**. that secreted by a venomous spider. **taipan venom**. the venom of *Oxyuranus scutellatus*. It is one of the most toxic of snake venoms, it is otherwise similar in most respects to krait venom. It is hepatotoxic and contains a clotting factor that usually brings death to a human victim within minutes unless treated with "Taipan" antivenin. **tarantula venom**. any venom produced by spiders of the family Theraposidae. Such venoms contain a number of proteins, at least one of which is cardiotoxic. The bite of a single tarantula is rarely seriously harmful to a human. Some tarantulas are dangerous to humans but those native to the United States are not considered dangerous. **toad venom**. the venom of toads of the genus *Bufo*; it contains bufotenin. **viper venom**. viperid venom. **viperid venom** (viper venom). any venom produced by snakes of the family Viperidae. These venoms may cause extensive focal and systemic tissue damage, defective coagulation, hemotoxic effects, or blood vessel damage. Damage to the heart, kidneys, and lungs may occur in some cases. **yellow jacket venom**. a hymenopteran toxin that may induce a possibly fatal (type I hypersensitivity) anaphylactic shock. The liberation of vasoactive amines may cause urticaria, chills, fever, a tightness in the chest, dyspnea, and even cardiovascular collapse and death.

venom gland. any gland that secretes (or excretes) a material toxic or irritating to another animal.

venter. **1a:** abdomen, belly, *q.v.*; the part opposite the dorsum or back of a bilaterally symmetric animal. **1b:** the surface of a chordate that is furthest from the notochord. **2:** the surface of a prostrate organism or structure, that is nearest to the substrate (e.g., in flatworms and fern prothalli). **3:** the part of an archegonium that contains the oosphere.

ventifact. a pebble with flat facets that meet at sharp angles due to wind action. Such pebbles occur most commonly in desert or polar regions.

ventilation. **1a:** breathing. **1b:** the process of moving air or other gasses into and out of the lungs or of water across gills. **2:** the blowing of fresh air into an interior space in order to replace stale or polluted, or otherwise contaminated air. *Cf.* suction.

ventrad. toward the venter (belly).

ventral. **1a:** pertaining to or nearer the venter, *q.v.* **2a:** pertaining to the underside of a leaf or other plant structure. **2b:** nearer to the surface of a petal that faces the axis of a flower.

Veratrum (false hellebore). a genus of nearly 50 species of hardy and showy perennial toxic herbs (Family Lilaceae). They occur in moist pastures and woodlands throughout North America and are cultivated in wild gardens around ponds or streams. All parts of the plant are toxic. Some species are teratogenic to sheep. The roots, which are especially toxic, are sometimes used to make an insecticide. Poisoning of humans is rare, however, and grazing animals generally avoid this plant. ***V. album*** (European white hellebore), *V. californicus*, and *V. viride* are sources of antihypertensive alkaloids. *Cf. Helleborus*. *V. album* (European false hellebore; European white hellebore). a species with greenish-white flowers, growing to a height of about 1.3 m. Its toxicity is similar to that of *V. viride*. *See* jervine. ***V. californicum*** (false hellebore; corn-lily; "skunk cabbage"). an attractive herb of the western United States that has white flowers marked with green and grows to a height of about 2 m. If parts of this plant are ingested during the second and third weeks of pregnancy by ewes, craniofacial malformations form in the newborn lambs. Affected lambs cannot breathe or feed properly and usually die within a few hours of birth. ***V. nigrum***. a plant that grows to a meter or more and bears crowded purple flowers. ***V. viride*** (American hellebore; American veratrum; white hellebore; American white hellebore; Indian poke; itchweed;

earth gall; duck ratten; devil's tobacco; Tabac du diable; devil's bit; poor Annie; bear corn). a coarse, erect, sparingly leafy herb that grows to a height of 1 to 2.5 m or more. The flowers are white or yellowish-green. It occurs in low or swampy grounds from Quebec, Minnesota, New Brunswick, and New England, south to Maryland, and in uplands as far south as Georgia and Tennessee. All parts of the plant are toxic. Human intoxication often results from confusion with other herbs. ***V. woodii*** (western false hellebore). a stout leafy plant that grows to a height of about 1.3 m or more. It occurs mainly in woodlands or on bluffs and hillsides in the western United States from Oregon to Iowa, south to Missouri and Oklahoma.

Vermaxan. a trade name for lindane.

vermifugal. anthelmintic (def. 1).

vermifuge. anthelmintic.

vermin. a noun in common usage that includes all noxious, parasitic, and nuisance animals such as lice, Norway rats, house mice, flies, and bedbugs. The term is, however, subjective and animals such as crows, snakes, and prairie dogs may be considered to be vermin by some people at some locations but not by others or at other locations.

vernal pheasant's eye. *Adonis vernalis*.

verruculogen (TR-1). a neurotoxic, tremorigenic indole mycotoxin produced by *Aspergillus caespitosus* and *Penicillium verruculosum*. It is extremely toxic to laboratory mice and domestic fowl, eliciting hypersensitivity to sound, ataxia, and tetanus spasmus.

verrugosa. *Lachesis mutus*.

Vertebrata (Craniata). the principal subphylum of the phylum Chordata. Included are cartilaginous and bony fishes, amphibians, reptiles, birds, and mammals. Representatives of this class typically have an endoskeleton of cartilage or bone, a dorsal vertebral column (backbone) that encloses a tubular nerve cord (spinal cord), and a complex nervous system with a well-developed brain in the cranium (skull). The blood and digestive systems are situated ventral to the vertebral column. All living vertebrates except the Agnatha have jaws, formed from the anterior pair of visceral arches. There are nine classes, including those known only from the fossil record: Agnatha, Placodermi, Acanthodii, Chondrichthyes, Osteichthyes, Amphibia, Reptilia, Aves, and Mammalia.

vertebrate. 1: having vertebrae. 2: a vertebrate animal; an animal of the subphylum Vertebrata, *q.v.*

vetch. a common name of any plant of the genus *Vicia*.

vetchling. a plant of the genus *Lathyrus*.

veterinary medicine. the medical science or practice that treats the diseases of animals including domestic species, wildlife, and zoo and aquarium animals.

veterinary science. a biomedical science concerned with animals. A science that broadly applies biological and veterinary knowledge to the problems to the care of domestic and wild animals and to the improvement of veterinary practice. *See* biomedical.

Vibrio cholerae (*Vibrio comma*; cholera vibrio). a Gram-negative bacterium (Family Spirallaceae) that is the etiologic agent of Asiatic cholera in humans. The proximate cause of the disease is the exotoxin, choleragen, *q.v.*

Vibrio comma. *Vibrio cholerae*.

Vicia (vetches). a genus of mostly weedy herbs (Family Fabaceae, formerly Leguminosae). Some species contain vicine, *q.v.* ***V. faba*** (fava bean; broad bean; horse bean; Windsor bean). the only species of *Vicia* normally cultivated for human food, it is indigenous to North Africa and southwestern Asia where it has been cultivated since ancient times. The stalk is strong, upright, with small leaflets, and edible seeds born in pods. The fava bean is rarely grown in the United States. ***V. sativa*** (spring vetch). the seed is cyanogenic and the foliage is photosensitizing to livestock. *See also* favism.

vicine (2,5-diamino-4,6-diketopyrimidine-3-β-D-glucoside). a white, crystalline, mononucleoside glycoside,

$$C_{10}H_{16}N_4O_8$$

It occurs in *Vicia sativa* and related vetches. *V. sativa* are often present among *Lathyrus sativus* seeds used in livestock feeds and may be responsible for some of the symptoms of lathyrism.

Vienna green. cupric acetoarsenite.

Vinca (periwinkle). a genus of erect or trailing herbs or small shrubs (Family Apocynaceae). *V. rosea* (= *Catharanthus roseus*) is believed to cause defects in grazing calves similar to those of crooked calf disease.

vinegar acid. acetic acid.

vinegar bacteria. *Acetobacter*.

vinyl (ethenyl). the hydrocarbon radical, $CH_2{=}CH$-.

vinyl chloride (chlorethylene; chloroethylene; chlorethene; chloroethene; ethylene monochloride; monochloroethene; monochloroethylene; vinyl chloride monomer; vinyl c monomer; VC). a colorless, flammable liquid, C_2H_3Cl, or gas (at room temperature) with a faint ether-like odor. Vinyl chloride is a very dangerous fire hazard. It is an irritant and is moderately toxic if ingested or inhaled. While it is a gas at room temperature (b.p. -13.9°), it is usually handled as a refrigerated liquid. Occupational exposures are almost exclusively via inhalation or contact. The public may be exposed to vinyl chloride via drinking water and in consumer products. The central nervous system, respiratory system, liver, and blood and lymph systems are all affected by exposure to vinyl chloride. Among the symptoms of poisoning are fatigue, weakness, cardiac arrhythmias, and abdominal pain. Cyanosis may occur. Chronic poisoning with vinyl chloride produces angiosarcoma of the liver and probably lung cancer in humans. It causes liver, lung, kidney, and mammary cancers and cancers of the blood vessels in laboratory mice and rats. It is also mutagenic and has reproductive effects (e.g., on spermatogenesis). It is used in large quantities as the monomer in the manufacture of polyvinyl chloride. It has also been used as an intermediate in other synthetic reactions, as a solvent, and as a propellent in household aerosol products. Many of them are being or have been restricted, and the last mentioned has been discontinued. Acute poisoning causes CNS depression, with symptoms similar to those of alcoholic intoxication, and hepatotoxicity. The carcinogenicity and mutagenicity of vinyl chloride is apparently due to its oxidation by the cytochrome P-45 monooxygenase enzyme system in the liver. The chloroethylene oxide so formed has a strong tendency to bond covalently to macromolecules (protein, DNA, and RNA) and to undergo transformation to chloroacetaldehyde, an established mutagen. Both of these inter mediates can be inactivated and eliminated from the body by conjugation with glutathione. During the above oxidation, the activity of cytochrome P-45 is destroyed.

vinyl chloride monomer. vinyl chloride.

vinyl cyanide. acrylonitrile.

L-5-vinyl-2-thiooxazolidone. a thyrotoxic substance that causes hyperplastic enlargement of the thyroid and symptoms of hypothyroidism. Thiooxazolidone glycosides have been isolated from plants which have caused livestock mortality especially among lambs from ewes fed on plants or plant products that contain these substances; many lambs are stillborn. Those that live have enlarged thyroids, are poorly developed, apathetic, and may fail to nurse. The ewes usually show few signs, although some enlargement of the thyroid may be observed on autopsy. Thiooxazolidone is the goitrogenic factor in various species of *Brassica*, including broccoli (*B. oleracea* var. *botrytis*), brussels sprouts (*B. o.* var. *gemmifera*), kale (*B. o.* var. *acephala*), kohlrabi (*B. o.* var. *gongylodes*, synonym, *B. o. caulorapa*), and rape (*B. napus*).

viosterol. vitamin D_2.

virgin. **1a:** pristine; unpolluted; untouched. **2a:** girl; maiden; **2b:** chaste; innocent. **3:** a sexually mature animal that has not yet coupled or mated another such animal.

Virginia pine. *Pinus virginiana*.

viricidal. able to destroy or neutralize a virus.

viricide (virucide). **1:** destructive of viruses. **2:** an agent that is destructive to viruses.

virucide. viricide.

virulence. **1a:** deadliness; lethality; toxicity; the capacity to be virulent. **1b:** the ability to harm, especially the ability of a pathogen (especially a microorganism) to rapidly cause serious disease. Virulence is often measured by the case fatality rate. **1c:** the quality of being toxic; the degree of toxicity of a substance. **2a:** the ability of a microorganism or virus to overcome host resistance. **2b:** the capacity of an organism (especially a microorganism) or a virus to cause disease in a given host, as indicated by the median lethal dose, LD_{50}, or the median infective dose, ID_{50}. *See*, for example, toxin (microbial toxin). **2c:** the degree of destructiveness of a pathogenic organism (especially a microorganism) or virus, as indicated (usually) by the median lethal dose, LD_{50}, the median infective dose, ID_{50}, or the case fatality rate.

virulent. **1:** severe, malignant, or rapidly progressing; referring especially to any microbial or viral infection marked by rapid onset of severe symptoms. **2:** pertaining to or characterized by virulence; extremely toxic, pathogenic, or harmful. **3:** pertaining to any infectious disease or poisoning marked by rapid onset of severe symptoms. **4:** of or pertaining to a highly infectious or extremely pathogenic microorganism.

viscera. the internal organs of an animal collectively.

visceral. of, relating to, or pertaining to the viscera.

visha. *Aconitum ferox*.

vitamin B_1. thiamine.

vitamin B_x. *p*-aminobenzoic acid.

vitamin C. ascorbic acid.

vitamin D_2 (9,10-secoergosta-5,7,10(19),22-tetraen-3-ol; calciferol; ergocalciferol; oleovitamin D_2; activated ergosterol; viosterol). an antirachitic vitamin and ro denticide.

Viton. a trade name for lindane.

vivaparous. giving birth to living young.

volatile oil of mustard. allyl isothiocyanate.

volcanic dust. *See* ash (volcanic ash).

vomit. **1:** to voluntarily or involuntarily disgorge (expel) stomach contents through the mouth. *See also* vomiting, regurgitate. **2:** vomitus, the matter so expelled. **3:** an emetic.

vomiting (emesis; vomition). retch; throw up; the voluntary or involuntary forcible ejection of partial or complete stomach contents through the esophagus and mouth. Vomiting provides protection against numerous noxious agents. Prolonged vomiting, however, may result in dehydration or the loss of essential electrolytes and can be fatal. Humans vomit readily when noxious agents are ingested or even tasted or smelled. Many species of livestock (e.g., cattle, horses, goats, and sheep) and wildlife do not readily vomit and are therefore more susceptible to many toxicants than are animals that vomit readily. **pernicious vomiting**. continuing, uncontrollable vomiting. **projectile vomiting**. vomiting with great force.

vomition. vomiting.

vomitive. emetic.

vomitory. emetic.

Vulpes fulva (red fox). a small, predatory, doglike animal (Family Canidae) that is usually harmless to poultry and other farm animals.

W

warfarin (4-hydroxy-3-(3-oxo-1-phenylbutyl)-2*H*-1-benzopyran-2-one; 3-(α-acetonylbenzyl)-4-hydroxycoumarin; 1-(4′-hydroxy-3′-coumarinyl)-1-phenyl-3-butanone;3-α-phenyl-β-acetylethyl-4-hydroxycoumarin; compound 42). a synthetic coumarin anticoagulant,

$$C_{19}H_{16}O_4$$

A vitamin K antagonist, it depresses the formation of prothrombin and increases capillary fragility, with increased hemorrhaging. It is used primarily as a rodenticide, usually applied as the sodium salt (warfarin sodium) or the potassium salt (warfarin potassium). Warfarin is named after the Wisconsin Alumni Research Foundation, where it was developed. **warfarin potassium**. a white, crystalline powder,

$$C_{19}H_{15}KO_4$$

Its anticoagulant activity and uses are similar to those of warfarin sodium. **warfarin sodium**. a white, crystalline or amorphous powder,

$$C_{19}H_{15}NaO_4$$

It is a cumulative anticoagulant with an action of intermediate duration, used as an oral rodenticide, although it may be effectively administered intravenously or intramuscularly. It has the same action as dicumarol and is extremely toxic to female laboratory rats, very toxic to male laboratory rats and mice but less toxic to laboratory rabbits. Symptoms of intoxication include hematuria, bloody feces, shock, coma, and death.

warm-blooded. *See* endotherm.

warm-blooded animal. *See* homeotherm.

wart hog. *See Phacochoerus*.

wart hog disease. African swine fever.

wart weed. *Euphorbia helioscopia*.

wasp stinger. *See* stinger.

wasting. **1:** squandering. **2a:** enfeebling, emaciating. **2b:** withering; fading; declining; losing strength or an excessive among of body mass (especially of muscle and vital tissue). **2c:** to cause wasting (defs. 1a, 1b). *See also* phthisis.

water. **1a:** a clear, tasteless, odorless, polar-covalent liquid, H_2O; f.p.= 0°C (32°F), b.p.= 100°C (212°F). It is a universal component of living systems and is critical to life. It is often referred as the "universal solvent" because of its solvent power. Exogenous water is a vital nutrient to living systems. In addition to its critical importance to the survival of individual organisms, water and water flow (influx, efflux, and internal cycling) are vital to the continuing existence and structure of ecosystems. It is a major vehicle whereby chemical pollutants are transported to and within ecosystems. It is a major medium of chemical transport and the major medium in which chemical reactions take place. **1b:** the liquid phase of H_2O; the other phases are solid (ice; crystalline water) and vapor (steam). **2:** a euphemism for urine. **available water** (plant-available water). That part of the water in the soil that can be absorbed into plant roots. It is defined as the difference between field capacity and the permanent wilting point. This usually includes a large portion of the capillary water. **bound water**. una vailable water. **brackish water**. a mixture of fresh and salt water. **capillary water**. **1:** water in soil above the water table due to capillary attraction. **2:** water retained in capillary pores of a soil; water with a surface tension greater than 60 cm (23.63 in.). **connate water**. water trapped in the interstices of an extrusive igneous or sedimentary rock at the time it was deposited. **crystalline water**. ice. **drainage water**. gravitational water (def. 1). **drinking water**. water from any source, treated (i.e., that from a municipal water supply) or untreated (e.g., that from a well) that is (1) sufficiently pure or unpolluted such that a person can drink it without harm or (2) water that is normally imbibed by humans or other animals, especially livestock. **excess water**. gravitational water (def. 1). **free water**. gravitational water (def. 1). **fresh water**. **1:** water that generally contains less than 1000 milligrams per liter of dissolved solids. Such water may or may not be drinkable (potable). **2:** surface or groundwater bodies that are uncontaminated by saline sources and have a rate of outflow to prevent accumulation of salt by water evaporation. **gravitational water** (drainage water; excess water; free water; percolation water). **1:** water that flows through soils due to the force of gravity. **2:** surface water, groundwater, or storm water flowing through a drain **3:** water supplied by gravity as opposed to that supplied by a pump. **gray water**. domestic wastewater composed of washwater from sinks, kitchen sinks, bathroom sinks and tubs, and laundry tubs. **ground water**, **groundwater**. subsurface water; all water beneath the soil surface that has no direct connection to the surface. Groundwater bodies lie over impervious strata. In many areas, groundwater is the major or even sole source of drinking water. Some groundwater bodies can be contaminated by chemicals from the surface by leaching or percolation through the soil and other strata above or by leakage or flow from waste dumps or mines. **perched water**. water that forms a zone of saturation above a shelf or bed of impermeable or nearly impermeable that retards further flow downward. **percolation water**. gravitational water (def. 1). **plant-available water**. available water. **receiving water(s)**. a body of water that receives water from another water body, from runoff, or from industrial, agricultural, or other effluents. **red water**. **1:** red tide. **2:** any of several diseases of certain domestic animals (e.g., cattle tick fever) indicated by production of reddish urine. **running water**. **1:** a body of water (e.g., that in a canal or river) that flows continuously in one direction. **2:** tap water; water from a pump. **salt water**. water of high salinity. **sea water**. salt water that occurs in the oceans of the world. The salinity is taken as the total amount by weight (in parts per thousand) of dissolved solids following conversion of all bromide and iodide to chloride, of all carbonate to oxide, and the complete oxidation of all organic matter. The salinity of ocean water averages 35 ppt. **soil water**. water in soil, usually total water in soil. **surface water**. **1a:** water in the hydrological cycle that is naturally open to the environment for a time, or moves on the surface of the ground (e.g., ponds, lakes, rivers, streams, reservoirs, impoundments, swamps, estuaries, oceans). Such water is readily contaminated from direct exposure to precipitation, dust, materials in surface runoff, or the activities and decomposition of living organisms. Also included under this term are water bodies that are partially underground (e.g., springs, wells, collectors) that contain fresh water and are directly influenced by aboveground water. **1b:** waters directly influenced by surface water (def. 1a) are often included in this term (e.g., wells, springs, cisterns and various other collectors). **unavailable water** (bound water). water in the soil that cannot be taken up by plant roots because it is strongly bound (adsorbed) to soil particles.

water buffalo. *Bos bubalus*.

water gardening. hydroponics.

water hemlock. *See Cicutum* and *Conium*.

water lily. *Zigadenus venosus*.

water moccasin. *Agkistrodon piscivorus*.

water table. groundwater table.

wax tree. *Rhus succedanea*.

waythorn. *Rhamnus cathartica*.

weavers' broom. *Sparteum junceum*.

wedgeleaf ceanothus. *Ceanothus cuneatus*.

weed. any plant that is considered to be undesirable generally, or at certain locations or under certain conditions.

weediness. **1:** the appearance of being overgrown with weeds, as a recently abandoned meadow. **2:** the unwanted effects of a plant.

weedkiller. herbicide.

Weil's disease. human leptospirosis (*See* leptospirosis).

Welsch onion. *Allium fistulosum*.

western diamond-back rattlesnake. *Crotalus atrox*.

western hemlock. *Cicutum occidentalis*.

western monkshood. *Aconitum columbianum*.

Western flat-topped agaricus. *Agaricus meleagris*.

Western honeybee. *Apis mellifera*.

Western poison oak. *Rhus diversiloba*.

Western red cedar. *Thuja plicata*.

western yellow pine. *Pinus ponderosa*.

wet deposition. *See* acid deposition.

wet-milling. a milling operation in which seed-containing feed material is steeped in water, with or without sulfur dioxide, to soften seed kernels as an aid to separation of the contents. Thus, for example, wet-milling can separate a bushel of corn into more than 31 pounds of starch, approximately 15 pounds of animal feed, and nearly 2 pounds of corn oil.

wether. a male sheep castrated before reaching sexual maturity.

wetland pasture poisoning. grass tetany.

Wetlands Reserve Program (WRP). a program to provide long-term protection for wetlands under the FACT Act of 1990. Producers enrolling in the program agree to implement an approved wetlands restoration and protection plan. The participants may then receive payments based on the loss, if any, in the value of their land due to placement of an easement on a portion of it. Enrollment in the program is limited to 975,000 acres.

WRP. Wetlands Reserve Program.

wharf rat. *Rattus norvegicus*.

wheat. **1:** any of several species of cereal grasses of the genus *Triticum*. **2:** the grain of these plants. **durum wheat**. *Triticum durum*.

wheatgrass. any plant of the genus *Agropyron*.

whetstone. a stone used to sharpen cutting tools by friction. *Cf.* oilstone.

white arsenic. arsenic trioxide.

white baneberry. *Actaea alba.*

white camas. *Zigadenus elegans.*

white cedar. *Thuja occidentalis.*

white clover. *Trifolium repens.*

white cohosh. *Actaea alba.*

white heath. *Erica arborea.*

white hellebore. *Veratrum viride.*

white horse nettle. *Solanum eleagnifolium.*

white loco. *See Oxytropis.*

white mustard. *Brassica hirta.*

white of the eye. sclera.

white point loco. *See Oxytropis.*

white quebracho. *Aspidosperma quebracho-blanco.*

white pine. *Pinus strobus.*

white ragweed. *Franseria discolor.*

white rice. *Oryza rufipogon.*

white rot. **1:** a rotting of wood caused by lignase-producing fungi, especially *Phanerochaete chrysosporium.* **2:** a fungal disease of onions and related plants caused by *Sclerotinia cepivorum.* **3:** any of a number of fungi that cause white rot, especially *Phanerochaete chrysosporium.*

white sage. *See* greasewood.

white sanicle. *Eupatorium rugosum.*

white snakeroot. *Eupatorium rugosum.*

whorl. a group of three or more leaves or flowers arranged in a complete circle around a stem (e.g., the flowers of spearmint). *See also* whorled.

whorled milkweed. *See Asclepias.*

whorled. **1a:** pertaining to a whorl. **1b:** pertaining to an arrangement along a shoot or twig in groups of three or more arising from a single node or stem joint. *Cf.* alternate, opposite.

wild balsam apple. *Echinocystis lobata.*

wild black cherry. *Prunus serotina.*

wild celery. *Angelica archangelica.*

wild cherry. *Prunus serotina.*

wild garlic. *Allium vineale.*

wild hippo. *Euphorbia corollata.*

wild horse. feral domestic horses, *Equus caballus.*

wild indigo. *See Baptisia.*

wild ipecac. *Euphorbia ipecacuanha.*

wild monkshood. *Aconitum uncinatum.*

wild morning glory. *See Convolvulus.*

wild mustard. *Brassica kaber.*

wild oats. *Avena fatua.*

wild onion. *Zigadenus venosus.*

wild pea. a plant of the genus *Lathyrus.*

wild radish. *See Raphanus*

wild rye. a grass of the genus *Elymus.*

wild sage. *Lantana camara*.

wild sunflower. *Helianthus annus*.

wild type. **1a:** an organ as it occurs in nature. **2:** an organism (or any aspect of an organism) that is considered typical, normal, or standard for the natural (wild) species or population from which it is drawn. Such an organism is used as a norm or standard against which aberrant or mutant individuals or alleles are compared. *Cf*. wild type allele (*See* allele). **3:** the condition of an organism before it is genetically engineered.

wilderness. land that retains its primeval character; any region or ecosystem where evidence of human presence is negligible.

Wildlife Services Program (WS). an animal and plant health inspection service for the purpose of protecting agriculture, natural resources, property, public health, and endangered species from undesirable and potentially harmful effects of wildlife. Included is the prevention of wildlife/airplane collisions at airports. Formerly named the Animal Damage Control Program.

wildlife. **1:** wild animals of types. **2:** game animals and other species of (usually terrestrial) vertebrate animals designated as wildlife by law or regulation (e.g., animals that are legally hunted, raptors, threatened and endangered species).

wildling. a seedling growing in nature as opposed to a nursery.

Wilson's disease. exfoliative dermatitis.

wilting point. the percentage by weight of water remaining in the soil when the plant begins to wilt. **permanent wilting point**. the percentage by weight of water remaining in the soil when the plant wilts permanently (does not recover).

wind gauge. anemometer.

Windsor bean. *Vicia faba*.

windthrow. blowdown

winter aconite. *Helleborus viridis*.

winter bluegrass. *Poa bulbosa*.

winter grass. common name of *Poa annua*.

Winton disease. Pictou disease.

witch grass. *Agropyron repens*.

witchgrass. *Panicum capillare*.

with calf. said of a pregnant cow.

witherite. *See* barium carbonate.

wolf's-bane. a plant of the genus *Aconitum*.

wolf's milk. *Euphorbia esula*.

wolfbane. *See* aconite, *Aconitum*.

wolfsbane. aconite.

wonderberry. *Solanum intrusum*.

wood tick. *See Dermacentor*.

wood. secondary xylem; the hard, usually non-living material of a tree or woody bush. *See also* xylem, lignification, annual ring.

wooden tongue. a common name of either of two diseases with similar signs: actinobacillosis and actinpomycosis.

woody aster. any plant of the genus *Xylorhiza*.

wool. the soft, curly hair or coat of domesticated sheep, goats, certain other ruminants, and rabbits. *See also* alpaca, Astrakhan, braid. **belly wool**. the wool that grows on the belly of a sheep. It is usually shorter and less desirable than body fleece. **black wool**. any black fleece. Even the presence of a few black fibers will be rejected as white, as it is not possible to separate the black fibers during manufacture. Such wool will be placed in lots to be dyed black. **black top wool**. that which contains a large amount of wool grease together with dirt at the tip of wool staples. It is most commonly produced by Merino sheep. It is usually a desirable wool of fine quality and good character. Shrinkage is high, however. **botany wool**. **1:** In Great Britain and the U.S,, any fine wool from Australia. **2:** any class of fine wool. **bright wool**. light-colored farm wool, as opposed to semibright, *q.v.* **britch wool**. wool from the thigh and twist region of the sheep; the coarsest and poorest wool on the entire fleece; usually manure-encrusted and urine-stained. **dead wool**. wool pulled from dead sheep, by means other than slaughter. **defective wool**. any wool that is faulty or that contains unusual amounts of extraneous materials such as burrs or seeds. **Delaine wool**. combing wool, produced mostly in the state of Ohio. **demiluster wool** (semiluster wool). wool with less luster than that classified as luster wool. **eastern pulled wool**. Wool pulled from sheep and lambs slaughtered in the east. In this process, the wool is first loosened from the skin, usually by using a depilatory and then pulled. **free wool**. wool free of defects. **luster wool**. a lustrous wool. *Cf.* demiluster wool. **semiluster wool**. demiluster wool. **semibright wool**. wool that is darker than bright wool usually due to soiling. However, it canbe scrubbed out to a good white color.

wool fat (degras; "yolk"). a greasy substance secreted by the skin of a sheep; it adheres to the wool. Wool fat is refined and sold as lanolin.

wooly-pod milkweed. *Asclepias eriocarpa*

wormseed mustard. *Eryusimum cheiranthoides*.

wormwood. *Artemisia absinthium*.

WS. Wildlife Services Program.

X

X-linked disease (sex-linked disease). a sex-linked genetic disease due to a mutant allele carried on an X chromosome.

xanthein. the part of the coloring matter in yellow flowers that is soluble in water. *Cf*. xanthin. (def. 1).

xanthin. **1:** the water-insoluble pigment in yellow flowers. *Cf*. xanthein. **2:** a yellow pigment in madder.

Xanthium (cocklebur; clotbur). a genus of toxic, semicosmopolitan, coarse, weedy, annual herbs (Family Compositae). All species have similar toxicities, and in many reports of poisoning, it is difficult to determine which of the approximately 20 species were involved, although *X. spinosum* and *X. strumarium* have often been implicated in North America. One or the other occurs in fields and wastelands throughout most of North America. Poisoning of livestock is common throughout much of the United States. The seeds are very toxic due to the presence of hydroquinone but are rarely eaten because they are contained in the burrs. The hydroquinone disperses throughout the plant, however, when the seed germinates. Thus, while the very young plants have a pleasant taste, they are toxic. All known cases of poisoning by this plant are due to the ingestion of seedlings in the cotyledonary stage of growth. Symptoms are similar in all classes of livestock and usually include anorexia, depression, nausea, vomiting, muscular weakness, weakened cardiac contractions, dyspnea, and prostration. Emprosthotonos and convulsions sometimes occur in severe cases. Vomiting may not occur in cattle. Swine are frequent victims.

Xanthocephalum lucidum. *Gutierrezia microcephala*.

Xanthocephalum sarothrae. *Gutierrezia microcephala*.

xeno-. a prefix meaning strange, foreign.

xenobiotic. **1a:** foreign to a living system. **1b:** a foreign compound; any natural or synthetic compound that is foreign to a living system. **1c:** any substance that enters and interacts with an organism and is neither endogenously produced or part of a normal metabolic pathway of the organism. **2:** a foreign (allochthonous) organic chemical. **3:** xenogeneic.

xenogenic (xenogenous). **1a:** exogenous; allochthonous; originating from outside the organism or system that is the subject of concern. **1b:** produced by a foreign compound within the organism or system of concern. **2:** xenogeneic.

xenogenous. xenogenic.

xerophyte. a plant that is adapted to dry or and habitats.

xylem. the principal fluid-conducting tissue of vascular plants. It conducts water and nutrients throughout the entire plant and serves also as a food storage organ; secondary xylem provides support. Xylem is comprised of elongate, hollow cells that are joined end-to-end, forming a continuous tube. The cells have thick, usually lignified walls and lack protoplasts when mature. *See also* phloem. **primary xylem**. that derived from procambium. **secondary xylem**. that derived from the vascular cambium. *See* wood.

Xylorrhiza (woody asters). a genus of plants (Family Compositaceae) closely allied with *Aster*; some authorities do not consider woody asters sufficiently distinct to rank as a separate genus. All species are selenium accumulators and most are reliable indicators of selenium-rich soils. The selenium content of woody asters varies with soil content and perhaps with species. Thus, consumption of these plants bylivestock may produce chronic or acute by selenium poisoning depending upon the selenium content and the proportion of forage comprised by *Xylorrhiza*. Even though plants of this genus are generally unpalatable to sheep, extensive losses have occurred as a result of consuming woody aster, especially *X. parryi*, when other forage is limited.

Y

yak. *Bos grunniens*.

yang tao. *Actinidia chinensis*.

yellow teat disease. malignant (bovine) anaplasmosis (*See* anaplasmosis).

yard. **1:** a piece of land surrounding or adjacent to a home or some other building or group of buildings. **2:** a tract of land that is used for a specific activity or business (lumberyard). **3:** a winter pasture or feeding area for herbivores such as deer or elk. **4:** an enclosed area where animals, such as chickens or pigs, are kept. **5:** an area that includes tracks and sidings, where railroad trains are serviced, made up, and switched. **6:** to put into a yard (e.g. to yard lumber) **7:** to be enclosed in, or gathered into a yard (or as if into a yard).

yard waste. waste due to the care of a lawn including, for example, grass clippings, leaves, weeds, and prunings less than six inches in diameter.

yean. to give birth; said of a sheep or goat.

yeanling. a newborn infant, especially the young of a sheep or goat, i.e., a lamb or kid.

yearling. a male or female farm animal, especially a cow or horse, during the first year of life.

yellow arsenic sulfide. arsenic trisulfide.

yellow foot agaricus. *Agaricus xanthodermus*.

yellow jasmine. *Gelsemium sempervirens*.

yellow jessamine. *Gelsemium*.

yellow lupin bean. *Lupinus luteus*.

yellow monkshood. *Aconitum lutescens*.

yellow oat grass. *Trisetum flavescens*

yellow oats. *Trisetum flavescens*.

yellow pine flax. *Linum neomexicanum*.

linuron(*N'*-(3,4-dichlorophenyl)-*N*-methoxy-*N*-methylurea; methoxydiuron). a substituted urea used as a selective pre- and postemergence herbicide. Linuron is moderately toxic to laboratory rats.

yellow trefoil. *Medicago lupulina*.

yellow turnip. *Brassica napus* var. *napobrassica*.

yellow-foot agaricus. *Agaricus xanthodermus*.

yellowsedge. *Andropogon virginicus*.

Yetrameres americana (chicken stomach worm). an internal parasite proventricular gland. Infested fowl may be asymptomatic but young birds especially may become emaciated and die.

yew family. Taxaceae.

yew poisoning. poisoning, usually of livestock, by ingesting the needles or berries of certain taxine-containing, woody, evergreen shrubs and trees of the genus *Taxus*, *q.*v. Symptoms of acute yew poisoning in humans are those of gastroenteritis. Death from cardiac or respiratory failure may intervene without further symptoms.

yew. **1:** any plant of the family Taxaceae. **2:** any tree or shrub of the genus *Taxus*. **3:** ewe.

yield curve. in forestry, a tabular or graphic depiction of stand volume, usually as a function of age of the stand.

yogurt. a fermented, semisolid, fermented product of skim or lowfat milk, sometimes augmented with protein. *Lactobacillus bulgaricus* and *Streptococcus thermophilous* are commonly-used fermenting bacteria. The characteristic mildly acid taste is due to lactic acid that is produced by the action of the bacteria on milk proteins. Flavors, flavoring, and sugar may be added. Bulgarian buttermilk is a type of yogurt.

yolk. food storage material, high in protein and fat, that is present in the eggs of most animals.

yolk sac. a membraneous sac with a rich blood supply that develops about the yolk of the vertebrate egg. It is attached to the growing embryo makes possible the conveyance of the nutrients of the yolk to the embryo. No yolk is present in the yolk sac of mammals; in marsupials it forms a temporary placenta.

Z

Z. the symbol for proton number.

Zamia. a genus of perennial tropical and subtropical plants (Family Zamiaceae) that superficially resemble both palms and ferns. ***Z. integrifolia*** (coontie; Florida arrowroot). a woody, superficially fern-like plant with palm-like leaves that arise from a thick, underground stem. They occur in dry, sandy soils throughout peninsular Florida. The foliage is decorative and used by florists. This and other species of *Zamia* cause wobbles in cattle.

Zea mays (maize; corn; Indian corn). an important grain, forage, and fodder plant (Family Cyperaceae) that presumably originated in Mexico and now consists of countless varieties. It is known to cause a number of illnesses in livestock. Problems may result, for example, from consumption of moldy corn or of forage or silage that contains unusually high concentrations of nitrate; *Zea mays* also has a cyanogenic potential. Poisoning of cattle by forage with high levels of nitrate can be fatal. Corn stalks may contain nitrate burdens, KNO_3, as high as 8%. Silage may contain high concentrations of nitrogen dioxide and nitrogen tetroxide. *See also* silo-filler's disease. ***Zea mays indentata*** (dent corn). a variety that produces grains with a dent in the top. It is the most widely cultivated corn.

zearalenone (F_2 toxin; 3,4,5,6,9,10-hexahydro-14,16-dihydroxy-3-methyl-1*H*-2-benzoxacyclotetradecin-1,7(8H)-dione; 6-(10-hydroxy-6-oxo-*trans*-1-undecenyl)-β-resorcylic acid lactone; compound F-2; FES). a mycotoxin, $C_{18}H_{21}O_5$, produced by various strains of *Fusarium* that infest corn (e.g., *F. roseum*, *F. graminearum*). It is the only known mycotoxin whose effects are primarily estrogenic. female swine that feed on infested corn suffer hyperfolliculinia with a high rate of abortion; males develop female sexual characteristics. Zearalenone is known to infest corn, oats, barley, wetland, and sorghum (both fresh and stored), in compounded cattle and swine rations, in corn ensiled at the green stage, and in hay. Infested plants sampled from pastures in temperate climates occasionally contain zearalenone at levels that are probably sufficient to cause reproductive failure of grazing herbivores. Zearalenone is carcinogenic to some laboratory animals.

zearalenone hyperestrogenism. fusarium estrogenism.

zebra. common name of a number of species of *Equus*, *q.v.* They have distinctive patterns of stripes on the head and body. **common African zebra**. *E. zebra*. **Grevy's zebra**. *E. grevyii*. **mountain zebra**. *E. zebra*. **plains zebra**. *E. burchelli*.

zebu. *Bos indicus*.

zephyr lily. *See Zephyranthes*.

Zephyranthes (zephyr lily). a genus of bulbous, usually attractive plants (Family Amaryllidaceae, *q.v.*) most of which are native to the temperate and tropical areas of the Americas. ***Z. atamasco*** (atamasco lily; rain lily). the common white, spring-blooming species found in wooded areas of the southern United States from Virginia to Florida, west to the coastal plains of Mississippi. It causes "staggers" in horses. The bulb is the most toxic part of the plant. It is extremely toxic to livestock. Symptoms appear within two days of foraging on the bulbs or leaves of this plant. They include staggering, soft feces with bloody mucus, collapse, and death. Most poisonings occur during the spring when other forage is in short supply.

zero tolerance. *See* tolerance.

Zigadenus (*Zygadenus*; black snake root; death camas). a genus of North American and Asian perennial herbs (Family Liliaceae) with grass-like leaves, whitish to yellow, greenish, or bronze flowers in clusters, and rhizomes or bulbs. The entire plant of most species, including the seeds, contains several toxic veratrum alkaloids and are among the most important poisonous plants of western North America. Death camas probably cause greater mortality among sheep on the spring ranges of the Western United States than any other plant. Toxicity varies considerably among the various species of death camas but symptoms are exceptionally similar. The minimum lethal dose of *Z. nuttallii* in sheep is less than one-hundreth that of *Z. elegans*. All species should be considered hazardous as individual species are difficult to identify. Early signs of poisoning in sheep include persistent excessive salivation, nausea, and sometimes vomiting. These are followed by muscular weakness, a slowing of the heart rate, low blood pressure, ataxia, tremors, and prostration. Labored breathing (dyspnea) is common and the animals may become cyanotic and, in severe cases, may become comatose and die. Cattle, horses, and domestic fowl may be poisoned as well as sheep. Swine are susceptible (as demonstrated in the laboratory) but natural poisonings do not occur because swine quickly expel the ingested material by vomiting. Humans have been poisoned when they mistook the bulb for that of an edible plant. Symptoms in humans are those of gastrointestinal irritation and cardiovascular collapse. ***Z. coloradensis***. *Z. elegans*. ***Z. densus*** (black snake root; crow poison). a bulbous plant that occurs in damp pinelands and bogs in the southeastern United States. It is very toxic and is just as poisonous to livestock as, and strongly resembles, *Amianthium muscitoxicum*. ***Z. elegans*** (*Z. coloradensis*; white camas; alkali grass). a highly toxic species with very narrow leaves and greenish flowers in clusters up to 0.3 m long. It is much more toxic than *Z. nuttallii*. ***Z. nuttallii*** (nuttall death camas; death camas; death-camass; poison camas; poison camass; merryhearts). a species with rather large flowers, nearly 1.3 cm across in loose clusters. It is extremely dangerous to humans and livestock. ***Z. paniculatus*** (foothill death camas; sandcorn). a species found in the western United States. It has yellow flowers and an extremely toxic bulb. ***Z. venenosus***. *Z. venosus*. ***Z. venosus***. (*Z. venenosus*; death camas; alkali grass; black snake root; soap plant; poison sego; water lily; wild onion; squirrel food; hog's potato). an herb that is widely distributed in North America, excluding the extreme southwest of Canada and Alaska. All parts of the plant are poisonous but the seeds are especially so; ingestion of the bulb can be fatal. The leaves are long, narrow, grass-like, and gathered about the base of the stem. The stem bears a terminal branched cluster of greenish-white to yellow-white flowers. The bulb is onion-like but with a dark coat and lacking the onion odor. Ingestion may cause increased salivation, muscular weakness with staggering, or even complete prostration, dyspnea, coma, and death.

zinc (symbol, Zn; C.I. 77945; C.I. pigment Black 16; C.I. pigment Metal 6)[Z = 30; A_r = 65.38]. a bluish-white, lustrous transition metal occurring naturally as the sulfide in zinc blende, or as the carbonate in smithsonite. It is a heavy metal of relatively low oral toxicity. Zinc is an essential trace element that normally occurs in animals in metalloenzymes, peptidases (e.g., carbonic anhydrase, alkaline phosphatase and dehydrogenase), or as cofactors (e.g., arginase and histamine diaminase) that catalyze the synthesis of DNA, proteins, and insulin.

zinc chloride. a white, granular or powdery crystalline solid, $ZnCl_2$, that is soluble in water, ethanol, glycerol, and ether. It is used, for example, in fireproofing, as a wood preservative, as a food preservative, in soldering fluxes, and in dry cell batteries. It is one of the most dangerous and toxic compounds of zinc. It is a potent irritant that can damage the eyes, skin, and mucous membranes on contact. Both the solid and fumes are toxic; inhalation of the fumes may cause severe pain and damage to the eyes and upper respiratory tract as well as pulmonary edema.

zinc oxide (Chinese white; C.I. 77947; C.I. pigment white 4; zinc white). an odorless, white, grayish, or yellowish powder, ZnO, with a bitter taste. It has a variety of uses including that of a seed dressing, a feed additive, a healing agent, and a pigment and mold-growth inhibitor in paints. It is an irritant of the skin and eye and is poisonous if given intravenously.

zinc phosphide. a toxic, ethanol-insoluble, dark gray, granular powder, Zn_3P_2, used as a rodenticide. It reacts violently with oxidizing agents and decomposes in water. When ingested, it is rapidly fatal to laboratory rats, rabbits, and various avian species; dogs and cats are more resistant. Zinc phosphide itself is not very toxic but it releases highly poisonous phosphine gas when it comes into contact with mineral acids such as HCl. Susceptible species release gastric HCl continuously, whereas the more resistant species secrete HCl intermittently.

zinc white. zinc oxide.

zincophyte. a plant that tolerates or is adapted to high levels of zinc in the soil. Such plants are often used as indicators of soils with high levels of zinc.

zineb. the zinc salt of dimethyldithiocarbamate, it is a fungicide of low toxicity to livestock. *Cf.* maneb. *See also* dithiocarbamate fungicide.

ziram. a fungicide of low toxicity to animals; it is, however, carcinogenic in some laboratory animals. Ziram is the zinc salt of ethylenebisdithiocarbamate. *See also* dithiocarbamate fungicide.

zoa. plural of zoon.

zooanthroponosis. anthropozoonosis.

zoogenous. originating, derived, or acquired from animals.

vaginal gland. any of the mucous glands in the mucous membrane of the vagina.

variance analysis. analysis of variance.

variegated thistle. *Silybum marianum*.

variety. a subdivision of a species, often loosely applied to a number of a groups within the species. In agriculture, this term is most often applied to a cultivated form of a plant and occasionally to a breed of animal. Cultivated varieties of horticultural plants are termed cultivars. Cultivars do not have latinized names.

zoology. The natural science concerned with animals.

zoon (plural, zoa). an animal.

zoonoses. plural of zoonosis.

zoonosis (plural, zoonoses; zoonotic disease). a disease that is transferrable from animals to humans under natural conditions. Among these are rabies, tuberculosis, and brucellosis (which becomes undulant fever in humans).

zoonotic disease. *See* zoonosis.

zoopathogenic. pathogenic to animals; of or pertaining to a substance or an organism that is pathogenic to or productive of disease in animals.

zoopathology. the science of pathology applied to animals.

zoophobic. of or pertaining to plants that animals avoid. Such plants are usually potentially injurious, harmful, or repugnant (e.g., because of spines, hairs, poisons).

zooplankton. plankton comprised of animals or that component of plankton comprised of animals.

zoosporangium. a sporangium that gives rise to zoospores.

zoospore. a motile, flagellate spore that sometimes behaves as a gamete, produced by many algae and some fungi from a zoosporangium.

zooteratology. the science of teratology, *q.v.*, as applied to animals.

zootoxic. toxic to animals.

zootoxicologist. a scientist concerned with poisonous and venomous animals; one who contributes to the body of knowledge; one who applies the findings of zootoxicology to problems of other fields of endeaver. *Cf.* phytotoxicologist, biotoxicologist.

zootoxicology. a science that deals with poisonous and venomous animals.

zootoxin (animal toxin). a toxin of animal origin (e.g., snake venoms, venoms of wasps and bees). All animal phyla contain species that produce toxins. Zootoxins are often complex mixtures that may include toxic enzymes, other proteins, peptides, and small organic compounds such as alkaloids, biogenic amines, glycosides, and terpenes.

zootrophic. heterotrophic.

zygomorphic. a botanic term denoting bilateral symmetry, especially that exhibited flowers whose parts are arranged bilaterally on the receptacle. *Cf.* actinomorphic.

zygote. **1a:** the diploid cell (fertilized egg) produced at fertilization by the union of male and female gametes. **1b:** individual derived by development from the fertilized egg.

zygotene. a stage of prophase of the first meiotic division.

zymogen. an inactive protein precursor of an enzyme.